W9-COR-544

ELECTRONIC DEVICES
A Top-Down Systems Approach

Michael Sanderson

DeVry Institute of Technology, Columbus, Ohio

PRENTICE HALL, Englewood Cliffs, New Jersey 07632

LIBRARY OF CONGRESS
Library of Congress Cataloging-in-Publication Data

Sanderson, Michael R.
Electronic devices : a top-down systems approach / Michael R. Sanderson.
p. cm.
Includes index.
ISBN 0-13-250879-6
1. Electronic circuits. 2. Semiconductors. I. Title.
TK7867.S25 1988
621.381--dc19 87-30447
CIP

Editorial/production supervision: Eileen M. O'Sullivan
Interior design and cover design: Jayne Conte/Maureen Eide
Manufacturing buyer: Peter Havens
Cover photograph: © John Giannicchi, Photo Researchers

A Division of Simon & Schuster
Englewood Cliffs, New Jersey 07632

Printed in the United States of America

10 9 8 7 6 5 4 3 2 1

ISBN 0-13-250879-6 025

Prentice-Hall International (UK) Limited, *London*
Prentice-Hall of Australia Pty. Limited, *Sydney*
Prentice-Hall Canada Inc., *Toronto*
Prentice-Hall Hispanoamericana, S.A., *Mexico*
Prentice-Hall of India Private Limited, *New Delhi*
Prentice-Hall of Japan, Inc., *Tokyo*
Simon & Schuster Asia Pte. Ltd., *Singapore*
Editora Prentice-Hall do Brasil, Ltda., *Rio de Janeiro*

In memory of my mother, Patsy, for her inspiration to write, and to Ginny, Rachel, and David for their love, support, and patience.

Contents

3

Fundamentals of Amplification 59

4

Ideal Amplifiers 88

5

Introduction to Solid-State Devices 126

Functional Aspects of Discrete Active Devices 159

Reference to Troubleshooting Topics

Portions noted as being (extensive) include detailed analysis or diagnostic steps, and often include examples.

It should also be noted that at the end of each chapter there are specific troubleshooting work problems to assist in providing additional practice.

Reference to Data Sheets

Preface

The purpose of this book is to provide clear and concise coverage of the fundamental aspects of semiconductor devices and the process involved in the logical development of electronic circuits. The objective is to provide the basis for a one- or two-semester study of electronic circuit development. The student is expected to have prior experience in dc circuit analysis and either prior or concurrent work in ac analysis. In as many cases as possible, basic algebra is used to resolve all calculated values.

This book deviates dramatically from a more traditional approach to semiconductor devices. The development of this text was prompted by three major factors. First, with the demand to incorporate more technical information in the preparation of a student for a career in electronics, the amount of time available to spend on solid-state-device technology was becoming a premium. Second, there were major stumbling blocks encountered by students who first approached these devices, ones that could be factored out without lessening the students' ability to deal with these devices. Finally, the role of the electronics professional is going through a major metamorphosis.

Students today are required to deal with an almost overwhelming quantity of electronic devices, circuits, and systems. The trend toward the incorporation of digital technology into all facets of electronics requires an expanded coverage of that field of electronics in a technical curriculum. In years past, knowledge of five or six distinct active devices was sufficient to comprehend the majority of the field of electronics. Now, it is not uncommon for students to be familiar with literally hundreds of devices and subsystems. Therefore, the amount of time that can be dedicated to any single topic area is limited. It is therefore necessary to approach any topic in a time-efficient manner.

In previous textbooks and courses dealing with solid-state devices, the topic was addressed by first studying the detailed physics of the devices. Then, all of the most fundamental dc bias arrangements were explored. Finally, the details of ac analysis were presented. In many cases, dc and ac circuits were treated as being totally separate, almost devoid of each other. Students seldom had a concept of an overall goal for the study of these topics. Therefore, students commonly learned what to do, via formula, to a device to make it work. Often though, they did not conceptually see what the overall objective was that all the formulas were designed to achieve. They lacked a set of conceptual goals. This made their study of the devices more complex. Often, if asked to reapply the concepts learned for one device to another device, they sought a formula rather than a logical pattern. In all too many cases, the memorization of a formula took on greater importance than the understanding of a principle.

The modern role of the electronics professional has changed dramatically. Not only is there a highly visible trend toward more and more digital electronic systems, but many of the systems, analog or digital, are either in integrated form or simply considered subassemblies of a whole system. More and more, the job of the electronics professional involves dealing with functional blocks of a system (or subassemblies) rather than with the performance or function of one discrete part.

In developing this book these concepts were factored into the essentials that needed to be known about discrete devices. Rather than starting with complex (and often confusing) semiconductor physics, this book starts by explaining the methods used to define the function of a system. The objective is to have the student be able to describe how a task will be achieved and what role electronics will play in achieving that task.

The student is then led through the process of using multiple stages to produce a function. Knowledge of the interrelationship of stages to each other will help establish goals for the development of any one single stage. Then the concept, usually missed by all traditional approaches, that amplification is simply the manipulation and control of one form of energy using another is presented. The objective is to have the student looking for the control function as basic circuits are developed.

The process of amplification, as characterized by ideal amplifiers, is used as a launching point for students to explore individual device functions. Device performance is related to the ideal model; circuit modifications to create performance closer to the ideal are more easily understood when such a relationship is understood.

The initial discussion of discrete semiconductors concentrates on the most popular circuitry used for amplification. The coverage of less popular variations (e.g., common-base BJT amplifiers) is saved for presentation in chapters earmarked for "special" circuits. Dc analysis is treated as a means to an end, not an end itself. It is presented as a necessary step to make the device functional for ac signal processing. As soon as possible, the processing of ac signals is introduced, since it is the main function of the basic amplifier. The student is reminded that the reason for different dc and ac analysis is simply based on traditional superposition analysis. Composite analysis is done to show the effect of both conditions existing in the same circuit.

Once the fundamentals of device manipulation is studied, the analysis of special-purpose circuits is done using simple amplifier models. The philosophy is that since the correct form of amplifier can be had, it is unnecessary to "reinvent" the amplifier each time a new application is seen. The objective is to make the student more flexible, more conceptually able to comprehend the process of achieving the effect of the special-purpose circuit. When it is helpful to use an amplifier circuit, the op-amp is applied in every possible case. Thus the amplification function is highlighted, not a complex string of dc bias elements mixed in with the description of some special effect.

Diodes and rectification are treated as separate concepts, not as primers to amplification devices. They are dealt with in a separate section of the text and are treated as distinct "other" semiconductors. So, too, control circuits and special-purpose devices are treated in their own right.

Introduced at key points in the text are the concepts of linear integrated circuits and methods for evaluating future device developments. Students must recognize that many of the circuits, subsystems, and complete systems that are and will be described in their studies are likely to be produced in integrated form, either now or in the near future. Also, it is presumptious to assume that device technology will stop with the BJT or FET form of semiconductor device. Students need to be armed with tools to evaluate future developments in this rapidly changing technology.

This repackaging of the topic of semiconductor devices and circuits will require changes in the way this material is presented to the student. This more conceptual approach to device analysis and application will, in most cases, call for a revision of the outlines, notes, and procedures used in a traditional approach. The improved comprehension of the material by the students and their greater flexibility to work with new material is a benefit worth the revision.

To ease the transition to this new form of presentation, an Instructor's Guide has been prepared. This guide gives a chapter-by-chapter rundown of highlights and embellishments for use by the course instructor. It also proposes timing and sequencing of the material, with optional suggestions.

This new approach to device analysis is the core of a restructuring proposed by this author to the DeVry Institutes of Technology for use in their electronics technician curriculum study of semiconductor devices. This system (currently eight schools nationally) has accepted this proposal, and in a modified form will be applying it nationally in that curriculum. The reported success of students in this course is proving exceptional results. Employers are finding that the ability of graduates to adapt to new and different systems is vastly improved due a lot to their ability to logically analyze system functions rather than search for a formula.

There are a few ideas about specific chapters that would be valuable to share at this time; again, further elaborations are given in the Instructor's Guide:

Chapters 1 and 2. These chapters, and the conceptual process they are intended to develop in the student, should not be treated in a hasty manner. Students should be encouraged to learn to define form and function as cooperative partners in achieving a task. Many of the basic concepts may be new to the way students have been trained to think in the past. It is suggested that participatory exercises be used to develop problem-solving skills early on.

Chapters 3 through 9. These chapters cover more of the traditional devices. It should be noted that less common circuits are omitted here for clarity (e.g., common-base BJTs, emitter follower, etc.). These are covered later under special applications.

In conclusion, the phrase "top-down approach" has been used to describe this system of study. It follows a process that everyone is more familiar with than the more traditional approaches. It is likened to assembling a jigsaw puzzle. We first look at the overall picture, the cover of the box. Once we have a clear concept of what the final product should look like, we begin assembly of the puzzle. First the border is established, the primary boundaries, and general pattern. Then as individual pieces are encountered, they are seen to fit the pattern, to have a role in the overall picture. When a section is completed, a sense of achievement, based on the boundaries and overall picture, can be felt. The sections then interlock to fit the finished product. Occasional odd pieces, the difficult concepts, can be dealt with more easily, if by nothing more than a process of elimination, they just do not fit anywhere else. It is hoped that using a "big-picture-down-to-small-part" process can improve the ability not only to understand today's puzzles, but to provide a means of analyzing tomorrow's yet-to-be-defined puzzles.

Michael R. Sanderson

SANDERSON OFFERS A COMPREHENSIVE TEACHING/ LEARNING PACKAGE

Helpful Learning Tools Include:

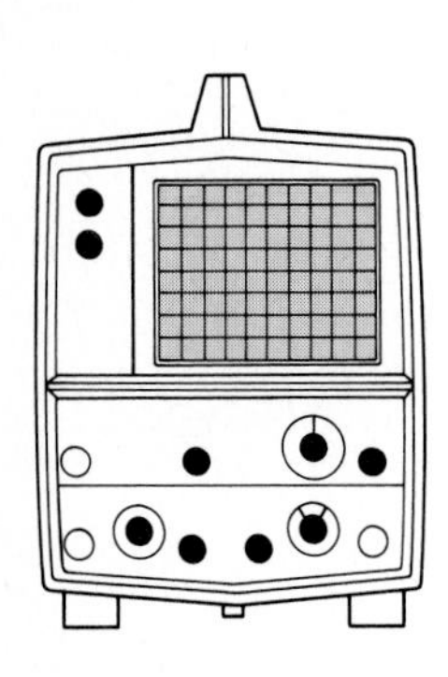

- Functional use of 2-color
- Data sheets are interspersed throughout each chapter when referenced
- Numerous examples are highlighted within each chapter
- Troubleshooting applications are incorporated throughout each chapter
- To distinguish a "troubleshooting section," an oscilloscope symbol is placed in the text margin
- End-of-chapter problems are grouped by sections
- Opening chapter objectives appear before each chapter
- End-of-chapter summaries are included

Teaching/Learning Supplements Include:

- Lab Manual to Accompany the text (25092-8)
- Instructor's Manual to the Lab Manual (25093-6)
- Instructor's Transparency Masters (25089-4)
- Instructor's Resource Manual (25091-0)
- Software disk Free upon adoption
- Shrinkwrapped FREE to every text is a Troubleshooting Supplement (includes additional troubleshooting applications and problems)

1 The Organization of Electronic Systems

Objectives

Upon completing this chapter, the reader should be able to:

- Identify basic functions of electronic devices.
- Identify the role of active and passive devices.
- Utilize block diagrams.
- Identify basic electronic processes.

Introduction

The application of electronics to the world around us is nearly limitless. In the past 10 years, the world has seen electronics become a major influence in every aspect of daily life. It is impossible to think of doing daily tasks without relying on electronics.

At the same time that the application of electronics is becoming more and more complex, the function of the electronic devices used still follows basic principles. It is first necessary for us to explore the general function of electronic circuits. Once we have a general understanding of how such circuits work, we will explore the specifics of how each task is achieved by each part within these circuits.

The control of electrical energy by electronic devices is the major effect we explore. In this chapter we see that such control is essential to achieve the functions desired.

Electronic functions will be categorized in order to ease their identification. The role of active devices (transistors, for example) will be contrasted with the role of conditioning elements (such as resistors). How each of these elements relates to the overall function of the circuit or system is explored in general terms. Several key definitions related to these devices and corresponding functions are developed. ■

1.1 THE CONCEPT OF CONTROL FUNCTIONS

Electronic systems are similar to many other physical systems. Most electronic processes, except those related to computer electronics, involve energy control and/or conversion. We are familiar with such energy control processes in our daily lives.

The gasoline engine is an example of energy control and conversion (Figure 1.1). The gasoline serves as a source of energy. Conversion from chemical energy to mechanical energy is achieved when the gasoline is burned in the engine. Special controls are required in this conversion process; we cannot just strike a match to a tank of gas.

Control is provided for this conversion by two basic elements. The first produces the overall means of converting from chemical energy to mechanical energy. Conditioning elements, the cylinders, spark plugs, and crankshaft (the raw engine parts), define the method of conversion. An active element, the carburetor, provides the second process. It uses an external input to provide control over the entire system's function.

The carburetor blends the external input and raw gasoline energy. Based on the information given the carburetor (from the gas pedal) it controls the flow of raw energy (gasoline) through the engine. Without this active control element, the conversion and use of energy is uncontrolled and inflexible.

An electronic system follows such a pattern (Figure 1.2). The energy source used in electronic systems is a battery or dc (direct current) power supply which simulates a battery. An information input is also provided. Control is achieved by both active and conditioning devices in the circuit. The active devices accept this information input and impress direct control over the energy supplied in response to this information. The conditioning devices in the circuit impress a constant level of control over energy flow either to set maximum limits or to support the function of an active device. The output of the system, voltage variations and/or current flow, is a controlled response to the information input.

In comparison to the systems in a car, the battery supplies energy much like the gasoline does. The active devices act much like the carburetor and the passive devices act like the basic engine parts. Just as we provide information to the engine which determines its operation, information is delivered

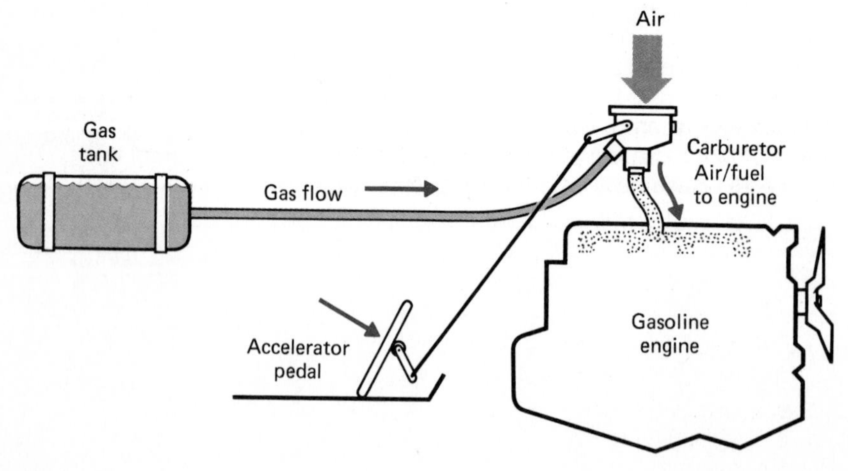

FIGURE 1.1 Engine Control System

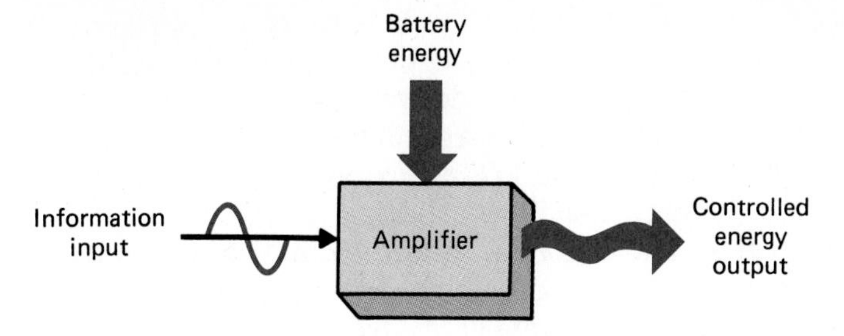

FIGURE 1.2 Simple Amplifier Function

to the electronic circuit which directs it to meter out the battery's energy in a specific way.

Electronic circuits and systems covered in this book will be shown to provide active control over current and voltage, and therefore, power. This differs from the use of passive devices (resistors, capacitors, and inductors), which provide a fixed control over energy flow. Series and parallel circuits involving passive elements will be used, as noted before, simply to provide limits or to support the actions of the active devices. The end result of this blend of active and passive devices will be a circuit that accepts information input(s) to provide control over the delivery of electrical energy supplied by a separate source (a battery).

REVIEW PROBLEMS

For the following descriptions, reproduce Figure 1.3 but label each section (block) with the specific terms shown in the question.

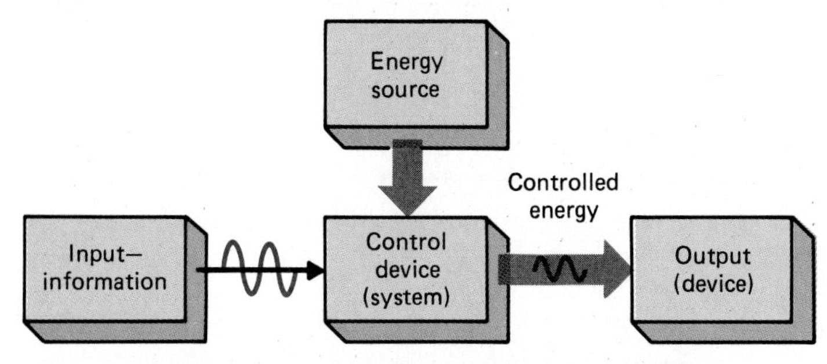

FIGURE 1.3 Energy Control in an Amplifier

(1) A backhoe operates in the following general manner: Pressurized fluid serves as a source of energy. The operator provides information to the control valves which directs the flow of fluid through the system. Hoses, cylinders, and the metal structure determine the way in which the controlled energy will be used. The end result is the movement of a load of dirt. (*Terms to use:* pressurized fluid; operator; control valves; hoses, cylinders, metal structure; movement.)

(2) A record player works in the following general manner: Batteries serve as a source of energy. The needle senses vibrations on the record and the cartridge converts these into audio signals (electrical variations) which are sent to the amplifier. Transistors in the am-

plifier use these signals to control the flow of battery-supplied energy. Resistors and capacitors condition this flow within usable boundaries. Electrical energy is then delivered to the speakers, which convert this energy into sound waves, which are in synchronization with the audio signal input. [*Terms to use:* Batteries; audio signal; transistors; resistors; capacitors; speaker (sound waves).]

(3) Sketch a simple diagram like Figure 1.3 for a controlled system of your choice. First describe it in general terms, then identify the major elements. Be certain to separate active from conditioning elements. Label each part of the system.

1.2 BLOCK DIAGRAMS AS AN AID TO SYSTEM DESCRIPTIONS

Most electronic systems start out as block diagrams. A *block diagram* is a drawing that shows the function of a system and how that system is to operate. A block diagram does not describe each small part in the system. Since the objective of the diagram is to illustrate how a system is to achieve a final goal, the designer of the system can quickly "block-out" the logical steps to take in far less time than it would take to compose a written description of the necessary steps. Later a part-by-part layout of the circuit will be based on the general information contained in this diagram.

As an example of the application of a block diagram in generating many circuits that achieve a certain task, we can look at the simple AM radio. There is a common link between early floor-standing radios that used vacuum tubes, to transistor radios that fit into your pocket, and finally to integrated-circuit radios that can fit on your wrist. Each shares the same common block diagram which describes how they work. There is one standard block diagram for the AM radio which describes how it is to function. Even the AM radio of the future will, in all likelihood, follow this standard block diagram.

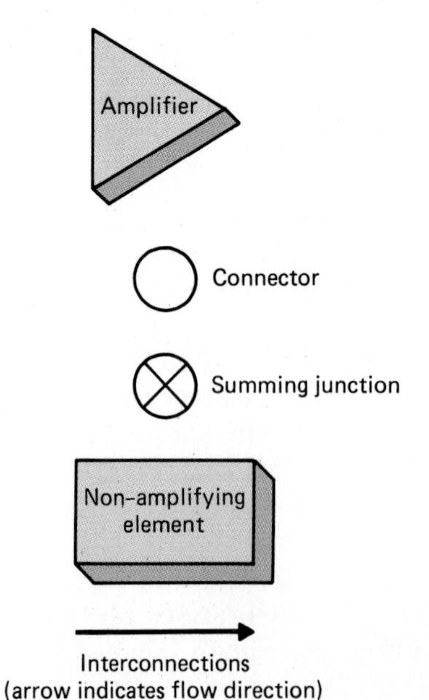

FIGURE 1.4 Basic Block Diagram Symbols

There are certain standard symbols and practices used when developing block diagrams. By knowing and understanding these it will be easier to understand the function of block diagrams. In Figure 1.4 several of the standard symbols are illustrated.

A block diagram can be expanded or modified as the process for achieving a task is refined. Often, a system starts out as a single block (idea) with just a definition of the overall task inside. Then as the process for achieving the task become clearer to the designer, more blocks are added, revisions are made, and the final diagram is developed. If the system is to be modified to do more or different tasks, the first step is to change the block diagram to fit the new modification. This will then lead to the changes needed in the actual circuitry used. Even though the diagram may include many blocks to describe the systems function, each block describes a distinct step to take in completing this function.

REVIEW PROBLEMS

(1) In your own words, briefly describe the use of block diagrams.

(2) Which of the following statements best describes what the role of each block in a block diagram should be? (*choose one*)

- **(a)** It should name each functional part (nuts and bolts) in the system.
- **(b)** It should name functional steps to take to achieve a task.
- **(c)** One block should include multiple steps bulked together into one complex block that describes the entire system in one step.

(3) *True or False:* A block diagram can be expanded once greater detail is known about each step needed to achieve a task.

1.3 DESCRIBING A SYSTEM USING BLOCK DIAGRAMS

At this point we are going to spend some time in describing the use and application of block diagrams. Since it is important to understand how electronic systems will function on a general basis, we will explore the use of block diagrams first before we tackle individual circuits.

Typically, a block diagram coordinates with a written or verbal description of a process. Each major step identified in this description corresponds with a block in the diagram. The individual part or parts used to achieve the necessary functions usually do not appear in the diagram.

When laying out a diagram it is necessary to identify all input and output functions needed. Inputs for electronic systems can range from simple on/off switch positions to complex ac waveforms. Outputs can be as simple as powering a light bulb to producing a variety of complex waveforms at high power levels. Inputs are then illustrated on the left of the diagram and outputs are shown on the right. There are occasional exceptions to this initial setup.

The steps necessary to achieve the task are usually placed between these two boundaries. How the information input finally controls the output is then described in the pattern shown in this central section. A first-step-to-last-step progression is then illustrated. Reading from left to right, the diagram shows

EXAMPLE 1.1

From the following written description, compose a block diagram.

> A furnace's setback thermostat will automatically reduce the operating temperature of the furnace at certain preset times of the day. In order to operate, this thermostat needs two inputs: time of day and the actual room temperature. Naturally, the output of this system is the signal that turns on the furnace's heating element. Between these blocks is the actual control circuitry. The information from the temperature sensor is sent to a temperature comparison block. Here it is compared with the desired "normal" temperature. The time of day is fed to a setback controller. Here, when the correct time of day occurs, the setback controller sends an additional signal to the temperature comparison section. This signal tells it to reduce its "normal" temperature. Finally, the output of temperature-comparison section is fed to the heating element on the furnace.

Solution:

The block diagram is shown in Figure 1.5.

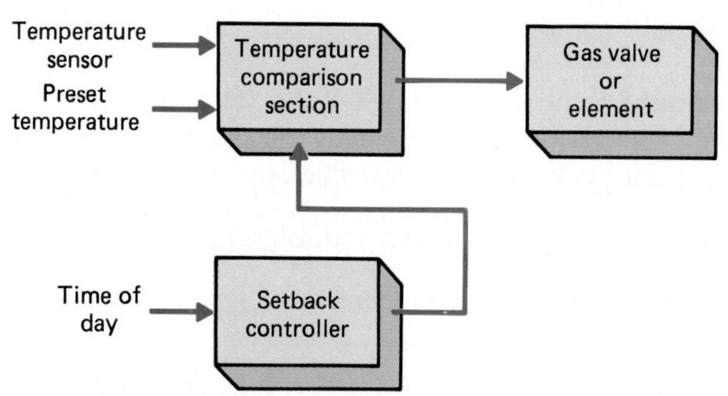

FIGURE 1.5 Simple Diagram of Temperature Control System

EXAMPLE 1.2

From the block diagram shown in Figure 1.6, compose a brief description of the process.

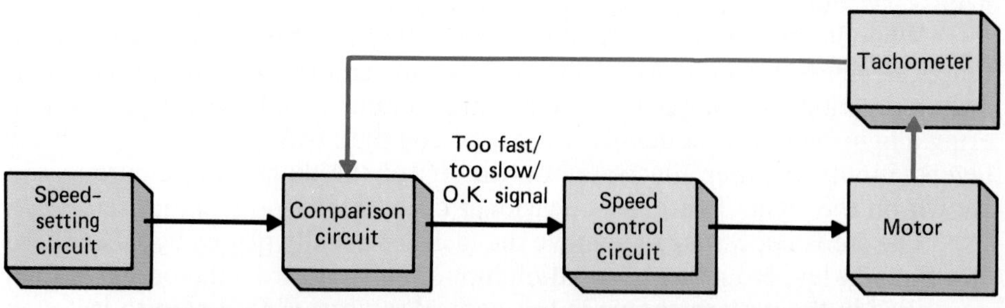

FIGURE 1.6 Simple Diagram of Speed Control System

Solution:

Description—motor speed controller: The speed of the motor is set by the speed control circuit. This speed control circuit is supplied with a TOO FAST–TOO SLOW–O.K. signal from the comparison circuit. This circuit receives its information from the tachometer (hooked to the engine) and from the speed-setting circuit. The speed-setting circuit is where the preset value of speed is initially given as an input.

how to start with the initial input and through a series of steps, how the final output is produced. Since many electronic systems typically are designed to control an energy source, these steps between input and output usually show how progressively control is exercised over this source of energy. Digital electronic systems (computers, calculators, etc.) are the exception to this process of control. These systems are designed to manipulate numbers and not

REVIEW PROBLEMS

(1) Fill in the blocks in the block diagram shown in Figure 1.7, using the description as a guide.

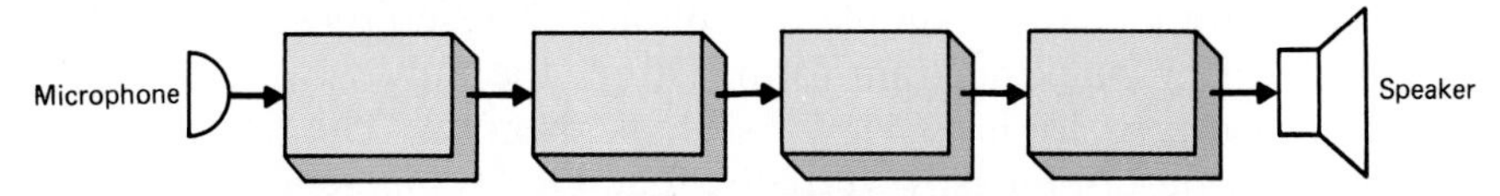

FIGURE 1.7 Basic Input to Output Organization

The microphone (an input) feeds directly to a preamplifier. Following this circuit are the tone and volume control sections. The output of these circuits supplies the power amplifier. The power amplifier finally feeds to a speaker.

(2) Write a brief but complete description of the block diagram shown in Figure 1.8.

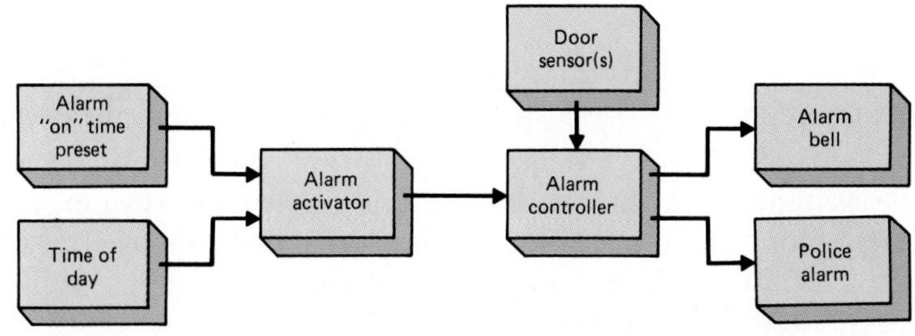

FIGURE 1.8 More Complex Block Diagram Organization

(3) Locate a process (not necessarily electronic) with which you are familiar. Write a brief description of this item or process and illustrate it with a block diagram. (For example, illustrate how to do a job that you are familiar with.)

provide control over some power supply's energy delivery. The application of digital electronic systems is not covered in this book but is left to texts that deal with them specifically.

It must be kept in mind that each important step in producing the final output must be illustrated by a block in the diagram. If not broken down into enough separate steps, the diagram lacks important clarity and precision. In most cases the dc power source is omitted in an electronic block diagram. Although not essential, it is so common to circuits that everyone assumes that it is to be used in the process.

From Examples 1.1 and 1.2 it can be seen that it is often much easier to illustrate a system's function using block diagrams than with longhand descriptions.

1.4 SAMPLE DESCRIPTION OF AN ELECTRONIC PROCESS

Electronic processes can usually be described clearly by the use of block diagrams. Electronic inputs take a variety of forms, such as simple on/off voltages from switches, voltage variations on a microphone, or complex radio signals. Outputs also take a variety of forms, from simple flashing lights to complex sound waves.

The steps through which these inputs are interpreted, processed, and presented as outputs can be simple or complex. These processes can be logically described in block diagram form. Often, the only way to interpret an electronic process easily is to view its block diagram. The process may be too complex to describe easily verbally. The electronic schematic, wherein each part is identified, may be too complicated to use easily when trying to understand how a circuit is to function.

In the following examples note that with a little explanation, the way in which each circuit functions is clearly visible in the block diagram. Once you have read the brief discussion, review the block diagram and see if you can understand the functions done as shown in the diagrams.

EXAMPLE 1.3

Figure 1.9 shows a sample diagram of a public address amplifier. Note that raw electrical energy is used in three of the four blocks (the dashed arrows). Typically, such a diagram would not show these because it would be assumed that they were a natural part of the system. (This is true of most block diagrams.) The one block that does not use this energy supply is a simple passive control circuit that uses only conditioning elements (no active elements).

At this point the reader is not expected to understand how the system shown in this diagram works. Simply follow the flow of the diagram. The microphones provide input "information" in the form of electrical variations (voltage and/or current). The mixer/preamplifier blends the variations from each microphone and increases the energy level of the blended signal. The level is not high enough yet to operate the speaker.

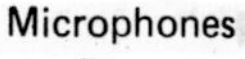

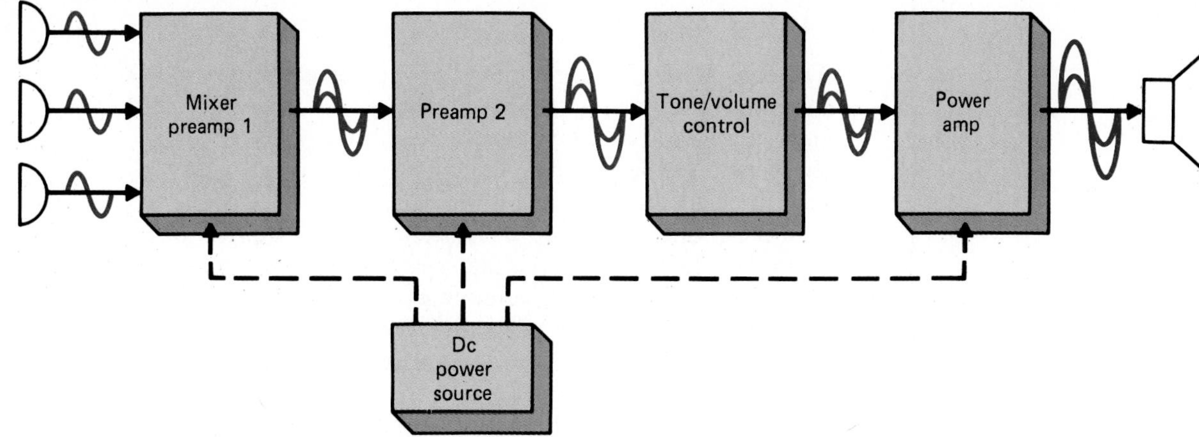

FIGURE 1.9 Power Supply Implied in Each Block

Preamplifier 2 and the power amplifier provide enough energy to operate the speaker. The tone and volume controls simply control the average level of the signal that flows through the system. The volume control adjusts the average level of all signals and the tone control (basically) sets the level that passes through the system at either the low or high frequencies.

To provide the energy necessary to operate the speaker, each amplifier has dc energy supplied from a separate power source (e.g., a battery). This source supplies the necessary voltage and current to each amplifier so that they can, in turn, provide an increase in the input signal's energy level.

Of major interest in any (linear) electronic system is the manipulation or control of the available energy source. In Example 1.3, each time information signals pass through a block (except the volume/tone control) its energy level is increased. The objective of the system is to take low-energy inputs and boost their energy level sufficiently to produce a high-energy output. The apparent gain in energy is due to the ability of the active devices (contained in each block's circuitry) to manipulate the electrical energy supplied by the dc source.

Blocks in other diagrams may produce other desired effects. Although we usually apply the concept of energy control somewhere in electronic systems, other steps not related to control may be involved. Some parts of the system (e.g., the tone control in Example 1.3) may need to be frequency selective. Others may simply sense voltage levels, while another may be used to blend different information signals. In computer systems, for example, energy control may be of little importance. Such a system is designed to make mathematical calculations or to make decisions based on a sequence of instructions, not to manipulate the electrical levels of inputs.

REVIEW PROBLEMS

(1) The easiest way to comprehend how an electronic system functions often is by reviewing its (*choose one*)
  (a) Schematic parts layout
  (b) Written description
  (c) Block diagrams

(2) The main energy source in an electronic system is often omitted from the block diagram of that system because (*choose one*)
  (a) It plays little role in achieving an output.
  (b) It is so common to use it that we assume it is there.
  (c) The block diagram is not supposed to provide a clear picture of what is happening.

(3) *True or False:* Every block in a block diagram must use the main energy source to function.

(4) *True or False:* In computer systems, energy control is of prime importance.

1.5 CATEGORIES OF ELECTRONIC OPERATIONS

Before an attempt is made to describe electronic systems, it will be helpful to categorize electronic circuits. The functions of electronic circuits can generally be grouped into five major categories.

Amplification A majority of electronic circuits provide amplification. Amplification involves the control by a low-energy input of the energy available from a high-energy source. The output of an amplifier is a high-energy duplication of the values supplied at the input. An example of amplification can be seen in the typical public address amplifier. A low-energy input from the microphone is used to control the delivery of high levels of energy to the speaker. The electrical signal delivered to the speaker is a duplication of the signal produced by the microphone.

Power Conversion Power conversion, as the name may imply, is the conversion from one form of electrical power to another form of electrical power. Often, such conversion involves both a change in the form of the energy (ac to dc) and a change in electrical levels (110 V to 12 V). A battery eliminator, like the one often used with electronic toys, fits in the category of power conversion. This device plugs into a 110-V 60-Hz ac outlet and converts this supply of energy to 9 V dc.

Electronic Control Electronic control is similar to amplification. An information input is accepted and the output follows the commands of the input. Unlike amplification, though, control circuits do not duplicate the form of the input (as a speaker duplicates the sounds at the microphone). Control circuits use inputs to direct the actions of other devices (such as motors or lights). The output is often not similar in form to the input. An example of a control circuit can be found in an electronic light dimmer. This circuit accepts the position of the control knob as an input but produces a proportional level of light as a final output.

Digital Logic Digital logic circuits are classified separately from most other electronic circuits. Digital circuits are designed to manipulate equations and logical statements. The input to a digital system is a string of numerical expressions. The output, although also a string of expressions, is seldom the same as the input. Computers and calculators are just two examples of digital logic circuits. Digital circuits often use the same forms of components that would appear in other electronic circuits. The way in which these components are used to produce an output differs dramatically from how they may be used in the other categories noted here.

Special Function Circuits Some electronic circuits or devices do not fit into the other categories; thus they are identified simply as special function circuits. Examples of these will follow later.

Even though these categories can be defined separately, they are not absolute divisions in terms of electronic circuits. With only slight modifications, a single type of electronic circuit could be usable in any of these categories. Often a complex system will incorporate several forms to achieve an overall task. An example of this can be seen in a digitally tuned, remote-controlled television set. Such a system includes circuitry that would fit into each of these classifications.

REVIEW PROBLEMS

(1) List, from memory, the five basic types of circuit functions, and describe each category in your own words.

(2) Place each of the following electronic circuits (or systems) into one of the five categories listed in this section.
   (a) Electronic digital clock
   (b) Stereo amplifier
   (c) Electronic speed control on an electric drill
   (d) Electronic smoke detector/alarm

(3) Make a list of at least 10 electronic circuits (or systems) that you have seen or used recently. Try to identify which of the five categories in this section best describes these items. Naturally, some may have to fit into more than one category. [As an example of this overlap, the rotary-dial telephone system fits both the digital (for the number coding) and the amplifier (for the voice transmission) catetories].

1.6 THE USE OF ELECTRONIC ELEMENTS

Very few systems can be described clearly by one block alone. To define an electronic system clearly it is necessary to describe each essential step taken to achieve the conversion from input to output. For most block diagrams, each block coordinates with a single electronic circuit that achieves the specified task. The exception to this is seen when using integrated circuits. One integrated circuit may be able to perform the tasks shown in several blocks. In fact, this is precisely the objective of using integrated circuits: to perform several tasks in a single element.

In the beginning of this chapter, conditioning and active elements were identified. As noted, **conditioning elements** are resistors, capacitors, and inductors. **Active elements** are diodes, transistors, vacuum tubes, and other special-purpose solid-state devices. Integrated circuits, which are complete circuits built into one package, use both conditioning and active devices to achieve their functions.

One role played by passive devices is as support for an active device. Another is to modify or condition the form of the information flowing within the circuit. Passive devices are also used to set certain electrical limits in the circuit. Their function may be blended with the function of an active device, thus making a more complex circuit. That circuit then corresponds to one block in a block diagram. We explore next the overall functions of electronic systems, concentrating on the use of block diagrams to establish building blocks. These blocks are used to describe how electronic systems produce an overall function. In later chapters we explore individual devices and circuits that will achieve these tasks.

Summary of Active and Conditioning Devices

Type	Form	Application
Active devices	Transistors, vacuum tubes, operational amplifiers	Provide electronically variable control over the flow of electrical energy; these devices respond to electrical variations from an input and provide control of energy in response to this input
Passive devices	Resistors, capacitors, inductors	Used for fixed control of currents and voltages within the circuit; unlike active devices, passive devices do not accept one input and control a flow of energy based on this input; they can work in cooperation with an active device to achieve an overall circuit function

REVIEW PROBLEMS

(1) *True or False:* In many cases, a block in a block diagram will represent one distinct circuit in a schematic drawing of a system.

(2) Achieving the overall function described in a block of a block diagram will usually require the use of (*choose one*)
  (a) One active device alone
  (b) A few conditioning (passive) devices
  (c) A blend of active and passive devices

(3) In your own words, write out a brief definition of the function of active and passive devices in a normal circuit.

1.7 EXAMPLES OF ELECTRONIC SYSTEMS

So far in this chapter, we have only gone over general concepts of electronic systems and circuit operations. To make electronic processes more clearly understood, we now review a few examples of electronic systems. To simplify our investigation of these circuits, we will use the standard symbols used in block diagrams (Figure 1.10).

Amplifier Block An amplifier is a circuit that provides "electronic gain." **Gain** is said to occur when a circuit produces an output that is identical to an input but which is much larger in electrical magnitude. This type of circuit obtains energy to produce this gain from a source separate from the information input.

General Function Block This block (Figure 1.11) may be used to describe almost any function in a block diagram. If it is used where amplifiers are specifically shown with the symbol of Figure 1.10, these blocks often show nonamplifying functions.

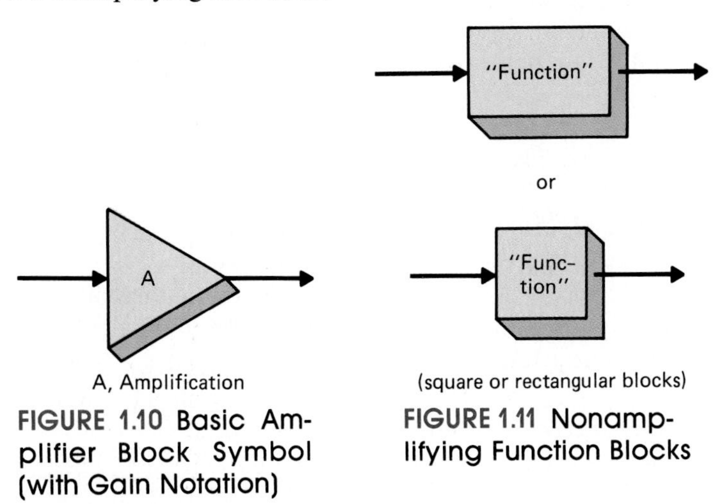

FIGURE 1.10 Basic Amplifier Block Symbol (with Gain Notation)

FIGURE 1.11 Nonamplifying Function Blocks

In the following examples several common electronic systems are investigated. Each system is developed from simple definitions of the desired functions needed. We expand from this general definition to more detailed block diagrams. You are not expected to be able to develop the resulting circuits, but simply to be able to carefully follow the system's development.

EXAMPLE 1.4

A simple block diagram for a public address amplifier is shown in Figure 1.12. Here a basic definition of the system is seen. Two microphones

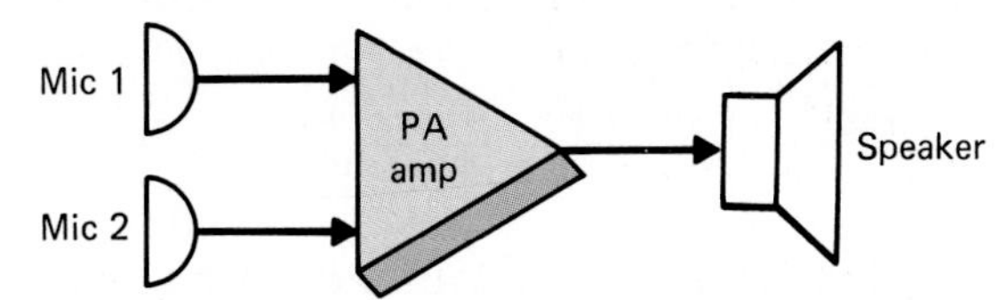

FIGURE 1.12 Simple Amplifier Diagram

produce the input. The amplifier must produce an output to the speaker which represents this input but at a much higher power level.

Figure 1.13 shows the first expansion to the diagram. Expansion is needed for two reasons. First, it is necessary to define more clearly all the functions desired of the amplifier system (tone and volume controls, for example). Second, it is known that the amount of amplification needed cannot be achieved in a single electronic circuit.

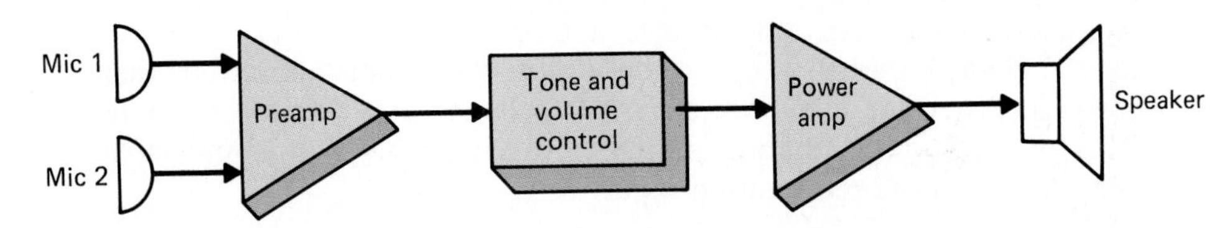

FIGURE 1.13 More Descriptive Amplifier Diagram (versus Figure 1.12)

A diagram that more closely represents the resulting circuit is shown in Figure 1.14. In this block diagram more detail is given as to how each step is to be achieved. We can now see the use of both amplifiers and passive circuits. A general overview of each section (or block) function is as follows:

Passive mixer. Here the two microphone inputs are electrically blended.

Preamplifier 1. Preamplifying simply means to provide amplification prior to providing the final output amplification.

Amplifier coupler. Because of the electrical nature of each of the two preamplifiers, it is necessary to isolate them electrically with a special circuit (again passive).

Preamplifier 2. This circuit will continue to provide amplification started by the first preamplifier.

Bass and treble tone control. In these two circuits special control is exercised of the high and low frequencies that pass through the system.

Volume control. In this passive circuit we are able to control the level of electrical signal that passes on to other circuits. Controlling this level will allow us to control the overall volume of the output to the speaker.

Power amplifier. This final amplifier produces the last gain in electrical signal level. It produces the highest level of electrical power of all the amplifiers in the system.

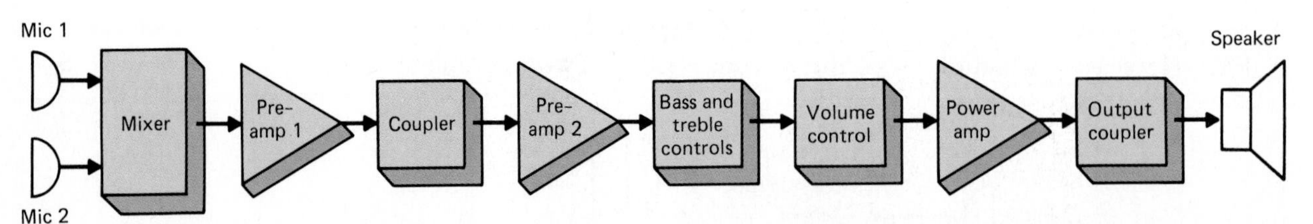

FIGURE 1.14 Complete Amplifier System Diagram Showing Multiple Functions

Output coupler. This circuit helps to connect the power amplifier to the speaker.

Review these blocks and the development of this system carefully. Be especially aware of the general functions indicated by each block. Again, as in other blocks shown, the power supply has been omitted for the sake of clarity. For each amplifier shown, one power source can be used to supply the energy needed to obtain gain.

EXAMPLE 1.5

The simple block diagram of Figure 1.15 shows the general steps needed to produce an automatic light dimmer. This dimmer will be designed to control the lights in a room automatically based on a preset level. (Unlike an ordinary dimmer, this one will compensate for sunlight or other lights in the room.) In this simple diagram only the basic elements are identified. Light-level sensing, level presetting, and the actual dimmer system compose the essential parts of the system.

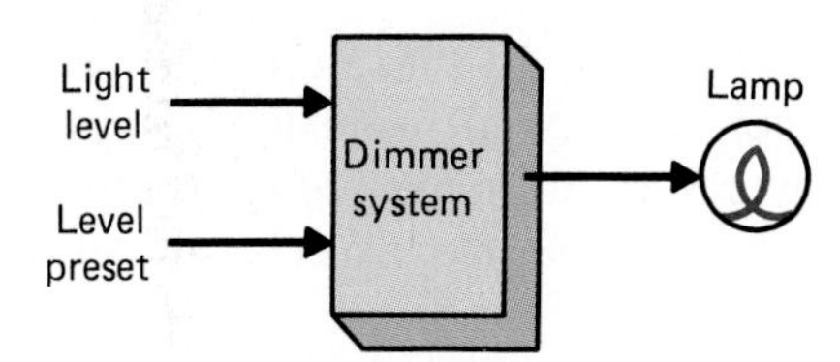

FIGURE 1.15 Simple Dimmer Control Function

Figure 1.16 shows the expanded block diagrams. In this expanded diagram, more detail is shown about how the process will be achieved. It was known that neither the level sensor nor the level preset device were able to control directly the energy flow to the light bulbs. Also, the problem of comparison of the two values, actual level and desired level setting, needed to be handled. Therefore, we see preamplification of the two input signals, a specific comparator to sense how they match, and a separate lamp controller.

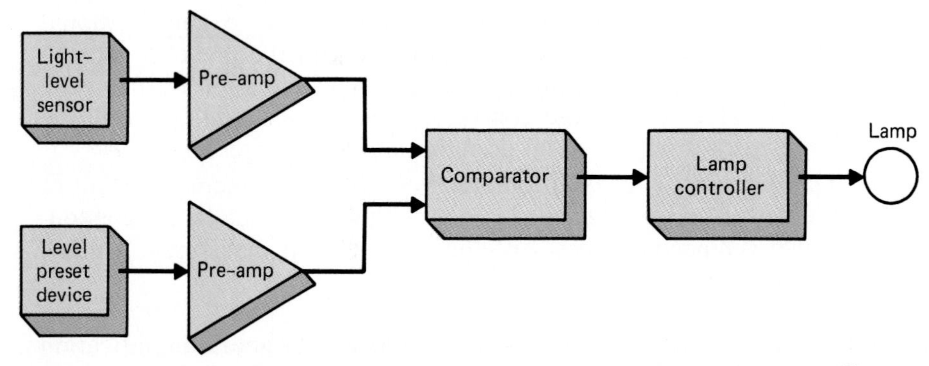

FIGURE 1.16 More Descriptive Dimmer Control Diagram (versus Figure 1.15)

The final block diagram that can lead to the development of the circuit is shown in Figure 1.17. Much more detail is now shown in this

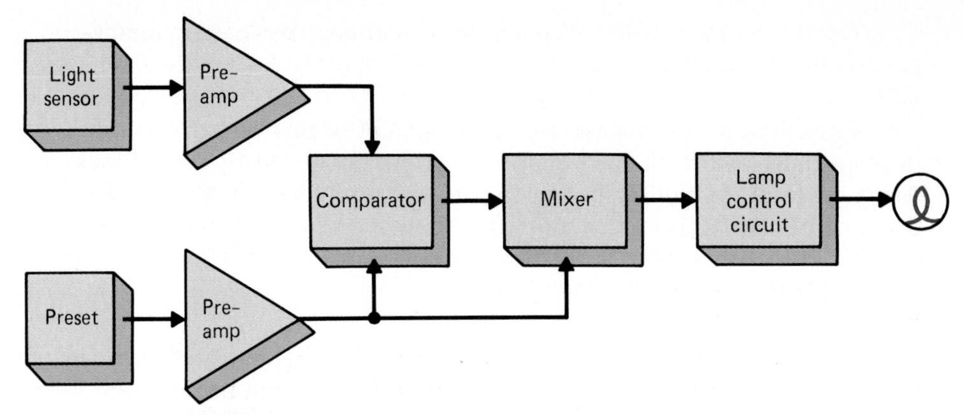

FIGURE 1.17 Complete Dimmer System Diagram

diagram. Here we see the steps necessary to control the lights. The following generally outlines each of the processes done in the blocks shown.

Light sensor/level preset. In these circuits, information about the desired light level and actual room light level is gathered. In the case of these sensors, they change resistance to reflect their settings. (A potentiometer could be used for the preset device, a photoresistive cell for the sensor.)

Preamplifiers. In the preamplifiers, the electrical values associated with the sensors' resistance changes (voltage drops or current flows) are boosted to levels usable by the circuits that follow.

Comparator. To determine if the actual light level matches the preset level, a circuit called a comparator is used. The output of this circuit represents the electrical difference between the two inputs. The output is either a positive level or a negative level, based on which input is higher than the other.

Mixer. The mixer is used to blend the value of the preset level with the output of the comparator. Here the comparator level will be either added to or subtracted from the preset value, depending on the comparator's output polarity (+ or −). (For example, should the room level be too bright compared with the preset level, the comparator would produce a value that would subtract from the preset value sent to the control circuit. Thus the lamp would be dimmed to reset the room brightness to the correct level.)

Lamp control circuit. In the lamp control circuit actual control of the power delivered to the lamp is achieved. Typically, such a controller is placed in series with the lamp and sets the average voltage supplied to the lamp.

(Note than an amplifier block was not used here since actual amplification was not done. Unlike the previous example, the output of this circuit does not reproduce the input but is simply controlled by it.)

Carefully review the development shown in each of these examples. As can be seen, the initial concept of these circuits starts as a basic plan of how the process should work overall. The steps that follow then refine the actual processes needed to achieve each step. The final diagrams would lead to the development of actual electronic circuits that achieve each task. For each block in the final diagram, a separate circuit would be produced. The circuits would then be connected together to produce the final results.

Some interesting facts can be noted as we compare these diagrams. None of the diagrams specify what individual components are to be used. In most cases, a variety of devices could be used to achieve the desired task. (Even the sensors noted in the dimmer circuit do not have to be resistance variable. This type was chosen for the sake of example only.) Also, even though each of the systems does a very different job, some of the circuits used in one may be nearly the same as circuits used in another. For example, the preamplifiers used in each may be identical circuits containing the same components. It is possible that the preamplifiers used in the public address amplifier could be used, without modification, in the dimmer circuit.

Block diagrams will be very useful when attempting to service an electronic system. By looking at these diagrams often the defect in the system can be isolated. For example, if the dimmer circuit failed to respond to room light but could be preset, we could almost be sure that the level preset control, its preamplifier, and the lamp control circuits were working correctly. We would then only have to concentrate on the other four circuits. In addition, we could speed up our testing simply by checking if the outputs of each block, as measured on the circuit, matched the listed values that would appear with the circuit's schematic. We would simply jump from block to block (coordinated with the circuit) until we found a section that failed to function. This section would then be targeted for careful tests and measurements.

REVIEW PROBLEMS

(1) Copy the final block diagram from Examples 1.4 and 1.5. On these copies indicate where another energy source would be an additional input. Use a heavy arrow (as shown in Figure 1.18) to indicate this additional input. Be prepared to discuss your reasons for adding this input and reasons for leaving it out of any block.

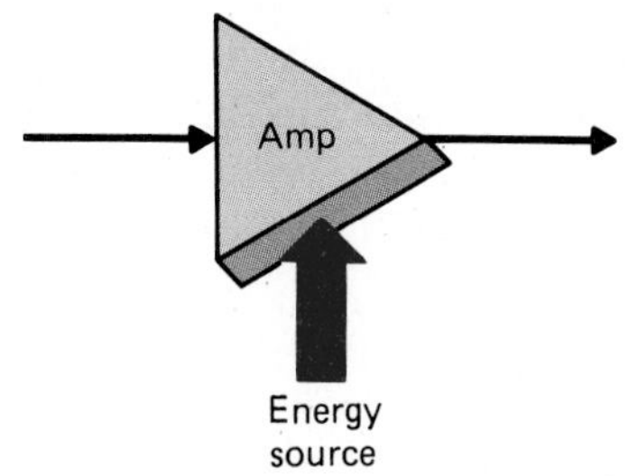

FIGURE 1.18 Introduction of Controlled Energy Source

(2) For Examples 1.4 and 1.5, write a brief description of the steps taken to achieve the final output. Be careful not to omit an important step, but also try to be brief in your writing. Use the final diagrams shown in each example as a guide.
(3) Select a process or operation in your daily life with which you are thoroughly familiar. Use a development process similar to the one shown in Examples 1.4 and 1.5 and prepare a block diagram for the process or operation. The process or operation you choose does not have to be electronic or even "scientific"; just select one with which you are familiar. (You could even diagram a daily routine or a how-to process.) Be sure to describe the process in a few sentences as a starting point.

SUMMARY

Electronic systems work much as other scientific systems do. Most electronic circuits (except digital circuits) deal with the manipulation of a major energy source's current and voltage delivery based on information supplied at the input.

Two categories of electronic elements are applied in this control process. Passive devices are used primarily to establish either fixed values of voltage and current within circuits or to set certain limitations on circuit values. Resistors, capacitors, and inductors are the typical passive elements in a circuit. Active devices complete the process, as they exercise control over the flow of energy based on information inputs. Transistors are examples of active devices.

To detail the methods in which energy will be manipulated by an electronic circuit, block diagrams are often used. These diagrams simply list the necessary steps to take to achieve a specific task. In most cases, for every block in the detailed diagram a corresponding electronic circuit appears in the system's schematic. The block diagram also indicates the general flow of information through the system.

Electronic circuits can be classified into one of five categories. An amplifier circuit will manipulate energy in coordination with the current and/or voltage values presented at the input to the amplifier. Control circuits also manipulate energy, but their output is usually not a duplication of the form of the input.

Power-conversion systems differ dramatically from either amplifier or control systems. In power conversion, the objective is to convert from one form of electrical energy to another. Unlike amplification and control circuits, power conversion circuits do not usually have an information input. Digital circuits deal only with the manipulation of numbers and logical decisions; they neither amplify, control, nor convert. Finally, some functions do not fit into a convenient category. These systems are simply defined by their exclusive function.

It has been shown that the functions of electronic systems can be illustrated by block diagrams. These diagrams are major stepping stones in the development of an electronic system. These diagrams are also used to understand more readily the function of the system they describe. Often, block diagrams are the key element used to begin troubleshooting a defective system.

Section 1.1

1.1. Draw a simple block diagram for the following description.

> The signal from two microphones enters a mixer. The single output of the mixer feeds into a preamplifier. This amplifier's output feeds to a power amplifier, then to a single speaker.

1.2. Draw a simple block diagram for the following description.

> An electric motor drives a gearbox which reduces its speed. The output of the gear box are two turning shafts. Shaft 1 feeds a fan blade that cools the area. Shaft 2 goes to a pulley system that runs a power saw. There is an on/off clutch between shaft 2 and the pulley system.

1.3. (Be inventive.) Make up your own type of system: mechanical, electrical, or any other. It should have at least four major functional features. First describe it in a short paragraph and then use block diagrams.

Section 1.2

1.4. Rearrange the block diagram shown in Figure 1.19 to make it easier to read and properly fit the basic procedures noted in Section 1.2.

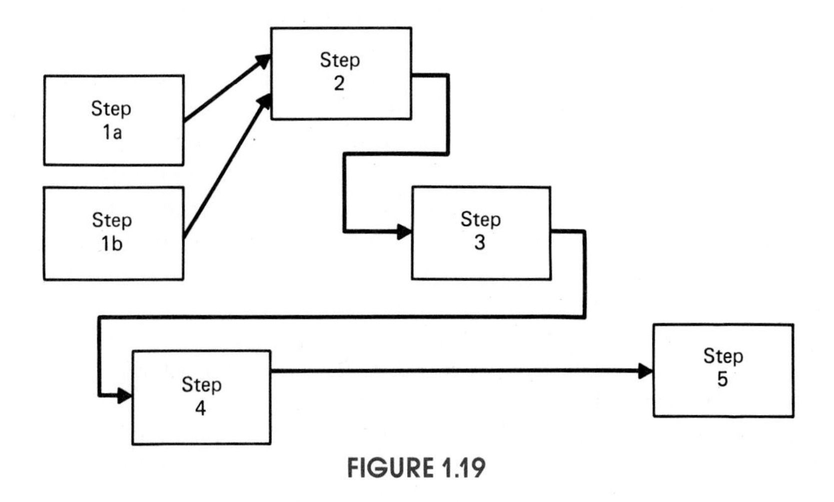

FIGURE 1.19

Section 1.3

1.5. Look ahead in later chapters and find one block diagram and its coordinated description in the text. Copy both the diagram and the description. Make any additional comments about what was shown that you feel are necessary.

1.6. Repeat problem 1.5 for a different diagram or one from another book.

Section 1.4

1.7. Locate, in a later chapter, a block diagram and electronic schematic which coordinate. Indicate which is easiest to comprehend and why. Indicate which illustration was used, and describe briefly which sections of the block diagram you feel you can (at least partially) comprehend. Which sections of the schematic can you understand?

1.8. Often when describing a system it is not necessary to specify exactly how a task (such as amplification) is to be done by indicating the exact parts to be used. It is usually only necessary to describe the function to be done (amplification). In the Index look up the number of references to *Amplifiers* or *Amplification* and list these as possible circuitry to use to achieve a single function.

Section 1.5

1.9. List the five basic categories of electronic circuits as presented in this chapter.

1.10. Write a brief description of the five categories of electronic circuits as noted in problem 1.9.

Section 1.6

1.11. Prepare a list of active and passive devices and a brief statement which indicates the difference between these two categories.

1.12. Describe why a resistor cannot be used as an active device. Include how it handles voltage and current in contrast to how an active device handles the same quantities.

1.13. If a new electronic device were invented, what features would you look for to determine if it was an active or passive device.

1.14. The following is a description of two electronic products. Determine if they are classified as active or passive, and why.

Product X. Product X is designed primarily to control the flow of ac current. It will produce an on/off control of that current flow based on the input of a triggering signal produced by a second circuit.

Product Y. Product Y is designed to limit the amount of either dc or ac current flow in a circuit. The amount of current permitted to flow is based directly on the level of voltage supplied to the device. Greater voltage produces greater current, and vice versa.

1.15. In a later chapter locate an electronic circuit schematic which contains at least eight individual elements. Copy this schematic and then indicate which elements are active or passive. (*Note:* Although you may be unfamiliar with the symbols used for active devices, you should be able to identify those which are passive. By a process of elimination, those that remain should be the active elements.)

SUGGESTED REPORT TOPICS

The following topics may be used as subjects for brief technical reports.

1.1. Write a brief report that discusses the control of energy in a system with which you are familiar. Be sure to point out those items which exercise active and passive control over certain quantities in the system. (*Possible topics:* a steam engine, a crane, an elevator, etc.)

1.2. In your own words, write a brief set of instructions that detail how to describe an operation using block diagrams. Be sure to point out the need to include the input elements, each functional step, and any special blocks that may be needed.

1.3. Using reference books or articles found in a library, prepare a general list of the proper steps to take when drawing block diagrams. Electronic drafting textbooks may be of assistance.

1.4. Prepare a short report that discusses, in your own words, the five categories of electronic systems discussed in this chapter. With each category present a list of at least five types of systems that could fit into these categories.

ANSWERS TO REVIEW PROBLEMS

Section 1.1

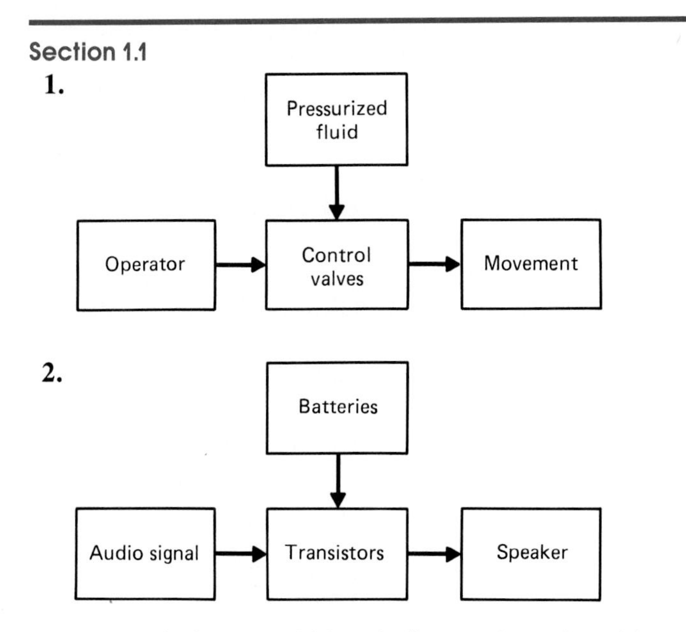

3. The solution should be similar to that of problems 1 and 2.

Section 1.2

1. Block diagrams illustrate the process by which a specific function is to be achieved.
2. b
3. True

Section 1.3

1.

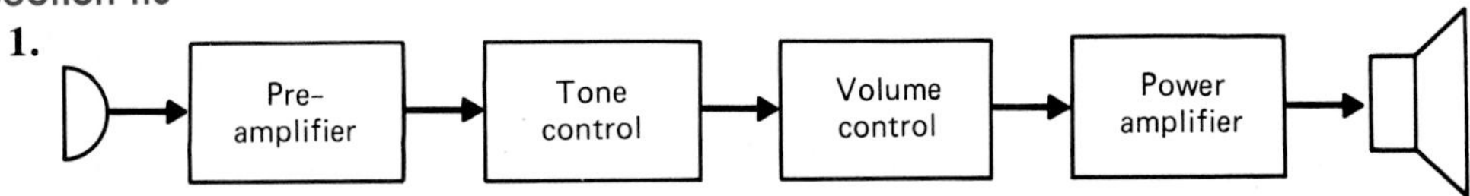

2. The alarm system is triggered by a door sensor to ring an alarm bell and a police alarm. The system is activated by an "on" time preset and the time of day.
3. The description and diagram should clearly match and both should be easily readable.

Section 1.4

1. a
2. b
3. False
4. False

Section 1.5

1. See Section 1.5 for details.
2. (a) Digital logic
   (b) Amplification
   (c) Electronic control
   (d) Electronic control
3. Solution is based on the electronic circuits selected.

Section 1.6

1. True
2. c
3. See "Summary of Active and Conditioning Devices" for details.

Section 1.7

1. For Example 1.4, add external energy source to preamp 1, preamp 2, and power amplifier; for Example 1.5, add external energy source to preamps, comparator, and lamp control.
2. See Section 1.7 for a description and diagrams.
3. Solution is based on the process or operation selected.

2 Expansion of the Block Diagram Process

Objectives

Upon completing this chapter, the reader should be able to:

- Identify the function of multiple-stage amplifiers.
- Recognize the use of special functions to produce special effects.
- Describe the process of feedback.
- Utilize proper block diagram methods.
- Follow a basic system's development.

Introduction

In this chapter we expand on the concept of block diagram processes developed in Chapter 1. The need for and function of multiple-stage amplifiers is explored. It is also shown that a basic amplifier's function can be modified by the addition of special function "external" components. Feedback is defined and shown to be a method of applying control to the amplifier's overall function.

Added details about the development of block diagrams are presented, as well as basic guidelines for preparing these diagrams. Finally, the process by which a system is developed, using block diagrams to assist in describing the system's function, is illustrated. The development begins with simple block diagrams and results in a final electronic circuit. Knowledge of the manipulation of amplifiers and the process of system development is a stepping-stone to understanding the role played by active devices in electronic circuitry. ■

2.1 THE CONCEPT OF AMPLIFICATION IN MULTIPLE STAGES

In Chapter 1 the general idea of amplification in multiple steps was introduced. It was noted that it was not possible to produce the gain of an entire amplifier system in a single circuit. It is often necessary to deal with input values that are in the microwatt range and from these produce outputs in the tens or hundreds of watts. It will be necessary to use a series of amplifiers, each contributing some amount of amplification to achieve the complete output of such a system.

It will be necessary to describe the requirements of an amplifier system by a simple block diagram, an expanded diagram, and then a mathematical formula that describes this system. Figure 2.1 shows a simple diagram of an amplifier. In this figure an input of 500 μW is supplied to the amplifier. The amplifier is then required to produce 5 W of output power. (Such an amplifier may be used in a portable cassette recorder.)

To describe the requirements placed on this amplifier, we will need to use the following equation, which shows just how much amplification, called **gain**, is provided by this amplifier:

$$\text{gain} = \frac{\text{output}}{\text{input}} \qquad \textbf{(2.1)}$$

This equation is in a general form. If the gain is to be rated as power gain, the input and output values must be in terms of power. The same is true if voltage gain or current gain are to be calculated; input and output units must match.

EXAMPLE 2.1

For the amplifier system shown in Figure 2.1, the gain needed would be:

$$\begin{aligned}\text{power gain} &= \frac{5\ \text{W}}{500\ \mu\text{W}} \\ &= 10{,}000\end{aligned}$$

What this means is that the amplifier must be able to produce an output power level that is 10,000 times as great as the input power. Remember, though, that the amplifier does not just create this output power; it controls the power from another energy source to produce this power. As you will learn in future chapters, to be able to provide this much control in one single electronic circuit is nearly impossible. Naturally, such control is ultimately possible, but it must be broken down into a series of steps. In Figure 2.2 the

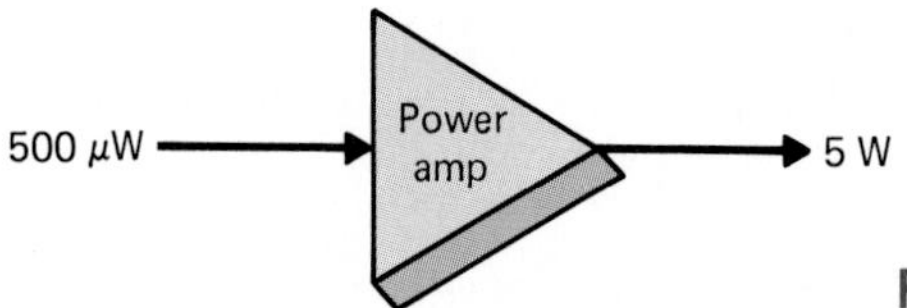

FIGURE 2.1 Simple Amplifier Diagram

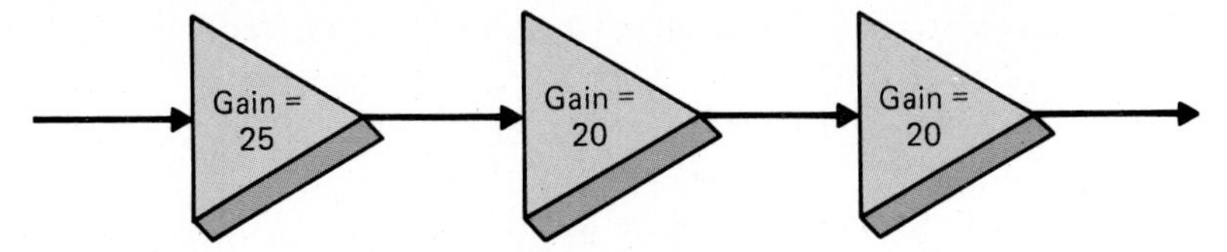

FIGURE 2.2 Multiple-Stage Amplifier Gains

system has been expanded into several steps, called **stages**, which each contribute to the final necessary gain of the system, as shown in Figure 2.1.

In Figure 2.2 we have taken the required system and broken it down into three amplifiers, called "stages." Each stage feeds its output to the next stage until the final output is achieved. It may seem unusual that each stage seems to have a rather low gain. Note that the first stage has a gain of only 25 and the next stages only 20. These seem to be a far cry from the desired gain of 10,000 for the system. Here, though, is where the real effect of using several stages together, called **multiple stages**, is seen. The following equation shows how we can go from gains of 20 or 25 to an overall gain of 10,000.

$$\text{system gain} = \text{gain 1} \times \text{gain 2} \times \text{gain 3} \times \cdots \qquad \textbf{(2.2)}$$

Equation 2.2 illustrates that gain from stage to stage of a multiple-stage system is the *product* of the respective gains—not the simple addition of gains as may first be suspected. For the system we have been using, Example 2.2 shows the math involved in producing this system gain.

EXAMPLE 2.2

For the system shown in Figure 2.2, the overall gain will be

$$\begin{aligned} \text{system gain} &= 25 \times 20 \times 20 \\ &= 10{,}000 \end{aligned}$$

In Example 2.2 we see that the overall gain is 10,000, as proposed in the beginning. But the system is broken down into three steps or stages instead of one massive stage. The gain of 10,000 is the same but is achieved in more realistic steps. Just to verify that this process is the same, remember that the input was to be 500 μW and the output was to be 5 W. We saw that this produced the demand of a gain of 10,000 from the system. Now let's use the following equation, which is a rearrangement of equation 2.1, to prove step by step that the input will produce the desired output:

$$\text{output} = \text{input} \times \text{gain} \qquad \textbf{(2.3)}$$

EXAMPLE 2.3

In this example we track the actual amount of gain through the system shown in Figure 2.2, using equation 2.3.

$$\text{Output for stage 1: } 500\ \mu\text{W} \times 25 = 12{,}500\ \mu\text{W} = 12.5\ \text{mW}$$

We then use the output of the first stage as the input of the second:

Output for stage 2: 12.5 mW $\times$ 20 = 250 mW

Finally, we use this output as the input to the last stage:

Output for stage 3: 250 mW $\times$ 20 = 5000 mW = 5 W

Thus we see that we can start with 500 μW and produce 5 W as an output from this system, exactly as was done in the simple system shown in Example 2.1.

We have thus far developed several equations that help to describe the overall function of a multistage system. We see that the overall gain of the system is the product of each individual gain value. Let's explore the use of these equations in the following examples.

EXAMPLE 2.4

Figure 2.3 illustrates the use of a power booster for a car stereo system. Such a system is used to increase the output power of a car stereo system from a typical 5-W output to 100 W, which would be used to run a higher-power speaker system.

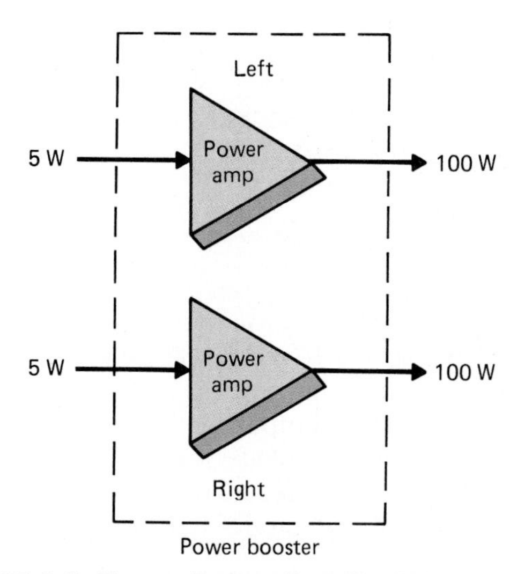

FIGURE 2.3 Stereo Power Booster Block Diagram

First, note that amplifiers are detailed for both the left and right radio output channels and for the booster output. In such a system, the left and right signals (music) are handled through two separate but identical amplifiers.

Now, using equation 2.1 we can see the actual gain requirement of the booster:

$$\text{booster gain} = \frac{\text{power out}}{\text{power in}}$$

$$= \frac{100\text{ W}}{5\text{ W}}$$

$$= 20$$

Thus the booster must produce a gain of 20. Naturally, this must be for each channel (left and right). Remember that the actual source of the 100 W of electrical power for each channel was the car's electrical system, not the radio.

EXAMPLE 2.5

Figure 2.4 illustrates a simple public address amplifier system. For the system we calculate the power gain and then trace a simple signal through each stage. For now we ignore any effect the volume control and tone control have on the value of the signals that pass through the system.

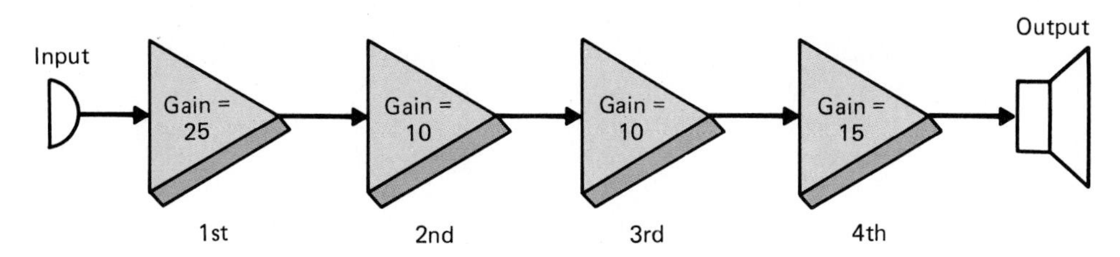

FIGURE 2.4 Four-Stage Amplifier

First, let's calculate the power gain of the system. For this we use the gains seen in each amplifier. From equation 2.2,

$$\text{system gain} = \text{gain 1} \times \text{gain 2} \times \text{gain 3} \times \text{gain 4}$$

$$= 25 \times 10 \times 10 \times 15$$

$$= 37{,}500$$

Now, let's assume an input power of 5 mW and see what the actual power output will be:

$$\text{output power} = \text{input power} \times \text{system gain}$$

$$= 5\text{ mW} \times 37{,}500$$

$$= 187.5\text{ W}$$

Assuming the same input, let's track this gain from stage to stage. We will assume the output level of one stage is the input level of the next stage:

Output power for stage 1: 5 mW input $\times$ gain 1 $=$ 5 mW $\times$ 25 $=$ 125 mW

Output power for stage 2: 125 mW $\times$ 10 $=$ 1.25 W

Output power for stage 3: 1.25 W $\times$ 10 = 12.5 W
Output power for stage 4: 12.5 W $\times$ 15 = 187.5 W

Thus we can see that the final output power of the system is 187.5 W by using either total gain or individual gains.

In this section we have explored the concept of multiple stages of amplification. It was noted that a system that has a large gain may have to be composed of several amplifiers tied together. These stages produce the overall gain of the amplifier desired. The gain of the system is calculated either by observing the ratio of output to input or by multiplying the gain for each stage. This information will enable us to observe typical electronic systems which employ multiple stages to achieve an overall gain. (Again, the reader is cautioned to note that the actual energy used to produce the gain comes from another source, a battery or power supply.)

REVIEW PROBLEMS

(1) If an amplifier has an output of 20 W and an input of 10 mW, what is the power gain of the system?
(2) For the amplifier system shown in Figure 2.5, calculate the total gain for the system based on each stages gain.

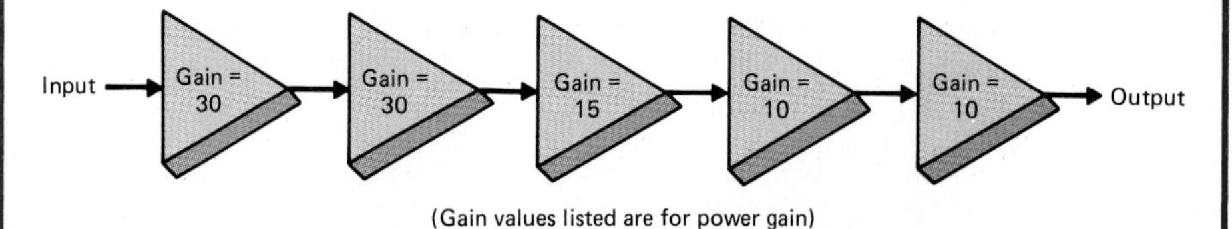

FIGURE 2.5 Five-Stage Amplifier

(3) Using the amplifier shown in Figure 2.5, assume that an input of 40 μW is applied. First calculate the output of the system based on the system gain calculated in problem 2, then calculate each stage's output, finally resulting in a system output. Compare these two answers.

2.2 A COMMON ADDITION TO AN AMPLIFIER CIRCUIT: THE VOLUME CONTROL

So far we have looked at simple amplifier circuits. In these circuits we have seen the need for multiple stages in order to achieve one overall gain value. Not all systems use just a series of amplifiers simply to boost a single input to a usable level. Although the use of amplifiers is seen in quite a few circuits, simple amplifiers are not able to achieve all the tasks needed in electronic circuits. We will find that with modifications, the general amplifier can be used to achieve quite a variety of more advanced tasks. Blending these special circuits with other basic amplifiers will allow for the development of more advanced systems.

First, let's review what the amplifier does. An amplifier is able to accept a (usually) low energy input in the form of voltage and/or current. It uses these variations to control the metering of electrical energy available from another energy source. The output then is a high-energy duplication of the input. Keeping this in mind, the challenge will be to investigate using other special circuitry in cooperation with an amplifier to produce not only a gain but also some electronic "special effect."

One item that was not discussed in Section 2.1 was the use of volume controls. Investigating the application of volume controls will start us on the road toward our study of the more sophisticated amplifier systems. Most amplifier circuits use some form of volume control to set the level of the output signal. There are several ways in which volume control could be achieved. As shown in Figure 2.6, the output level could simply be divided across a variable resistor prior to reaching the appropriate output device. If this were done, there would be a simple voltage-divider circuit established by the potentiometer. By adjusting the potentiometer, more or less voltage (and current) would be delivered to the output device; thus the power delivered to this device would be increased or decreased. Such a system would be highly inefficient. All of the power that was not used by the output device but was available from the amplifier may simply be wasted in heating the variable resistor. This would be similar to using the brakes on a car to control its speed on the road while allowing the engine to run at full speed.

If we use a circuit like the one shown in Figure 2.7, where the volume control (still a variable resistor) is now in control of the input to an amplifier, greater efficiency is achieved. This arrangement, where the control of volume is seen as a part of the amplifying process, wastes far less energy than the

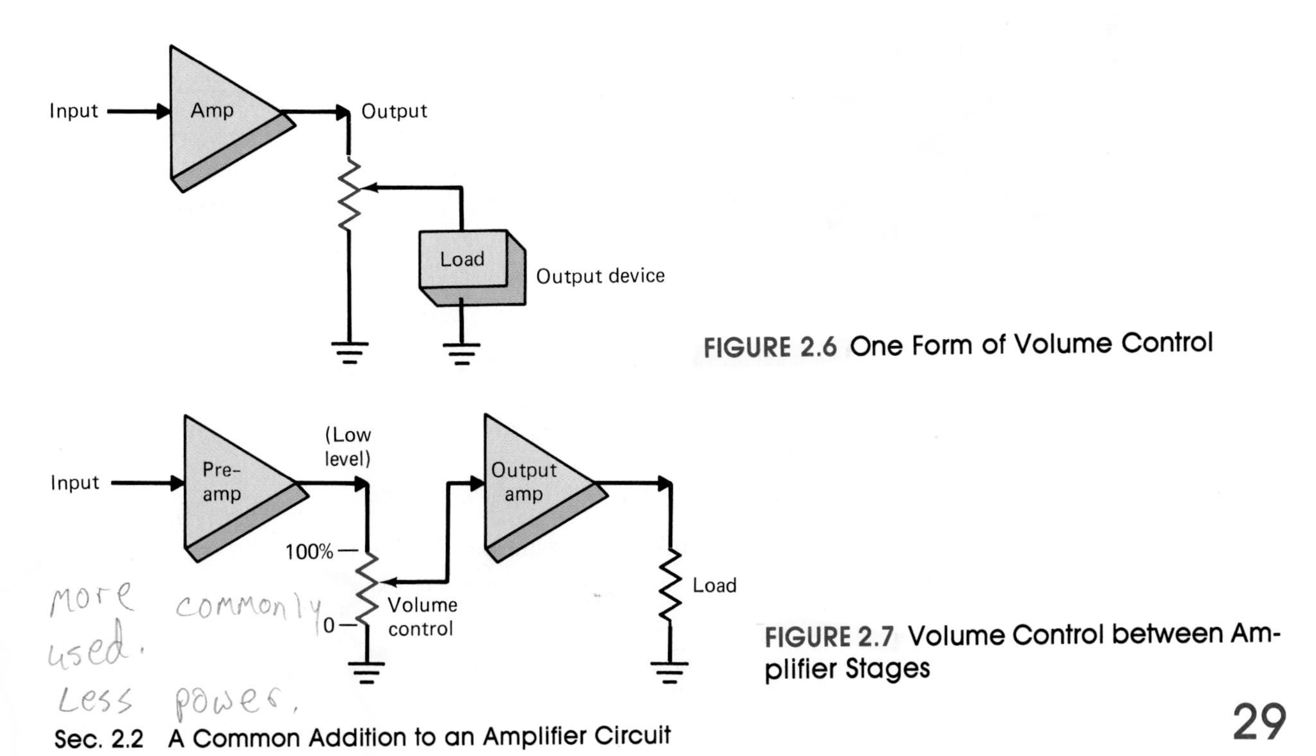

FIGURE 2.6 One Form of Volume Control

FIGURE 2.7 Volume Control between Amplifier Stages

previous circuit would. Here there still is a voltage-divider circuit but the actual energy losses are quite small compared to the losses seen when applying control at the final output. The principle is about the same, reducing the voltage and current delivered to the following element, but in this case the amount of power lost in the control process is far less than when control is done at the output.

An interesting effect has occurred. The function of the volume control, its reduction in these low power levels, has been magnified by the next amplifier in line. The following example should clarify this effect by illustrating it with actual numerical values.

EXAMPLE 2.6

For the diagram shown in Figure 2.8, we calculate the output level given that the control potentiometer is set to deliver (a) 90%, (b) 50%, and (c) 10% of the power available from amplifier 1 to amplifier 2.

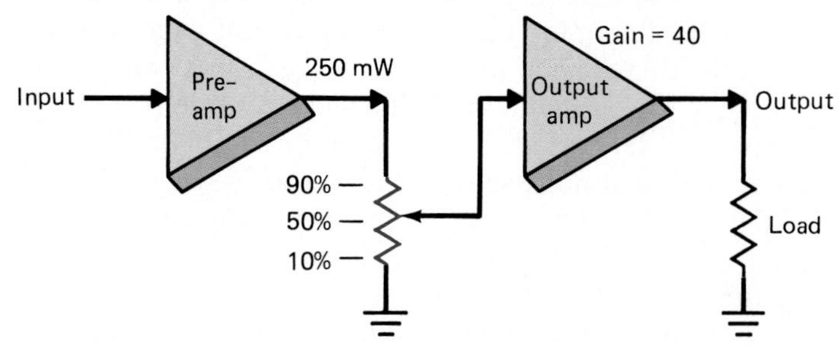

FIGURE 2.8 Volume Control with Functional Values

(a) When the control is set to deliver 90% of the power from amplifier 1 to amplifier 2, we see:

$$250 \text{ mW} \times 90\% = 225 \text{ mW}$$

$$225 \text{ mW} \times 40 \text{ (gain of amplifier 2)} = 9 \text{ W (output)}$$

The power lost in controlling the output is 250 mW − 225 mW = 25 mW.

(b) When the control is set to deliver 50% of the power from amplifier 1 to amplifier 2, we see:

$$250 \text{ mW} \times 50\% = 125 \text{ mW}$$

$$125 \text{ mW} \times 40 \text{ (gain)} = 5 \text{ W (output)}$$

The power lost in controlling the output is 250 mW − 125 mW = 125 mW.

(c) When the control is set to deliver 10% of the power from amplifier 1 to amplifier 2, we see:

$$250 \text{ mW} \times 10\% = 25 \text{ mW}$$

$$25 \text{ mW} \times 40 \text{ (gain)} = 1 \text{ W (output)}$$

The power lost in controlling the output is 250 mW − 25 mW = 225 mW.

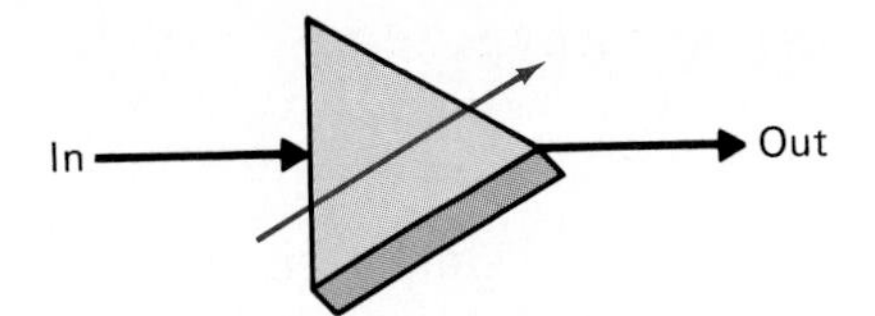

FIGURE 2.9 Variable Gain Amplifier

From Example 2.6 it can be seen that control of the final output is determined by controlling the input to the second amplifier. Notice especially that the amount of power lost in the control process is less than $\frac{1}{4}$ W in every instance. Were the control device placed at the output of amplifier 2 as proposed in Figure 2.6, the potentiometer may have to dissipate up to 10 W of power to exercise the same type of control.

The interesting thing to note about this circuit is that the effect of the potentiometer has been virtually amplified along with the actual information. In other words, the potentiometer appears to be controlling up to 10 W when it is really exercising control only over, at best, $\frac{1}{4}$ W. By coupling this special effect, volume control, with the process of amplification, two processes are developed in one simple step.

There is one other popular means by which volume control is achieved. If the control device is properly placed within the actual circuitry for the amplifier, we can exercise control over the gain for that amplifier. Were we to use this process in amplifier 2 of the circuit in Example 2.6, the actual gain of that amplifier would be adjustable. The gain could be changed from the value of 40 as listed, down to as low as nearly zero (0). Thus the output could still be varied, but the input values to the amplifier would remain the same. In block diagram form, this type of amplifier could be illustrated as shown in Figure 2.9.

In later chapters we explore how to construct variable-gain amplifiers. This form of volume control often uses even less power from the system than do the other methods. Such a circuit requires an understanding of the internal workings of an amplifier. The general application of such a circuit is easy to understand. If the gain of the amplifier can be made adjustable, then as changes are made to the gain it follows that the output levels also change.

REVIEW PROBLEMS

(1) List the two preferred methods of volume control discussed in this section. Briefly, in your own words, write a simple description of each type of control.

(2) Repeat the calculations in Example 2.6 but place the potentiometer control circuit at the input to amplifier 1. Here the same percentage of control is exercised but it is done at the immediate input to the circuit, not partway between. Compare your findings to those shown in Example 2.6.

(3) Return to the circuit of Example 2.6 but instead of using the potentiometer control circuit, redraw it with a variable-gain amplifier for amplifier 2. Repeat the output calculations for gain values for this amplifier of 4, 20, and 36 (the same percentage of change as used in the example). Again compare your results to those found in Example 2.6.

2.3 OTHER CIRCUITRY CONNECTED TO AMPLIFIER STAGES

In Section 2.2 the application of a volume control in a simple circuit was explored. There we found that the volume control was able to cause the same effect when placed between amplifier stages as it would do directly connected to the output. Now we explore connecting other passive circuit elements between stages and see what effect they have on the circuit's overall operation.

Another common application of passive elements used in conjunction with amplifier stages is seen when a filter circuit* is added to a typical amplifier system. First, let's review what a filter circuit is. A **filter circuit** is a combination of resistors, capacitors, and/or inductors. The purpose of the filter circuit is to allow certain signals (information) at specific frequencies to be passed through the filter while other undesirable signals are eliminated by the filter. Therefore, a filter designed to pass signals of about 10,000 Hz would allow these signals through while eliminating signals at other frequencies (Figure 2.10).

Now, if this filter were added between the stages of a multiple-stage amplifier, the effect of the filter would act on the information passing through the system. In Figure 2.11 such a combination has been made. In this illustration the filter is added between amplifier stages 2 and 3. In this position it has no effect on the information entering the system, only on the information passed from amplifier 2 to amplifier 3. Much like the effect of the volume control seen in Section 2.2, this filter affects the information flow from stage

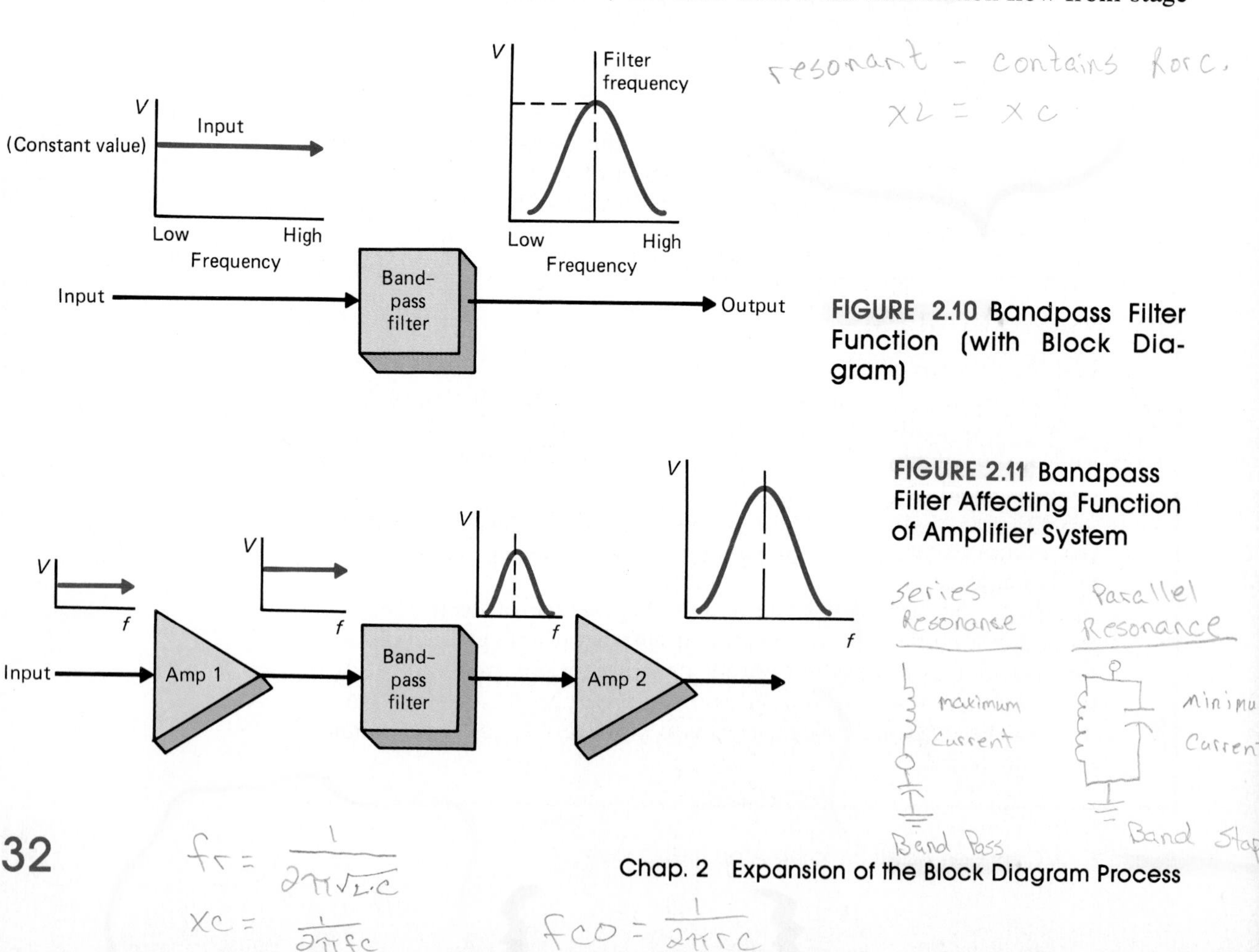

FIGURE 2.10 Bandpass Filter Function (with Block Diagram)

FIGURE 2.11 Bandpass Filter Affecting Function of Amplifier System

2 to stage 3 and on the final delivery of energy to the load. Only signals that will pass through this filter will be passed on to the following sections of the amplifier. Now, in effect, we have created a "filtering amplifier" or an amplifier that will respond (give an output) only when the correct input frequency is applied; other frequencies are eliminated by the filter.

Such a circuit can be quite useful. If the filter is designed only to pass frequencies above some relatively low frequency (say, 20 Hz)—a high-pass filter—then signals below this will not be passed on from stage 2 to stage 3. Such an application is seen in many amplifier circuits. Since each amplifier uses a dc source for energy it is typical for some of this dc voltage to be available at the output of an amplifier. It would be a waste of energy for a following circuit to try to amplify this level, since it is not a part of any information. Therefore, as shown in Figure 2.12, a simple filter is often used when joining one amplifier with another. The filter is selected to block dc levels while passing most of the ac information signal on to the next stage of amplification.

Another application of this form of filtering may be seen in the simple bass and treble controls used in a portable radio. Bass (low frequency) and treble (high frequency) signals may be reduced by using simple filters as shown in Figure 2.13. In this circuit, both the bass and treble levels are controlled by the filters shown. These filters are adjustable; thus the amount of bass or treble passed on to the next stage is varied by the use of simple potentiometers. In both this and the previous illustration, note that the effect of the filter was "amplified" by the system.

Another typical use of such a filter is in a radio. If, as shown in Figure 2.14, a filter is used at the input of an amplifier circuit, and if that filter is able to pass a single frequency (a bandpass filter) on to the first amplifier, the system will be able to respond to that frequency and only that frequency. If the filter is adjustable, the actual frequency to which it is sensitive can be changed. The circuit will follow the setting of the filter and will reproduce only the information related to the chosen frequency. An actual radio circuit requires several more specialized circuits than are shown in this simple diagram, but the effect of tuning the radio to a specific frequency, thus selecting a specific station, follows this basic principle. In this application we see an amplifier circuit that will amplify signals of only one specific frequency, the one selected by the filter circuit.

In this section we have expanded on the idea of adding control or "special effect" circuits to amplifier circuits. We have seen the use of filters in connection with amplifiers. In this application, the filter's effect on the information signals is carried through the circuit and is produced as an overall effect on the final output. Actually, any circuit or other device that affects the information signal will affect the final output. We see that not only is the original information input to the system amplified, but along with it is an "amplification" of any effects that cause a change in this signal.

FIGURE 2.12 DC Blocking between Amplifier Stages

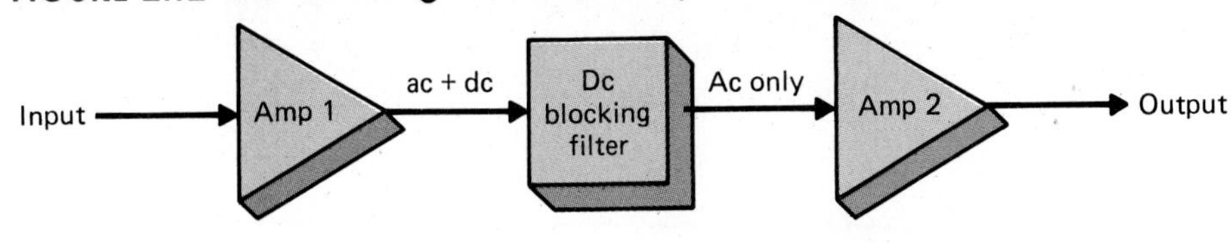

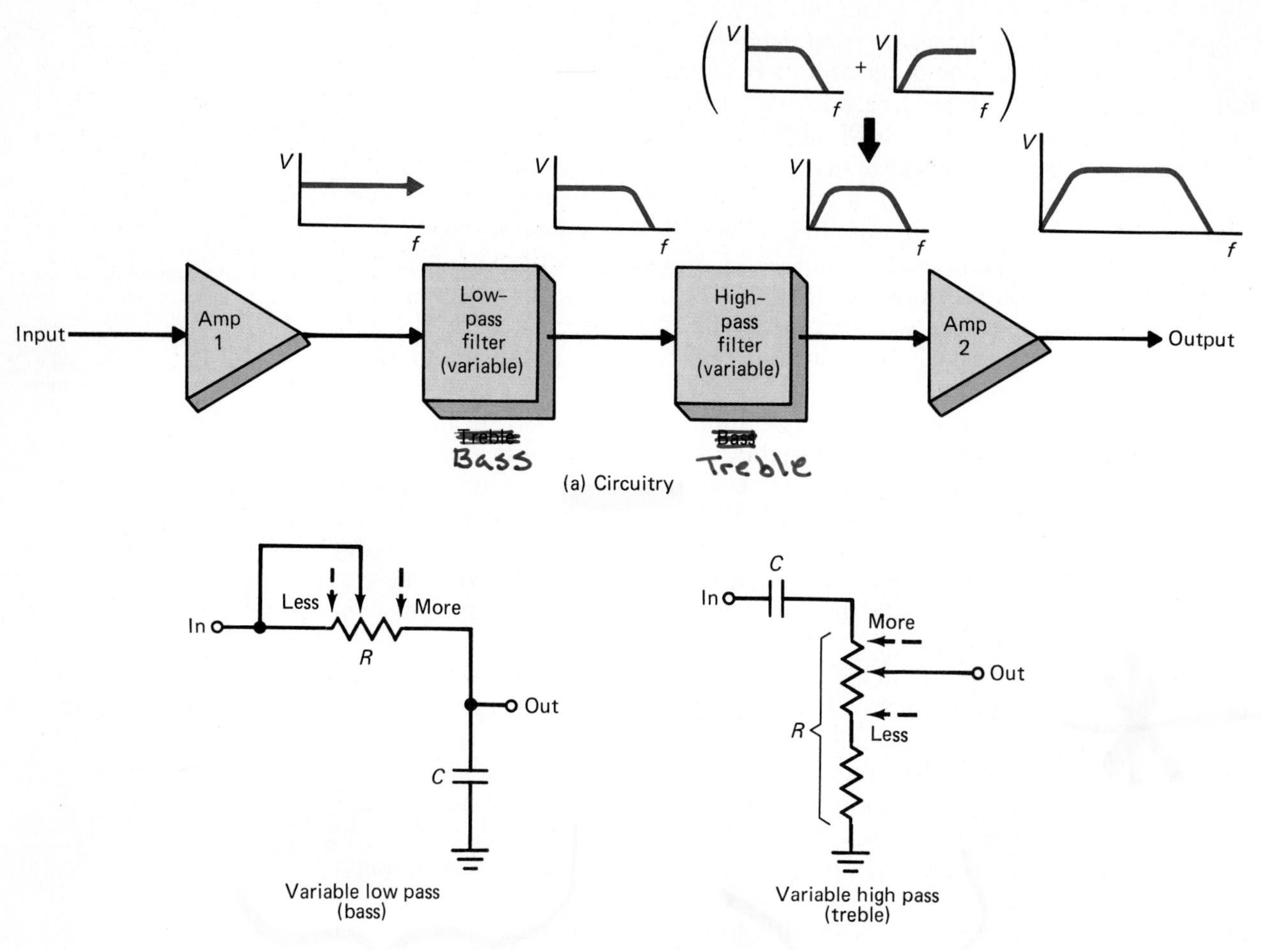

(a) Circuitry

(b) Composite of variable filters

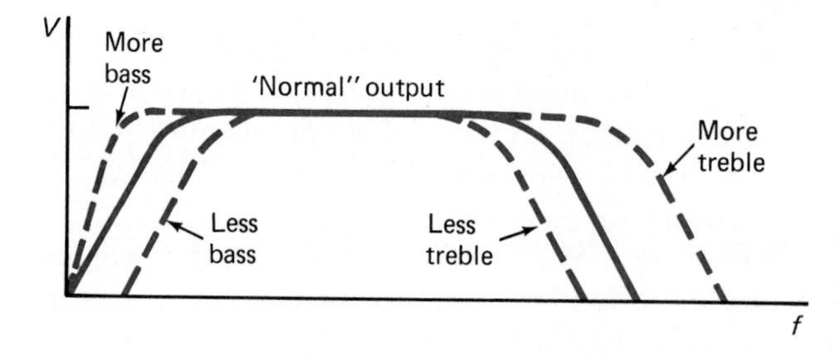

(c) Effect on frequency response

FIGURE 2.13 Bass and Treble Control

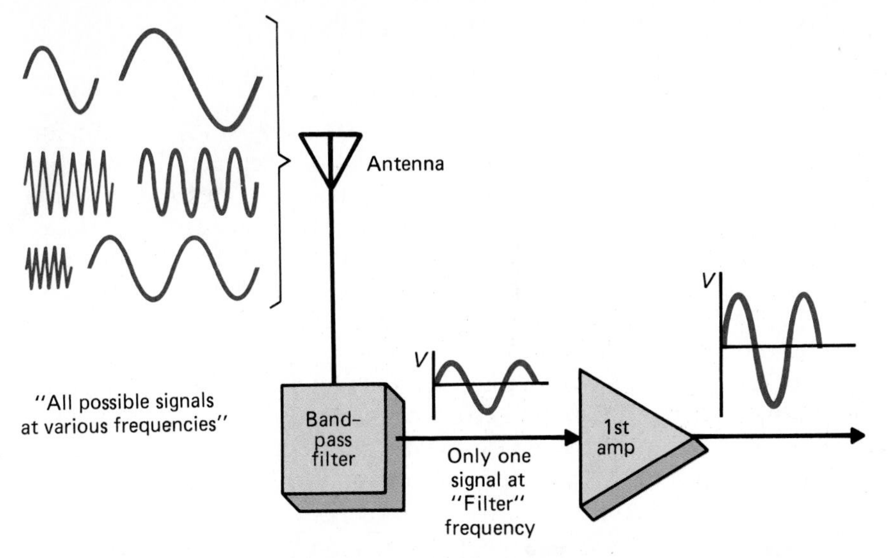

FIGURE 2.14 Bandpass Filtering of Radio Signals

REVIEW PROBLEMS

(1) In the diagram shown in Figure 2.15, a filter is shown attached to a multistage amplifier. Discuss the effect of this filter on the final output if the filter is a bandpass type which is tuned to 100 kHz (only allows signals at 100 kHz to pass through it). Discuss the effect the circuit would have on signals of 10, 100, and 200 kHz, both with and without the filter in place.

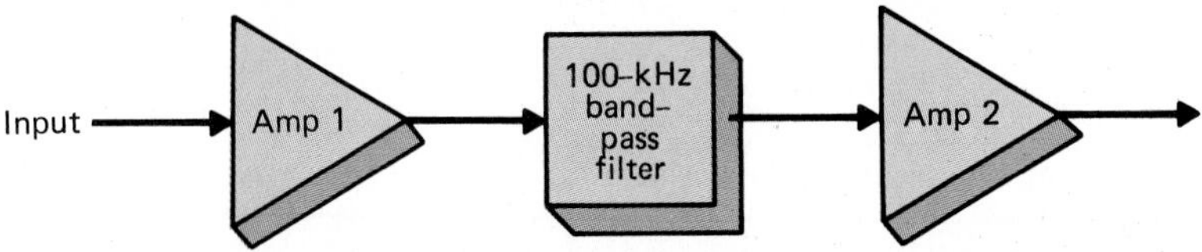

FIGURE 2.15 Bandpass Filter Used Between Amplifier Stages

(2) Refer to later chapters to find further applications of filter circuits. Illustrate the block diagrams that describes two such applications.

(3) In the diagram in Figure 2.16 two bandpass filters are used in a multistage amplifier. One filter is tuned to 1000 Hz and the other is tuned to 20 kHz. Using sample inputs of 500 Hz, 1000 Hz, 10 kHz, 20 kHz, and 50 kHz, discuss what the output will look like from this configuration. Assume that each filter passes *only* the bandpass frequency to which it is set. (The results may be surprising.)

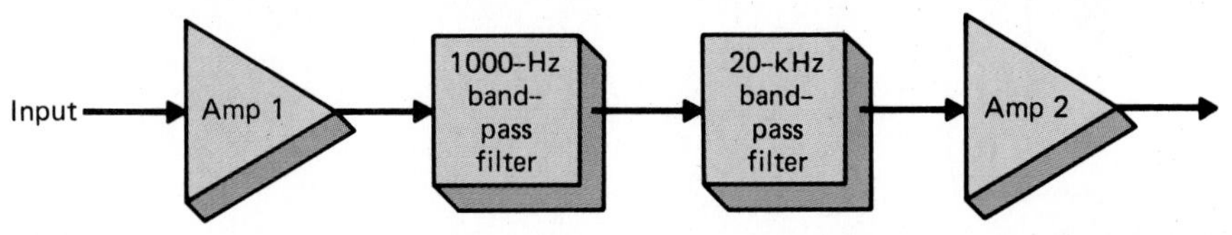

FIGURE 2.16 Two Bandpass Filters Used between Stages (Multiple Effects)

2.4 GENERAL INTRODUCTION TO FEEDBACK

So far in this chapter we have looked at multiple-stage amplifiers and have explored ways to affect the operation of these systems by studying their forward (input to output) flow of energy. The output of one amplifier was tied to the input of the next amplifier. As the signal passed through the system it appeared to increase in magnitude as it went from stage to stage. This process, where output feeds input from stage to stage, is sometimes called **feedforward**. There are occasions, though, where it is desirable to sample output values and return them to input locations. This process, where output samples are returned to inputs, is called **feedback**.

In Figure 2.17 the two most common forms of feedback are shown: single stage and multiple stage. The overall effect of the feedback process is nearly the same whether used in one stage or spanning several stages. The general discussion of feedback in this chapter will be applicable to either single- or multiple-stage amplifier forms.

First, we detail what generally happens in a system that uses feedback. Using Figure 2.18, let's track the flow of information through the system. The forward flow of information at first glance appears to be just like that in a

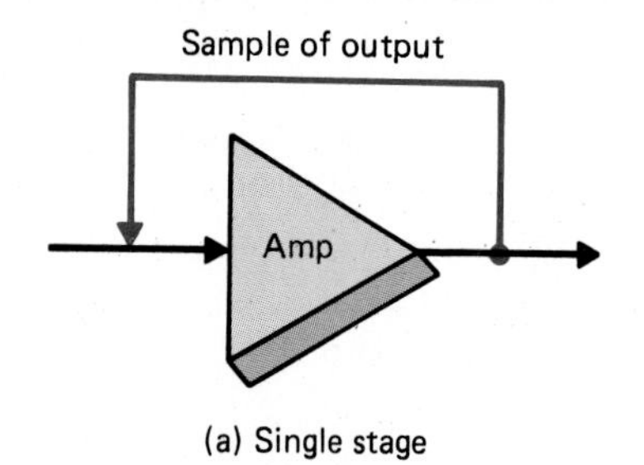

(a) Single stage

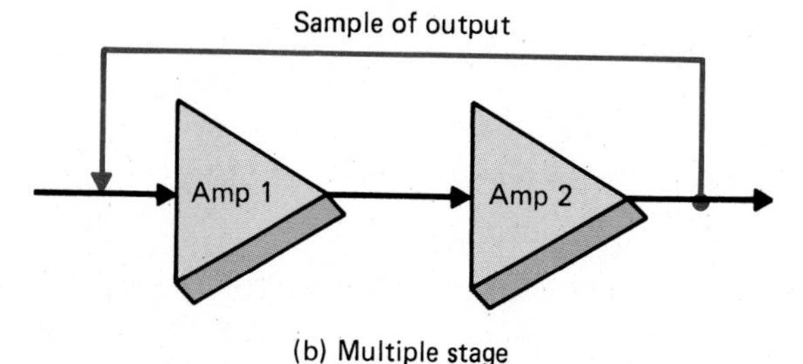

(b) Multiple stage

FIGURE 2.17 Feedback

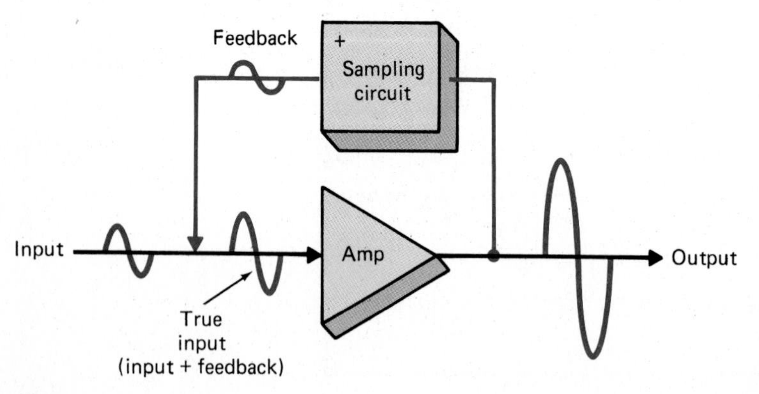

FIGURE 2.18 Process of Positive Feedback

standard amplifier. An input should create a higher-level output. Initially, this is true. But at the instant an output begins to be produced, a sample of this output is returned to the input terminal and is blended with the original input signal. Let's now think of what may happen.

If this feedback signal is simply added to the original input signal, the actual value of the input is mathematically increased. The true input to the system is then:

$$\text{Positive-feedback equation: true input} = \text{original input} + \text{feedback} \quad \textbf{(2.4)}$$

Therefore, the value of input to the amplifier increases. If the input increases, the output would be expected to increase as well. Now—follow this closely—if output increases, feedback samples will increase, too. With the feedback increasing the **true input** will also increase, and the output will go through another increase. This process can continue, everything going up, until the amplifier is destroyed by massive flows of energy. This does not seem like a wise application. Later in this chapter we see how this form of feedback can be used without destroying the amplifier.

This previous form of feedback is called **positive (+) feedback** since the feedback sample is added to the input signal. A second form of feedback is called **negative (−) feedback**. As may be guessed, negative feedback is subtracted from the input signal, not added to it (Figure 2.19).

$$\text{Negative-feedback equation: true input} = \text{original input} - \text{feedback} \quad \textbf{(2.5)}$$

Let's follow the input signal through an amplifier that uses negative feedback. As in the other system, let's assume that the initial signal begins passing through the amplifier as expected, and produces an output. As this output begins to be produced, a sample of it is fed back to the input of the amplifier. This signal is now subtracted from the original input. Therefore, the true input is lower in level than the original input. It therefore follows that the output will be reduced. Now, since the feedback signal is simply a sample of the output, the value of feedback signal is also reduced. Repeating this in a shorter form: *input* creates *output*, which produces *feedback*; *feedback* reduces *true input*; lower *true input* creates a lower *output*; lower *output* produces less *feedback*.

Now to carry this one step further: If feedback is reduced, then the true input is reduced less (less subtracted from the original input) and thus becomes closer to the original input level. If you continue to follow this progression through the system several times, it seems that output, feedback, and true input seem to be jumping all over the scale—first up, then down, then up

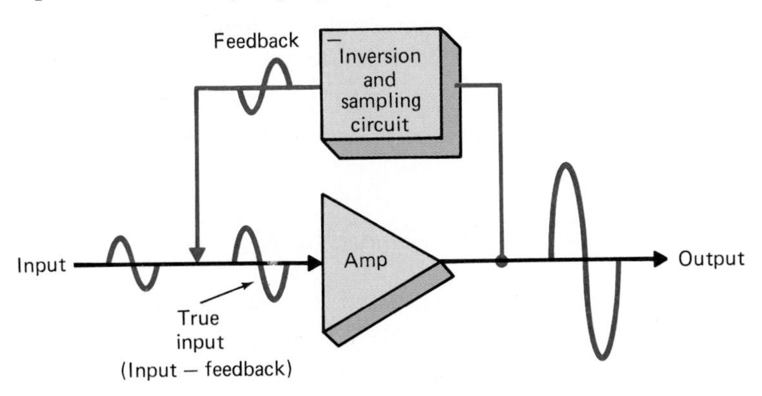

FIGURE 2.19 Process of Negative Feedback

again. In a properly structured negative-feedback system, such a jumping around of values does not occur.

What actually happens is that the system, almost immediately, balances itself. The true input, output, and feedback levels will automatically adjust to individual levels that will allow for a stable (nonfluctuating) output. Thus the output will look quite normal in comparison with other nonfeedback amplifiers except that the gain value for the system will be somewhat less than if no negative feedback were applied.

It does not seem as if either of these circuits, positive feedback or negative feedback, has any practical application. If positive feedback is used, the system appears to increase wildly in output until it is destroyed. If negative feedback is used, the actual output seems to be much like that of a nonfeedback amplifier but with reduced gain. We have to explore the possible applications of these amplifiers carefully before their usefulness will be discovered.

REVIEW PROBLEMS

(1) A real-life example of positive feedback can be seen in a standard public address system. Often, if the volume control is set too high, a squeal is produced from the speakers. Here the feedback is produced mechanically as the microphone picks up some of the sound produced by the speakers. Write a brief description of this system's input-to-output-to-feedback cycling. Try to pattern your description after the discussion of positive feedback in this section. Also try to explain why the feedback stops if the microphone is covered up. (It is not necessary to try to explain why the system produces a specific tone or ''squeal''; just discuss the overall process.)

(2) A real-life example of negative feedback can be seen in a standard lawn-mower engine. This engine runs at a rather constant speed regardless of the amount of ''load'' (heavy versus light grass) it sees. The feedback element is a control valve that senses engine speed and adjusts the flow of gasoline to the carburetor in response to speed changes. In a few sentences, roughly describe this negative feedback system. As a guide, use the description of negative-feedback amplifiers presented in this section. (Note that engine speed is the controlled output of this system.)

(3) A thermostat that controls the temperature of a room is a classic example of the application of feedback. Try to classify this as a positive- or negative-feedback system. Explain your reasons for making this choice, then discuss what would happen if the system used the other form of feedback. (Note that temperature is the controlled output of this system.)

2.5 GENERAL FEEDBACK APPLICATIONS

In Section 2.4 the topic of feedback was presented. In this section the general function of positive and negative feedback was discussed. It was found that positive feedback produced an output that was ever increasing whenever an input was applied. If left unchecked, this output would potentially destroy the amplifier with high power levels. Negative feedback, on the other hand, reduced output to a level that would maintain a balance among output, feedback, and input. Both forms of feedback do have practical applications. In this section we explore some general applications of both negative and positive feedback.

Even though it seems that negative feedback would not serve any purpose in an amplifier other than reduction of the output, it is used in quite a few amplifiers. To discover a practical application for negative feedback, it is necessary to look at a nonideal amplifier system. So far in this book we have used ideal systems when discussing amplifiers. All of the examples used have assumed that input levels were constant, that the amplifiers used would perform exactly as specified under all kinds of operating conditions, and that the output could be any level wanted.

In Figure 2.20 a comparison of ideal to nonideal circuit conditions is shown. In this illustration it is shown that the input may fluctuate from very high to very low levels. Such changes may create functional problems in the amplifier itself or may be undesirable at the output of the system.

An illustration of how negative feedback can reduce these problems will be demonstrated using an AM radio as an example. Since the radio stations that may be received are at variable distances from the radio receiver, the level of power from them that reaches the radio antenna is also variable. It would be very inconvenient if the output volume level dropped just because a distant station were tuned in. This is an example of where we would want a relatively constant output level if the average input levels were to change.

In the same radio circuit, the electronic components used can change value if they are heated or cooled excessively. (More details on this effect are presented in later chapters.) Due to these changes, the actual gain for each amplifier could change. An individual amplifier stage with an expected gain of, say, 50 could actually fluctuate from a gain slightly greater than 1 to a gain that went over 100, due only to temperature extremes. Again such changes would not be satisfactory. The radio volume could get louder or softer just

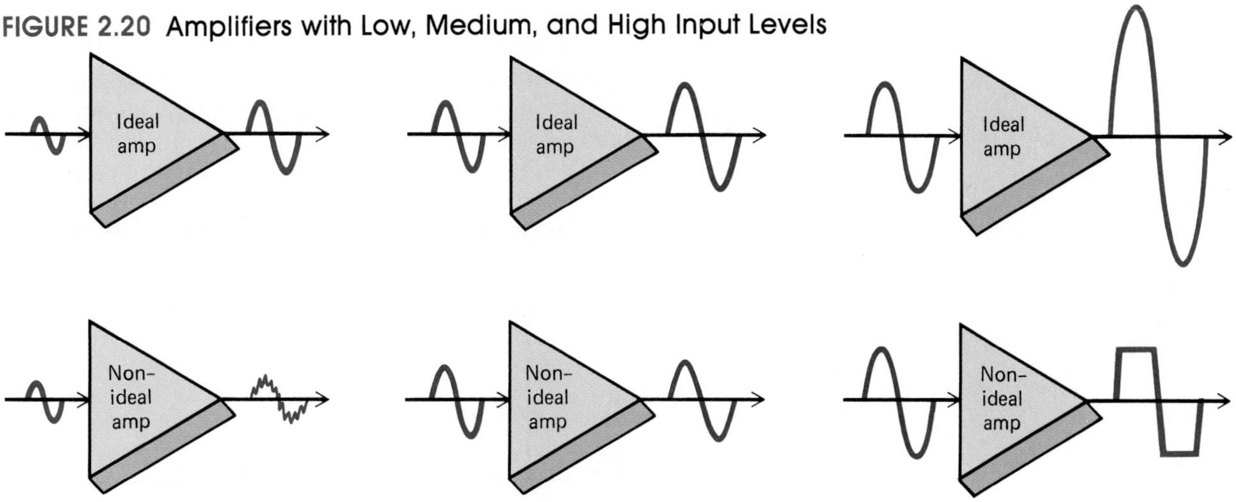

FIGURE 2.20 Amplifiers with Low, Medium, and High Input Levels

by warming or cooling the radio. Naturally, neither of these effects is desirable. First, there is no reasonable way to control the causes of these changes in output levels. Therefore, it would be necessary to try to control the system so that such variations would be compensated for and would not show up as changes in the output.

If negative feedback were to be used, such control would be had. In Figure 2.21 negative feedback has been added to a sample block diagram of a radio (simplified). (In this figure the process of radio reception has been simplified into two major sections. One section deals with the tuning of the radio station and with amplifying and converting the radio signals into audio signals (sound frequencies). The second section uses standard amplifiers to boost this audio signal to a level that can be used by speakers. This simplified example will be used to illustrate a feedback application.)

The negative feedback provides a balancing effect to help counter the types of problems that may occur as noted earlier. The feedback level (a sample of the output from the receiver circuit) is set in such a way that under normal operating conditions [Figure 2.22(a)] the overall system gain is at a value below its maximum potential. This is done so that there is room for the gain to rise or fall in response to the feedback control signal. Should some change in the input cause the output level to drop [Figure 2.22(b)] the feedback level would drop as well. A drop in this negative feedback would result in an *increase* in the true input to the amplifier section. (Remember, negative feedback subtracts from the actual input, and in this case we subtract less than before.) If this input were to go up, the output would be expected to rise since greater input produces more output. Remembering that what triggered these changes was a *drop* in output level, the resulting changes in feedback now push the output level back to where it was! The system would therefore appear to increase its overall gain value since the output would be stabilized to its previous level even though the input had dropped below its previous level.

Just the opposite would happen if the input had increased in value [Figure 2.22(c)]. If a stronger signal on the antenna would tend to drive the output above its normal level, feedback would increase. This increase would then drive the true input lower and thus stabilize the output to its previous "normal" level.

Several points need to be noted about this example. First, for the sake of this example the use of feedback in a radio circuit has been oversimplified. Typically, such feedback occurs in several locations in a circuit and is achieved using several electronic techniques. The overall effect, though, is relatively the same. Also, notice in the circuit that the audio (sound) amplifiers are not

FIGURE 2.21 Maintaining Constant Amplitude Using Negative Feedback

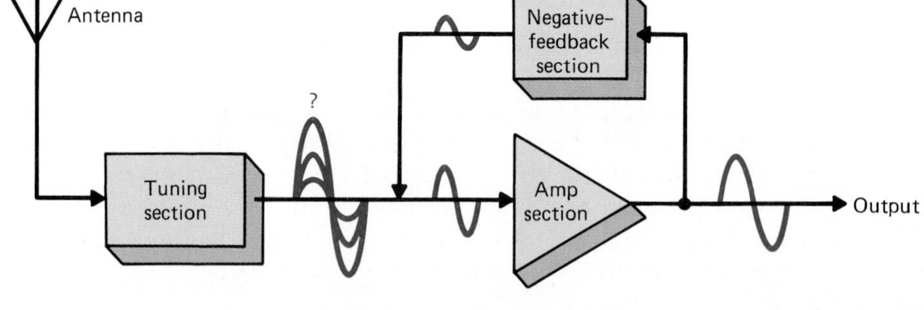

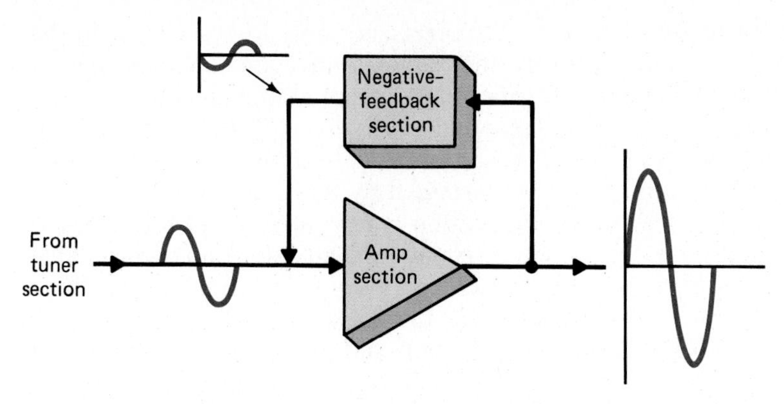

(a) Normal operation

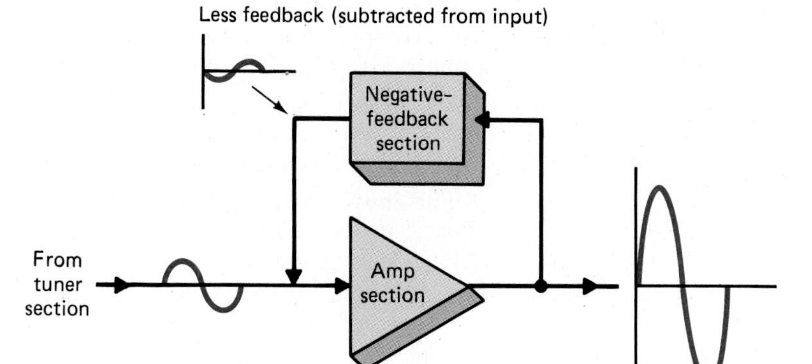

(b) Lowered input

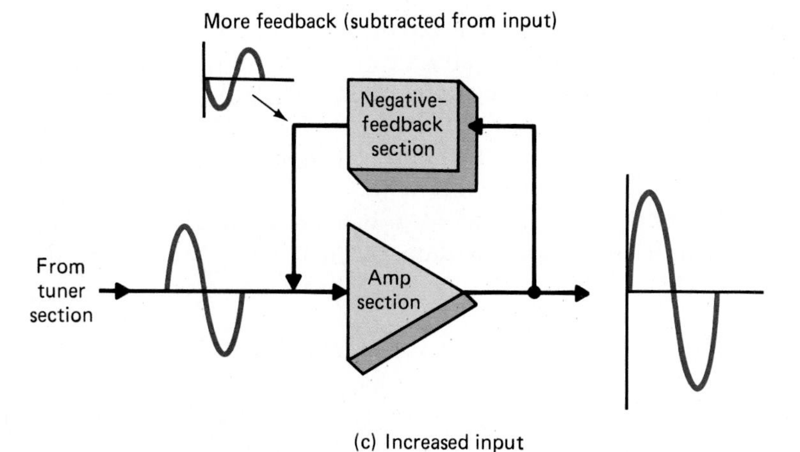

(c) Increased input

FIGURE 2.22 Feedback Circuit Operation

involved in this feedback process. The control of the gain of these initial circuits needs to be independent of the volume out of the speaker. Finally, although the discussion of this process implied that feedback control happens in a step-by-step process, the actual control produced by this feedback is instantaneous. Thus any cause/effect relation as described here would be difficult, if not impossible, to actually observe in a real circuit.

This negative-feedback effect will correct not only for variations in the input signal, but will, using exactly the same procedures, compensate for changes in the individual amplifiers. Should their gain change, higher or lower, the feedback level will immediately respond and adjust the true input to a level that will ultimately compensate for these errors. Naturally, there will be limits to how much compensation can be determined by this process.

There are quite a few other applications of negative feedback. In fact, most popular circuits use some form of negative feedback to provide reliable output levels. It has been noted in these descriptions that negative feedback produces less than the best possible gain from a circuit when operating at normal conditions. This loss is easily compensated for by the addition of one more stage of amplification. Now that an application of negative feedback has been shown, let's explore an application of positive feedback.

Although positive feedback seems to be quite destructive, as it causes an amplifier to rise in output to its maximum level and hold, there are several practical examples of its use. One of these examples can be seen in a typical alarm system. An alarm system is a simple control circuit that receives an input signal, an opened or closed alarm sensor switch, and responds by causing an alarm (i.e., bell) to be sounded. Within this system there must be what is called a **latching circuit**. The objective of the latching circuit is to hold the output on after it is triggered, even if the input should return to normal. For example, if an intruder were to enter through a door and then close it right after using it, the alarm would not only start to ring when the door's alarm switch was interrupted but would continue ringing after the switch was reset by the closed door.

In Figure 2.23 a simple latch circuit's diagram is illustrated. Here positive feedback is used to ''simulate'' the input signal after that signal is removed. In this circuit the output is simply a dc level that will operate the alarm bell. A sample of this is fed back to the output as positive feedback. In this case, if the bell circuit is on, the level of the feedback is the same as the level of the triggering input. The feedback level is zero if the bell circuit is off.

If there is an input from the triggering device, the circuit responds by allowing a large level of voltage to be passed on to the bell circuit. At the same time the positive feedback is activated and is added to the trigger input. Whether there is just the trigger input or the trigger input plus the feedback, the bell circuit will still operate in the same way.

Such a circuit is constructed internally so that the electrical conditions that are produced when it latches do not damage the electronic components

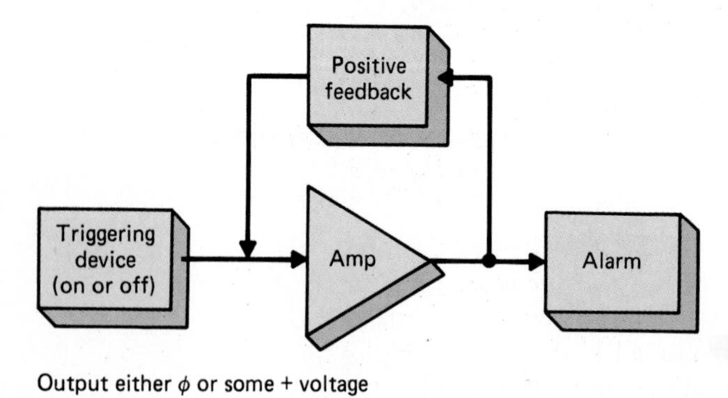

FIGURE 2.23 "Latching" Circuit Using Positive Feedback

used in that circuit. This circuit demonstrates one of several applications of positive feedback.

In this section we have explored an application of negative and then positive feedback. The reason for our exploring these topics was to provide greater background for the coverage of more specific circuitry and applications in future chapters. In these chapters we will see quite a bit of use for the positive- and negative-feedback concepts we have developed in these last few sections.

REVIEW PROBLEMS

(1) In the discussion of negative feedback used in the radio circuit it was noted that this feedback was also effective for compensating for a moderate change in the gain of the system's amplifiers. In a brief paragraph or outline, discuss how this would happen.

(2) For the alarm circuit presented in the discussion of positive feedback, show where a switch could be placed to reset the alarm circuit. Do not make it a simple on/off switch for the battery supplying all of the circuits.

(3) A typical AM radio uses negative feedback not unlike that discussed in this section. In a short paragraph, describe why, when such a radio is tuned between stations (no available broadcast signal), radio interference (noise) signals, which are usually of a very low level, can often be heard on the speaker.

2.6 OTHER BLOCK DIAGRAM CONCEPTS

Up to this point in the chapter we have introduced the concept of multiple stages, of special additions to these stages and of feedback. In each of these discussions, the amplifiers have been arranged in serial fashion. The output of one stage is fed the input of the following stage. Although many systems use this arrangement, quite a few other systems do not use such a straight-line arrangement of the stages. There are quite a few ways in which stages can be arranged. In fact, nearly every type of electronic circuit has a somewhat unique arrangement of its internal stages. There are several common variations of the simple series, often called **in-line**, arrangement of amplifiers. In this section we explore a few of these variations.

In some systems, circuits are connected in a parallel-like mode. The most common example of this can be seen in a typical stereo amplifier. In Figure 2.24 a simplified stereo amplifier system is shown. In this illustration it can be seen that the amplifier blocks used for both the left and right channels are identical. Any function that appears in one circuit is duplicated in the other circuit. In fact, it is common to use the same individual electronic components in each circuit. For example, if a 10,000-Ω potentiometer is used for volume control in one, a separate but identical 10,000-Ω potentiometer will be used

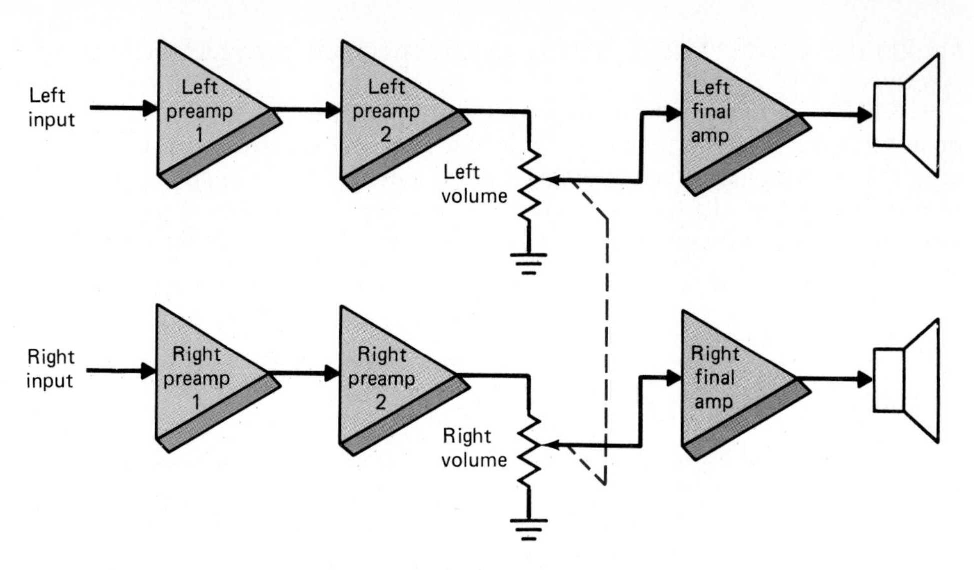

FIGURE 2.24 Stereo Amplifier (Simple Diagram)

in the other. It is often the case, in a stereo system, that each amplifier can function as an independent amplifier. Should one fail, the other can continue to operate without interruption.

These two (nearly) independent circuits are built on the same chassis both for convenience of operation (all controls in one location) and so that they can share one common power supply. Even though there may be only one volume control or one tone control on the front panel of the amplifier, the single "knob" used to adjust these factors often is attached to two independent controls inside the cabinet, one for each circuit. Dashed lines like those shown between certain control sections in Figure 2.24 are typically used to indicate a mechanical (not electrical) connection between circuits (i.e., a common "knob" which adjusts each circuit's volume control potentiometer).

It was noted in the last paragraph that these circuits are *nearly* independent. One can operate regardless of how the other one is functioning, to a point. Since these two systems share some common elements and circuitry, should one of these shared elements become defective, the defect will be reflected in each amplifier. For example, should the power supply for this system become defective, both circuits will fail to operate. On the other hand, if the preamplifier for the left channel were to fail, which would result in no output from that channel, the right channel could be expected to operate normally.

Such parallel-like functions can be seen in quite a few systems. It is not necessary for such functions to be exact duplicates of each other. Often, a system will have alternate functions that are either done automatically or can be selected separately by the operator. In many cases these independent circuits will share some common circuitry. The diagrams discussed in Example 2.7 demonstrate possible configurations that are not simple in-line circuitry. Although these do not illustrate specific circuits, they demonstrate the fact that it is not necessary to use simple in-line layouts to produce desired functions.

EXAMPLE 2.7

Consider the three block diagrams presented in Figure 2.25. Circuit (a) will allow several different input functions to be selected. Once one function is selected, the output of the circuit related to that function then feeds into circuits that are commonly used by all of the other input circuits. Here, the input circuits may be quite different from each other, but the sharing of a common output circuit makes it logical to place these all into one common package.

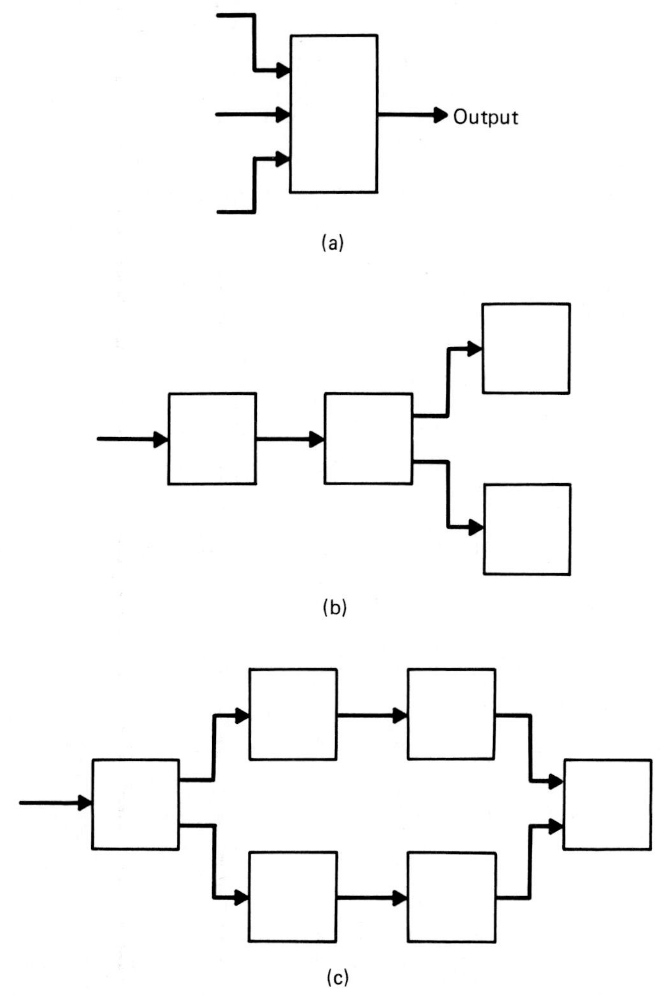

FIGURE 2.25 Possible Block Diagram Arrangements

In circuit (b), there is one input. This input passes through some initial in-line circuits before being fed to several output circuits. These outputs do not have to be similar in nature. Since these outputs share a common input circuitry, it is again logical to blend them into one complete system.

In circuit (c), although there is a single input and a single output, in the center of the system the information is passed to several different blocks. The outputs of these blocks combine to produce the final output. Since there are several common circuits within this system, it makes sense to have these circuits presented in one complete diagram.

For each of the circuits in Example 2.7, there is an additional reason why the individual circuits would be blended into one common assembly. In most cases it is no great challenge to design the individual circuits so that they can be operated from one common power supply. With each system using a central power source, the overall cost of the system can be much less than if each circuit were to be operated independently with a separate supply.

There is one more special arrangement of the block diagram illustration that should be noted at this point. In a block diagram it is customary to indicate all inputs at the far left-hand side and all outputs at the right-hand side. A simple example of such an arrangement can be found in a common stereo amplifier. Many amplifiers allow either speakers or headphones to be used as the output device. The speakers are connected to the final power amplifier in the system so that maximum volume can be delivered. But as shown in Figure 2.26, the output for the headphones is tapped prior to this final stage. The headphone output is gathered at this location in the system since headphones would be damaged if too much power (as would be found at the final output) is applied to them.

The output that would be fed to a tape recorder is obtained even farther back from the final output. This location is used for two reasons. First, the

FIGURE 2.26 Multiple Input/Multiple Output Block Diagram (Stereo)

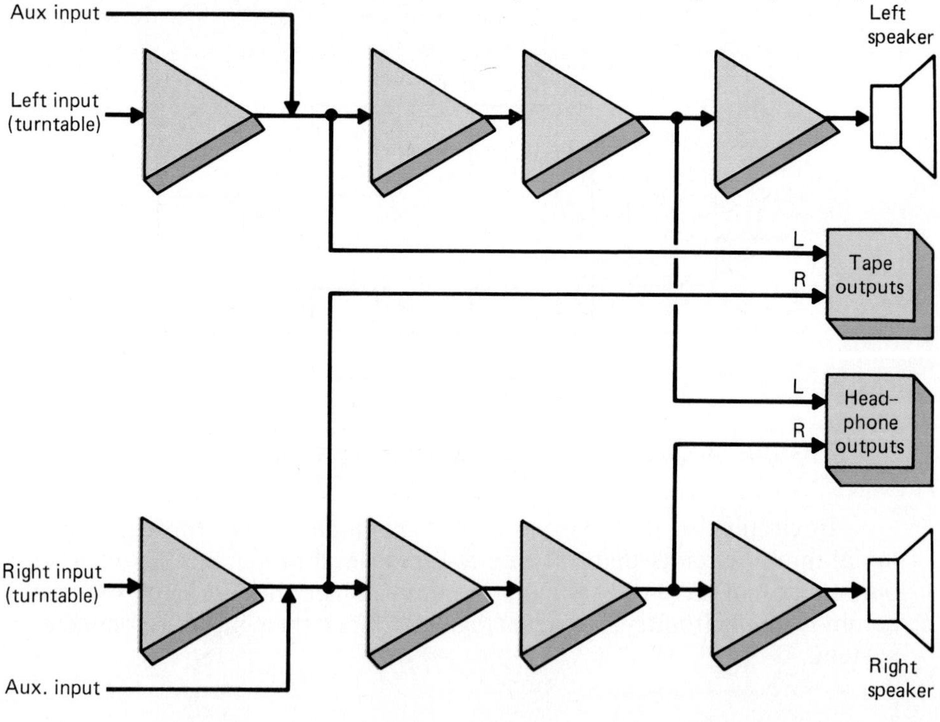

tape recorder is designed to operate using lower input levels than those of headphones. Thus fewer stages of amplification are needed. Even more significant, though, is that at this location in the circuit, the signal has not been affected by either the tone control circuits or the volume control. Thus what is recorded will not be subject to these settings, but will be a pure representation of the information input to the amplifier.

It was indicated at the beginning of this section that there can be quite a variety of arrangements within a block diagram. So far we have investigated some of the major forms of these configurations. The simple in-line diagram shows a direct flow of information from input to output. Other forms, including parallel-function types, can be more complex to interpret but still follow a logical pattern. Being able to follow the form and flow indicated in a block diagram can be a major aid in being able to understand the overall function of complex electronic systems. Future sections of this text will use block

REVIEW PROBLEMS

(1) A graphic equalizer is a collection of circuits that divides the audio-frequency range into specific bands and allows the user to select the volume level desired for each band. An equalizer can be built as an integral part of an amplifier system. Typically, the equalizer circuits are placed in parallel with each other and are located somewhere prior to the final power amplifier stage in such a system. They are fed by one common signal from a preamplifier. Based on this description, identify which diagram presented in Figure 2.25 would most likely represent an amplifier with a built-in graphic equalizer.

(2) A television set has one source of information, the signal received on the antenna. There are several forms of output for a television set: the picture, the picture's colors, and the sound. The separation of this information is done automatically within the set's circuitry. Identify which diagram in Figure 2.25 most closely matches this brief description of a television set's operation.

(3) Using the stereo amplifier diagram shown in Figure 2.26, discuss briefly what section (block) of the system may be suspected of being defective should the following conditions be observed (for each set of conditions, describe a possible defect).

  (a) There is no sound out of the speaker connected to the right channel. Using the headphones, both the right and left channels work. (The speakers have been tested and are operating correctly.)

  (b) The system works okay when using the auxiliary input, but there is no left-channel output to the tape, headphones, or speakers when using the turntable input (the turntable itself checks out okay).

  (c) The system has no output from either the left or the right channel, not even to the tape recorder. It is plugged in and turned on, but still there is no output. (In your discussion, try to isolate the one most probable defect.)

diagrams to describe circuit functions before an attempt is made to explore the complexities of individual circuit elements.

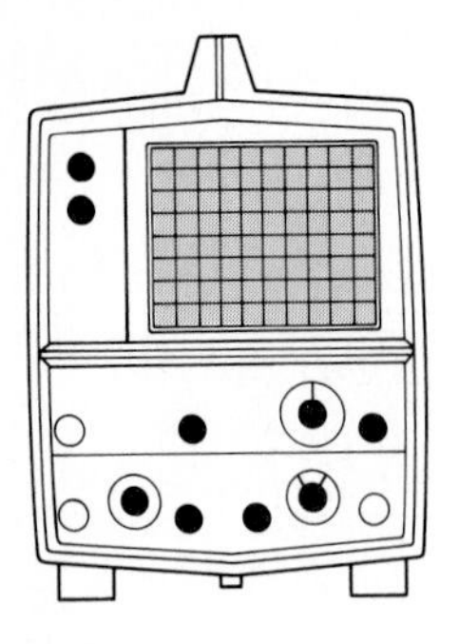

2.7 AN EXAMPLE OF THE DEVELOPMENT OF A SYSTEM USING BLOCK DIAGRAMS

This section is dedicated to the development of a single system, a light dimmer. The reader is not expected, at this time, to be able to duplicate this development process—rather to work to follow the logical progression of the system's development. The important point to know from this description is the *logical progress* taken to develop this system. The reader is encouraged to take notes regarding the steps taken and to attempt to think ahead and predict the next step in the development of this product.

The concepts of block diagram usage and application have been presented in previous sections of this chapter. In this section we trace the development of circuits using block diagrams. Understanding how circuits are developed this way will afford a greater understanding of the overall function of electronic circuitry. It is not our objective to learn all the details of circuit design or to become prepared to design systems, but rather to have a general concept of how complex circuitry is developed.

Initially, the design of an electronic system starts with an objective that describes the specific task that needs to be achieved. For example, if we wanted to develop a light dimmer, our objective may be stated as follows:

> The circuit will need to provide a variable output voltage level given a standard ac line voltage. The level of output should be controlled by a simple "dial" setting. This input should be able to control the light level of several standard lamps totaling a maximum of 400 W of power consumption. Such control should be done in a more desirable manner than is currently available.

Notice in this statement that we have carefully specified the type of input, the type of output, and the means of providing the desired output (with the dial setting). Also, this description implies that there are methods available but they are not desirable. (Prior to the development of modern light dimmers, the only practical ways to dim lights either used bulky transformers or highly inefficient rheostats, neither of which was desirable for home use.)

Although quite simple in comparison to other block diagrams, the one illustrated in Figure 2.27 shows the basic functional parts of this dimmer circuit. In this diagram all the initial parts of the functional circuit are shown. The input of line voltage that is to be controlled, the other control input—"dial"—position, and the output to a typical load (lamp) are clearly seen.

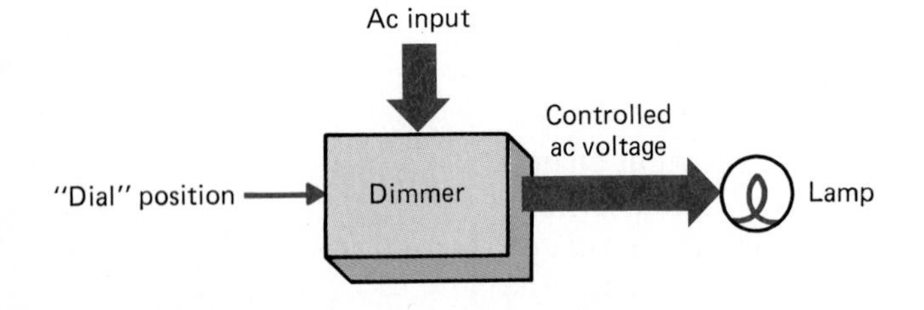

FIGURE 2.27 Simple Dimmer Diagram

One of the next steps taken would be to develop a more specific statement relative to how the process would be achieved. For a more complex system the initial block would be expanded into more descriptive blocks. For the dimmer, though, the designer would have found that there was one electronic component that could provide the desired control, if supported by a few other simple components.

The details of this support process for the main control element would then dictate the final steps needed to develop the dimmer. All electronic systems need such support elements. Their description is most often left for the final steps in the development process.

The block diagram shown in Figure 2.28 includes the final details that will lead to the development of an electronic schematic for the light dimmer. Many of the steps described are there because the control element chosen needs such circuitry to operate. The following is a brief description of each of the blocks noted.

AC Voltage Sampling Block This block samples the input of ac voltage. This sample is needed as a low-level control voltage for the electronic device used.

Phase-Shift Bridge Block This block is necessary not because of the function of the overall system, but is tied to the device chosen to control the output. The designer would have added this block only after investigating the operation of the control device that was to be used. In another design using some other device, such a circuit may not be necessary. Notice that attached to this block is the actual control dial. It has taken the form of a potentiometer. This controller is not attached to the actual control device used, but is associated with another function which feeds this device.

Triac Block This block indicates the use of a triac (an electronic control device). This is the component the designer chose to use. Its function now dictates some of the other blocks that are in place in this figure.

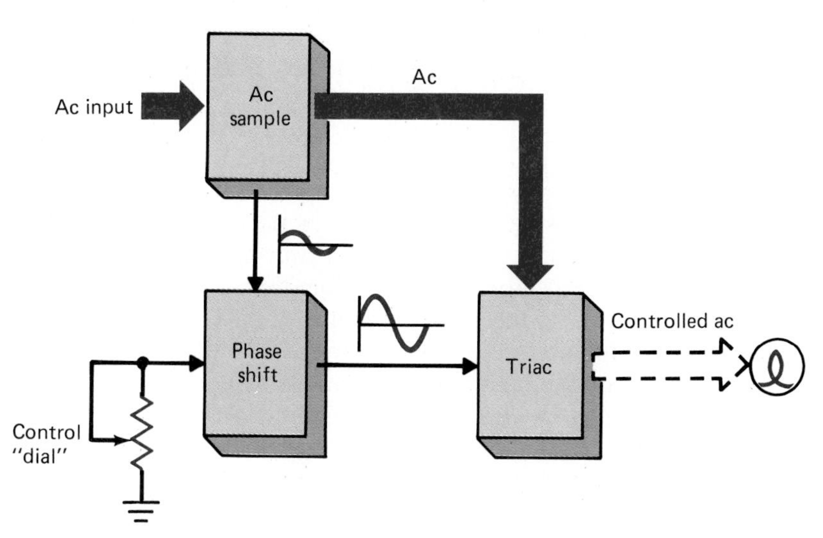

FIGURE 2.28 More Functional Dimmer Diagram

Notice that actual component values or part numbers are not used even in this final diagram. Even though the designer selected a triac as the control device, a specific part number for the triac is left out. These details will be included in a final schematic. Only the basic steps required to operate a triac for this application are indicated.

The next step would involve tracing the process described by this flowchart. Refer to Figure 2.28 and follow the steps as outlined.

1. Electrical power is input to the system (ac input).
2. A sample of this ac is taken (a low-level voltage).
3. This sample is then phase-shifted from the input phase based on the setting of the control device (dial). Simplified, this phase shifting presents a time delay between the peaks of the voltage applied to the control lead of the triac and the peaks in the original supply voltage.
4. The triac then controls the average voltage delivered to the load. (This description is based on how a triac functions. For a more detailed explanation of the triac's operation, see later chapters.
5. Since the average voltage delivered to the lamp is controlled by the triac, the average amount of brightness from the lamp is variable.

The next step that would be taken in this circuit's development would be to prepare a schematic drawing. This drawing would show in detail the specific parts used in the electronic circuit. Actual part numbers would be selected, and equations would be used to calculate exact operating conditions. The block diagram would be used to guide the designer in establishing the exact layout of all the functional parts.

Figure 2.29 shows the final electronic schematic for the light dimmer circuit. From this composite illustration, the actual function of specific chosen parts can generally be surmised. The resistors outlined by the voltage sampling block can be seen to compose a rather simple voltage divider. The resistor/capacitor combination highlighted by the phase-shift block will provide a phase shift of the voltage supplied to the triac. As the potentiometer is adjusted, the ratio of resistance (a 0° shift) to capacitive reactance (a 90° shift) changes, thus adjusting the composite shift from nearly zero (0) to close to 90°. Finally, the triac (a solid-state control device) controls the voltage delivered to the lamp based on phase-shifted voltage supplied by the phase-shift bridge section. (Again, how or why the triac responds to this input signal will be described clearly in later chapters.)

As can be seen, though, there is a rather direct relationship between the final schematic diagram and the block diagram from which it was developed. Even though we have not yet discussed the exact electrical properties of the triac, we can generally analyze the function of the circuit, and what is more important, be able to relate form more clearly to function.

From this simple analysis of the development of a light-dimmer circuit, we can see how more complex circuitry is developed. By more clearly understanding how circuitry is developed to meet a set of objectives detailed in a block diagram, we can have a clearer understanding of how a circuit is expected to operate. Even if the actual components specified in the schematic diagram were to change, due to the selection of a different triac, for example, we would still be able to break it down into specific tasks done by certain parts if we understood the objectives spelled out in the block diagram.

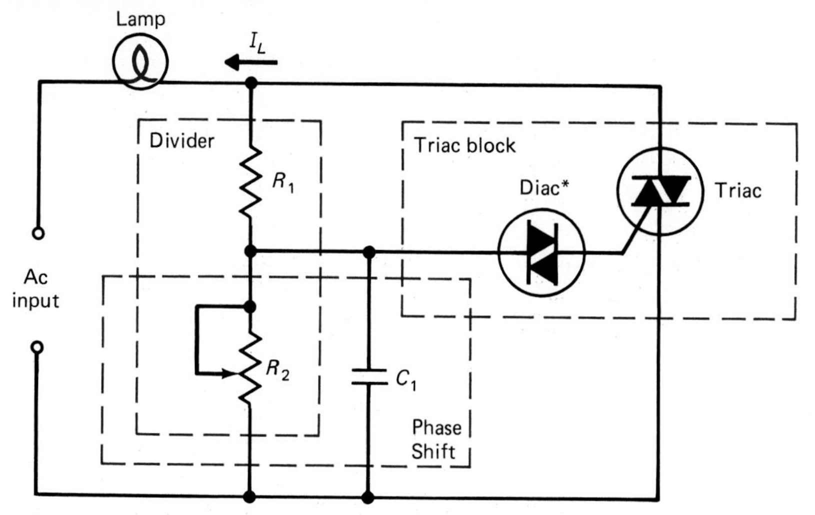

FIGURE 2.29 Dimmer Schematic Showing Block Diagram Coordination

REVIEW PROBLEM

(1) As practice in developing a complex process from a simple block diagram, start with a single block that specifies the task of going from your home to the school you are attending. Now expand that block into a series of blocks that specify simply north, south, east, and west directions based on the routes you normally take—omit the actual route numbers (e.g., go east 10 miles, north 50 miles, etc.). Now actually sketch a simple map showing the path taken; use exact road and route names. When finished, write a brief discussion of which of the directions could be used more universally, and why.

SUMMARY

In this chapter we have expanded the process of using block diagrams. The use of multiple stages to provide adequate amplification for a system has been demonstrated. In using multiple stages it was shown that several steps are needed to provide adequate gain for a low-energy input to be able to produce a high-energy output. Additionally, several "special effects," such as feedback, volume control, and filtering, have been added to basic amplification circuitry. Several amplifier applications have also been described to aid in understanding the overall concept of gain. Finally, an example of the development of a control system was shown to illustrate the block diagram process.

KEY EQUATIONS

2.1 $\text{gain} = \dfrac{\text{output}}{\text{input}}$

2.2 system gain = gain 1 × gain 2 × gain 3 × . . .

2.3 output = input × gain

2.4 Positive-feedback equation: true input = original input + feedback

2.5 Negative-feedback equation: true input = original input − feedback

PROBLEMS

Section 2.1

2.1. If the input to an amplifier was 3 mW and the output was 15 W, what is the value of its power gain?

2.2. If an amplifier system had a power gain of 15,000, with an input of 8 mW what would its output be?

2.3. If four amplifiers in a row had a power gain of 20 each, what would be the total power gain for the system?

2.4. If a system was to have a total gain of 900 and was composed of three amplifiers and the gain of two of the three amplifiers was 15 each, what must be the gain of the third amplifier?

2.5. If the amplifier system in problem 2.3 had input voltage 2 V and input current 1 mA, what would its output power be?

Section 2.2

2.6. Illustrate two possible locations for a volume control in the block diagram shown in Figure 2.30.

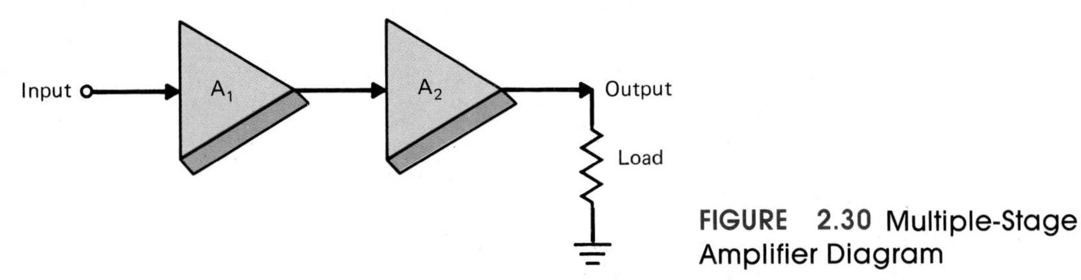

FIGURE 2.30 Multiple-Stage Amplifier Diagram

2.7. Define the general application of a variable-gain amplifier.

2.8. Indicate the total gain of the system shown in Figure 2.31 if the variable-gain amplifier was set to the following gains:
(a) 5 (b) 13 (c) 25 (d) 100

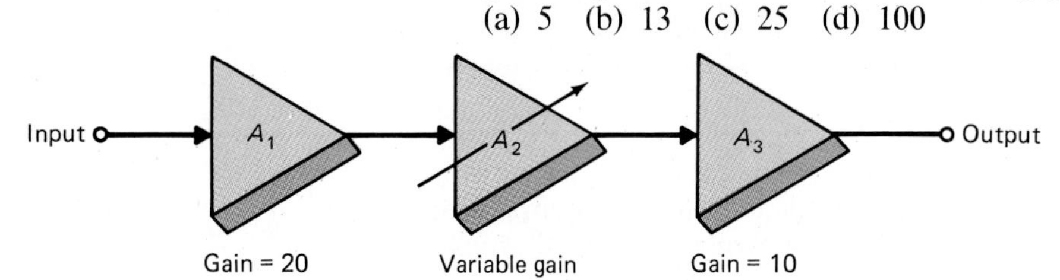

FIGURE 2.31 Variable-Gain Amplifier Used in Multiple-Stage Diagram

2.9. For the system in problem 2.8, if the amplifiers were rearranged so that amplifier 2 was first and then rearranged so that it was last, what change in the gain statistics would result, using the four gain values shown in problem 2.8? Verify your argument by calculating the gains under this new arrangement using the four gains in problem 2.8.

Section 2.3

2.10. If the filter shown in Figure 2.32 were a bandstop type (eliminating one specific frequency) tuned to 15 kHz, discuss the effects the circuit would have on the following signals.
(a) 1 kHz (b) 8 kHz (c) 15 kHz (d) 25 kHz

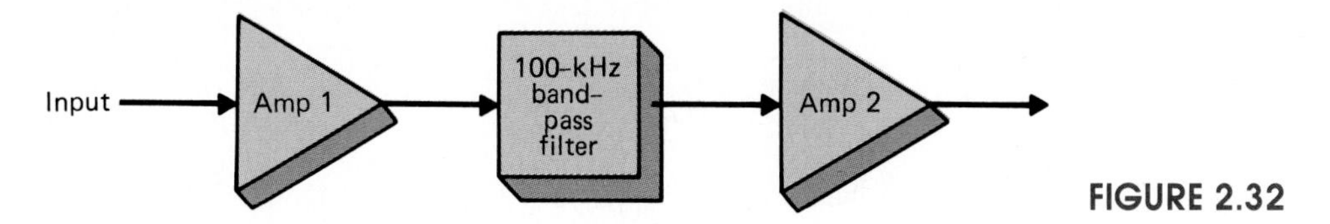

FIGURE 2.32

2.11. Using the system shown in Figure 2.33, assume that one filter is tuned to 200 Hz and the other tuned to 15 kHz. Describe what the output will look like for inputs ranging from 100 Hz to 25 kHz. (Sketch this output.)

Section 2.4

2.12 Illustrate an amplifier with negative and then one with positive feedback applied. Discuss the differences in their outputs.

2.13. If an amplifier has an input of 1 V, a voltage gain of 10 (for a 10-V output), and 2.5% of the output is returned to the input as negative feedback, what is the true input to the amplifier after the feedback?

2.14. Identify a real-life illustration of negative or positive feedback and discuss briefly the process involved. (You may use the review problems at the end of Section 2.4 as a guide; just do not copy them as your example.)

Section 2.5

2.15. Illustrate an ideal amplifier, then a nonideal amplifier. Briefly discuss their differences.

2.16. Write a brief "step 1, step 2, . . ." outline of the function of feedback shown in Figure 2.34.

2.17. Repeat problem 2.16 for Figure 2.35.

2.18. Although negative feedback has several advantages, there is one major disadvantage, a loss in potential output. Discuss how this loss occurs.

2.19. It is possible to have too much negative feedback in a circuit. Estimate what effects would be seen for a simple negative-feedback amplifier if the negative-feedback quantity approached or even exceeded the input level.

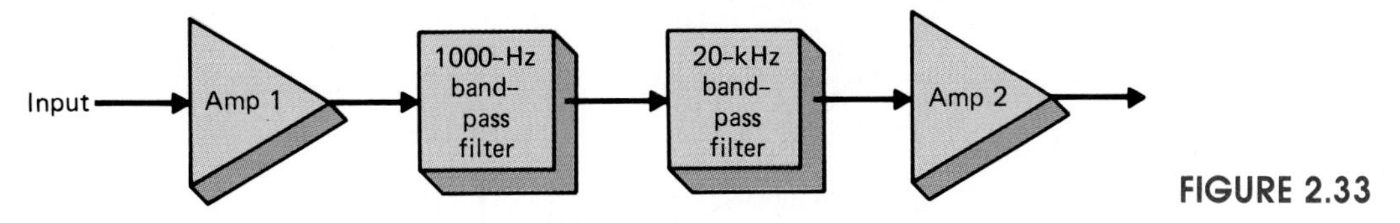

FIGURE 2.33

FIGURE 2.34

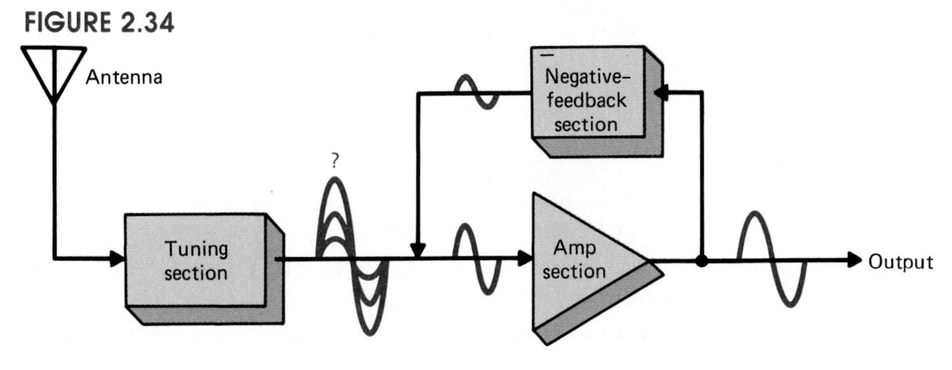

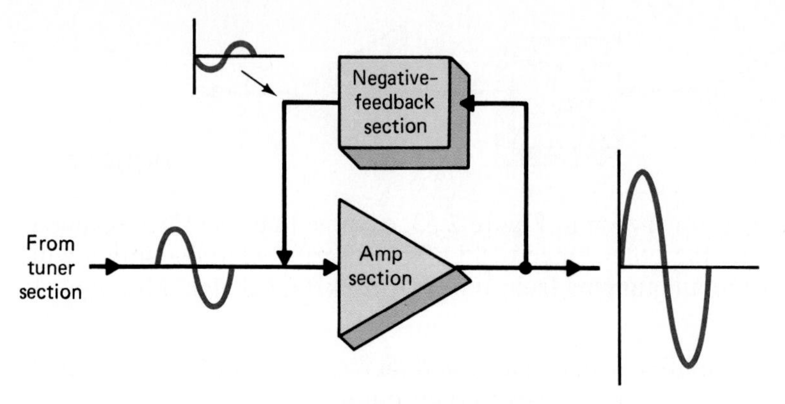

(a) Normal operation

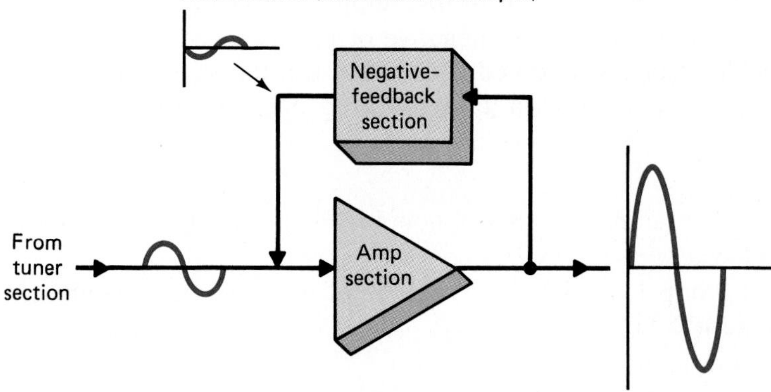

(b) Lowered input

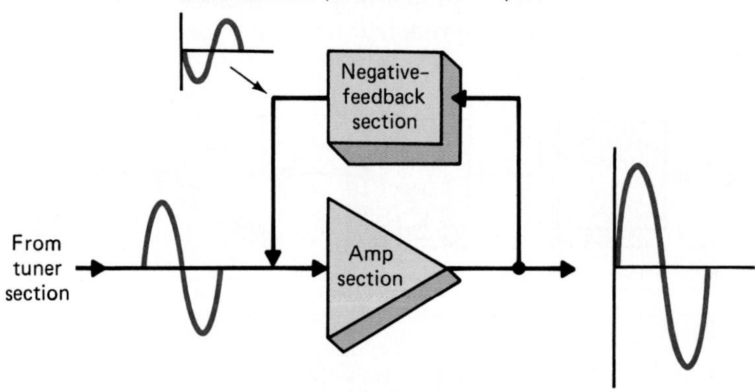

(c) Increased input

FIGURE 2.35

Section 2.6

2.20. Illustrate the three basic amplifier configurations presented in Section 2.6, then briefly describe their differences.

2.21. A videocassette recorder has one input, the antenna signal. This information is then divided into video and audio signals when recorded on the tape. When taking the signals off the tape (again separate video and

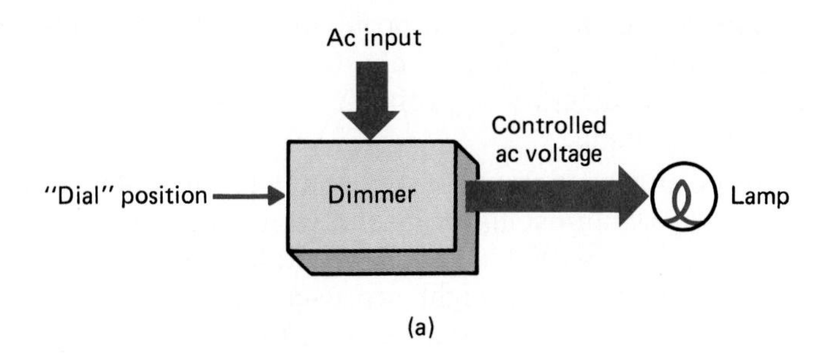

(a)

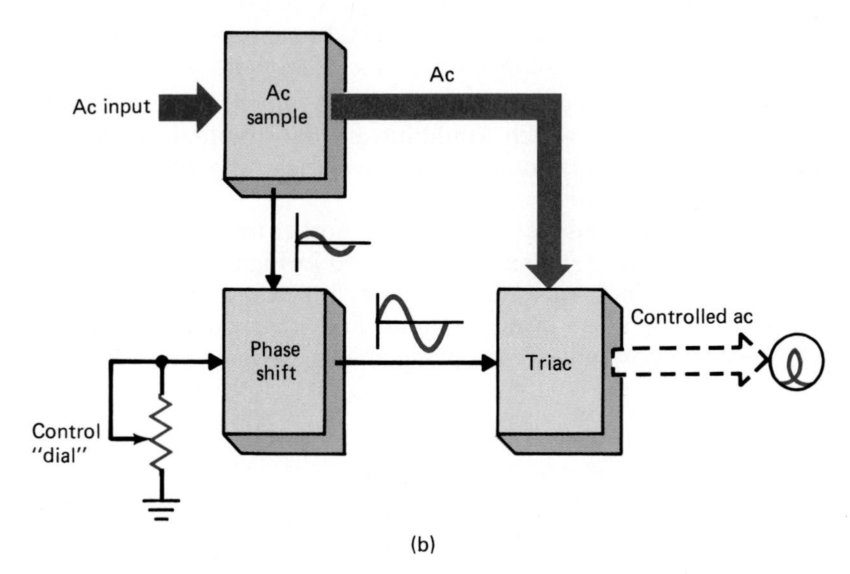

(b)

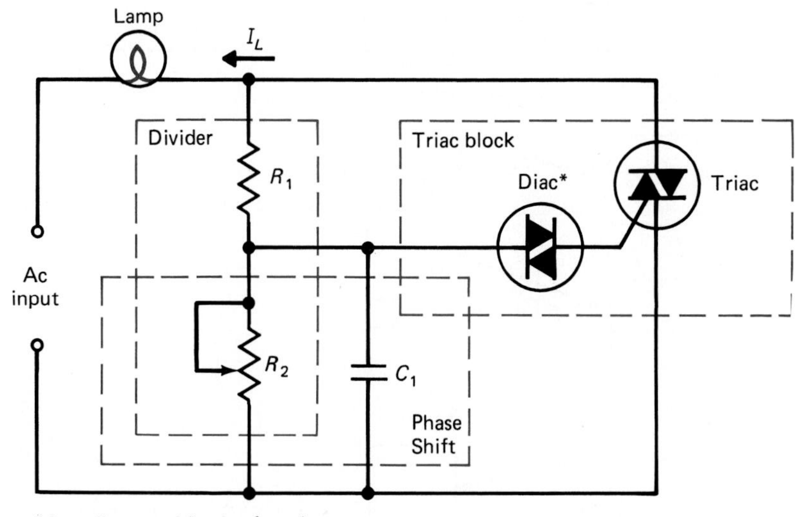

(c)

FIGURE 2.36

audio), they are recombined into one signal for feeding to a TV set. Illustrate these functions in block form.

2.22. Often video recorders (as described in problem 2.21) have a plug for a camera and microphone plus output jacks for direct video and audio information. Modify the diagram of problem 2.22 to show where these connections would be made.

2.23. A simple electronic organ operates in the following manner: When a key is struck, it turns on an oscillator circuit (frequency generator) set to a frequency coordinated with that key. The output of all the generators (for simplicity assume that only eight are used) is fed to a common amplifier. The output of this amplifier feeds a tone control circuit (to create various sounds) which feeds, through a volume control, to the final power amplifier. The speaker is attached to this power amplifier. From this brief description, construct a block diagram.

2.24. Figure 2.36 illustrates a light dimmer. Indicate which of the diagrams would be unchanged and which would have to be modified for each of the following changes. (It is not necessary to actually show the changes; just support your decisions.)

(a) The specific triac used was changed to another model but operated in basically the same way as the one used in Figure 2.36(c).

(b) In actual operation it was found that when the triac operated it created undesirable noise in the ac input lines that would tend to interfere with radios and TV sets plugged into the same line. To solve

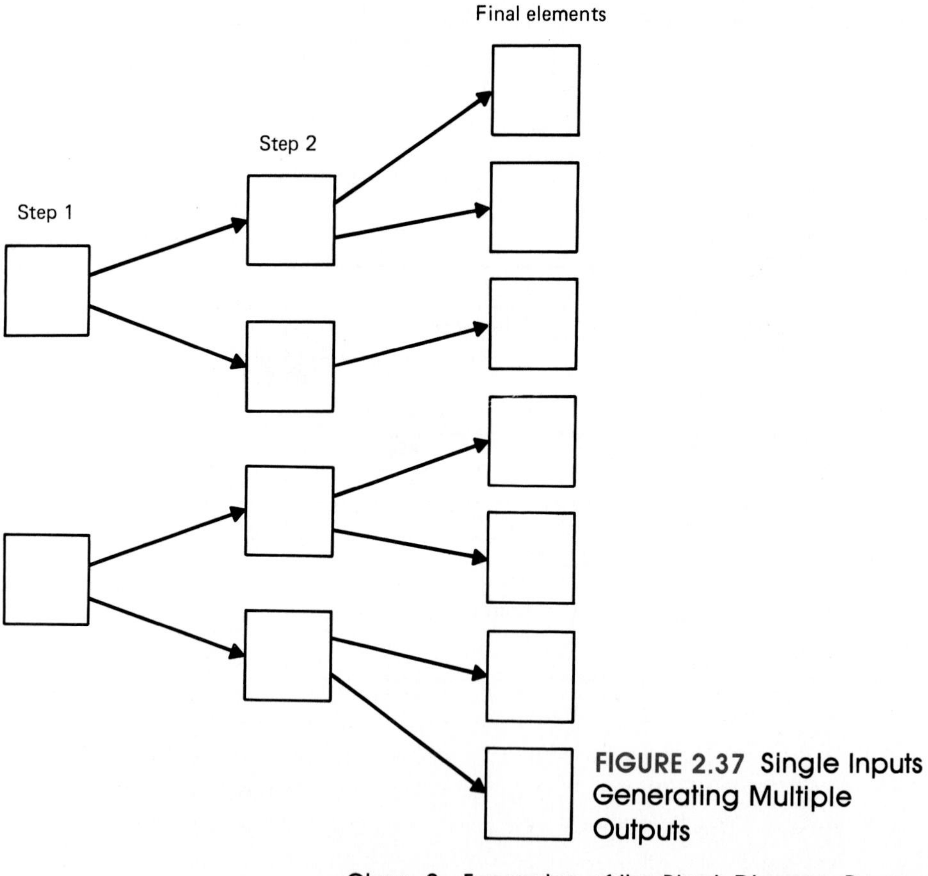

FIGURE 2.37 Single Inputs Generating Multiple Outputs

the problem, a simple noise filter was added across the ac input lines.

(c) After this circuit was in use, an integrated circuit (IC) was developed which incorporated all the schematic parts found in Figure 2.36(c). It was therefore only necessary to connect it to the ac lines and attach a load and potentiometer as a controller. (In an IC, it is not possible to modify or repair any of the internal parts; therefore, it usually is not necessary to specify their function in a separate drawing.)

2.25. On a single sheet of paper, show the expansion of each segment of this system as it was developed. On the left side (about halfway down the page) show the initial blocks used to start the project [see Figure 2.36(a)], vertically. Then in the center, illustrate the blocks that these expanded into [see Figure 2.36(b)], again vertically. Finally, on the right side of the page, show (just by name) the final elements that were a result of the development [see Figure 2.36(c)]. Connect these blocks, left to right, to show how the system expanded from a basic idea to a finished system. (Use Figure 2.37 as a guide to how to lay out this problem.)

ANSWERS TO REVIEW PROBLEMS

Section 2.1

1. 200
2. 1,350,000
3. 54 W

Section 2.2

1. Control between stages and variable-gain amplifier (see the text for a description).
2. (a) 9 W
   (b) 5 W
   (c) 1 W
   (The same values)
3. The illustration would be the same as Figure 2.8 except the control potentiometer would be omitted and a variable gain amplifier would replace the amplifier shown. Output values would be identical to those in example 2.6.

Section 2.3

1. The system output would exist only for input signals at a frequency of 100 kHz. The 10- and 200-kHz inputs would not be present at the output, whereas the 100-kHz input would be amplified.
2. The answer depends on the system located in the text.
3. Regardless of the input frequency, no output would be had since there is no single frequency that would pass both filters.

Section 2.4

1. Noise is input to the microphone; this is amplified and outputted from the speaker. The microphone receives the sound from the speaker and uses it as an input, thus producing a repeated cycle (positive feedback). Placing a hand over the microphone blocks the sound input, thus breaking the output-to-input part of the cycle.

2. As engine speed is reduced (perhaps due to higher grass), the control valve senses this and increases the gas flow to the engine, thus increasing its speed back to a "normal" value. Higher engine speed, in a similar manner, causes a reduction in gas flow, thus reducing speed.
3. Negative feedback, since the furnace control is inversely proportional to the room temperature

Section 2.5

1. If the amplifier's gain were to change, negative feedback could compensate for the changes, thus maintaining a relatively constant output. If gain went up, more feedback would be applied, which would reduce the effective input. This lower input, multiplied by the higher gain, would produce roughly the same level of output as seen before the change in gain. Should the gain go down, feedback would decrease and the effective input would go up. This action would also maintain a relatively constant output.
2. The switch could be used to break the positive-feedback path. Breaking this path, if there is no new input, will reset the circuit.
3. Since the radio is between stations and thus no usable signal is found, the feedback is nearly zero. With such low feedback, even the very small noise signals are amplified and reproduced as an output.

Section 2.6

1. Figure 2.25(c)
2. Figure 2.25(b)
3. (a) The last (right-hand) amplifier of the right-channel circuit is probably defective.
   (b) The first (left-hand) amplifier of the left-channel circuit is probably defective.
   (c) Since a power supply is connected to each amplifier and since nothing seems to be operational, it is quite probable that the power supply is at fault.

Section 2.7

1. There is no one correct solution, but it is suggested that the reader carefully review (if not "walk through") each set of diagrams, or have a partner follow them to cross-check their accuracy.

3 Fundamentals of Amplification

Objectives

Upon completing this chapter, the reader should be able to:

- Identify the role of active and passive devices.
- Recognize the effect of polarity-sensitive devices.
- Name specific active and passive devices.
- Coordinate active devices to block diagrams.
- Identify the role of the power supply in amplification.
- Use block diagrams for troubleshooting.

Introduction

In this chapter we deal with the basic functional elements of amplification and the typical components used to achieve amplification. Once the process of amplification is defined, these components, the active devices, are described. Their basic form and function are presented. The role of active devices is contrasted to passive, or conditioning, devices. Both their individual function and their interrelationship within circuitry are explored. How these circuits coordinate to block diagrams is another important topic of this chapter. Finally, the role of the energy source, the dc supply, is described as the control function of the amplifier is illustrated. Once all the primary parts of an amplifier are specified, the process of troubleshooting, based on the functions of these parts in the circuit, is presented. These details form an essential primer to understanding amplifier circuits of any type. ■

3.1 AMPLIFICATION AND AMPLIFIERS DEFINED

In Chapters 1 and 2 the concept of amplification has been used. We have used the terminology for gain and even looked at multistage amplifiers with large values of cumulative gain. In this section we define and describe gain and relate it to the electrical effects that go on in an amplifier. In addition, we investigate some of the areas where there are limitations to amplifier use.

As presented in Chapters 1 and 2, gain involves inputting a low-energy signal and achieving a high-energy output which is a reproduction of that signal. By observing that input-to-output relationship for an amplifier, we can define its gain. (Remember that gain is simply the ratio of output to input, in the form of voltage, current, or power.) There is no system, either electronic or otherwise, which can directly increase an energy level without using more energy to do so.

All electronic amplifier systems require another energy input in addition to the information signal. They require electrical power from a simple dc source. This dc source is not a part of the signal that is being amplified but supplies the energy necessary to produce a high-energy output. An amplifier is simply providing a control function. The amplifier is designed to control the flow of electrical power from the battery to the output load. Amplifiers provide this control in response to changes in the input signal. If the input drops to a very low level, the amplifier meters out less of the battery energy to the load. Should the input's level rise higher than average, the amplifier passes a higher level of battery energy to the load.

One other characteristic of the amplifier is its ability to respond to rapid input changes. As an ac input goes through hundreds, thousands, even millions of changes in amplitude every second, the amplifier will keep up with these changes and present an output that changes in perfect synchronization. Therefore, an ac input will produce an identical ac output, only at a higher energy level (higher voltage and/or current).

Although amplifiers do produce a relatively simple control function, it is important to note that individual amplifiers do have certain limitations. All electronic devices have voltage, current, and power limitations which become limits for the amplifier circuit in which they are used. In addition, devices have frequency-response limits. A device used in an audio amplifier (low frequency), for example, may not work in a circuit used for higher radio frequencies.

Many amplification devices can permit current to flow in only one direction. Since by definition, ac, the typical form of an input signal, has an alternating direction of current flow, it would therefore be severely affected by such a device. In fact, without taking special preventive measures, such a device would render an ac input useless and may be severely damaged by this input. In fact, much of this book is dedicated to dealing with and overcoming the problems noted above.

3.2 THE ROLE OF ACTIVE AND PASSIVE ELEMENTS IN AMPLIFIER CIRCUITS

Active devices are those elements that can produce the control function called gain. Passive elements are elements that cannot gain but are used to establish levels of or limits to current, voltage, and/or power. Typical elements that fall into the passive category are resistors, capacitors, and inductors. Transistors and integrated circuits are the most commonly used active devices. (Even though an integrated circuit may contain many other "devices" inside, it usually functions as a distinct element in a complete circuit. Since the function of the internal devices in an integrated circuit cannot be split apart from the whole circuit's function, it will often be referred to as a singular "device.")

Although the active device may be the central gain-producing element in an amplifier, passive elements are equally essential to the amplifier's function. As pointed out in Section 3.1, amplifiers have electronic limitations. Often these limitations are due to the active device, which is used when building an amplifier. Often, passive devices are selected specifically to overcome or to reduce the effects of these limits or restrictions. In other cases, the passive devices in a circuit are selected to place even greater control over the amplifier's overall function.

As an example of the roles played by active and passive devices, let's review the highlights of a typical amplifier's design. At this time we will not worry about the exact equations and statistics used to select specific device values. First, let's concentrate on the amplifier's overall requirements.

Using Figure 3.1 as a starting point, it can be seen that the amplifier must produce a gain value. The output level is to be higher than the input level. Therefore, an active device that can provide this high-level output will be needed. The energy delivered at the output is provided from the power supply under the control of the active device. Let's assume that an active device has been selected which can provide the required gain value. If we can find such a device, it seems that our problem has been solved. In some cases, where an integrated circuit has been selected, we may need to do little more than "plug it in." It would be necessary to use passive devices to assist in making the active device function exactly as required.

One of the more common uses for passive devices, even when using an integrated circuit, is to make the amplifier selective in the type of signal that it will amplify. In Figure 3.2 input and output filters (discussed in Chapter 2) have been added to the basic amplifier. The objective of these simple high-pass filters is to block the flow of dc current from the input element to the amplifier and from the amplifier to the output device.

Should dc flow into the input line of this amplifier, several problems may be produced. This dc level may interfere with the dc conditions produced by the power supply connected to this amplifier. Another reason for not allowing dc levels in at the input terminal would be that this portion of the input may be totally useless. Since most signals to be amplified are ac, the dc that may be present at the input may not represent any valuable information. (There are some applications where this is not the case.) In this case the dc from the input circuit would be treated by the device just like any other information

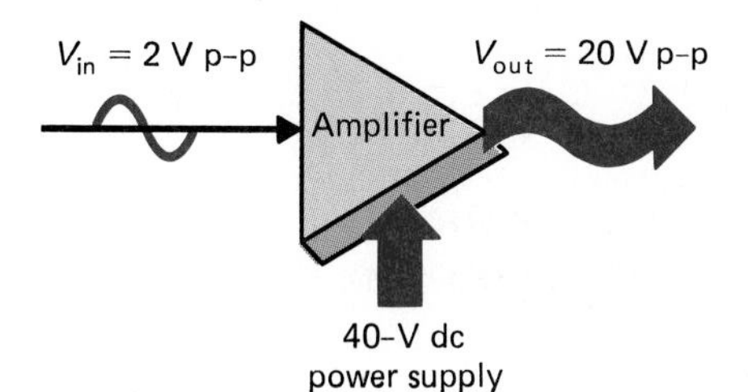

FIGURE 3.1 Control of Energy in Amplifier System

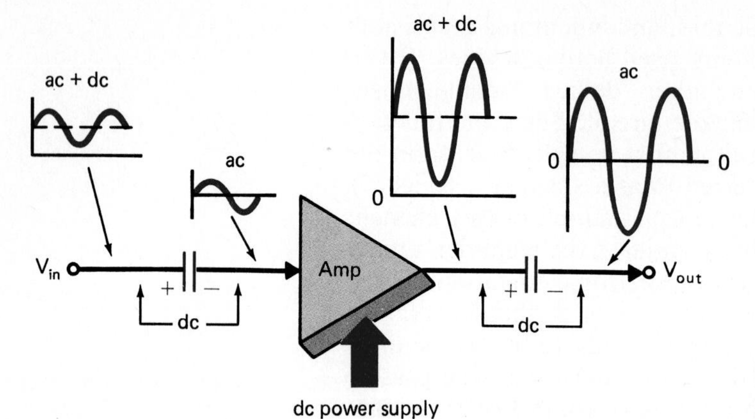

FIGURE 3.2 Use of Capacitors to "Contain" DC Levels

signal and it would attempt to amplify it. The active device would therefore be using power supply energy to amplify a useless quantity.

For similar reasons, the output line contains a high-pass filter. The only value that should be forwarded to the output should be the amplified ac information signal. Thus we have added the passive-device filters to this amplifier to isolate ac signals from unwanted dc values. Such filters exist in nearly every form of amplifier, even with sophisticated integrated-circuit amplifiers.

Another application of passive devices to such an amplifier is to use them to assist in matching the input and output impedance (resistance) of this amplifier to either the impedance of the circuit connected to its input line or the one connected to its output. If the transfer of electrical power into the amplifier in Figure 3.3 is to be efficient, the amplifier's input impedance (resistance) should closely match the impedance of the circuit connected to its input. Similarly, then, the output impedance of the amplifier should match the impedance of the load.

Often, the actual input and output impedance of the active device selected for use in this amplifier does not meet these requirements. Therefore, it may be necessary to use passive devices, often resistors, in series or parallel with the active device to achieve the desired impedance value.

So far, we have seen the need to add passive elements to the amplifier to filter unwanted dc and to provide impedance matching. Another active-device characteristic that is not desirable but can be overcome by the use of passive elements is the single-polarity problem noted earlier. Most amplifying devices can pass current in only one direction. The flow of current through the device can vary in level but cannot reverse direction. Therefore, it is

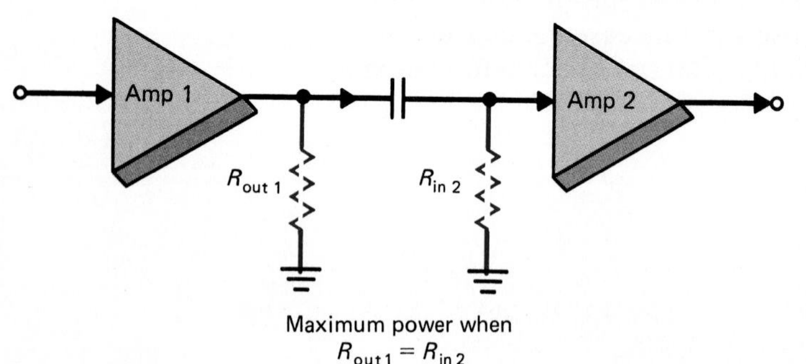

FIGURE 3.3 Impedance Matching between Amplifiers

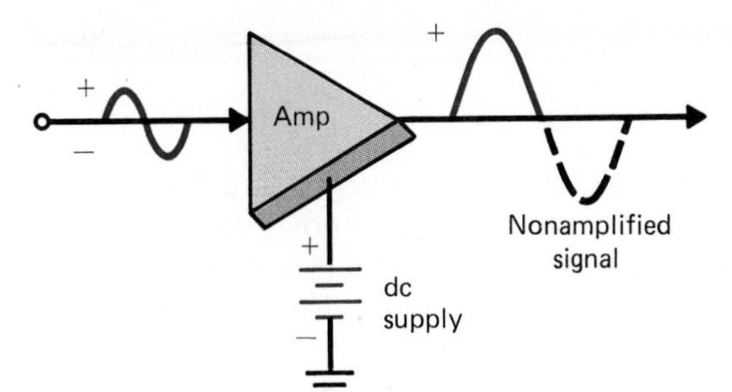

FIGURE 3.4 Loss of Input Signal Due to Single-Polarity Function of Output

possible that either the positive (+) or negative (−) portion of an ac input may not be amplified. Furthermore, it is possible that the device used may be damaged if the input attempts to force it to work in an unacceptable direction.

This polarity sensitivity can also be demonstrated by investigating the power supply from which the amplifier obtains the energy used for amplification. Most power supplies are of one polarity in reference to ground. It would therefore be impossible to produce an output polarity (in respect to the same ground) that was of a different polarity than that of this supply (see Figure 3.4).

This problem, the conflict between the ac input and the single-polarity requirement of the active device, can also be solved by the use of passive circuit elements, again typically resistors. If this amplifier's circuitry provided a preset dc voltage level at the input terminal of the active device, any ac input would be added to or subtracted from this level, depending on polarity. When properly done, the blend of these voltages could result in an actual input that had a varying level but was of one polarity (see Figure 3.5). For example, should an active device be unable to respond to the negative portion of an input sine wave (see Figure 3.4 again), this problem could be resolved by using a simple resistive voltage-divider circuit operating off the dc supply [thus providing a constant positive (+) dc level at the device's input; see Figure 3.5]. Such a voltage would add to the input sine wave. The composite of these two values would be a waveform which was always positive (+) in polarity, like that shown in Figure 3.5. Since the voltage presented to the active device is now *all* positive, it can respond to all values of the original input waveform.

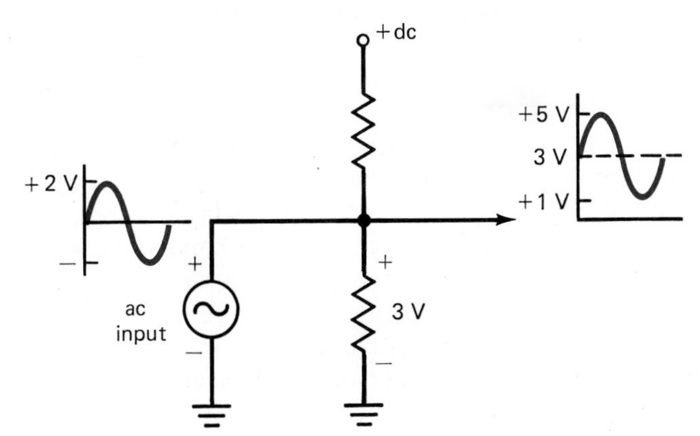

FIGURE 3.5 Mixing AC with DC to Obtain All Positive (+) Polarity Signal

EXAMPLE 3.1

As shown in Figure 3.6, if the input peaked at both a positive 5 V (+5 V) and negative 5 V (−5 V) and the dc was a constant positive 7 V (+7 V), the sum of these two still results in a positive level that ranges from +12 V to +2 V. Thus the input presented to the amplifier could always be positive in polarity, guaranteeing that current will flow only in one direction. (This is true as long as the input voltage's negative level never exceeds the positive dc level set by the voltage-divider circuit.)

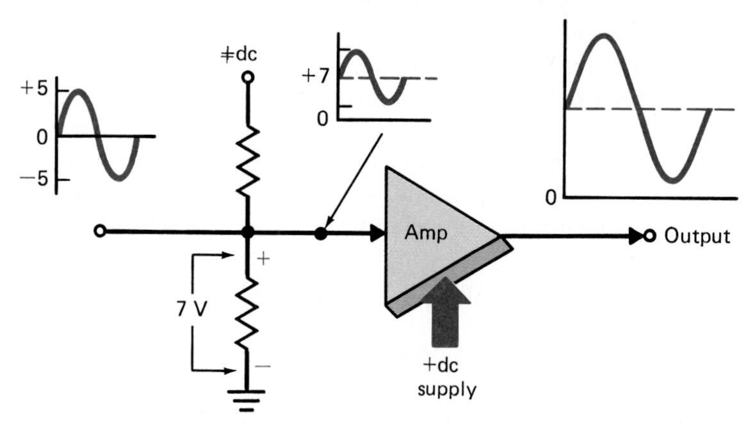

FIGURE 3.6 Amplifier Output with Single-Polarity Input

A very similar effect can be seen at the output. Since this device will be controlling a positive-polarity dc source (in reference to ground), it stands to reason that the output will also be all positive in polarity. It will still vary, but will remain positive in polarity.

Another reason for the need for additional passive devices is to provide system stability. Many electronic devices are quite sensitive to temperature variations and other changes in operating conditions. Also, most are manufactured to meet average standards, but on an individual basis may not meet these standards precisely. Active devices also have a certain tolerance or degree of error, much like resistor tolerances.

Because the exact values of characteristics such as gain are not certain from one device to the other, and since these values can change due to temperature and other affects, it is necessary to construct circuitry that will allow for such variations. If the circuit used must depend on an exact value of gain from the active device, failure or damage could result if variations from this exact value occur. With some preplanning, though, a circuit can be made that will accept these variations but still produce the desired output.

The actual process of obtaining stability is usually unique to the specific active device used and the circuit layout that is chosen. Again, it is common to use passive devices in coordination with the active device to act as monitors to sense instabilities in the circuit (changes from the desired gain, for example).

Once the errors are detected, automatic corrections can be made to the circuit, which will compensate or make up for the errors.

One active "element" that does not require a large amount of support by external passive devices is the integrated circuit (IC). Inside an IC are active devices such as transistors used to produce gain. In addition to these active devices, passive devices are built into the integrated circuit. (The term "integrated circuit" implies many elements, both active and passive, connected together in one package.) The advantage is, of course, that since these passive devices are built into the IC, they do not have to be provided externally. The more sophisticated the IC, the fewer are the number of external elements needed to support its operation.

So far in this section, we have reviewed the role of both the active devices and the passive devices used in an amplifier circuit. We have seen that the active devices provide the gain or control function desired of an amplifier system. The passive devices, which are added to the active device, modify the input and output voltages to help control the flow of energy and to stabilize the device should it vary from normal standards. Thus, for most active devices, passive elements provide support to assure the function of the overall system.

REVIEW PROBLEMS

(1) In this section, several active device types were mentioned. Prepare a list of these and other active devices from simple types to the more complex integrated circuits. In this list include the specific names of various types of transistors, integrated circuits, and other forms of active devices. Use both this book and others in the library to prepare this list.

(2) In certain applications it is necessary to have an amplifier which is selective of the frequencies that it will amplify. Such an amplifier fails to provide an output unless the input is at (or quite near) the desired frequency. Examine Figure 3.2 and, referring to a circuit analysis text that discusses resonant circuits (frequency selective), prepare a sketch for a circuit that uses such a filter instead of the simple dc filter illustrated.

(3) If a multistage amplifier circuit were to be used (as discussed in Chapters 1 and 2), and the average gain of each of four stages was to be 50, discuss the impact on the overall gain of this system should the amplifiers become unstable and their gain vary high or low by as much as 20%. Calculate the effect of such instability on the output. Write a brief discussion as to why such instability would not be desirable.

3.3 GENERAL IDENTIFICATION OF ACTIVE DEVICES

In this section we attempt to identify generally the various types of active devices. From very sophisticated integrated circuits to simple transistors, we attempt to define what each device can and cannot do and, in general, how to deal with such devices when applying them to an amplifier-based process.

Integrated circuits (see Figures 3.7 and 3.8), rapidly becoming the most popular form of active device, have a distinct advantage over other forms of active devices. Since ICs contain complete circuits in one compact package, they require fewer external components to achieve their electronic function. About the only elements that cannot be built into an IC are those whose physical size, mechanical needs, or power consumption prohibit them from fitting in the small package of an integrated circuit. Some components are intentionally omitted from the IC so they can be selected by the user, thus allowing the IC to be custom applied to a task.

From a repair standpoint, ICs simplify the job of diagnosing a system's malfunctions and getting a piece of equipment back into operation. If such equipment contains ICs, it is often necessary simply to identify the defective block in a block diagram which coordinates with one IC in the circuit. Replace that IC and the system is back in operation.

One special form of IC that will be used extensively in this text is the operational amplifier (op-amp; Figure 3.9). The op-amp is one of the smaller scale ICs. It is actually somewhat of a bridge between the discrete active device, which requires many passive devices to support its operation, and the large-scale integrated circuit, which can achieve a complete function by itself. Unlike the larger-scale ICs though, it requires the use of some external passive devices to become fully functional. It is far more stable than discrete active devices and has characteristics which are usually far better than those found in discrete devices.

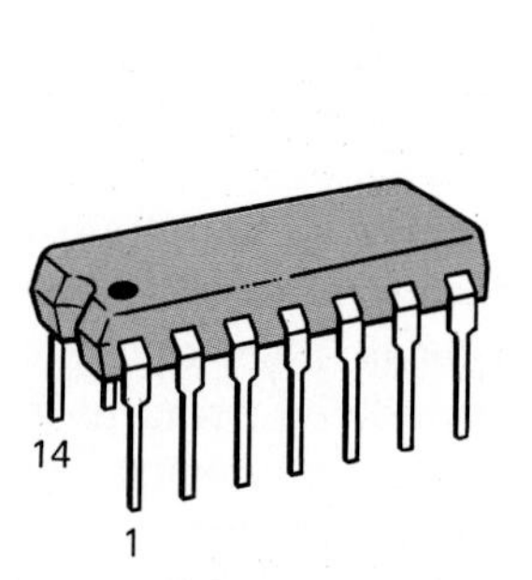

FIGURE 3.7 Integrated Circuit Dual-In-line Package (DIP)

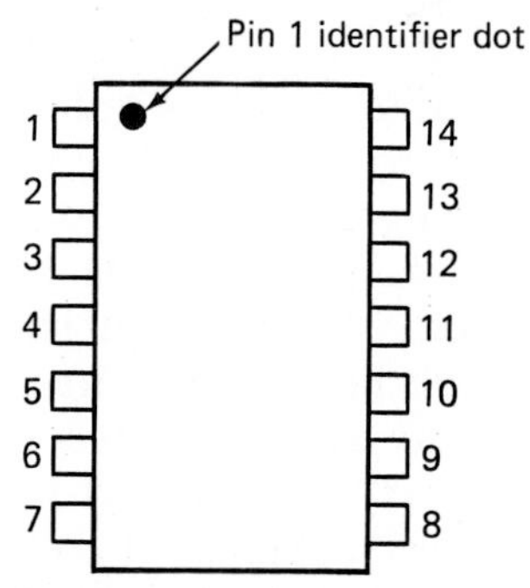

FIGURE 3.8 Pin Numbering for a 14-Pin IC (DIP)

FIGURE 3.9 Operational Amplifier

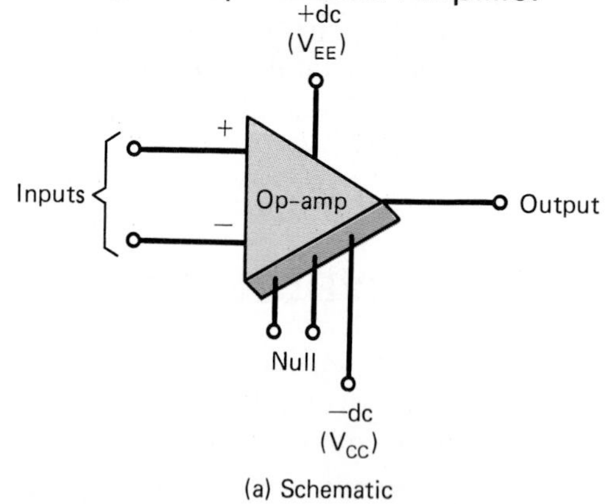

(a) Schematic

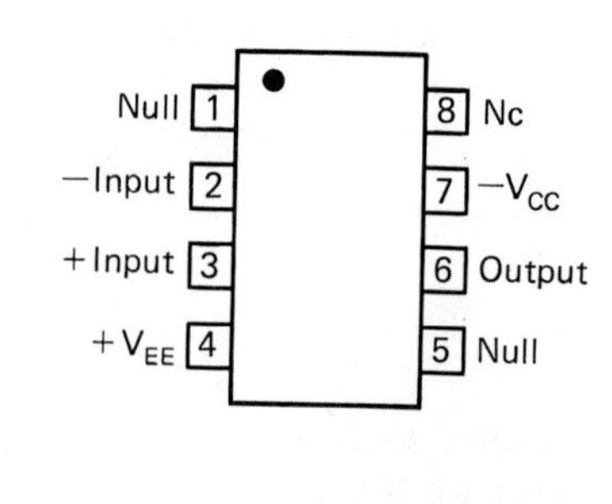

(b) Pin-out (8-pin mini-dual-in-line package)

Since the larger-scale ICs can often achieve the task of several amplifiers in a circuit, it would be difficult to describe how a circuit functions by using them in an example. But since the op-amp can usually be applied only as a single amplifier, it will serve quite well as we discover how systems work on a stage-by-stage basis. Additionally, many of the special-purpose passive devices that are added to an op-amp circuit will also fit into the design of a circuit that uses discrete active devices.

Individual nonintegrated devices used to produce gain are called discrete active devices. These devices require the most support from passive devices. Since they achieve only the function of gain and are single elements, they are quite basic in form and function. The most popular discrete active device is the semiconductor transistor. (Prior to transistors, electron tubes were used. They have such a limited application in modern electronics, although their function will not be discussed in this book.)

The term ''transistor'' is an identification of a large category of discrete active devices. Transistors are any of a group of three-terminal semiconductor devices that have the ability to control current flow.

There are two distinct versions of the transistor. First there is the **bipolar junction transistor**, more commonly called the BJT. The BJT was the first form of the commercially available transistor. In later chapters we explore how the BJT operates, but generally it is a current-controlled device; that is, the amount of current applied to the input of the device controls the amount of output current (or voltage) from the device. Figure 3.10 illustrates typical shapes for these devices. Figure 3.11 shows the schematic symbols used for the two available forms of BJTs (NPN and PNP types).

FIGURE 3.10 Typical Transistor Packaging

The second version of the transistor is the **field-effect transistor** (commonly abbreviated FET; Figure 3.12). Unlike the BJT, the FET responds only to input voltage levels. The amount of output voltage or current is proportional to the level of input voltage. Thus the FET is a voltage-controlled device. As can be seen from these two descriptions, the BJT and FET are quite different in the form of input to which they will respond.

FIGURE 3.11 BJT Symbols

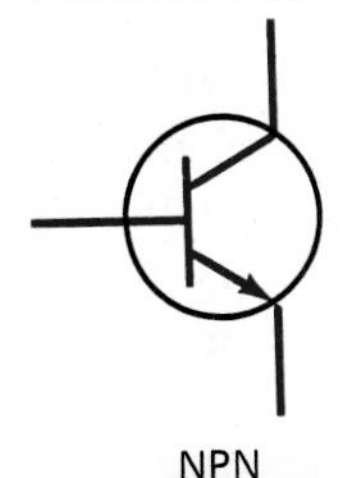

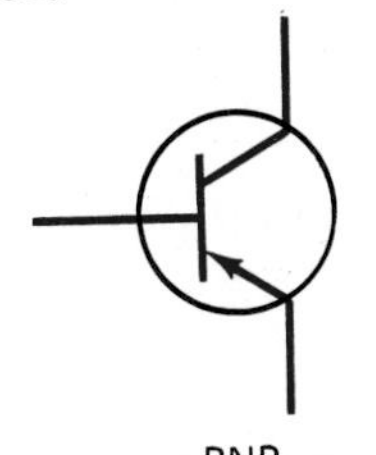

FIGURE 3.12 Field-Effect-Transistor Symbols

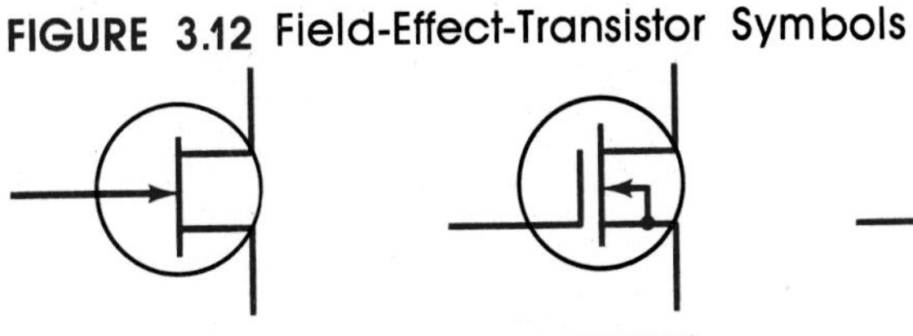

BJTs and FETs are the two major categories of transistors that are currently available. Although these two differ quite a lot in how they react to inputs and are mechanically quite different on the inside, they are both made up of similar semiconductor material and are often packaged similarly. There are quite a few versions of both BJTs and FETs available for use in electronic circuits. These versions have different specifications related to voltage, current, power, and so on. In subsequent chapters we explore these various specifications and the varieties of BJTs and FETs that are available. There are several other categories of discrete semiconductor devices that are not in the category of transistors. These are special-purpose devices and are discussed in special sections of this book.

EXAMPLE 3.2

The schematics for three versions of the same circuit are shown in Figure 3.13. Schematic (a) uses one large-scale IC to produce the desired output. Schematic (b) employs the smaller-scale op-amp ICs. Schematic (c) uses discrete BJT transistors to achieve the same output. As can be seen, the circuit becomes most complex when BJTs are used. These three circuits are presented to illustrate the amount of passive-device (resistor, capacitor, inductor) support needed by each type of device.

REVIEW PROBLEMS

(1) To gain more knowledge of the types of large-scale integrated circuits that are available, use a manufacturer's or distributor's catalog to locate at least five different integrated circuits (do not use digital types). List the manufacturer's part number, the function of the integrated circuit and basic power supply, and the input and output specifications for these devices.

(2) Using Figure 3.13, prepare a list of possible devices (both active and passive) that could be defective if a fault were found related to the second step in the block diagram. Make a separate list for each of the three circuit versions. Then prepare a brief conclusion as to how much time would be involved in repairing each version (troubleshooting and replacement time together; assume all parts soldered to a circuit board).

(3) Refer again to Figure 3.13 and list the elements that appear to be common to both the large-scale IC and the op-amp version of this circuit, then compare the op-amp version to the discrete transistor type. Where possible, try to identify the purpose of these elements and attempt to describe their effect on the overall function of each circuit.

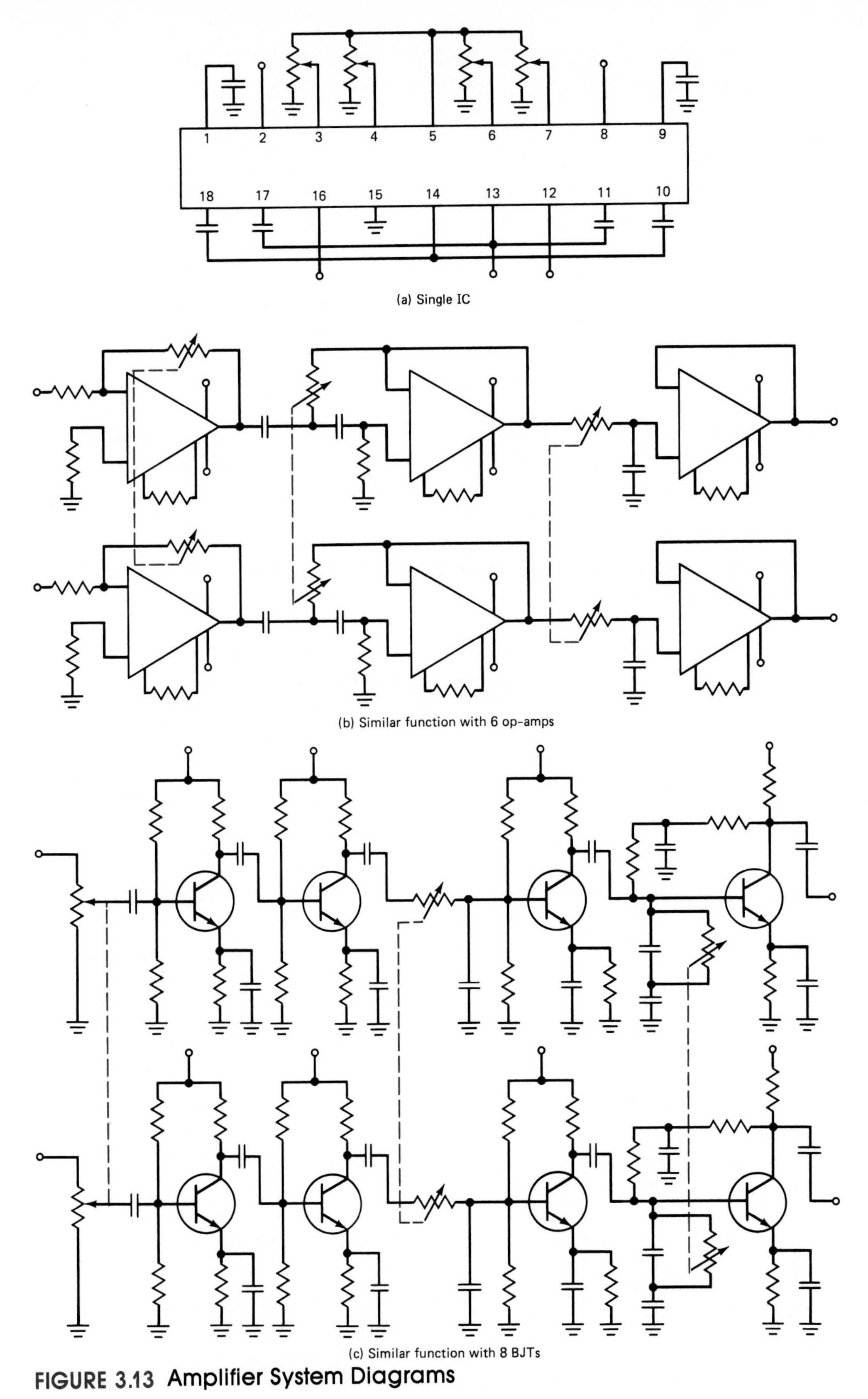

(a) Single IC

(b) Similar function with 6 op-amps

(c) Similar function with 8 BJTs

FIGURE 3.13 Amplifier System Diagrams

3.4 COORDINATION BETWEEN ACTIVE DEVICES IN A SCHEMATIC AND BLOCK DIAGRAM

In Figure 3.13 a set of schematics were shown to coordinate both with each other and with a master block diagram. Each of the schematics achieved the same function as described by the block diagram, but they used different active devices and supporting passive devices to achieve the function. In this section several examples of the relationship of schematic drawings (especially the active devices used) to block diagrams showing an overall function will be given. Being able to make such comparisons will prove to be quite helpful both in reading the following chapters and when actually working with these diagrams in an on-the-job situation. The basic assumptions we will use as we analyze these circuits are as follows:

1. Active devices, either transistors, ICs, or op-amps, are used to obtain gain. Their function will be to accept an input (either voltage or current) and to produce an amplified output (again either voltage or current).
2. Passive devices (resistors, capacitors, inductors, transformers, etc.) will be used either to condition the signal flowing through the circuit (as a filter does) or to support the special needs or correct for the faults of the active device (such as stability).

When we analyze these circuits, we attempt first to describe the role of the active device in terms of the function detailed by the block diagram. Then a very general description of the role of the passive elements will be made. In this investigation we will not spend time justifying the specific values used in each circuit; that will be left to other chapters in this text.

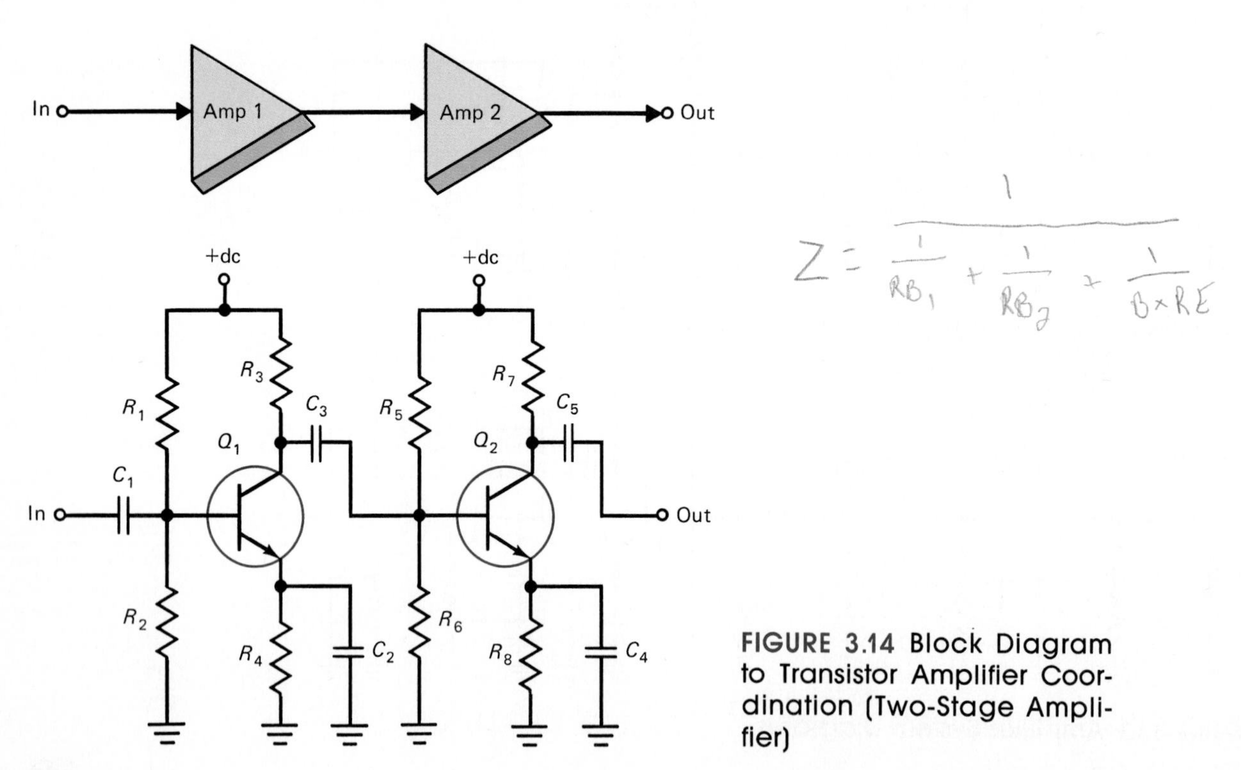

FIGURE 3.14 Block Diagram to Transistor Amplifier Coordination (Two-Stage Amplifier)

EXAMPLE 3.3

In Figure 3.14 the two BJTs (Q_1 and Q_2) relate to the basic function defined by amp 1 and amp 2, respectively. They provide the gain from input to output. Since BJTs are polarity sensitive, the two resistors on each amplifier's input (R_1 and R_2, R_5 and R_6) provide a dc level which is added to the ac signal to produce a single-polarity input for each transistor. Output resistors (R_3 and R_4, R_7 and R_8) function with the device to provide the correct signal output level. These resistors also help determine the gain and stability of the amplifier. Capacitors are used in the traditional role of blocking dc voltages and passing ac voltages.

EXAMPLE 3.4

In Figure 3.15 the op-amps achieve the function of amplification indicated in the block diagram. The resistors in the circuit help to establish the gain of each op-amp. Since op-amps do not have dc input problems, capacitors to block dc are not required.

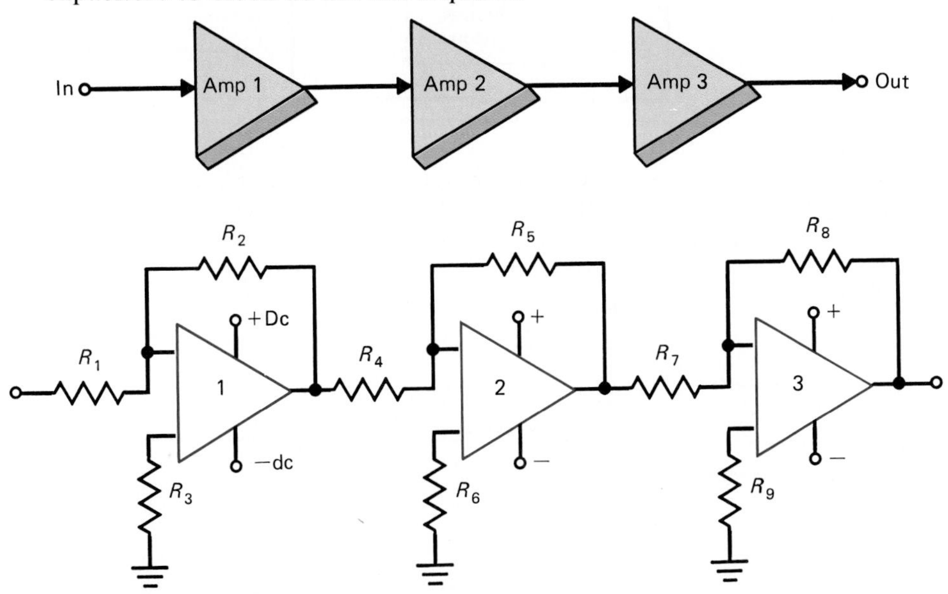

FIGURE 3.15 Block Diagram to Op-Amp Amplifier Coordination (Three-Stage Amplifier)

EXAMPLE 3.5

Unlike the previous examples, a complete function on one integrated circuit is presented in Figure 3.16. The block diagram shown is for the *internal* function of the IC. To summarize its function would involve just listing one amplifier block with a large gain characteristic. As can be seen, the IC amplifier incorporates three amplifiers in one package. As usual, capacitors are used to pass ac and block dc. The function of the resistors is not to establish bias conditions but to produce operating specific to that IC.

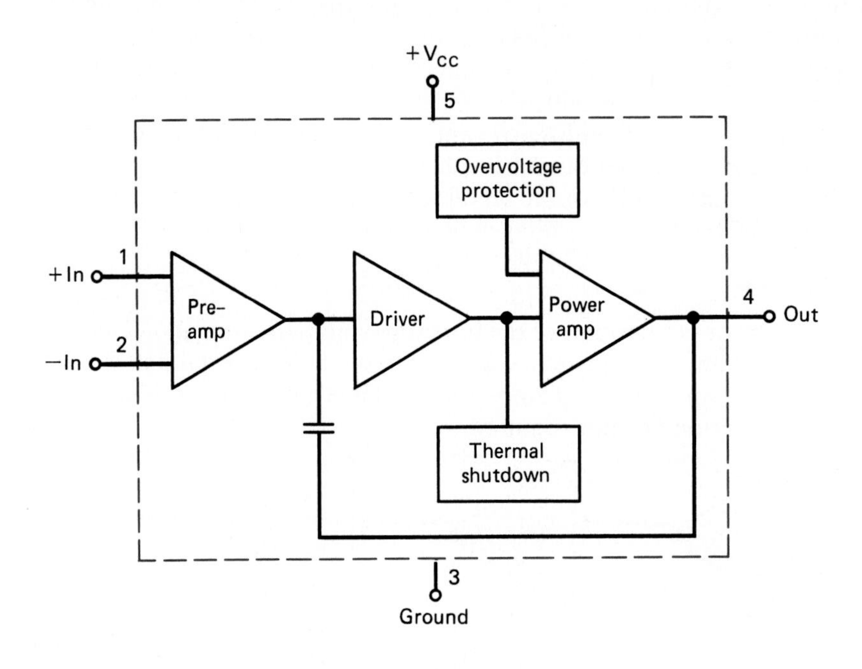

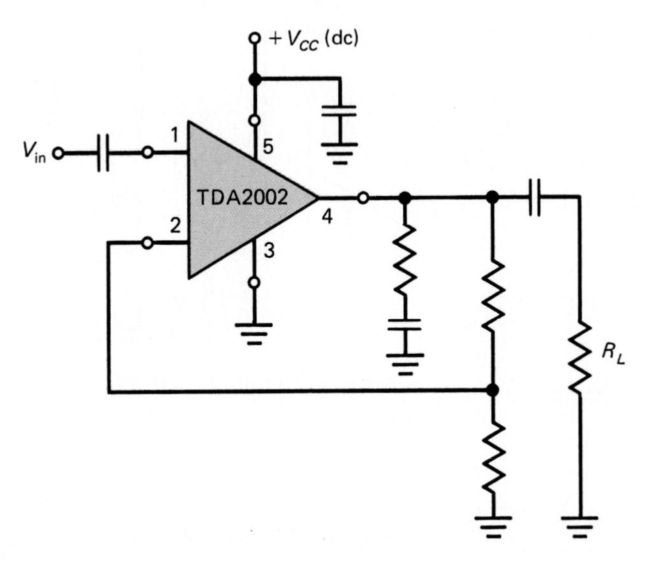

FIGURE 3.16 Amplifier System in a Single IC Package

REVIEW PROBLEM

(1) As was done in Examples 3.3 to 3.5, prepare a general overview for the circuit and block diagram shown in Figure. 3.17.

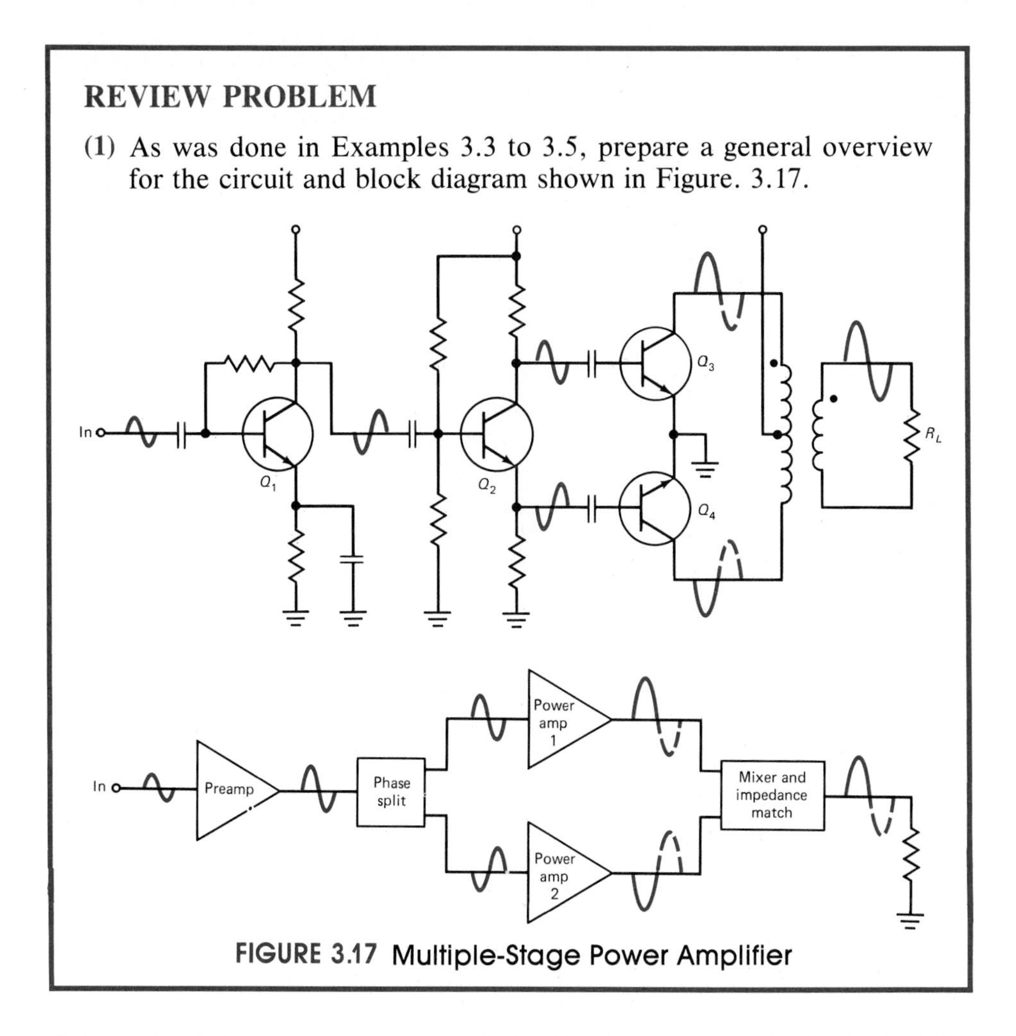

FIGURE 3.17 Multiple-Stage Power Amplifier

3.5 CONTRAST OF ENERGY SOURCES TO INFORMATION SOURCES

In the preceding sections we have discussed the roles of active and passive devices. In this section we investigate the forms and types of energy sources used in these circuits. We will see that there are basically two types of sources, the information source and a source used to supply operational power to the circuit.

Looking at the simple amplifier diagram shown in Figure 3.18, we can see the two forms of energy sources that this circuit needs. In all of our previous examples we have simply assumed that the input value was a given

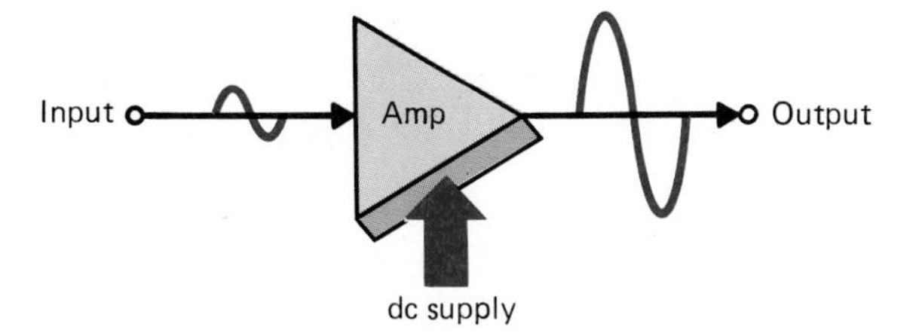

FIGURE 3.18 Simple Amplifier Function

figure. Even though we initially discussed the power supply as a special part of the amplifier, we did not specifically detail its role in amplification. We now pay special attention to each of these sources of energy.

It is a common error to assume that the energy applied at the input creates the energy that is finally delivered to the output of the circuit. If this misconception was true, there would be no need for amplifiers. But this factor, where input energy creates output energy, is applicable only in systems as simplistic as a child's tin-can-and-string telephone. The actual amount of energy that is consumed at the input of an amplifier is usually very small when compared to the energy delivered to the output device. We have described the term "gain" and have shown that it is a measure of how an amplifier can produce *more* output than is available at the input terminal.

When doing an analysis of a system that contains several amplifier stages, it is common to use these source symbols to represent the internal production of signals. As the input passes through the first initial stage of an amplifier, it is subjected to the gain of that amplifier, and a higher-energy signal is produced at the output. To easily analyze the following stage of the amplifier, we can simply use the symbol for a source as the input of this second stage (Figure 3.19). Often, each stage of a system can be isolated in this way and thus simplify the analysis by breaking it up into easy-to-work-with pieces.

The other source in the system is called the *power supply*. The power supply provides the energy that is to be controlled by the active device. When an amplifier is operating, both active and passive components use up electrical energy in a nonproductive manner. Many components will heat up as current passes through them. This is already a known fact for resistors, and as we explore active devices will also be shown to be true for them. This loss of energy in the circuit is not useful. The heating will not aid in the production of the higher-level output expected. It will simply be a waste of energy. This energy must come from somewhere and in fact is provided by the circuit's power supply. Thus it can be seen that this second energy source must provide

FIGURE 3.19 Isolation of Single Amplifier from Multiple-Stage Circuit (Showing Use of Input Signal Source to Represent Previous Stage)

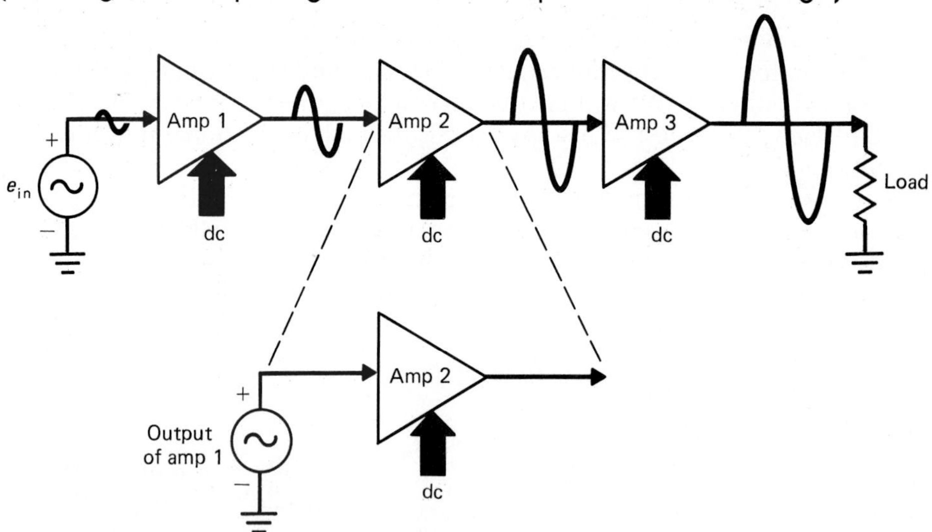

the energy used to produce an amplified output and to supply the energy normally lost due to heating. Additionally, this supply is called on to provide the voltage and/or current needed to support the active devices.

The term used for all of the special needs of the active device is **bias**. To bias the active device is to establish the correct electrical conditions (voltage or current) around the device so that it will react in a predictable and controlled manner. These bias conditions are usually set by the designer of the circuit, in respect to the needs of the active devices used. In some circuits bias is established simply by attaching a power supply. In other circuits, a network of passive elements working in coordination with the power supply is needed to establish bias conditions.

It is interesting to note that quite a large amount of the energy consumed by a system from the power supply is lost energy. Only a small percentage of the energy (power) consumed from the power supply actually results in output power. Most systems only deliver, to the output, from 10 to 40% of the energy used from the power supply. Thus it is possible for a system that delivers 10 W to a load to actually consume 40 W from the power supply.

The use of dc rather than ac as a power supply voltage is, in nearly every case, mandatory. Since the input to an amplifier is in the form of amplitude and frequency variations, if the power supply attached to this amplifier were also to have amplitude and frequency variations, the output would be a mix of these two signals (one desirable, one undesirable). If these two signals were to be blended electronically it would be nearly impossible to separate them later on. The only form of voltage (current) which carries no information in the form of frequency or amplitude changes is dc. Thus, even if the dc is mixed with the information, there is no interference caused by it. Dc can often be separated from ac quite easily and effectively. Simple low-pass or high-pass filters can achieve this separation. One popular form of power supply is the battery. Outside of the battery's long-term loss of energy capacity, it can provide a reasonably constant level of voltage (dc).

Another form of supply is one that can be attached to an electronic system but will not loose power over time as a battery does. Later we investigate these power supplies. Actually, they are power converters. They have as an input the ac power from a wall outlet and convert this energy to the constant dc level needed by an electronic system.

In most cases either the battery or the conversion type of supply can be

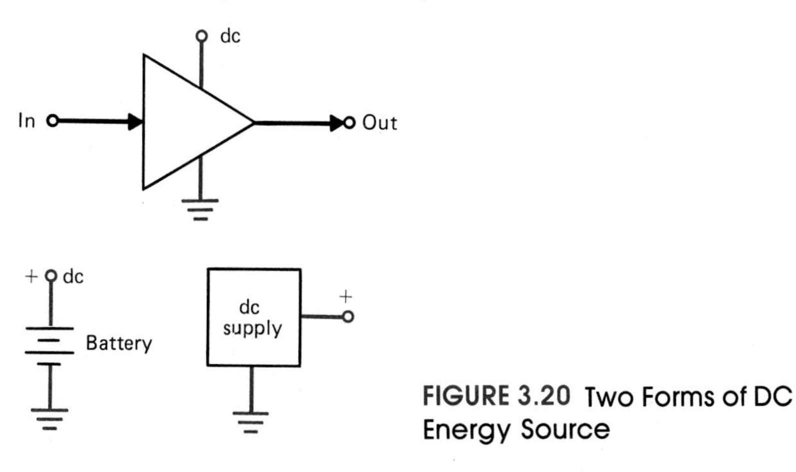

FIGURE 3.20 Two Forms of DC Energy Source

used with any circuit. Therefore, it is customary simply to specify the need for a power supply without specifying exactly which type is to be used. In Figure 3.20 the typical schematic symbols used for power supply elements are shown. In the first illustration the standard symbol for a battery is shown.

REVIEW PROBLEMS

(1) Figure 3.21 shows three amplifiers that produce gain for a final output. Redraw each amplifier individually, substituting the symbol for a signal source at each input terminal. Assume that each is designed to produce a voltage output based on the gain listed for each. Be sure to list the actual value of this source with each new drawing.

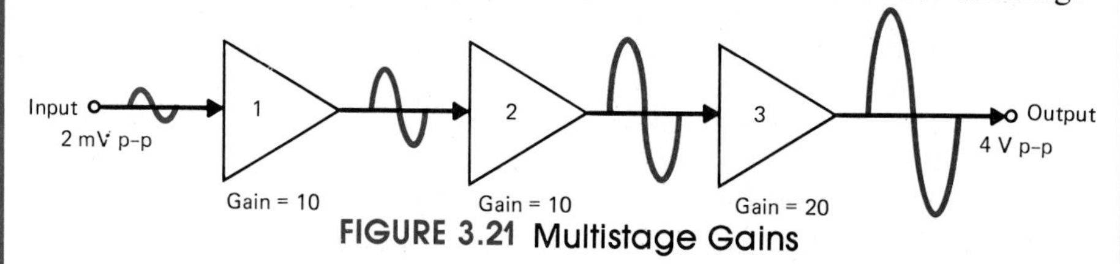

FIGURE 3.21 Multistage Gains

(2) An electronic circuit was found to have an excessive amount of "hum." This defect produced a constant 60-Hz noise (hum) in the output of the amplifier. The input to the amplifier was tested and no hum was detected at that location; in fact, the hum existed even when the input was disconnected. Based on discussions in this section, speculate as to the source of this undesirable noise. Suggest a possible way to test the system to find the source of the hum.

(3) Figure 3.22 could be simplified by the use of connector symbols for a common power supply. Redraw the schematic using such symbols and a common power supply symbol (substitute it for the schematic shown in the power supply block).

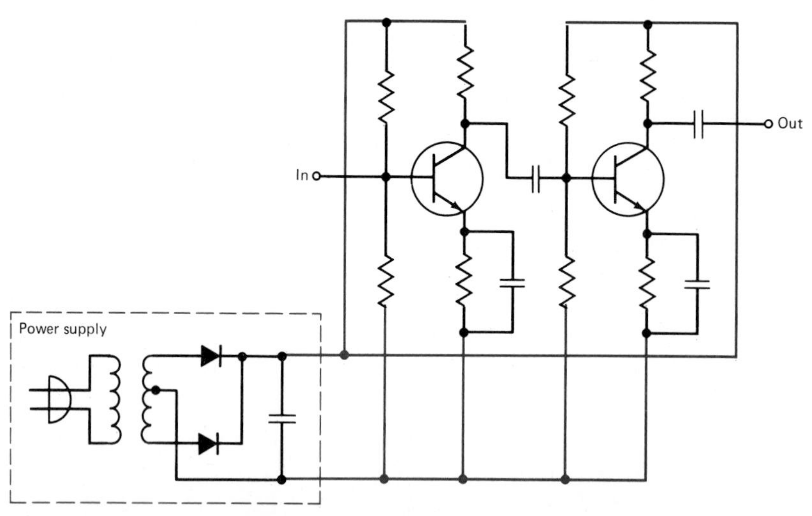

FIGURE 3.22 Amplifier Showing Power Supply Connections

A more common symbolism is shown in the second illustration. In this illustration simple connector symbols are used to label the connection points for the supply voltage. In the adjacent block, the exact statistics for the supply are given. This illustration is most often used since, as noted above, either a battery- or converter-type supply could be used.

In this section we have explored the use of two sources in an electronic circuit. First, the information source was shown to provide the signal that was to be amplified by the system. This input is typically illustrated using the common ac source symbol. The main energy supply, from which each active element obtains energy to produce gain, is a dc energy source. Both of these sources are needed if a system is to produce the desired amplification.

3.6 TROUBLESHOOTING BASED ON THE GENERAL CONCEPTS OF FUNCTIONAL PARTS

In this section we take the concepts of active- and passive-device applications and of energy sources used and apply them to a general troubleshooting process. The failure of a system to operate correctly can often be traced to the failure of some small segment of the overall system. In most cases, this failure is due to the breakdown of one specific component.

Many modern circuits are designed with integrated circuits and modular subsystems. The task of repair is often related to isolation of the defective subsystem. In some cases, this may involve locating and replacing a defective circuit board that contains the appropriate subsystem. Or it may require the isolation of one defective integrated circuit. In some cases, the isolation of one or more discrete devices (active or passive) is desirable. More often than not, though, the actual cost of replacing a single component exceeds the value of the entire circuit board on which it is mounted. Therefore, from an economical standpoint, it is not worth the effort to do more than isolate subsystem defects and remove the circuit board that contained the defective component.

In some cases, the defective circuit board may be repaired or remanufactured. This still requires isolation of the defective section and on-the-spot replacement of the complete subsystem. The defective subsystem would then be repaired at a later time, either at the manufacturing facility or when the repair technician was not under pressure to return the entire system to operation.

Of course, there are cases where the replacement of single defective elements is desirable. When the defective element is an integrated circuit, it is often treated as a subsystem, and repair involves the replacement of this device. (ICs are not subject to internal repair; they are usually discarded when found to be defective.) Some systems are custom built and thus subassemblies may be one of a kind. Naturally, a replacement for the defective subsystem may not be possible in this case. To return such a system to operation, it would be necessary to isolate the single defective element and replace it.

Therefore, isolation of a defective subsystem is often the preferred method of solving the problem of repair. In Chapter 2 we reviewed the use of block diagrams to find a defective section of a system. Once such isolation has been done, a few additional steps need to be taken before we can be assured that the defect is specifically due to that block's failure.

First, it may be necessary to locate the probable source of the defect. If we assume that a specific segment of a circuit is at fault (specific block in the block diagram), it may be worthwhile to be able to describe the form of the defect. Was it due to the failure of an active device? Is the fault related to defects in one or more of the supporting passive devices? Can the failure be traced to an error in the signal source? Did the circuit's power supply fail? Without generally trying to indicate the source of the failure, it may be very difficult to be certain if a simple replacement of that one subassembly will fix the whole problem. The problem may show up again, either immediately or in the future.

In future chapters we detail the precise functions of active devices and their associated passive devices. For now, though, we review logically how defects in these and other elements may become obvious, and how to isolate the proper corrective action to take to get the system functioning.

In the following paragraphs it is assumed that (as presented in Chapter 2) work has been done to isolate the defect to a specific block in a system's block diagram. Each heading identifies the type of defect that, in general, could have created the subsystem's failure.

Power Supply Faults

A defect in the power supplied to a circuit can result not only in that circuit's malfunction, but could damage some or all of the active and passive devices in that circuit. One of the most common tests of a circuit is first to measure the value of the power supply voltage delivered to the circuit. Any major deviation from the specified level will cause the entire circuit to function abnormally. Usually, with the defective circuit in place, the level of dc voltage supplied to the circuit is measured. If this does not meet the specified level, further tests of the supply are made before proceeding.

The symptoms that are produced by a defective power supply range from rather simple to major. If a power supply is not able to provide enough voltage and current to support a circuit's operation (such as weak batteries in a battery-operated system), it is possible for the system to function, but not up to specifications. Such a simple defect may not damage the circuit, and repair of the power supply functions may be all that is needed.

Some power supply defects cannot easily be detected, because they occur only for brief moments. There may be a brief drop in the output of the power supply. In this case the drop may be severe enough that it will damage sensitive active devices. There is also the possibility that the supply may produce a surge of higher-than-normal voltage. In this case, damage to active devices is quite possible. It may be difficult to detect such power supply defects. It is not uncommon, if the source of a defect is not found, to assume that the supply may have these transient problems. (In later chapters details on how to compensate for these defects are presented.)

One symptom of a defective supply is that if a supply feeds several circuits, a defect in the supply will show up as a defect in all these circuits. It is unusual to have several circuits malfunction without having the power supply be indicated as the source of the defect. Therefore, if more than one block of a block diagram was found to be defective, one would first suspect the attached power supply before attacking each of the individual circuits.

Signal Source Faults

If the source of the signal that is to be amplified is defective, this defect will show up in the output of that section. In fact, the process of isolation of a faulty section is often done simply by tracing the flow of the signal from section to section. If the output of a section duplicates the input, that section is often assumed to be good, even if the input is not in the form in which it is supposed to be. Typically, one would trace the signal back from stage to stage until a defective signal was produced by a circuit that had a proper input signal. Thus the defect must have been produced in that stage.

The defects seen in an input signal to a system (or subassembly) fit one of about three categories. First, the signal may not be at all in the form expected. It may be too low or too high in amplitude or it may contain other signals that were not intended to be present. Fortunately, most input signal defects do not produce effects that could damage the components in a circuit. Usually, the results of an error in the input signal produce an enhanced error in the output of the system.

The most common method of testing for defective signal sources is by substitution. First, the suspected source is disconnected from the circuit. This could mean a direct disconnection of the signal source (e.g., a microphone) or may require removing the interconnection between two stages, since one stage is the signal source for the other. Then a source with a known value (such as a signal generator) is connected in place of that source. If this signal is handled correctly, the defect must lie in the disconnected source.

Active-Device Faults

One of the most probable causes of a malfunction is the active device in a circuit. Since it is commonly the least stable device in the circuit, it is often the source of malfunctions. In most cases, faults in the active device will produce noticeable errors in the circuit. Many voltage levels, both dc and ac, will not match the normal values listed with the schematic if the active device is defective. Assuming that the power supply and signal source have been verified to be correct, the active device is usually targeted for replacement if the circuit is malfunctioning. The method for testing specific active devices to see if they are defective will be covered in later chapters, since the test procedure is often specialized to the type of device being used.

The causes for device failure often fall into categories that are common among all devices. An active device will usually fail if either the electrical limits set for the device are exceeded, if it is subjected to environmental conditions that destroy its characteristics, or if it is applied in a circuit that produces long-term "fatigue" on the device. Modern solid-state devices do not have a typical "life span" after which they can no longer be used. Therefore, an active-device failure is usually caused by some other defect, in the system (e.g., power supply surges), in its application in that circuit (e.g., design error), or in the operating environment (e.g., operating in a high-temperature environment).

The point of this argument is that once an active device is found to be defective, the balance of the system needs to be inspected to attempt to isolate the cause of this defect. In some cases, an active-device failure is caused by a true device failure or by an unusual transient in the circuit (e.g., accidental short circuit that produced a current surge in the device). In many other cases, the source of the failure of that device is traceable to some other defect. The simple replacement of that device will not necessarily eliminate the source of future defects in the system.

Passive-Device Faults

It is possible for a passive device to fail in a circuit. Passive devices can deteriorate with age. Also, passive devices may either fail or deviate from their expected values if subjected to the same type of abuse as that noted previously for active devices. A passive device's failure may cause a chain reaction in the circuit, which could cause several devices, most often its related active device, to become damaged.

Resistors typically fail either by totally opening (infinite resistance) or by drifting way off value. Capacitors can either drift off value, become shorted, or develop excessive leakage currents. Inductors usually fail by opening or by drifting off value. Usually, if one of these components drifts off value, simple voltage-level measurements can detect such a problem since a drift off value will produce an out-of-proportion voltage level on the device. Total defects can usually be seen as major deviations in expected voltage levels.

In some cases, a simple replacement of the defective device will correct the problem. In other cases, the fault was caused by some intolerable operating condition. If that condition is not eliminated, the failure can be expected to occur in the future. As well, a defect in a passive device could probably cause either damage or fatigue in other devices in a related circuit. Thus once the passive device is replaced, further tests should be made to determine if more damage resulted from that device's failure and to attempt to isolate the source of that device's failure.

It can be seen that the failure of one circuit could be the result of one or more defects elsewhere. Although the objective of troubleshooting a circuit is to return the circuit to a functional state, it is necessary to attempt to isolate the cause of the circuit's failure. The causes could be related to power supply defects, input signal defects, active-device failures, and/or passive-device faults. Knowing the source of the failure is a major step in the right direction to solving the overall problem of getting the system back to long-term operation. Solving operational problems, though, could involved anything from the simple replacement of a defective resistor to a complete redesign of the circuit to compensate for defects that would occur again if not eliminated by a restructuring of the circuit.

REVIEW PROBLEMS

(1) In this section a basic procedure for steps to take in isolating a defect were presented. Prepare a simple description of a step-by-step procedure to take to detect the cause of a malfunction. You may assume that the defective block has been isolated.

(2) Prepare a checklist of the type of test equipment that would help a technician isolate defects in an audio amplifier. Along with each type of equipment listed, include a list of possible uses for that equipment in the testing of the amplifier.
(3) An electronic control system is defective. Upon testing, a transistor was found to be defective. When reviewing a set of repair reports on the system, it seems that the transistor was replaced three times before. The last time it was replaced was just two weeks ago. Suggest other action to take besides just replacing the defective device to help keep this defect from happening again.

SUMMARY

In this chapter the relationship of amplifier circuit elements to achieving the task of amplification (as seen in block diagrams) has been presented. Knowledge of this relationship is not only essential for understanding how present-day amplifiers function, but is a general process for analysis of amplifiers which will be encountered in the future.

The reader is encouraged to return to this chapter as new amplification systems are presented in later chapters in order to review these basic principles. Having a clear concept of the role of active and passive elements in a system will be a primary aid to using these systems and their internal parts more effectively.

In this chapter it was shown that active devices come in many different forms. The role of each, though, was seen to be the production of gain in one form or another. These devices produced outputs that were greater in magnitude than the input by manipulating the energy available from a power supply. Passive elements did not have the capacity to produce gain but instead produced a basic manipulation of voltage and current values. They aided the active device by providing necessary bias conditions.

PROBLEMS

Section 3.1

3.1. Prepare a brief list of the applications and any special limitations for amplifiers as noted in Section 3.1.

3.2. Illustrate an amplifier symbol incorporating both the signal input and output, the power supply, and input/output filtering.

3.3. List the basic reasons for using dc filtering at the input and output of an amplifier. Indicate a drawback to using this filtering.

3.4. If the input resistance of one amplifier is 1 kΩ and the output resistance of an amplifier that feeds this one is 2 kΩ, what problems may result because of these resistance statistics?

3.5. Indicate the expected output of the polarity-sensitive amplifier illustrated in Figure 3.23 if the waveforms shown in Figure 3.24 were applied to the input.

3.6. If the waveforms in Figure 3.24 were applied to an amplifier that functioned on a negative supply voltage, indicate what the output from that amplifier would be.

3.7. Using the waveforms in Figure 3.24, indicate what dc level could be

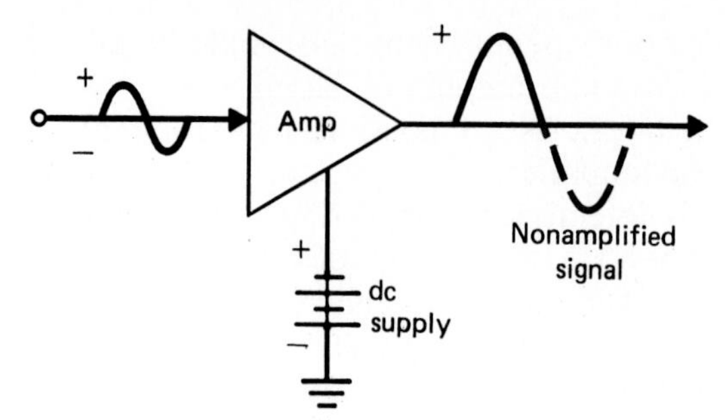

FIGURE 3.23

added to these to produce an undistorted output and illustrate the wave with the dc added. Do this for the amplifiers indicated in problems 3.5 and 3.6.

3.8. If an amplifier had an input voltage of 8 mV p-p (ac) and an output voltage of 1.6 V p-p (ac), find its voltage gain statistic. If this amplifier became unstable and its gain dropped by 15% from this value, find the new output (input is unchanged).

3.9. The data sheets for several active devices are presented in Appendix A. Look through these sheets and select two devices which, based on the brief description included for that device, could be used for amplification. (Look especially for information that relates to gain.) List the device part number, its name, and a brief quote of the data that lead to the decision to include the device in your selection.

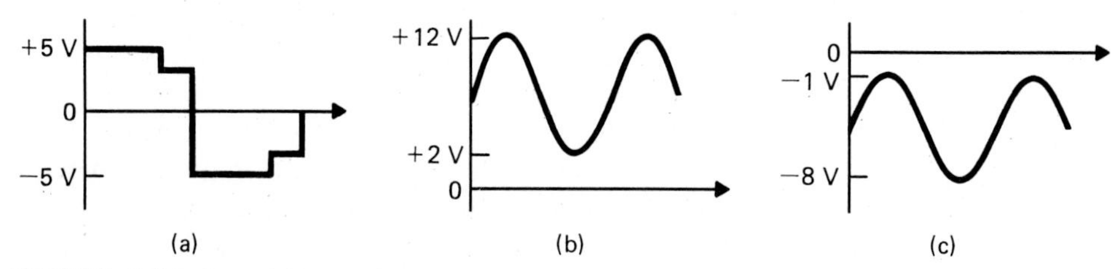

FIGURE 3.24 Input Waveforms

Section 3.3

3.10. Refer again to Appendix A and categorize the devices listed as either BJT, FET, or op-amp types. If a type does not fit, simply skip it for now. List each part number in the proper category.

3.11. Briefly describe the difference between a BJT device and an FET device.

3.12. Briefly discuss the advantages of using an op-amp over using a BJT or FET.

3.13. Locate the specification sheet for the TDA2002A device in Appendix A. Copy its block diagram and write a brief description of the number of functions it achieves.

Section 3.5

3.14. Reproduce the block diagram and schematic from Figure 3.25. On the block diagram indicate the application of dc power to each block symbol. Highlight the dc source connections in the schematic. Complete this drawing by showing a 20-V battery connected to these terminals and to ground. Also, interconnect these grounds using one common "wire."

3.15. Repeat problem 3.14 for the diagrams and schematics shown in Figure

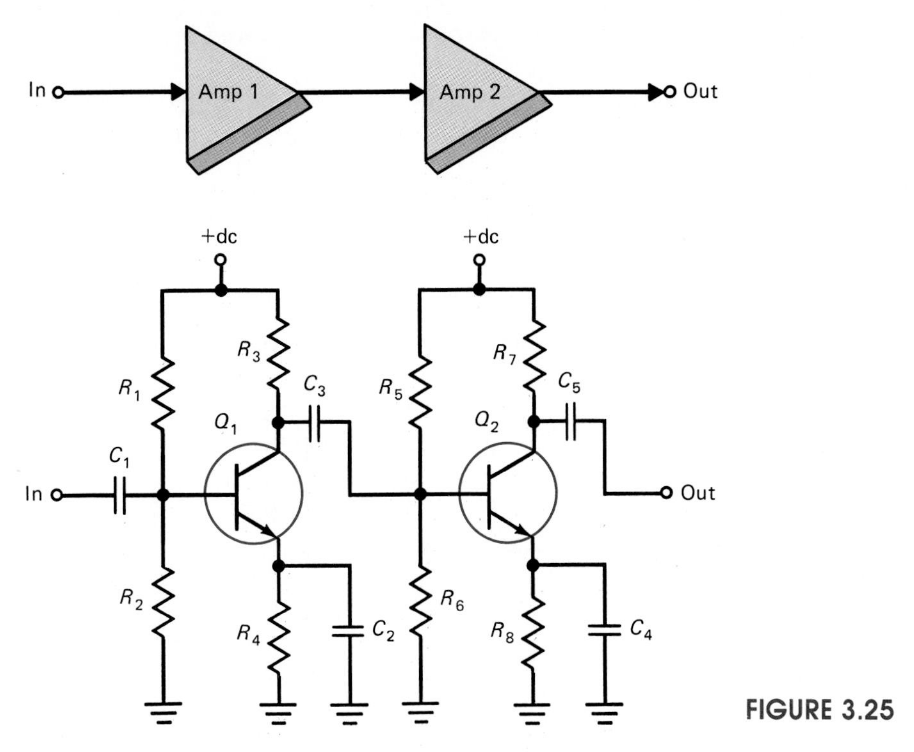

FIGURE 3.25

3.26. Note that two sources are used in the schematic and thus two dc inputs should be shown in the block diagrams. Use two separate 9-V battery symbols for the sources in this schematic, each sharing a common ground.

Section 3.6

3.16. If the first transistor in the schematic shown in Figure 3.25 were to fail to amplify at all, what effect would that have in the output? Indicate in

FIGURE 3.26

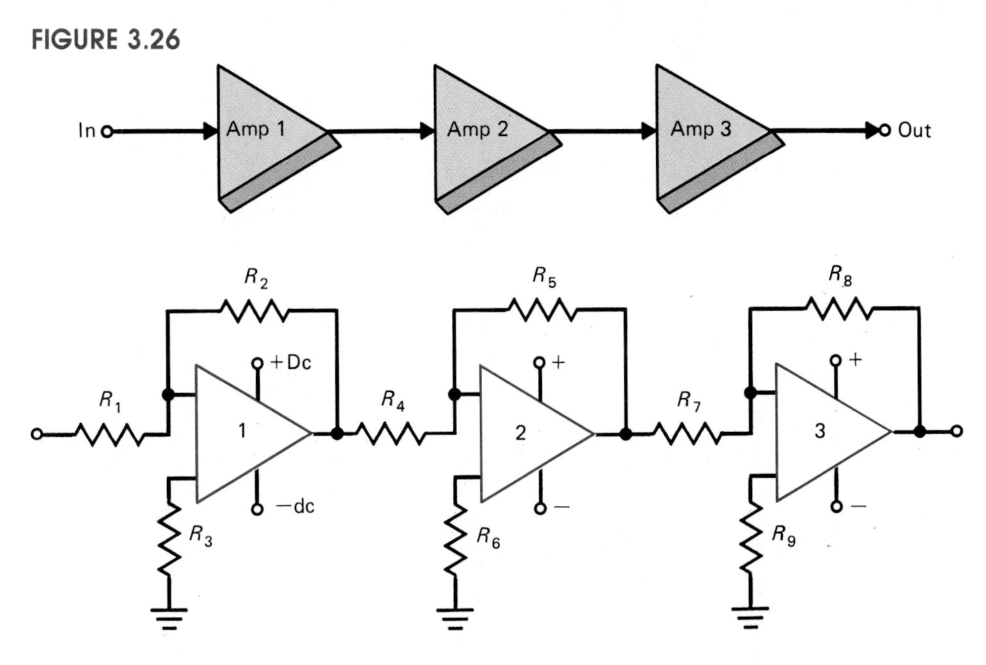

the block diagram where the flow of information would be lost due to this defect.

3.17. A defect developed in the driver amplifier in the TDA2002 amplifier shown in Figure 3.27. Indicate what can be done to correct for this fault.

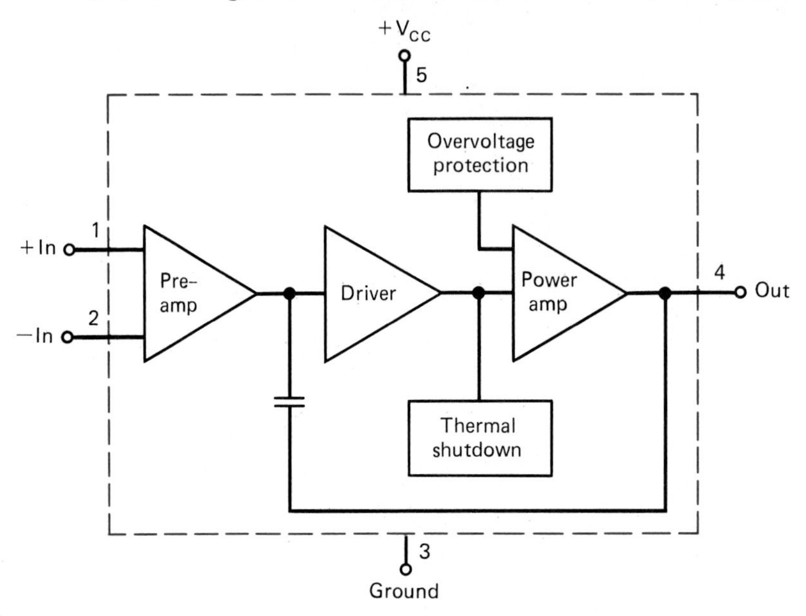

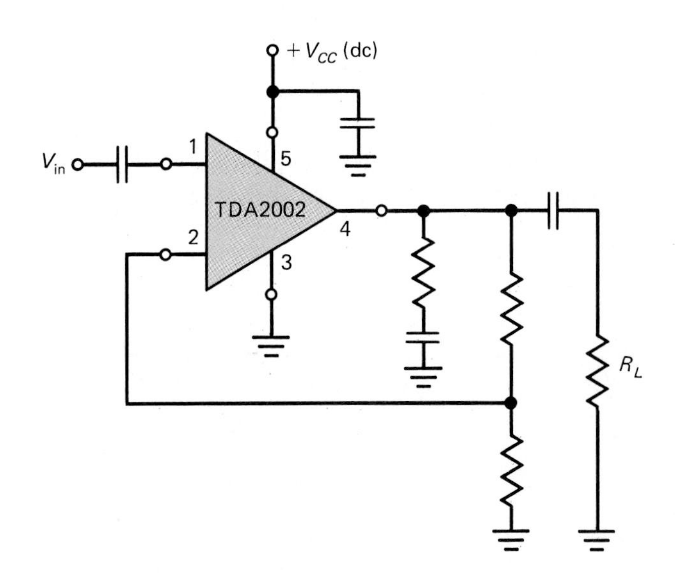

FIGURE 3.27

3.18. In Figure 3.26 the amplifier circuits shown are built on small plug-in modules. A test was run using the block diagram and it was found that the input signal failed to come out of amp 2. Which module (1, 2, or 3) should be suspected of being defective? If the module costs $2.44 to replace but may take 1 hour to repair, what is the best decision to make to get the system back in operation (repair or replace)?

3.19. The circuit shown in Figure 3.28 is defective. When using the block diagram it was found that the positive segment of the output signal (out

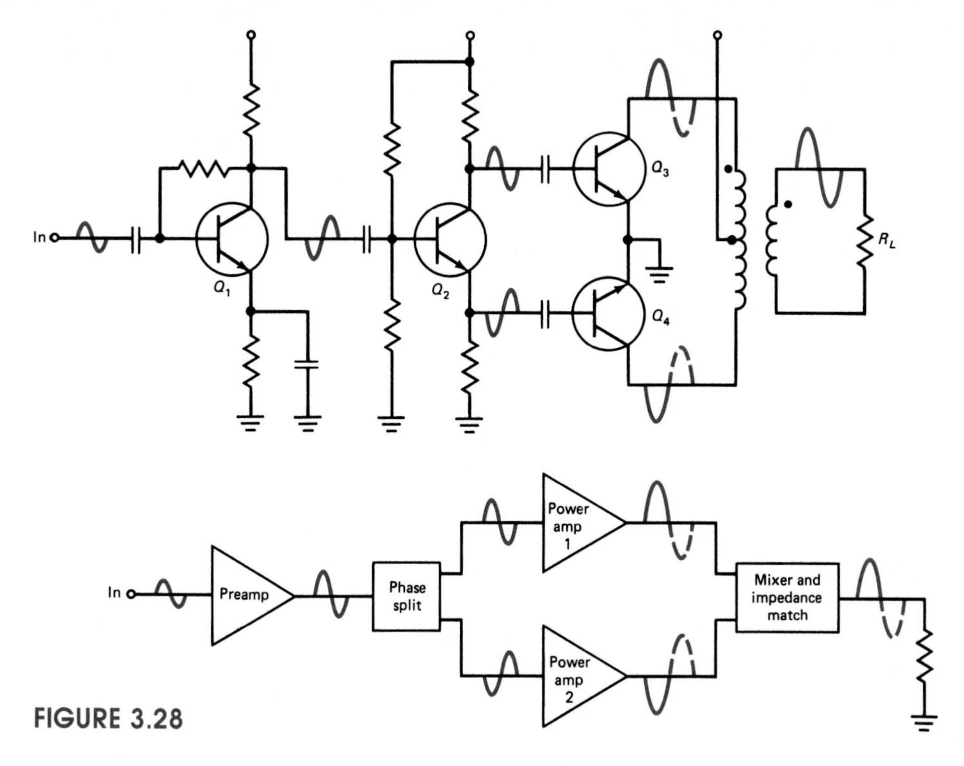

FIGURE 3.28

of power amp 1) was not present. Which device in the schematic is probably defective? If that device seemed to be functioning correctly when tested, what other device(s) could be at fault? (The negative portion of the wave was found to be fine.)

3.20. Knowing that capacitors are commonly used to couple amplifiers together, break the two amplifiers shown in Figure 3.25 down into two distinct amplifiers, each with an input source and an output load and a distinct dc power source. The load resistor used in the separate illustration of amplifier 1 actually represents what resistive value?

ANSWERS TO REVIEW PROBLEMS

Section 3.2

1. See the text and other books for listings.
2.

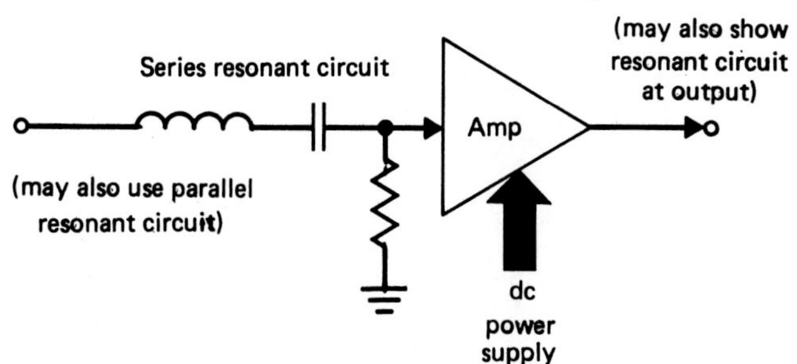

3. Normal gain = 6,250,000, gain when all are 20% low = 2,560,000, gain when all are 20% high = 12,960,000. The output therefore varies from less than half the normal value to over twice the normal value. This could

cause serious problems when trying to maintain a relatively "normal" output function.

Section 3.3

1. Answer based on research of ICs.
2. (a) The integrated circuit could be defective. Repair could be simply the replacement of the IC. This could probably be done in about 5 minutes.
   (b) The defect could be in the second op-amp or support components. Repair could involve testing and/or replacement of each of the (roughly six) components. This could probably be done in about 10 to 15 minutes.
   (c) The defect could be in the second transistor or support components. Repair could involve testing and/or replacement of each of the (roughly eight) components. This could probably take over 15 minutes to do.
3. Between the IC and the op-amp it appears that variable resistors and possibly capacitors are similar. These are probably used to provide variable gain or some other variable function. Between the op-amp and discrete transistor versions it appears that both the items listed above and a few more capacitors and resistors are common. The use of these added elements is not certain at this point.

Section 3.4

1. This written overview should closely follow the flow indicated in the block diagram and should be certain to include each element.

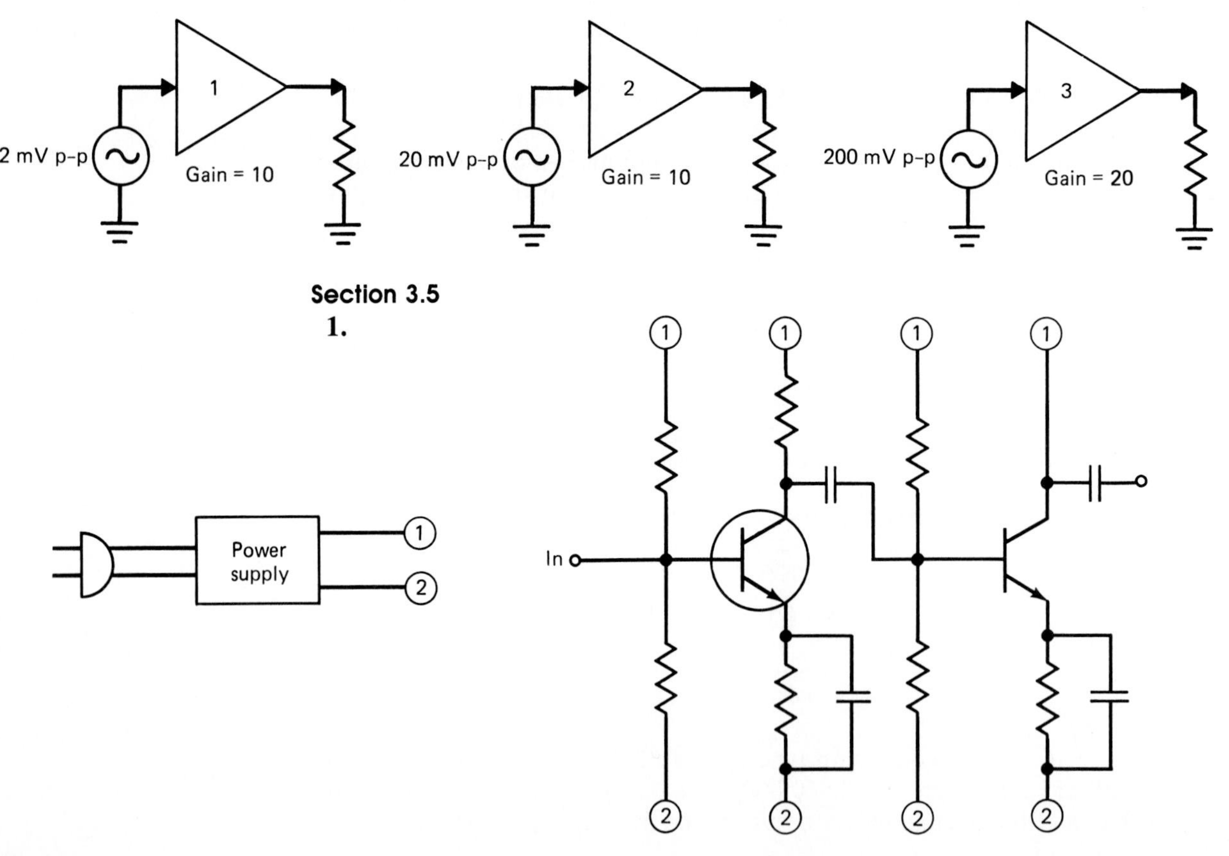

Section 3.5

1.

2. The hum could be being produced in the power supply. To test for this, the supply could be checked with an oscilloscope to see if there was an ac output in addition to the expected dc output.
3.

Section 3.6

1. Discussion should match the main points in Section 3.6.
2. VOM (or DVOM): To measure dc voltages and currents; signal generator: to input a known signal to the amplifier; oscilloscope: to monitor the flow of the signal through the amplifier; tester used for specific active devices: to check their function
3. The technician should attempt to locate the cause of the transistor's failure. Other circuitry should be inspected to see if there are voltage or current levels which exceed the ratings listed for that device. Other defects, such as off-tolerance resistors, should also be inspected. Also, it should be determined if the circuit is being applied in the correct manner; is it being used properly? Have the operating environment or conditions been changed? Finally, is it possible that the transistor that was selected for use does not meet the demands of the circuit (e.g., redesign is required)?

4 Ideal Amplifiers

Objectives

Upon completing this chapter, the reader should be able to:

- Recognize typical amplifier characteristics.
- Identify special forms of gain.
- Identify four forms of ideal amplifiers and recognize their application and related gain equations.
- Identify the function of negative feedback and relate its use to ideal amplifiers.
- Contrast nonideal amplifiers to ideal models.
- Relate ideal models to troubleshooting.

Introduction

In this chapter ideal amplifiers are defined. By understanding the role and function of the ideal amplifier, a clearer concept of the goals of all amplifying systems can be more easily had. Since real-life amplifiers do not meet all of the conditions of ideal amplification, contrasting them to the ideal will help to define faults and to investigate means to compensate for these faults.

Along with this introduction to ideal amplifiers, we take a general look at how multistage amplification is achieved. Concepts of feedback as they relate to amplification are also explored, as is how feedback relates to the amplifier definitions. Several popular feedback models are presented. Once these concepts are presented, the real-life (or nonideal) amplifier is contrasted to the ideal. By making these comparisons, a clearer concept of the function of these nonideal amplifiers will be had.

Finally, the use of the ideal model is demonstrated as troubleshooting is done within a system. By using these models it will be seen that troubleshooting can be simplified as each part of a system can be related to these models. ■

4.1 CHARACTERISTICS OF AMPLIFIERS

When we used amplifiers in previous chapters, we worked with rather simple definitions of the amplifier. For the most part, we considered it to have little more than a basic gain figure. As such, we were dealing with these amplifiers as if they were *ideal amplifiers*. What we will find in this chapter is that ideal amplifiers take on a variety of characteristics, not just a simple gain figure. Thus there will be several forms of the ideal amplifier. To identify these amplifier forms properly, we need to identify three primary features of any amplifier: gain, input impedance, and output impedance.

The term "gain" was used quite liberally in earlier chapters. We have defined gain in terms of output-to-input ratios. If the output of a system was 500 times larger than the input, we identified the gain as being 500. This fact has not changed. What we must concentrate on, though, is what form the gain takes.

It is necessary to be specific as to the type of gain we are working with: voltage, current, or power gain. Voltage gain has been used in many places in previous chapters. Another form of gain figure is an amplifier's current gain. If there is a ratio of input current to the output current delivered to a load, there must be a statistic for current gain. Finally, if there is a ratio for both voltage gain and current gain, there must be a notation for power gain. A certain amount of power will be dissipated at the input of the amplifier and the amplifier will deliver some amount of power to its load. Therefore, the ratio of output power to input power is its power gain.

EXAMPLE 4.1

In Figure 4.1 an amplifier is shown with input voltage and current and output voltage and current. The voltage-gain calculation is simply the ratio of output voltage to input voltage. The current gain is calculated similarly. Finally, the power gain is calculated as the ratio of input power to output power.

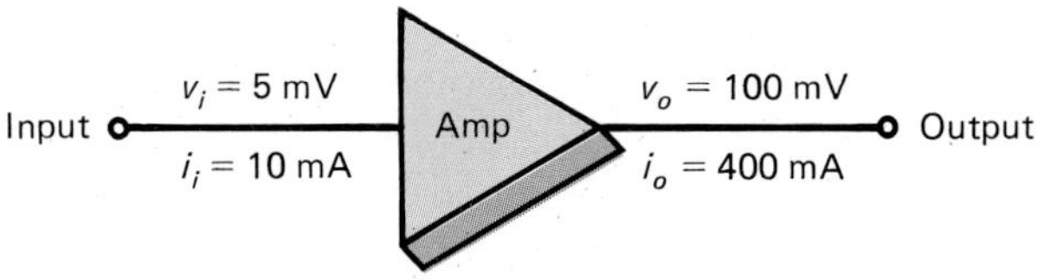

Input $v_i = 5$ mV, $i_i = 10$ mA — Amp — $v_o = 100$ mV, $i_o = 400$ mA Output

$$A_v = \frac{v_o}{v_i} = \frac{100\text{ mV}}{5\text{ mV}} = 20$$

$$A_i = \frac{i_o}{i_i} = \frac{400\text{ mA}}{10\text{ mA}} = 40$$

$$A_p = \frac{p_o}{p_i} = \frac{100\text{ mV} \times 400\text{ mA}}{5\text{ mV} \times 10\text{ mA}} = 800$$

FIGURE 4.1 Voltage, Current, and Power Gain

In Example 4.1 all the gains of the amplifier were calculated. It can be seen that an amplifier has not one but three gain figures. This is true of all

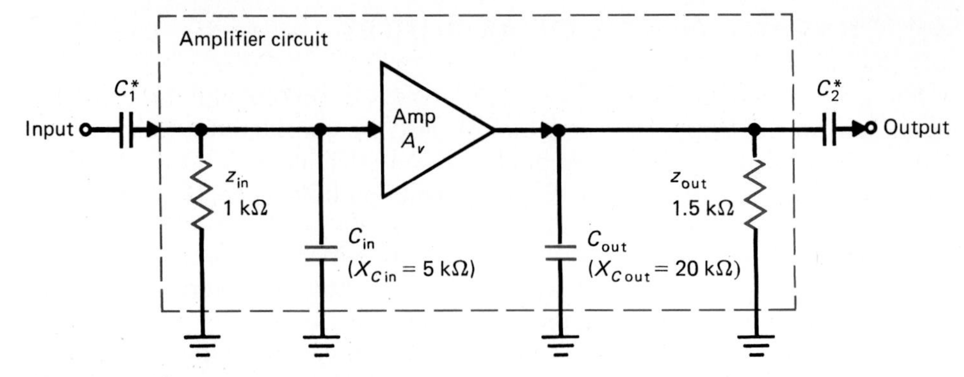

*Typical input/output coupling capacitors.

(The values of X_{Cin} and X_{Cout} would be based on the frequency of the input signal.)

FIGURE 4.2 Equivalent Resistance and Capacitance Associated with Amplifier Circuitry

amplifiers. In many cases, though, one of these characteristics will be singled out as being most important, either because of the application of that amplifier or because of the active device used in the amplifier.

In the beginning of this section it was noted that there were three major features to look for in any amplifier. Gain, either voltage, current or power, is but one of these features. The other two are input impedance (resistance) and output impedance (resistance).

In order to work with either of these features, we first need to define what impedance or resistance of the amplifier actually is. There are two forms of this characteristic. One of the forms is simple resistance. By using the specific term **resistance**, we are referring to ohmic resistance to dc current. In many cases an amplifier exhibits infinite dc resistance as viewed from the input or output connections. This is due to the use of coupling capacitors that were chosen to prevent the flow of dc current in or out of these connections. (Use of these capacitors was discussed in an earlier chapter.)

The more general term **impedance** relates to the complex resistance seen by the flow of ac through the amplifier. Since coupling capacitors are usually designed to pass the desired ac signal into and out of the amplifier, that flow of current will see some value of impedance as it enters the amplifier. When the amplified flow exits the amplifier, it will pass through another value of impedance. These input and output impedances are probably complex. That means that the impedance will be a composite of a dc resistance term and ac impedance terms. In Figure 4.2 typical values for both the dc resistance and ac impedance related to the input and output terminals of a normal amplifier are shown.

To see clearly gain and impedance together in one general overview of an amplifier, let us take the amplifier shown in Figure 4.2 and construct a Thévenin equivalent* model for the amplifier. By viewing this model, we can

* The Thévenin equivalent of any circuit or device is a simple "looks like" model of the circuit or device. This model shows the input and/or output as acting like a voltage source in series with a resistance. The source represents the voltage produced by the device or circuit, and the resistance represents the series/parallel combination of "internal" resistances. In some cases, especially at inputs, voltage is not produced by the circuit or device and therefore the voltage source is eliminated, leaving only an equivalent resistance.

have a greater appreciation for the use of these three figures: gain, input impedance, and output impedance. We will use this sort of modeling in later sections of the book to help us understand many types of amplifiers. (It should be noted that a Norton equivalent† could have been used, but for this example, the Thévenin equivalent was selected.)‡

EXAMPLE 4.2

From Figure 4.2 we can see that the dc input resistance of this amplifier would be infinite since a capacitor is shown that blocks dc current. In addition to this capacitor we can see a real resistance that represents the input of the amplifier after that capacitor. To characterize the input impedance, then, we need to include both elements in our model.

Since this is an input terminal and would not be expected to supply current back to the source, we would not include any Thévenin equivalent source in the model of this input terminal. (The equivalent is simply the input impedance, both R and C; see Figure 4.3 for the development of this model.)

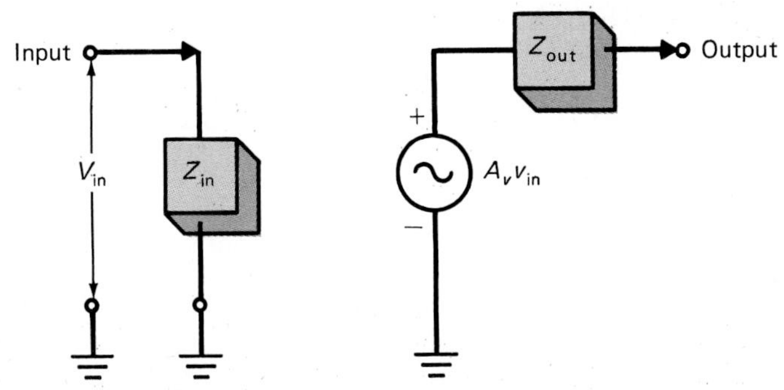

FIGURE 4.3 Thévenin Equivalent of a Voltage Amplifier

To model the output terminal, then, we would again look for the impedance seen from the output connections. We would look for any series or parallel combinations seen from these points, looking backward (in relationship to the normal flow of the signal). In this case we again see a basic capacitance/resistance combination. For the same reasons as obtained at the input terminal, this combination represents ac impedance only.

Finally, we need to identify the source value of our model. In the case of the output, we will see a source. That source is related to the voltage gain of the amplifier as it reacts to an input value. We can simply illustrate it as a voltage source with a value of $A_V \times V_{in}$. This is nearly the same type of source that we would see in a simple circuit analysis problem. The only unusual factor is that it is not listed with a fixed

† The Norton equivalent of a circuit or device is also a "looks like" model of that circuit or device. The Norton model uses current sources and paralleling resistances to represent the circuit/device function.

‡ The reader is not expected to be able to establish the values for Thévenin or Norton equivalents. For further details regarding these models, consult a textbook dealing with electronic circuit analysis topics.

(constant) output. The actual voltage output is left unspecified since it is a direct function of the input voltage, which is not given with this circuit. (Should an actual input voltage be given, a numerical value can be used for this source.)

The point of Example 4.2 is to illustrate that an amplifier can be represented by the combination of its input impedance, output impedance, and gain (as a voltage source). In fact, if an input signal were to be applied to the original amplifier (Figure 4.2) and the model of it (Figure 4.3), exactly the same statistics could be obtained for input current, input power, output current, and output power using either form.

The statistics that were shown for this amplifier were for a nonideal amplifier. In the following sections the ideal conditions for such a voltage amplifier are presented. If we can carefully describe what an ideal voltage (or current, or power) amplifier should look like, we can then relate actual, nonideal amplifiers to these models.

REVIEW PROBLEMS

(1) If an amplifier has the following statistics, what are the values of A_v (voltage gain), A_i (current gain), and A_p (power gain). (Remember that power is simply the product of voltage and current.)

$$V_{in} = 2\text{ V} \qquad V_{out} = 22\text{ V}$$
$$I_{in} = 5\text{ mA} \qquad I_{out} = 500\text{ mA}$$

(2) For the values shown in problem 1, calculate a simple value for input and output resistance. (Note that in this problem, no complex impedance terms are listed.)

(3) For the circuit shown in Figure 4.2 and the equivalent shown in Figure 4.3, assuming a 0.2-V input voltage, calculate the output voltage for both examples (assume an infinite load or open circuit at the output terminals). Also calculate the input current (ac) for both examples. Compare your results from both circuits.

SPECIAL INFORMATION ABOUT GAIN STATISTICS

Before we proceed to discuss voltage, current, or power gain there are three special forms of the gain statistics that we need to review. So far in this text we have looked at gain as a positive number that was greater than 1. We need to investigate three other options for gain: negative gain, gain between 0 and 1 (gain less than "unity"), and undefinable gain.

Negative Gain

Initially, the concept of negative gain (say $A_V = -35$) may seem to represent a loss. This is not the case. Negative gain simply means that there is a value for the gain statistic but that the input signal suffers a 180° phase shift by the

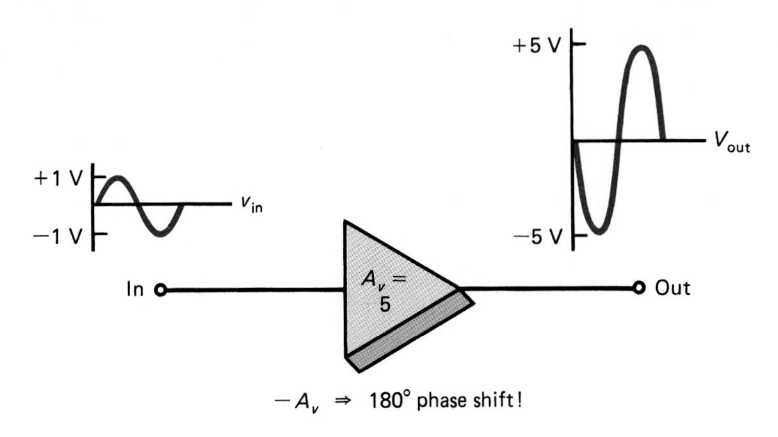

FIGURE 4.4 Negative Gain Amplifier (−180° Phase Shift)

time it reaches the output. That means that if the input begins to rise in value, the output will fall, and vice versa.

Looking at Figure 4.4, it can be seen that the output is larger than the input by a factor of the gain value. But the negative sign of the gain value indicates that the amplifier causes a 180° phase shift of the signal as it passes through the amplifier.

Many amplifiers will produce negative gain characteristics. Since such a phase shift does not normally produce any complications in the amplification process, usually no special treatment is given to the negative-gain amplifier. It is usually handled just like a positive-gain amplifier, as long as the sign of the output value is kept intact.

Gain Between 0 and 1

If an amplifier does not produce an increase in output signal over input signal (output $<$ input), its gain value will range between 0 and 1 (Figure 4.5).

It is possible for an amplifier to have a gain for either voltage or current in this range. If an amplifier's primary role is to produce a current gain, it is possible that the actual voltage gain falls below unity (1). There are even cases where all gains fall below unity; these will be discussed later when we look at special-purpose amplifiers.

For the sake of comparison, a simple voltage-divider circuit (not an amplifier) produces a gain of less than 1; that is, the output is less than the input. Normally, we do not represent such circuits with this sort of gain value, but it does illustrate the concept of gain less than 1.

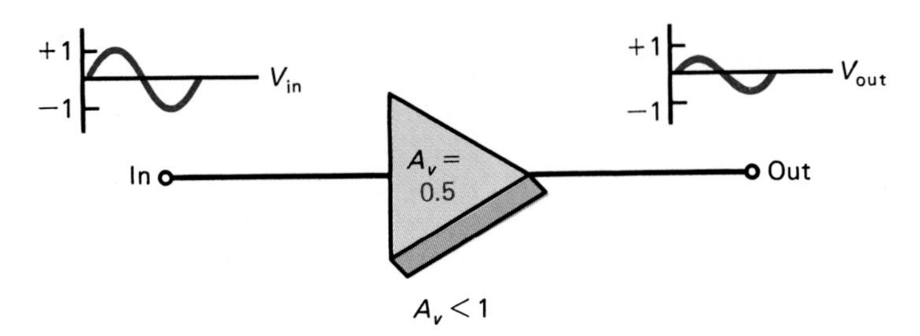

FIGURE 4.5 Amplifier with Gain Less Than Unity

When dealing with either negative gain or gain less than 1, we treat these the same way that we treat positive gains greater than unity. If we are composing a total system gain, we simply multiply by these gains as we encounter them. There are two factors to be aware of, however.

1. If there are two (or an even number of) negative gains, they cancel, since a negative multiplied by a negative results in a positive answer. This is simply verified by the fact that if one amplifier shifts phase by 180° and the next shifts by an additional 180°, the composite shift is 360°, or is back in phase.
2. In most amplifier systems, if a decimal gain is encountered for current or voltage in one amplifier, it can usually be compensated for by another stage which produces a gain greater than unity. For example, if one amplifier has a voltage gain of 0.8 and the following has a gain of 20, the composite gain is 16 (0.8×20).

Finally, it is not uncommon to encounter an amplifier system that has a mixture of these forms of gain. One stage may produce a negative gain, another with a gain less than 1, and yet another with a positive gain greater than unity. In any case, though, the composite system gain is simply the product of all gains.

Undefinable Gain

There are certain situations where there may not be a clear way to define the voltage or current gain for an amplifier. This can happen when it is not possible to obtain characteristics for an input or output value of voltage or current.

As an example of this problem, there are some active devices that respond only to the level of input voltage. They draw an almost undetectable amount of current (nanoamperes) at their input terminal. Yet they produce a reasonable level of output current.

To attempt to calculate the current gain of such a device, dividing output by input, the gain would be in the millions. This is unrealistic. In such a case, then, the value for current gain would be considered undefined. In other words, no realistic value of current gain can be had for the circuit, since it is impossible to force any realistic value of input current into the circuit. Should such a condition exist and this undefined value of gain encountered, the gain for that value (voltage or current) must be ignored for that stage.

For any circuit where one of the gain characteristics is undefined, the other characteristic must be defined by a reasonable numerical value (e.g., if current gain is undefined for a circuit, voltage gain must be a reasonable value). It should be noted that if the characteristic for either current gain or voltage gain is undefined, power gain will also be undefined for that stage.

It can therefore be seen that gain can be represented in one of several ways, either for voltage, current, or power. In future examples and problems, these varieties of gain will be used. For some of the ideal amplifiers that are described in future sections, we will relate to these special forms of gain.

4.2 THE IDEAL VOLTAGE AMPLIFIER

In Section 4.1 it was noted that gain, input impedance, and output impedance were special characteristics of a common amplifier. In this section we look at each of these characteristics as they relate to an ideal voltage amplifier. Although any amplifier can produce some form of voltage gain, there are cases where an amplifier is required to produce nearly ideal voltage gain characteristics. The more we know about these ideal conditions, the easier it will be to investigate the normal characteristics of a typical amplifier. In your circuit analysis courses, you have probably studied ideal voltage sources. An ideal voltage amplifier will have characteristics much like this ideal source.

Following are the requirements for an ideal voltage amplifier.

1. An ideal voltage amplifier displays a constant value for voltage gain regardless of input or output conditions.
2. An ideal voltage amplifier produces a constant output voltage level (given a constant input level) for any load conditions.
3. An ideal voltage amplifier has an infinite input impedance.
4. An ideal voltage amplifier has zero output impedance.

Each of these features must be had to assure that an amplifier is an ideal voltage amplifier. The constant-voltage-gain figure must be known to assure voltage amplification. The amplifier must produce its output regardless of the current it delivers to the load, even if the current is unusually high. If the source that drives this amplifier has any internal impedance, the ideal amplifier must not produce a voltage division between itself and the source. Therefore, it must have an infinite input impedance.

Finally, to be able to produce a constant voltage at the output regardless of the current demands of the load, it may not present any output impedance of its own. Were it to have an output impedance that was not zero, it would form a voltage divider with the load. Were such a divider to exist, the true output to the load would then depend on the ratio of load resistance to the output resistance of the amplifier. The output voltage then would vary depending on the load. An ideal voltage source must, by definition, produce a constant output voltage. Figure 4.6 shows a basic model for this ideal voltage amplifier. Since this amplifier has an infinite input impedance, it would not draw any current from the input source. Also, the output has no limit to current delivered to the load. Therefore, the value for current gain is undefined. If current gain is undefined, power gain is also undefined.

To contrast this ideal voltage amplifier with a real amplifier (Figure 4.7), the following considerations need to be made:

1. A real amplifier will produce voltage gain, but only under certain limited conditions. First, the value for this gain is not a fixed number. Although a real voltage amplifier could have a calculated gain value, it would not be constant for any input voltage or for any output load condition. For example, if an amplifier had a calculated gain of 100, along with that value might be the restriction that it is valid for inputs below 100 mV with loads of 50 Ω or more. If one of these restrictions is exceeded, the actual gain may go below the figure given.

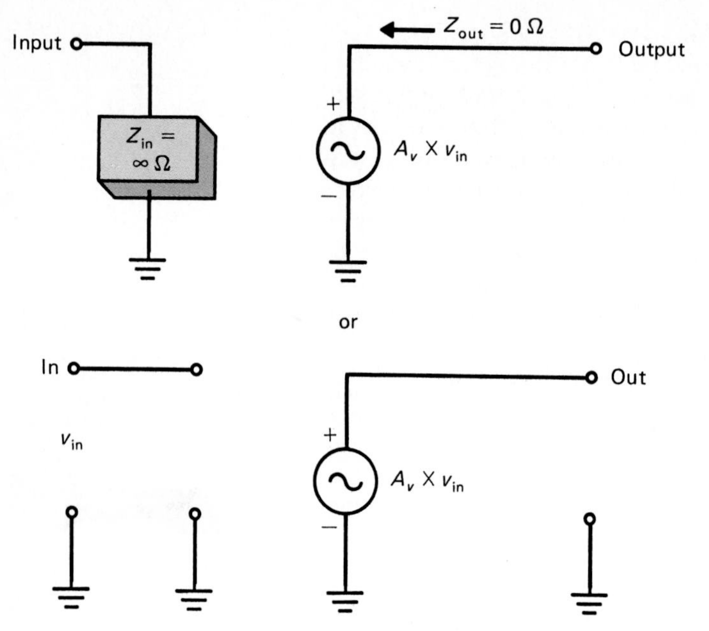

FIGURE 4.6 Ideal Voltage Amplifier Input and Output Equivalents

2. The input impedance of a real voltage amplifier may come quite close the the ideal (infinite). It is not unusual to have an amplifier with an input impedance in the megohm range. Compared to the impedance of a source for the input signal, impedances in this range may be so high that we could *assume* it to be infinite. But should the source's impedance also be in the megohm range, a voltage divider may be produced, thus defeating the infinite-impedance assumption.
3. Probably one of conditions of the voltage amplifier that is most difficult to obtain is that of zero output impedance. Since an amplifier is constructed of components that exhibit resistance, expecting zero output impedance is not realistic. In certain circuits, the output impedance of such an amplifier can be reduced to a few ohms. (In some very specialized circuits, an output resistance of less than 1 Ω can be found, but this is not the common case for most voltage amplifiers.)

If this real output impedance is dramatically less than the impedance of the load connected to the output, the current-limiting effects of that impedance

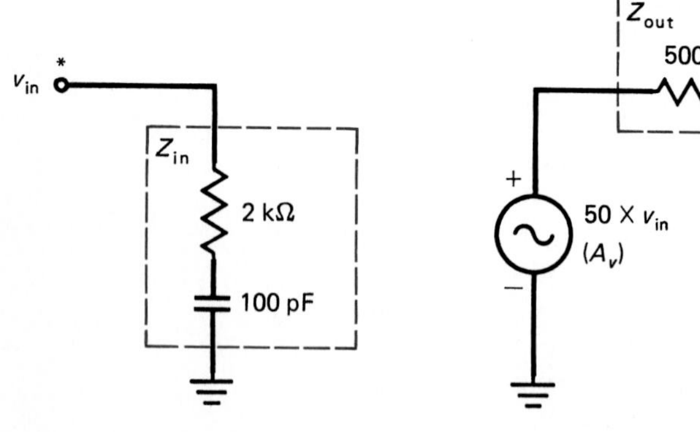

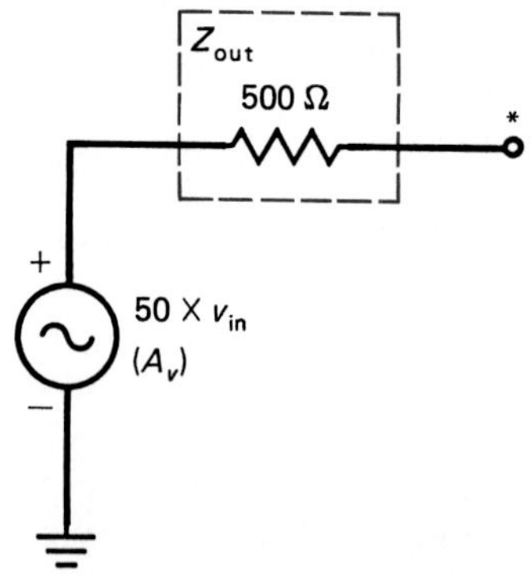

FIGURE 4.7 Practical Values for a (Nonideal) Voltage Amplifier

are sufficiently small that we can *assume* that it is nearly ideal. The basic relationship needed to make this assumption is that the circuit impedance should be one-tenth (or less than) that of the load.

The ideal voltage amplifier can, therefore, only be approximated under real conditions. Even though ideal voltage amplifiers do not exist, it is important to know their characteristics. The closer that a real voltage amplifier can approximate these characteristics, the more we can depend on it to act like the ideal. Also, if we are working with a voltage amplifier and we find that it is not responding as expected, we can examine gain and input and output impedances to learn if and how that amplifier is nonideal. This may help determine how to correct the amplifier's faults.

REVIEW PROBLEMS

(1) Using Figures 4.6 and 4.7, calculate the final output voltage from the following input and load conditions. (Concentrate primarily on the effect of input and output voltage dividers.)
(a) $V_{in} = 0.1$ V, source resistance $= 10$ kΩ, $R_l = 100$ Ω
(b) $V_{in} = 0.1$ V, source resistance $= 40$ Ω, $R_l = 2$ kΩ
Write a brief comparison of the results of these calculations.

(2) An ideal voltage amplifier should produce its voltage-gain figure for any input. Discuss briefly the problem of applying this to a circuit where the input is 4 V, the voltage gain is 80, but the dc voltage source is only 12 V.

4.3 THE IDEAL CURRENT AMPLIFIER

In Section 4.2 the ideal voltage amplifier was presented. It was noted that the amplifier produced a constant voltage gain for any input or output condition. In contrast to this, the *ideal current amplifier* will have constant current gain. Each of the features of the ideal voltage amplifier—input and output impedance, along with gain—will be changed when looking at the ideal current amplifier. The output of the ideal current amplifier will take on the characteristics of the ideal current source used in circuit analysis.

Following are the requirements for an ideal current amplifier.

1. An ideal current amplifier displays a constant value for current gain regardless of input or output conditions.
2. An ideal current amplifier produces a constant output current level (given a constant input current) for any load condition.
3. An ideal current amplifier has zero input impedance.
4. An ideal current amplifier has infinite output impedance.

Each of these features must be achieved to assure that an amplifier is in fact an ideal current amplifier. Since its objective is to provide an unchanging current gain value, it must produce current gain independent of other variables, such as voltage and impedance. Zero input impedance is necessary because if it were a nonzero value, it may present a restriction to the input current fed to the amplifier. Such a restriction would go against the concept of "constant gain for *any* input."

Just like an ideal constant current source, this ideal current amplifier must have infinite output impedance. In Figure 4.8 the basic illustration of an ideal current amplifier is shown. Note that the output impedance is in parallel with the load impedance (just as for a current source in circuit analysis). If this impedance were not infinite, it could form a current divider with the load and thus have less than ideal current delivered to the load. Since this amplifier has zero input impedance, there would be no voltage drop at its input terminals. Therefore, a value for voltage gain is meaningless. In addition, the amount of voltage delivered to the load is strictly dependent on the value of the load (via Ohm's law).

Much like an ideal current source used in circuit analysis, the ideal current amplifier is not achievable in real-life applications. First, the concept of a fixed current gain for any operating condition is not possible. Let's assume, for example, that we had an ideal current amplifier with a current gain of 50. If the input were 1 A and the load resistance 10 kΩ, the output into that load would be 50 A. So far that is quite possible, statistically. But if 50 A were forced to pass through that load, the voltage drop on the resistor would have to be 500,000 V. That is where the concept becomes unrealistic! Just imagine what would happen if we were to open the output terminals (near-infinite load resistance). What is realistic is an amplifier which displays characteristics that are close to those of the ideal current amplifier, under very restricted operating conditions.

1. A real current amplifier will be able to maintain a relatively constant current gain. But this gain can be depended upon only if the input level and resulting output level are within a restricted range. These restrictions will be demonstrated when we study specific amplifiers.
2. All circuits must exhibit some realistic value for impedance. Therefore, the input impedance will not be zero but may be relatively low, especially when compared to the impedance of the source that produces the input to this amplifier. In a similar manner, output impedance will be quite high in respect to the load impedance but will not be infinite.
3. There will always be a restriction to the amount of voltage that can be produced at the output terminals. Usually, this restriction is set by the power supply used to operate that circuit.

In Figure 4.9 more realistic conditions for a current amplifier are shown. Each of the nonideal values shown represents a more realistic approach to

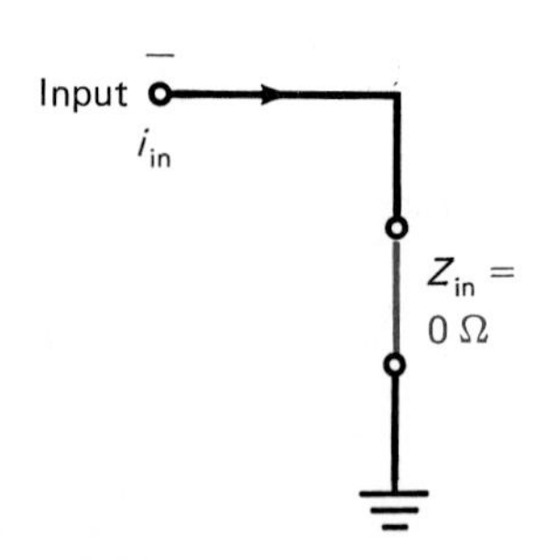

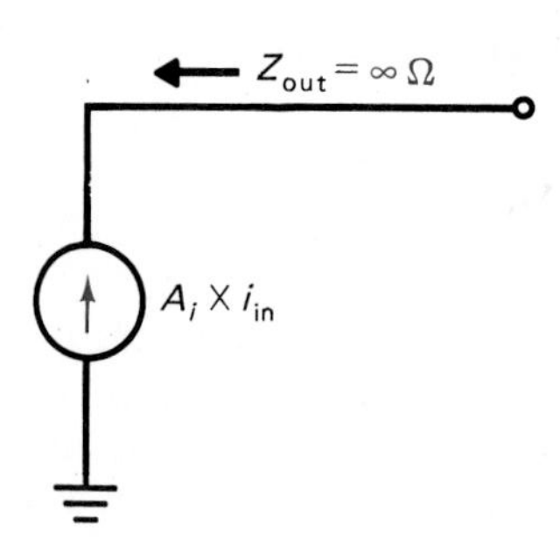

FIGURE 4.8 Thévenin Equivalent of a Current Amplifier

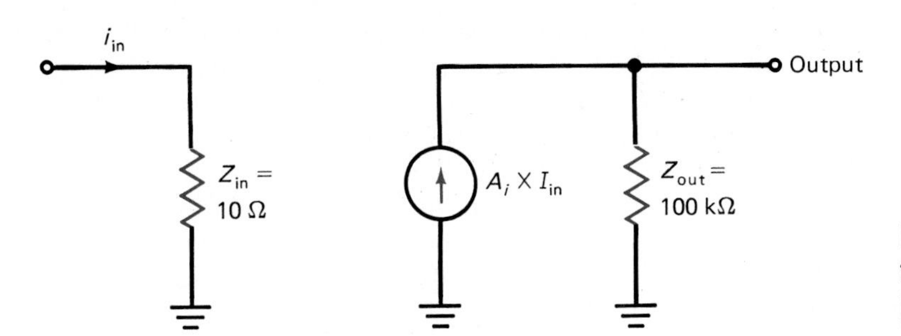

FIGURE 4.9 Practical Values for a (Nonideal) Current Amplifier

the concept of current amplifiers. It should be noted, though, that when we encounter the application of a current amplifier, the closer its characteristics match those of the ideal current amplifier, the more likely it is to produce the ideally constant current gain.

REVIEW PROBLEMS

(1) Using Figures 4.8 and 4.9, calculate the output current delivered by both amplifiers based on these two sets of data.
   (a) I_{in} = 50 mA, source resistance = 20 Ω, load resistance = 50 kΩ
   (b) I_{in} = 50 mA, source resistance = 150 Ω, load resistance = 100 Ω

   Write a brief comparison of the results of these calculations.
(2) An ideal current amplifier should be able to produce output current regardless of voltage conditions. Briefly discuss the problem of supplying 100 mA to a current amplifier with a current gain of 100 and a load resistance of 1 kΩ. Assume that the power supplv for this circuit is 40 V and able to supply up to 200 A of current. (Also assume that input and output impedances are such that current dividers are not a problem.) Under such conditions, the actual gain of the amplifier will not meet the gain expected. The amplifier will not just shut off but will now have another gain value; calculate that value.
(3) Using the data given in problem 2, calculate what the limit to input current would be based on trying to maintain a gain value of 100 with the other restrictions noted in this problem.

4.4 THE OPTIMUM POWER AMPLIFIER

In preceding sections, ideal voltage and current amplifiers were presented. Their characteristics were quite distinct. The ideal conditions for input and output impedances were quite specific, along with conditions for gain. An ideal power amplifier is not quite as clearly defined. Yes, it is to provide power gain, but we need to define what that means. The factors of input and output impedance cannot be given exact values (such as zero or infinite) that are usable in all circuits. Thus the concept of an optimum (application based) rather than an ideal power amplifier will be explored in this section.

First, power gain is a factor in most amplifiers. If voltage and current are present at the output of the circuit, there is a value for output power. But in the case of ideal voltage amplifiers, no current is drawn at the input terminal. Therefore, a value for input power is nonexistent, and power gain is undefined (infinite). The value of power gain for an ideal current amplifier is also infinite.

If we step away, even slightly, from the ideal circuits and make a few simple assumptions, power gain is a real characteristic. If the voltage amplifier draws even a slight amount of current at its input terminals, there will be a figure for power gain based on

$$\text{Power in: } P_{\text{in}} = V_{\text{in}} \times I_{\text{in}} \tag{4.1}$$

$$\text{Power out: } P_{\text{out}} = V_{\text{out}} \times I_{\text{out}} \tag{4.2}$$

$$\text{Power gain: } A_p = \frac{P_{\text{out}}}{P_{\text{in}}} \tag{4.3}$$

For this amplifier and a current amplifier that has some input voltage drop, there will be values for input and output voltages and currents. Therefore, some power gain can be expected. All nonideal amplifiers have a value for power gain. In certain cases, though, the amplifier is not designed to produce an output power level that is substantial. In such a case, the value for power gain may not be important.

Some amplifiers are designed to produce large power levels in a load. For example, in a stereo amplifier, the last amplifier in either channel (left or right) must be a power amplifier. This amplifier delivers large quantities of electrical power to a speaker, which then converts this to mechanical power (vibration of the air).

An optimum power amplifier can only be defined in terms of the actual circuit and load to which it is connected. Two factors are always true of an optimum power amplifier. To provide the maximum transfer of power, the input impedance must match the impedance of the amplifier or source connected to its input terminal; in a like manner, this power amplifier's output impedance should match the impedance of the load (Figure 4.10). Therefore, the input and output impedance values are not fixed (as zero or infinite) but must be matched to the circuitry to which this amplifier is connected. The actual value of power gain is dependent on the circuit in which this amplifier is being used.

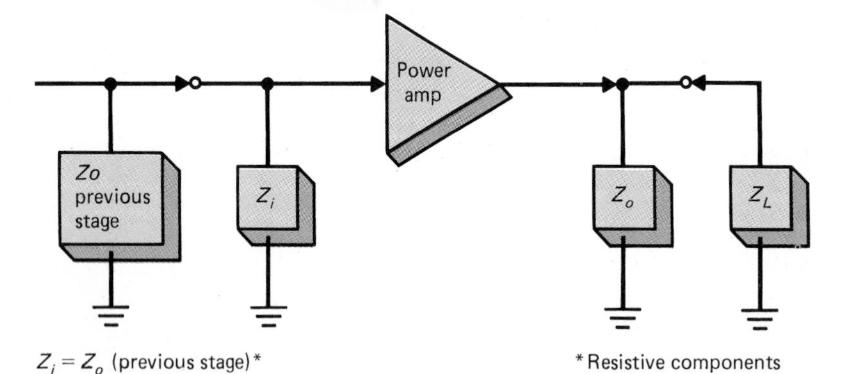

FIGURE 4.10 Impedance Matching for Maximum Power Transfer

In some cases, the power amplifier may be used mainly to match impedances. In this application, when input and output power values are compared, the actual gain figure may be small (in the range 1 to 10). For some other circuits, the power gain may be quite large. In any case, the values for impedances and gain are very specific to the circuit in which the power amplifier is being used. Therefore, unlike the voltage and current ideals, one system's power amplifier cannot be moved from that application to another and maintain the same functional values.

Fortunately, a power amplifier is generally designed using the same process as that used for a simple voltage or current amplifier. The only modification would be in the impedance-matching requirements at the input and output terminals. The only other variation seen in the power amplifier is that since it often produces large amounts of voltage and current and therefore large amounts of power, it is often necessary to construct it out of heavier-duty components than are used in a simple voltage or current amplifier.

The power amplifier is often called a *large-signal amplifier*. That means that the signal (information) passing through it has high voltage and current levels. Amplifiers that do not handle such large levels are called *small-signal amplifiers*.

The following is a summary of the major characteristics of power amplifiers:

1. Input impedance must match the impedance of the input producing circuit or source.
2. Output impedance must match the impedance of the load connected to the output.
3. Power gain values (A_p) are specific to the circuitry to which this amplifier is connected.
4. Unlike ideal voltage or current amplifiers, the optimum power amplifier for one circuit cannot be placed in another circuit and be expected to have the same operating characteristics.

Special Computation of Power Gain

In equation 4.3 power gain was calculated based on input power and output power. An alternative way in which power gain can be calculated for a circuit is based on the fact that since

$$V_{out} = V_{in} \times A_v \text{ (voltage gain)}$$

and

$$I_{out} = I_{in} \times A_i \text{ (current gain)}$$

and if we factor this into the power gain equation, which states that

$$A_p = \frac{V_{out}}{V_{in}} \times \frac{I_{out}}{I_{in}} = \frac{P_{out}}{P_{in}}$$

we obtain

$$A_P = \frac{(V_{in} \times A_v) \times (I_{in} \times A_i)}{V_{in} \times I_{in}}$$

which can be simplified to

$$A_p = A_v \times A_i \qquad \textbf{(4.4)}$$

Therefore, the value of power gain can be found by multiplying the voltage gain of the amplifier by its current gain. This calculation can be used as a quick means of finding the power gain of an amplifier.

REVIEW PROBLEMS

(1) For the circuit shown in Figure 4.11, find the value of ideal input impedance, output impedance, and power gain for the power amplifier.

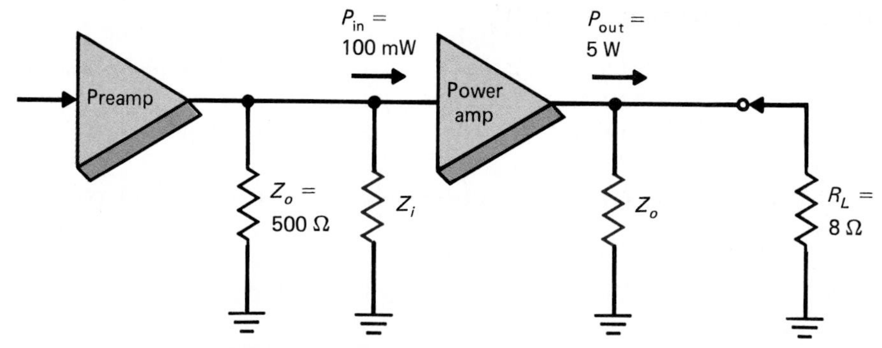

FIGURE 4.11 Power Amplifier Functions

(2) It was stated that impedance matching was a required function of the power amplifier. Briefly discuss what will happen to output power levels if in the circuit of Figure 4.11, the output impedance of the power amplifier is much larger than the impedance of the load. (If you need assistance, refer to the topic of maximum power transfer in a circuit analysis textbook.)

(3) Given the following statistics for a power amplifier, calculate the power gain, first using equations 4.1 to 4.3, and then using equation 4.4. Compare these solutions.

Input: $V_{in} = 0.5$ V, $I_{in} = 400$ mA
Output: $V_{out} = 5$ V, $I_{out} = 6$ A
Gains: $A_v = 10$, $A_i = 15$

4.5 OTHER FORMS OF IDEAL AMPLIFIERS

In earlier sections, we looked at amplifiers that produced either voltage gain, current gain, or power gain. There are two other forms of amplifiers that should be investigated at this time. The previous amplifiers operated with a straightforward representation of gain in the form of voltage, current, or power. The two remaining forms of amplifiers do not deal with such simple relationships of input to output. These amplifiers, in a manner of speaking, translate one input form into a different output form. In one of these amplifiers, input current produces constant output voltage and input voltage produces constant output current in the other. Although this sounds like an odd combination, these amplifier forms are quite common in electronics.

The ideal **transconductance amplifier** produces a constant output current based on input voltage levels (not on input current, as in the current amplifier). The opposite of this, the ideal **transresistance amplifier**, produces a constant output voltage based on input current. Both of these amplifiers have impedance and gain values which differ greatly from the current or voltage amplifiers covered earlier.

Transconductance Amplifier

To understand the workings of the transconductance amplifier, let us first break down the word used to describe it. The term "trans" is used to indicate a relationship of input to output. (We could have called a voltage amplifier a transvoltage amplifier.) The "conductance" part of the term is what actually defines the fact that voltage-in produces current-out. If we go back to the circuit analysis definition of conductance, we can see that the output-to-input ratio actually forms a conductance-like unit. From circuit analysis,

$$G \text{ (conductance)} = \frac{1}{R \text{ (resistance)}} \tag{4.5}$$

If we substitute the voltage-to-current relationship that defines resistance into this equation, we get

$$G = \frac{1}{E/I}$$

which yields

$$G = \frac{I}{E} \tag{4.6}$$

From this equation we see that conductance is the ratio of current to voltage. Thus the term "conductance" in the word "transconductance" refers to the ratio of current to voltage. Now remember that gain is defined as the ratio of output to input. So, in equation form,

$$\text{Gain} = \frac{\text{output}}{\text{input}} = \frac{\text{current}}{\text{voltage}} = \text{conductance} \tag{4.7}$$

(*Note:* Since $R = E/I$ and conductance $= 1/R$, then conductance $= I/E$.)

For the first time, we have a unit related to gain. (Gain in voltage, current, or power amplifiers was unitless since the terms that were in the gain ratio contained exactly the same units.) This unit is the same as the unit used in circuit analysis for conductance, the siemens. As an alternative, the transconductance gain is simply stated as an ampere/volt ratio (no unit named). Also, special terms are used to identify this "gain." The term A_g or the alternate term g_m (used later in this text) is used to identify the transconductance gain value.

EXAMPLE 4.3

This example is used to show how the units in a transconductance amplifier are manipulated. Using Figure 4.12, calculate the gain value for the transconductance amplifier given the following statistics:

$$v_{in} = 1.2\text{ V} \qquad i_{out} = 312\text{ mA}$$

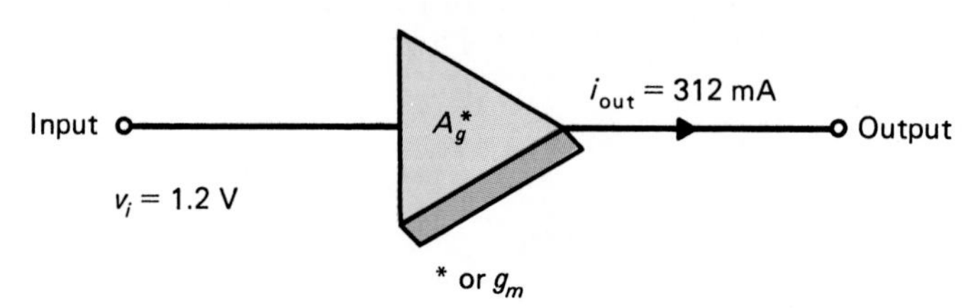

FIGURE 4.12 Transconductance Amplifier

Solution:

$$v_g = \frac{i_{out}}{v_{in}} = \frac{312\text{ mA}}{1.2\text{ V}} = 0.26\text{ S or }260\text{ mA/V}$$

or we could say that $g_m = A_g = 260$ mA/V.

Since we are describing the *ideal* transconductance amplifier, the value of output current is assumed to be constant (given a fixed input voltage) under any operating conditions. There are ideal values for input and output impedance for the ideal transconductance amplifier. If the input is voltage sensitive, as was seen in the voltage amplifier, the input impedance must be infinite (see Figure 4.13).

The output impedance of the ideal transconductance amplifier must be infinite as well. This is just like the output impedance of the current amplifier. In some ways, this amplifier is like a blend of the voltage amplifier (input) and the current amplifier (output) models.

Transresistance Amplifier

Another form of amplifier is the transresistance amplifier. This amplifier is the opposite of the transconductance amplifier, as the name may imply. Again, the term "trans" indicates an input-to-output relationship. The term "resis-

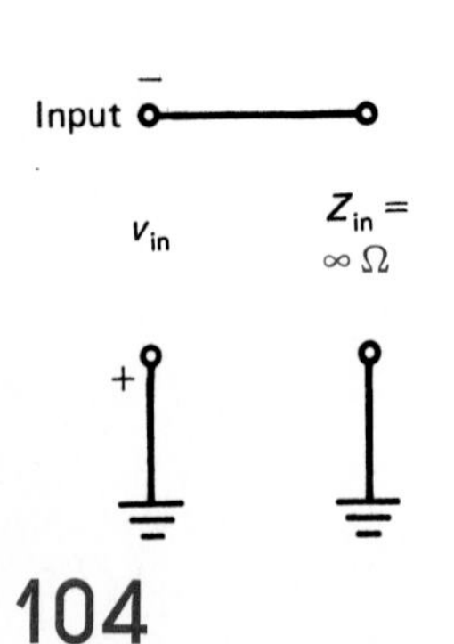

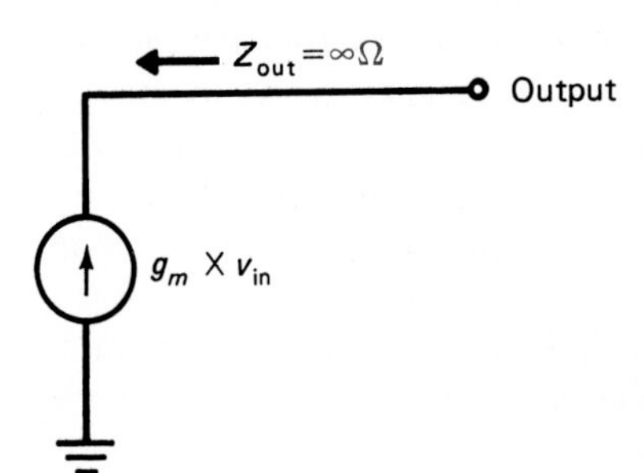

FIGURE 4.13 Ideal Transconductance Amplifier Input and Output Equivalents

tance'' relates the output to input quantities. In this amplifier, input current produces an output voltage. Since resistance is defined as the ratio of voltage to current ($R = E/I$) and this amplifier's gain would be a ratio of v_{out} to i_{in}, it assumes the name transresistance to imply this relationship. [It should be noted that the transresistance amplifier does not amplify resistance (e.g., output resistance to input resistance ratio).]

The units used to describe this gain are, as in resistance, ohms. It it also common to use the unit ''volts per ampere,'' so as not to confuse this amplifier with simple resistors. The symbol used to represent this gain is A_r or r_m. More details on the uses of these symbols will be presented in later chapters when applications are shown. (See Figure 4.14 for further details.) Gain for a transresistance amplifier:

$$A_r = \frac{v_{out}}{i_{in}} \quad \text{ohms} \tag{4.8}$$

EXAMPLE 4.4

Using the following statistics with a transresistance amplifier like that shown in Figure 4.14, calculate the gain figure A_r.

$$v_{out} = 5\text{ V} \qquad i_{in} = 40\text{ mA}$$

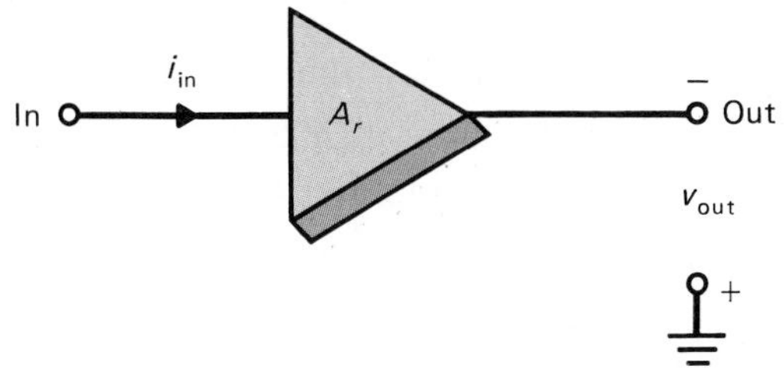

FIGURE 4.14 Transresistance Amplifier

Solution:

$$A_r = \frac{v_{out}}{i_{in}} = \frac{5\text{ V}}{40\text{ mA}} = 125\ \Omega\ (\text{V/A})$$

The impedance terms for the ideal transresistance amplifier are the opposite of those for an ideal transconductance amplifier. In order to accept any current source at the input and not form a current divider, the ideal transresistance amplifier must display zero input impedance. At the output, the impedance must also be zero. The *ideal* transresistance amplifier is expected to produce a constant output voltage, given a constant input current, for any load or circuit conditions.

Figure 4.15 summarizes the characteristics of the four forms of amplifiers that have been presented in this chapter. It can be useful when comparing each form's characteristics.

Type	Gain	Gain Ratio	Z_{in}	Z_{out}
Voltage	A_v	$\frac{v_{out}}{v_{in}}$	∞	0
Current	A_i	$\frac{i_{out}}{i_{in}}$	0	∞
Transconductance	$A_g (g_m)$	$\frac{i_{out}}{v_{in}}$	∞	∞
Transresistance	A_r	$\frac{v_{out}}{i_{in}}$	0	0

FIGURE 4.15 Comparison of Ideal Amplifier Values

REVIEW PROBLEMS

(1) Write a brief comparison of the transconductance amplifier to the transresistance amplifier. Discuss input and output quantities and impedances in this comparison.
(2) Given a transresistance amplifier with $A_r = 200\ \Omega$ and an input current of 240 mA, calculate the value of output voltage.
(3) If an amplifier has the following characteristics, identify it as being most like a transconductance amplifier or transresistance amplifier. (It is not ideal, but it will look more like one than the other.) Then calculate the gain figure for that amplifier (either A_g or A_r).

$$v_{in} = 1.5\text{ V} \qquad v_{out} = 3\text{ V}$$
$$i_{in} = 3\ \mu\text{A} \qquad i_{out} = 6\ \mu\text{A}$$

(*Hint:* Using Ohm's law, you may need to look at input/output impedances to determine the type of amplifier being shown.)

4.6 MULTISTAGE AND INTEGRATED IDEAL AMPLIFIERS

The previous sections concentrated on the definitions of certain forms of ideal amplifiers. In this section we explore how ideal amplifier concepts can be applied to multistage systems. It should be remembered that it is often necessary to use multiple stages of amplification since one amplifier may not be able to supply all of the gain needed to produce a final output. Integrated-circuit amplifiers will be presented as well, since their function may simulate an ideal amplifier.

Whether or not a multistage system is built of ideal amplifiers, the system may be described in terms of ideal characteristics. In fact, one of the initial ways to inspect a complex system is simply to identify the input and output conditions for the entire system. When that is done, a model for the system, no matter how complex inside, is shown as a simple "equivalent" amplifier. This equivalent is not based so much on all of the individual parts that are inside the system, but on what the system looks like from the outside.

In Figure 4.16 a complex system is reduced to a rather simple equivalent amplifier. Even the most complicated system can be reduced to some form of equivalent model. (If the system is multifunctional, several models, one for

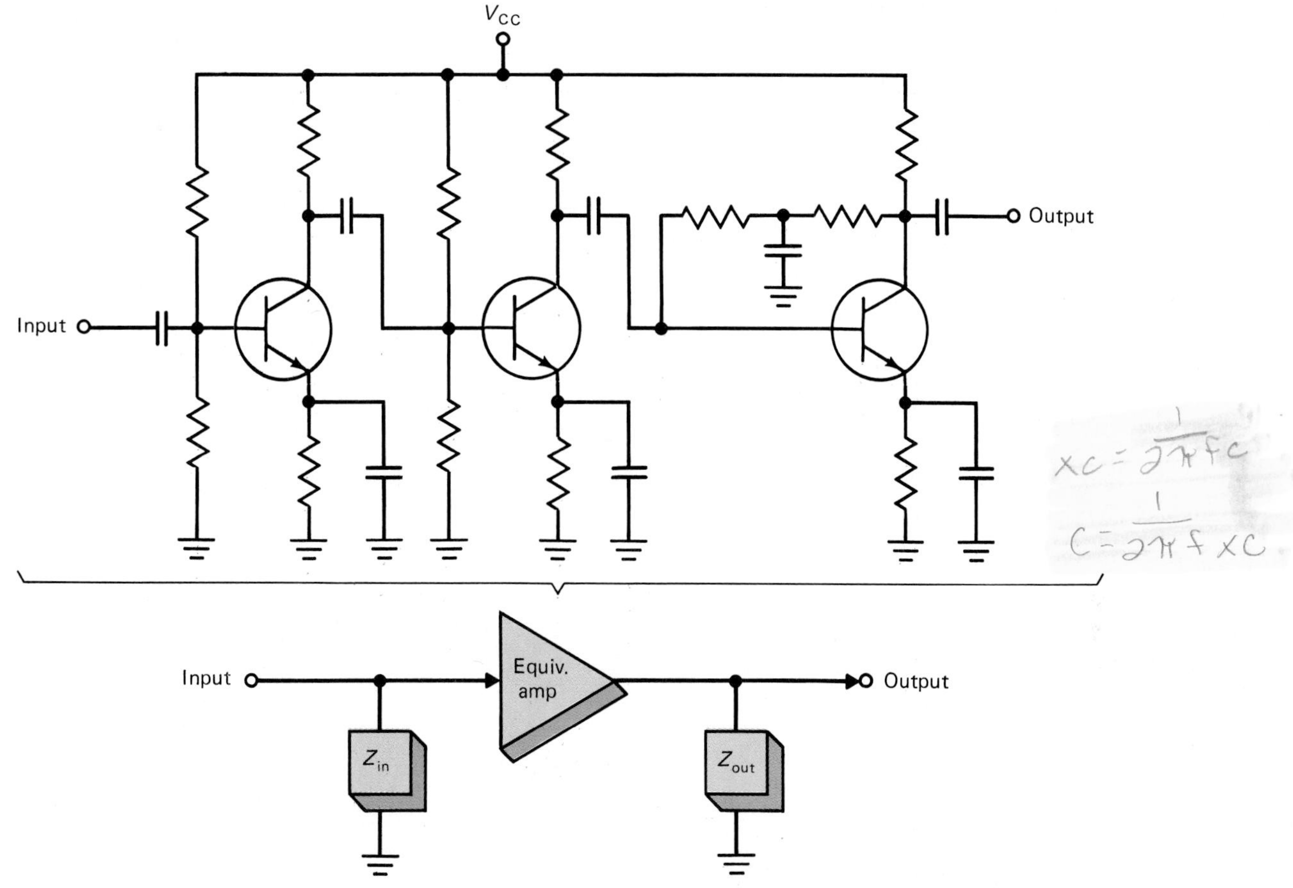

FIGURE 4.16 Characterizing Complex Amplifier System Using a Single Equivalent Amplifier

each function, may be required.) Often, troubleshooting can be aided by first investigating the ideal model(s) associated with such a system.

EXAMPLE 4.5

Let's assume that a malfunction in a product could be due to any one of four amplifier systems within the product (Figure 4.17). Each of the systems was on a separate circuit board, and each contained nine amplifiers (a multistage system). There would be two possible ways to identify the actual malfunction.

One way would involve making measurements within all 36 amplifiers (nine in each of four systems). This could be quite time consuming. The other method would be first to test each of the four systems using ideal models. If careful comparisons were made, one system would show to be obviously defective. This system would have the poorest match to the ideal of the four that were suspected. In many cases, the circuit board that contained the defective system would simply be pulled from the product and replaced with one that functioned.

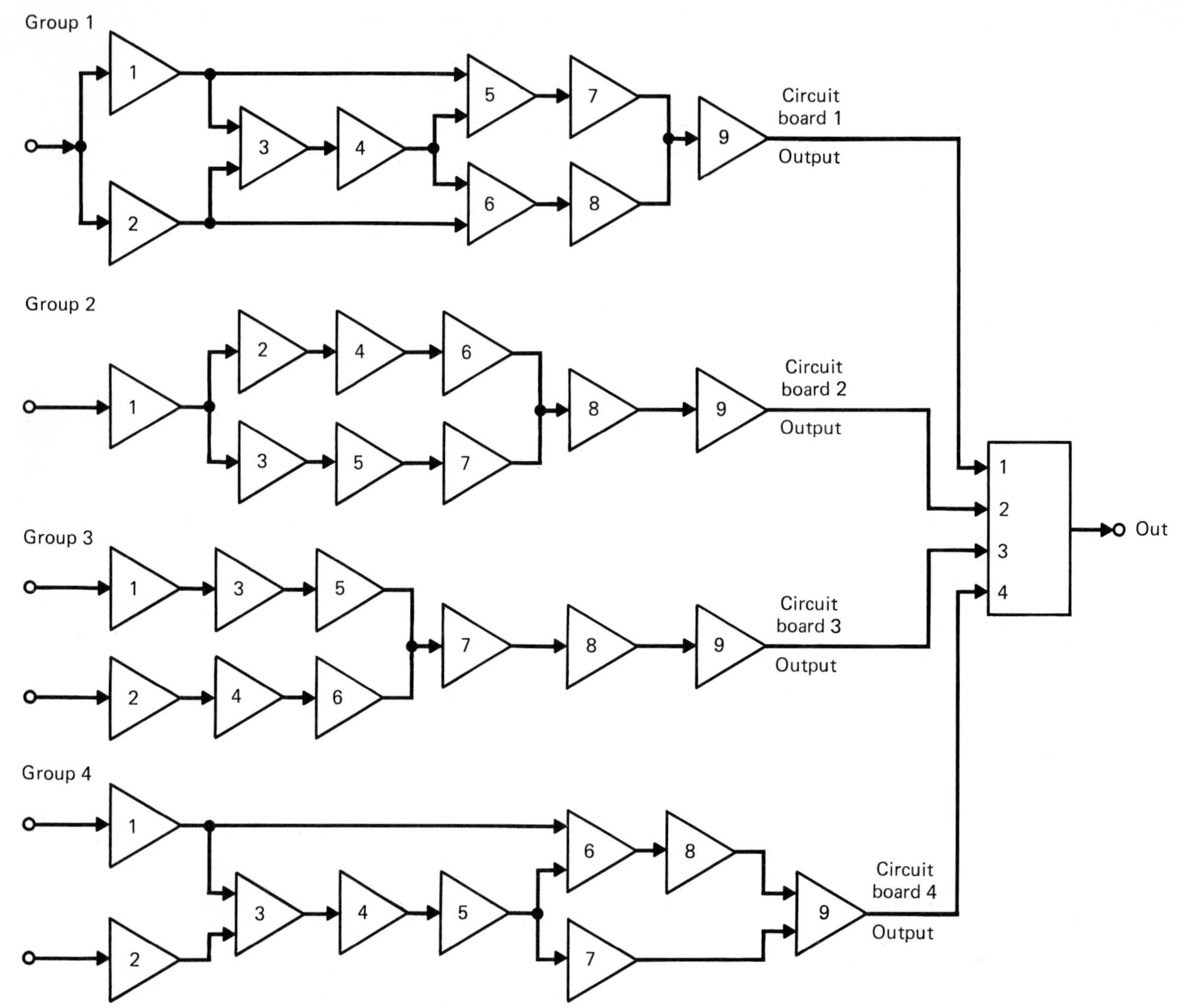

FIGURE 4.17 Complex Multiple Amplifier System

As can be seen in Example 4.5, a multistage system can be investigated simply by looking at its input and output terminals and determining its match to an ideal model. Such a comparison is not isolated to a single amplifier stage.

When integrated circuits, especially large-scale integrated circuits (multicircuit, multifunction integrated circuits), are used in a system to complete a whole function, it is necessary to use the same form of modeling. In such a case the model of the IC is used to determine if it, as a system, is functional.

In this section we have explored the use of ideal amplifier models to describe systems that incorporate more than one amplifier circuit. The multistage amplifier can, where appropriate, be described simply by its input and output characteristics. Certain ICs can also be described by input-to-output functions. When these descriptions are coordinated with the ideal amplifier model tests, measurements and expanded applications of these systems can be made with a minimum amount of effort.

EXAMPLE 4.6

One example of this modeling (coordination of real to ideal based on input/output) can be shown using the common public address amplifier shown in Figure 4.18. The input characteristics of this amplifier shows a high impedance, low current is drawn from the source (microphone) with a low voltage input. The output, on the other hand, shows only 8 Ω of impedance, with relatively high current and voltage delivered to the load. How this circuit achieves this function may not be important in describing its overall function.

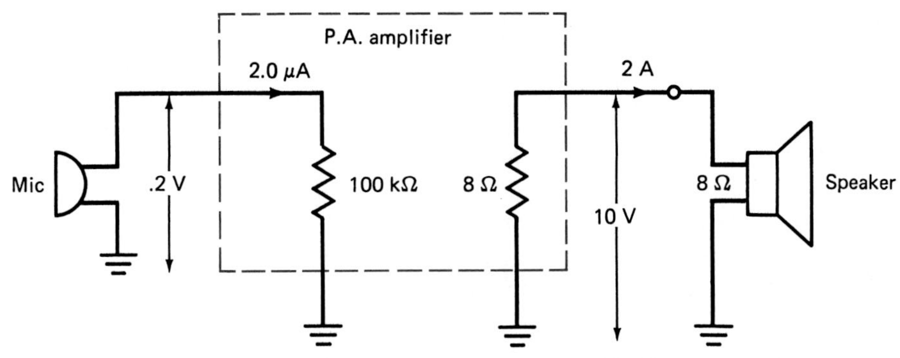

FIGURE 4.18 Input/Output Equivalent of Multistage Public Address Amplifier System

If we compare this circuit to one of the ideal models shown in the preceding sections, we see that it matches most closely the description of the ideal voltage amplifier. Often, it is not necessary to break this system down into more complex parts than are shown in this simple model. If we were interested to see if this amplifier can be used to amplify a signal other than that from a microphone, or to operate a load other than a speaker, we need only to see how these new conditions would coordinate with the ideal model. If the system were felt to be malfunctioning, we could first test the input and output conditions used on the ideal model. (It should be noted here that there are occasions where more extensive testing may be needed than implied in this section.)

REVIEW PROBLEMS

(1) Given a seven-stage amplifier with the input and output characteristics noted below, indicate which form of ideal amplifier most closely matches it as a possible model.

input impedance = 4 Ω   output impedance = 50 kΩ

$$A_i = 50 \qquad A_v = 6 \qquad A_p = 300$$

(2) An integrated circuit is being used as a voltage amplifier in a public address amplifier. First, sketch a simple model of this circuit using its coordination to an ideal model. Then indicate if this circuit could be used as a voltage amplifier in a cassette tape recorder. Briefly discuss why this application is possible (use a very general argument).

(3) In a product, there are four multistage systems which are designed to act like voltage amplifiers. Suggest how, measuring either voltage or current (not resistance), one could determine which of these systems were defective. Use the ideal voltage amplifier as a model.

4.7 FEEDBACK APPLICATIONS TO IDEAL AMPLIFIERS

In many cases, feedback can be used with nonideal amplifiers to obtain nearly ideal functions. To understand how feedback will affect the operation of a real amplifier, in this section we discuss the process of feedback in relationship to both real and ideal amplifiers. Ideal amplifiers will be used when matching feedback functions to amplifier forms. Real amplifiers and their shortcomings will be used to demonstrate the effects of feedback. Feedback will be considered a functional section of an amplifier system, rather than a separate system.

Feedback Feedback is the process of sampling a portion of the output of a system or amplifier and transferring that sample back to the input. Feedback is used to provide control over the amplification process and over the flow of information.

By using negative feedback in the proper way, we can gain more control over an amplifier's operation. An amplifier that does not function in an ideal manner, for example, can be controlled by feedback in such a way that it will more closely duplicate the desired ideal functions.

How Negative Feedback Works

Negative (subtractive) feedback is used to obtain greater control over the operating characteristics of an amplifier. One of the highly desirable features of the ideal amplifier is that its gain remains constant for any variation in load conditions, or any other operational conditions. Real amplifiers cannot maintain such a constant (stable) gain quantity. By using negative feedback, greater "stability" can be obtained.

The amount of voltage or current that is returned to the input from the output using feedback is usually rather small. Since it is subtracted from the input signal, it must be less than the actual input quantity or it would overcome the input. Thus the actual level of input supplied to the amplifier is less than the original input (actual input = original input − feedback value). So far this appears to be a useless loss of valuable input!

If the feedback quantity did not produce the stability we were looking for, this loss to the input signal would be highly undesirable (see Figure 4.19).

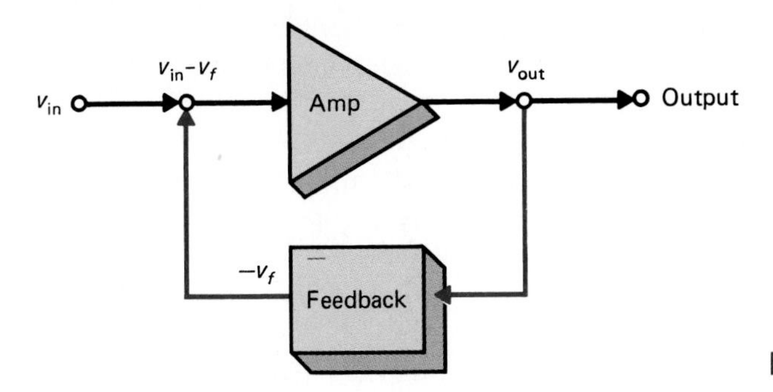

FIGURE 4.19 Negative Feedback

It must be remembered that the feedback quantity is a direct sample of the output of the amplifier.

To illustrate the role of feedback to stability, let's make the assumption that the feedback quantity will reduce the actual input to 70% of the original level. This would be done for average operating conditions. (The output, of course, is not as large as it could be, but that will not be a significant problem.) Now, let's assume that the output were to drop below its "normal" level due to some amplifier problem.

At the instant that this drop happened, the level of the feedback quantity would also drop. Since less is now being fed back (subtracted) at the input, the actual input would increase. (Note that less is subtracted from the original input.) Now the actual input may have gone up to 80% of the maximum (original) level. Since there is more actual input for the amplifier, the output level will rise. Remember that the problem was that the output had dropped. Now, instead of the lower output level that we started with, the level is nearly back to what it was before the instability problem happened in the amplifier. (Such corrections are not 100% perfect, but a major improvement in stability can be achieved.)

This counterbalancing process, where the output sample is used to reduce or increase the actual input to the amplifier, in a direction opposite to the changes seen at the output due to amplifier errors, creates a much more stable amplifier. This process occurs almost instantly, even for very minor changes in the amplifier's output. Since such errors are compensated for instantly, the actual output of this system remains relatively constant.

EXAMPLE 4.7

It may be helpful to look at some characteristics of such a feedback amplifier. First, let's look at an amplifier without negative feedback.

The amplifier is a current type that produces a normal gain of 100 with an input of 10 mA. Let's assume that the amplifier has one fault that causes its gain to go down to 90, then a second fault that causes it to jump to 110.

Normal output: 10 mA $\times$ 100 = 1 A
Fault 1 output: 10 mA $\times$ 90 = 0.9 A (10% loss)
Fault 2 output: 10 mA $\times$ 110 = 1.1 A (10% increase)

If a negative feedback of just 5% of the output were applied to this circuit, the amount of change in output due to these faults would be:

Normal output: 200 mA (note loss in overall gain)
Fault 1 output: 195.7 mA (2.2% loss from 200 mA normal)
Fault 2 output: 203.7 mA (1.9% increase from 200 mA normal)

Even though the second circuit has a lower overall output, (200 mA versus 1 A) the percentage of error is greatly reduced. From our study of multistage amplifiers in earlier chapters, to recover from this loss in output we need add only a following amplifier stage with a gain of 5 to compensate (5×200 mA $= 1$ A). The following is a detailed algebraic solution of the second part of this example:

$$\text{actual input} = \text{initial input} - \text{feedback} \qquad \textbf{(4.9)}$$

$$\text{feedback} = 5\% \times \text{output} = 0.04 \times \text{output} \qquad \textbf{(4.10)}$$

$$\text{output} = \text{actual input} \times \text{gain} \qquad \textbf{(4.11)}$$

Therefore, factoring equation 4.10 into 4.9, we get

$$\text{actual input} = \text{initial input} - 0.04 \times \text{output} \qquad \textbf{(4.12)}$$

Finally, factoring equation 4.12 into equation 4.11 yields

$$\text{output} = (\text{initial input} - 0.04 \times \text{output}) \times \text{gain} \qquad \textbf{(4.13a)}$$

Rearranging yields **(4.13b)**

$$\text{output} = \frac{\text{gain} \times \text{input}}{1 + (\text{feedback} \times \text{gain})}$$

If this equation is solved for each of the conditions noted in the solution, we get answers of

Normal output: (10 mA $-$ 0.04 $\times$ normal output) $\times$ 100 or

$$\text{Normal output} = \frac{100 \times 10 \text{ mA}}{1 + (0.04 \times 100)} = 200 \text{ mA}$$

$$\text{Fault 1 output: } \frac{90 \times 10 \text{ mA}}{1 + (0.04 \times 90)} = 195.65 \text{ mA [error} = \text{(approx.) } 2.2\%]$$

$$\text{Fault 2 output: } \frac{110 \times 10 \text{ mA}}{1 + (0.04 \times 110)} = 203.7 \text{ mA [error} = \text{(approx.) } 1.9\%]$$

Methods of Negative-Feedback Production

There are four basic methods of producing feedback. Each of these methods relates to how voltage and current are handled through the feedback process and by the amplifier that uses the feedback. First we review the handling of voltage and current at the input and output terminals of an amplifier (Figure 4.20). Then we investigate how these quantities (V and I) are handled in the feedback process.

First, let's consider the form that input to an amplifier can take. The input could be a voltage or current level. Although a real amplifier may have both voltage and current at its input terminals (not having ideal input impedance), the form of the amplifier or components used will make it respond primarily to only one of these quantities.

Next, we can look at output. An amplifier will have either voltage or current as its primary output. The primary output, then, is that quantity which the amplifier will attempt to hold constant given a variable load on the output. Much as we divided amplifiers into four types before (voltage, current, transconductance, and transresistance), we will divide feedback into four forms. Each of these forms corresponds to a combination of input and output quantities (V and/or I). Each of the feedback types will be shown to relate to one of the four basic amplifier types listed previously.

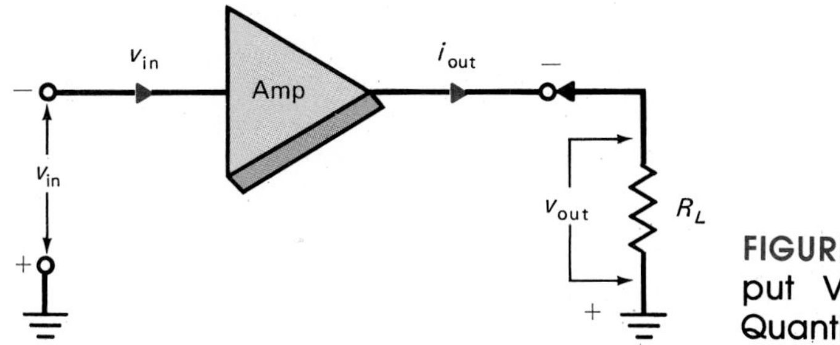

FIGURE 4.20 Input and Output Voltage and Current Quantities

Figure 4.21 illustrates each type of feedback we will discuss and the corresponding amplifier to which that type is related. It will be necessary to take note of the type of amplifier used, the form that the feedback takes, and the fact that all of the feedback is negative (subtractive) in relationship to the main input value.

Note: The term **shunt**, when related to feedback, indicates a parallel connection. For example, to place two resistors in shunt is to place them in parallel. In feedback, to sample the output in shunt means to attach the sample points across the load, therefore in parallel with it. To feed back to the input in shunt means to return the signal in parallel with the input signal.

The term **series**, when related to feedback, indicates a series type of connection. To sample the output in series means to interrupt the path between the amplifier and the load and extract the sample. To feed back a series signal means to place that signal (like a source) in series with the actual input source.

The following is a discussion of each type of feedback and how its affects the amplifiers with which it is associated.

Voltage Sampling/Series Mixing (Series/Shunt) This form of feedback is related to the voltage amplifier. The output quantity of this amplifier is voltage. A sample of this output (taken in shunt) is inverted (negative feedback) and returned to the input as a voltage quantity. This voltage is reapplied to the input in series with the input voltage. Since the feedback is negative, it is subtracted from the input quantity.

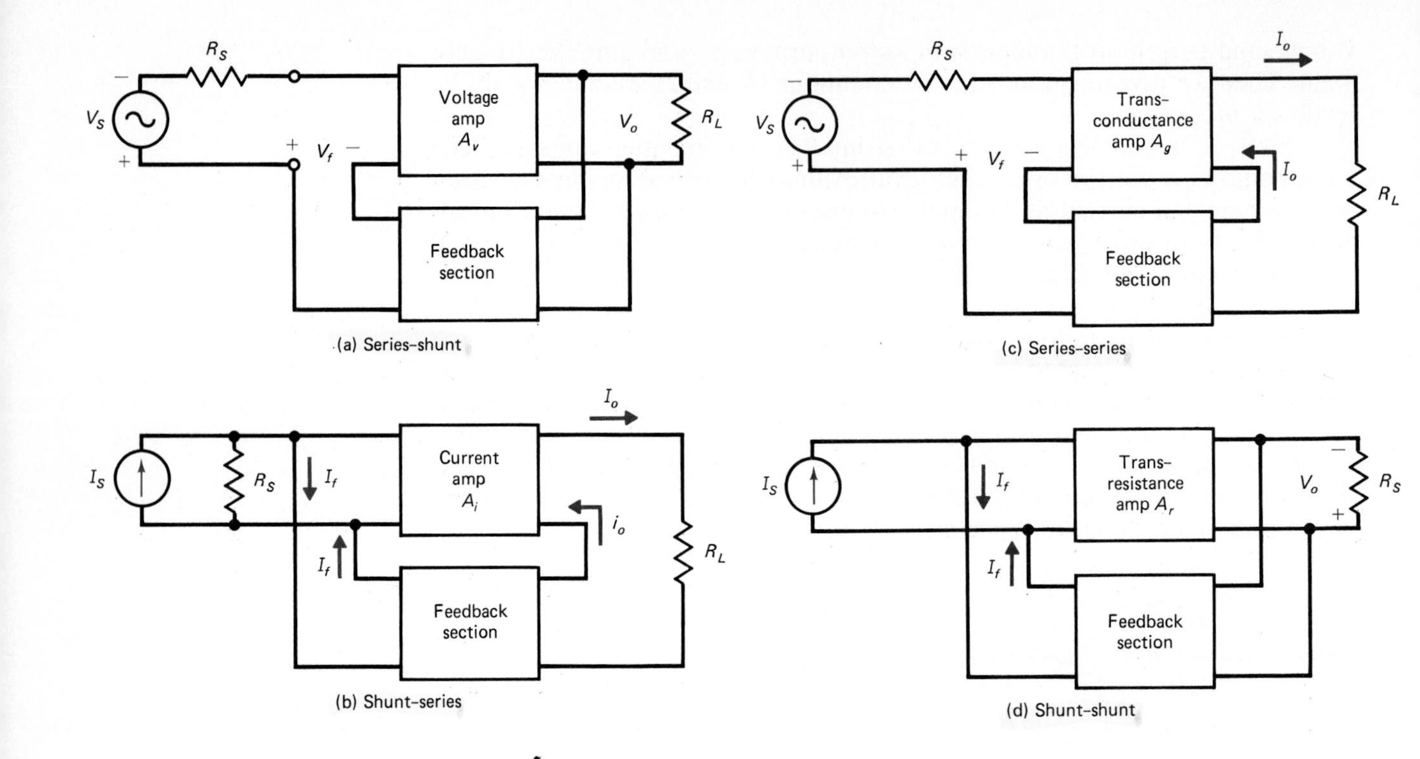

FIGURE 4.21 Feedback Configurations for the Four Amplifier Types

Current Sampling/Shunt Mixing (Shunt/Series) A basic current amplifier uses this form of feedback. Both the input and output quantities are current. The output flow is interrupted and sampled in the feedback section. It is inverted (negative) and added in shunt with the input current source.

Current Sampling/Series Mixing (Series/Series) In this feedback method the output current is sampled. The feedback section both converts current to voltage and supplies this voltage as a feedback quantity which is opposite in polarity from that of the input. This form of feedback is used on amplifiers that accept input voltage and produce output current (the transconductance type).

Voltage Sampling/Shunt Mixing (Shunt/Shunt) The last of the four forms samples the output voltage from the amplifier and supplies an inverted current back to the input terminals. Thus this feedback section must also convert an output voltage sample to current. This form of feedback is to be used with the transresistance amplifier.

In the previous examples and discussions it was shown that with negative feedback the actual gain of the complete system was less when feedback was used. When looking at an amplifier with feedback, the actual gain value to use is the **effective gain**, not the potential gain of just the amplifier. In using negative feedback the amplifier achieves greater stability, but with a lower gain value.

If an amplifier, for example, has a potential gain of 100, if negative feedback is used that gain could well drop to only 30 or 40. There are two major reasons for this feature. The actual process of negative feedback, the subtraction of values from the input, presents a lower actual input to the amplifier than is originally supplied. Second, for feedback to function properly it must be able to subtract either more or less value (voltage or current) from the supplied input. Therefore, the "normal" value of actual input (after feedback) must be below the maximum possible input.

In these discussions negative feedback was used to compensate for faults in the amplifier circuit itself. Actually, this feedback could also be used to compensate for other errors. If, for example, the original input to the system were to vary, negative feedback could be used to compensate for these errors, too.

There are limits to how much stability negative feedback can produce. Negative feedback can reduce the amount of instability in a system, but it cannot eliminate instability. Negative feedback, for example, could reduce potential errors of 50% to possibly only 15%, but never to 0%. Also, feedback cannot compensate for severe instability. For example, if an amplifier's gain were to drop below the effective gain of the system (with feedback), the feedback could not make up for the amplifier's inability to produce an acceptable output. Therefore, the effective gain of the system using negative feedback is the minimum amount of gain the amplifier inside the system must be able to produce.

It should be noted here that feedback *does not* have to be restricted to use with a single amplifier stage but can be applied to multiple-stage circuits as well. In multiple-stage circuits, feedback could be applied to the individual stages or could as well be used to span the complete set of amplifiers in the group. It is not uncommon to use one main feedback signal tapping the multistage circuit's output and returning a tiny sample to the first stage's input. The effects of feedback and the calculations are the same (or only slightly modified) whether using single amplifier stages or multiple stages.

In this section on feedback, we have used the example of feeding back information signals. This form of feedback is the easiest to observe and understand initially. In any circuit, though, it is equally common to see the dc source levels within the circuit being sampled and fed back for control. The effect is nearly the same—control of the amplifier to compensate for instability. In any circuit or system, either source levels or signal levels, or even both, could be used in the feedback process. Which form is used is often determined by the type of active elements used to make up the amplifier system.

There also are special forms of feedback control. Some feedback circuits are frequency selective and therefore produce a frequency-selective control (feed back more voltage or current at one frequency than at another). Others will sample a signal output and convert it to a dc level to provide control not of the information input but of the dc source levels at an input amplifier. Some are even used as simple on/off controllers that switch on (or off) certain circuits, depending on the sample seen at the output. When specific applications for electronic circuits are explored later in this book, it will be shown that in many cases the form the feedback takes governs more of the final function of the circuit than does the selection of any other circuit element.

The objective of this section was to introduce the reader to the application of negative feedback. The function of negative feedback was shown to be the production of a more stable output from an amplifier which could be internally unstable (changing gain). In future sections, more detailed applications and calculations involving feedback will be presented. At this point it is sufficient to know the overall role played by feedback in amplifiers.

As a final note on feedback, most feedback is done through the use of passive elements. Voltage or current dividers are commonly used to reduce the quantity of the output to a level usable at the input. Resistors are often used as the sensor that will convert current to voltage (Ohm's law) or voltage to current flow. Often, the negative polarity (or current direction) that is necessary for negative feedback is obtained through the normal function of the amplifier itself (examples of this are shown later).

REVIEW PROBLEMS

(1) If a current amplifier fails to produce a stable output, feedback could be used to help maintain stability. Sketch such a system and prepare a list of the advantages of feedback (and any disadvantages).

(2) Even with feedback, an amplifier cannot produce ideal output for any conditions. Discuss conditions under which even a strong feedback system would not be able to force an acceptable output from an amplifier.

(3) Calculate the effects seen (and percent error) for an amplifier with the following statistics, based on feedback equations 4.9 through 4.13.

Input: 0.02 V
Feedback: 1% of output
Normal gain of amplifier: 150
Fault 1 gain: 110
Fault 2 gain: 180

[Normal output = 3.0 V, fault 1 output (no feedback) = 2.2 V (26.7% error), fault 2 output (no feedback) = 3.6 V (20% error).] Once you have finished this, calculate the effective gain of the system under all three operating conditions: normal, fault 1, and fault 2.

4.8 INVESTIGATION OF NONIDEAL AMPLIFIERS

In this chapter both ideal and nonideal amplifiers have been used in examples and discussions. Before proceeding and using either form, it is important generally to contrast the two and to investigate what factors would make an amplifier not fit the ideal model. Once this is done, we can be at liberty simply to show an amplifier function and assume the ideal model, knowing in reality that real-life amplifiers will not be ideal.

As has been noted, ideal amplifiers are expected to produce their specific function of voltage, current, or power amplification under any operational conditions. In reality, no amplifier can be expected to meet this objective. Even amplifiers that are nearly ideal in function have some ultimate limits.

Normal Amplification Limits

Every amplifier has maximum limits as to how much output it can produce. This is the most obvious limitation. Usually, these maximum limits are set either by the specific parts used in the amplifier or by the power supply for that amplifier.

All electronic devices—resistors, transistors, integrated circuits—have a maximum amount of power they can dissipate (handle) before they will overheat and be destroyed. If such elements have a limit, the amplifier that uses them will have a limit as well. Therefore, it is not possible to assume that if, for example, a voltage amplifier has a gain of 100 and we attempt to input 50 V, the individual components can tolerate the 5000-V level at the output while handling any amount of current at the same time. An amplifier could be built to handle such a level, but components would have to be specially selected to handle what could be several hundred or even thousands of watts of power. Therefore, maximum power limits for individual elements may set operational limits for even the best amplifiers.

Another ultimate limit would be the capacity of the power supply for an amplifier to produce the desired output levels. All power supplies are limited to a maximum voltage and current they can produce. If an amplifier were to attempt to exceed these limits, the power supply would either be damaged or would simply impose its limit on the amplifier. If in the previous example a power supply were used that could produce no more than 100 V, then no matter what the "gain × input" calculation was, the output would not go above 100 V. If the circuit continued to "overload" the power supply, the supply might be damaged.

There are often input restrictions associated with an amplifier. If the output cannot go above a certain level due to potential component failure, this is reflected back to the input as a limit there as well. An example of this can be seen if we review the preceding paragraph. The system was limited to a maximum output voltage of 100 V, due to the power supply. If the amplifier has a gain of 100, we can use this gain figure in reverse to see a limit in input voltage. Thus the maximum input voltage would be 1 V (maximum output)/ 100 (gain), or an input limit of 1 V. If 1 V is exceeded at the input, the amplifier will attempt to produce an output greater than 100 V, which cannot happen.

Most amplifiers also have a minimum input-level restriction. Naturally, this is not based on the same factors that set the maximum limitations. Usually, this limit is based on the physical properties of the amplifier. Most active devices will not produce a dependable output if the input is below a specified minimum. Such "low-level" outputs may either be nonexistent or so distorted (changed from the input) that they are unusable.

Therefore, there will be both a maximum and a minimum limit to the level of voltage, current, or power that an amplifier can handle. In many cases, these limitations are often the basis used to select the active device used in the amplifier.

Specialized Amplifier Limitations

Beyond the normal maximum and minimum limitations noted, there are other limitations that will restrict an amplifier's function. These features are specialized since in some applications they can be ignored. One of these limitations deals with frequency response. There is an upper limit to the frequencies that an amplifier can be expected to amplify. In most cases, these are set by the active device used to provide gain. Some devices are unable to amplify an input frequency above 40 kHz. If a frequency above that limit is applied at the input the output may either be nonexistent or will be distorted. Quite a few amplifier circuits cannot amplify a dc input level. This is often due to the use of capacitors connected directly to the input terminal of the amplifier. (Capacitors will block dc from entering the circuit because they will simply charge to that level and act like an open circuit to current flow.)

When an amplifier is used to switch a system on or off, such as in a motor control circuit, time delays may be produced by nearly any of the components used in the circuit. In fact, about the only element that does not typically produce a delay is a resistor. Such a delay may, however, create a serious problem in controlling the operation of the motor. Therefore, what appears to be a nearly ideal amplifier may become totally useless because of the time delay it produces.

Most devices are sensitive to temperature and in some cases to humidity. One of the most frequent causes of instability in a device or circuit is temperature variation. An amplifier that has a voltage gain of 100 at room temperature may have a gain of 150 in a heated environment and a gain of 50 at zero degrees. If nothing is done to stabilize the amplifier (such as negative feedback), its operation would be undependable since an exact temperature would be difficult to maintain continuously. Such a circuit may be acceptable to use in a system that is operated only in a temperature-controlled room. It would never be able to be used in a portable radio, for example, since the temperature could not be properly controlled.

There are other reasons for not being able to obtain amplifiers that will perform ideally under any situation or application. Just the simple tolerance of the components used, both active and passive, can create problems. It would be impossible to duplicate ideal conditions if the elements used to build a circuit were not precise. An additional problem would exist when replacing a defective component. If the replacement were not exactly the value of the element it was replacing, this ideal amplifier could become unstable and wind up being totally useless.

Knowing that circuits will not initially be ideal and that instability will probably occur, the selection of circuits used for particular applications can be done in such a way as to overcome these potential problems. This, in fact, is the prime reason for applying negative feedback in an amplifier system.

One of the big advantages gained when using integrated circuits rather than discrete devices is the overall stability of the system. Integrated circuits usually apply feedback and other internal controls so that from the outside they are very stable and act nearly like an ideal model. This simplifies the task of developing, calibrating, and troubleshooting such a system. For example, a transistor radio using discrete devices may have 20% or more of the internal components dedicated to maintaining stability. Any one of these elements (or others) could fail or change in value enough to create a problem.

The same basic radio circuit, made from one or two integrated circuits, would have the problem of instability solved internally by the integrated circuit(s).

REVIEW PROBLEMS

(1) Describe, in your own words, what could happen if the circuits inside a car radio were adversely affected by temperature. What if the gain of the amplifiers were to increase when the radio got warm and decreased when it got cold?

(2) In Appendix A are several "specification sheets" for solid-state devices. Select any one of the transistor (device) sheets. Go over that sheet and list any of the characteristics which are indicated as being approximate, an average, having a range (say, from 40 to 60), or stated to be temperature dependent. You will probably not know what each of these characteristics means, but this list does summarize some of the features that could lead to instability.

4.9 USING IDEAL MODELING TO ASSIST IN TROUBLESHOOTING

In previous chapters, the block diagram was shown to be of assistance when troubleshooting a defective circuit. Once the block diagram is reviewed and the potential source of the problem is found, it will be necessary to investigate the circuitry associated with the problem block. When doing this, it will be necessary to use schematics and the actual circuit.

Ideal models will help when working with schematics and circuits. In most applications, the circuit used was designed to produce an output that simulated an ideal circuit's function. As long as the input(s) and other conditions are normal, the circuit would be expected to maintain an output similar to that expected of an ideal counterpart.

If, when working with the circuits, we can relate their function to that of ideal models, tests can often be made to determine where the fault in the system exists. For example, if a circuit should maintain the constant current output of a current amplifier, and when tested, that current was ever changing, the circuit could be suspected of being defective. Similarly, if a transconductance amplifier did maintain constant current at the output for a variety of load conditions, it is probably operating as expected.

Since we also know that negative feedback is often used to make amplifiers function more like ideal systems by providing stability, if a circuit seems to be unstable, the fault may be in the feedback circuit (or with a lack of enough feedback to maintain stability). Therefore, if we can keep in mind the ideal model of the circuit that we are troubleshooting, we can use that model to assist in judging if the circuit is meeting the proper objectives. In future chapters many of the devices will be contrasted to their ideal model counterparts. Much of the circuit development will be related to producing more ideal circuits.

REVIEW PROBLEM

(1) In preceding sections of this chapter, models of the following ideal amplifiers were presented. For each type, write a short procedure for testing each to see if it is functioning as expected (voltage and/or current measurements). The first one has been done for you as an example.

(a) Voltage amplifier

To test the voltage amplifier, apply a fixed voltage to the input. Measure the output voltage to a variety of loads. The output should remain constant and be at a level of: gain × input. The input current should be checked. It should be zero since this amplifier's input resistance should be infinite.

(b) Current amplifier

(c) Transconductance amplifier

(d) Transresistance amplifier

SUMMARY

In this chapter the ideal forms of amplifiers were introduced. Each model had specific input, output, and resistance (impedance) characteristics. Each was designed to produce a specific function. The following is a list of these amplifiers and a summary of their operating conditions:

Type	Input Quantity	Output Quantity	Input Impedance	Output Impedance
Ideal voltage amplifier	Voltage	Voltage	Infinite	Zero
Ideal current amplifier	Current	Current	Zero	Infinite
Ideal transconductance amplifier	Voltage	Current	Infinite	Infinite
Ideal transresistance amplifier	Current	Voltage	Zero	Zero

(Since the ideal power amplifier has variable specifications, it was omitted from this table.)

Becoming more familiar with the features of these ideal models will assist in the understanding of all forms of amplifier circuits, since they can usually be related to one of these models.

Feedback was shown to be used to assist in producing a more ideal model from a nonideal amplifier circuit. Negative feedback was shown to help eliminate stability problems. By using feedback, an amplifier with unpredictable gain and a variable output could be made more stable and predictable. One disadvantage of negative feedback was that the maximum gain of the amplifier could not be used, but this could be compensated for by the addition of more amplifier stages.

It was noted that all amplifiers have limitations. Most cannot produce ideal or even nearly ideal characteristics for every possible operating condition. Therefore, these limitations must be considered carefully when dealing with any amplifier.

KEY EQUATIONS

4.1 Power in (P_{in}) $= V_{in} \times I_{in}$

4.2 Power out (P_{out}) $= V_{out} \times I_{out}$

4.3 A_p (Power gain) $= P_{out}/P_{in}$

4.4 $A_p = A_v \times A_i$

4.5 G (conductance) $= 1/R$ (resistance)

4.6 $G = \dfrac{I}{E}$

4.7 Gain $= \dfrac{\text{output}}{\text{input}} = \dfrac{\text{current}}{\text{voltage}} = \text{conductance}$

4.8 Gain for a transresistance amplifier: $A_r = \dfrac{v_{out}}{i_{in}}$ (ohms)

PROBLEMS

Section 4.1

4.1. If an amplifier had the following input and output statistics, find A_v, A_i, and A_p.

$$V_{in} = 0.02\text{ V} \qquad V_{out} = 4\text{ V}$$
$$I_{in} = 0.25\text{ mA} \qquad I_{out} = 100\text{ mA}$$

4.2. A voltage amplifier has the following statistics. Illustrate a simple Thévenin equivalent of that amplifier (input and output).

$$R_{in} = 1.6\text{ k}\Omega \qquad A_v = 60 \qquad R_{out} = 2.5\text{ k}\Omega$$

4.3. For the amplifier described in problem 4.1, construct a Thévenin equivalent model of that amplifier.

Section 4.2

4.4. List the requirements of an ideal voltage amplifier.

4.5. An ideal voltage amplifier is (since it is ideal) to produce its listed voltage gain under any operating conditions. Even if an amplifier was built that duplicated an ideal amplifier, discuss why this gain factor is still unrealistic. (Be sure to note such factors as power supply limits and power consumption.)

Section 4.3

4.6. List the requirements for an ideal current amplifier.

4.7. If an ideal current amplifier has an input of 10 mA and a current gain

of 100, it should produce 1 A output current. Even if the amplifier is nearly ideal, this becomes unrealistic when load resistances become high. Indicate the types of voltages that would be required to support this current for the following load resistances.
(a) 10 Ω (b) 1 kΩ (c) 5 MΩ (d) Open circuit

4.8. Contrast the ideal current amplifier to the ideal voltage amplifier. Compare their input and output conditions.

4.9. The ideal power amplifier is defined in terms of the circuitry that surrounds it. Basically explain why this is the case. (Why isn't there one form of ideal power amplifier?)

4.10. Given an amplifier with the following characteristics, find the power gain first by using equation 4.4, then by using equation 4.3. Compare the results.

$$V_{in} = 0.2\text{ V} \qquad V_{out} = 11\text{ V} \qquad I_{in} = 5\text{ mA} \qquad I_{out} = 40\text{ mA}$$

(Assume that the output current is delivered into a nearly zero ohms load, thus is due exclusively to the output resistance of this amplifier.)

4.11. For the amplifier statistics listed in problem 4.10, find the input and output resistance for the amplifier. If a load were attached to this amplifier that matched the output resistance, find the value of power delivered to that load resistor.

Section 4.5

4.12. List the characteristics of both the ideal transconductance amplifier and the transresistance amplifier.

4.13. Contrast the transresistance amplifier to an ideal voltage amplifier. How are they similar? How do they differ?

4.14. A transconductance amplifier has the following characteristics. Find its gain value (A_g).

$$V_{in} = 0.4\text{ V} \qquad I_{out} = 38.5\text{ mA}$$

4.15. Although it is not specifically a transconductance amplifier, find the transconductance gain figure for the amplifier listed in problem 4.10.

4.16. Find the transresistance gain for the following amplifier statistics. Also, indicate what value the input and output resistance of this amplifier should be to approximate an ideal transresistance amplifier.

$$I_{in} = 340\text{ mA} \qquad V_{out} = 12.4\text{ V}$$

4.17. Any nonideal amplifier will have characteristics that simultate any of the four forms of ideal amplifiers. Although some of the statistics may rate it as a poor voltage amplifier, others may classify it as an adequate transconductance type. Given the following characteristics for a basic amplifier, rate it in terms of all four forms of amplifiers (look at input and output impedances and gain characteristics for each of the four forms). Finally, decide which ideal form it is most like and defend your decision.

$$V_{in} = 20\text{ mV} \qquad V_{out} = 1.1\text{ V} \qquad I_{in} = 0.5\text{ mA} \qquad I_{out} = 1.8\text{ mA}$$

Section 4.6

4.18. An op-amp is an integrated-circuit amplifier system with a high input impedance (potentially), a low output impedance, and a high ratio of input voltage to output voltage. Classify it as one of the ideal forms and defend your selection.

4.19. In some electronic applications, a typical current amplifier (close to the ideal model) is applied in a circuit that depends on voltage gain statistics. Propose, with the addition of series or parallel resistances, how such a change in characteristics could be obtained. (You cannot change the basic current amplifier but may add extra components to make it "look and act" like a voltage amplifier.)

4.20. Ideal amplifier models were presented as a basis on which to compare other amplifier systems. In troubleshooting, these comparisons can aid in speeding the process. Prepare a brief summary of the quantities to be measured around each of the four models to determine if it is functioning properly.

Section 4.7

4.21. Indicate the four forms of feedback and relate these to the four forms of amplifiers.

4.22. List the advantages and disadvantages of negative feedback.

4.23. Too much negative feedback can destroy the output of an amplifier. Verify this statement by indicating what would happen to a negative-feedback amplifier as the feedback value approached the input value.

4.24. A governor on a lawn-mower engine produces a negative-feedback function. If the engine speeds up, it reduces the gasoline flow; if it slows down, it increases the flow. Relate this simple use of negative feedback to that used in an amplifier.

4.25. Using the feedback equation 4.13b, calculate the effects seen (and final percent error) for an amplifier with the following characteristics. Also find the output and error without feedback and compare these characteristics.

Input: 12 mV
Feedback: 0.6% of the output
Normal gain of the amplifier: 100
Fault 1 gain: 135
Fault 2 gain: 82

4.26. Using the same characteristics (and equation) as shown in problem 4.25 but with a 3% feedback, recalculate the effects and errors for this system.

Section 4.8

4.27. For each of the four types of ideal amplifiers, list those specifications that would change when using a "real" equivalent of those models.

4.28. List the factors within real amplifiers that could fluctuate during the course of operation. Also list circuit or environmental conditions that could cause these changes.

ANSWERS TO REVIEW PROBLEMS

Section 4.1

1. $A_v = 11$, $A_i = 100$, $A_p = 1100$
2. $R_{in} = 400\ \Omega$, $R_{out} = 44\ \Omega$

3. $V_{out} = 40 \times 0.2 = 8$ V, $I_{in} = 0.2$ V/1 kΩ = 200 μA (Values are the same for both forms of the circuit.)

Section 4.2

1. (a) For Figure 4.6, $V_{out} = 5$ V; for Figure 4.7, $V_{out} = 0.139$ V.
   (b) For Figure 4.6, $V_{out} = 5$ V; for Figure 4.7, $V_{out} = 3.92$ V.
2. Even though the output should be 320 V ($V_{in} \times A_v$), the dc source prevents the output from going above 12 V; thus this function is unrealistic.

Section 4.3

1. (a) For Figure 4.8, $I_{out} = 2.5$ A; for Figure 4.9, $I_{out} = 1.11$ A.
   (b) For Figure 4.8, $I_{out} = 2.5$ A; for Figure 4.9, $I_{out} = 2.34$ A.
2. The output should be 10 A to the 1-kΩ load. This would produce 10,000 V. Since the supply can produce only 40 V, that would be the maximum possible drop on the load. 40 V/1 kΩ = maximum current to the load then (or 40 mA). The output is limited to less than the input (100 mA) and the gain would be 0.4 (a loss).
3. The maximum input current would be 400 μA.

Section 4.4

1. $Z_{in} = 500\ \Omega$, $Z_{out} = 8\ \Omega$, $A_p = 50$
2. The voltage output of the amplifier would be dropped across its internal resistance and not efficiently passed on to the load, thus reducing the output power.
3. $P_{in} = 200$ mW, $P_{out} = 30$ W, $A_p = 150$
   (equation 4.4) $A_p = A_v \times A_i = 10 \times 15 = 150$
   They are identical values.

Section 4.5

1. See Figure 4.15 for details.
2. 48 V
3. Transconductance; $A_g = 4$ μS

Section 4.6

1. Current amplifier
2.

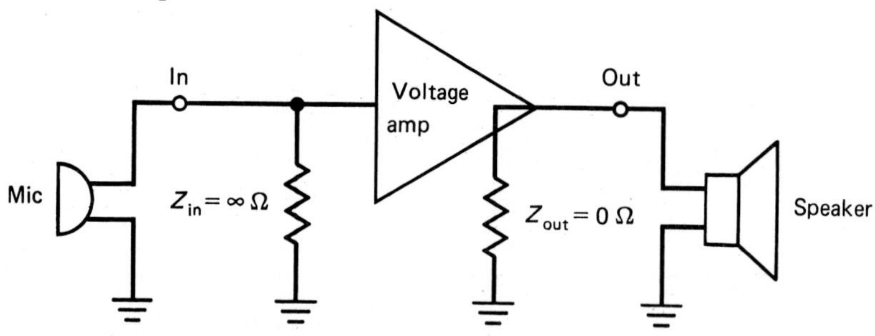

   Yes, it could also be used with the cassette since either application would require voltage gain, high input impedance, and low output impedance, the characteristics of any voltage amplifier.
3. Measure both input and output voltages to determine the voltage gain. Then if additional data are needed, measure input and output current to determine input and output impedance. The defective amplifier system will show an unacceptable statistic for one of these values.

Section 4.7

1.

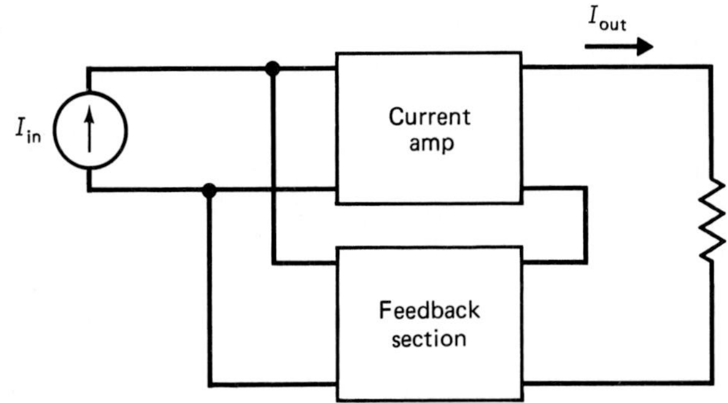

For advantages and disadvantages, see Section 4.7.

2. If the output would attempt to produce an output higher than that deliverable by the dc source; if the internal circuit were defective, not able to produce gain; if the feedback signal were so high as to eliminate the input level; if the input signal were virtually undetectable
3. Output with feedback = 1.2 V; fault 1 output with feedback = 1.05 V (12.7% error); fault 2 output with feedback = 1.29 V (7.5% error); effective gain, no fault = 60; effective gain, fault 1 = 52.5; effective gain, fault 2 = 64.5

Section 4.8

1. The radio's volume would change as temperature in the car changed: louder for hotter temperatures, softer for cold temperatures.
2. The answer is dependent on the device selected.

Section 4.9

1. (b) To test a current amplifier, apply a fixed current input. Measure the output current to a variety of loads. The output current should remain relatively constant. The input voltage should be checked and should be near zero ($R_{in} \simeq 0$).
   (c) To test the transconductance amplifier, apply a fixed input voltage. Measure the output current to a variety of loads. The output current should remain relatively constant. The input current should be checked and should be near zero ($R_{in} \simeq$ infinite).
   (d) To test the transresistance amplifier, apply a fixed input current. Measure the output voltage to a variety of loads. The output voltage should remain relatively constant. The input voltage should be checked and should be near zero ($R_{in} \simeq 0$).

5 Introduction to Solid-State Devices

Objectives

Upon completing this chapter, the reader should be able to:

- Identify semiconductor types and their characteristics.
- Recognize the forward-bias conditions needed by semiconductors.
- Calculate basic gain figures for active devices.
- Relate devices again to ideal modeling.

Introduction

In this chapter we deal with the functional characteristics of semiconductors. It is a primer for future chapters which deal with specific device applications. The devices introduced in this chapter are:

The semiconductor diode
The bipolar junction transistor (BJT)
The field-effect transistor (FET)
The operational amplifier (op-amp)
Large-scale integrated circuits (LSI)

The basic function, electrical operation, and relationship to ideal amplifier models is presented for each of these elements. A comparison is made of each classification, contrasting their characteristics. ■

5.1 CATEGORIES OF SOLID-STATE DEVICES

The term "solid-state" refers to a broad classification of electronic devices. Within that classification is a wide variety of types and forms of devices. In this section we explore generally each of the primary forms of solid-state devices.

Solid-state devices are manufactured using a special process and special materials. Another term used for solid-state devices is "semiconductor." However, this term is related to the application of these solid-state devices. Solid-state devices are manufactured primarily using silicon or germanium. Both of these materials are elements that are found in nature. Although germanium is used to manufacture solid-state devices, most devices are made of silicon. This book will deal mainly with the applications of silicon-based semiconductors.

To produce a solid-state device, silicon is used in a highly refined and purified crystalline form. Once this chemically pure silicon is manufactured, it is treated with a variety of other chemicals, called impurities, to produce specialized electrical properties. The chemicals that are blended with the silicon change its characteristics in such a way as to produce the semiconductor material.

By selecting the correct combination of impurities and physical arrangements of semiconductor materials, a semiconductor device can be made that will change its conduction based on a variety of stimuli. Some semiconductor devices change their conduction when stimulated by electrical quantities (either voltage or current). Others are especially sensitive to light or heat or other physical stimuli. By changing the chemical composition and physical arrangements of semiconductor materials, a variety of devices can be produced. The term "semiconductor" therefore relates to solid-state devices that change electrical conduction due to some form of external stimuli.

Semiconductor devices are produced in a wide variety of configurations (internal arrangement of semiconductor materials). Semiconductor diodes, transistors, operational amplifiers, and integrated circuits are some of the forms of these configurations.

5.2 THE SEMICONDUCTOR DIODE

Inside the diode is a special configuration of semiconductor material. This arrangement produces what is called a *semiconductor junction*. This junction is what causes this solid-state device to have the special characteristics of a diode. The diode is a two-terminal device (two leads).

The semiconductor diode is polarity (+ or −) sensitive. If an attempt to pass current through the diode in one direction is made in (due to +/− connections), the diode will permit this flow and act like a relatively good conductor. An attempt to pass current in the other direction (reversing the +/− connections) will be blocked by the semiconductor material, which now produces very high resistance to current flow. Thus the diode is like a one-way valve, passing current in only one direction.

A diode responds to the polarity (+ or −) of voltage applied to each of its leads. Application of a voltage polarity that coordinates with the direction in which the diode will pass current is called **forward biasing** the diode. Applying a polarity that is opposite this, thus causing the diode not to conduct current, is called **reverse biasing** the diode. This response to polarity is found in many other semiconductors.

In Figure 5.1 the schematic symbol for the diode is illustrated. Also illustrated are three popular forms of physical diodes. The diode lead that coordinates with the arrow end of the symbol is called the **anode** (A), and the lead that coordinates with the bar shaped end is called the **cathode** (K). (These names and the "K" abbreviation are a holdover from older vacuum-tube technology.)

The proper way to "forward bias" the diode and thus have it allow current flow is to make the anode lead most positive (+) and the cathode negative (−). Reversing this polarity causes the diode to be reverse biased and therefore not conduct. This singular direction of current flow through the diode has a wide variety of applications. A diode cannot produce gain since it does not have the capacity to control a second flow of energy in a circuit based on the level of an input or control signal.

There are a few other semiconductor effects exhibited by the diode that do need to be pointed out before we proceed to active semiconductor devices. First, when the diode is in the reverse-bias mode (nonconducting), there is always some small amount of leakage current that flows through it. This means that the diode presents a very high resistance in the reverse direction, but not a perfect open circuit. In the reverse direction it is possible to apply enough force (voltage) to cause a breakdown in the semiconductor and ultimately cause it to conduct. This "reverse breakdown," in a normal diode, can destroy the device through overheating due to high current flow.

When operated in the forward direction, the diode will allow current to flow. Even though the diode will conduct when properly biased (+ anode, − cathode) there is a minimum amount of voltage required to maintain conduction. This minimum voltage is called the "forward drop" or "junction potential" for the diode. This potential is on the order of 0.2 to 0.75 V, depending on the type of semiconductor used (germanium or silicon) and other operational conditions, such as temperature. Since silicon diodes are most often used, in this book a value of 0.7 V will be assumed to be the "normal" forward drop on the diode. Actually, in most applications the real drop is less than this, but by assuming 0.7 V, we can cover for the "worst-case" condition.

The forward drop is more than a minimum voltage needed to create conduction. That voltage needs to be applied to the semiconductor junction on a continuous basis to produce conduction. If 20 V is applied in the forward-bias polarity, 0.7 V is dropped across the device terminals to maintain conduction, leaving only 19.3 V for use by a connecting resistance. Unlike a resistor, though, the drop on the diode is not in proportion to current flow. Once the minimum drop (0.7 V) is satisfied, that voltage remains relatively

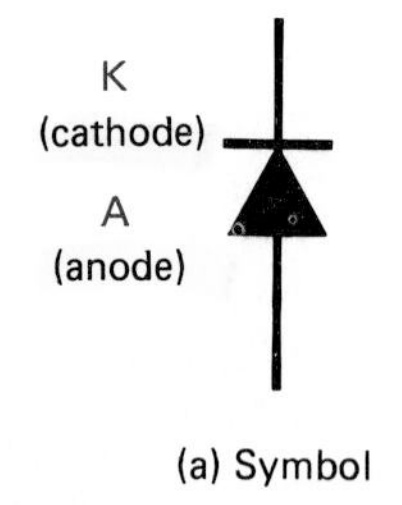

(a) Symbol

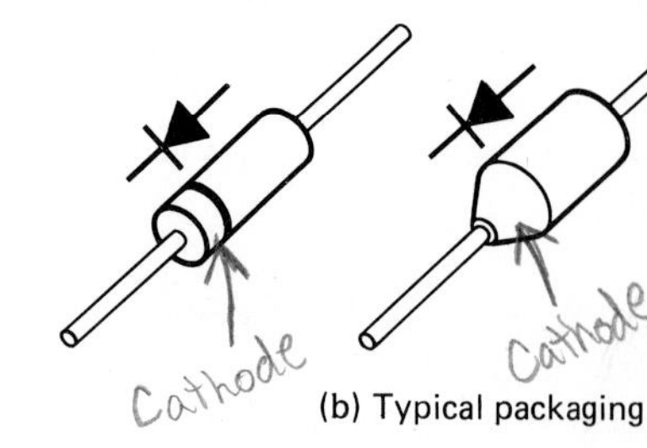

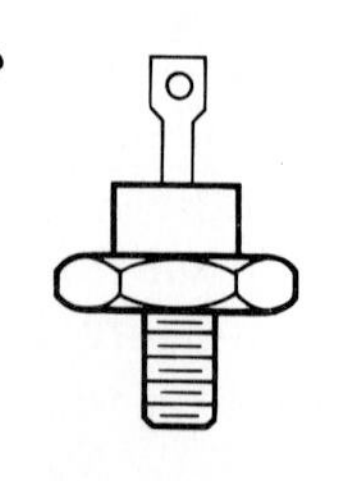

(b) Typical packaging

FIGURE 5.1 Diode

constant for a wide variety of forward current flows. Figure 5.2 illustrates the function of the diode in both the forward and reverse directions. [Another term used to indicate a voltage level on a device is to call that value the "bias" on the device. Thus there is a forward bias (when polarity will produce conduction) and a reverse bias.]

In Figure 5.2 both the forward- and reverse-bias conditions on the device are shown. This graph shows the voltage-to-current relation for a typical diode. Note that voltages on the right side (positive) are in 0.1-V steps and those on the left (negative) are in tens of volts. The vertical current above the axis is scaled in milliamperes, whereas below the axis the scale is in microamperes. The curve on the left side of the center axis is for reverse-bias conditions. Note when the curve does have current flow, it bends in the negative current direction. This simply means that current flows opposite the direction shown for forward bias. The right side of the curve is for forward-bias conditions. Superimposed on this graph is the graph for a simple resistor.

First let's investigate the forward-conduction side of the curve. Note that when the applied voltage is well below the 0.7-V minimum, the amount of current flow is extremely low. Once the minimum forward voltage is reached, the device "turns on" and permits a high amount of current flow. This curve can be contrasted to the straight-line ratio of voltage to current for the resistor. As voltage on the resistor increases above 0.7 V, it continues to limit current flow ($I = E/R$, Ohm's law). After the 0.7-V drop is satisfied, current through the diode can be nearly any value (there is a maximum limit above which the diode can be damaged due to overheating).

Reversing the polarity on the diode will cause it to limit conduction to a few microamperes. Therefore, when reverse biased, the diode fails to permit any substantial current flow. There is a maximum limit to the level of reverse-polarity voltage that the diode can tolerate; when that level is exceeded, the diode "breaks down" and conducts. Typically, the high flow of current at this high reverse voltage will usually destroy the diode. (It will cause the diode to be open in both the forward and reverse directions.) Certain diodes (called

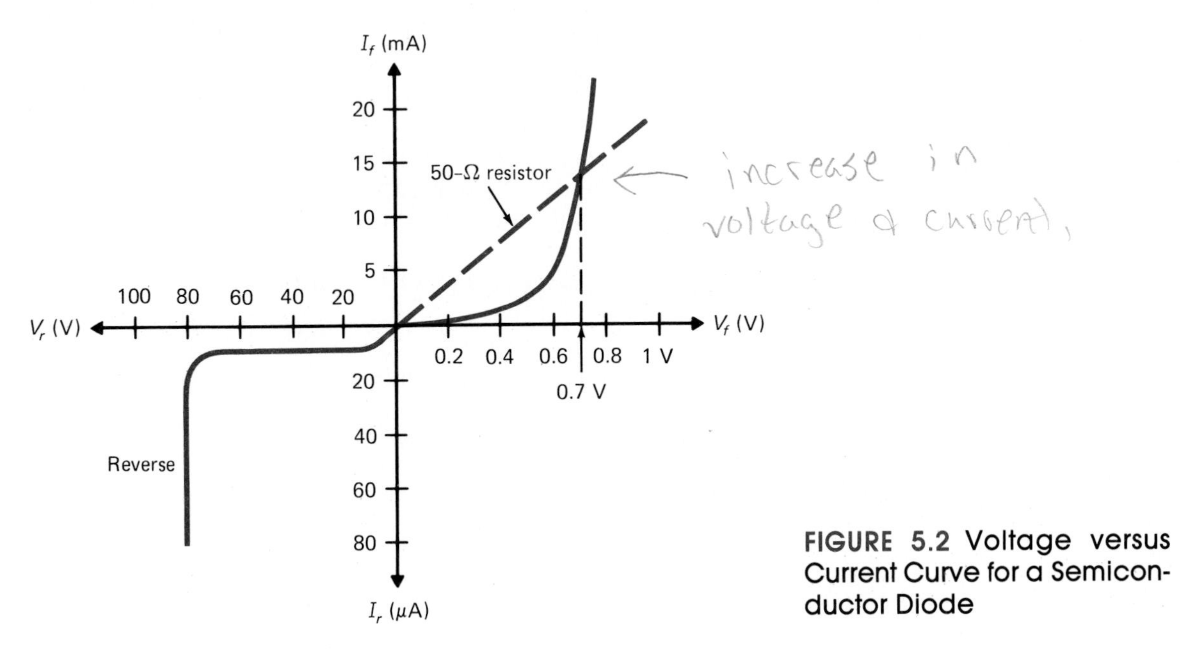

FIGURE 5.2 Voltage versus Current Curve for a Semiconductor Diode

zener diodes) are made to be used in this reverse direction and are therefore designed to handle the current flow produced at these high breakdown voltage levels. Although the diode is expected to act like an open circuit in the reverse direction, there is a very small amount of reverse current flow. This small amount of flow, called "reverse leakage current," is usually so low in value that it does not create any adverse effects in circuits attached to the diode. The following example illustrates a basic use of the diode in a circuit. The germanium diode exhibits characteristics quite similar to the silicon diode. The major difference (related to application) is that the germanium diode produces a 0.3-V forward drop instead of the 0.7-V approximation for a silicon diode.

EXAMPLE 5.1

In this example the diode will be applied first in a forward direction [Figure 5.3(a)] and then in reverse [Figure 5.3(b)]. A simple dc supply and resistor will be connected to the diode. The objective will be to see how the diode affects current flow and voltage distribution.

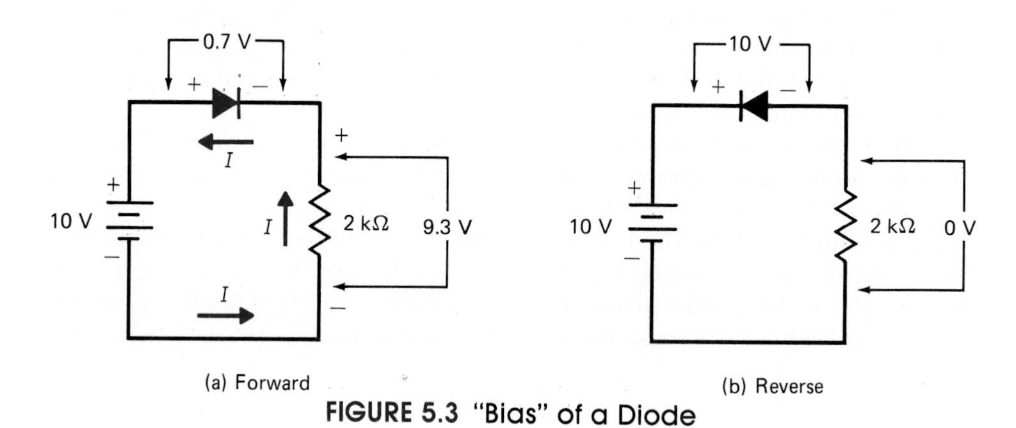

FIGURE 5.3 "Bias" of a Diode

Solution:

(a) In the first circuit, where the diode is forward biased, find the voltage on the diode and resistor and the circuit current flow.

Diode voltage: $V_d = 0.7\text{ V}$ (standard forward drop)

Using Kirchhoff's voltage law, we can find:

Resistor voltage: $V_r = 10\text{ V} - 0.7\text{ V} = 9.3\text{ V}$

Circuit current: $I = \dfrac{V_r}{R} = \dfrac{9.3\text{ V}}{2\text{ k}\Omega} = 4.65\text{ mA}$

(Note that current is set by the resistor, not the diode!)

(b) In the second circuit, where the diode is reverse biased, find the voltage on the diode and resistor and the circuit current.

Diode voltage: $V_d = 10\text{ V}$ (open circuit in reverse bias)
Resistor voltage: $V_r = 10\text{ V} - 10\text{ V} = 0\text{ V}$
Circuit current: $I = \dfrac{V_r}{R} = 0\text{ A}$ (no flow)

In exploring the diode, there were no factors of gain when calculating how it manipulated energy. The diode does not produce a gain figure. Other semiconductor devices, such as transistors, do produce gain. The discussions in Section 5.3 center around gain-producing devices.

REVIEW PROBLEMS

(1) Using Figure 5.3, estimate the value of voltage on the resistor for the following dc input voltages.
(a) 0.2 V **(b)** 0.6 V **(c)** 2.0 V **(d)** 35 V

(2) In this section dc voltages were used to forward and reverse bias a diode. For a simple series circuit shown in Figure 5.3, change the dc source to an ac sine-wave source. Sketch what will happen when the ac cycle is positive and when it is negative. (The diode will not change how it reacts to polarities!)

(3) Redraw the circuit in Figure 5.3 and show the proper voltage distribution, but in this case use a germanium diode. (The symbol is the same as for silicon.)

5.3 TRANSISTORS (A BROAD CATEGORY)

The term **transistor** refers to a wide variety of gain-producing solid-state devices. Each is made from silicon (or possibly, germanium), but there their similarity ends. Each device has specific characteristics and coordinates with one of the four ideal amplifier models noted in earlier chapters. We will break the category of transistors down into two major divisions: the bipolar junction transistor (BJT) and the field-effect transistor (FET). These two forms operate quite differently from each other. By understanding how each functions, we can understand how individual devices classified in that division will operate.

5.4 THE BIPOLAR JUNCTION TRANSISTOR

The bipolar junction transistor is a gain-producing device. That is, it has the ability to control one flow of current based on the value of another (independent) flow. The BJT is a three-terminal device (three leads). Unlike the diode, the BJT can have two paths for current flow. To observe this possible flow pattern, Figure 5.4 illustrates the standard BJT schematic symbol, a popular shape for the BJT and an illustration of two potential paths for current flow. Initially, the BJT does not appear to provide gain or control, just two paths for current flow. But the special arrangement of the impurities applied to the silicon inside the device allows the BJT to provide control of one flow based on the value of the other.

First, we need to establish a common set of terms to use when referring to the BJT. Note that there are two forms of the BJT illustrated in Figure 5.5. Also note the specific labeling of the three leads on each illustration. One lead is called the **base** (*B*). The **emitter** (*E*) and **collector** (*C*) are the other two leads. Later we investigate each of the three names and coordinate them to device functions. The form on the left is called an **NPN** type and the other is called **PNP**. These notations relate to two distinct arrangement of the doped silicon inside the device. Be certain to note the difference between the NPN symbol and the PNP symbol.

Like the diode, the BJT is polarity sensitive. To produce any conduction, the proper polarity of a source must be connected to the correct lead of the device. Since this is a three-terminal device, two sources could be connected to it at once. The NPN and PNP markings assist in coordinating the proper polarity to connect to produce a "forward conduction" (like that of the diode; see Figure 5.6).

When a supply is connected with the correct polarity between the *base and emitter* leads, forward conduction occurs. The current passes through the device much like the flow through a diode. The NPN transistor has forward electron current flowing from the base to the emitter; the PNP has flow from emitter to base. Just like a diode, there will be an approximate forward drop of 0.7 V. If either of these connections is made with the opposite polarity, a reverse-bias condition exists and no flow will be seen.

When observing the base-to-emitter connection, this forward drop is called V_{BE}, or voltage drop—base to emitter. In a like manner, the voltage from the *base to collector* is termed V_{BC}. Actually, any voltage around the circuit can be defined using a two-letter (from–to) combination. In addition, the polarity of the voltage is given as seen by the lead represented by the first letter in reference to the second lead. Therefore, a $V_{BE} = +0.7$ V indicates a positive base in reference to the emitter.

So far, though, we have observed only diode-like actions between these leads. The one lead combination that has not yet been discussed is the *collector-to-emitter* combination. [Upon initial observation, it does not seem possible to produce a flow from the collector to the emitter, since both are labeled

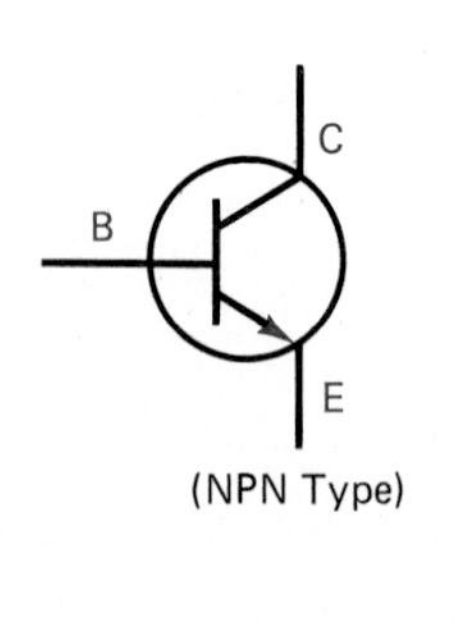

(a) Symbol

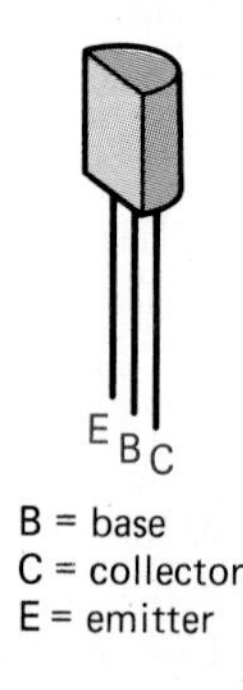

(b) Case with leads identified

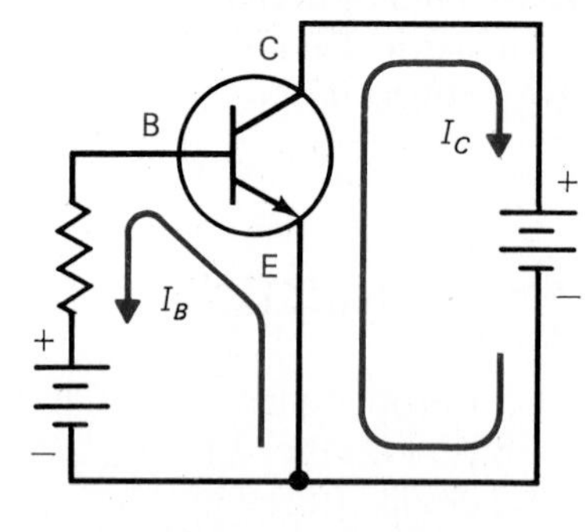

(c) Forward biasing

FIGURE 5.4 NPN BJT transistor

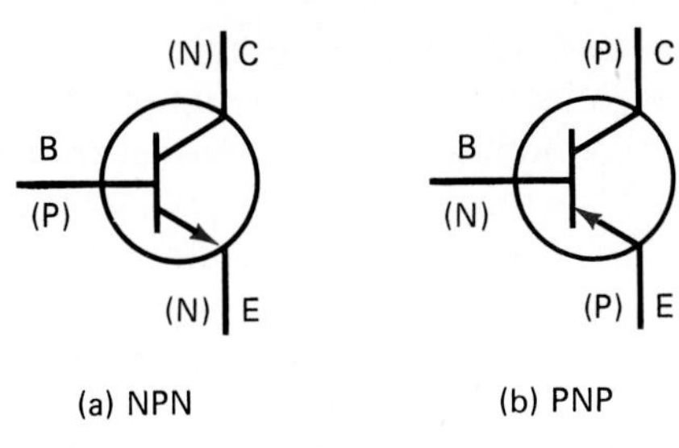

FIGURE 5.5 Lead Identification for BJT transistors

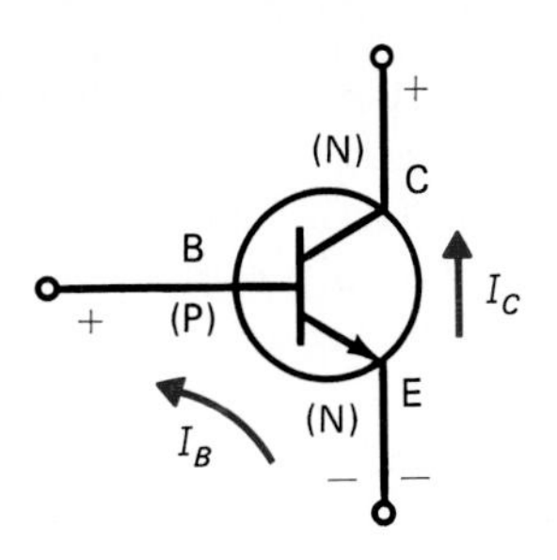

FIGURE 5.6 Current Paths through an NPN Transistor

with the same polarity markings. But it is here where control of one flow based on the value of another (gain) is observable.]

In Figure 5.7 a very simple arrangement of sources around an NPN transistor has been made. (PNP will be shown to be nearly the same.) At first observation, the only conduction that can be seen is that from emitter to base. Since the source there is connected with the correct polarity, we can assume that forward conduction occurs. Therefore, one could guess that the voltage and current values in this circuit would be:

V_{BE}: 0.7 V
V_R: 5.3 V
V_{CE}: 10 V

Current at the base: $I_B = \dfrac{5.3\text{ V}}{2\text{ k}\Omega} = 2.65\text{ mA}$

Current at the collector: $I_C = 0\text{ A}$

But this would not be the case. In fact, there would be a strong flow of current at the collector lead. When base current is present, the emitter-to-collector path becomes conductive in proportion to the amount of base current that is flowing. Therefore, the base current flow produces a conductive path from the emitter to the collector. This path will then permit a flow of current from the collector to emitter. Thus the current at the collector is not zero but is some predictable value.

The amount of conductivity between the emitter and collector and thus the potential amount of collector current flow is directly proportional to the amount of base current. The larger the value of base current, the larger is the

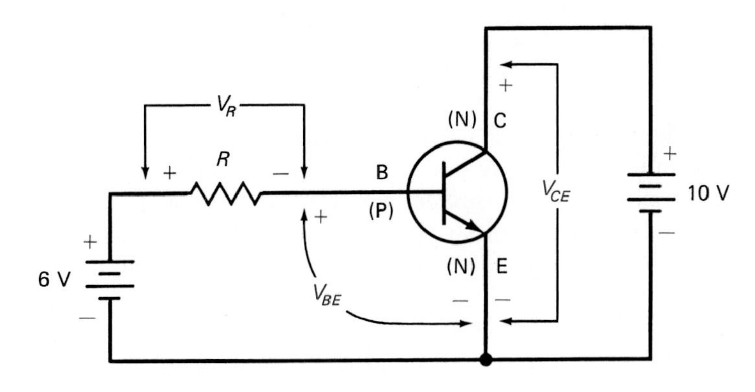

FIGURE 5.7 Forward-Biased NPN Transistor Showing Identification of Voltage Potentials

value of conductivity and thus the larger the collector current. If base current is lowered, conductivity drops and so does collector current. Different versions of the BJT have different proportions of conduction versus base current and thus different relationships between the current into the base and the collector current it permits. This relation is given a special name, **beta** (β).

$$\beta = \frac{I_C}{I_B} \tag{5.1}$$

or, rearranged to represent I_C,

$$I_C = \beta \times I_B \tag{5.2}$$

The value for the term β is a manufacturer-supplied specification for every BJT. The value of beta can range from 10 to 200 or higher. With a beta of 100, for example, a transistor could have a base current of 50 μA and would be able to control a collector current of 5 mA. Beta is therefore a gain figure for the device. The beta indicates how the base current (the input) can control the collector current (the output).

Since the current at the base controls the flow of current through the collector, the BJT is therefore a current amplifier. It meets the basic requirements of the current amplifier. First it is sensitive to current at the base, where inputs are commonly applied. It also produces an output current. The input (base) side is relatively insensitive to voltage changes once the 0.7-V forward bias on the base-to-emitter junction is satisfied. The collector (output) side is, within limits, insensitive to the voltage supplied. Collector current will flow at the rate determined by the gain figure (beta) and will not be affected by the level of the voltage supplied to the collector. In this respect, the device is nearly an ideal current amplifier. It does not, however, meet all the specifications of the ideal current amplifier.

Some of the specifications of the BJT make it less than an ideal current amplifier. The BJT does not have zero input resistance (impedance), although the value is usually low. It also does not have infinite output resistance (impedance). In fact, the output resistance, which is directly related to the conductance of the collector portion of the device, is variable. The BJT is also quite unstable, not a good factor to have in an amplifier if we desire ideal characteristics. The gain of the device is quite sensitive to temperature and is also not fixed for various levels of base current flow. That means that as temperature or the level of base current changes, different values for gain may occur. In later sections specific circuit elements will be added to the BJT to provide greater stability and higher reliability.

In this section dc supplies have been used to illustrate the general function of the BJT. The BJT is capable of responding to a variety of input signals: dc, pulsed dc, and ac. More of these applications are described in later sections of the book.

All of the analysis done on the NPN transistor can be applied to a PNP device. The only variation is that voltage polarities and current directions are exactly opposite those seen in the NPN device. The PNP transistor (BJT) will not be emphasized in this book due to its correlation to the NPN transistor. Also, the NPN transistor is used more often in electronic circuits than is the PNP type.

BJTs have a wider variety of characteristics than have been presented in this introduction to that device group. In later sections more of the characteristics will be explored as they relate to circuit applications. The important concept is that the BJT can provide current gain in an electronic circuit by the process of base current controlling the flow of collector current. There are other applications for the BJT, which are described in later chapters.

REVIEW PROBLEMS

(1) In Figure 5.6, if the polarity of the supply connected to the base of the BJT were reversed, it would become reverse biased (the same as a reverse-bias diode). Redraw this figure using a reverse-biased V_{BE}. Then discuss briefly what would happen to the value of current at the base, the value of V_{BE}, the voltage on the base resistor, and then the resulting collector current. You may wish to refer to equation 5.2.

(2) Redraw Figure 5.6 using a PNP transistor (BJT). Be certain to connect the power supply using the correct polarity and to indicate the paths for current flow. (Be sure to forward bias the base–emitter section.)

(3) Using equations 5.1 and 5.2, solve for the following values.

(a) A transistor has a beta of 120 and a base current of 65 μA. What is the value of the resulting collector current I_C?

(b) The same transistor ($\beta = 120$) now has the following base currents applied.

(1) $I_B = 100\ \mu A$

(2) $I_B = 30\ \mu A$

Find the new collector currents.

(c) Discuss if under all three conditions of parts (a) and (b) there would be a change in V_{BE}. What element in the circuit could have caused the change in I_B? (See Figure 5.7 for assistance.)

5.5 THE FIELD-EFFECT TRANSISTOR

The field-effect transistor (FET) is another form of solid-state device. Like the BJT, it is a silicon-based semiconductor device, and also like the BJT, it is a three-terminal device. It can use an input quantity to control an alternate output quantity; therefore, it can produce gain. Otherwise, the FET is dramatically different from the BJT.

The internal structure of the FET differs from that of the BJT. This different arrangement of materials creates a completely different set of electrical properties. Also, there are various forms of FETs. In addition to having polarity sensitivity (there are "N" and "P" versions of FETs much as there are PNP and NPN versions of the BJT), the FET is produced in three internally and functionally different forms. In Figure 5.8 the various names and schematic symbols for these versions are presented in chart form.

Even though there seems to be quite a variety of FET devices, they all function in basically the same way. Notice, too, that there are only three primary divisions of the FET, with each having an "N" and a "P" version. These three types are the JFET (junction field-effect transistor), the E-MOS-

Type	N-Type	P-Type
Junction (JFET)		
Depletion-metal-oxide semiconductor (DE-MOSFET)		
Enhancement-metal-oxide semiconductor (E-MOSFET)		

FIGURE 5.8 FET (Field-Effect Transistor) Versions

FET [enhancement (type)-metal-oxide semiconductor FET)] and the DE-MOSFET [depletion/enhancement (type)-MOSFET]. In this section we concentrate on the JFET. Since the other forms are specialized variations on the basic function of the JFET, we reserve coverage of their specific characteristics and applications to later chapters.

Note: Usually, JFETs are referred to simply by the term FET. When using MOS types, the complete notation is applied. For the sake of simplicity the term FET will be used from now on to refer to the standard JFET. Of course, N or P types will be delineated. When speaking of the FET, either the letters F.E.T. can be named or it may be pronounced as a word ("fet"). The use of the name "transistor" when referring to an electronic device implies a reference to the BJT only.

Not only is the symbol for the FET different from that of the BJT, but the labeling of its terminals is also unique. In Figure 5.9 the three terminals of the FET are identified as the **gate**, the **source**, and the **drain**. In a very general sense, the gate lead coordinates with the base lead of the BJT and the drain and source coordinate roughly with the collector and emitter leads. Also, in Figure 5.9 it can be seen that the basic package of the FET does not differ from the packaging of the BJT. The only way to delineate the FET from the BJT's physical package is by looking up the part number of the device. (These are printed on the device.)

The FET (either N or P types) does not function like the BJT either internally or externally. The FET is a transconductance device. That is, its input accepts a voltage level and uses that level to produce an output current. This voltage-in-to-current-out process differs dramatically from the BJT cur-

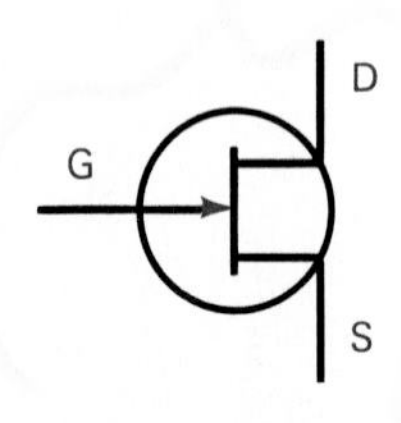

(a) (N) JFET

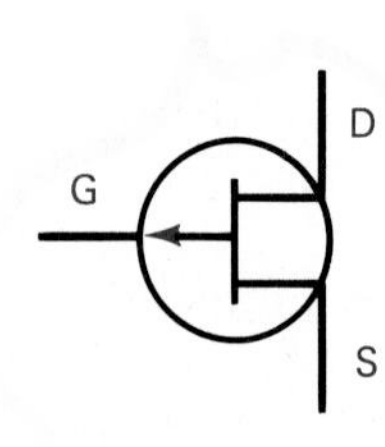

(b) (P) JFET

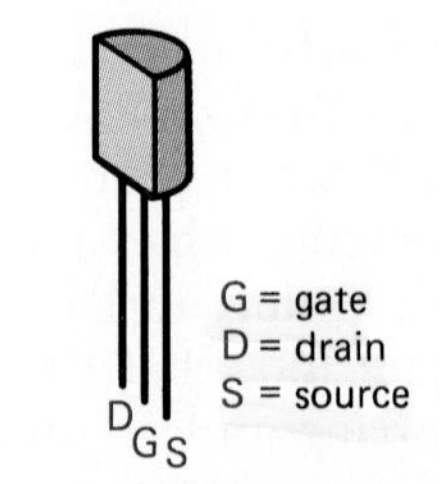

(c) Lead identification

FIGURE 5.9 JFET's

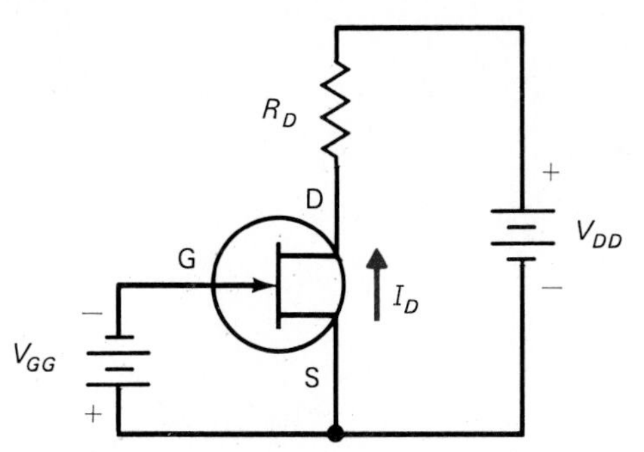

FIGURE 5.10 Initial Bias Connections for the JFET

rent-amplifying device. In Figure 5.10 the FET is shown in a simple circuit arrangement. An input voltage source is connected to the gate lead and another source is connected to the drain lead. The source is used as a common tie point for both the gate and drain suplies. It is important to notice the polarity of these connections. For the N-type FET, the gate lead must be connected to the negative (−) terminal of the gate supply; the positive terminal is connected to the source lead, thus completing the circuit at the input. The polarity of the supply on the drain side has its *positive* lead connected to the drain and its *negative* lead connected to the source.

Like a transconductance amplifier, the FET is sensitive to voltage levels at its input (not current). The flow of current in the drain lead (output side) is controlled by the level of voltage, both dc bias and ac signal at the gate lead (input side). Changes in input voltage will produce coordinated variations in the drain current. An increase in gate voltage produces a decrease in drain current! With zero gate voltage, the maximum amount of drain current will be flowing. This function is illustrated in Figure 5.11. This function, where output current is inversely proportional to gate voltage, does not create a problem when applying the device to a functional circuit. The important feature is that output current *is* proportional to gate voltage. The inverse function can simply be factored into the method in which the FET is applied to a system.

> *Note:* The term "channel" is often used with FETs to identify the source-to-drain semiconductor material. This channel's conductivity is affected by the gate voltage potential. An N-type FET is commonly called "N-channel" and a P-type is referred to as a "P-channel" FET.

FIGURE 5.11 JFET Output Current

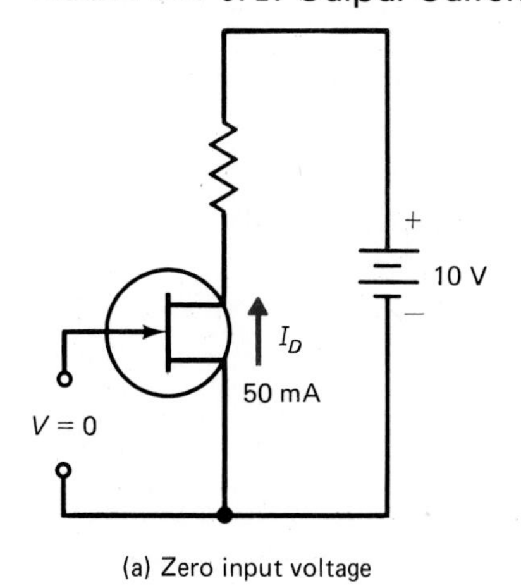

(a) Zero input voltage

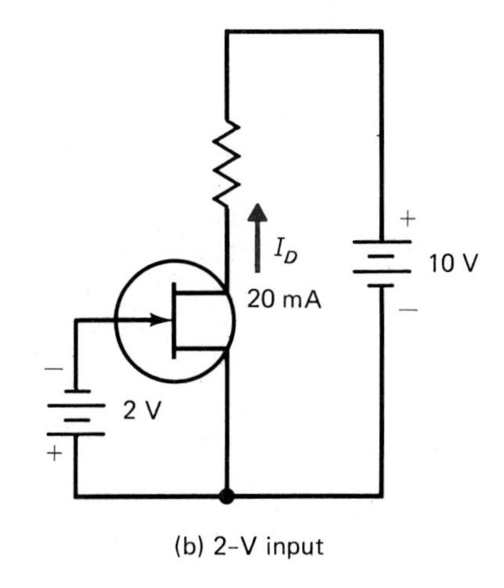

(b) 2-V input

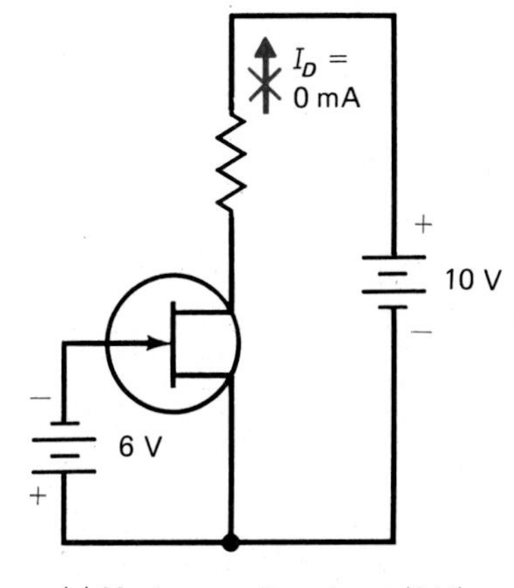

(c) Maximum voltage input (6 V)

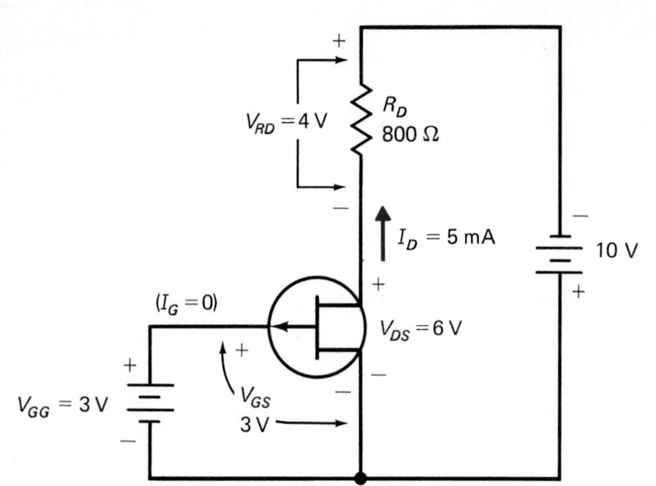

FIGURE 5.12 Typical Voltages and Currents for the JFET Circuit

Let's now investigate the other functions of a FET and compare them to the ideal transconductance amplifier. A transconductance amplifier should have infinite input resistance (impedance). The FET does have an extremely high input resistance, usually in the megohm range. Therefore, like the transconductance amplifier, it does not draw (appreciable) current from the input source. The input is therefore a static (no current flow) voltage level that is a blend of dc bias and signal voltages. The voltage on the gate of the FET is the same as the level produced by the sources (dc + signal) connected to it (see Figure 5.12).

The output of the FET is a current that is inversely proportional to this gate voltage. Therefore, a voltage at the gate lead will control the level of conduction at the output. The output resistance (impedance) of the FET does not coordinate with that specified by the ideal model. The ideal transconductance amplifier has infinite output resistance. The FET has a value that is simply related to the proportion of current and voltage associated with the gate and drain leads.

There is an amplification (gain) factor for the FET. Unlike beta for a transistor, though, the gain figure for an FET is a ratio of output current to input voltage. Such a factor is not as easily understood as beta is. The following example should help to illustrate this function.

EXAMPLE 5.2

An FET has an input voltage of 1.5 V and an output current of 30 mA. Find the gain for this device.

Solution:

$$A_g = \frac{I_{\text{out}}}{V_{\text{in}}} \tag{5.3}$$

(A_g is the standard notation for gain in a transconductance amplifier.)

For this example,

$$A_g = \frac{30\text{ mA}}{1.5\text{ V}}$$

$$= 20\text{ mS}$$

[The division of voltage by current results in a conductance term (opposite of resistance "E/I"), which is given the units of siemens.]

In Example 5.2 the calculation resulted in a conductance term. This is often called the amplifier's transconductance (here used to express gain). An alternative notation used in FET data sheets to represent this gain is the symbol g_m. There are a variety of JFETs available from semiconductor manufacturers. Each has its own value of g_m and specific values for other operating parameters. The method of interpreting these data and selecting an FET for a specific application will be presented later. This general survey of the FET used the N-type FET. The P-type FET will have a similar response, but all source polarities and the output current direction will be reversed.

It should be noted here that like the BJT, the FET has a wide variety of electrical limitations and restrictions. As with the BJT, a wide variety of support circuitry for the FET will be developed in future sections.

The FET is a voltage-sensitive device (at the gate lead) which produces an output current in response to that voltage. Although this response is an inverse relation (increased input causes decreased output), the FET still fits in the category of transconductance amplifiers. Its input resistance is very high; therefore, it draws negligible current from the input (gate) supply. Although its output resistance is not high, it does provide a reasonably stable output current. In future applications, the specific parameters of the FET are explored in more detail. Other forms of this device, the E-MOSFET and the DE-MOSFET, have characteristics similar to the JFET and are explored in later sections.

REVIEW PROBLEMS

(1) Refer to Figure 5.8 and write a simple description of how the "N" or "P" JFET schematic symbols differ and how to identify them.

(2) If a JFET had a gate voltage of 2 V and a resulting drain current of 400 μA, what would be the value of A_g or g_m (transconductance)? (Be sure to include the proper units.) Then, if that voltage were to double to 4 V, what value of drain current would potentially result? (Remember the inverse relationship of input to output for a FET!)

(3) Prepare a simple comparison of the BJT and the FET. In this comparison, include typical input and output quantities and impedances. Contrast each with its coordinated ideal amplifier models.

(4) In a simple BJT circuit (see Figure 5.7), a resistor was used in the base (input) circuit. That resistor was used to control base current and to be a place where the remaining supply voltage in excess of 0.7 V (forward drop on the transistor) could be distributed. In the FET circuit (Figure 5.9), no input (gate) resistor was used. Does gate current in a FET have to be controlled? Is there a standard forward drop on the gate lead? Does a voltage divider need to be present in the input circuit? In answering these questions, contrast the BJT and FET input circuits (or their ideal models).

5.6 THE OPERATIONAL AMPLIFIER

So far our discussion of solid-state devices has centered on discrete semiconductors. These devices, BJTs and FETs, are singular (discrete) elements that can be used to produce a single amplifier stage. The operational amplifier, more commonly called the **op-amp**, is an integrated-circuit amplifier. The integrated circuit is a complete circuit built on a single piece of silicon. It contains several semiconductor devices and the necessary resistors and other support elements to create a complete circuit system.

A typical op-amp contains 20 BJT-type transistors, 12 or more resistors, and other passive components. All of these elements are built on a single silicon "chip" in one complete manufacturing process. Together, these elements constitute a complete amplifier system. This system can be used to complete the process described by a block in a block diagram.

The op-amp is designed internally in such a way as to produce a nearly ideal amplifier. One of the most important features of the op-amp is its stability. It was noted that both BJTs and FETs suffered from problems of instability. Special circuits need to be constructed to compensate (correct) for instability. The op-amp usually does not require external components to compensate for such instabilities. The necessary compensation is built into the amplifier system. Figure 5.13 illustrates the op-amp schematic symbol and typical packaging for the op-amp. The op-amp's symbol matches the block diagram symbol used to indicate any amplifier. The op-amp is so flexible that it can be used in many amplifier applications. Unlike discrete amplifier elements, op-amps can have a usable gain value in the hundreds or thousands.

Although the op-amp is a self-contained amplifier system, it does require external components to prescribe its operational characteristics. The majority of op-amps available are voltage amplifiers. Values of input voltage produce

FIGURE 5.13 Op-Amp

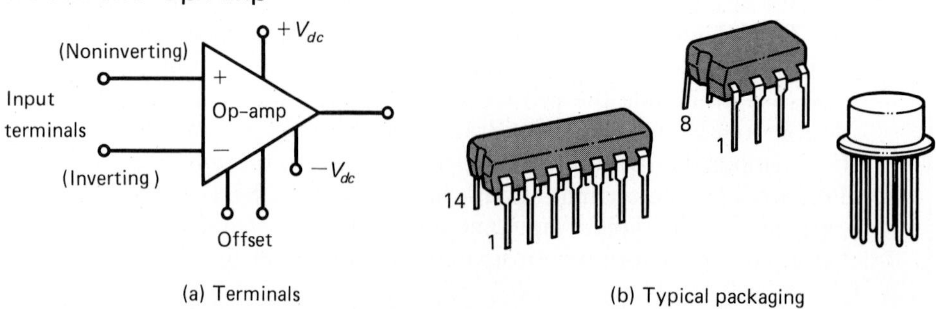

(a) Terminals (b) Typical packaging

control over the level of output voltage. As a voltage amplifier, the op-amp could be thought of as a sort of "generic" amplifier. The gain value of a typical op-amp is controllable through the choice of various external components. Its input impedance is typically in the 250-kΩ range and its output impedance starts around a few hundred ohms. Both impedances can be changed (higher or lower), again by the addition of external components. It will typically handle frequencies from 0 Hz (dc) up to 1 MHz, and it will operate on a wide variety of power supply values. Unlike the BJT and JFET, the op-amp will accept either positive or negative input levels and produce an output.

By adding the correct combinations of external components, the op-amp can be made to nearly duplicate any amplifying function. It is the addition of these external components, at either the input or the output, or as feedback, that will allow its user to prescribe its functional characteristics exactly.

Unlike the BJT and FET, the op-amp is not a three-terminal device. Being a complete system, the op-amp has a different orientation of its leads. Most op-amps are packaged in an 8-, 14-, or 16-pin dual-in-line package (DIP). Often, two op-amps are contained in one 16-pin package. The 8-pin package is called a mini-dip (mini-dual-in-line package). Figure 5.14 shows a sketch of the mini-DIP as seen from the top. This illustration, called a **pin-out diagram**, indicates what quantities need to be connected to what pins (numerically organized) on the packaging. These pins coordinate with the connections made to the op-amp's schematic symbol. (Note on the symbol that the pin numbers are indicated.)

As can be seen, the op-amp has more connections to be made than does a discrete semiconductor. First, most op-amps require both a positive and a negative supply voltage, in reference to a circuit ground (associated with one of the input leads; see Figure 5.15). There are two input leads, the (+) noninverting and the (−) inverting. Connection of an input to the noninverting terminal will produce an output that is of the same polarity as that input (+in = +out, −in = −out). An input applied to the inverting terminal (−) will result in an opposite polarity output (+in = −out, −in = +out; again, see Figure 5.15). The remaining connections will be discussed when we explore op-amp applications.

The op-amp has certain minimum and maximum ratings. Even though it is quite versatile and can be applied in many circuits, it also has certain undesirable characteristics. There are a variety of op-amps available, each having a distinct set of characteristics.

FIGURE 5.14 Op-Amp Mini-DIP

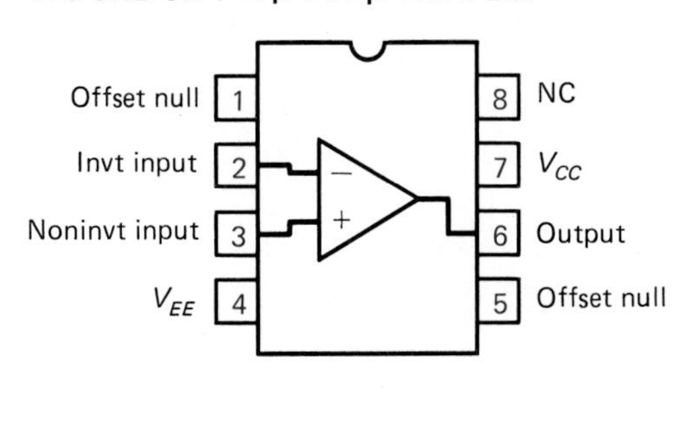

(a) Pin-out

+Dc (V_{EE})
+In
3
+
7
6
Out
−In
2
−
4
5
1
−Dc (V_{CC})
Offset

(b) Schematic with pin numbers

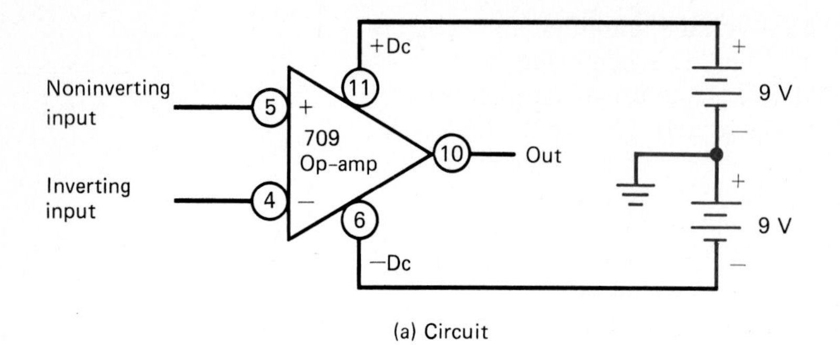

(a) Circuit

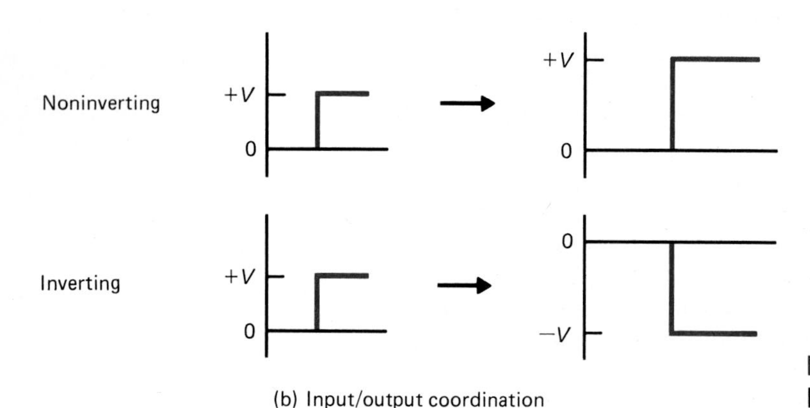

(b) Input/output coordination

FIGURE 5.15 Use of Dual Polarity Power Supply (Op-Amp)

REVIEW PROBLEMS

(1) Turn to Appendix A and locate the specification sheet for the MC34071 operational amplifier. List the following information found on that sheet.
  (a) The number of internal components (you will have to count BJTs, resistors, etc.)
  (b) The typical large-signal voltage gain [two listings may be given, one for room temperature (T_A = 25°C) and another for extreme temperatures; use the ones for room temperature]

(2) There are both inverting and noninverting terminals on the op-amp. Predict what the output would look like if an ac sine wave (both + and − polarity) were applied to either of these terminals, inverting and noninverting. [Simply assume that the gain is 5 (output = 5 × input).]

(3) Copy the pin-out diagram from Figure 5.14 and then sketch a simple wiring diagram to illustrate the batteries (V_{EE} and V_{CC}), noninverting input, and output terminal connections. Be certain to include a ground terminal between the two supply voltages.

5.7 LARGE-SCALE INTEGRATED CIRCUITS

The final category of semiconductors that will be discussed is that of large-scale integrated circuits. Large-scale integration (LSI) means that large numbers of circuits, and therefore many components, are contained in one integrated-circuit package. Even though the op-amp had a number of components inside, it does not compare to the large-scale integrated circuit. Literally hundreds to thousands of components are included on one silicon chip about $\frac{1}{4}$ in. by $\frac{1}{4}$ in.

Nearly all large-scale integrated circuits are designed to produce a singular function. Unlike the op-amp, whose final function can be tailored to a specific need, most linear (amplification) LSI circuits cannot. LSI-ICs are available in two distinct forms: linear (also sometimes called analog) or digital. A **linear** device has the capacity to handle continuously variable electrical signals (such as sine waves) and produce an output which is usually proportional to these variations. **Digital** devices deal with numerical codes and produce a logical output (yes/no type), again in a numerical code. In this text we deal only with the linear type of integrated circuit.

Most LSI-ICs do not require many external components to operate. Usually, only components that are not easily duplicated on the silicon chip (typically, inductive elements), consume large amounts of power, or are specifically designed to be external to a system (such as a volume control for an amplifier) are left out of the IC package. A complete stereo amplifier could be produced by using one or two LSI-ICs and appropriate volume and tone control potentiometers.

Often the connections to LSI-ICs, external components to be used, and the applications for that circuit are specified by the manufacturer. Currently, many popular electronic circuit functions are available on LSI-ICs. Figure 5.16 illustrates one of the many LSI-ICs available.

Since many functions are available on single ICs, many of the circuits that previously used discrete devices are no longer being built. It is quite possible that in the near future the only need for discrete devices will be either for very high power applications or when circuits are nearly one of a kind, which would make them too expensive to be manufactured as a specially built IC.

FIGURE 5.16 Black-and-White TV Subsystem (Single IC)

Monomax Black-and-White TV Subsystem

Improved versions of previously introduced MC13001P/MC13002P circuits which perform all electronic functions of a monochrome TV receiver with the exception of the tuner, sound channel and power output stages. These improved drop-in replacements require fewer external IF components and feature a wider AGC range, lower video output impedance, and increased horizontal drive current capability.

P SUFFIX
Plastic Package
Case 710-02

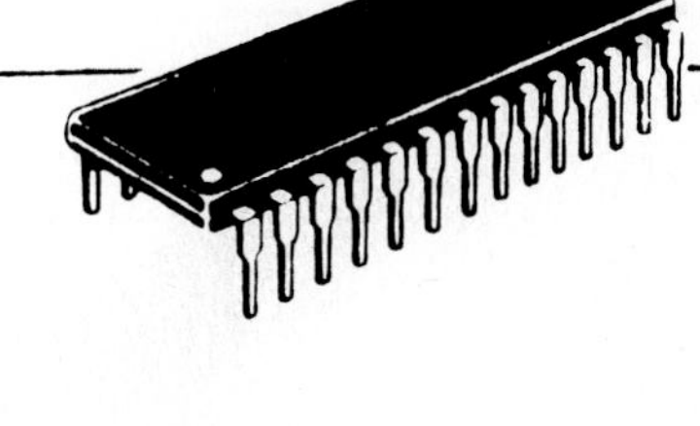

DEVICE	FEATURE
MC13001XP	525 Line NTSC
MC13002XP	625 LINE CCIR

REVIEW PROBLEM

(1) Several types of LSI-ICs are listed in Appendix A. Locate these and list their part numbers, names, and a brief description of their use. (Further information on linear LSI-ICs can be found in a technical library, either listed in special IC reference books or in distributors' catalogs.)

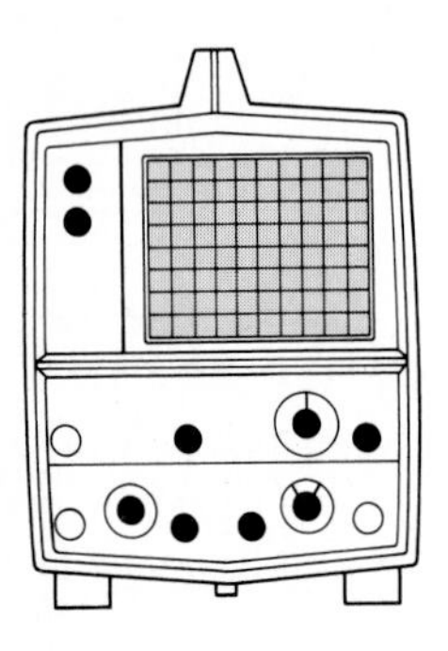

5.8 COMPARATIVE REVIEW OF SEMICONDUCTOR DEVICES

In preceding sections we introduced several forms of semiconductor devices. In this section we review their specifications and make some comparisons between the devices. By being able to compare and contrast these elements, one can understand more completely how they will be applied in circuits and why one form may be chosen over another for use in a specific application.

Since the semiconductor diode is not an amplifying device, but simply a two-terminal device sensitive to polarity, it cannot be compared to other amplifying devices. Its role in electronic circuits is not amplification or control but simply one of polarity or current sensitivity. Although the BJT and potentially other devices will display this sensitivity, that sensitivity is not their prime function as it is for the diode. In fact, the polarity sensitivity of amplifying devices is a drawback, not an advantage. The applications for a diode are totally distinct from those of amplifying devices and will be elaborated in future sections dedicated specifically to these applications. Also, there are several varieties of diodes, each with distinct applications; these, too, will be investigated later.

Active (Amplifying)-Device Comparisons

Each of the active devices discussed in this section (BJTs, FETs, op-amps, and integrated-circuit amplifiers) has distinct characteristics and thus distinct applications. BJTs and FETs are categories of discrete active devices. Each must have special support circuitry to operate. They are used to produce a single amplification (or special-purpose) stage in a complete system. Op-amps are integrated-circuit amplifiers that do not need the same type of support circuitry. Op-amps can produce gain values equivalent to several stages of discrete device amplifiers. Also, the op-amp can be tailored to simulate many forms of amplifiers. Finally, integrated-circuit amplifiers (more complex than the op-amp) are usually designed to perform one specific task, with the need for few, if any, external components. In the following discussions, integrated-circuit amplifiers will not be included, since their operation is quite different from that of other amplifying elements.

Matching Devices to Ideal Models

First it is important to match each of these devices to its ideal amplifier counterpart. By being able to relate these devices to a model, comparisons and applications can be more easily understood.

1. *The BJT.* Since the BJT is a current-amplifying device (input current produces output current), it can be related to the ideal current amplifier (Figure 5.17). It deviates from this model, though, in several ways:
   (a) It does not have an ideal input impedance. The value may be low, but is not zero.
   (b) It does not have an ideal output impedance. Instead of the ideal of infinite impedance, the BJT has a variable, and potentially low, output impedance.
   (c) Its current gain figure (beta) is not constant; the device is too unstable to be considered ideal.
2. *The FET.* The FET is a transconductance device. It produces output current under the control of the input voltage. It, too, deviates from the model (Figure 5.18), but not as drastically as the BJT does from its model.

FIGURE 5.17 Coordination of BJT to Ideal Current Amplifier

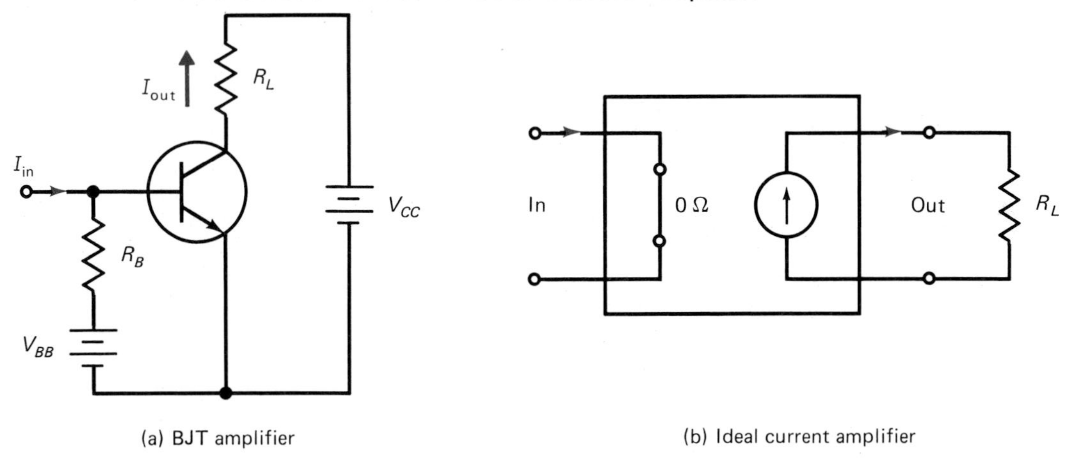

(a) BJT amplifier (b) Ideal current amplifier

FIGURE 5.18 Coordination of JFET to Ideal Transconductance Amplifier

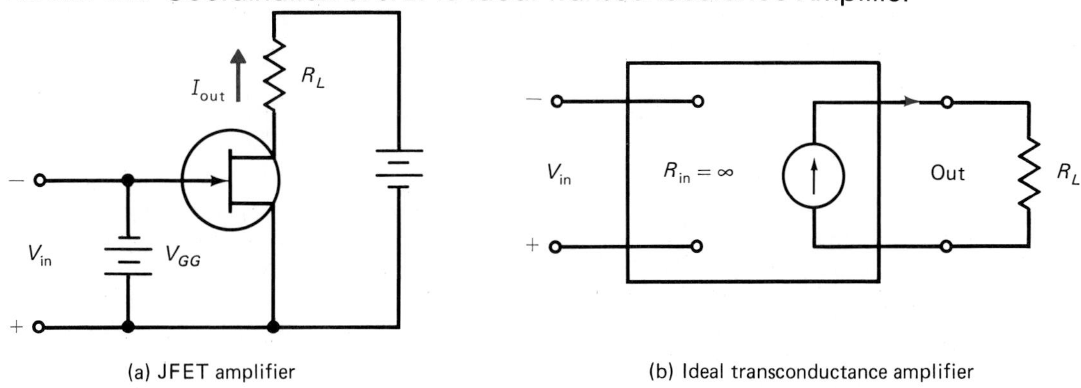

(a) JFET amplifier (b) Ideal transconductance amplifier

(a) The input impedance of the FET *is* quite close to the ideal model. Megohms of input impedance are possible with FETs.
(b) The output impedance of the FET, which should be high to match the model, is variable, much like that of the BJT.
(c) The gain (g_m) is not constant and can, on occasion, be unstable.

3. *The op-amp*. Initially, the op-amp is a voltage amplifier. With the addition of external parts, though, it can be made to duplicate many of the characteristics of other amplifier forms. As such, it is quite flexible and has a wide variety of applications.

Understanding the function of each of these semiconductor devices will allow us to make some general comparisons among them and to suggest types of applications. First, the BJT is the least stable of the three and duplicates its ideal model less than the others duplicate their models. It will be necessary to add quite a few external components to the BJT to provide both stability and closer duplication of the ideal model. Often, special control circuits which include feedback are required to make a dependable amplifier using a BJT.

The FET will also require some support circuitry, since it, too, does not duplicate its ideal model. In some cases, feedback will be used to improve stability. Also, the FET is polarity sensitive at its input terminal. If either device is to be used to amplify a dual-polarity signal (such as a sine wave), special circuitry will be needed to compensate for the "wrong-polarity" portion of the signal.

The op-amp's function in a circuit is controlled by the external components that are tied to it. Although by itself it is a high-gain voltage amplifier, its characteristics are such that it is normally not usable without external support components.

Now that the problems of each active device have been noted, we can focus on their application. The BJT can be applied when an amplifier is needed to provide current gain. Beyond this obvious application, there are quite a few other applications for the BJT. If current gain can be produced, power gain should be present. Current levels cannot be increased, assuming some consistency between input and output voltage, without having a resulting power gain. Input current must be a by-product of some level of voltage present on a resistance. If output current is produced, a voltage drop on the load attached to that amplifier must exist. By considering both input and output voltage levels, it is possible to investigate a factor for voltage gain for the BJT. Therefore, the BJT can have realistic ratings for current gain, power gain, and voltage gain. The BJT, although it is primarily a current-gain device, can be placed into a circuit where figures for voltage and power gain are usable (see Figure 5.19).

By the correct manipulation of external elements (usually resistors), the BJT can be made to function as a voltage amplifier, power amplifier, or even transconductance or transresistance amplifier. It should be noted that for these "simulated" amplifiers, the BJT circuit is far less ideal than the models it is duplicating. Often, this "simulation" is valid for only a very limited range of operating conditions. Regardless of the circuit design and the model being simulated, the BJT itself is unchanged as a current-amplifying device.

The FET is less flexible than the BJT. Since the FET has a nearly infinite input impedance, the potential of current gain for that device is relatively undefined. It will permit such low levels of current to flow into its input (gate)

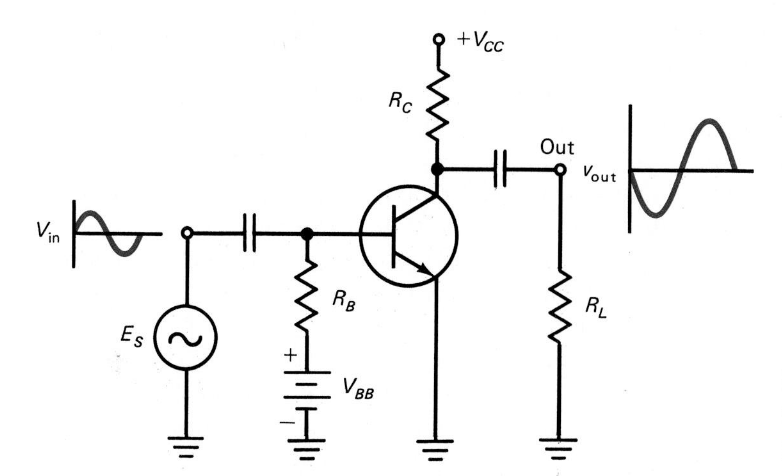

FIGURE 5.19 BJT Amplifier with Signal and DC Bias Connections

lead that comparing that flow to its typically strong flow of output current is unrealistic. Since the BJT can be used to produce current gain easily, it is unnecessary, in most cases, to structure an FET circuit so that it can simulate current gain.

Since the FET is voltage sensitive at its input and since its output current could be used to produce a voltage drop on a resistive load, the FET could be placed in a circuit that would produce an output voltage that was proportional to the input voltage, thus indicating voltage gain (Figure 5.20). It must be remembered that the FET, by itself, is a transconductance device.

Finally, reviewing the op-amp, we find that it is highly flexible when it comes to what form of ideal amplifier we wish to simulate. By manipulating the elements used external to the op-amp, quite a few amplifier configurations can be formed. It is for this reason that in future chapters, where special applications of amplifying circuits are presented, the op-amp will often be used to represent the central amplifier in the circuit.

Evaluating Other Amplifying Devices

There are presently a few other forms of amplifying devices which are not discussed in this book. It can be expected that in the future even more forms will be developed. To be able to understand the application of these devices, it is necessary to look for certain primary characteristics. The following check-

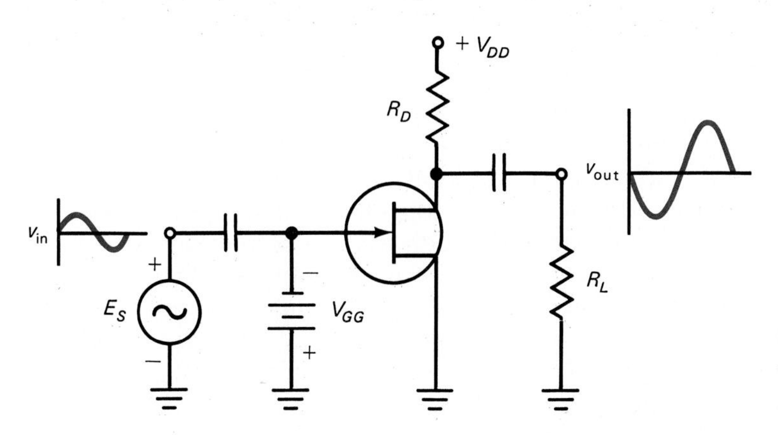

FIGURE 5.20 JFET Amplifier with Signal and DC Bias Connections

list when applied to a device will assist its user in classifying just how it would be applied in an electronic circuit.

1. Any amplifying device should first be classified as fitting one of the four basic model types (voltage, current, transconductance, or transresistance). The device input and output impedance value, coupled with its gain relationship, can be used to classify the device.
2. Next, how the device deviates from the ideal model should be determined. By comparing the device to its ideal counterpart, one can discover how it will or will not be limited in application.
3. Finally, the device specifications can be reviewed to see how it could be used to simulate other models. Some specifications may prohibit it from easily fitting a model (such as the FET being a current amplifier).

By making these observations and comparisons, one can then fit that device into categories which will help the user to determine applications for the device.

Thus the important characteristics of a device to look for initially are:

Input impedance (resistance)
Output impedance (resistance)
Gain figure and its statistical value

By investigating these, the device can be classified more clearly as approximating one of the four ideal models. Understanding how the device can be applied to a variety of uses becomes much easier when it is matched to an ideal counterpart.

REVIEW PROBLEMS

(1) One of the devices not discussed in this book is the vacuum-tube triode. Its basic characteristics are:

High input impedance
Moderately high (but variable) output impedance
A gain figure that relates input to output voltage

Conduct the three-part comparison noted in this section to help identify applications for this device.

(2) A photovoltaic cell (solar cell) has the capacity to produce voltage levels in proportion to light levels. But (for certain types of cells) it does not have the capacity to produce much current output. If I wished to operate a current meter with the output of this device, thus indicating light level (like the light meter on a camera), which device—BJT or FET—would be my best choice to use as an amplifier between the cell and meter? Justify your answer by discussing the general function of the device you choose.

(3) It was noted that the FET would not be very good at duplicating a current amplifier since it had such a high input impedance (little if any input current flow permitted). Which of the other ideal models might be difficult to duplicate with an FET? (Again, justify your answer.)

SUMMARY

In this chapter several of the semiconductors that will be used in the rest of this book have been presented. It was found that the diode, BJT, FET, op-amp, and integrated-circuit amplifiers all had different schematic symbols and different operating characteristics. Figure 5.21 illustrates each device's schematic symbols, a brief description of its function, and coordination to an ideal model.

The diode is a simple two-terminal device that does not amplify. It is sensitive to polarity, conducting with the proper "forward-bias" polarity and blocking current (not conducting) when the "reverse-bias" polarity is applied. Further discussion of the diode and its applications, together with various forms of diodes, is reserved for later chapters.

Device	Symbol	Function	Ideal model
Diode		Polarity sensitive conduction	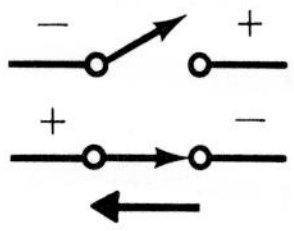 − + / + −
BJT	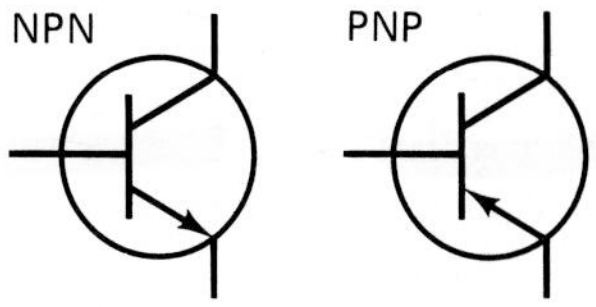 NPN PNP	Input current controls output current	Current amplifier
JFET*	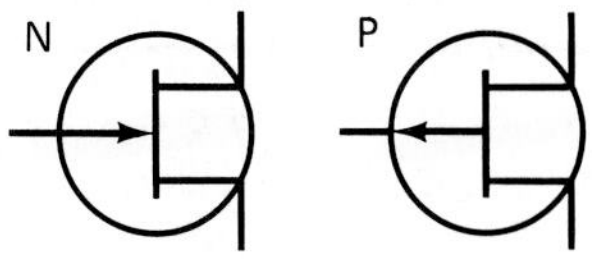 N P	Input voltage controls output current	Trans-conductance amplifier
Op-amp	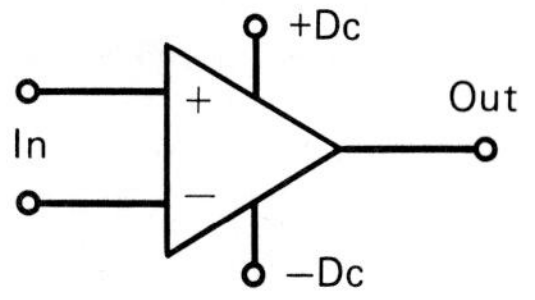 +Dc, In, Out, −Dc	Input voltage controls output voltage	Voltage amplifier
IC (system)	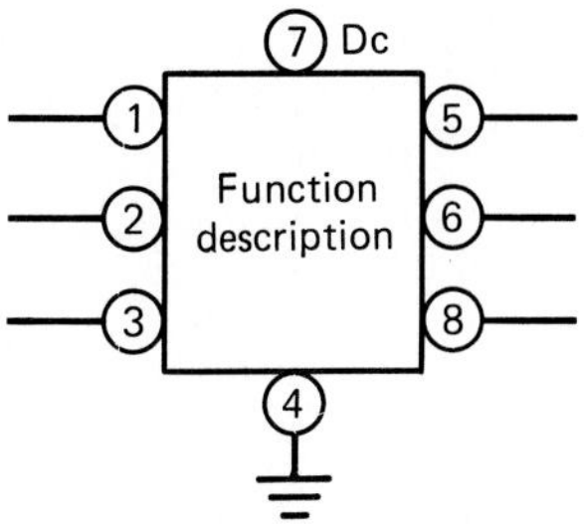 7 Dc, 1, 2, 3, 5, 6, 8, 4, Function description	Individual to each device	Based on function

FIGURE 5.21 Comparison of Semiconductor Devices

Transistors, both BJTs and FETs, are discrete amplifying devices. They can accept an input value and control an output level in response to that input. The BJT is a current amplifier, accepting a current level at the input and permitting a proportional level of current flow at the output. The FET, a transconductance device, accepts a voltage input and responds to that level by controlling the flow of output current. In the case of the FET, an inversion of the basic control function is seen where increases in input voltage produce corresponding decreases in output current. Neither of these devices is ideal; both instability and nonideal impedance terms create less-than-ideal operating functions.

The op-amp does come close to meeting ideal amplifier conditions. It also can be tailored, by the use of external components, to simulate a variety of amplifier systems. The op-amp has its limitations. The op-amp will be used later in the book to demonstrate amplifier applications, since it does have relatively universal amplification characteristics.

KEY EQUATIONS

5.1 Beta $(\beta) = \dfrac{I_C}{I_B}$

5.2 $I_C = (\beta) \times I_B$

5.3 $A_g = \dfrac{I_{\text{out}}}{V_{\text{in}}}$

PROBLEMS

Section 5.2

5.1. Using the circuit in Figure 5.22, find the forward and reverse current levels on the diode for the following load resistances.
(a) 56 Ω (b) 1 kΩ (c) 20 kΩ

5.2. A delicate piece of test equipment is battery operated. If the batteries are inserted the wrong way (+ for − and − for +), the instrument will be damaged. Illustrate a circuit that will protect the instrument in the event that the batteries are inserted incorrectly. (Use a diode. Also, simply show a battery symbol for the batteries and a "black box" for the instrument.)

5.3. Using the diode curves in Figure 5.22, explain briefly what the diode

FIGURE 5.22

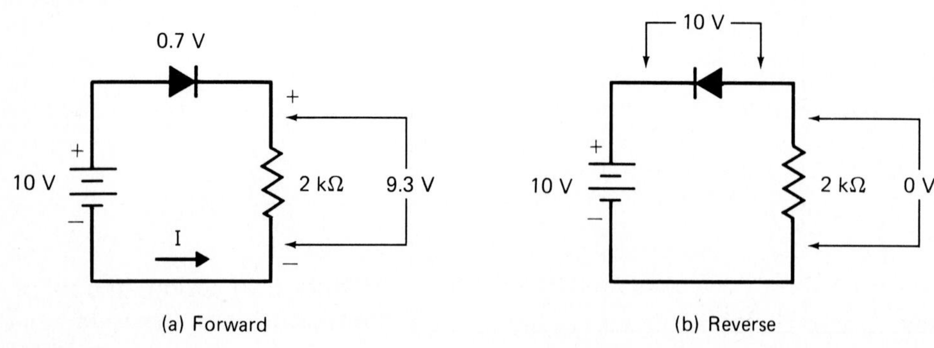

looks like when the forward voltage is between 0.3 and 0.4 V. (An open? A good conductor? Resistive?)

Section 5.3

5.4. Illustrate the symbols for both an NPN and a PNP transistor (BJT) and label the base, emitter, and collector leads.

5.5. Illustrate the correct bias polarities and current paths for an NPN and a PNP transistor.

5.6. Define the term ''beta.'' Indicate by an equation how it relates currents passing through the BJT.

5.7. If a transistor had a beta of 100 and a base current of 0.5 mA, what would be the resulting collector current?

5.8. What factors of a BJT make it less than an ideal current amplifier?

Section 5.5

5.9. Illustrate the schematic symbols and label the leads for both ''N''- and ''P''-type JFETs.

5.10. With which ideal model does the JFET coordinate? How does it differ from that model; how is it similar to it?

5.11. In Appendix A, look up the specification sheets for both BJT- and FET-type devices. List (separately) the part numbers for each type and identify their polarity markings (NPN, PNP, etc.).

5.12. Illustrate the proper polarity connections for an ''N''- and a ''P''-type JFET.

5.13. Briefly discuss the relationship of gate voltage to drain current for a JFET. Indicate what a change in gate voltage will do to drain current.

5.14. If an FET has a gain value of 470 mS, what will be the output level (drain current) for the following input voltage level of 2.4 V?

Section 5.6

5.15. Illustrate the schematic symbol for an op-amp.

5.16. Describe the basic difference between a noninverting and an inverting input to an op-amp.

5.17. Illustrate what output would result if the op-amp in Figure 5.23 had negative-going pulses (opposite those shown in Figure 5.23).

Section 5.7

5.18. What does the term ''LSI'' mean?

5.19. (If possible) Investigate several LSI linear ICs (not digital) in a technical library or through another resource. Find at least five different devices, list their part numbers, and provide a brief description of their application. Use either manufacturers' or distributors' catalogs, composite IC index(es), magazines, or circuit application books.

Section 5.8

5.20. Prepare a chart that lists the BJT, FET, and op-amp and their comparable ideal model schematics (general).

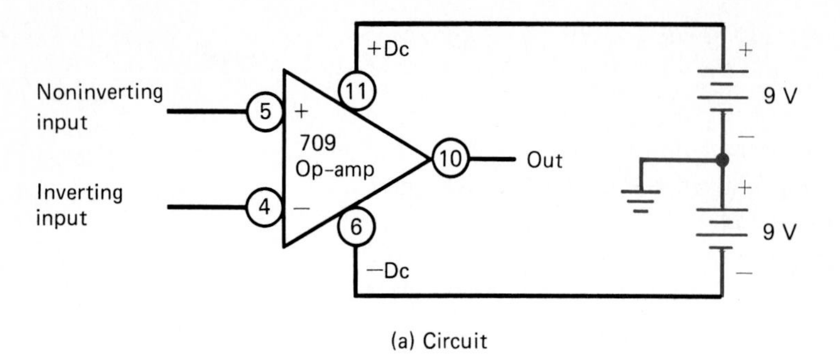

(a) Circuit

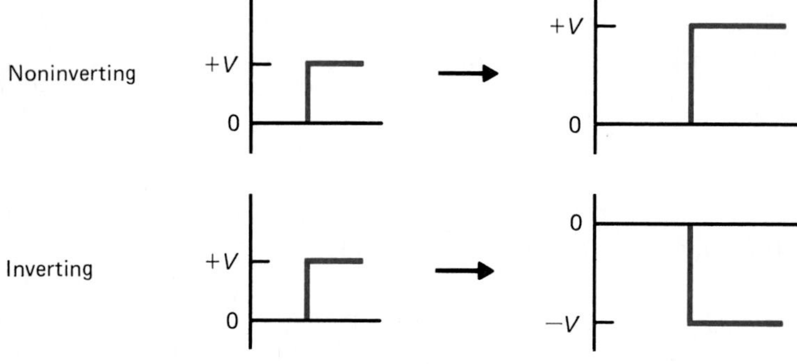

(b) Input/output coordination

FIGURE 5.23

ANSWERS TO REVIEW PROBLEMS

Section 5.2

1. (a) 0 V
(b) 0 V
(c) 1.3 V
(d) 34.3 V

2.

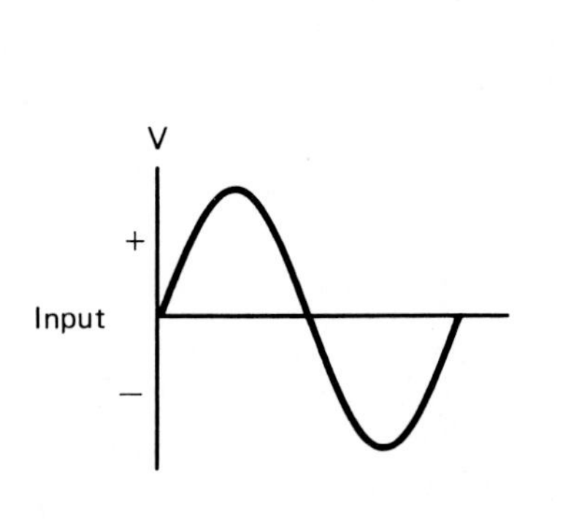

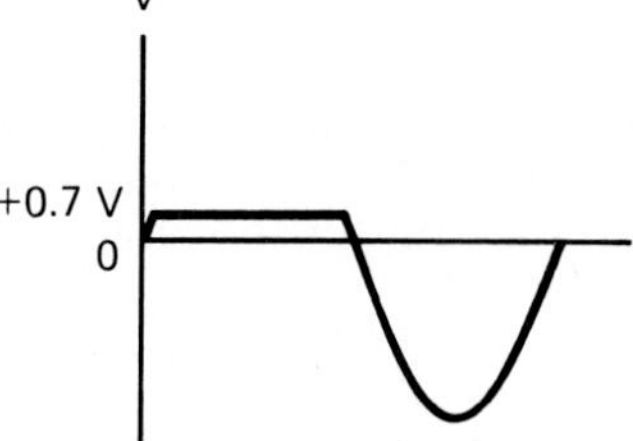

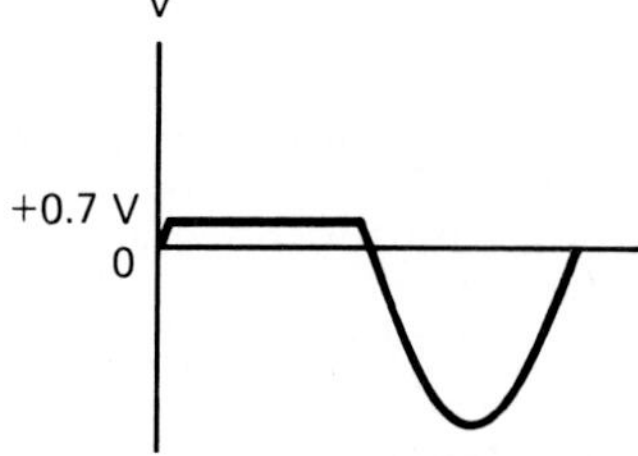

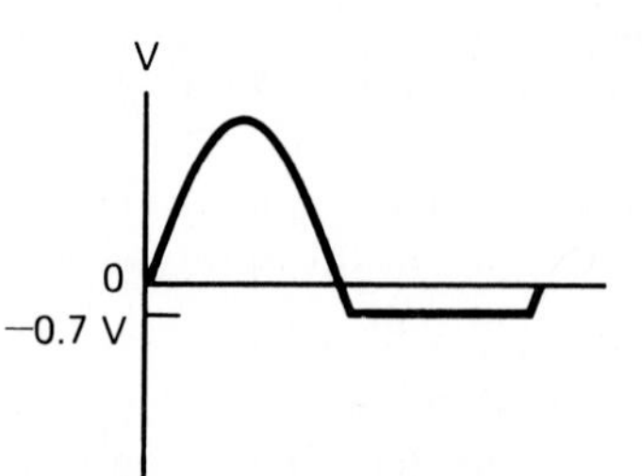

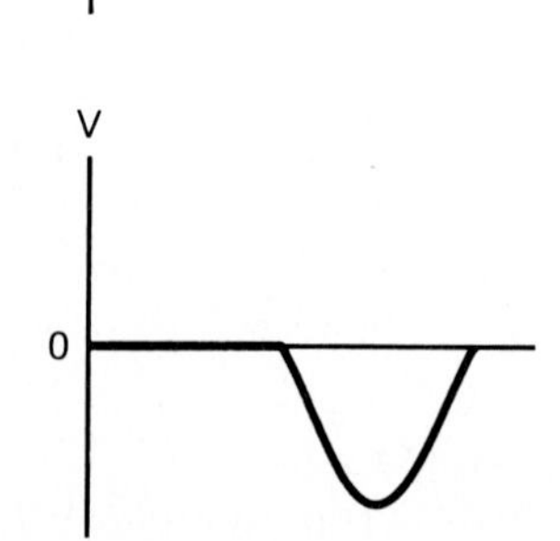

3.

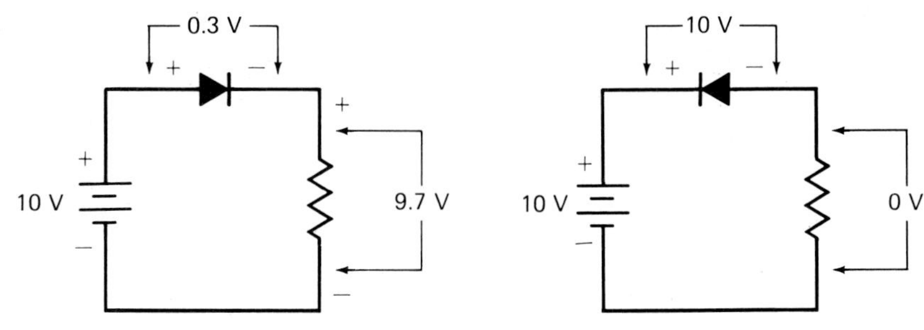

Section 5.4

1.

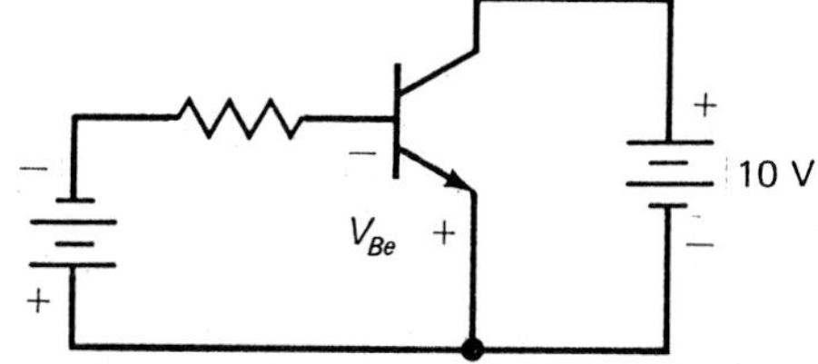

The following voltages and currents would exist: $I_B = 0$, $V_{BE} = V_{BB}$, $V_R = 0$ V, $I_C = 0$

2.

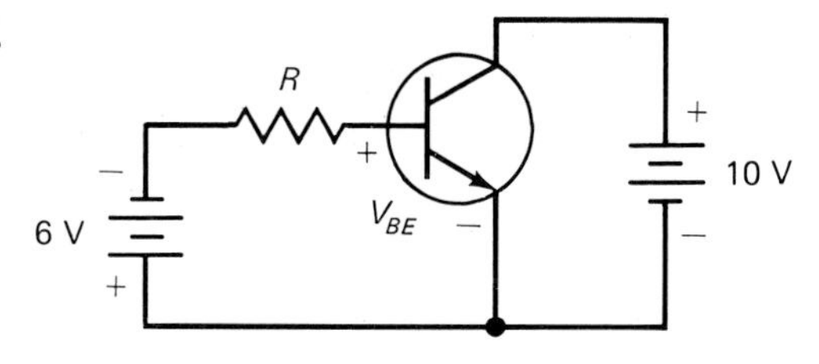

3. (a) $I_C = 7.8$ mA
   (b) (1) 12 mA
       (2) 3.6 mA
   (c) V_{BE} would remain constant. The change in I_B could result in a change in the base resistor value or power supply voltage.

Section 5.5

1. The N and P polarities are indicated by the direction of the arrow on the gate lead.
2. 200 μS; drain current at 4-V gate voltage ≃ 200 μA.
3. The BJT operates similarly to an ideal current amplifier. It has relative low input impedance, relatively high output impedance, and its gain is the ratio of input current to output current. The FET operates similarly to an ideal transconductance amplifier. It has very high input impedance, relatively high output impedance, and its gain is a ratio of output current to input voltage.
4. Since the FET has nearly infinite input impedance, it does not draw input current. Thus a current-limiting resistor is not needed. The total value of input voltage is simply dropped at the gate lead of the device (to ground). Since the BJT is a current amplifier, the input current must be controlled, unlike the FET.

Section 5.6

1. (a) 48 components
   (b) $A_{VOL} = 100$
2. The output from a noninverting input would be five times the input and in phase. The output from an inverting input would also be five times the input but would be 180° out of phase.
3.

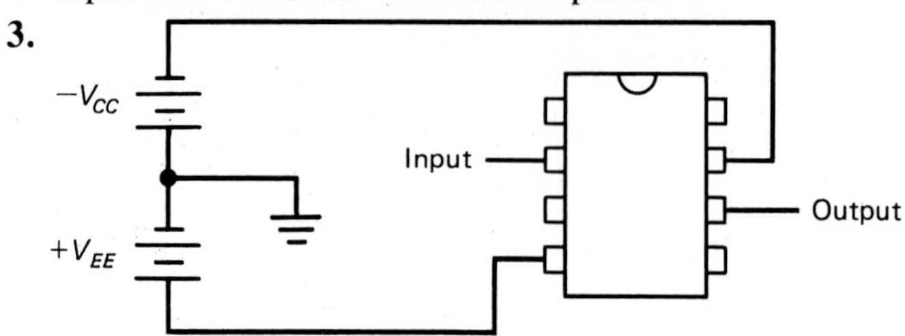

Section 5.7

1. The answer depends on the devices selected.

Section 5.8

1. This device is most like a voltage amplifier, but the output impedance deviates quite a bit from the ideal (0 Ω). Since it has high input impedance, it may be more difficult to fit it into a current or transresistance amplifier application.
2. An FET; since the input would be voltage from the cell and the desired output is current, the FET would be best able to produce this function.
3. A transresistance amplifier; again because of the need for low input resistance.

MC34071, MC34072
MC35071, MC35072
MC33071, MC33072

Advance Information

HIGH PERFORMANCE SINGLE SUPPLY OPERATIONAL AMPLIFIERS

SILICON MONOLITHIC INTEGRATED CIRCUIT

HIGH SLEW RATE, WIDE BANDWIDTH, SINGLE SUPPLY OPERATIONAL AMPLIFIERS

A standard low-cost Bipolar technology with innovative design concepts are employed for the MC34071/MC34072 series of monolithic operational amplifiers. These devices offer 4.5 MHz of gain bandwidth product, 13 V/μs slew rate, and fast settling time without the use of JFET device technology. In addition, low input offset voltage can economically be achieved. Although these devices can be operated from split supplies, they are particularly suited for single supply operation, since the common mode input voltage range includes ground potential (V_{EE}). The all NPN output stage, characterized by no deadband crossover distortion and large output voltage swing, also provides high capacitive drive capability, excellent phase and gain margins, low open-loop high frequency output impedance and symmetrical source/sink ac frequency response.

The MC34071/MC34072 series of devices are available in standard or prime performance (A Suffix) grades and specified over commercial, industrial/vehicular or military temperature ranges.

- Wide Bandwidth: 4.5 MHz
- High Slew Rate: 13 V/μs
- Fast Settling Time: 1.1 μs to 0.10%
- Wide Single Supply Operating Range: 3.0 to 44 Volts
- Wide Input Common Mode Range Including Ground (V_{EE})
- Low Input Offset Voltage: 1.5 mV Maximum (A Suffix)
- Large Output Voltage Swing: −14.7 V to +14.0 V for $V_S = \pm 15$ V
- Large Capacitance Drive Capability: 0 to 10,000 pF
- Low T.H.D. Distortion: 0.02%
- Excellent Phase Margins: 60°
- Excellent Gain Margin: 12 dB

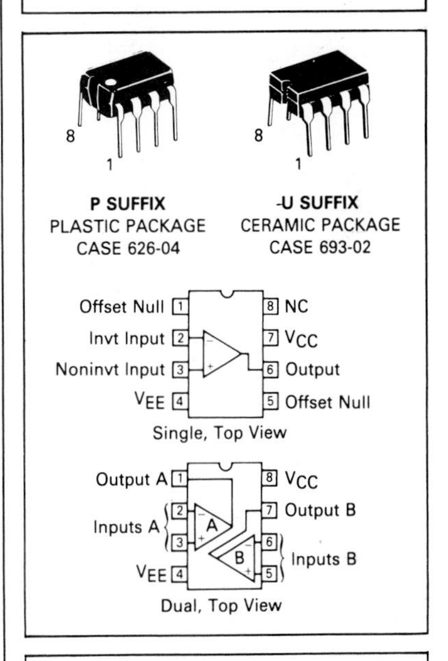

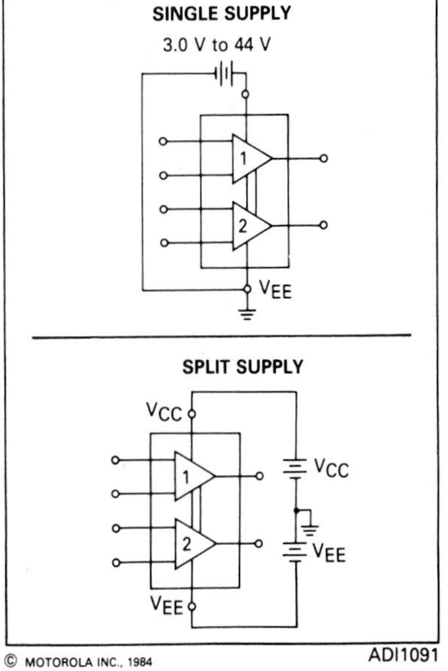

ORDERING INFORMATION

Op Amp Function	Device	Temperature Range	Package
Single	MC35071U,AU	−55 to +125°C	Ceramic DIP
	MC33071U,AU	−40 to +85°C	Ceramic DIP
	MC33071P,AP	−40 to +85°C	Plastic DIP
	MC34071U,AU	0 to +70°C	Ceramic DIP
	MC34071P,AP	0 to +70°C	Plastic DIP
Dual	MC35072U,AU	−55 to +125°C	Ceramic DIP
	MC33072U,AU	−40 to +85°C	Ceramic DIP
	MC33072P,AP	−40 to +85°C	Plastic DIP
	MC34072U,AU	0 to +70°C	Ceramic DIP
	MC34072P,AP	0 to +70°C	Plastic DIP
Quad	MC34074 Series	Refer to MC34074 Data Sheet	

This document contains information on a new product. Specifications and information herein are subject to change without notice.

ADI1091

DC ELECTRICAL CHARACTERISTICS (V_{CC} = +15 V, V_{EE} = −15 V, R_L connected to ground, T_A = T_{low} to T_{high} [Note 3] unless otherwise noted)

Characteristic	Symbol	MC3507_A/MC3407_A/ MC3307_A			MC3507_/MC3407_/ MC3307_			Unit
		Min	Typ	Max	Min	Typ	Max	
Input Offset Voltage (V_{CM} = 0)	V_{IO}							mV
V_{CC} = +15 V, V_{EE} = −15 V, T_A = +25°C		—	0.5	1.5	—	1.0	3.5	
V_{CC} = +5.0 V, V_{EE} = 0 V, T_A = +25°C		—	0.5	2.0	—	1.5	4.0	
V_{CC} = +15 V, V_{EE} = −15 V, T_A = T_{low} to T_{high}		—	—	3.5	—	—	5.5	
Average Temperature Coefficient of Offset Voltage	$\Delta V_{IO}/\Delta T$	—	10	—	—	10	—	μV/°C
Input Bias Current (V_{CM} = 0)	I_{IB}							nA
T_A = +25°C		—	100	500	—	100	500	
T_A = T_{low} to T_{high}		—	—	700	—	—	700	
Input Offset Current (V_{CM} = 0)	I_{IO}							nA
T_A = +25°C		—	6.0	50	—	6.0	75	
T_A = T_{low} to T_{high}		—	—	300	—	—	300	
Large Signal Voltage Gain V_O = ±10 V, R_L = 2.0 k	A_{VOL}	50	100	—	25	100	—	V/mV
Output Voltage Swing	V_{OH}							V
V_{CC} = +5.0 V, V_{EE} = 0 V, R_L = 2.0 k, T_A = +25°C		3.7	4.0	—	3.7	4.0	—	
V_{CC} = +15 V, V_{EE} = −15 V, R_L = 10 k, T_A = +25°C		13.7	14	—	13.7	14	—	
V_{CC} = +15 V, V_{EE} −15 V, R_L = 2.0 k, T_A = T_{low} to T_{high}		13.5	—	—	13.5	—	—	
	V_{OL}							
V_{CC} = +5.0 V, V_{EE} = 0 V, R_L = 2.0 k, T_A = +25°C		—	0.1	0.2	—	0.1	0.2	
V_{CC} = +15 V, V_{EE} + −15 V, R_L = 10 k, T_A = +25°C		—	−14.7	−14.4	—	−14.7	−14.4	
V_{CC} = +15 V, V_{EE} = −15 V, R_L = 2.0 k, T_A = T_{low} to T_{high}		—	—	−13.8	—	—	−13.8	
Output Short-Circuit Current (T_A = +25°C) Input Overdrive = 1.0 V, Output to Ground	I_{SC}							mA
Source		10	30	—	10	30	—	
Sink		20	47	—	20	47	—	
Input Common Mode Voltage Range	V_{ICR}							V
T_A = +25°C		V_{EE} to (V_{CC} − 1.8)			V_{EE} to (V_{CC} − 1.8)			
T_A = T_{low} to T_{high}		V_{EE} to (V_{CC} − 2.2)			V_{EE} to (V_{CC} − 2.2)			
Common Mode Rejection Ratio (R_S ≤ 10 k)	CMRR	80	97	—	70	97	—	dB
Power Supply Rejection Ratio (R_S = 100 Ω)	PSRR	80	97	—	70	97	—	dB
Power Supply Current (Per Amplifier)	I_D							mA
V_{CC} = +5.0 V, V_{EE} = 0 V, T_A = +25°C		—	1.6	2.0	—	1.6	2.0	
V_{CC} = +15 V, V_{EE} = −15 V, T_A = +25°C		—	1.9	2.5	—	1.9	2.5	
V_{CC} = +15 V, V_{EE} = −15 V, T_A = T_{low} to T_{high}		—	—	2.8	—	—	2.8	

NOTES: (continued)

3. T_{low} = −55°C for MC35071,A/MC35072,A
   = −40°C for MC33071,A/MC33072,A
   = 0°C for MC34071,A/MC34072,A

   T_{high} = +125°C for MC35071,A/35072,A
   = +85°C for MC33071,A/33072,A
   = +70°C for MC34071,A/34072,A

MC34071,A • MC34072,A Series

MAXIMUM RATINGS

Rating	Symbol	Value	Unit
Supply Voltage (from V_{CC} to V_{EE})	V_S	+44	Volts
Input Differential Voltage Range	V_{IDR}	Note 1	Volts
Input Voltage Range	V_{IR}	Note 1	Volts
Output Short-Circuit Duration (Note 2)	t_S	Indefinite	Seconds
Operating Ambient Temperature Range MC35071,A/MC35072,A MC33071,A/MC33072,A MC34071,A/MC34072,A	T_A	 −55 to +125 −40 to +85 0 to +70	°C
Operating Junction Temperature	T_J	+150	°C
Storage Temperature Range Ceramic Package Plastic Package	T_{stg}	 −65 to +150 −55 to +125	°C

NOTES:

1. Either or both input voltages must not exceed the magnitude of V_{CC} or V_{EE}.
2. Power dissipation must be considered to ensure maximum junction temperature (T_J) is not exceeded.

EQUIVALENT CIRCUIT SCHEMATIC (EACH AMPLIFIER)

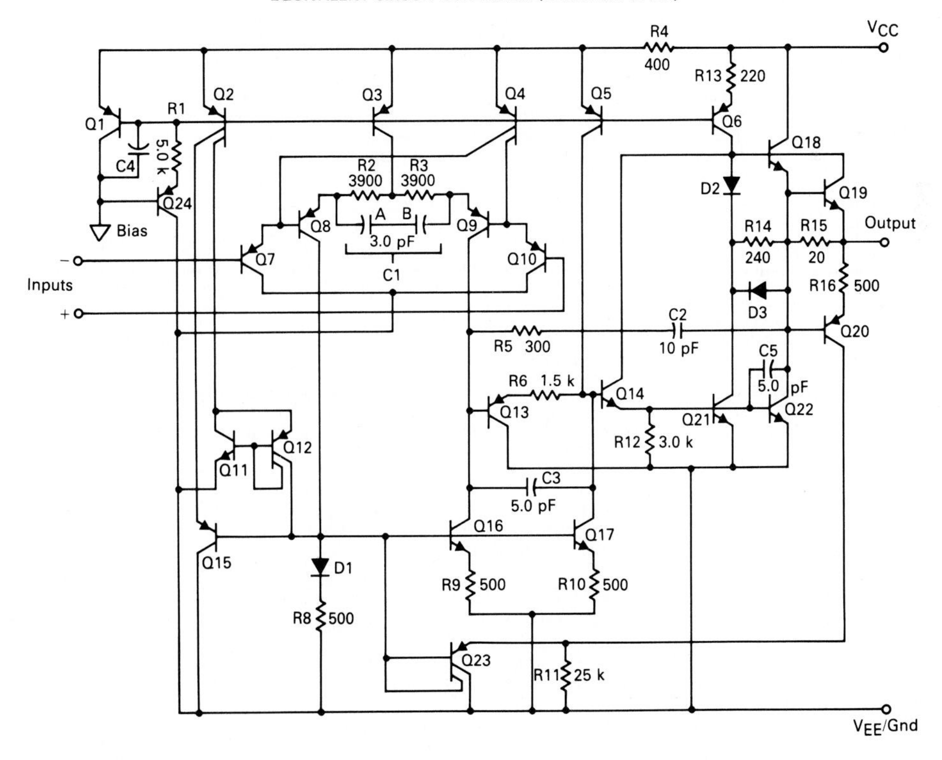

MC34071,A • MC34072,A Series

AC ELECTRICAL CHARACTERISTICS (V_{CC} = +15 V, V_{EE} = −15 V, R_L connected to ground, T_A = +25°C unless otherwise noted)

Characteristic	Symbol	MC3507_A/MC3407_A/ MC3307_A Min	Typ	Max	MC3507_/MC3407_/ MC3307__ Min	Typ	Max	Unit
Slew Rate (V_{in} = −10 V to +10 V, R_L = 2.0 k, C_L = 500 pF)	SR							V/μs
A_V + 1.0		8.0	10	—	—	10	—	
A_V − 1.0		—	13	—	—	13	—	
Settling Time (10 V Step, A_V = −1.0)	t_s							μs
To 0.10% (±1/2 LSB of 9-Bits)		—	1.1	—	—	1.1	—	
To 0.01% (±1/2 LSB of 12-Bits)		—	2.2	—	—	2.2	—	
Gain Bandwidth Product (f = 100 kHz)	GBW	3.5	4.5	—		4.5	—	MHz
Power Bandwidth A_V = +1.0, R_L = 2.0 k, V_O = 20 V_{p-p}, THD = 5.0%	BWp	—	200	—	—	200	—	kHz
Phase Margin	ϕm							Degrees
R_L = 2.0 k		—	60	—	—	60	—	
R_L = 2.0 k, C_L = 300 pF		—	40	—	—	40	—	
Gain Margin	A_m							dB
R_L = 2.0 k		—	12	—	—	12	—	
R_L = 2.0 k, C_L = 300 pF		—	4.0	—	—	4.0	—	
Equivalent Input Noise Voltage R_S = 100 Ω, f = 1.0 kHz	e_n	—	32	—	—	32	—	nV/ $\sqrt{Hz}$
Equivalent Input Noise Current (f = 1.0 kHz)	I_n	—	0.22	—	—	0.22	—	pA/ $\sqrt{Hz}$
Input Capacitance	C_i	—	0.8	—	—	0.8	—	pF
Total Harmonic Distortion A_V = +10, R_L = 2.0 k, 2.0 ≤ V_O ≤ 20 V_{p-p}, f = 10 kHz	THD	—	0.02	—	—	0.02	—	%
Channel Separation (f = 10 kHz, MC34072,A Only)	—	—	120	—	—	120	—	dB
Open-Loop Output Impedance (f = 1.0 MHz)	z_o	—	30	—	—	30	—	Ω

For typical performance curves and applications information refer to MC34074 series data sheet.

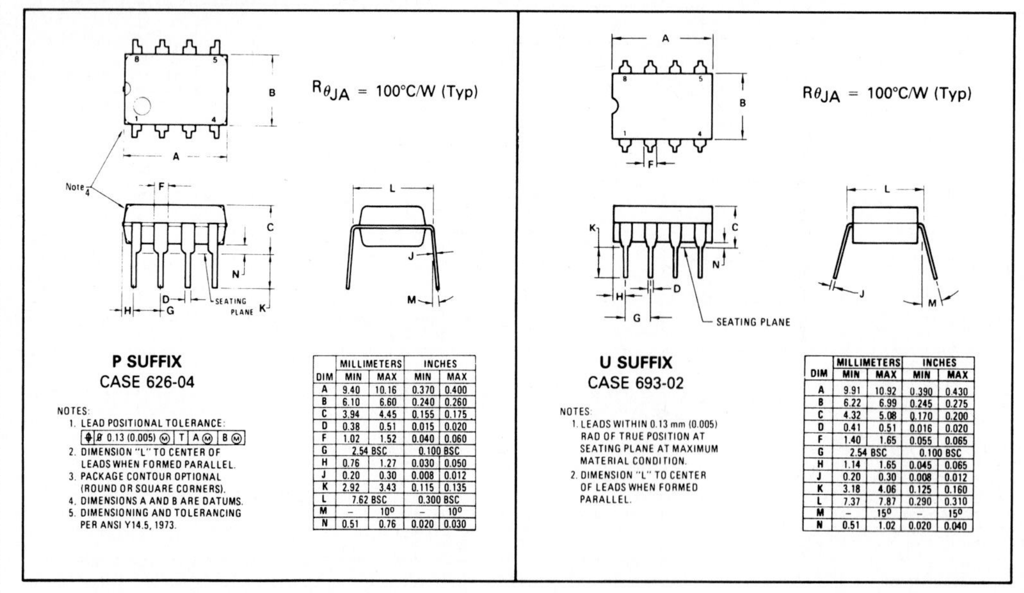

P SUFFIX
CASE 626-04

NOTES:
1. LEAD POSITIONAL TOLERANCE: ⌖ Ø 0.13 (0.005) Ⓜ | T | A Ⓜ | B Ⓜ
2. DIMENSION "L" TO CENTER OF LEADS WHEN FORMED PARALLEL.
3. PACKAGE CONTOUR OPTIONAL (ROUND OR SQUARE CORNERS).
4. DIMENSIONS A AND B ARE DATUMS.
5. DIMENSIONING AND TOLERANCING PER ANSI Y14.5, 1973.

DIM	MILLIMETERS MIN	MILLIMETERS MAX	INCHES MIN	INCHES MAX
A	9.40	10.16	0.370	0.400
B	6.10	6.60	0.240	0.260
C	3.94	4.45	0.155	0.175
D	0.38	0.51	0.015	0.020
F	1.02	1.52	0.040	0.060
G	2.54 BSC		0.100 BSC	
H	0.76	1.27	0.030	0.050
J	0.20	0.30	0.008	0.012
K	2.92	3.43	0.115	0.135
L	7.62 BSC		0.300 BSC	
M	–	10°	–	10°
N	0.51	0.76	0.020	0.030

U SUFFIX
CASE 693-02

NOTES:
1. LEADS WITHIN 0.13 mm (0.005) RAD OF TRUE POSITION AT SEATING PLANE AT MAXIMUM MATERIAL CONDITION.
2. DIMENSION "L" TO CENTER OF LEADS WHEN FORMED PARALLEL.

DIM	MILLIMETERS MIN	MILLIMETERS MAX	INCHES MIN	INCHES MAX
A	9.91	10.92	0.390	0.430
B	6.22	6.99	0.245	0.275
C	4.32	5.08	0.170	0.200
D	0.41	0.51	0.016	0.020
F	1.40	1.65	0.055	0.065
G	2.54 BSC		0.100 BSC	
H	1.14	1.65	0.045	0.065
J	0.20	0.30	0.008	0.012
K	3.18	4.06	0.125	0.160
L	7.37	7.87	0.290	0.310
M	–	15°	–	15°
N	0.51	1.02	0.020	0.040

MOTOROLA *Semiconductor Products Inc.*

BOX 20912 • PHOENIX, ARIZONA 85036 • A SUBSIDIARY OF MOTOROLA INC.

17857 PRINTED IN USA 5-84 IMPERIAL LITHO C22510 12,000

ADI1091

6 Functional Aspects of Discrete Active Devices

Objectives

Upon completing this chapter, the reader should be able to:

- Identify the need for and process of compensation.
- Calculate basic bias values for BJT and FET circuits.
- Identify the use of output load resistors.
- Manipulate both dc and ac equations for BJTs and JFETs.
- Identify the need for stability and the forms of stabilization circuits.
- Calculate gains and impedances for stabilization circuits.
- Troubleshoot stabilization circuits.
- Identify the application of bypass capacitors.

Introduction

In this chapter more specific details of how to work with the individual forms of discrete active devices (BJT and FET types) are presented. The process of compensation for nonideal characteristics and device limitations is introduced and methods of evaluating the compensation effects are shown. The concept of stability biasing and the problems of stabilization are presented as well. Finally, the use of bypass capacitors, used to modify circuits for ac applications, is shown. The solution of the bias and signal-processing circuitry is shown to result in a composite solution of current and voltage levels around the device. ■

6.1 THE GENERAL CONCEPT OF COMPENSATION AND INPUT/OUTPUT CONDITIONING

In previous chapters it was pointed out that discrete devices do not meet the specifications of the ideal model that coordinates to their function. Both the BJT and the JFET are polarity sensitive at their input terminals and instability makes them somewhat undependable. These problems need to be resolved if the devices are to be used in electronic circuits. Since each device has electrical limitations, both maximum and minimum boundaries, circuits in which they are used must provide electrical restrictions so that these boundaries are not overstepped.

When components are added to a circuit to help overcome both device limitations and boundary limitations, this is called **compensation**. The added components are selected to compensate (make up) for the shortcomings of the device. Resistors can provide voltage dividers, limit current, and have voltage drops proportional to current flow. For these reasons, resistors are most often used to provide compensation. In addition, negative feedback can be used to assist in the compensation process.

There are two electrical quantities that any amplifier deals with. First, there is the information signal that is to be amplified within the active element. Then there is the dc energy source which is present to provide the energy necessary to produce an amplified output. (Remember, the process of amplification is simply the control of one source's potential energy by another, weaker information source.) We will use this dc source not only to provide raw energy for an output but also to provide the proper compensation for the active device.

Once compensation has been provided by the dc source in the circuit, the amplification process can proceed without the need for concern about the active device. The energy used to compensate for device shortcomings is usually wasted within the circuit. Therefore, the dc source must supply more energy (or power) to the circuit than the circuit uses directly to produce an output. For example, a circuit may consume 500 mW of power from the dc source to be able to produce 200 mW of usable output. The remaining 300 mW is used to operate other components that compensate for the device's inequities.

In this and the following chapter, each circuit is divided into two forms: one for dc compensation (also called bias) and the other for ac signal processing.

6.2 INITIAL INPUT BIAS CONDITIONS FOR THE BJT

In Chapter 5 it was noted that the discrete semiconductor devices, BJTs and FETs, required special bias arrangements. In this section these special conditions will be introduced and using some of the circuit analysis methods noted in Section 6.1, equations will be developed to calculate actual operating conditions (voltages and currents). First, we need to identify some of the special terminology used with these devices.

From the past chapters a simple identification of the BJT's leads is already known (base, emitter, collector). Current flow and voltage potentials for an NPN transistor are identified as follows (see Figure 6.1):

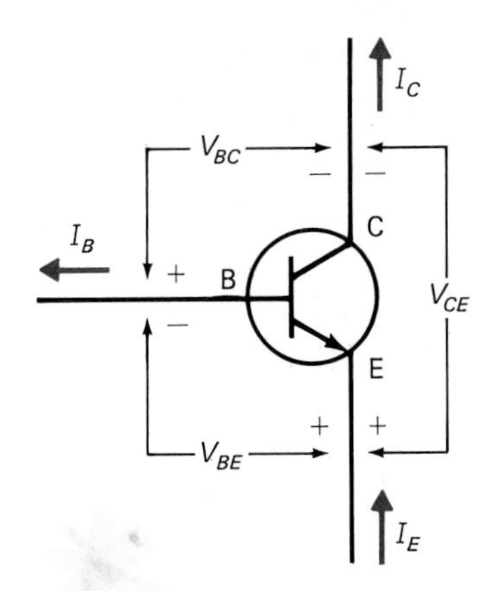

FIGURE 6.1 Voltages and Current Identification for an NPN BJT

V_{CC} Typical notation used for the "collector" dc voltage
I_B Current flow at the base lead
I_C Current flow at the collector
I_E Current flow at the emitter
V_{BE} Base-to-emitter voltage (typically, 0.7 V)
V_{CE} Voltage from the collector to the emitter
V_{BC} Voltage from the base to the collector

The most common configuration (arrangement) of the BJT is called the **common-emitter** configuration. In such an arrangement, the input (or controlling) value is applied to the base lead, with the emitter being at the ground potential for this input. The controlled source (power supply) is applied to the collector lead, again with the emitter being connected to the ground potential of this source. Thus the input current flows through the base and emitter leads and the controlled current flows through the collector and also the emitter. Since both input and output currents flow through the emitter, it is said to be "common" to both. Figure 6.2 shows the common-emitter configuration using a dc input source. There are other ways in which to arrange the leads of the BJT in reference to input and output, but these will be shown later.

It is now necessary to deal with the first operational problem encountered with the BJT, its polarity sensitivity. The BJT, as discovered earlier, can have only base current (I_B) flowing in a single direction. The base lead must be

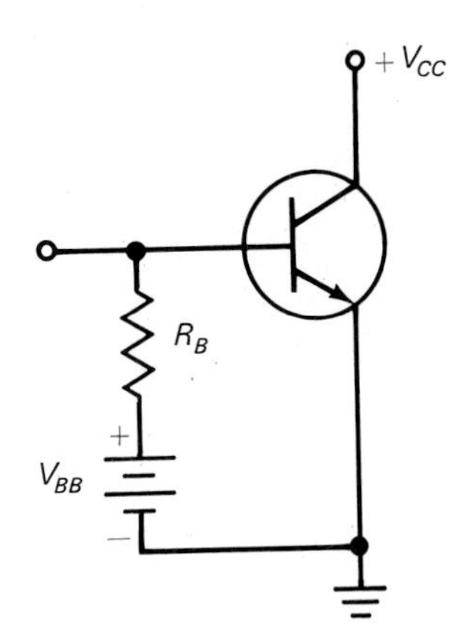

FIGURE 6.2 Simple Input Control Bias for an NPN BJT

forward biased in order to have base current flow permitted by the base–emitter junction (inside the device). When dealing with an NPN transistor, this means that the base lead must be more positive (+) than the emitter lead. In fact, it must exceed the potential at the emitter by +0.7 V (the typical forward drop for silicon devices). If a PNP device were to be used, the base must be more negative (−) than the emitter, again by 0.7 V. Stated another way, in an NPN device, base current must flow from the emitter to the base (negative to positive), and in a PNP device, from the base to the emitter. Figure 6.3 summarizes this process.

In Figure 6.3 the direction of collector current is also shown. To have base current flow encourage collector flow, the collector voltage polarity must match that of the base. If this collector voltage were to be either opposite to the base polarity or even drop below the 0.7-V base potential, the collector-to-base junction could become forward biased just like the emitter–base junction should be. Should this happen in this common-emitter configuration, the circuit would fail to operate as expected and might become damaged.

One final fact needs to be observed in this illustration. The actual value of the emitter current I_E is a sum of base and collector currents. Since both currents flow from ground through the emitter, the total amount of emitter current is the sum of both I_C and I_B. As a reminder, the base resistor is used to limit base current since the base–emitter junction will not restrict the base current.

$$I_E = I_C + I_B \tag{6.1}$$

In Chapter 5 it was shown that I_C is related to I_B via the term "beta." Soon we will use this factor to observe more device functions.

A problem exists when the BJT is used to amplify an ac input. If a sine wave such as that shown in Figure 6.4 is applied to the base of the device, not all of that signal will cause the device to be forward biased. That portion of the input wave that is positive with respect to ground (the emitter) and above 0.7 V will forward bias the device. Current will then flow through the base lead and will encourage flow through the collector lead. *But* any portion of the wave that reverse biases the base–emitter junction (V_{BE} = −voltage) or simply falls below the mandatory 0.7 V potential drop *will not produce base current*. Without base current, there will not be a responding collector current.

This principle is also demonstrated by the fact that if the current flow to the base were to reverse, it would be logical to assume that collector flow

FIGURE 6.3 Comparison of Voltage Polarities and Currents

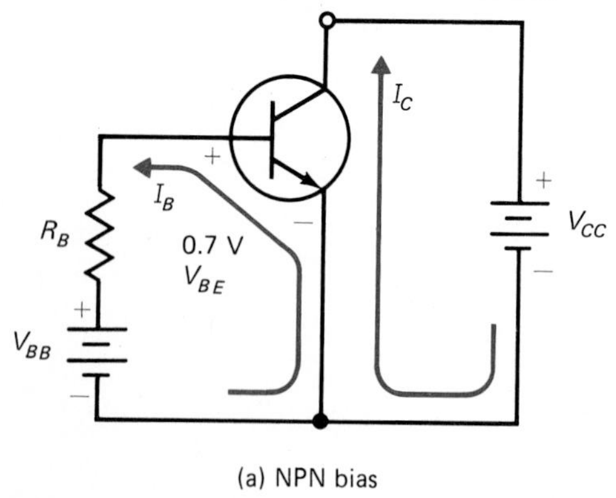

(a) NPN bias

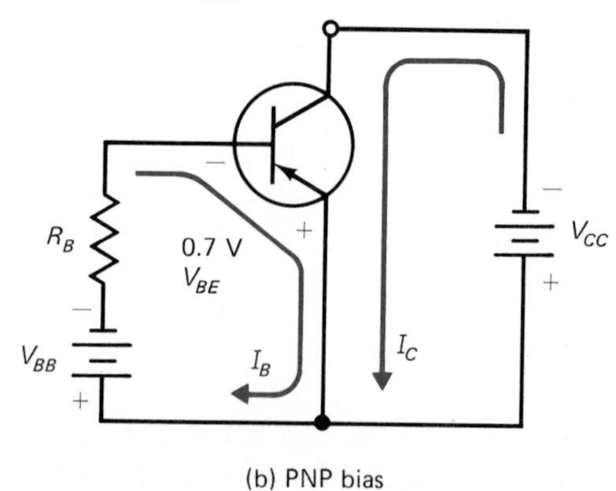

(b) PNP bias

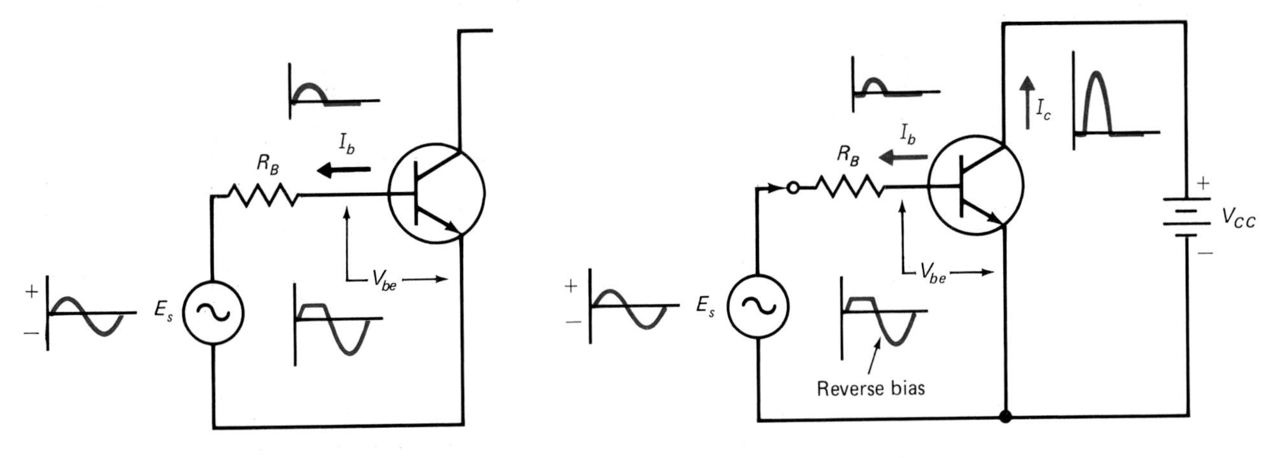

(a) Loss of negative portion of input signal

(b) Failure to produce output current

FIGURE 6.4 Failures in BJTs Caused by Bias Effects

would reverse in response to this change. But since the collector is supplied with a dc source only, a reversal of collector current is impossible.

The condition where the base current would be zero is given the special name **cutoff**. When cutoff occurs, the collector current also drops to zero and the collector-to-emitter section of the device looks like an open circuit. In Figure 6.4 the waveform of collector current has been added to the illustration. Even though the positive input segment of the wave (above 0.7 V) is amplified, none of the portion that reverse biased the device is reproduced. Naturally, this is an unacceptable output for a typical amplifier.

Actually, physical damage can be done to a transistor if it is continually forced into this ''hard'' reverse-bias condition (in this context, ''hard'' means forcing a large reverse-bias potential on the base–emitter junction). Therefore, it is necessary to overcome this problem. To do this, we introduce our first dc bias condition for the BJT. If the base of the transistor is connected to a source of continuous forward-bias potential, and if that potential, when added to the input, always produces a composite voltage that is of the correct polarity to forward bias the device, base current will always flow. We can therefore avoid a condition where the transistor is in the cutoff mode.

In Figure 6.5 a **base bias resistor** has been added to the circuit. This

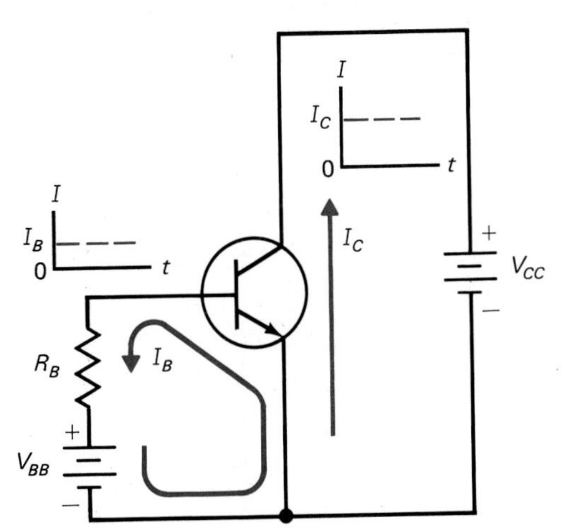

FIGURE 6.5 DC Bias for an NPN BJT

resistor, called R_B, maintains a constant flow of base current, even if there is no usable input applied to the circuit. This process can be related to maintaining a middle-range idle speed on a lawn-mower engine. The engine can speed up or slow down from that idle value but will never be allowed to stall (or reverse) due to "backing off" on the throttle.

The actual current flow to the base is maintained at a midlevel by the addition of the base resistor. Without an input signal, base current flows due to this resistor, and resulting collector current flow is also produced (beta × base current). This flow does not produce a usable or practical output; it simply consumes energy from the dc source.

In Figure 6.6 an ac source has been added to the circuit. The capacitor was used in series with this source so that dc current from the power supply would not flow through the ac source's low output impedance to ground. Now there are two sources of base current in the circuit, the dc bias current and the input signal current. This input sine wave will produce a current through the base of the device which will be added (algebraically) to the dc current. The composite input wave form illustrates such this process. (The addition of these potentials was illustrated in the circuit analysis review section of this chapter.) Observe the input waveform carefully. Notice that at no time does it drop below the minimum forward-bias voltage required by the device. Nor does the polarity at the base ever become negative (−) and attempt to reverse bias it. Therefore, all the ac information that is to be amplified is of the correct polarity (current direction).

Figure 6.7 illustrates the composite output collector current flow that will be produced. Notice that it represents all the input waveform. This ac output is riding on a corresponding dc level, just as seen at the input. Both the dc level at the base and the ac input signal are affected by the gain—beta. Every point on the input curve is "amplified" by beta to produce collector current. Now, though, the transistor produces a varying current flow from the dc source in synchronization with the input signal. (When dc is added to ac as done here, the dc level is often called an **offset**. Thus if a signal has 4 V dc added to it, it could be referred to as a "4-V offset.")

This process, where a dc level is added to an ac input, is a very common practice when dealing with polarity-sensitive devices. The dc level needed to make the device fully functional depends on several factors. First, the dc level

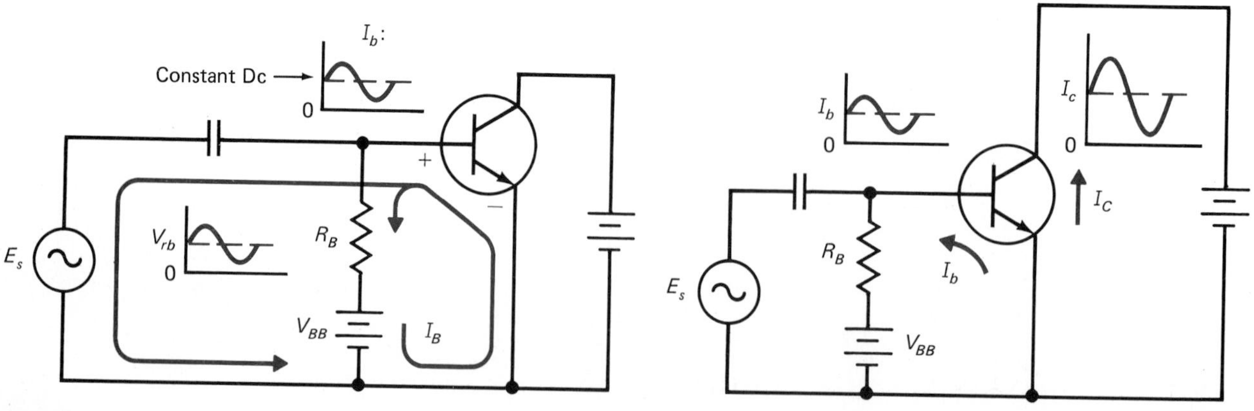

FIGURE 6.6 Amplification of Complete Input Waveform by the Addition of DC Bias

FIGURE 6.7 Output Current Waveform for a Biased BJT

must be sufficient to cause all of the input to be of one correct polarity. If the device being used can accept only a positive (+) input polarity, the dc level must be of a high enough positive voltage to counter the negative peak of the input signal. If the device must have a negative polarity at the input, the dc offset must be sufficiently negative to negate the positive peaks of the input signal.

EXAMPLE 6.1

(a) If an NPN transistor (requiring positive base polarities for forward bias) had an input signal current that had a peak-to-peak level of 20 mA, the offset (dc bias) would have to be at least 10 mA from a positive voltage source in reference to a grounded emitter. With that amount of offset, the negative peak of the input current, when added to the offset, would result in a minimum current of 0 mA [see Figure 6.8(a)].

(b) Consider a PNP transistor with a 14-mA p-p input current. In that case, 7 mA of offset current from a negative supply would be required [see Figure 6.8(b)].

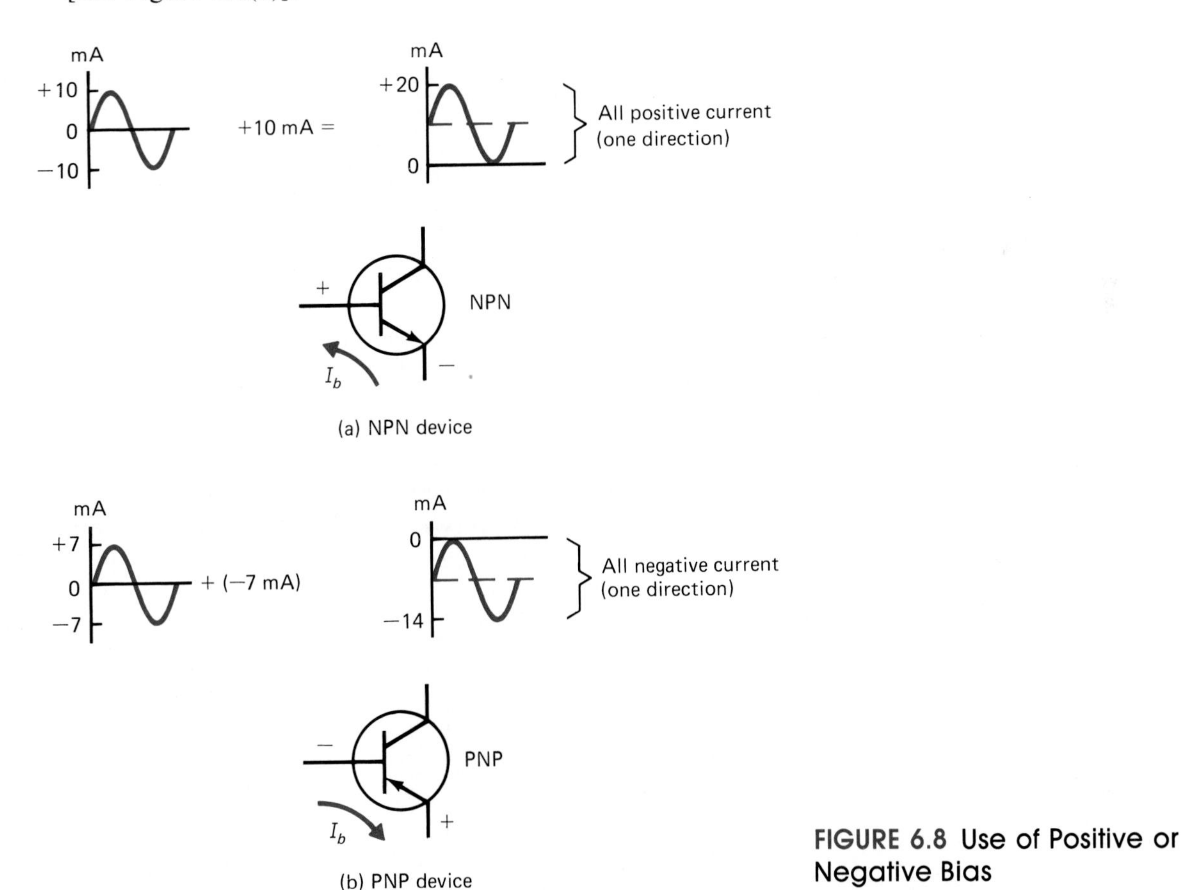

FIGURE 6.8 Use of Positive or Negative Bias

The need to offset the ac input is not the only reason for the base bias circuitry. Just raising the ac to all positive levels (as for an NPN transistor) is often not sufficient. First, some base bias voltage is used to forward bias the device, the standard 0.7-V potential drop on the base–emitter junction. This is also supplied by the dc source. (Once forward bias is achieved, it is not necessary to redo this bias with the information signal.)

Finally, all discrete devices have some minimum input levels that must be exceeded if operation is expected. If the input level falls below this amount, the output will not properly duplicate the input. Such a condition often results in distortion of the input wave. There is a wide selection of BJTs; for example, one of the factors that makes them different is that of the minimum input specifications. For example, when examining the specifications for a specific NPN transistor, it may be found impractical to operate it with base currents below 1 mA. Therefore, the minimum current (dc plus ac) should not drop below that 1-mA level. If it does fall below that level, the output may fail to represent the input accurately. In Chapter 8 these specifications are explored in greater depth.

REVIEW PROBLEMS

(1) If a BJT in the common-emitter arrangement (see Figure 6.3) has a dc base current of 40 μA and $\beta = 120$, what will be the level of the collector and emitter currents?

(2) Sketch a common-emitter PNP transistor circuit, adding the base bias resistor. Indicate the paths for current (and label them) as well as voltages V_{BE}, V_{CE}, and V_{BC}.

6.3 INITIAL INPUT BIAS CONDITIONS FOR THE FET

The FET has a different polarity arrangement than the BJT. There are N-type and P-type FETs, but the FET does not have a forward bias condition as does the BJT. The FET uses a voltage potential at the input (gate lead) to control the output current. An increase in the input voltage creates a reduced conductivity of the source-to-drain semiconductor material.

To bias the FET properly, the polarity of the gate bias voltage should be negative for N-type and positive for P-type. The typical configuration for the FET is the **common-source** mode. In this mode the source lead is at the most ground potential in reference to the drain lead (see Figure 6.9). It will

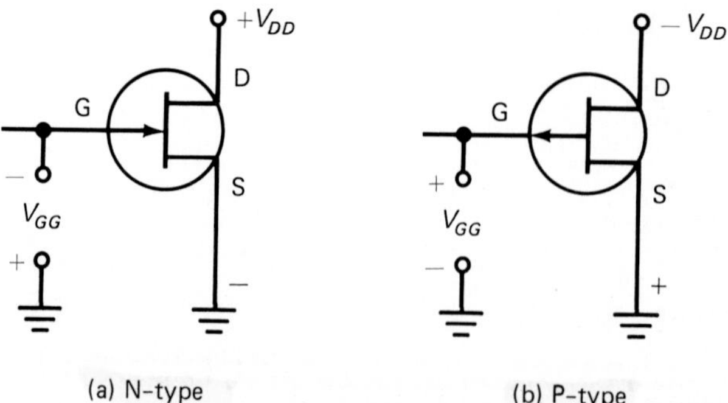

FIGURE 6.9 Bias of JFETs

be noted that the drain lead for the N-type FET is actually at a *positive* potential, opposite that of the gate. For the P-type, the drain is *negative* and the gate is positive. This could present a problem if we attempt simply to bias the gate lead with the supply connected to the drain lead, using the source as a common ground for both. Before we go any further with this analysis, we need to establish some common names for the voltages and currents associated with the FET. The following terminology is illustrated in Figure 6.10.

- V_{DD} Voltage source attached to the drain (using the common-source mode)
- V_{DS} Voltage across the device, the drain in reference to the source
- V_{GS} Voltage at the gate in reference to the source
- I_D Drain current
- I_S Source current
- I_G Gate current (usually not an appreciable amount; ignored in most cases

The problem then is:

1. For the P-type FET, the gate voltage V_{GS} must be positive, but the source V_{DD} must be negative to support the need for negative drain voltage V_{DS} and the resulting current I_D and I_S flowing from source to drain.
2. For the N-type FET, the gate voltage V_{GS} must be negative, but the source V_{DD} must be positive to support the need for positive drain voltage V_{DS}, and the resulting currents I_D and I_S flowing from drain to source.

These are summarized in Figure 6.11. What this all means is that the gate

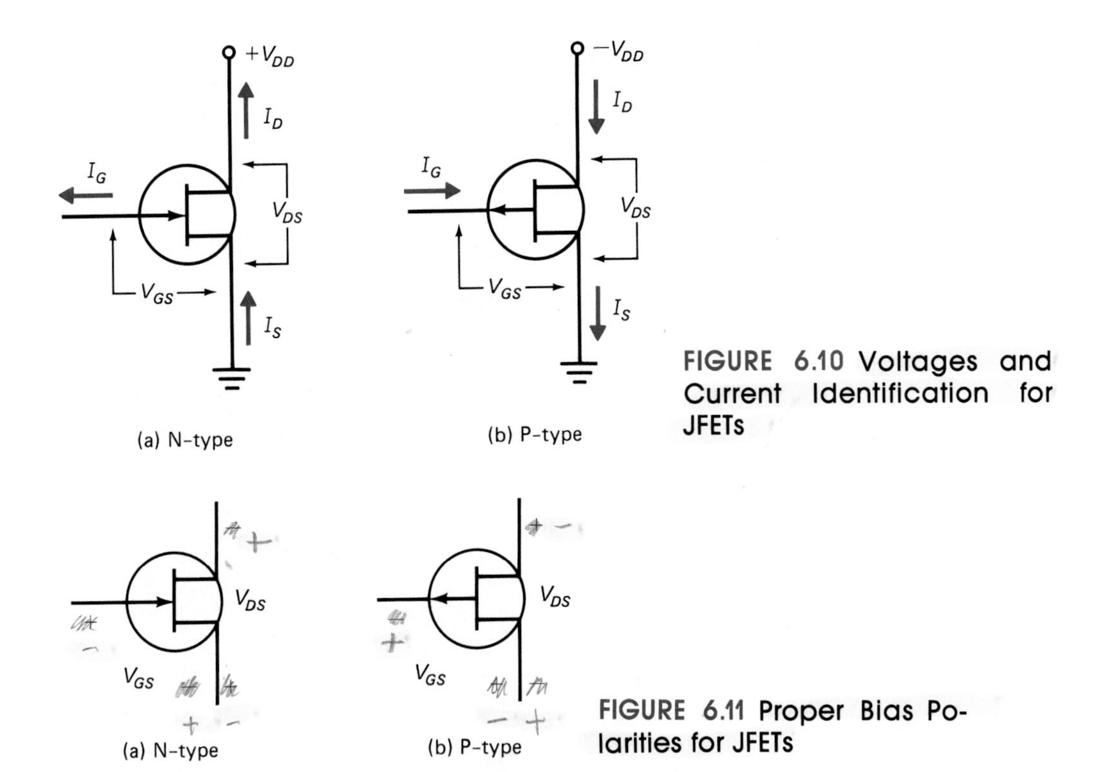

FIGURE 6.10 Voltages and Current Identification for JFETs

FIGURE 6.11 Proper Bias Polarities for JFETs

lead of the FET cannot be biased with the same voltage polarity as that used at the drain (in the common-source mode). Since this is the case, another source, or a means of getting the proper voltage and polarity on the gate, is needed.

First, the most simple solution would be to add another power supply at the gate: a V_{GG} supply. Figure 6.12 shows that possible solution. Such a source would provide the correct amount of bias and the right polarity. The problem would be that such an arrangement would be quite costly and such a source at the gate would make the total resistance seen at the gate be nearly zero (due to the internal resistance of the voltage source). This would defeat the advantage of having the high input impedance of the FET. Also, such a source, if it nearly matched the ideal form of a voltage source, would force the gate lead to stay at that source's voltage, even if a variable input voltage were applied to the circuit. Needless to say, such an arrangement, using a V_{GG} supply, is not acceptable.

Obtaining the Correct Gate Bias Using a Source Resistor

Odd as it may seem initially, by placing a resistor between the source and the ground, the correct *gate* bias can be obtained. First, remember that we need to establish a constant bias potential at the gate for the same reason that we maintained a bias current through the base of a BJT. We need to have the input potential be all of one polarity if it is to affect the drain current totally.

The circuit of Figure 6.13 will be able to achieve the task of having the gate voltage be the correct polarity and still maintain the correct direction for drain current. (There is one element left out of this circuit, a capacitor across the new R_S; it will be added and explained later.) The first element that needs to be explained is the new resistor at the gate, R_G.

Actually, R_G serves no purpose for the actual production of gate bias voltage. Since there is no gate current drawn by the FET, there is no drop

FIGURE 6.12 Application of DC Bias Potentials (JFET)

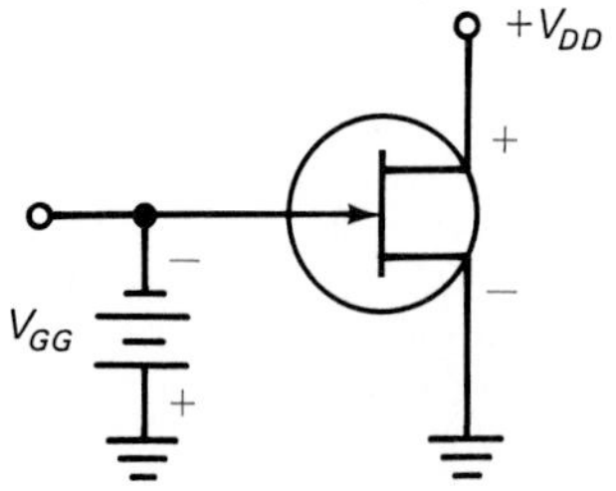

$+V_{DD}$ V_{GG}

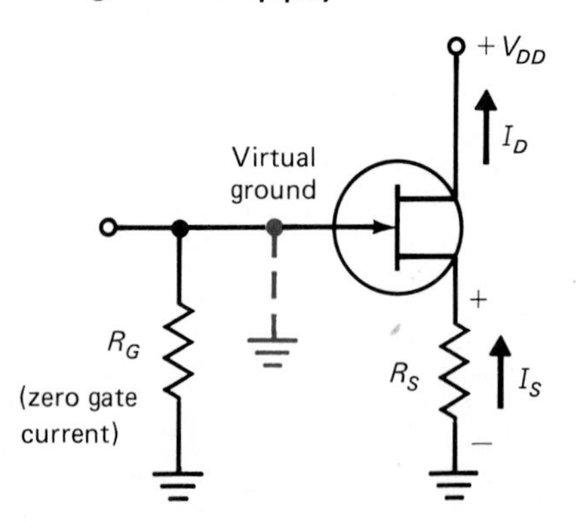

FIGURE 6.13 Establishing Gate Bias Potential Using a Single dc Supply

on R_G; therefore, it is the same as if the gate were connected to ground [Kirchhoff's voltage law: ground + 0 V (drop on R_G) = ground potential]. This resistor needs to be in the circuit for the input signal. If the gate were directly connected to ground, which would be fine for the dc bias condition, any input signal would be grounded through that connection. Naturally, that cannot be done. The R_G resistor, typically in the megohm range, is therefore used to present a high resistance to the input source, instead of ground.

But since that resistor cannot have a current flow from the gate lead of the FET, the gate is at what is called **virtual ground**. Thus to the FET it appears that the gate lead is grounded. Virtual ground is important to the analysis of the gate lead when viewing it from the source lead, since V_{GS}, the potential difference between the gate and the source, is the value of the gate bias (Figure 6.14).

If current flows through the source lead of the FET, a voltage drop will have to be present on the source resistor R_S. That would make the source voltage slightly more *positive* in reference to ground. [For example, if the drop on R_S were 2 V, it would be of a polarity (+) at the source in respect to the (−) ground. Thus the source would be at a voltage of +2 V, in reference to ground.]

Now that the source is at a positive potential to ground, it is also at a positive potential in reference to the *gate lead*. This must be true since the gate lead was just shown to be at virtual ground. Now if we inspect the value of V_{GS}, it is *negative*, since it is a comparison of voltage *from* the gate *to* the source. Therefore, by raising the gate to a more positive voltage than when it was at ground, it makes the gate appear to be more negative (still at the old ground potential). By this subtle manipulation of the voltage potential at the source lead, we have created the same effect as if we had used the separate source at the gate V_{GG}.

The application of this gate bias voltage has the same effect as when we applied base bias current to the BJT. This voltage adds to the input voltage and maintains a singular polarity (either all negative for an N-type FET or all positive for a P-type FET). The exact type of waveform is created at the gate lead that was seen at the BJT's base lead. The waveform produced at the drain lead of the FET is quite similar to that seen on the collector of the transistor. It must be remembered at this point, though, that the FET has an inverse relationship between input levels and output levels. As input voltage increases, output current decreases, and vice versa.

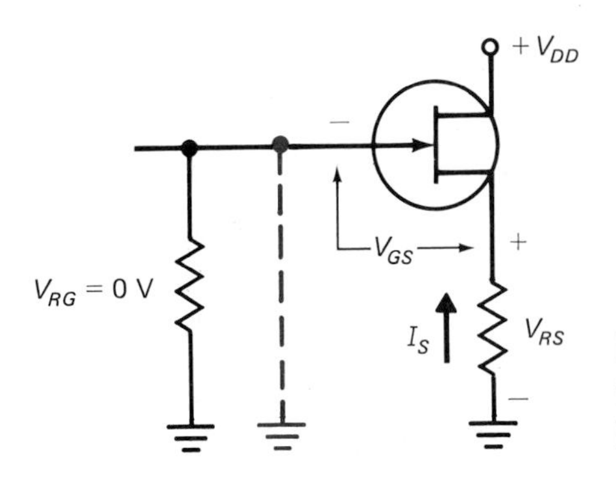

FIGURE 6.14 Establishing V_{GS} through the Use of a Source Resistor

This inversion process simply produces a 180° shift in the phase of the output signal in reference to the output. As the input voltage increases to its maximum peak value, the output current is reduced to its minimum peak value. As input drops to the minimum, output current rises to its highest level. Normally, such an inversion will create no problem in dealing with the output. The exact shape of the input wave is reproduced, but in an inverted form (see Figure 6.15).

One last point must be made about this special FET circuit. As shown in Figure 6.15, a capacitor must be added in parallel with the source resistor R_S. This capacitor C_S is called a **bypass capacitor**. Its purpose is to *prevent* the ac portion of the source current from passing through the resistor R_S. If this ac current passed through R_S, it would create a changing voltage drop on that resistor. If that drop were to rise and fall with the signal, so would the bias on the gate lead. The effect this would have on the circuit would be that some, if not all, of the input level would be canceled by these variations. (This is a case of an undesirable negative feedback—negative due to the inversion through the FET.)

This capacitor would be selected with a sufficiently low reactance to the ac current that a portion of the output (ac + dc) would pass through it to ground without creating a voltage drop that would affect the V_{GS} bias potential. Figure 6.16 illustrates the split of source current—dc through R_S and ac through C_S.

In summary, the FET also needs to have the input "conditioned" to a singular polarity. Input signals that reach the gate lead and are of the wrong

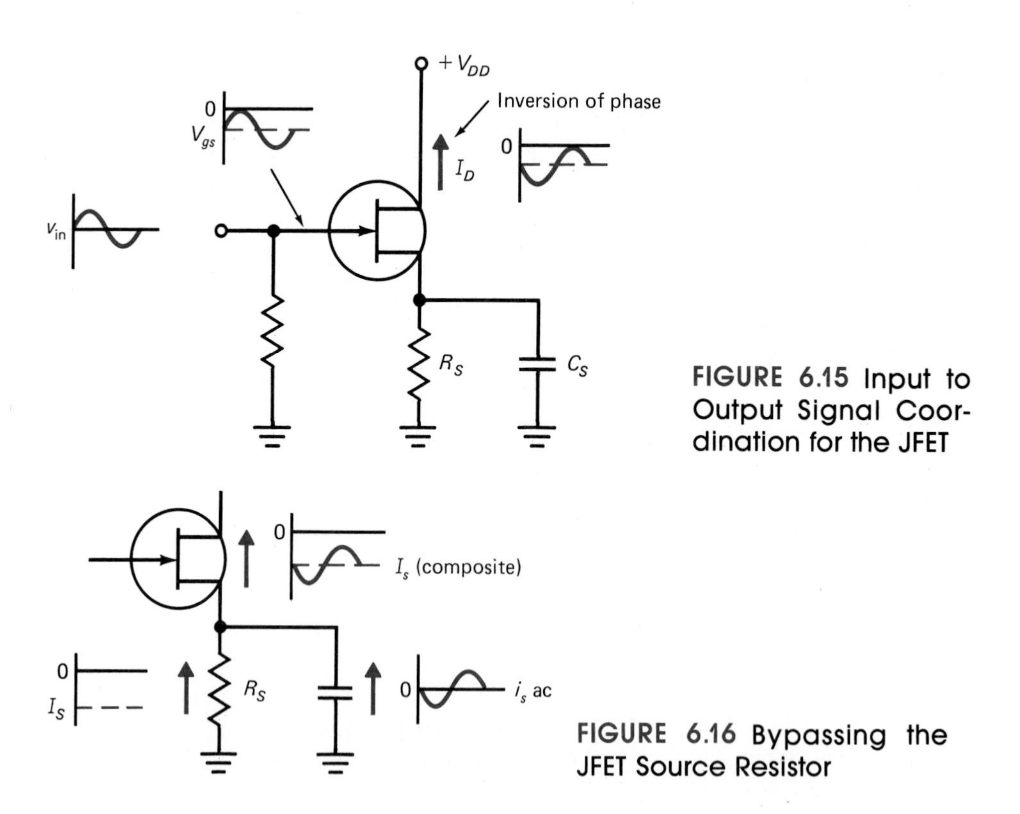

FIGURE 6.15 Input to Output Signal Coordination for the JFET

FIGURE 6.16 Bypassing the JFET Source Resistor

polarity will not be "amplified." Since there is a need to supply one polarity to the gate and a different one to the drain, the drain power supply (V_{DD}) cannot be connected directly to the gate lead. The correct gate-to-source polarity is achieved by manipulating the source lead while keeping the gate at virtual ground. Thus the correct polarity of V_{GS} is determined using only one power supply for the circuit.

REVIEW PROBLEMS

(1) Redraw and label Figures 6.12, 6.14, and 6.15 using P-type FETs instead of the N-types shown.

(2) The same types of conditions will help to determine the correct amount of gate bias potential (1 V, 2 V, etc.) as would be used to determine the correct base bias for a BJT (except the minimum forward voltage on the BJT). Review Section 6.2 and in your own words, restate those conditions for an FET.

(3) It was stated that without the bypass capacitor across R_S (see Figure 6.15) the input signal could be canceled by the changes in voltage on R_S. Illustrate this possible effect by using the following data (be certain to include the inversion through the device in your analysis):

input voltage = 4 V p-p   dc voltage on R_S = 3 V

R_S = 100 Ω

output current $i_D = i_S$ = 50 mA max., 30 mA min.

6.4 OUTPUT RESISTORS

So far we have concentrated on the input side of both the BJT and the FET. Now we will take a look at the output side of these devices. Since the BJT and FET produce the same form of output and current in response to an input, we can deal with the use and manipulation of these outputs in a similar manner. The one major difference between these two devices' outputs is that the BJT's output current flow is directly proportional to the input (as input rises, so does output); in contrast, the FET has an inverse relationship between input and output (increased input produces *decreased* output). This difference will create no real problem when discussing how to handle the output.

Up to this point we have not used the output of these devices to do any productive work. All that has been observed is that these devices produce a variable current flow from the power supply. Unless these variations are used by some other device (resistor, speaker, another amplifier), they serve no productive purpose. Therefore, it will be necessary to attach a **load** of some type to these amplifiers. A load is any device that can use the output for some productive means. Since there are a wide variety of types of loads that could be used, we will use a resistor to represent "any" type of load device, since most loads will have a voltage-to-current relationship much like a resistor. When a load is attached to these amplifiers, we see several effects occurring.

First, we need to determine where the load is to be attached. Typically, the load would be attached in such a way as to be able to sense the output quantity (current in the case of the BJT and FET). Therefore, it will be necessary to place the load resistor in the path of output current flow. For the BJT that could be in series with either the collector or the emitter; for the FET, in series with either the drain or the source lead. Figure 6.17 illustrates the most popular output connections for these two devices.

In Figure 6.17 the output current for the BJT, I_C, must pass through the load resistor as it passes through the transistor. For the FET, drain current passes through its load resistor in much the same way. Since the resistor on the BJT is attached to the collector, it is typically called a collector load resistor and is given the notation R_C. For the FET, the drain resistor is called R_D. In either case, the output current flows through this resistor as it is controlled by the device.

At this time it may be thought that the addition of load resistors to these devices would cause a change in the expected output current. In an ordinary series circuit, the addition of a resistor should cause a decrease in current flow through that circuit. That is not the case with the BJT and FET. They control the value of output current. Remember, their internal resistance (or conductance) is not a fixed value but is variable. They compensate for the added resistance by lowering their output resistance to maintain a constant output current flow. Figure 6.18 illustrates this point.

What does change in the circuit with the addition (or change in value) of a load resistance is the voltage dropped on the transistor (BJT or FET). Without the load resistor, the devices dropped all the supply voltage. With the load resistor attached, the supply voltage is divided between the device and the load. The amount of voltage on either element is easily calculated.

Simple Ohm's law calculation where I is the output current:

$$V_{\text{load}} = R_{\text{load}} \times I \tag{6.2}$$

Simple Kirchhoff's voltage law calculation where V_{CC} is the dc source in the output circuit:

$$V_{\text{device}} = V_{CC} - V_{\text{load}} \tag{6.3}$$

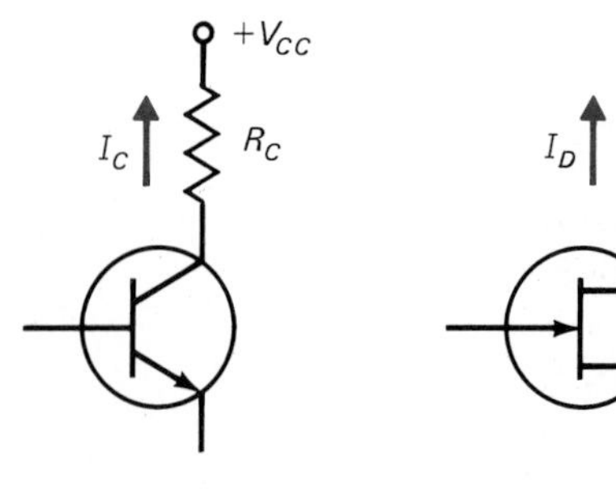

(a) BJT

+V_{DD} I_D R_D

(b) JFET

FIGURE 6.17 Use of Output Load Resistors

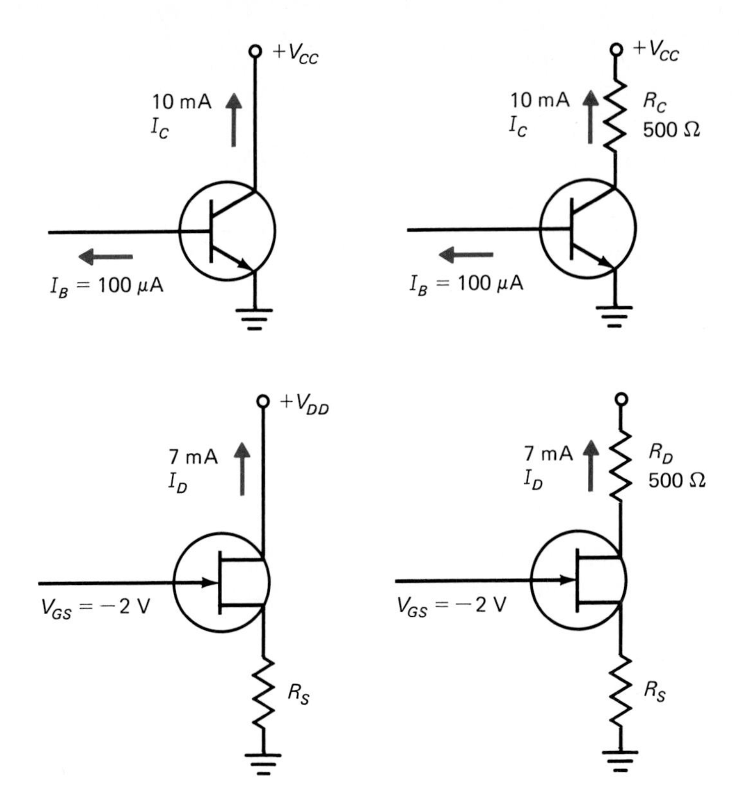

FIGURE 6.18 Output Current is Controlled by the Active Device, Not the Load

What is important to remember is that the BJT or FET is in control of setting the value of output current. Output current is therefore *not* calculated by using the traditional V_{CC}/R_{total} equation. In fact, since these devices adjust their internal conductivity with changing input levels, there is no one fixed value of internal resistance for them; thus a calculation of $R_{total} = R_{device} + R_{load}$ would not be possible.

Were the load resistor to change in value in either of these circuits, the device (BJT or FET) would readjust internally so that the same level of current was flowing at the output as had been flowing before the change. What will change is the voltage divider between the device and the load. Figure 6.19 illustrates such a change using a BJT and three different load resistors. As can be seen in Figure 6.19, the BJT has set the current at a constant level regardless of the loads that were attached. This current, through the three different load resistors, produced three different divisions of the supply voltage between the resistors and device.

Constant Output Current Effect It should be remembered at this point that the BJT was matched with the ideal current amplifier and the FET with the ideal transconductance amplifier. Each of these ideal models would produce a constant current output, regardless of voltage levels.

There is one case where the load resistor will affect the output current.

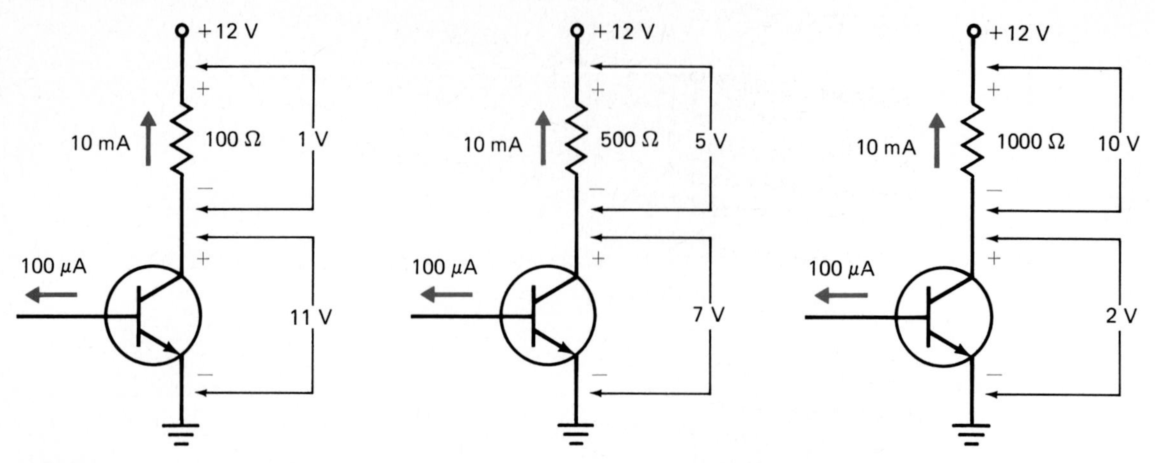

FIGURE 6.19 Voltage Division between the BJT and Collector Resistor

If the transistor (BJT/FET) attempts to produce a current flow that is too high for that resistor to support in coordination with the power supply, that resistor will limit current flow. Basically, the limit to the flow is

$$I_{\max} = \frac{V_{CC}}{R_L} \tag{6.4}$$

where $I_{\max}$ is the maximum output current and R_L the load on the BJT/FET.

The "worse-case" voltage divider for the system would be where all the supply voltage was dropped on the load resistor and none was left for the BJT/FET. In such a case, the transistor would display maximum conduction (lowest resistance). Thus the resistance in this circuit would be (approximately) the value of R_L. Current would then be limited to the value given in equation 6.4. Figure 6.20 illustrates the condition where a BJT's output current is being limited by a load resistor.

One other noticeable effect can be observed when adding a load resistor to each of these devices. The voltage drop on the load resistor is directly proportional to the current passing through the transistor (BJT/FET). If current increases (as long as it is below the limit set by the resistor), the drop on the

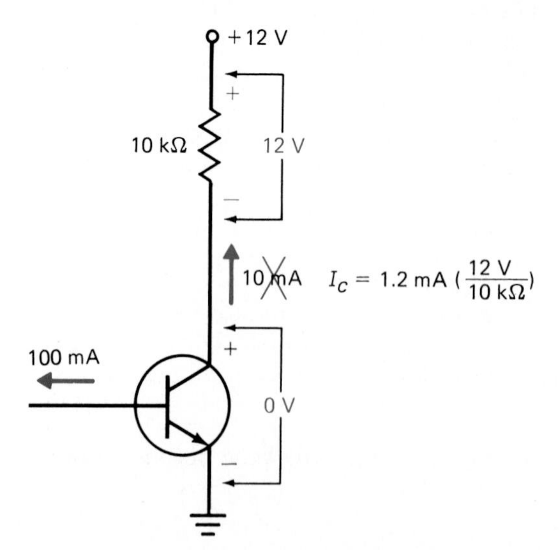

FIGURE 6.20 Collector Resistor Placing a Limit on the Maximum Output Current

load resistor will increase. Decreases in current will create a lower voltage on the load. Since this voltage is now proportional to the current, the current level can be indirectly observed by monitoring the voltage on that resistor. In essense what has happened is that a device that produces current output has now been placed in a circuit where both voltage and current can be seen as output quantities.

Figure 6.21 illustrates this point. If the input current to the BJT were to be an offset ac voltage, the output current would also vary. With these changes in current, the voltage on the collector resistor would also vary. These waveforms are shown in the illustration. As a result of the changes in voltage on the collector resistor, there would be changes in the voltage across the device's collector-to-emitter leads. Basically, this is verified by Kirchhoff's voltage law, which (in one form) states that the total voltage (V_{CC}) applied must equal the sum of the two drops ($V_{RC} + V_{CE}$).

It is important to note that there is a special phase relationship between all of these quantities (for a BJT):

1. Output current I_C is in phase with input current I_B.
2. The voltage on R_C is in phase with the current through it, I_C, which makes it in phase with the input current I_B.
3. The voltage on the device, V_{CE}, is 180 ° out of phase (inverted) in comparison to the voltage on R_C, since $V_{CE} = V_{CC} - V_{RC}$. Thus it is 180° out of phase with the input I_B. Thus the collector-to-ground (circuit output) voltage will follow this phase shift.

One special point should be noted. Only the changing (ac) value of the input is inverted. The dc level at the collector resistor (R_C) and across the device (V_{CE}) are *always* of the same polarity as the source V_{CC}, since that source supplies the operational voltage for the output.

For an FET, the relationship is about the same, but the initial inversion seen through the device must be taken into account. It should be remembered that if input increases on the FET, the output decreases, and vice versa. Thus the output current is initially 180° out of phase with the input voltage. The basic breakdown of the output is as follows, for an FET:

1. The output current is 180° out of phase with the input voltage.
2. The voltage on the drain resistor, V_{RD}, coordinates with the output current, thus making it also 180° out of phase with the input voltage.

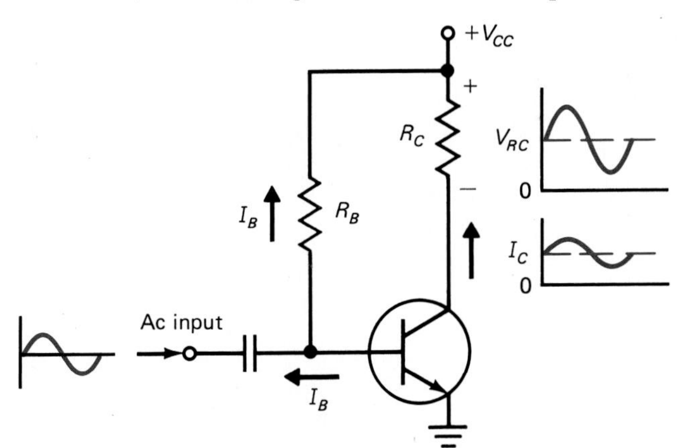

FIGURE 6.21 Production of AC Output Voltage on the Collector Resistor

3. The voltage on the device, V_{DS}, is also the result of supply voltage minus the drop on R_D. Therefore, if the voltage on R_D is low, the voltage on the device is high. This creates a device voltage V_{DS} that is back in phase with the input. (Input peaks *high*—creating a low output current—lowering the drop on R_D—creating a *higher* drop on the device!)

Note: To assist in understanding this process, observe the waveforms illustrated in Figure 6.22.

Note: The dc voltages at the output, just as for the BJT, are not affected by the inversion. They will match the polarity of the controlled source V_{DD}.

One interesting point can be observed from this summary. The actual output of such a circuit can be monitored in *three* ways. First, the flow of output current (the true controlled output of the device) can be monitored. Second, the voltage on the collector (BJT) or drain (FET) resistor can be seen to follow the output quantity. Finally, the voltage across the device (V_{CE} for a BJT and V_{DS} for an FET) will represent this output as well.

At this point it is interesting to note that these two amplifiers, BJT (common emitter) and FET (common source), can now take on the appearance of two other amplifier models. If the output current of the BJT is used as the prime output, it is a current-in-current-out amplifier, which matches the current amplifier model. If the output voltage (on either R_C or the device) is monitored, it is a current-in-voltage-out amplifier, which matches the transresistance amplifier model.

For the FET, it is initially a transconductance amplifier (voltage-in-current-out), but if its output voltage is monitored and a voltage-in-voltage-out ratio is observed, it becomes a voltage amplifier.

Thus both amplifiers can take on either of two forms:

		Type of Amplifier if Monitoring:	
Device	Input Quantity	Current Output	Voltage Output
BJT	Current	Current	Transresistance
FET	Voltage	Transconductance	Voltage

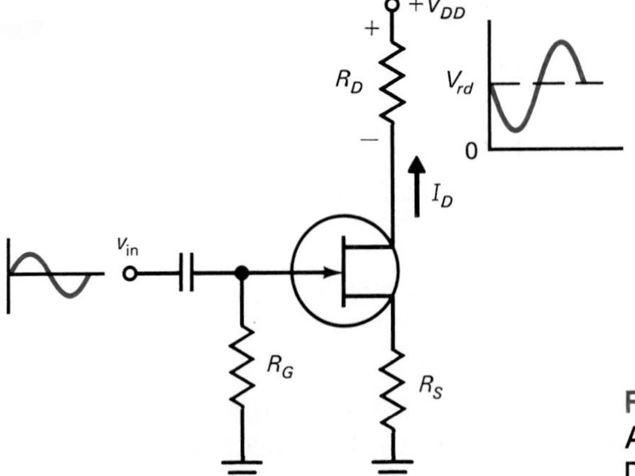

FIGURE 6.22 Production of AC Output Voltage on the Drain Resistor

Neither of these derived forms of amplifiers matches the ideal model, but simply approximates its function. The voltage output of these devices is *not* fixed or constant. It will always depend on the value of the load resistor. Therefore, they will not meet the ideal conditions of having constant voltage outputs under varying load conditions. The actual value of output voltage will not be set by the device, but by an Ohm's law calculation based on the load and output current. Therefore, from each device, with proper manipulation of the output quantity, we can duplicate two of the four forms of amplifier models. By the addition of an R_C on a BJT or an R_D on an FET, output voltage now varies in relationship to the input quantity.

These resistors, R_D and R_C, serve another purpose besides acting as voltage dividers. As noted before, they place a limit on the maximum current through the device. In many cases, this is a necessary limitation to prevent the device from overheating and becoming destroyed. The BJT especially is sensitive to this "overcurrent" problem.

REVIEW PROBLEMS

(1) Given a BJT circuit with $\beta = 100$, $I_B = 40\ \mu A$, $V_{CC} = 12$ V, and $R_C = 500\ \Omega$, find the maximum limit to collector current flow, $I_{C(\max)}$.

(2) For the same system as in problem 1, if the base current (dc) were to be the following values, find the value of collector current that would result. (Note your calculation for problem 1.)
(a) $I_B = 100\ \mu A$ **(b)** $I_B = 200\ \mu A$ **(c)** $I_B = 500\ \mu A$
(d) $I_B = 1$ mA

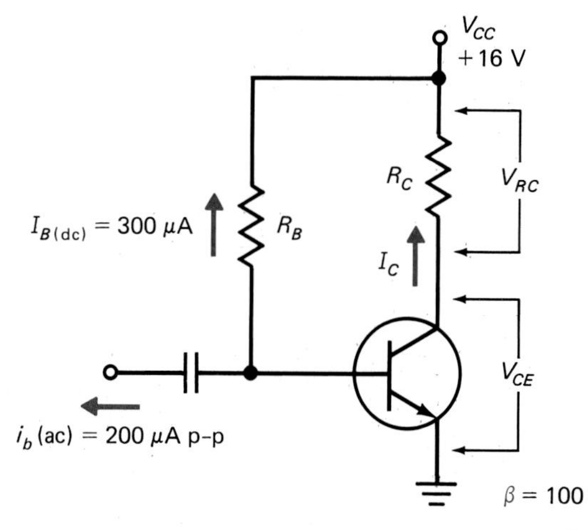

FIGURE 6.23 Coordination of AC and DC Voltages for a Simple BJT Circuit

(3) Using the BJT circuit shown in Figure 6.23 with the data shown in that figure, find the following values.
(a) Output currents: I_C (dc) and I_C (ac) p-p
(b) Voltage on R_C: V_{RC} (dc) and v_{RC} (ac) p-p
(c) Voltage on the device: V_{CE} (dc) and v_{CE} (ac) p-p

Inside the BJT, there is no basic limit to the amount of collector-to-emitter current that could flow except that if too much passes through the device, it overheats and is destroyed. Therefore, if there is a high base current, the collector current could be pushed so high as actually to destroy the device.

Each model of the BJT has, as one of its specifications, an upper limit to current that should not be exceeded. Often, a collector resistor is selected so that the current flow in that side of the circuit is restricted at or below that limit. In an FET, such a limit normally does not exist. Maximum current flow will be limited by the FET itself. But for both devices, the R_D and R_C resistors are usually kept as a functional part of the basic amplifier circuit. They are often not the actual loads that use the output to perform a function. They serve as artificial loads on the amplifiers, simply performing the role of voltage division and, for the BJT, current limiting.

An actual load, such as a speaker or another stage of an amplifier, is usually attached to the circuit *in addition* to these resistors. Since this section of text deals with just single stages of amplification and the handling of bias and signal inside such amplifiers, the process of attaching an *external* load will be presented in a later section.

6.5 STABILIZATION OF DISCRETE DEVICES

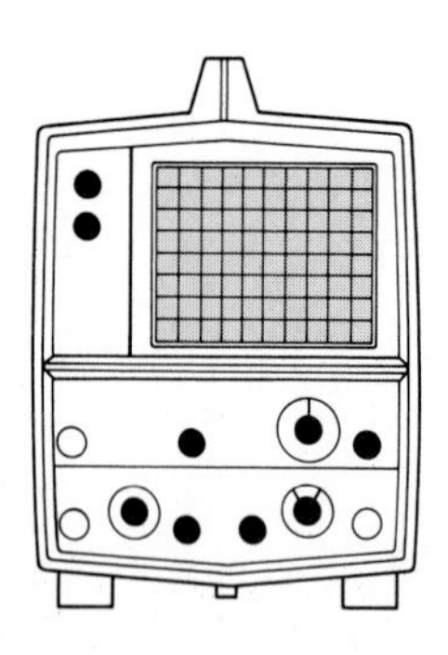

In past chapters it was noted that both the BJT and the FET had problems of instability. The most common cause of circuit instability is temperature. The BJT is especially sensitive to temperature variations. When the internal temperature of the device increases, the gain (beta) of the device, together with certain other parameters, will also increase. This creates a serious problem since increased values of beta will cause collector current to increase. Instability is defined in terms of these undesirable (and often unpredictable) changes in output current.

If collector current were to increase, further heating of the device would occur. This heating is a by-product of the power consumption of the device. (Greater collector current produces greater device power, since power = $E \times I$.) This process could continue—rising current producing rising heat, which produces more current—until the device is destroyed. In some cases, this wild increase in collector current and temperature is totally out of control and device damage is certain; such a process is often called **thermal runaway**. Even under conditions where collector current would be limited by the use of a collector resistor (R_C), this change in current due to heat would tend to distort the output waveform. Therefore, some level of control over the device is needed.

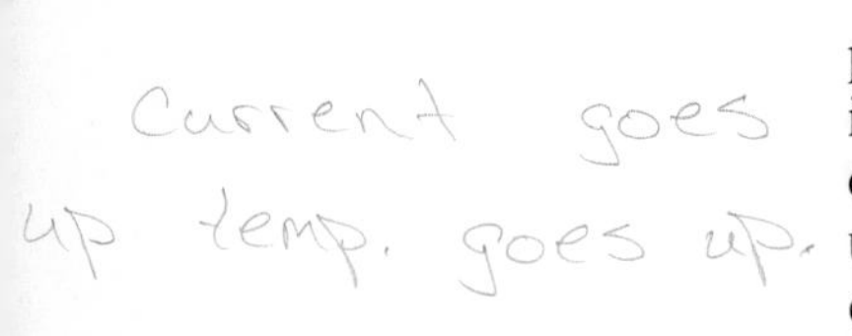

In the FET the relationship of temperature to heat is opposite that of the BJT. As temperature increases, current tends to decrease. Therefore, the FET is less likely to become damaged by moderate overheating. Even though the variations will not cause thermal runaway, they could still tend to distort the output waveform.

Another problem exists that is solved when stabilization is added to a BJT or FET circuit. Just like resistors, capacitors, and inductors, transistors (BJT and FET) are manufactured with tolerance values. For example, the beta of two BJTs with the same part number may be quite different (say, 100 for one and 80 for the other). These devices could not be considered interchangeable in a circuit that depended totally on the value of beta for operation. A stabilized circuit would accept such a substitution and still maintain normal operating conditions.

In this section we explore popular circuits used to produce a stable output, one that will not depend on minor variations in device parameters. Since the BJT is more susceptible to instability and because later we will see that the basic FET circuit has some built-in stability factors, we concentrate on stabilization of the BJT circuit.

Before we can see a solution for instability, we need to investigate some of the results of instability. (From now on the term "instability" will represent both the type of instability caused by overheating and the problem created by device substitution. Since both forms involve a changed gain figure, solving for one problem will also solve for the other.) When a BJT becomes unstable, a radical change in the output current results. This change makes the device undependable and not match the ideal models used to represent the devices function. Since the dc bias current at the input is held at a constant value, the input is insensitive to the errors that show up only at the output.

If some form of feedback could be employed, where the changes in output current would be reflected back to the input and adjust the input current to compensate for the error, greater stability could be achieved. Negative feedback from the output can be applied to the input to provide greater stability.

In Figure 6.24 an emitter resistor R_E has been added to the simple BJT circuit. This resistor provides stability against variations in the device's beta. It provides negative feedback of the output current via the voltage drop on the new emitter resistor. This causes inverse variations of base current which counteract the changes in collector current caused by the undesirable change in device beta. (For a more detailed analysis of this process, see Chapter 7.) This circuit therefore provides a simple form of stability through the use of negative feedback. This circuit form is a basic building block for more sophisticated stabilization circuits and is the simplest form of common-emitter BJT circuitry in popular use.

A variation on this circuit, one that produces even greater stability, is called **beta-independent** stability biasing. In such a circuit, the values of voltage and current are forced to be specific values through the design of the support

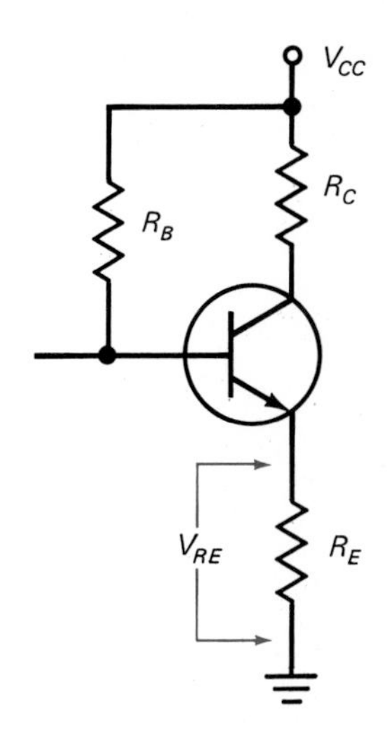

FIGURE 6.24 Voltage Drop on an Emitter Resistor

circuitry (resistors and capacitors). This second form of stability biasing is one of the most popular ways of biasing the BJT.

The circuit shown in Figure 6.25 illustrates a beta-independent stabilized circuit. This circuit forces the value of I_C to be one specific value. The I_B is actually not controlled by the base resistor but is set by the demand for the production of I_C by the device.

It may be better to view this beta-independent circuit more as a beta "forcing" circuit. The two base resistors (R_{B1} and R_{B2}) form a simple voltage divider. The voltage produced on R_{B2} due to this divider produces a corresponding voltage on the emitter resistor (R_E), minus the transistor's forward drop V_{BE}. This voltage on the emitter cuases a fixed value of emitter current via the emitter resistor (R_E). Since collector current is assumed to be equal to emitter current, the value of I_C is thus preset. Therefore, independent of the device's beta (or other operational characteristics) the dc collector current is predetermined by the R_{B1}–R_{B2} divider circuit. Because of these effects, the equation for current gain becomes

$$A_i = \frac{R_{B2}}{R_E} \quad \text{(for ac input signals)} \tag{6.5}$$

Although the device's beta does not set the value of current gain, beta does play a role in the function of the circuit. For this circuit to operate properly, the value of the beta for the transistor must be greater than the demand for gain placed on it by the rest of the circuit (R_{B2} and R_E). Should the beta of the transistor be less than the value predicted by the current-gain equation, the circuit will malfunction by adopting this lower gain value.

Therefore, there is a major disadvantage in using this stabilized circuit. The selection of R_{B2} and R_E must be such that the gain of the circuit is always *below* the worst-case gain of the transistor in use. If the circuit gain must be below that of the transistor, we are not getting the most potential out of that device. If, for example, a transistor with an average beta of 120 was used in a stabilized circuit, the total gain for that system may be only 40 or 60, not even close to the 120 potential of the device by itself.

In addition to this drawback, there is the question of efficiency. An amount of energy from the power supply (V_{CC}) is used to operate the R_{B1}/R_{B2} voltage-divider circuit. This energy consumption is not used to produce an output but is lost (as heat) in order to provide stability. In addition to the

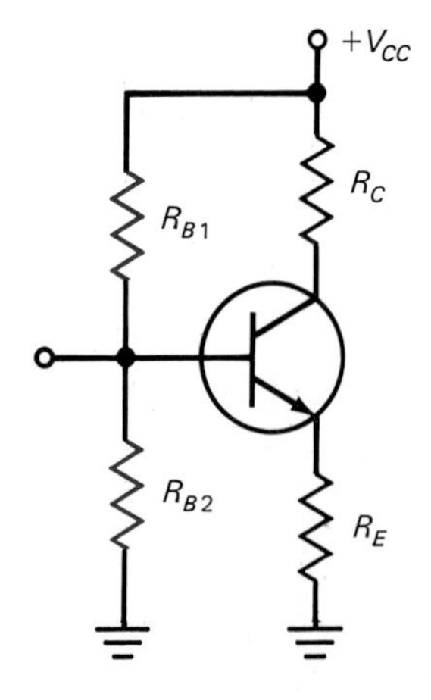

FIGURE 6.25 BJT Bias Circuit

individual circuit losses, additional stages of amplification will have to be added to the total system (multistage amplifiers) to make up for the reduction in potential gain ($A_i \ll$ device beta). Even with these drawbacks, this is one of the most popular bias arrangements, mainly because it is so stable and accepting of devices with different betas.

Although current gain (A_i) is primary to the BJT, which is a current-gain device, three other quantities are important to the application of the BJT circuit. Voltage gain, input impedance, and output impedance are very desirable ac parameters of these circuits. In many cases, an amplifier is selected for its voltage-gain (A_v) characteristic since often input quantities are easily measured as voltage.

If the amplifier is to be matched to input and output elements which display values of resistance/ac impedance in order to increase efficiency through maximum power transfer, the input and output impedance of the amplifier are also important quantities. Fortunately, in their simplified forms, the equations for these three quantities are based on resistance quantities within the circuit and are therefore quite simple to calculate. These equations are:

$$\text{Voltage gain: } A_v \simeq \frac{R_C}{R_E} \tag{6.6}$$

$$\text{Output impedance: } Z_o \simeq R_C \tag{6.7}$$

$$\text{Input impedance: } Z_i \simeq R_{B1} \parallel R_{B2} \parallel \beta \times R_E \tag{6.8}$$

Note: In the equation for input impedance (Z_i) it can be seen that the value of R_E, as seen from the input side of the circuit is "magnified" by the beta of the device. This is due to that resistor's effect on emitter current, which, by *division* of beta, affects base current. In many cases this parallel equation for input impedance is simply $Z_i \simeq R_{B2}$, since the value of R_{B1} and $\beta \times R_E$ are often dramatically larger than R_{B2}.

EXAMPLE 6.2

Given the following circuit values for a beta-independent circuit, calculate A_i, A_v, Z_i, and Z_o.

$$R_{B1} = 20\ \text{k}\Omega \qquad R_{B2} = 2\ \text{k}\Omega \qquad R_E = 100\ \Omega$$

$$R_C = 1\ \text{k}\Omega \qquad \text{transistor's beta} = 120$$

Solution:

$$A_i \simeq \frac{R_{B2}}{R_E} = \frac{2\ \text{k}\Omega}{100\ \Omega} = 20$$

$$A_v \simeq \frac{R_C}{R_E} = \frac{1\ \text{k}\Omega}{100\ \Omega} = 10$$

$$Z_i \simeq R_{B1} \parallel R_{B2} \parallel (\beta \times R_E) = 20\ \text{k}\Omega \parallel 2\ \text{k}\Omega \parallel (120 \times 100\ \Omega) = 1579\ \Omega$$

(An assumption of $Z_i \simeq R_{B2} = 2\ \text{k}\Omega$ would be reasonable.)

$$Z_o \simeq R_C = 1\ \text{k}\Omega$$

Thus it can be seen that calculation of all four significant quantities for these circuits is quite simplified when using the beta-independent circuit. These calculations can also be applied to the simpler emitter resistor feedback circuit by substituting the value of the single base resistor in equations that use R_{B2} or R_{B1} and R_{B2}.

In the FET circuit, it was noted that stabilization may not be as much of a problem as it is for the BJT. There are two reasons for this. First, the FET will tend to be a bit more self-stabilizing since the FET will tend to reduce output current if it overheats, thus avoiding thermal runaway. Also, in establishing the special gate bias voltage, using R_S, feedback for stability is already achieved. This duplicates the stabilization seen in the BJT circuit, which uses the simple emitter resistor feedback connection.

Just as stability is achieved using an emitter resistor in the BJT circuit, the source resistor that establishes gate voltage potential creates a feedback path for the FET (see Figure 6.26). Since this special arrangement has been explained before, only the negative-feedback process to compensate for stability problems will be discussed here. For most applications, this is the extent to which the FET needs to have dc stabilization.

Special note for FET circuits: It is possible to see an arrangement of gate resistors (an R_{G1} and R_{G2}) that may look a lot like this bias arrangement. If this combination is used with a MOSFET device, the same form of stability biasing is being employed. But if these resistors are used with a JFET, this form of bias is *not* being employed. Such a divider on the JFET gate lead would potentially cause the gate to be biased with the wrong polarity. Instead, these two resistors are selected so that the divider still places the gate at a more ground-like potential than the source ($V_G < V_S$).

All of these circuit forms use negative feedback to stabilize the discrete device, the R_E for a BJT and the R_S for an FET, fit into one classification of feedback (as presented in a previous chapter). Each of these forms senses output current and feeds back a voltage that is proportional to that current. Each of these is of the series–series feedback form. As a review of this process:

1. Output current is sampled (R_E or R_S in series with output current).
2. That current is converted to voltage (Ohm's law for each resistor).

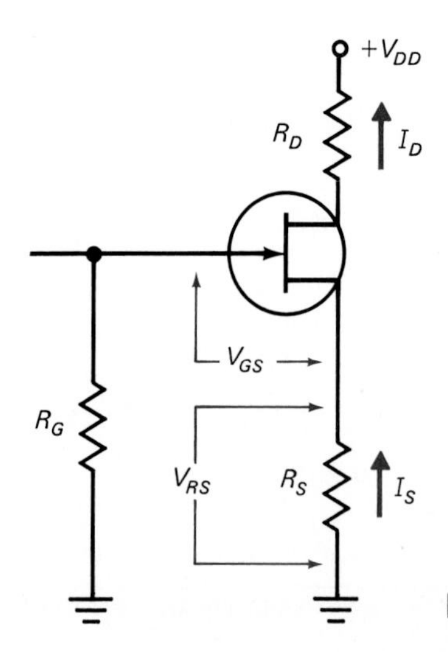

FIGURE 6.26 JFET Bias Circuit

3. The voltage is of an opposite polarity from the initial bias voltage (for the BJT by the subtraction of voltage on R_E from the voltage on R_B, for the FET by the inverse relationship of input voltage to output current), creating negative feedback.
4. This feedback is algebraically (by polarity) subtracted from the input as a series value of voltage.

The exception to this process is the use of the base voltage divider (R_{B1} and R_{B2}) for a BJT. These resistors do not constitute a feedback process, but instead create a forcing function that makes the output current constant. The device adjusts its draw of base current to compensate for the amount of collector current demanded.

There is one final popular negative-feedback form used for BJT transistors. (Actually, this form is also popular with op-amps, which will be shown later.) This method is called **voltage feedback**. In such a process, a sample of the voltage on the collector (to ground) voltage potential is fed back to the base of the device to actually set the value of base current. Such a circuit does not use a standard base bias resistor (R_B). In this process (see Figure 6.27), the voltage that is dropped on the collector of the transistor is used to provide the potential that, through the feedback resistor, creates base current I_B. With this configuration the value of current gain becomes (again) a ratio of resistances:

$$\text{current gain } A_i \simeq \frac{R_F}{R_C + R_E} \tag{6.9}$$

(Again omit R_E if it is not used in the circuit.)

The current gain of the circuit is therefore the ratio of the feedback resistor to the resistance seen on the output side of the circuit. In this process, dc output voltage is sampled and used to create the value of base current. Since variations in output bias voltage will create variations in the base bias current, this form of feedback fits in the category of **shunt–shunt feedback** (voltage sampling–current feedback). As in other stability biasing circuits, the potential gain of the device must exceed the gain forced by the controlling circuit.

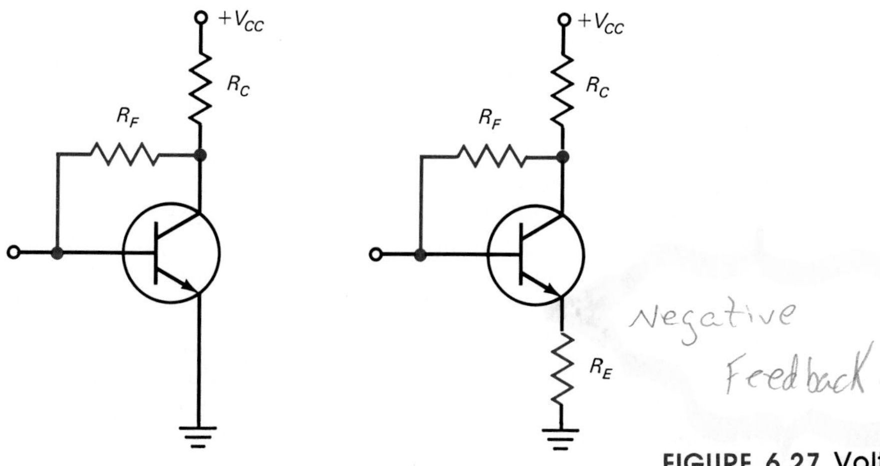

FIGURE 6.27 Voltage Feedback Resistor for a BJT Circuit

Concluding Remarks

In this section several popular forms of stabilization circuitry have been covered. The major concept about stabilization is that it is always necessary to sample the output bias conditions and use that sample to control the input bias quantity in order to produce stability. The loss of potential gain from the semiconductor device is traded for a more stable and thus more dependable device.

Later, when op-amps are explored, it will be seen that such stabilization is integrated into the op-amp circuitry. Inside the op-amp there are circuits much like the ones presented in this section which eliminate the need for external stabilization circuitry.

It is a rare case where one of the discrete transistors (BJT or FET) will be used without applying some form of stabilization. Most circuits used from this point on will include stabilization elements. Therefore, it is important to understand the objective of stabilization and just how it can be achieved within the discrete transistor amplifier circuit.

REVIEW PROBLEMS

(1) Prepare a chart that includes an illustration of the stabilization circuits shown in this section. With these illustrations, write a brief description of how stability is achieved.

(2) With all of the stabilized circuits it was noted that the BJT used had to have a gain value that always exceeded the circuit's "forced" gain. Write a brief discussion of what could happen, in any one of these circuits, if the BJT could not support the forced gain value.

(3) If too much feedback is used in any of these circuits (again the voltage-divider bias circuit is not included), the actual value of gain could be so low that the circuit will provide almost no amplification. Using an emitter resistor for feedback (like Figure 6.24), illustrate how too much feedback could actually reduce gain below unity (1).

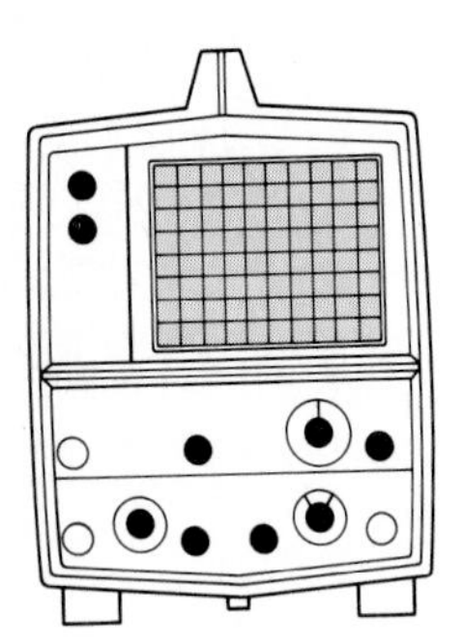

6.6 FAULT DIAGNOSIS AND TROUBLESHOOTING STABILIZATION CIRCUITS

The use of stabilization in BJT or FET circuits initially provides circuits which are less prone to fault and are more tolerant of device errors than circuits that do not use stabilization. One of the greatest advantages of these circuits from a repair point of view is that they are far more tolerant to the use of replacement or substitute devices. It is not necessary to find a replacement that precisely matches the device being replaced. Naturally, maximum limit specifications (such as maximum allowable collector current) must be matched or exceeded by the replacement device. The device must be able to produce the gain demanded by the circuit, not to match the typical (or maximum) gain of the transistor being replaced. Many manufacturers produce a device substitution list which can be used to "cross-match" one device (part number) which meets or exceeds the specifications of the device being replaced (potentially with a different part number). Using a stabilization circuit will then permit this substitution without affecting the overall gain of the system.

Since stabilization circuits provide more consistent gain and operational conditions, general device gain variations, such as would be caused by temperature changes, will usually not produce circuit malfunctions. The following is a list of the common faults that could occur in such circuitry:

1. *Device failure.* If the transistor fails to function, the circuit cannot compensate for this fault. An open base–emitter junction in a BJT or open source condition for a FET are two common forms of device failure.
2. *Error in minimum device gain.* If the device used or the environmental conditions in which it is used (temperature) produces a gain characteristic for the device that is below the demand of the stabilization circuit, the total gain will have to follow this lowered value. The use of a defective device, use of an improper replacement device, or application in an "unplanned" environment could create such a fault.
3. *Excessive device gain.* Although stability biasing can overcome most device gain variations, when the device's gain becomes abnormally high, instability may be introduced. This is due to the fact that stability circuits cannot produce 100% control. It is possible that if device gain was too high, the level of feedback may come close to exceeding a nominal bias level at the input (subtracting too much feedback from the input bias condition). Such a problem could create distorted output signals or ultimately damage the transistor. The causes for this form of fault are similar to those seen in item 2.
4. *Failure of stabilization elements.* Although simple resistors are used to provide stability, they are subject to failure. If the are either far off from their intended value or open/short due to failure, stabilization may become nonexistent, or they may create electrical conditions that could result in damage to the device. Although it is not too common to have resistors open or short, their connections (improper solder joints on printed circuit boards, for example) could produce these faults. Usually, there is enough flexibility in the initial circuit design that simple tolerance variations for these resistors will not create problems in the stabilization process.

Troubleshooting stability-biased circuits is usually not complicated. The objective of troubleshooting would be to isolate the source of the failure so that correction or replacement of defective components could be done. In most schematics, the normal operational values for dc bias voltages are stated for each amplifier circuit. The following descriptions show typical values for these voltages as a basic guide should they not be listed with the schematic.

Note: Most circuit voltages are measured in reference to circuit ground. This is done to save time, as one meter lead can be connected to ground and the other simply used as a probe to test each potential in reference to ground. Usually, voltages across resistors directly connected to the power supply are not measured. If it is necessary to measure these resistor voltages, Kirchhoff's voltage law can be used to determine the correct value for comparison.

Troubleshooting the Simple Stabilized BJT Circuit (Reused for Stability)

Refer to Figure 6.28(a) for details. The numerical values listed below relate to that circuit, assuming a supply voltage, V_{CC}, of 10 V.

V_E Emitter-to-ground voltage: this voltage is typically $\frac{1}{10}$ of the value of V_{CC} ($V_E = 1$ V).

V_C Collector-to-ground voltage: this voltage, which includes emitter voltage, is typically $\frac{1}{2}$ the value of V_{CC} ($V_C = 5$ V).

V_B Base-to-ground voltage: this voltage is $V_E + 0.7$ V (approx.) ($V_B = 1.7$ V). (If a germanium transistor is used, add 0.3 V instead of 0.7 V.)

Errors in These Voltage Potentials Figure 6.29 is a basic diagnosis of the circuit conditions that would most likely create the listed errors in these voltage levels. In this analysis, the three voltages are assumed to be measured and compared to listed (or estimated) values. To use the diagnostic chart shown, locate the initial base voltage defect seen. Follow it to examine the collector or emitter voltage fault to determine potential sources of error.

In Figure 6.29 values marked "low" are assumed to be nonzero amounts that vary significantly from listed values. Values marked "high" are assumed to be below V_{CC} but vary significantly from listed values. When defects are noted, they could refer either to actual component faults or to component interconnections (bad solder joints, for example). This chart represents the most common faults seen in the basic emitter-resistor stabilization type of circuit.

The last entry, where all dc voltages are listed as "O.K.," represents a circuit that demonstrates a fault in handling the amplification of an input signal but shows normal bias conditions. In such a case, one would suspect another circuit or other components to be at fault rather than this circuit. Usually a

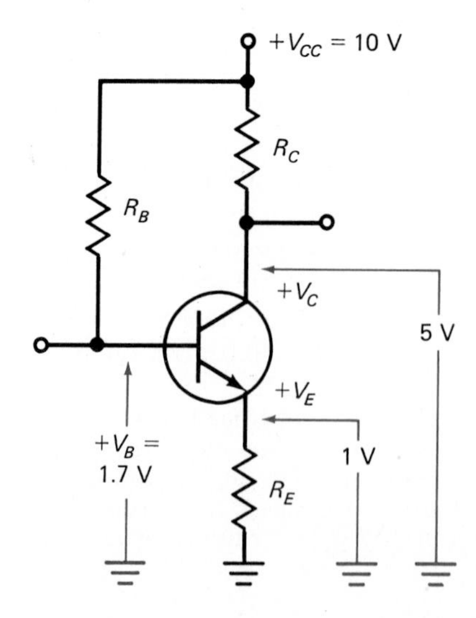

FIGURE 6.28 Typical BJT Voltage Levels (Simple Bias Circuit)

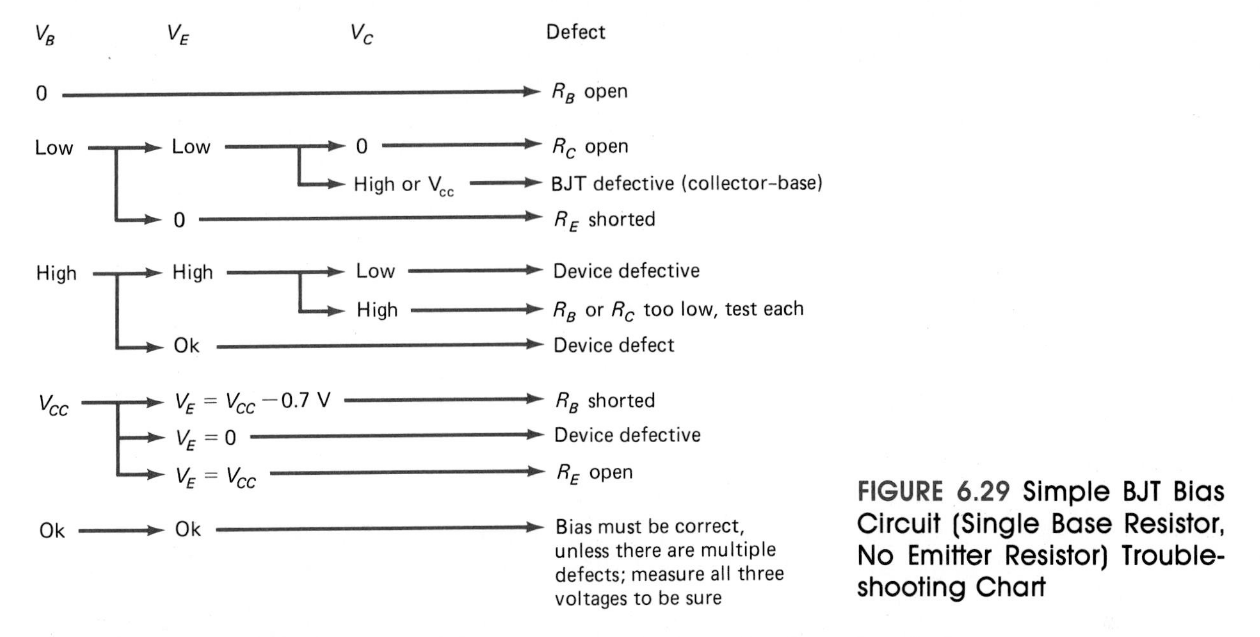

FIGURE 6.29 Simple BJT Bias Circuit (Single Base Resistor, No Emitter Resistor) Troubleshooting Chart

test of the dc bias conditions is sufficient to ascertain if the circuit on the whole is properly functional.

Troubleshooting the Beta Independent Circuit

The process of troubleshooting the beta independent circuit is quite similar to that of the simpler circuit used thus far. Tests of the dc bias voltages will usually indicate the condition of the circuit elements. Figure 6.30 shows the typical values used in such a circuit. A troubleshooting chart to assist in diagnosing faults in the circuit is shown in Figure 6.31. Again in this circuit the concept of low and high voltage levels relates to values that noticeably deviate from standard values listed on a reference schematic. If values are not listed

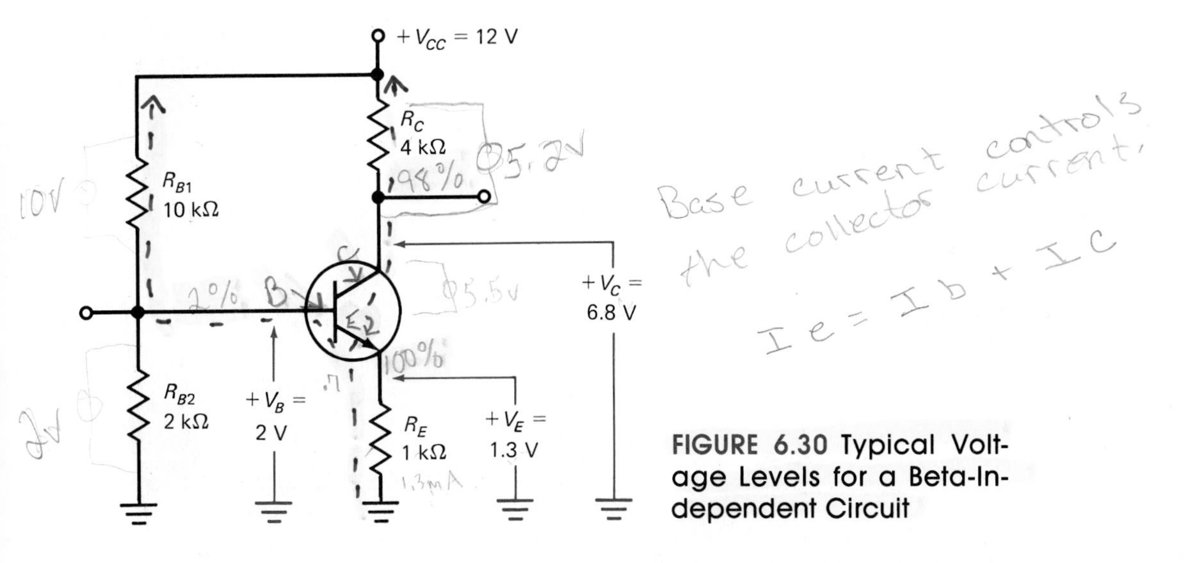

FIGURE 6.30 Typical Voltage Levels for a Beta-Independent Circuit

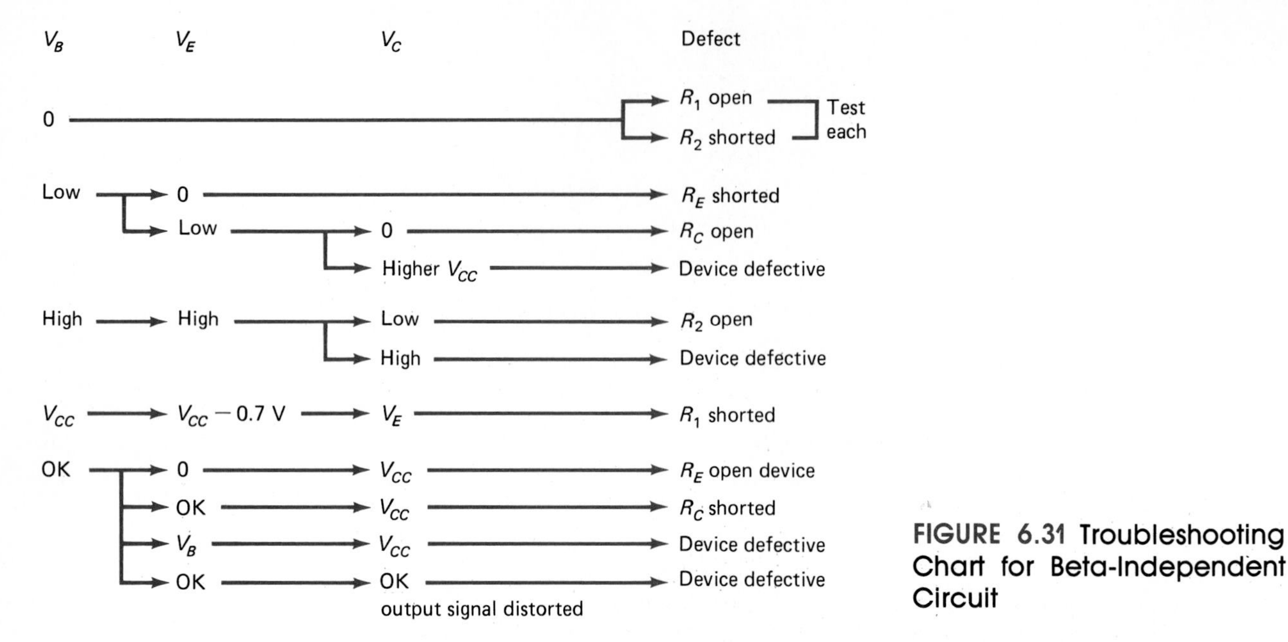

FIGURE 6.31 Troubleshooting Chart for Beta-Independent Circuit

on the schematic, the following guidelines should be used to approximate typical values:

- V_B Base-to-ground voltage: typically, $\frac{1}{10}$ to $\frac{1}{4}$ of V_{CC}
- V_E Emitter-to-ground voltage: 0.7 V less than V_B (approx.) (if a germanium transistor is used, 0.3 V less)
- V_C Collector-to-ground voltage: typically $\frac{1}{2}$ of V_{CC}

Troubleshooting the Voltage Feedback Circuit

Figures 6.32 and 6.33 will provide guidelines for troubleshooting the voltage feedback circuit. For this circuit, if voltage values are not supplied, the sche-

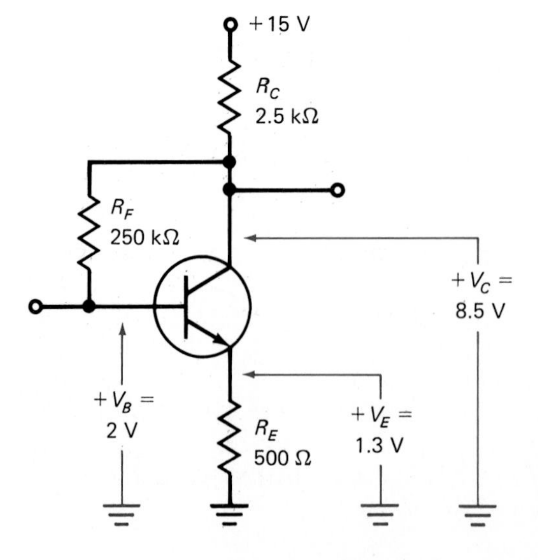

FIGURE 6.32 Typical Voltage Levels for a Voltage Feedback Circuit

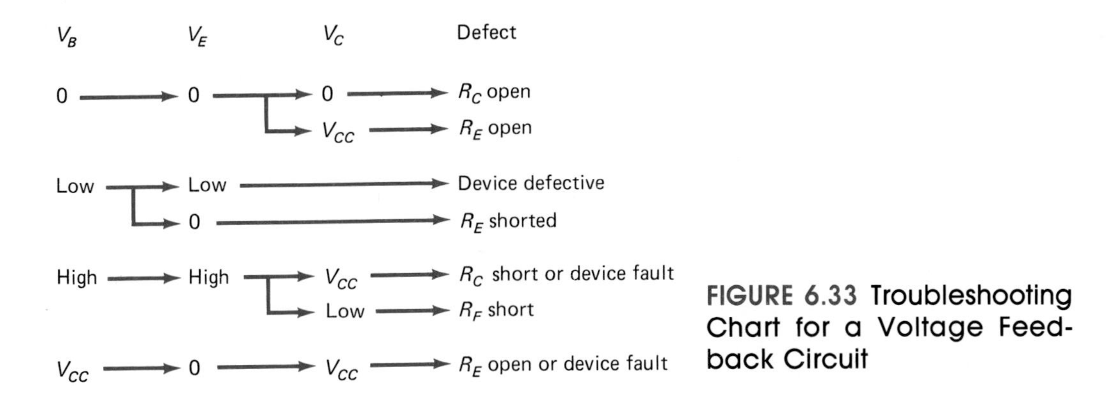

FIGURE 6.33 Troubleshooting Chart for a Voltage Feedback Circuit

matic estimates can be made using the guidelines presented for the simple emitter-resistor feedback circuit.

Each of these troubleshooting charts has been prepared based on potential component faults. In the case of resistors, typical open or short conditions were applied, as well as the case where the resistance is far off from the listed value. For the transistor, the effects of open conditions associated with emitter–base and collector–base lead combinations were investigated. Also, conditions where the device beta is either too low to support circuit gain or so high that the output is unstable were also considered. The only conditions not considered were short-circuit conditions associated with lead combinations or a collector–emitter lead paring.

Collector-to-emitter defects must, by the nature of the device's construction, affect the base lead, too. Such defects would show up in the same manner as the collector–base and collector–emitter problems. Were the collector–base lead to short circuit, either internally or due to a mechanical error (soldering connection), in nearly every case the base–emitter current would run so high as to destroy (open) that junction. Were the base–emitter junction to short (as the base–collector could), a simple test will detect this for all circuit arrangements. Base voltage and emitter voltage will be exactly equal and collector voltage will be equal to V_{CC} were this to happen. Thus should these voltage levels occur, it can be assumed that the device is defective or the base–emitter has been externally shorted.

Troubleshooting the JFET

For the JFET, similar troubleshooting practices can be used. Since there is only one typical JFET stability (and normal bias) circuit used, troubleshooting will be less complicated.

It can be seen that only the source and drain voltages are listed for the JFET shown in Figure 6.34. Since the gate lead is at virtual ground, zero voltage would be as seen at this lead when measured in reference to ground. Figure 6.35 summarizes the possible defective conditions for such a circuit. As for the BJT circuits, this chart summarizes the typical defects that could occur in such a circuit. Again, it was developed based on typical component failures (or mechanical faults).

To confirm some of the entries in these charts, it is suggested that the reader construct test circuits and duplicate some of these faults. To protect

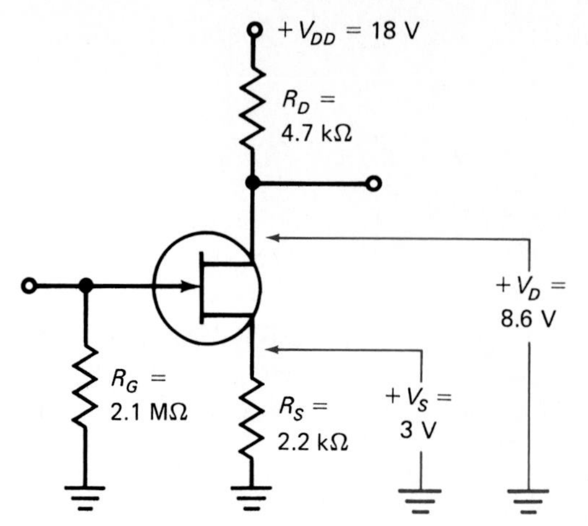

FIGURE 6.34 Typical Voltage Levels for a JFET Circuit

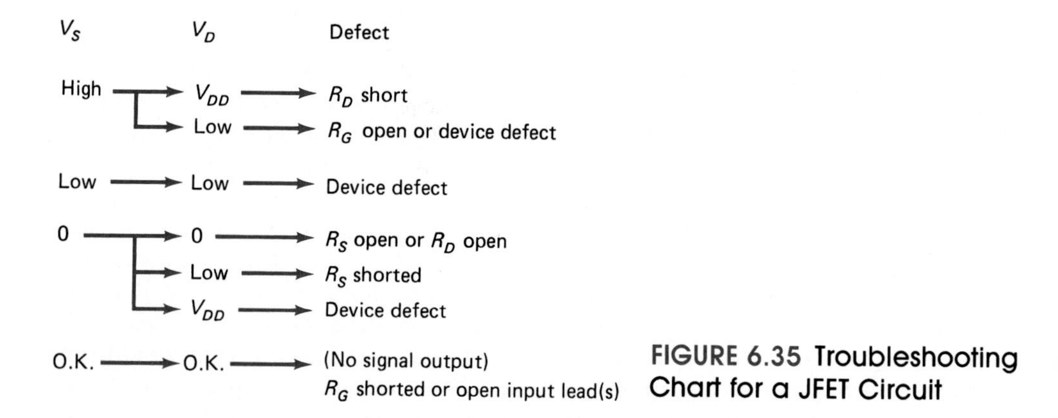

FIGURE 6.35 Troubleshooting Chart for a JFET Circuit

the active device used (BJT or JFET), short-circuit conditions for resistors or device lead combinations in the circuits should *not be duplicated*. Open components or device leads can be effectively tested without damaging the active device used.

REVIEW PROBLEM

(1) Copy each of the schematics presented in the troubleshooting portion of this section (include normal voltage levels). Then prepare two fault examples listing new voltage levels for those listed in the schematic. (Make up "good guess" values.) Exchange your circuits and listed defects with another student and attempt to establish the source of each other's faults. [Be cautious in your choice of faults, as some combinations will not exist due to a single fault (e.g., beta-independent circuit with emitter voltage at V_{CC}).]

6.7 BYPASSING DC BIAS RESISTORS FOR MORE EFFECTIVE SIGNAL PROCESSING

In the past sections of this chapter, the addition of several components provided special bias conditions or stability gain. In most applications, once stability is achieved through the dc biasing process it is not necessary to repeat the process for ac signals. In fact, in many cases, the components that create dc stability may actually interfere with the signal that is being amplified by the device.

The following is a basic analysis of how one of the stability elements causes a problem when processing ac. For the basic circuit of Figure 6.36 which uses negative feedback via R_E, it was found in Section 6.6 that stability was achieved since variations of the collector current would be fed back and subtracted from the base current. If collector current went up, base current was forced to go down. If collector current went down, base current was allowed to increase to compensate. Imagine what problem this could cause with a changing ac input.

If the input to this amplifier went up (as a peak in a sine wave), that rise would cause a rise in collector current; that rise would cause R_E to produce more negative feedback. This would then push the base current back down to the "normal" value. In the process of doing this, all (or most) of the ac signal would be *eliminated*. Feedback could therefore cancel the rises and falls in the cycling of the ac input, thus defeating the amplification process.

Such a process is not found in a normal amplifier. So how can there both be stability *and* amplification of an input signal? Back in the section that discussed the FET, a capacitor was used to bypass the ac signal around the source resistor which was used to create gate voltage potentials. Such a capacitor can be used again, only this time to bypass the signal around the stabilizing element.

First, a capacitor is chosen which will have very low reactance at the frequencies that are expected at the input of the circuit. Often, values from 5 to 100 μF are used in typical circuits. The objective is to have the capacitor

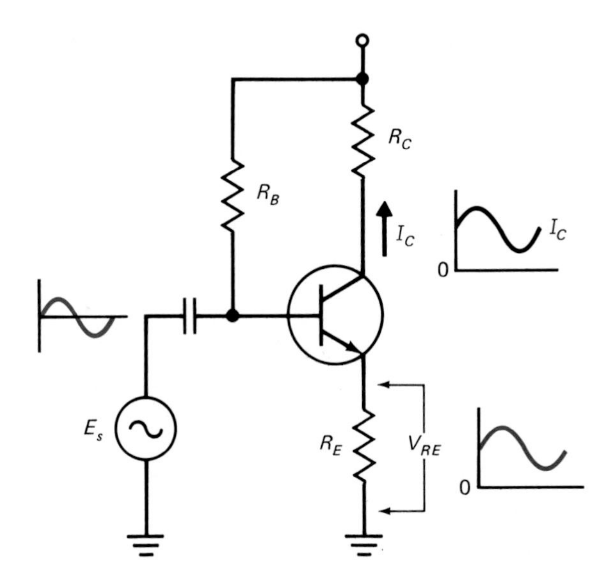

FIGURE 6.36 Problem Produced if the Emitter Resistor Is Unbypassed

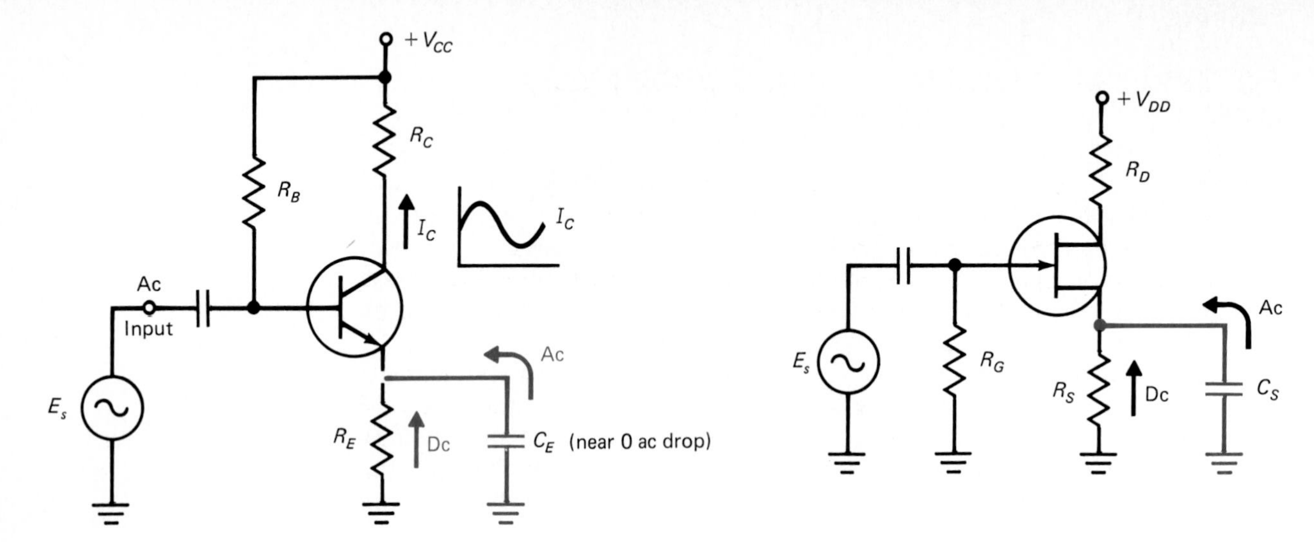

(a) Emitter capacitor in a BJT circuit (b) Source capacitor in a JFET circuit

FIGURE 6.37 Use of Bypass Capacitors

be very conductive to the ac portion of the combined signal and have very high resistance to any dc.

In Figure 6.37 the use of bypass capacitors is shown. Again, the objective is not to have the ac signal fed back to the input for control, just the dc bias potentials. The capacitor/resistor combination (R_E/C_E or R_S/C_S) actually presents two different circuits to the device: one for the dc bias and another for the ac signal. If we simply study the flow of current in each of these circuits, the process of bypassing will be clear.

In Figure 6.38 these two circuit forms are illustrated. Each is described below.

FIGURE 6.38 Coordination of DC and AC Levels When Using a Bypass Capacitor

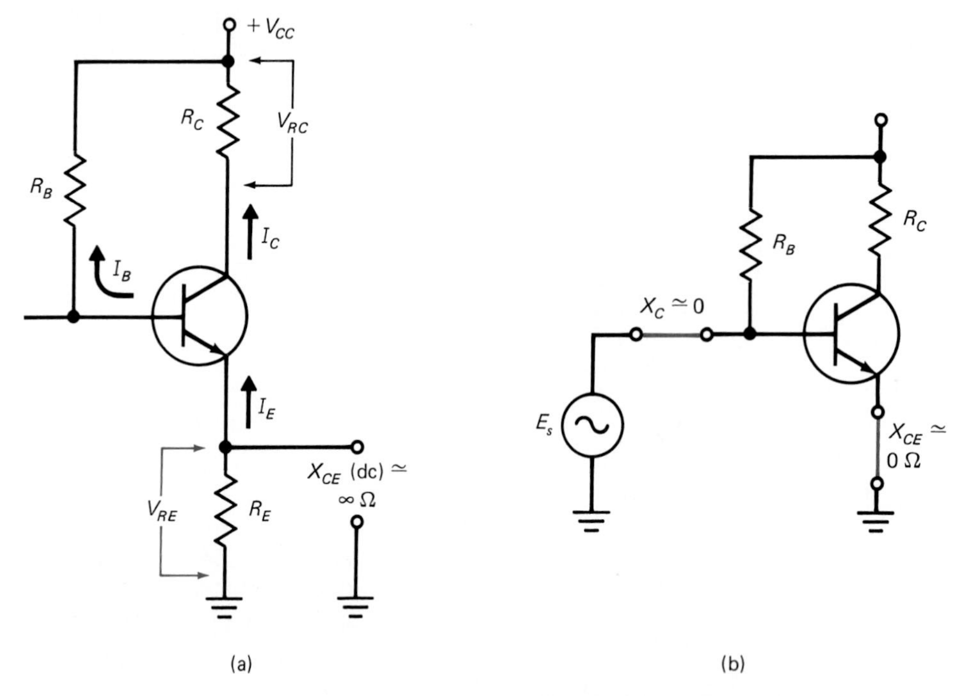

1. The dc circuit consists of the resistor in parallel with the dc reactance of the capacitor. This reactance is nearly infinite (only the leakage resistance of the capacitor). Therefore, the flow of dc current will be through the resistor R_E. This dc will produce a coordinated drop on the resistor, and feedback will be achieved using one of the forms noted in Section 6.5.
2. The ac circuit contains the same elements as the dc circuit, but the reactance of the capacitor is not infinite as it was in description 1. Now the reactance of the capacitor [$X_C = 1/(2\pi fC)$] is very low (a few ohms, based on the lowest frequency that will pass through the amplifier). Therefore, it will easily pass the ac emitter current to ground. This low ohmic value, in parallel with the emitter resistor (R_E), will divert all ac current around the resistor, through this capacitor. Thus no voltage drop is seen on the resistor due to the ac part of the composite output.

Figure 6.39 shows the division of the composite wave through a bypassed feedback resistor. The composite current enters the leg of the circuit connected to both components. The dc portion of this current flows through the resistor on the left, the ac portion through the capacitor on the right.

Equation Changes Due to Bypass Capacitors

Since the bypass capacitors eliminate the feedback resistor (or at least a portion of it), equations for ac conditions shown earlier in this chapter which included this resistor are no longer valid. Therefore, new equations will need to be developed that will aid in calculating the quantities for current and voltage gain and input and output impedance.

Since many bypass capacitors eliminate the emitter resistor R_E, a new means for establishing emitter resistance will be needed. The term r_E' will be used. This term represents what is called a "dynamic" or operational resis-

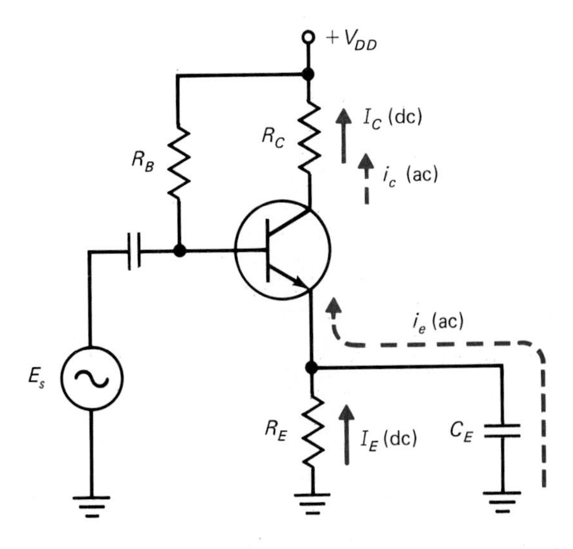

FIGURE 6.39 Blend of Collector Currents; Split of Emitter Currents in the Emitter Bypass Circuit

tance of the emitter junction inside the transistor. It will be necessary to use this resistance since the bypass capacitor places the emitter at ground for all ac functions. The value of this resistance is found by

$$r_E' = \frac{25 \text{ mV}}{I_E} \tag{6.10}$$

where r_E' is the "dynamic" (operational) resistance of the emitter and I_E the dc value of emitter current. (The 25 mV is a standard voltage used as a dynamic drop on the emitter junction inside the device.)

EXAMPLE 6.3

Find the dynamic emitter resistance of a transistor with a dc emitter current of (a) 5 mA; (b) 10 mA; (c) 20 mA.

Solution:

(a) $r_E' = \dfrac{26 \text{ mV}}{5 \text{ mA}} = 5.2\ \Omega$

(b) $r_E' = \dfrac{26 \text{ mV}}{10 \text{ mA}} = 2.6\ \Omega$

(c) $r_E' = \dfrac{26 \text{ mV}}{20 \text{ mA}} = 1.3\ \Omega$

This value of dynamic resistance can then be substituted for emitter resistance into equations that previously included R_E, when R_E is now bypassed. For the simple emitter resistor and beta-independent circuits:

Quantity	Unbypassed Equation	Bypassed Equation
A_i	$\dfrac{R_{B2}}{R_E}$	$\dfrac{R_{B2}}{r_E'}$
A_v	$\dfrac{R_C}{R_E}$	$\dfrac{R_C}{r_E'}$
Z_i	$R_{B1} \parallel R_{B2} \parallel \beta \times R_E$	$R_{B1} \parallel R_{B2} \parallel \beta \times r_E'$
Z_o	R_C	Unchanged

In a similar manner, any other BJT circuits that use the emitter resistor to calculate ac quantities will have an identical form of substitution.

There are a few cases where it may be desirable to feed back some of the ac signal. If it is necessary to control the level of the input so that it will not "overdrive" the amplifier (distort the output signal because the input signal was too strong), some ac feedback could be used. The amount of ac feedback usually must be less than the level of dc stability feedback, to maintain some reasonable gain for the amplifier. To achieve a "complete" stability feedback and have a "partial" signal feedback, a circuit like the one shown in Figure 6.40 could be used.

In the circuit of Figure 6.40 a simple split was made in the normal R_E used for dc stability feedback. The top resistor (R_{E1}) feeds back both dc and ac signals, since it is unbypassed. The bottom resistor (R_{E2}) is used to feed back the remaining portion of the necessary dc value. The total dc fed back is from $R_{E1} + R_{E2}$, the ac only from R_{E1}.

The last stability circuit discussed, the voltage feedback type, usually contains a bypass capacitor. As shown in Figure 6.41, the capacitor is placed in such a way as to eliminate ac from the feedback path. The placement of this capacitor is a little trickier than in the previous circuits. The feedback resistor R_F must be split (in value) with the bypass capacitor placed to ground from the junction of these two resistors (both add up to make the value of R_F). This arrangement is necessary so that neither input nor output signals would be passed to ground through this bypass capacitor. In this location only ac feedback voltages are eliminated from the circuit.

It should be noted here that bypassing either of the two base resistors in the popular beta-independent circuit is not done. Bypassing either of these two resistors would destroy the value of the ac input to the amplifier. It is necessary to bypass the output emitter resistor in this circuit just as in the simpler emitter feedback circuit.

Although it does not relate directly to the process of bypassing feedback devices, it is important at this point to illustrate the role of input and output capacitors. Their role is much like that of the bypass capacitor, presenting high resistance to dc and low impedance to ac. It is often very necessary to contain the dc levels used to bias active devices within the circuit. If these dc bias levels were to pass out of the circuit to other circuits, or potentially to delicate input and output devices, faults, distortion or failure of the attached device could result. Also, if these outside devices would be able to draw current from these circuits, the actual level of bias within the circuit would have to be calculated, *including* the outside element's voltage- or current-

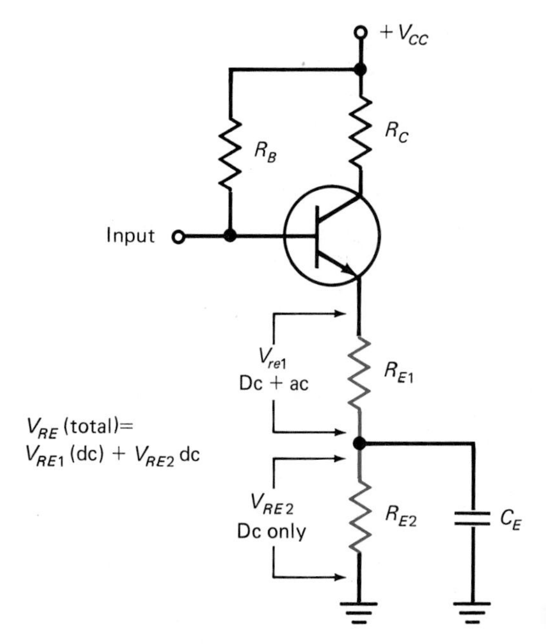

FIGURE 6.40 Partially Bypassed Emitter Resistor

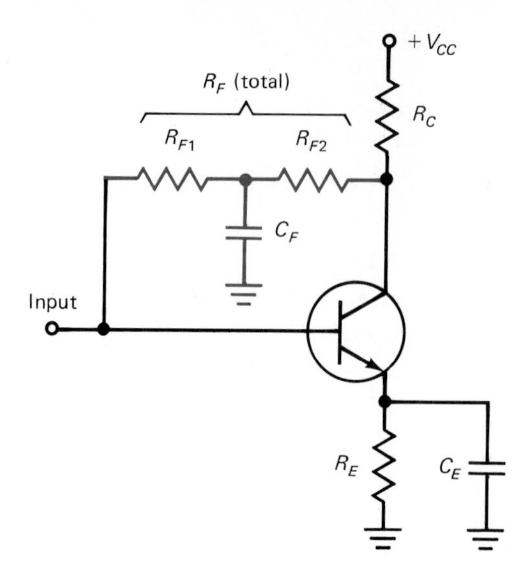

FIGURE 6.41 Bypassing the Voltage Feedback Circuit

divider effect on the circuit. Worse yet, if the other circuit (or device) could supply dc current or voltage to a circuit, it could damage the circuit or device being used.

Figure 6.42 illustrates what problem there could be if a circuit's bias levels were to flow out of the circuit and if external dc levels were allowed to pass into the circuit. Figure 6.43 illustrates the simple use of capacitors to "contain" the bias levels within the circuit used to support the BJT. As can be seen in the latter illustration, only the ac information signal can pass into the input of the BJT circuit, and only the ac output signal is allowed "out" of the circuit.

The process of **direct coupling**, where dc is allowed to pass (as in Figure 6.42) into and out of the amplifier, is used in some circuits. In later chapters, direct coupling will be illustrated. But most amplifiers use some form of **isolation**, to keep bias levels contained within the individual device's circuitry.

Since the bypass process involves the combination of resistors and capacitors, it may be speculated that an R/C time constant may be present. In

FIGURE 6.42 Two-Stage Amplifier Where DC Bias Levels Are Allowed to "Mix"

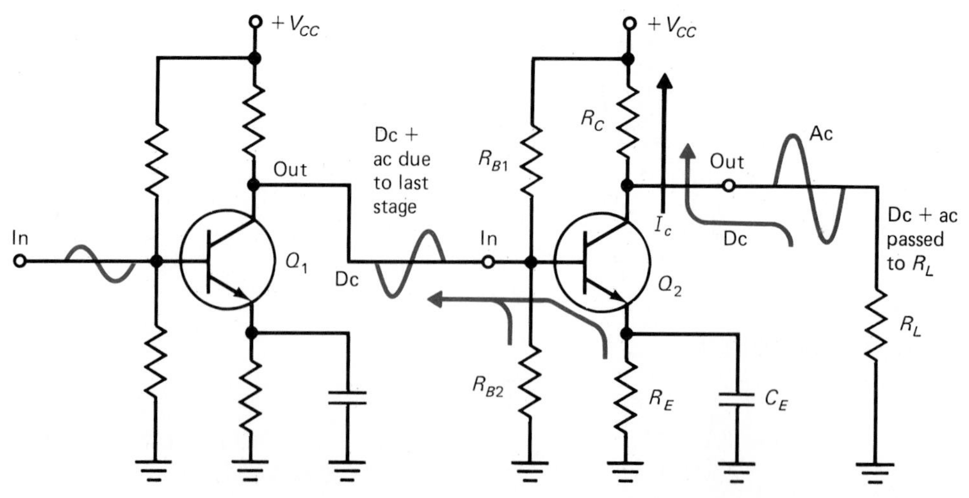

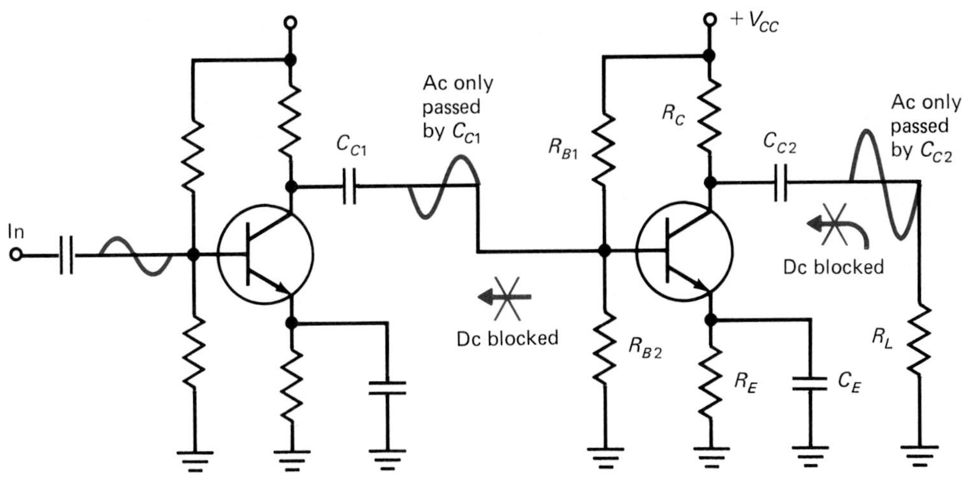

FIGURE 6.43 Use of Input/Output Capacitors to Block DC Current between Amplifier Stages

fact, such a time constant does exist. In most cases, though, this time constant creates no adverse effects in the circuit. The instant that the circuit is turned on, the capacitor does charge. It will charge up to the proper dc level to support the feedback resistor. During this brief period, while the capacitor charges, it may temporarily bypass the resistor even for dc voltage levels. Usually, this time constant is so brief, though, that no noticeable effect is seen in the overall function of the circuit.

In high-power circuits, a brief surge of dc charging current to these capacitors may prove to be the cause of circuit damage (or extreme fatigue on the transistor). Some high-power systems are designed with special limiters or controls to reduce the initial "surge" of current caused when all of the capacitors used for bypassing or coupling begin to charge when the system is initially turned on. If device fatigue or failure cannot be isolated to another cause, it may be necessary to investigate (usually in coordination with design engineers) capacitor "surge" currents as a possible cause for the active device's failure.

Troubleshooting Bypass Capacitor Problems

Bypass capacitors can produce circuit faults. Since capacitors (especially electrolytic types) are subject to defects due to age, they can be the source of circuit faults. It is important to note that capacitor failure can produce either short-circuit conditions or open-circuit conditions (for the bypassed ac signals) and "leakage," wherein the capacitor begins passing unacceptable amounts of dc current.

Open Capacitor If the bypass capacitor used to prevent ac feedback in a circuit (such as the emitter bypass capacitor in a BJT circuit) opens, then ac signals will be fed back to the input. This will cause a noticeable loss of ac gain for the circuit. This defect will not affect the dc bias levels around the device. It may be necessary to use an oscilloscope to test the bypass circuit.

Reading across the bypass capacitor should produce little, if any, ac signal voltages. The presence of ac across this capacitor indicates that it is open (or more likely, its connections are open). An open input or output capacitor will show an immediate loss of ac signal either into or out of the circuit.

Shorted Capacitor Should the bypass capacitor short, both ac and dc voltages normally seen across it will be shorted to ground. The effect is most commonly seen in a loss of dc levels, as ac is supposed to be shorted (bypassed) to ground. It will be hard to tell the difference between a shorted capacitor or a short in the resistor it is designed to bypass. Individual component testing often will be required to determine this fault.

Shorted input or output capacitors will produce a severe inbalance in the dc bias levels in the two stages which they connect. Usually, the amplifiers, if they function at all, will produce highly distorted ac voltages. A prime indicator of this defect is that the dc bias levels in both amplifiers which are coupled through this capacitor will be far off normal levels. This contrasts with most other problems, which are typically isolated to a single amplifier stage.

Leaky Capacitor A bypass capacitor that becomes leaky will often result in a gradual decline in the amplifier's function, since leakage is usually due to long-term wear. The leaky capacitor will cause two amplifier problems. First, since the capacitor now passes an unacceptable level of dc current, and since it is in parallel with a feedback resistor, the total value of feedback resistance (dc) is reduced. This will result in a lower dc feedback voltage and therefore less control on the amplifier's stability. Second, the actual value of capacitance will vary a lot from the original value. When this occurs, control of the ac signal will begin to falter, producing potentially lower gain. The most common sign of this fault is a reading of lower dc feedback voltage levels (e.g., lower emitter voltage for a BJT circuit) and a lessening of amplifier performance.

Leaky input or output capacitors will produce a varying degree of dc current to pass from amplifier stage to stage. This will produce bad readings for dc levels in both the output of one stage and the input of the next. Again for the leaky condition, this and the shorted condition are one of the few cases where the dc levels in both stages will be interrupted by one malfunction.

REVIEW PROBLEMS

(1) To confirm the bypassing effect of the capacitor/resistor combination noted in this section, calculate the value of X_c for a 5-μF capacitor using the following frequencies. Then place this in parallel with a 100-Ω resistor and calculate the total impedance of this combination. Note at approximately what frequency the value of the resistor becomes nearly meaningless. (You may wish to refer to a circuit analysis textbook to review these calculations. At this point, values of phase shift are not significant.)
(a) dc (0 Hz) **(b)** 5 Hz **(c)** 20 Hz **(d)** 200 Hz **(e)** 1000 Hz

(2) Illustrate the use of partial signal feedback (see Figure 6.37 as a guide) in an FET circuit, noting that R_S is the feedback resistor. (Specific values need not be shown in the illustration.)
(3) If the capacitor used to bypass the stability feedback resistor in a simple BJT circuit (see Figure 6.35) were to develop more and more leakage due to aging, discuss what effect this would have on the feedback process. (Concentrate on the dc bias levels only.)
(4) In a BJT beta-independent circuit, it was found that little, if any, ac gain was produced. Dc measurements of bias were taken and found to be quite close to normal. An oscilloscope was then used and it was found that the ac base voltage level was quite lower than expected and that there was ac voltage present on the emitter of the device. Discuss a defect that could cause this fault.

SUMMARY

This chapter has concentrated on the functional features of the more popular transistor biasing arrangements. It is necessary to become familiar with these various arrangements, their similarities and differences, and the general concepts of how they function. A clear understanding of each of these circuits will allow the reader to deal clearly with the many functional amplifier circuits encountered both in this text and in real-life situations. The role of each element and the possible problems produced when they fail is most important to anyone dealing with electronic circuitry.

The following chapter details many of the formulas used to design these special circuits. Although a knowledge of these design calculations can be helpful when working with these circuits, it is not necessary to know them when attempting to troubleshoot or build these circuits.

KEY EQUATIONS

6.1 $I_E = I_C + I_B$

6.2 $V_{\text{load}} = R_{\text{load}} \times I$

6.3 $V_{\text{device}} = V_{CC} - V_{\text{load}}$

6.4 $I_{\text{max}} = \dfrac{V_{CC}}{R_L}$

The following are for Beta independent circuits:

6.5 $A_i = \dfrac{R_{B2}}{R_E}$ (for ac input signals)

6.6 Voltage gain: $A_v \simeq \dfrac{R_C}{R_E}$

6.7 Output impedance: $Z_o \simeq R_C$

6.8 Input impedance: $Z_i \simeq R_{B1} \parallel R_{B2} \parallel \beta \times R_E$

For the voltage feedback circuit:

6.9 current gain: $A_i \simeq \dfrac{R_F}{R_C + R_E}$

Calculation of "dynamic" emitter resistance:

6.10 $r_E' = \dfrac{25 \text{ mV}}{I_E}$

PROBLEMS

6.1. Draw the schematic symbol for both an NPN and a PNP transistor (BJT) and label their terminals. Then indicate the directions for current flows (I_B, I_C, and I_E).

6.2. Draw the schematic of an NPN transistor with proper base and collector supplies, a base current-limiting resistor (R_B), and a sine-wave input (with capacitor). Label the proper battery polarities and sketch a normal base current and collector current waveform.

6.3. Repeat problem 6.2 using a PNP transistor.

6.4. For the circuit of Figure 6.44, find the value for I_B and I_C.

6.5. For the circuit in Figure 6.44, if an input current of 500 μA p-p ac were applied at the base, what would the composite waveforms for base current and collector current be? (Indicate values as well.)

6.6. Illustrate the schematic symbol for an ''N''- and a ''P''-type JFET and indicate the proper gate, drain, and source bias polarities on these devices.

6.7. Illustrate the common schematic for an ''N'' JFET in a common-source configuration: first with a separate gate supply voltage, then with gate voltage set by the potential difference across a source resistor.

6.8. State the basic relationship of drain current to gate voltage for a JFET.

6.9. Why is it necessary to provide a bypass capacitor around the source resistor in a JFET circuit?

6.10. What ideal amplifier model matches the BJT? Which matches the JFET?

6.11. Given a BJT circuit with $\beta = 150$, $I_B = 50$ μA, $V_{CC} = 20$ V, and $R_C = 1$ kΩ, find the following values.

(a) Collector current
(b) Voltage on the collector resistor and across the device
(c) Maximum limit to the collector current

6.12. Using the data from problem 6.11, find the collector current and output voltage distribution given the following base currents.

(a) 100 μA (b) 200 μA

6.13. Briefly define the problem of stability for a BJT, being certain to indicate cause/effect relationships.

6.14. Describe briefly how the addition of an emitter resistor to a BJT circuit aids in providing stability.

6.15. What effect does the R_{B1}/R_{B2} combination in a beta-independent circuit have on the values of voltage and current in that circuit?

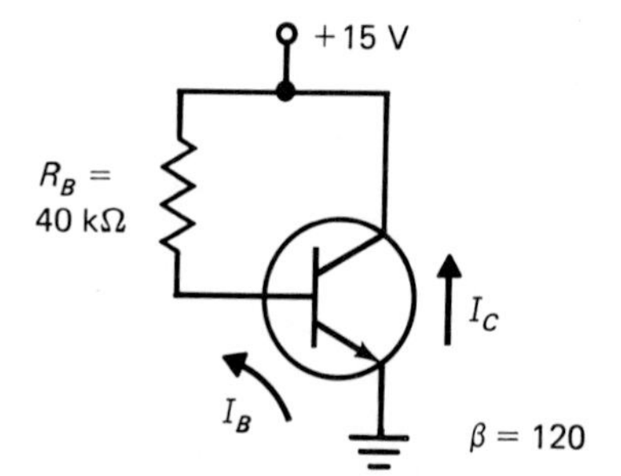

FIGURE 6.44 Simple BJT Bias Circuit

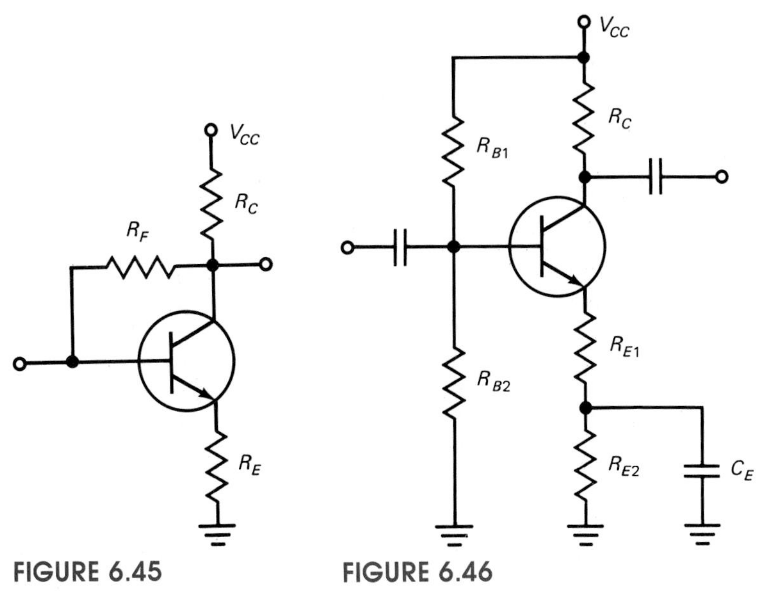

FIGURE 6.45 FIGURE 6.46

6.16. Define the basic purpose of bypassing the emitter resistor with a capacitor in a BJT circuit. Is the source resistor in a JFET circuit bypassed for the same reason?

6.17. Define briefly the purpose of only partially bypassing the emitter resistor in a circuit, such as in Figure 6.45.

6.18. Illustrate the placement of the bypass capacitor in a voltage feedback circuit (such as Figure 6.46).

6.19. Define the basic purpose of the input and output "coupling" capacitors used in BJT or JFET circuits.

ANSWERS TO REVIEW PROBLEMS

Section 6.2

1. $I_C = 4.8$ mA, $I_E = 4.84$ mA
2. See Figure 6.3(b).

Section 6.3

1. The drawings would be similar to those listed except for the substitution of the P-type JFET and a change in dc source polarities.
2. The gate bias potential for a JFET must be sufficiently high so that the input signal is properly converted to an "all-one-polarity" type of input.
3. With no bypass resistor on R_s and a 20-mA (ac) source current through the 100-Ω source resistor, a 2-V (ac) source voltage would exist. Since this would be out of phase (inverted) with the input, the value of gate-to-source voltage would be reduced by the 2 V to only 2 V (p-p). Thus only one-half of the original input signal would be used to produce an output. This reduction would be repeated until virtually no input level remained.

Section 6.4

1. Maximum collector current = 24 mA (due to R_c)
2. (a) 10 mA

(b) 20 mA
(c) 24 mA (due to R_c)
(d) 24 mA (due to R_c)

3. (a) I_C = 30 mA (dc), i_C = 20 mA p-p (ac)
(b) V_{RC} = 6 V dc, v_{RC} = 4 V p-p (ac)
(c) V_{CE} = 10 V dc, v_{CE} = 4 V p-p (ac)

Section 6.5

1. Based on data in Section 6.5.
2. If the gain of the BJT was lower than that "forced" by the circuit, the overall gain would have to be reduced to that value. It is possible that the device (and circuit) may not function properly in this case, and either damaged components or a distorted output would result.
3. Since voltage gain is R_c/R_e, if $R_e = R_c$, the gain would be unity (1) and the output would be the same (but inverted) as the input.

Section 6.6

1. Solutions are based on student selected faults.

Section 6.7

1. (a) X_c = infinite, total impedance = 100 Ω
(b) X_c = 6366 Ω, total impedance = 99.98 Ω
(c) X_c = 1592 Ω, total impedance = 99.8 Ω
(d) X_c = 159 Ω, total impedance = 84.6 Ω
(e) X_c = 31.8 Ω, total impedance = 30.3 Ω
At 1 kHz the resistor value is beginning not to affect the value of impedance.

2.

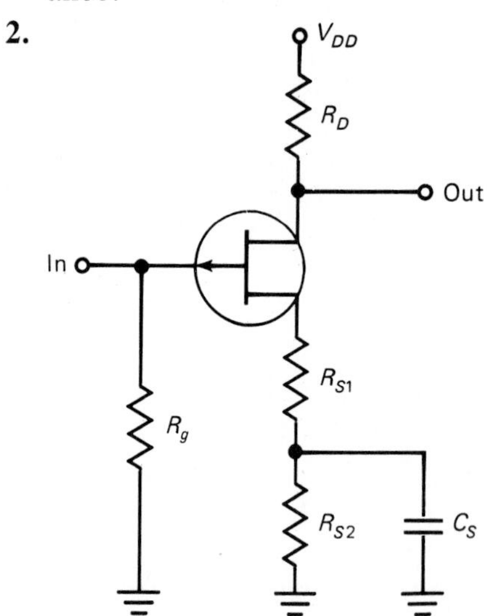

3. Leakage creates a lowering of the dc resistance (assumed to be infinite) for a capacitor. If this occurs, the total value of emitter resistance (dc) will also drop. A drop in this value will cause a loss of stability and a lowering of the voltage-gain figure.
4. The emitter bypass capacitor could either be defective (leaky), disconnected, or may be of an improper value to bypass the desired ac frequency (X_c too high at the selected frequency).

7 Detailed Analysis of Popular Amplifier Circuitry

Objectives

Upon completing this chapter, the reader should be able to:

- Present a more detailed analysis of BJT and FET circuits.
- Use hybrid parameters to describe ac equivalent circuits.
- Apply superposition analysis to amplifier circuits.
- Identify the affect of external loads on amplifiers and revise calculations accordingly.

Introduction

In Chapter 6 we presented the more popular circuit arrangements for BJT and FET amplifiers. It presented basic functional data and analysis tools, including essential equations. This chapter deals with the same circuitry, but provides a more detailed analysis of these circuits. ■

7.1 ANALYSIS OF COMPENSATION AND INPUT/OUTPUT CONDITIONING

In Chapter 6 we introduced the circuitry that is typically used to compensate for device inequities and to provide stability. In this chapter we concentrate on developing more detailed equations for calculating the effects these elements have on the amplifier circuit. Since this book is based on application-oriented uses for active devices, it is not essential for the reader to become proficient at doing the calculations in this chapter. Knowledge of these computations will, though, enhance the reader's ability to understand these applications more fully.

In the previous chapters it was pointed out that discrete devices did not meet the specifications of the ideal model that coordinated to their function. Also, both the BJT and the JFET were polarity sensitive at their input terminals. In addition, instability made them somewhat undependable. These problems need to be resolved if these devices are to be used in electronic circuits. Finally, since each device has electrical limitations, both maximum and minimum boundaries, circuits in which they are used must provide electrical restrictions so that these boundaries are not overstepped.

As a review of Chapter 6, components are selected to compensate (make up) for the shortcomings or minimum/maximum requirements of the active device. Resistors provide voltage dividers, limit current, and assist in providing negative feedback, which is a part of the compensation process.

Again, in this chapter, analysis will concentrate first on the dc bias process, then progress to analysis of the circuit with an ac signal applied. Control of the device will be achieved primarily with control of the dc bias conditions, with ac functions fitting into the controlled (compensated) function of the circuit.

This analysis can be viewed as a basic "superposition analysis" problem. First dc analysis will be conducted, then ac analysis will be done for the same circuit. A composite (dc + ac) solution will then be used to describe the complete effect. Since dc and ac analyses serve such distinct purposes, we will often not concentrate on the composite solution. We will simply say that if dc compensation is proper, a simple ac analysis will be adequate to describe the input-to-output relationship.

7.2 A SUMMARY OF SPECIAL CIRCUIT ANALYSIS TOPICS

To be able to evaluate bias and compensation circuitry, a little review of select circuit analysis topics is in order. These topics and examples have been expressly selected to prepare for the circuitry used with semiconductor devices. It is assumed that the exact methods of analysis of these circuits has been practiced by students in a complete course on circuit analysis. Only one semiconductor device concept is applied in this section. To illustrate the inclusion of a "forward voltage drop" as it would be encountered when using a BJT, the diode will be shown in some examples. Since this normal forward drop of 0.7 V will be seen in any BJT circuit, for simplicity, an ordinary diode will be used to represent this potential loss of voltage. It will be treated simply as a fixed loss of voltage potential (not an Ohm's law loss, $I \times R$).

Simple Current-Limiting Effects of Resistors

One of the basic principles of electronic circuits is the process of limiting current flow by using a resistor. The application of this to semiconductor circuits is rather common. Examples 7.1 and 7.2 illustrate the process of current limiting using resistors.

EXAMPLE 7.1

In the circuit shown in Figure 7.1, identify the current-limiting effect of the resistor.

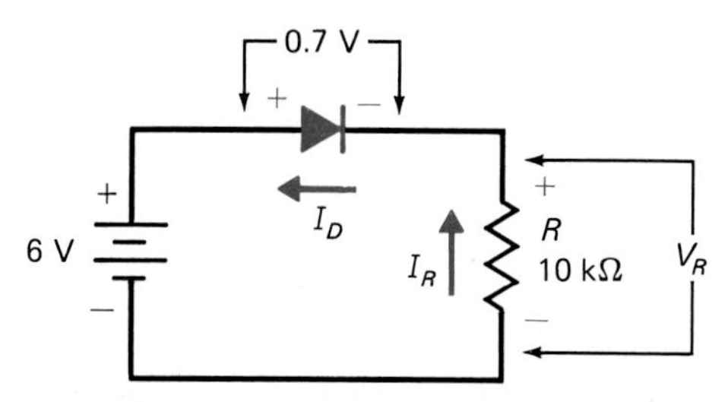

FIGURE 7.1 Forward-Biased Diode Circuit

Solution:

In this figure, note that current through the diode will be set by the value of the resistor (not by the diode) to the value shown by

$$V_R = E_s - 0.7 \text{ V} = 5.3 \text{ V}$$

$$I_R = \frac{V_R}{10 \text{ k}\Omega} = 0.53 \text{ mA}$$

$$I_D = I_R = 0.53 \text{ mA}$$

(Remember that basically, the diode does not limit current when forward biased; it just produces the standard 0.7-V drop.)

EXAMPLE 7.2

For the circuit shown in Figure 7.2, find the maximum limit to current flow set by R_1.

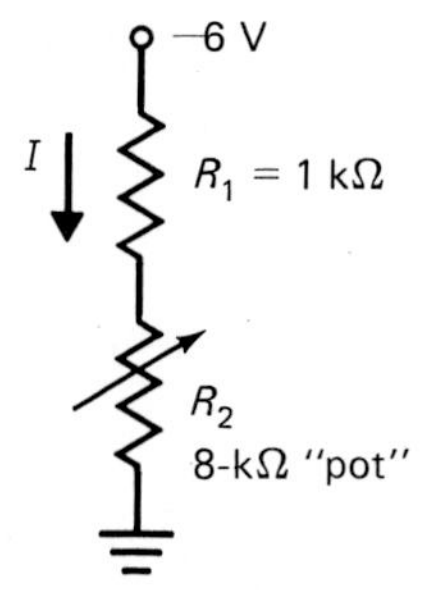

FIGURE 7.2 Series Circuit with Variable Resistor ("POT")

Solution:

In this circuit, the limit to maximum current flow is set exclusively by resistor R_1. Since resistor R_2 is variable, its value could range from the maximum of 5 kΩ to a minimum of 0 Ω. Thus total resistance could range from 6 kΩ to as low as 1 kΩ. The maximum current will flow when the minimum total resistance is seen. Therefore,

$$I_{\text{max}} = \frac{E_S}{R_{\text{min}}} = \frac{6\text{ V}}{1\text{ k}\Omega} = 6\text{ mA}$$

The current in this circuit could not exceed 6 mA.

Sharing a Common Source and Common Ground (Parallel Analysis)

The following example is used to illustrate both a blend of two parallel circuits and their resulting current flow and further to illustrate the use of common source/ground symbols.

EXAMPLE 7.3

In the circuit shown in Figure 7.3, find the current flow in the left and right legs of the circuit and the current flowing in the ground connector (labeled I_T).

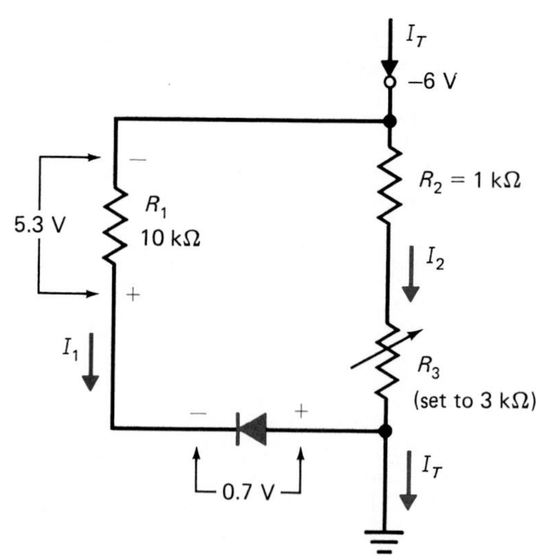

FIGURE 7.3 Series/Parallel Circuit with a Connecting Diode and Variable Resistor

Solution:

(Note first that the potentiometer is set to 3 kΩ.) The current in the left leg of the circuit is the same as was seen in Example 7.1.

$$I_1 = \frac{5.3\ \text{V}}{10\ \text{k}\Omega} = 0.53\ \text{mA}$$

And with the total resistance in the right leg being equal to

$$R' = R_2 + R_3 = 4\ \text{k}\Omega$$

The current on that side is

$$I_2 = \frac{E_S}{R'} = \frac{6\ \text{V}}{4\ \text{k}\Omega} = 1.5\ \text{mA}$$

Now the total current that flows from the source, through the circuit (branches), and then back through the ground connection to the negative (−) battery terminal is

$$I_T = I_1 + I_2 = 0.53\ \text{mA} + 1.5\ \text{mA} = 2.03\ \text{mA}$$

Note that the voltage divisions on each side can be verified by using Kirchhoff's voltage law:

$$V_{R1} + 0.7\ \text{V} = 6\ \text{V}$$

$$V_{R2} + V_{R3} = 6\ \text{V}$$

Solving a Circuit Using Current Ratios

In some circuits, all resistance values may not be known. Since we are working with semiconductors, their value of conductance in any one application (and therefore resistance) is not usually known. In such circuits, voltage or current ratios may be known even though resistance terms are not available. Using these relationships can be helpful when attempting to find the balance of the circuit values (voltages or other currents). Examples 7.4 is used to illustrate this point.

EXAMPLE 7.4

The circuit shown in Figure 7.4 is a duplication of the one shown for Example 7.3. In this circuit, though, the value of R_3 is not known, but the proportion of current in the right leg (where R_3 is) to the current in the left leg is known. Find all the voltages and currents around this circuit. As a final step, calculate the value of the unknown resistor, R_3.

Solution:

Calculating the current on the left side is identical to the way it was done in Example 7.3. Therefore,

$$I_1 = 0.53\ \text{mA}$$

Since R_3 is not known, we must use an indirect method of finding the

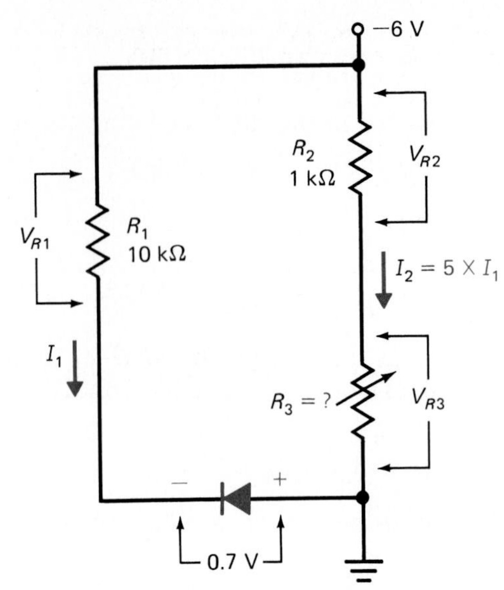

FIGURE 7.4 Series/Parallel Circuit with Unknown Resistance; Using Current Ratios as an Aid to Calculating Currents and Voltages

values on the right side. But it is stated that

$$I_2 = 8 \times I_1 = 4.24 \text{ mA}$$

With that amount of current flow on that side, we can calculate the value of the voltage drop on resistor R_2:

$$V_{R2} = I_2 \times R_2 = 4.24 \text{ mA} \times 1 \text{ k}\Omega = 4.24 \text{ V}$$

By now using a variation on Kirchhoff's voltage law, we find the voltage on R_3

$$V_{R3} = E_S - V_{R2} = 1.76 \text{ V}$$

and the total current flow from the battery (and to ground) is

$$I_1 + I_2 = 0.53 \text{ mA} + 4.24 \text{ mA} = 4.77 \text{ mA}$$

Finally, we find the unknown value (or setting) of R_3:

$$R_3 = \frac{V_{R3}}{I_2} \simeq 415 \ \Omega$$

More Complex Analysis

There are some situations where circuit values are solved progressively. First one side of the circuit is solved, then those solutions are used to resolve the other side. Following is an example of such a solution.

EXAMPLE 7.5

For the circuit shown in Figure 7.5, find the values of all voltages and current flows labeled.

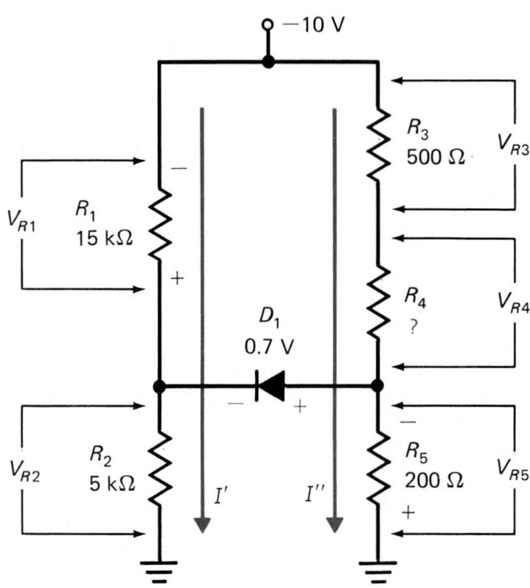

FIGURE 7.5 Complex Series/Parallel Circuit with Unknown Resistance

Special note: This circuit was selected to represent a special circuit used in transistor biasing. To simplify the analysis it will be assumed that although the diode shown is forward biased, the flow of current through it is so small in value that it will be ignored when calculating the remaining values of voltage and current.

Solution:

Since resistor R_4 is unknown, current and voltage on the right side of the circuit cannot be found directly. Therefore, the left side will be worked with first:

$$R' = 15\ \text{k}\Omega + 5\ \text{k}\Omega = 20\ \text{k}\Omega$$

$$I' = \frac{10\ \text{V}}{20\ \text{k}\Omega} = 0.5\ \text{mA}$$

(Since, unlike the other examples, the value of current I'' is not related to I', it will have to be found using other methods.) With I' flowing through R_1 and R_2, their drops can be found to be

$$V_{R1} = 7.5\ \text{V} \qquad \text{and} \qquad V_{R2} = 2.5\ \text{V}$$

Now the value of V_{R2} will be used to find a voltage on the right side of the circuit:

$$V_{R5} = V_{R2} - 0.7\ \text{V (diode drop)} = 2.5\ \text{V} - 0.7\ \text{V} = 1.8\ \text{V}$$

With 1.8 V o_n resistor R_5,

$$I_{R5} = \frac{1.8 \text{ V}}{200 \ \Omega} = 9 \text{ mA}$$

This is also the currents:

$$I'' = I_{R3} = I_{R4} = I_{R5} = 9 \text{ mA}$$

(Remember that it is assumed that the current through the diode connecting the left and right circuits is so small that it will not affect either I' or I''.)

To finish this solution,

$$V_{R3} = 500 \ \Omega \times 9 \text{ mA} = 4.5 \text{ V}$$

$$V_{R4} = 10 \text{ V} - 4.5 \text{ V} - 1.8 \text{ V} = 3.7 \text{ V}$$

Therefore, since the ratio of I'' to I' was not specified, the value of I' was dependent on voltages on the left side of the circuit, where all values were easily known.

Complex Analysis Using Simultaneous Solutions

There will be some cases where the values for current and voltage can be found only by simultaneous solutions. That is, one segment of the circuit is dependent on values in another segment, and vice versa. The following example illustrates this point.

EXAMPLE 7.6

In the circuit shown in Figure 7.6, the values of voltages and currents are dependent on each other. I'' is related to I' ($I'' = 20 \times I'$). The value of I' depends on the value of the voltage on R_2. The voltage on R_2 is caused by both I' and I'' flowing through it. To solve for voltage and current, some simultaneous equations will have to be used to find the one value for I' and I'' that will allow the circuit to be fully balanced.

Solution:

Since I' is not known, the following simple equations are written based on known values, Ohm's law, or Kirchhoff's current law:

$$I'' = 20 \times I'$$

$$I_{R1} = I' + I'' = I' + 20I' = 21I'$$

$$I' = \frac{V_{R1}}{R_1}$$

$$V_{R1} = 10 \text{ V} - V_{R2} - 0.7 \text{ V}$$

$$V_{R2} = I_{R2} \times R_2 = I_{R2} \times 500 \ \Omega$$

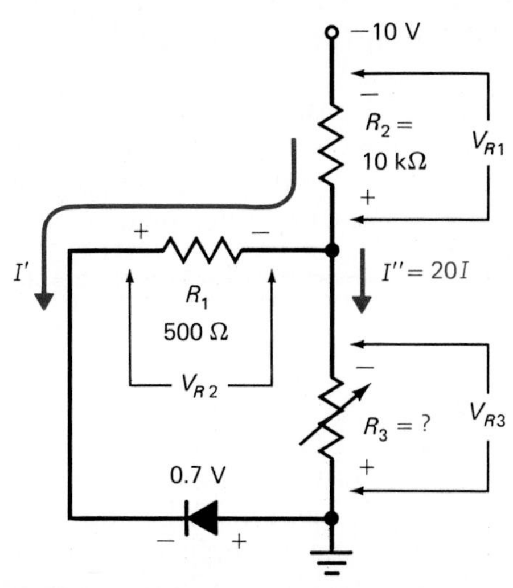

FIGURE 7.6 Another Complex Series/Parallel Circuit with Unknown Resistance; Using Simultaneous Equation to Solve for Currents and Voltages

But none of these equations has enough information to solve for a numerical answer. Therefore, we will have to work the problem using substitution of one equation into the other until number values are known (simultaneous solutions):

$$V_{R2} = 21I' \times 500\ \Omega$$

$$V_{R1} = 10\text{ V} - (21I' \times 500) - 0.7\text{ V} = 10\text{ V} - (10.5\text{ k}\Omega\ I' - 0.7\text{ V})$$

$$I' = \frac{10\text{ V} - 10.5\text{ k}\Omega\ I' - 0.7\text{ V}}{10\text{ k}\Omega}$$

$$10\text{ k}\Omega \times I' = (10\text{ V} - 10.5\text{ k}\Omega\ I' - 0.7\text{ V})$$

Adding 10.5 kΩ I' to both sides yields

$$20.5\text{ k}\Omega \times I' = 10\text{V} - 0.7\text{ V} = 9.3\text{ V}$$

$$I' = \frac{9.3\text{ V}}{20.5\text{ k}\Omega} = 454\ \mu\text{A}$$

and thus

$$I'' = 20(454\ \mu\text{A}) = 9.07\text{ mA}$$

To finish the analysis by finding voltage and current from the first equations used:

$$I_{R2} = 9.52\text{ mA}$$

$$V_{R2} = 4.76\text{ V} \qquad V_{R1} = 4.54\text{ V} \qquad V_{R3} = 10\text{ V} - 4.76\text{ V} = 5.24\text{ V}$$

[To check this solution for V_{R3}: V_{R3}(also) $= V_{R1} + 0.7\text{ V} = 5.24\text{ V}$, which checks with the calculation above.]

Analysis Using Superposition Methods

It is very common for a circuit to have two sources of energy; one may be dc and the other ac. Since these sources may be applied to the circuit at different points, the composite value of dc and ac voltage will require a two-part solution. Example 7.8 will illustrate this point.

EXAMPLE 7.7

The circuit shown in Figure 7.7 has two voltage sources. One is a dc source (E_{S1}) and the other is ac (E_{S2}). The objective will be to find the voltage on R_2 as a composite of these two voltages. The capacitor from E_{S2} is there to prevent any dc current from source E_{S1} from flowing through source E_{S2}. This capacitor will charge to the dc voltage on R_2 and hold that value. Its reactance (X_C) will be assumed to be very low in reference to the ac source, thus not dropping any (appreciable) amount of that voltage.

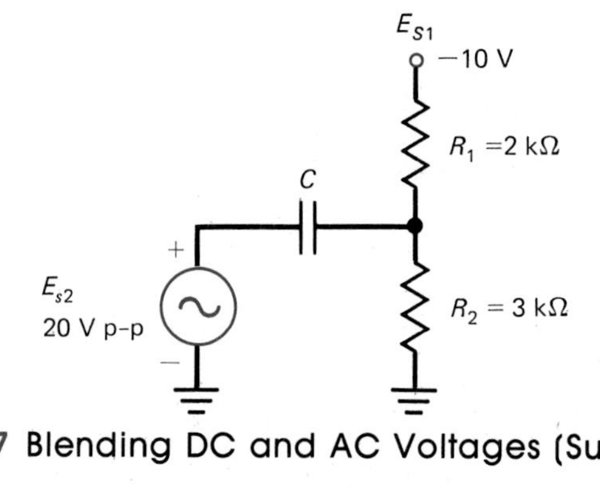

FIGURE 7.7 Blending DC and AC Voltages (Superposition Analysis)

Solution:

To resolve this composite answer, superposition analysis will be done. To do this, two independent circuits will be used. One will solve for the dc voltage on R_2 and the other will be used to find the ac voltage. These circuits are shown in Figure 7.8.

Circuit (a) will be used to solve for the dc voltage on R_2. Note in this circuit that the ground connection for the battery is indicated for greater clarity. Also note that the capacitor has been replaced with an open connection (how it will look once it charges). As well the source E_{S2} has been replaced with its internal impedance (0 Ω). This then becomes a rather simple series circuit:

$$R_{T1} = R_1 + R_2 = 5\ \text{k}\Omega \quad I_1 = \frac{10\ \text{V}}{5\ \text{k}\Omega}\ \text{dc} = 2\ \text{mA}$$

$$V_{R2}\ \text{dc} = 2\ \text{mA} \times 3\ \text{k}\Omega = 6\ \text{V (dc)}$$

In circuit (b), note the changes from the one used in the previous analysis. The capacitor (with X_C low) is now a short circuit and the battery

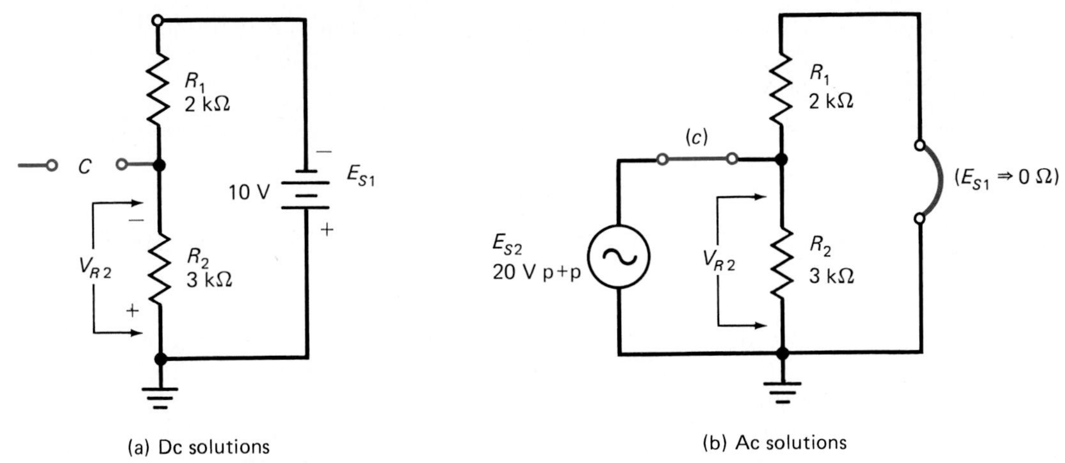

FIGURE 7.8 Solutions for Circuit of Figure 7.7

has been replaced with its internal impedance (0 Ω). The ground connection has also been completed. Upon careful observation, the two resistors, R_1 and R_2, are now in parallel, as seen by this source. Therefore,

$$V_{R2} = 20 \text{ V p-p (ac)}$$

The composite solution of these values results in a total voltage that includes both the ac and the dc values. Note the illustration of this composite value in Figure 7.9.

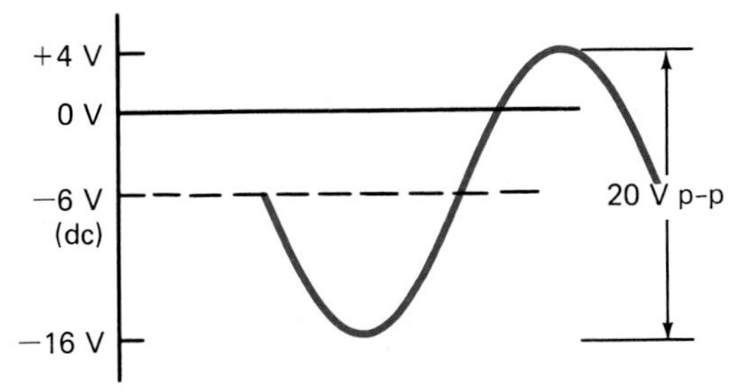

FIGURE 7.9 Composite of AC and DC Voltages

By no means have we covered all possible solutions in these examples. They are selected illustrations of the type of circuit solutions that will be needed when dealing with quite a few of the semiconductor circuits. Outside the basic form of the diode, these circuits did not employ semiconductors. It will be seen, though, that semiconductor circuits used in future sections of this book will be solved using much the same form of analysis as shown here in these examples.

REVIEW PROBLEM

(1) It is suggested that the reader copy each of the circuits shown in these examples on a separate sheet of paper. Then, without referring

to the examples, attempt to solve for all circuit values found in each problem. Compare your steps and answers to those shown in the examples.

7.3 A MORE-DETAILED SOLUTION FOR BJT INPUT BIAS CONDITIONS

In Chapter 6 we introduced the basic principles used to investigate the BJT common-emitter circuit's bias conditions. The following example expands the basic equations used there by producing a more detailed analysis of the circuit voltages, currents, and operational conditions. As well, ac levels at the input are translated into output values. These calculations verify the general analysis established in Chapter 6 by tying actual circuit values to the functions of bias and amplification.

EXAMPLE 7.8 (COMPOSITE CALCULATIONS)

In this example it will be assumed that the current gain (beta) for dc bias is equal to the current gain for ac signals. Although this is a common assumption, the actual values for dc and ac current gains are not usually the same value. If more precise calculations are desired, specific ac and dc current gain values would be used in the appropriate equations. (This process is demonstrated in a later example in this chapter.)

Consider the device shown in Figure 7.10. First, we observe the forward-bias condition on the device. It is of the correct polarity (+) and

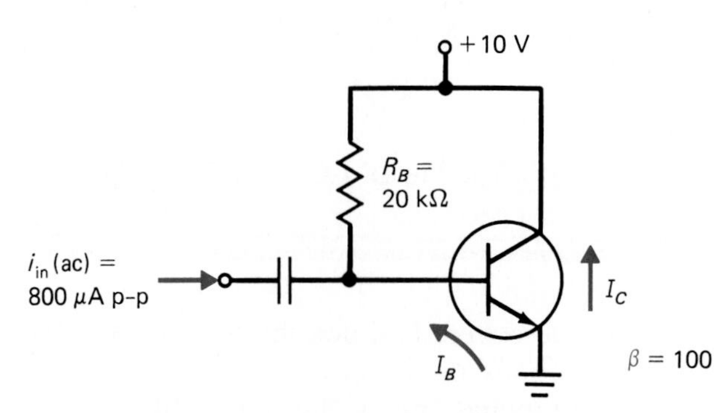

FIGURE 7.10 Simple DC Bias of a BJT

thus 0.7 V will be dropped:

$$V_{BE} = 0.7 \text{ V}$$

Therefore, the voltage left for the base resistor will be

$$V_{RB} = 10 \text{ V} - 0.7 \text{ V} = 9.3 \text{ V} \qquad \text{(Kirchhoff's voltage law)}$$

This can be used to calculate the dc ("idling") base current:

$$I_B = \frac{V_{RB}}{R_B} = \frac{9.3 \text{ V}}{20 \text{ k}\Omega} = 465 \ \mu\text{A}$$

The resulting dc collector current will be

$$I_C = \beta \times I_B = 100 \times 465 \ \mu\text{A} = 46.5 \text{ mA}$$

Note when observing the ac input that 800-μA p-p signal (+400 μA, −400 μA peaks) will ride on the 465-μA dc level set by R_B to yield peak base currents of

$$I_{B(\text{max})} = 465 \ \mu\text{A dc} + 400 \ \mu\text{A peak} = 865 \ \mu\text{A}$$

$$I_{B(\text{min})} = 465 \ \mu\text{A dc} - 400 \ \mu\text{A peak} = 65 \ \mu\text{A}$$

Now to find the peaks (maximum and minimum) at the collector side,

$$I_{C(\text{max})} = I_{B(\text{max})} \times \beta = 865 \ \mu\text{A} \times 100 = 86.5 \text{ mA}$$

$$I_{C(\text{min})} = I_{B(\text{min})} \times \beta = 65 \ \mu\text{A} \times 100 = 6.5 \text{ mA}$$

Finally, the amount of emitter current will be

$$I_E = I_C + I_B \qquad \text{(for maximum and minimum analysis)}$$

$$I_{E(\text{max})} = 86.5 \text{ mA} + 865 \ \mu\text{A} = 87.365 \text{ mA}$$

$$I_{E(\text{min})} = 6.5 \text{ mA} + 65 \ \mu\text{A} = 6.565 \text{ mA}$$

Thus, by using the factors noted in this section, all values associated with this simple transistor circuit can easily be calculated. Figure 7.11 illustrates the input and output waveforms for this circuit, based on these calculations.

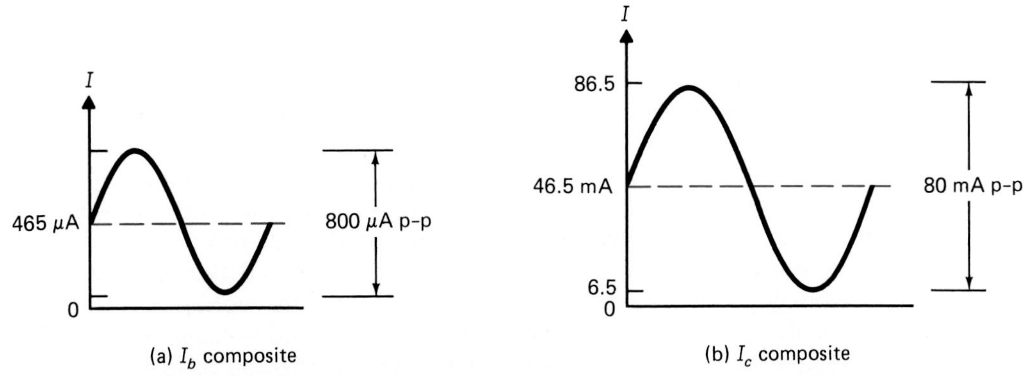

FIGURE 7.11 Composite of Input and Output AC and DC Currents for the BJT

REVIEW PROBLEMS

(1) If a common-emitter NPN transistor circuit was as shown in Figure 7.12, calculate the following values.
(a) I_B **(b)** I_C **(c)** I_E **(d)** V_{BE} **(e)** V_{CE}

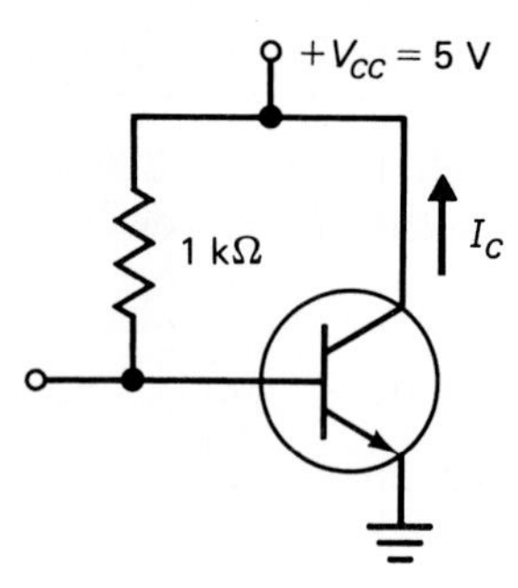

FIGURE 7.12 Simple BJT DC Biasing

(2) If the circuit in problem 1 had an ac input of 10 mA p-p, sketch the composite (ac + dc) input and the composite value of collector current.

7.4 STABILIZATION OF DISCRETE DEVICES

In this section a more detailed analysis will be made of the various stabilization circuits developed in Chapter 6. For the BJT, the simple emitter resistor stabilization circuit and the beta-independent circuit will be investigated in greater detail. Similar FET circuits will also be explored. For a general introduction to the functional aspects of these circuits, it is suggested that Section 6.7 be reviewed first.

Investigation of the Emitter Resistor Stabilization Circuit

The following is a more detailed analysis of the role played by the emitter resistor in the basis emitter resistor stabilization circuit (Figure 7.13). First, on the output side of this circuit there is a simple voltage divider of the V_{CC}. The voltage drop on the emitter resistor R_E is due to emitter current. The important feedback process is best seen when looking at the input circuit.

In past chapters it was seen that the base resistor R_B and the base–emitter section of the device formed a simple series circuit for the dc base current flow. The normal 0.7-V drop on the transistor simply left the balance of the supply voltage on the base resistor. Now, though, the voltage on the base resistor is also a by-product of the voltage on the emitter resistor.

The emitter resistor is now thrown in series with the base resistor and device. (Again the device creates the standard 0.7-V drop, that is, unaffected by the new emitter resistor.) What does happen is that there is a simple divider of the remaining voltage (V_{CC} − 0.7 V) between R_B and R_E. *But* the drop on the emitter resistor is due to the collector current through the device. Therefore, the voltage remaining for the base resistor is

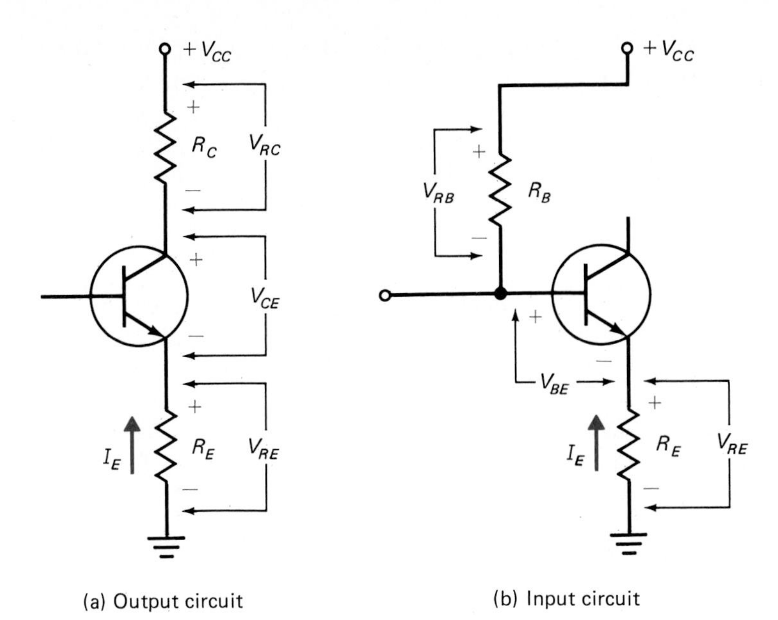

FIGURE 7.13 Division of the Input and Output Circuits for a BJT Amplifier

$$V_{RB} = V_{CC} - (V_{BE} + V_{RE}) \tag{7.1}$$

So far this is a simple Kirchhoff's law analysis which illustrates that the voltage on the base resistor in inversely proportional to the voltage on the emitter resistor. (If the voltage on R_E goes up, the voltage on R_B goes down, and vice versa.) The important factor is what R_B controls.

Since R_B sets the value of base current, it is a key element in determining the amount of collector current flow. If the voltage on R_B were to increase (from some reference value), the value of I_B would increase in proportion to this change. A drop in I_B will produce a drop in I_C through the beta of the transistor.

In Figure 7.14, some normal statistics relating these quantities are shown.

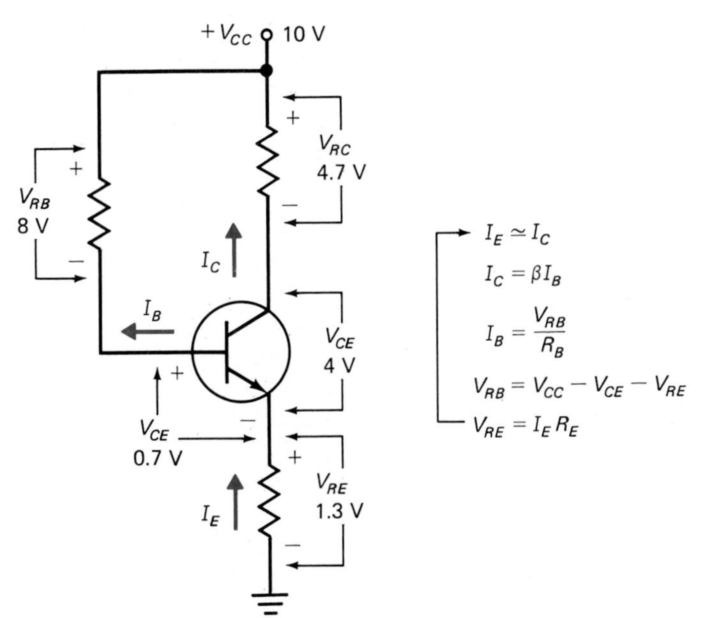

FIGURE 7.14 DC Bias Voltage Analysis for a BJT Amplifier Using an Emitter Resistor

First it is seen that I_B produces I_C through the device's actions. But it is seen that I_B is affected by the drop on R_E, caused by I_E (remember that $I_E = I_C + I_B$). What we have here is an interrelationship which creates stability in the circuit. The following analysis gives a basic rundown of what can happen if the emitter current of the device would shift due either to a change in I_C or an error in the value of I_B:

Initial Set of Conditions First it is assumed that the device has a stable flow of base and collector current, resulting in a stable emitter current.

Changing the Emitter Current If the emitter current of the device increased (for example) as a result of more base or collector current flow, the following events would occur:

1. The voltage on R_E would go up.
2. This increased voltage on R_E would result in a drop in the voltage on the base resistor R_B.
3. Decreased voltage on R_B will cause a drop in the base current I_B.
4. Decreased base current will cause collector current and thus emitter current to drop to a lower level.

System Regains Stability Remember that the problem was an increase in emitter current; now, due to this feedback, emitter current is reduced. The reduction places the value of I_E back (close) to the value it had prior to when the fault occurred initially.

The same type of comparison can be made if the emitter current were to decrease. The drop in this current would cause a lower voltage on R_E. This would produce more voltage on the base resistor R_B, therefore raising base current and ultimately emitter current. The circuit would be back to the value I_E had prior to the change. As can be seen from these descriptions, the actual value of I_E and therefore I_C was stabilized to a constant (or near-constant, depending on the total influence of the feedback on I_B).

Investigation of the Beta-Independent Circuit

The circuit shown in Figure 7.15 illustrates a beta-independent stabilized circuit. It is called beta independent because the value of the circuit's gain is not set by the beta of the device, but by simple resistor ratios. This circuit actually forces the value of I_C to be one specific value. The I_B is actually not controlled by the base resistor but is set by the demand for the circuit to produce a correct value of I_C. Naturally, how such a circuit is independent of beta and how it will achieve stability need further explanation.

We will take a look at the base circuit. We see first that there is a simple series combination of resistors R_{B1} and R_{B2}. These form a voltage divider of V_{CC}, independent of the rest of the transistor circuit. Figure 7.16 illustrates

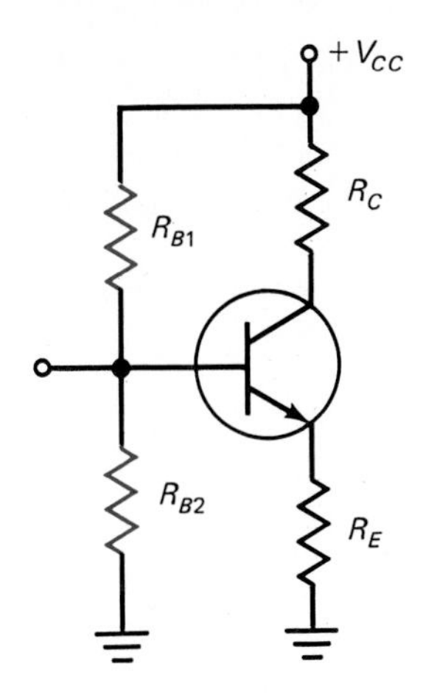

FIGURE 7.15 Beta-Independent BJT Amplifier

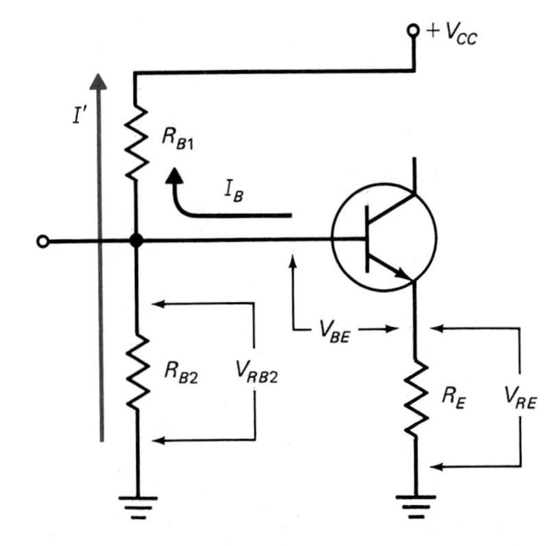

FIGURE 7.16 Input Circuit for the Beta-Independent BJT Amplifier

the input side of this circuit. Although the base–emitter junction and R_E are in parallel with the second base resistor (R_{B2}), they have little influence on the voltage dropped on that resistor. This is easily achieved if the current through R_{B1} and R_{B2} (shown as I' in Figure 7.16) is a value which is larger than the normal value of I_B used to bias the transistor. (If this current is a lot larger than the I_B current that splits off at the junction of R_{B1} and R_{B2}, then I_B will have little affect on the current drawn from V_{CC} by these two resistors.) Therefore, the voltage on R_{B2}, which is the voltage on the base of the transistor, is held to a value determined by the simple divider of R_{B1}/R_{B2}.

$$V_{RB2} = \frac{V_{CC} \times R_{B2}}{R_{B1} + R_{B2}} = V_B \text{ (base voltage on the BJT)} \tag{7.2}$$

If this voltage—the voltage on the base of the device—is held constant by the divider of R_{B1} and R_{B2}, the voltage at the emitter is a constant value. Emitter voltage is simply

$$V_{RE} = V_B - 0.7 \text{ V (normal forward drop)} \tag{7.3}$$

Now, if the voltage on the emitter is held to a constant level preset by the two base resistors, let's look at the influence on the output side of the device. Figure 7.17 illustrates the output side of this circuit, with the value of

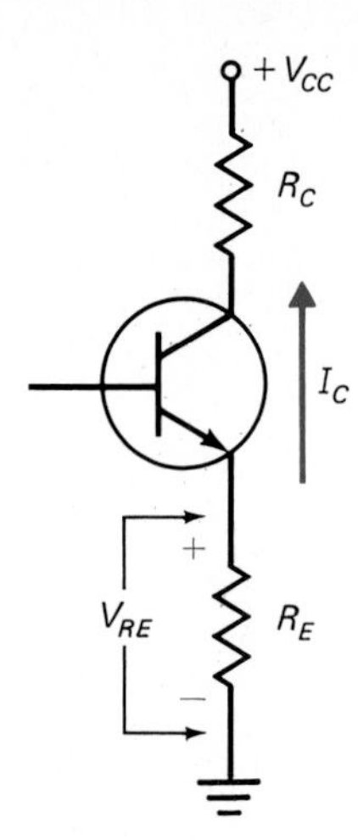

FIGURE 7.17 Output Circuit for the Beta-Independent BJT Amplifier

V_{RE} preset by the input side. It the voltage on R_E is preset, the current through it must also be preset. Since that current is I_E and since $I_E \simeq I_C$, the collector current must also be preset!

$$I_{RE} = \frac{V_{RE}}{R_E} = I_E \simeq I_C \qquad (7.4)$$

Therefore, the collector current is *not* set by the device, but is determined by the base divider circuit and the value of R_E. This results in an ac current gain equation (as shown in Chapter 6) which is expressed in terms of resistance ratios as:

$$A_i \text{ (current gain)} \simeq \frac{R_{B2}}{R_E} \qquad \text{(for ac inputs)} \qquad (7.5)$$

(Note again that this equation assumes that $R_{B1} \gg R_{B2}$, which is typically the case.)

Investigation of the FET Circuit Using R_S for Feedback

Just as stability is achieved using an emitter resistor in the BJT circuit, the source resistor that establishes gate voltage potential creates a feedback path for the FET (see Figure 7.18). Since this special arrangement has been explained before, only the negative-feedback process to compensate for stability problems will be discussed. The following is a simple step-by-step analysis of this process.

Initial Set of Conditions Under normal operation, one value of source current (equaling drain current) will flow. That value will create a drop on the resistor R_S which sets the value (by potential differencing) of gate bias voltage.

Changing the Source Current If there is a drop (for example) in the source current caused by some form of instability, the following would occur:

1. There would be a corresponding drop in voltage on the source resistor R_S.

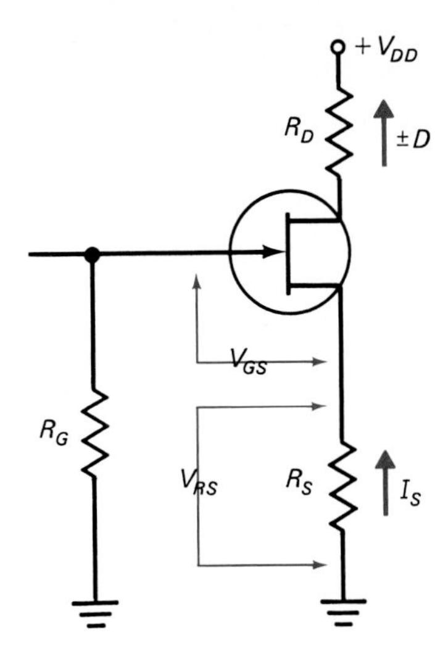

FIGURE 7.18 JFET Bias Resistors

2. This would cause a drop in the voltage potential difference between gate and source (V_{GS}).
3. This drop in gate voltage would cause an increase in the drain–source current.

Reminder: The output current of the FET is inversely proportional to the gate-to-source voltage!

Thus the device is restabilized using this source resistor as a feedback to the gate of the device. For most applications, this is the extent to which the JFET needs to have dc stabilization. This circuit is therefore the most common one used with the JFET.

Investigation of the Voltage Feedback Stabilization Circuit

In the voltage feedback circuit (see Figure 7.19), the voltage that is dropped on the collector of the transistor is used to provide the potential that, through the feedback resistor, creates base current I_B.

Initial Bias Arrangement Since this arrangement is new it will be necessary first to look at the initial bias arrangements, before stability is investigated. The base current that governs the amount of collector and emitter current is set by the voltage on R_F. That voltage, using Ohm's law, produces an amount of base current flow.

This voltage is set by the value of voltage that is dropped across the device (V_{CE}). The actual value of base current is therefore based on the drop on the device. That drop is dependent on the voltage across the collector resistor (R_C); $V_{CC} - V_{RC}$. [If there is an emitter resistor (R_E), its drop is also subtracted from V_{CC} to set the drop on the device.]

Since the drop on R_C (or R_C and R_E) depends on the value of I_C, which is set by I_B, the values of I_B and I_C are interrelated. Actually, to solve for either value a series of simultaneous equations must be developed. Without

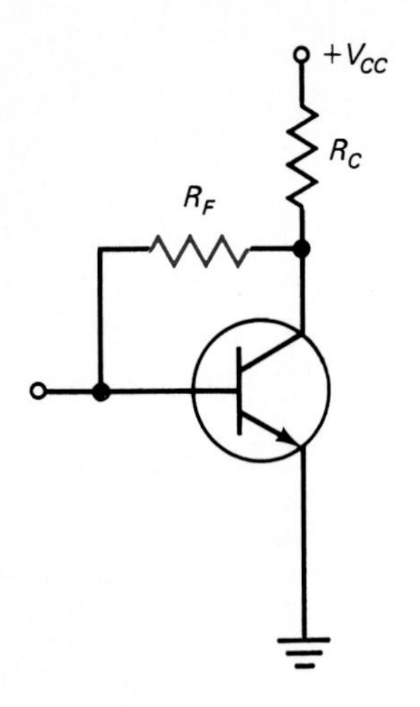

(a) Acting alone

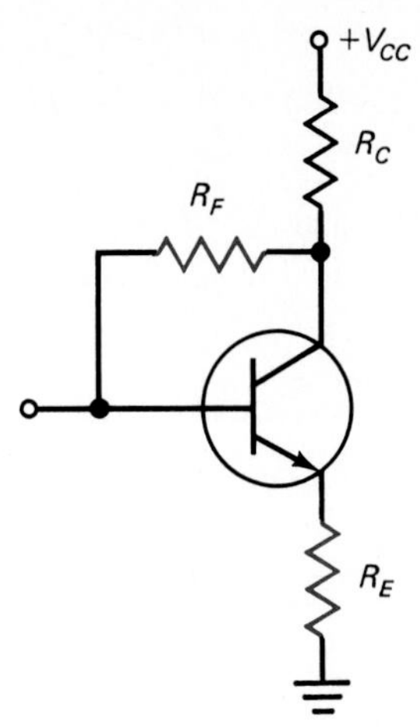

(b) With an emitter resistor

FIGURE 7.19 Voltage Feedback Bias Current for a BJT

going through the solution of these equations, the following gives an approximation of the equations for the initial bias levels:

$$I_B \simeq \frac{V_{CC}}{R_f + \beta(R_C + R_E)}$$

$$I_C = \beta I_B \tag{7.6}$$

$$V_{CE} = V_{CC} - I_C(R_C + R_E)$$

(If there is no R_E, simply omit it from these equations.)

Although the term β shows up in the equation for I_C, that does not mean that the system is still tied to the instability of β. Upon closer observation, note that β also shows up in the calculation for I_B. After solving these equations, the actual current gain of this circuit will become (approximately)

$$A_i \simeq \frac{R_F}{R_C + R_E} \tag{7.7}$$

(Again omit R_E if it is not used in the circuit.)

Concluding Remarks

These more detailed investigations of the functional role of stabilization/feedback resistors should assist in clarifying how stability is achieved in each of these popular circuits. There are many variations on these basic circuit forms. As these variations are encountered, most can be related to the primary forms shown in this section. Therefore, familiarity with both these equations and solutions and how they relate to Ohm's law and Kirchhoff's laws will assist the reader in evaluating these other circuit configurations.

Note: Equations used in this section, especially those for ac gain, are the more popular simplified forms of equations derived from complex analysis of the circuits used. Detailed derivations of these equations have been omitted to emphasize the application of them to the circuits involved.

REVIEW PROBLEMS

(1) Using the circuit shown in Figure 7.20, find the following values.
(a) V_{RE} **(b)** V_{RB} **(c)** V_C (collector voltage)

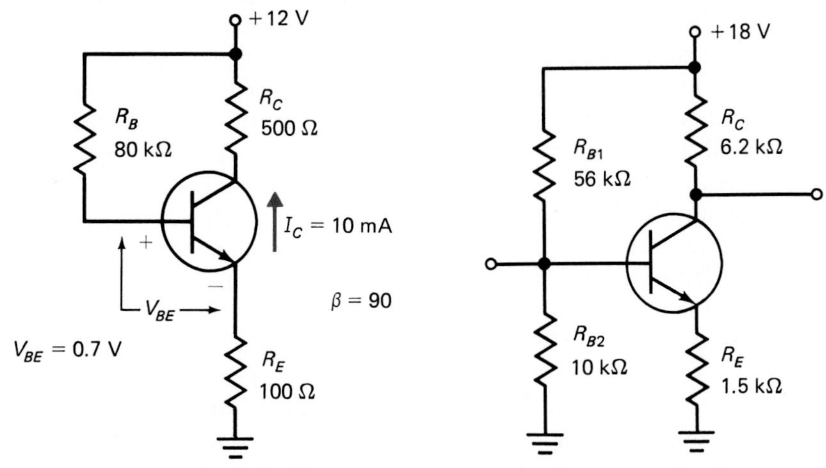

FIGURE 7.20 BJT Amplifier Using Emitter Resistor Stabilization

FIGURE 7.21 Beta-Independent Biasing for a BJT

(2) Using the circuit shown in Figure 7.21, find the following values.
(a) V_{RB2} **(b)** V_{RE} **(c)** I_C **(d)** A_i (current gain)

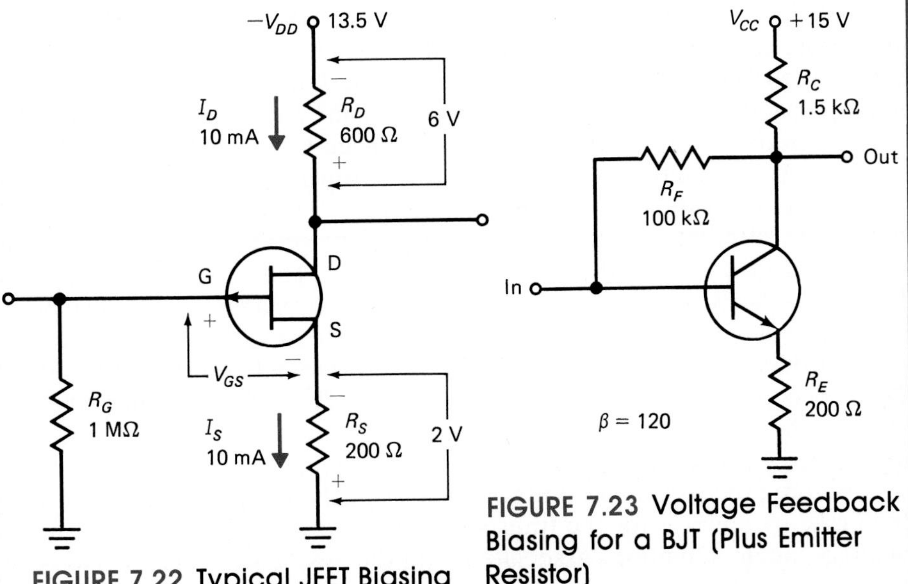

FIGURE 7.22 Typical JFET Biasing

FIGURE 7.23 Voltage Feedback Biasing for a BJT (Plus Emitter Resistor)

(3) Write a short description of how stability is achieved in the circuit shown in Figure 7.22 if there would be a momentary drop in the source current (I_S) through the JFET. Use a step 1, step 2, and so on, description.

(4) Using the voltage feedback circuit shown in Figure 7.23, find the following values.
(a) I_B **(b)** I_C **(c)** V_{CE} **(d)** A_i (current gain)

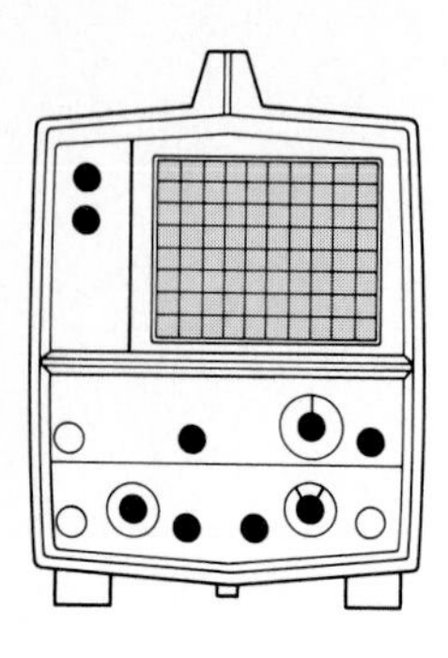

7.5 SUPERPOSITION ANALYSIS FOR AC AND DC CONDITIONS

In Chapter 6 capacitors were added to the basic dc bias circuits to bypass the feedback resistors or to isolate the input or output elements from the circuit's bias conditions. In this section we analyze each of these popular circuits to see that there are distinct dc and ac equivalents of each of them.

Outside of looking at ac current-gain equations, we have concentrated on the dc voltages and currents through the circuits. Now we will look at the flow of the signal (ac) through the circuit. Since the information signal enters these circuits from a different location than where the dc bias currents enter (from the power supply terminal), and since the output is often extracted across the device or in reference to the internal load resistor (R_C or R_D), the way the circuit looks from these different vantage points must differ from the way it looks when seen from the power supply.

To analyze each of these circuits, a two-part process will have to be used. The first is the normal dc bias analysis, done in previous sections. The second will be for the ac signal flow through the circuit. Basically, this is the second half of a superposition analysis. The composite solution, the dc added to the ac signal flow, will constitute a complete analysis of the system.

Analysis of BJT Circuits

Figure 7.24 illustrates this breakdown of a basic common-emitter circuit with a stabilization resistor R_E. In the dc model [part (b) of the figure] the input and output capacitors are opened and the capacitor C_E is opened. It can be seen that the dc conditions are isolated inside the circuit by the input and output capacitors. Therefore, this circuit remains the same as used before.

When observing the ac circuit, each capacitor is shorted (this presumes that their X_C is small). Also, the dc source is replaced with its internal resistance (0 Ω) from the V_{CC} to ground terminals. As can be seen, the circuit looks quite different from the dc model, especially since our point of reference for current flow is shifted to a new set of paths. This circuit is redrawn for the sake of clarity in Figure 7.25. Several noticeable changes have taken place. First, by replacing the dc source with 0 Ω (a short circuit) R_B is now from the base (through the short) to ground. R_C is also grounded through this short. R_E has been eliminated by the bypass capacitor. [It is of interest here that this circuit is the same as if R_E had not been used. R_E provides only dc stabilization and does not function (if bypassed) in the ac circuit.]

Before proceeding with the complete ac analysis of this circuit, there are certain ac parameters for the transistor that have yet to be explored. In past chapters the dc conditions associated with the device have been used. Now, it is necessary to describe certain special conditions associated with the BJT when operated with an ac signal. The following is a list of the (popular) *h* (for "hybrid") parameters for a BJT. These parameters are derived from statistical analysis of the device's performance using changing (ac) signals. These parameters are often supplied along with other specifications in manufacturers' data sheets which detail specific BJTs. (These parameters presume that the device is used on the common-emitter configuration.) The identification of these parameters is simply by calling out the three-letter name "*h-i-e*."

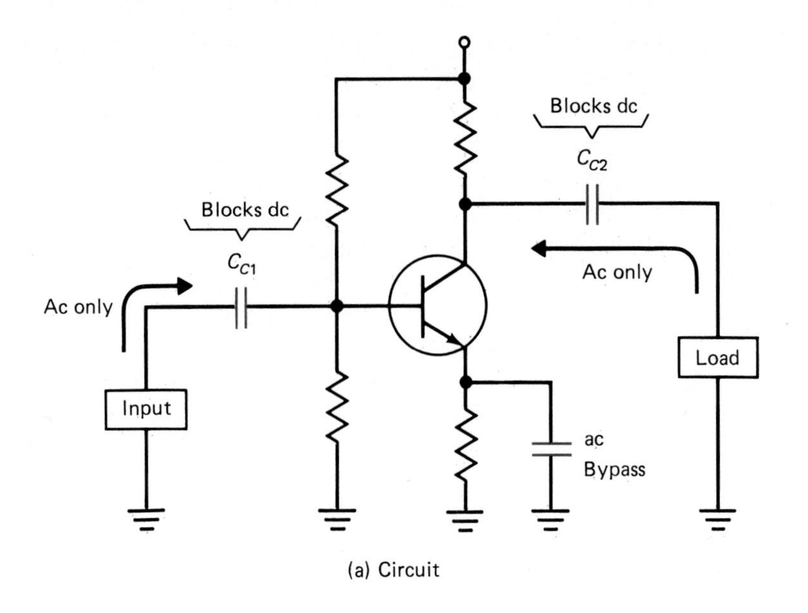

(a) Circuit

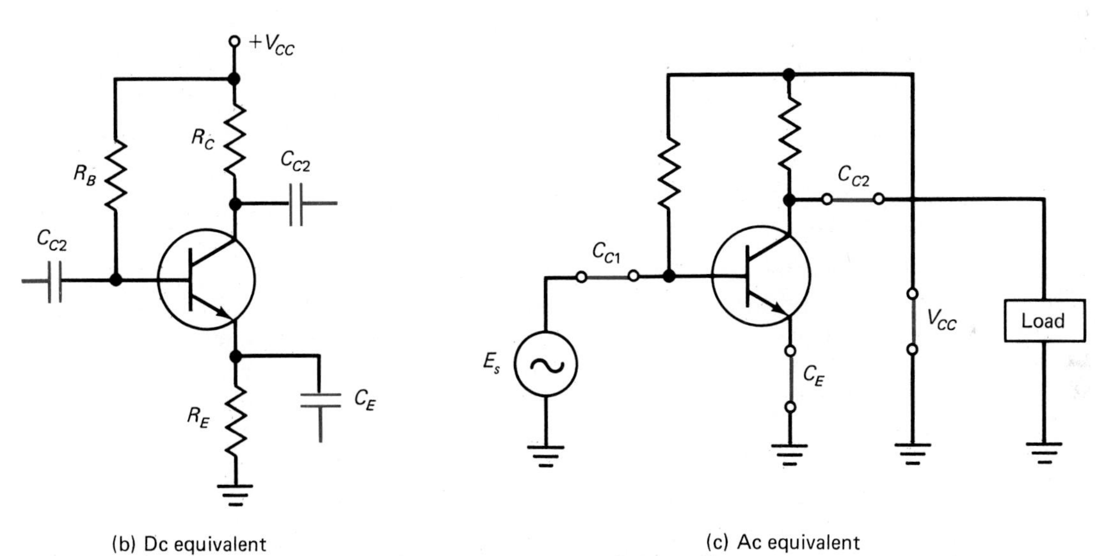

(b) Dc equivalent (c) Ac equivalent

FIGURE 7.24 Use of Input, Output, and Bypass Capacitors

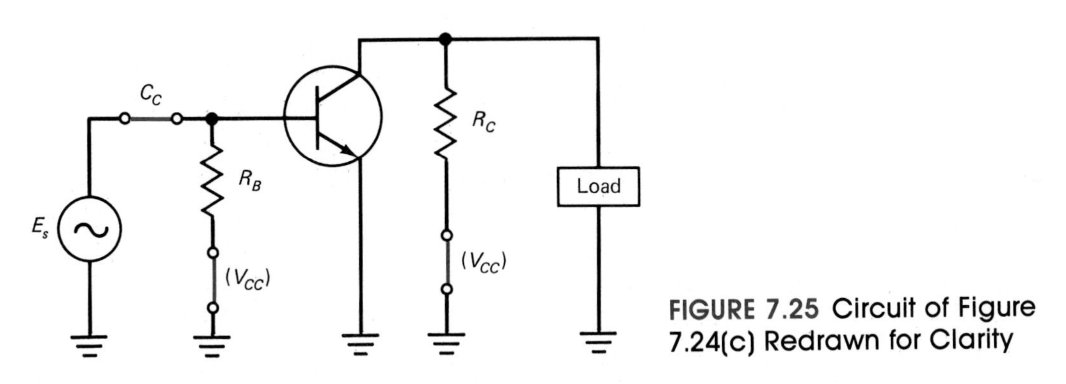

FIGURE 7.25 Circuit of Figure 7.24(c) Redrawn for Clarity

h_{ie} This parameter is the transistor's input impedance from the base to emitter. Typical values range from about 500 Ω to around 4 kΩ.

h_{re} This parameter is the transistor's reverse transfer voltage ratio. This parameter is not commonly used.

h_{fe} This parameter is the transistor's forward current transfer ratio, or ac current gain. Often the value for β is used in place of this term.

h_{oe} This is the output conductance of the transistor. Usually, for circuit analysis purposes, it is used to find the ac output resistance of the transistor by calculating $1/h_{oe}$. This resulting impedance is usually in the 100 kΩ range.

These parameters will be referred to as the transistor is placed into ac circuits. These parameters are generally not applicable to the dc bias conditions of the device.

Now it is necessary to determine the important characteristics for this version of the circuit. The following are the factors that would be important (generally) to the function of this circuit as seen by the input and load.

Special note: Since the following quantities are of importance to the ac analysis, they will be given standard ac notations: lowercase letters for voltages, currents, and so on, and the term "impedance" to refer to ac resistance factors.

Input Impedance (Z_i) It will be necessary to know the input impedance of the amplifier as seen from the input source. By knowing this, potential voltage drops, current limits, and maximum power transfer can be observed.

Output Impedance (Z_o) To adequately match a load to this circuit, output impedance must be determined for the amplifier. Matching the load to this value will assist in obtaining maximum power transfer to the load and in observing the current divider that exists between internal impedance and the load.

Current Gain (A_i) The ratio of ac output current to ac input current is the current gain (A_i). This gain value can vary from the value of dc current gain (β).

Voltage Gain (A_v) If there is a value of input current and impedance, there will be information about input voltage. If there are the same conditions associated with the output, the output voltage level can be found. The ratio of these two can be important, often more important than current gain alone.

Power Gain (A_p) If there is voltage gain and current gain, it follows that there must be a power gain. Unless otherwise specified, power gain will be the simple product of $A_i \times A_v$.

Note: In the following steps, the approximate equations for these quantities will be presented.

For the circuit shown in Figure 7.26 (a duplicate of Figure 7.25) the following are the equations for each of these quantities:

$$\text{Input impedance: } Z_i \simeq h_{ie} \parallel R_B \tag{7.8}$$

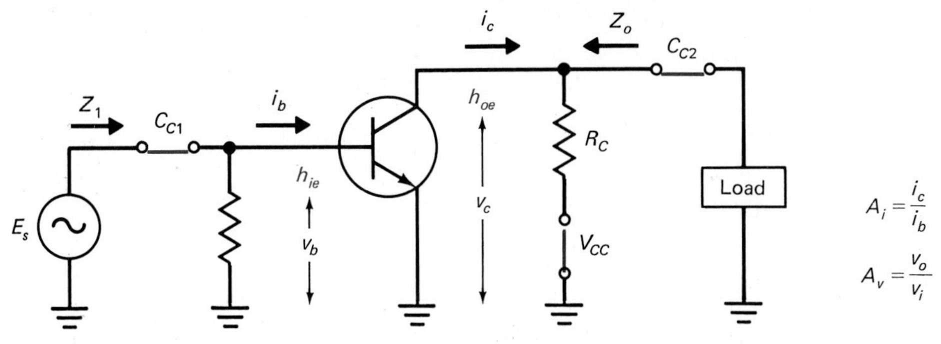

FIGURE 7.26 AC Equivalent Circuit Indicating Input and Output Impedances for the BJT and Identifying AC Voltages and Currents

[The base resistor is in parallel with the input resistance of the transistor (base to emitter).]

$$\text{Output impedance: } Z_o \simeq R_C \tag{7.9}$$

(This is generally true since the actual ac impedance of the device (h_{oe}) is usually quite high. Since R_c is in parallel with this device impedance, the total impedance is approximately $= R_c$.)

$$\text{Current gain: } A_i \simeq h_{fe}\beta \tag{7.10}$$

[Even though R_E provided stability and less dependence on β, it has been eliminated in this circuit. The gain is therefore for a simpler, nonstabilized circuit. The circuit's ac current gain is therefore simply the device's gain value (h_{fe}).]

$$\text{Voltage gain: } A_v \simeq \frac{(-1) \times h_{fe} \times R_c}{h_{ie}} \tag{7.11}$$

[This formula is basically one that relates input current and resistance (input voltage) to output current and resistance (output voltage). The h_{fe} term simply incorporates the ratio of output current to input current. The (-1) multiple simply implies a phase inversion (output voltage is 180° out of phase with input voltage).]

EXAMPLE 7.9

For the circuit shown in Figure 7.24, given the following circuit and device parameters, calculate Z_i, Z_o, A_i, and A_v. Also calculate input current (ac) output current (ac) and output voltage (ac).

$$R_C = 2\ \text{k}\Omega \qquad R_E = 200\ \Omega \qquad R_B = 200\ \text{k}\Omega$$

$$h_{ie} = 1\ \text{k}\Omega \qquad h_{re} = 2.5 \times 10^{-4} \qquad h_{fe} = 50\ h_{oe} = 25\ \mu\text{A/V}$$

$$V_{CC} = 20\ \text{V input voltage (ac)} = 100\ \text{mV p-p}$$

[h_{oe}, although rated as conductance, is usually stated as a current-to-voltage ratio (inverse of resistance!).]

Solution:

First, to obtain the device's output impedance,

$$\frac{1}{h_{oe}} = 40\ \text{k}\Omega$$

Now calculate the input impedance:

$$Z_i \simeq h_{ie} \parallel R_B = 1\ \text{k}\Omega \parallel 200\ \text{k}\Omega \simeq 1\ \text{k}\Omega$$

$$Z_o \simeq R_C = 2\ \text{k}\Omega$$

$$A_i \simeq h_{fe} = 50$$

$$A_v \simeq \frac{(-1) \times h_{fe} \times R_c}{h_{ie}} = \frac{-1 \times 50 \times 2\ \text{k}\Omega}{1\ \text{k}\Omega} = -100$$

$A_p = A_i \times A_v$ (standard equation for power gain) $= 50 \times 100 = 5000$

Then calculate the functional values for this circuit:

$$V_{\text{in}} = 100\ \text{mV p-p (above)}$$

$$I_{\text{in}} = \frac{V_{\text{in}}}{Z_i} = \frac{100\ \text{mV}}{1\ \text{k}\Omega} = 100\ \mu\text{A p-p}$$

$$I_{\text{out}} = I_{\text{in}} \times A_i = 100\ \mu\text{A} \times 50 = 5\ \text{mA p-p}$$

$$V_{\text{out}} = V_{\text{in}} \times A_v = 100\ \text{mV} \times (-100) = -10\ \text{V p-p}$$

[Since $V_{\text{out}}/I_{\text{out}} = Z_o$, these calculations can be checked to confirm that 10 V p-p/5 mA p-p = 2 kΩ. (Note that the sign on the V_{out} term was omitted since it only implies phase inversion, which does not affect impedance values.)]

Therefore, all of the important quantities, current gain, voltage gain, and input and output impedance are defined for the device. It should be noted that in the previous dc analysis, the factor of voltage gain was not considered. Since the BJT is a current-amplifying device, it was not expected to provide a stable voltage gain. Now, though, as long as the circuit remains stable with fixed values for the components used to support the BJT, voltage-in and volt-

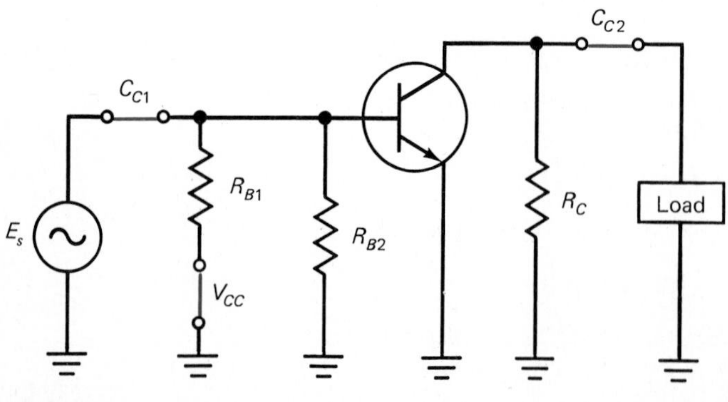

FIGURE 7.27 AC Equivalent Circuit for Beta-Independent BJT Biasing

age-out can be considered as being reasonably predictable *for the circuit*. Therefore, the gain value of output voltage to input can be estimated.

If the beta-independent circuit is used, only a slight modification to these equations need be made. The illustration of the ac equivalent circuit is shown in Figure 7.27. The only real change in the circuit is the addition of R_{B1} in parallel with R_{B2} instead of the single base resistor R_B. Factoring this new value of total base current into the previous equations will yield

$$Z_i \simeq R_{B1} \parallel R_{B2} \parallel h_{ie} \quad (7.12)$$

$$Z_o \simeq R_C \quad (7.9)$$

$$A_i \simeq h_{fe} \quad (7.10)$$

$$A_v \simeq \frac{(-1) \times h_{fe} \times R_c}{h_{ie}} \quad (7.11)$$

EXAMPLE 7.10

Using the following circuit elements and the same h parameters used in Example 7.9, find Z_i, Z_o, A_i, and A_v. Then, using the listed input voltage, find input current and output current and voltage.

$$R_{B1} = 50\ \text{k}\Omega \qquad R_{B2} = 5\ \text{k}\Omega \qquad R_C = 4\ \text{k}\Omega$$

$$R_E = 1\ \text{k}\Omega \qquad V_{CC} = 20\ \text{V with input voltage} = 20\ \text{mV p-p}$$

Solution:

$$Z_i = R_{B1} \parallel R_{B2} \parallel h_{ie} = 50\ \text{k}\Omega \parallel 5\ \text{k}\Omega \parallel 1\ \text{k}\Omega \simeq 820\ \Omega$$

$$Z_o \simeq R_C = 4\ \text{k}\Omega$$

$$A_i = h_{fe} = 50$$

$$A_v = \frac{(-1) \times h_{fe} \times R_C}{h_{ie}} = \frac{-1 \times 50 \times 4\ \text{k}\Omega}{1\ \text{k}\Omega} = -200$$

$$A_p = A_i \times A_v = 50 \times 200 = 10{,}000$$

Now figuring the input level that will produce an output level:

$$V_{\text{in}} = 20\ \text{mV p-p}$$

$$I_{\text{in}} = \frac{V_{\text{in}}}{Z_i} = \frac{20\ \text{mV}}{1\ \text{k}\Omega} = 20\ \mu\text{A p-p}$$

$$V_{\text{out}} = V_{\text{in}} \times A_v = 20\ \text{mV} \times (-200) = 4\ \text{V p-p}$$

$$I_{\text{out}} = I_{\text{in}} \times A_i = 20\ \mu\text{A} \times 50 = 1\ \text{mA p-p}$$

Thus it can be seen that the addition of components to the ac model will cause a change in these equations, but only by the series or parallel relationship of each of the added elements to those already in the circuit.

The one circuit that takes on radically different formulas for these quantities is the voltage feedback form of biasing. Since there is a unique connection between output and input through the feedback resistor, the ac model for that circuit will provide some unique relationships. The equations for the voltage feedback circuit shown in Figure 7.28 are as follows:

$$Z_i \simeq \frac{R_F}{A_v} \parallel (h_{fe} \times R_E) \tag{7.13}$$

$$Z_o \simeq R_c \tag{7.9}$$

$$A_v \simeq \frac{(-1) \times h_{fe} \times R_C}{h_{ie}} \tag{7.10}$$

$$A_i \simeq \frac{R_F}{R_C} \tag{7.14}$$

One may speculate that since this circuit differs noticeably from the previous ones shown, all the equations would be quite different as well. The equations for Z_o and A_v are the same as with the other circuits since R_F is typically so much larger than R_C and other resistances. This resistance indirectly in parallel with R_C will result in an equation nearly equal to R_C. In a similar manner, A_v is not noticeably affected by R_F as compared to these other circuits.

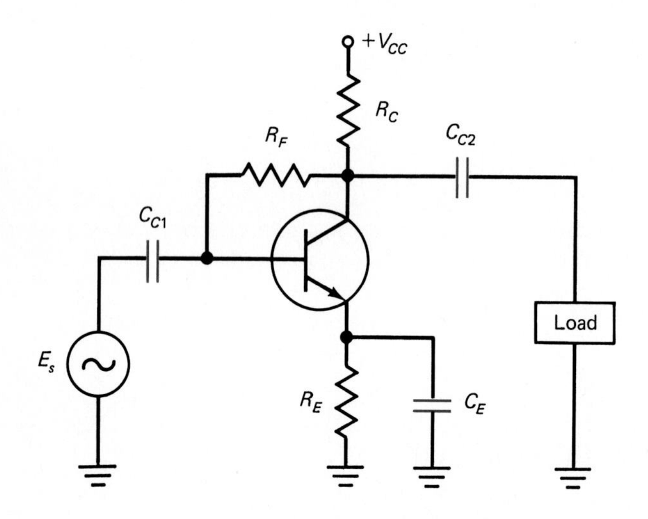

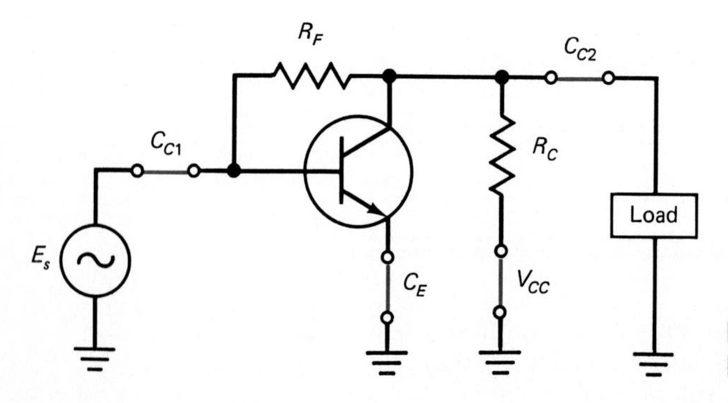

FIGURE 7.28 Voltage Feedback Biasing with AC Equivalent Circuit

EXAMPLE 7.11

For the voltage feedback circuit shown in Figure 7.28, using the following values, find Z_i, Z_o, A_v, and A_i. Also, given the input voltage listed, find the input current and the input and output current.

$$R_F = 40\ \text{k}\Omega \qquad R_C = 2\ \text{k}\Omega \qquad R_E = 200\ \Omega$$

$$V_{CC} = 20\ \text{V} \qquad V_{\text{in}} = 40\ \text{mV}$$

$$h_{ie} = 1\ \text{k}\Omega \qquad h_{fe} = 60 \qquad \frac{1}{h_{oe}} = 40\ \text{k}\Omega$$

Solution:

$$A_V = \frac{(-1) \times h_{fe} \times R_C}{h_{ie}} = \frac{-1 \times 60 \times 2\ \text{k}\Omega}{1\ \text{k}\Omega} = -120$$

$$A_i = \frac{R_F}{R_C} = \frac{40\ \text{k}\Omega}{2\ \text{k}\Omega} = 20$$

$$Z_i = \frac{R_E}{A_v} \parallel h_{fe} R_E = \frac{40\ \text{k}\Omega}{120} \parallel 60 \times 200\ \Omega \simeq 324\ \Omega$$

$$Z_o = R_C = 2\ \text{k}\Omega$$

$$A_p = A_i \times A_v = 20 \times 120 = 2400$$

Now for operational values:

$$V_{\text{in}} = 40\ \text{mV p-p}$$

$$V_{\text{out}} = 40\ \text{mV} \times (-120) = -4.8\ \text{V p-p}$$

$$I_{\text{in}} = \frac{40\ \text{mV}}{324\ \Omega} = 123\ \mu\text{A}$$

$$I_{\text{out}} = I_{\text{in}} \times A_i = 123\ \mu\text{A} \times 20 = 2.46\ \text{mA}$$

At this point a composite chart of each of these popular circuits, plus some of the variations used in previous chapters, will help to clarify all of these new equations (see Figure 7.29).

Final Superposition Composite Analysis In the beginning of this section it was proposed that the analysis of this circuit would involve a superposition process. So far we have looked at dc bias levels and at ac signal levels. To compose a final analysis, we simply need to blend both values into one composite solution. The circuit in Figure 7.30 shows both ac and dc voltages and currents that may result from a typical set of independent solutions.

Circuit | ac model | Equations

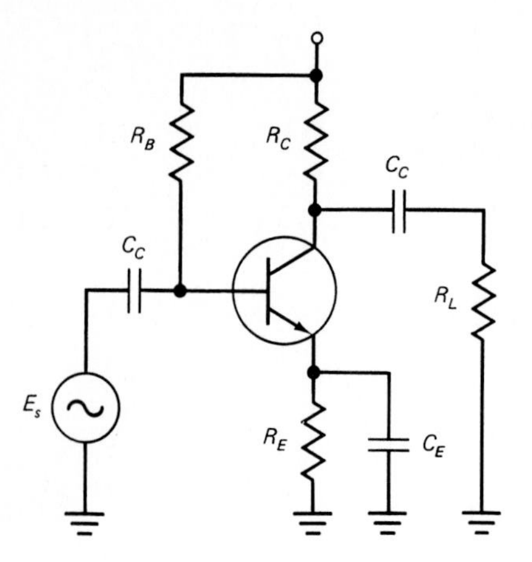

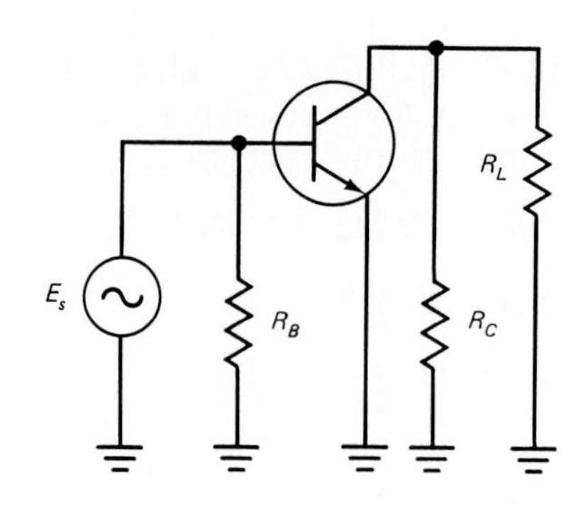

$Z_i = h_{ie} \| R_B$

$Z_o \simeq R_C$

$A_i \propto h_{fe}\ (\beta)$

$A_v \simeq \dfrac{-h_{ie} \times R_C}{h_{ie}}$

(a) Simple BJT biasing with emitter feedback resistor

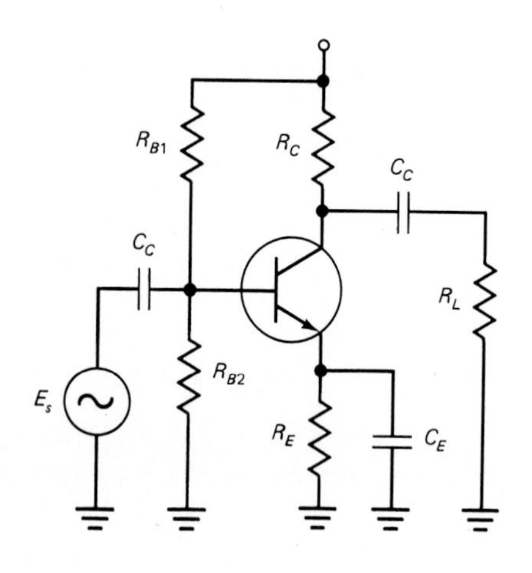

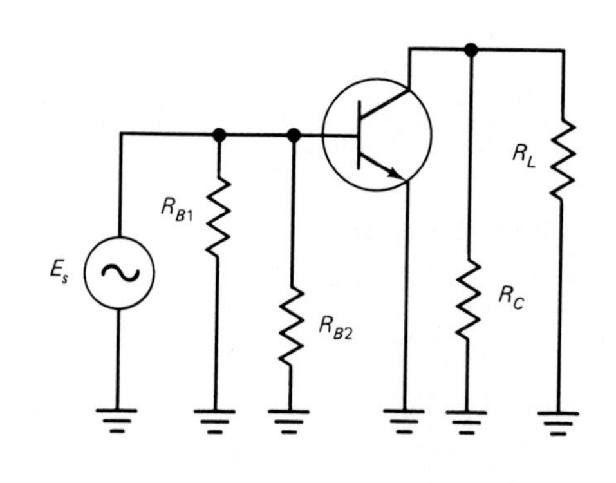

$Z_i = R_{B1} \| R_{B2} \| h_{ie}$

$Z_o \simeq R_C$

$A_i \simeq h_{fe}$

$A_v \simeq -\dfrac{h_{fe} R_C}{h_{ie}}$

(b) Beta-independent bias

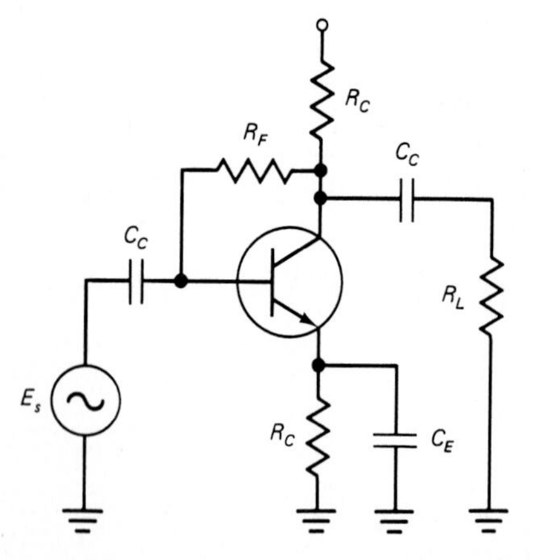

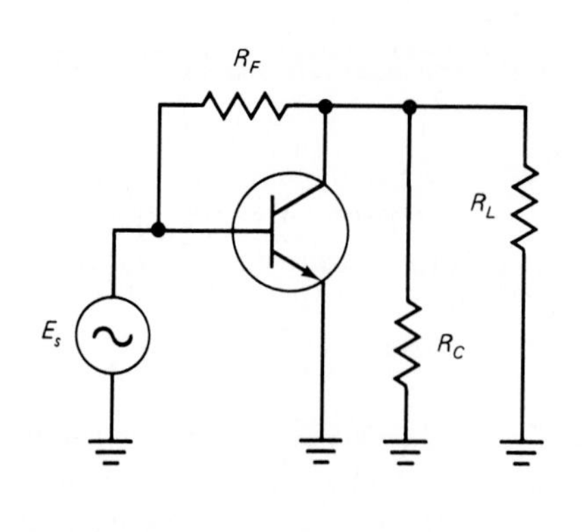

$Z_i \simeq \dfrac{R_E}{A_v} \| h_{fe} \times R_E$

$Z_o \simeq R_C$

$A_v \simeq \dfrac{-h_{fe} R_C}{h_{ie}}$

$A_i \simeq \dfrac{R_F}{R_C}$

(c) Voltage feedback bias with emitter resistor

FIGURE 7.29 Typical Circuitry, AC Model, and Typical Equations for Gains and Impedances

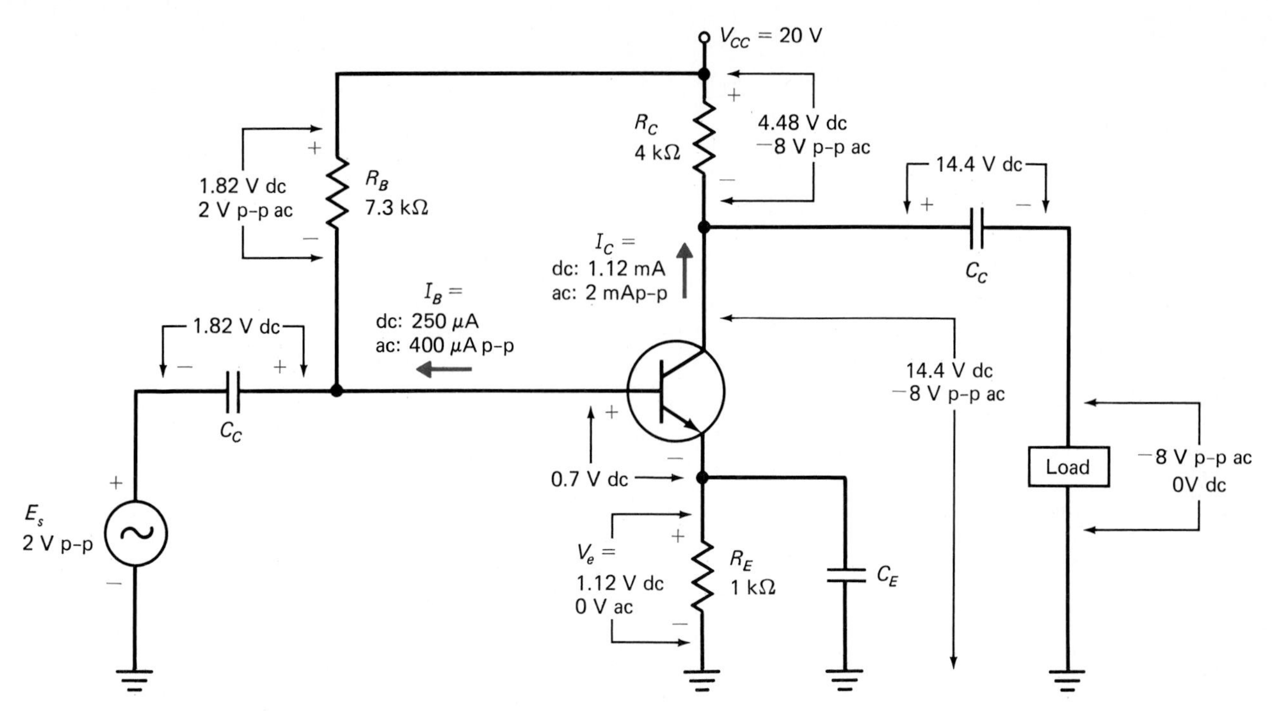

FIGURE 7.30 BJT Amplifier with All DC and AC Voltages Labeled

First compose the base voltage level and current level:

Total base voltage: 1.82 V dc with 2 V p-p ac
Total base current: 250 μA dc with 400 μA p-p ac

Now compose the levels on the output side:

Total collector current: 1.12 mA dc with 2 mA p-p ac
Collector resistor voltage: 4.48 V dc with −8 V p-p ac
Total emitter voltage (on R_E): 1.12 V dc with 0 V ac (bypassed by C_E)
Total emitter current: 1.12 mA dc with 2 mA p-p ac
Voltage from collector to ground: 14.4 V dc with −8 V p-p

Since the output capacitor would eliminate the dc level from the voltage on the collector resistor (and device), the output would simply be the −8 V p-p ac. (Note phase inversion!) Each of these voltages, when resolved by superposition, would appear (when measured on an oscilloscope) as shown in Figure 7.31.

Analysis of FET Circuits

The analysis of FET circuits is much easier than that for BJT circuits. There are not as many configurations of the FET circuit (not as many stability problems) and since the FET's gate lead is considered an open circuit, the input conditions are usually quite simple. As well, since the output side of the FET

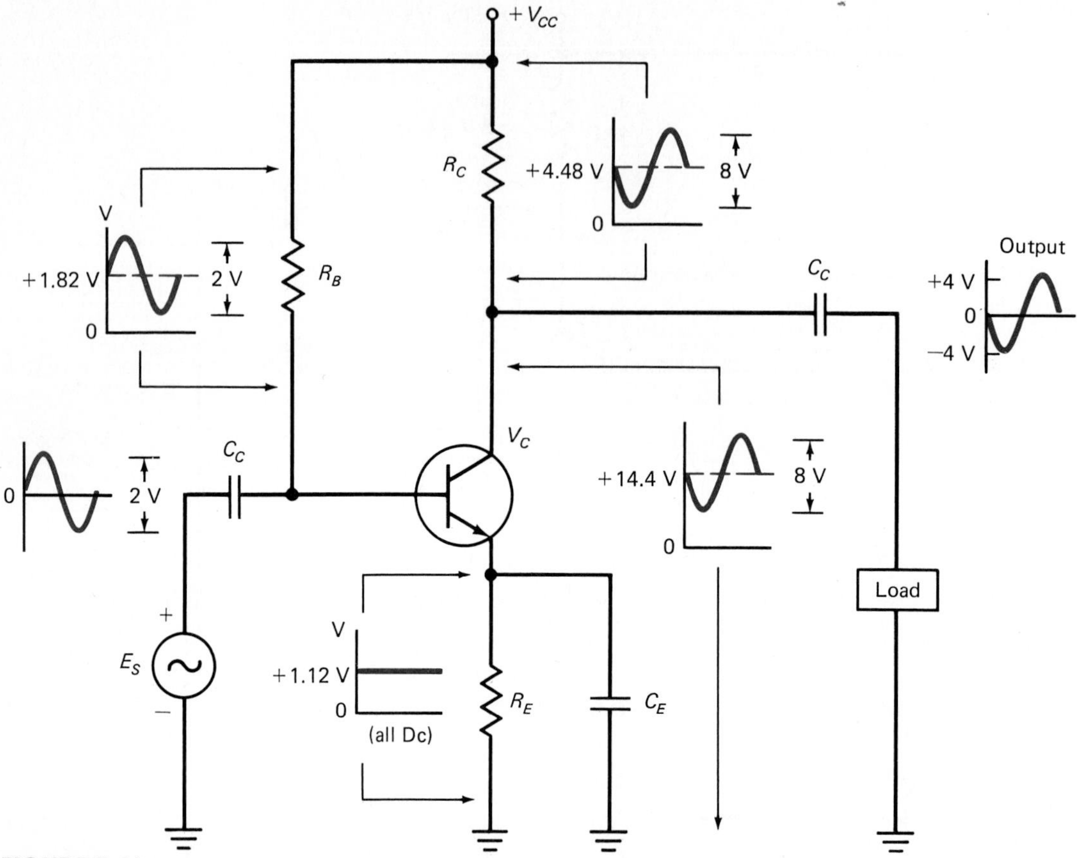

FIGURE 7.31 Reproduction of Figure 7.30 Showing Composite DC and AC Waveforms

is quite similar to that of the BJT, the equations will resemble those for the BJT.

Before we proceed, we need to describe the important ac parameters for the FET. Following is a summary of these (basic) parameters.

Input Impedance There is no specific term used to identify input impedance except Z_{in}. Typically, input impedance is much greater than 1 mΩ (100 mΩ is not uncommon).

Output Impedance (r_{ds}) Drain-to-source resistance, ac. This figure may be given as an output admittance—y_{ds} or y_{os}. Simply invert this value to obtain resistance: $r_{ds} = 1/y_{os}$. Also, this figure is not a fixed value (since the FET is a transconductance device. In Chapter 8 further details will be presented to elaborate on this value. Typical values range from about 10 kΩ to 1 MΩ.

Transconductance (g_m) The transconductance of an FET is also variable. It is affected by the voltage values on the device. There is one value of g_m which is typically given by manufacturers and that is listed as either

g_{mo} or y_{fs}. These values relate to a zero voltage condition on the gate of the device (unbiased). Actual operational values of g_m are then related to this base value, corrected by the operational value of gate voltage.

$$g_m \text{ (operational value of } g_m) = g_{mo}\left(1 - \frac{V_{GSQ}}{V_P}\right) \qquad (7.15)$$

where g_{mo} = data sheet value of g_m at 0 V_{GS}
V_{GSQ} = operational in-circuit value of gate-to-source voltage

$V_{GS(\text{off})}$ Sometimes called V_P; gate voltage value where conduction is reduced to zero on the drain/source side ($I_D \simeq 0$). This specification is explained in greater detail in Chapter 8.

For the sake of simplicity in this chapter, the active value of g_m (precalculated) will be used. There is no figure for current gain for the FET since the gate current is negligable.

Figure 7.32 illustrates a typical JFET circuit. The equations for input and output impedance, voltage gain, and current gain for the ac signal will be

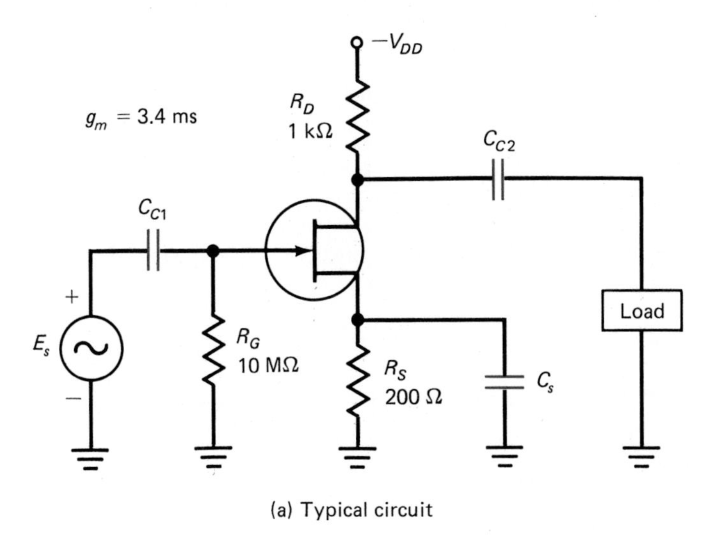

(a) Typical circuit

(b) Ac equivalent

FIGURE 7.32 Typical JFET Bias Circuit and AC Equivalent

shown. The basic ac circuit equations are as follows for Figure 7.32:

$Z_i \simeq R_G$ (since input impedance of the FET is assumed to be nearly infinite) (7.16)

$Z_o \simeq R_D$ (since the output impedance is much higher than R_D) (7.17)

$A_v \simeq (-1) \times g_m \times R_D$ (7.18)

g_m is precalculated from data sheet values of g_{mo} and the circuit's V_{GS}. A_i is undefined in terms of the device since there is no current drawn through the gate lead. If A_i is necessary, it must be calculated by a simple i_{out}/i_{in} ratio.

The following is the solution of the impedance and gain values for the circuit in Figure 7.32.

EXAMPLE 7.12

For the circuit shown in Figure 7.32, calculate the impedance and gain values.

$$Z_i = R_G = 10\ \text{M}\Omega$$

$$Z_o = R_D = 1\ \text{k}\Omega$$

$$A_v = (-1) \times g_m \times R_D = -1 \times 3.4\ \text{mS} \times 1\ \text{k}\Omega = -3.4$$

A_i is basically undefined.

Solution:

Compute actual values due to the ac and dc circuitry. Due to dc bias, the following levels exist:

$$I_D \simeq 5\ \text{mA} \qquad V_{GS} \simeq -1\ \text{V} \qquad V_{RD} \simeq 5\ \text{V}$$

For the ac conditions,

$$V_{out} = A_v \times V_{in} = -3.4 \times 15\ \text{mV} = -51\ \text{mV p-p}$$

To find output current ac, simply divide V_{out} by Z_O:

$$\frac{-51\ \text{mV}}{1\ \text{k}\Omega} = -51\ \mu\text{A p-p}$$

[To confirm the unrealistic value of current gain, a computation of ac input current would give $I_{in} = 15\ \text{mV}/10\ \text{M}\Omega = 1.5\ \text{pA}$ (picoamperes!). Therefore, the current gain would be

$$A_i = \frac{I_{out}}{I_{in}} = \frac{-51\ \mu\text{A}}{1.5\ \text{pA}} = -34{,}000$$

Further confirming the unrealistic value of power gain A_p,

$$A_p = A_v \times A_i = 115{,}600$$

Although these values are based on valid calculations, their unrealistically high value gives a false illusion of potential gain. That is why these values are not considered when evaluating an FET.] Figure 7.33 illustrates the composite (dc + ac) voltages for this FET circuit.

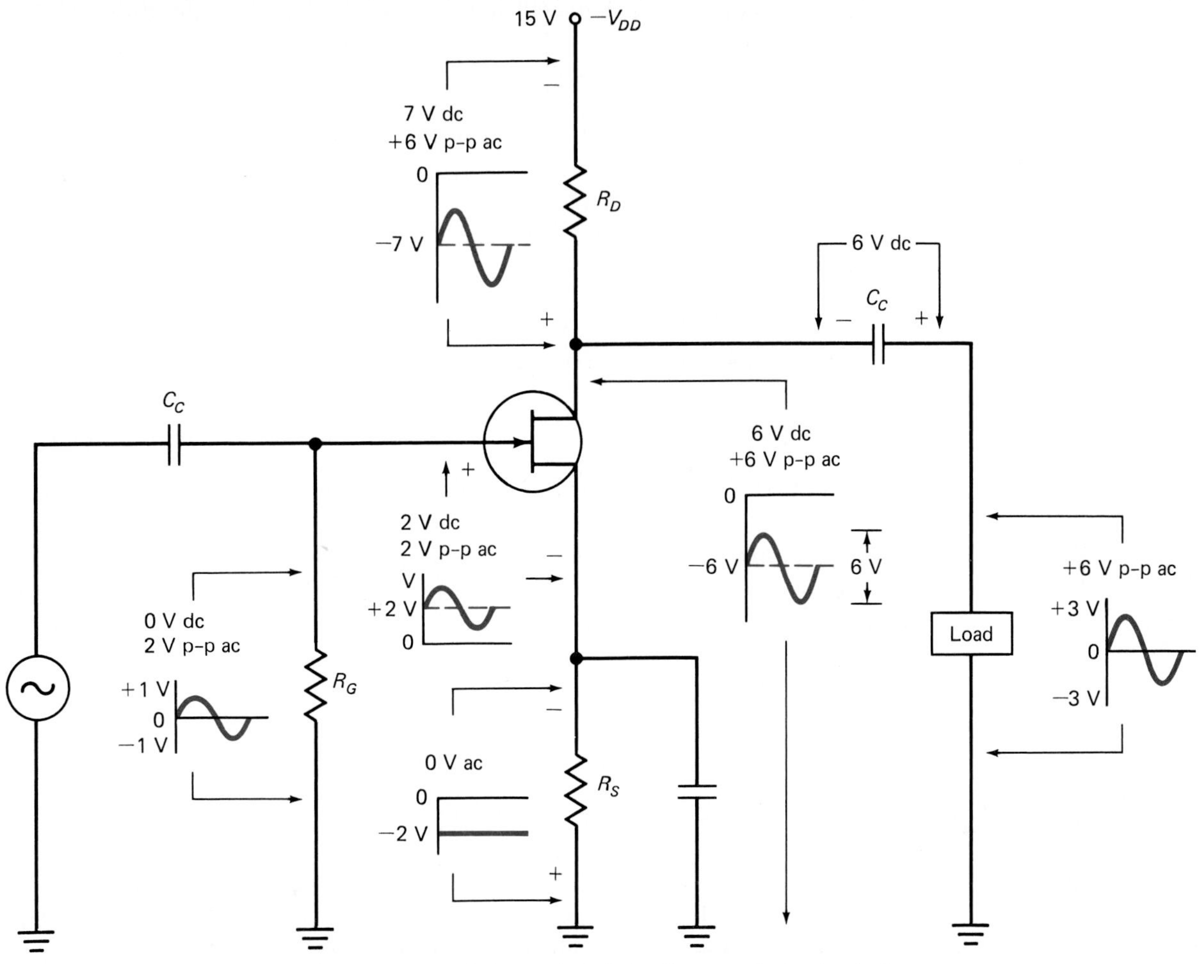

FIGURE 7.33 JFET Amplifier with All DC and AC Voltages Labeled, Also Showing Composite Waveforms

Final Superposition Composite Analysis In this section the equations that define the ac figures for discrete amplifiers have been presented. These are simply related to common circuit analysis practices, functional device characteristics, and the normal relationships for gain equations. The solutions of the dc and ac equivalent circuits are composed to form a composite solution, related to the composite solution obtained from any other superposition analysis.

Special Consideration Due to External Resistances

In all of the examples presented in this chapter, the only load listed for the amplifier was the collector resistor for the BJT circuit and the drain resistor for the FET. If the ac output of these circuits, through the coupling capacitor, is to be used by an eternal system (e.g., a speaker, another amplifier, etc.), the impedance of that system will have some effect on the circuit. In cases where the load (R_C for a BJT or R_D for an FET) is used to calculate an ac gain value, an external load will have an effect on these equations.

In Figure 7.34, the relationship of a load (here represented by R_L) to the internal load on the device (R_D or R_C) is shown. Simply, this load is in parallel with the internal load value and a normal parallel resistance equation can be used to find the "true" ac load on the amplifier. Because of the use of the capacitor to attach this load, the circuit's dc values are totally unaffected by this resistor.

$$R_{ac'} \text{ (true ac load on amplifier)} = R_{\text{internal}} \parallel R_L \qquad (7.19)$$

The formulas that the external load resistor will effect are as follows:

For a BJT:

Simple circuit, with or without RE: A_v (voltage gain) (assuming bypass capacitor on R_E)

Beta-independent circuit: A_v (voltage gain)

Voltage feedback: A_v (voltage gain), A_i (current gain), Z_i (input impedance) (only because it affects A_v)

For an FET:

Standard bias circuit: A_v (voltage gain)

In each of these equations, if a load resistance is known, the actual value of output load should be adjusted to reflect that the load resistor is in parallel with the internal load. It should be noted that the external load *does not* affect the value of Z_o for the circuit. That impedance is for the circuit alone; actually, it is its impedance as seen from the R_L position.

In most applications it is important to obtain the maximum power transfer out of the amplifier to the external load resistor R_L. Therefore, it is common practice to select a load resistance that will match the internal resistance of the circuit (Z_o). Or, if the load cannot be changed, an attempt would be made to change the output impedance of the amplifier (Z_o) to the intended load.

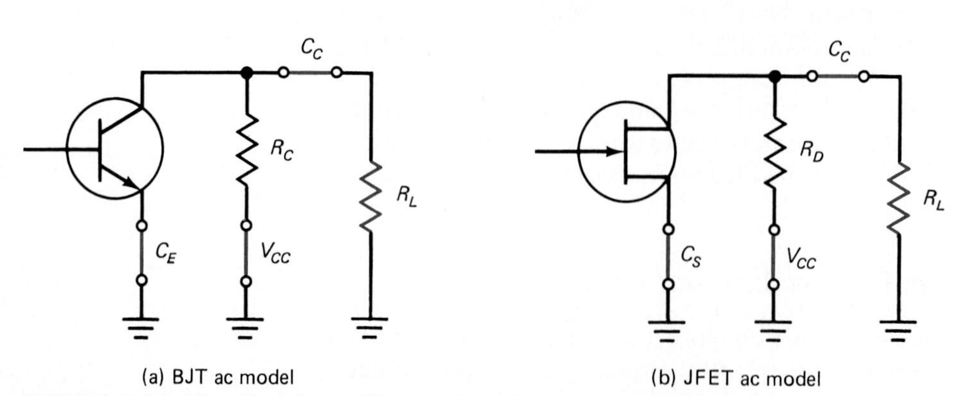

FIGURE 7.34 Parallel Condition of Load and Internal Load

Since the load resistor is in parallel with the internal resistance of the amplifier, they form a current divider for the ac output circuit (see Figure 7.34). If the load is properly matched ($R_L = Z_o$) to the internal impedance, they form a perfect 50/50 divider for the output current (they share the same voltage). As well, because of the influence of R_L on the voltage gain equations (again assuming that $R_L = Z_o$), the output voltage will be about one-half of that potential output if it were not there (R_L = open). The accumulated affect of this load then is (generally)

$$\text{output current to } R_L = \tfrac{1}{2} \text{ of the potential } I_{\text{out}}$$

$$\text{output voltage to } R_L = \tfrac{1}{2} \text{ of the potential } V_{\text{out}}$$

Therefore,

$$\text{output power to } R_L = \tfrac{1}{4} \text{ of the potential power output of the circuit}$$

(This does match the maximum power transfer formulas seen in standard circuit analysis problems.) Thus it can be seen that if an external load is attached to these circuits, it will have an effect on output quantities. That effect, though, is easily predicted by using standard circuit analysis techniques.

In a similar manner, if the input source to the circuit has an internal impedance, special circuit analysis may be required for the ac model. But unlike a load resistor, that impedance will not affect the operational equations for the circuit (Z_i, Z_o, A_v, A_i, or A_p). The only effect seen there would be a simple voltage divider of the input potential of the source between its internal resistance and the input impedance (Z_i) of the amplifier.

REVIEW PROBLEMS

(1) Prepare a chart that summarizes all of the basic ac and composite equations for the following circuits.
  (a) BJT beta-independent circuit
  (b) BJT voltage feedback circuit
  (c) FET basic circuit (as in Figure 7.32)
  This chart will help you summarize the methods used to find composite values.

(2) The voltage, current, and power that are present at the output of each of these circuits are larger in magnitude than those supplied by the input source. Write a brief description as to where these increased energy levels come from (the device, the ac source, power supply, etc.).

(3) The ac and dc voltages and currents associated with Figure 7.30 have been calculated. It was noted in some of the first chapters of this book that amplifier systems are not highly efficient. To verify this comment, total the amount of power supplied by the ac source (rms), the power supplied by the dc source, and the ac power supplied at the output of the circuit (rms).

$$\text{dc power input} \simeq V_{CC} \times (I_{RB1} + I_C)$$

$$\text{ac power input} = V_{\text{in(rms)}} \times I_{\text{in(rms)}}$$

$$\text{ac power output} = V_{\text{out(rms)}} \times I_{\text{out(rms)}}$$

Add up both the dc power and the ac source power. This value is the total power input to the circuit. Now calculate the system's efficiency (how much of the input power is used to create a usable output) by the following equation:

$$\text{efficiency (\%)} = \frac{P_{\text{out}}}{P_{\text{in(total)}}} \times 100\%$$

If this efficiency is less than 100%, more power was consumed from all the sources than was produced at the output. Write a brief description of how the excess power is being consumed.

(4) Using the equations given prior to all the examples in this section and copying down the values used in each of these examples, practice solving for each of the values without copying the procedures shown in the examples. Use the examples only to check your work.

(5) If a new device was invented which had the following parameters, compose a sample set of equations to define Z_i, Z_o, A_i, and A_i for the device, given that it was used in the circuit shown in Figure 7.35

$$d_{ix} = \text{input impedance}$$

$$d_{ox} = \text{output impedance}$$

$$d_{fx} = \text{forward voltage gain}$$

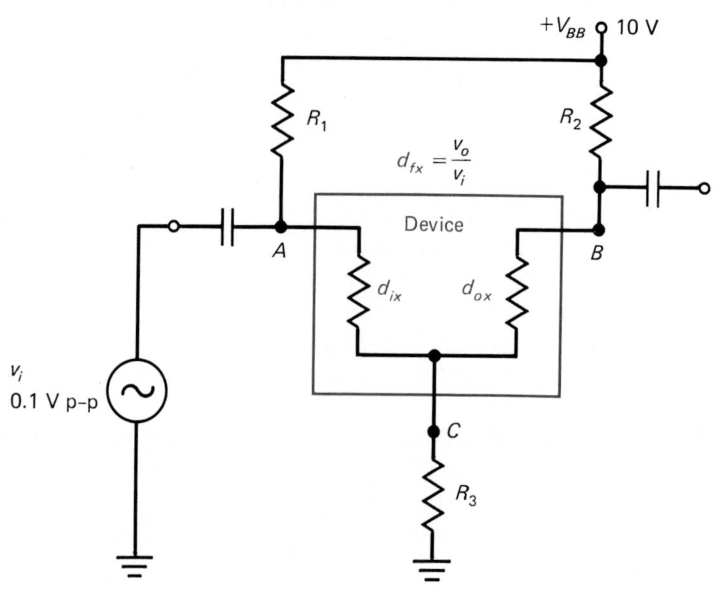

FIGURE 7.35 Ficticious Device Showing Ficticious Parameters

(To compose current gain, simply write and simplify a formula for the ratio of $I_{\text{out}}/I_{\text{in}}$, where each current is based on an Ohm's law equation. Be sure to substitute the formula for output voltage based on input before simplifying! This value will contain, when simplified: d_{fx}, d_{ix}, d_{ox}, R_1, R_2, and R_3.)

(6) For problem 5, if the following values were given, solve for output voltage and current—ac only:

$d_{ix} = 5\ \text{M}\Omega \qquad d_{ox} = 100\ \Omega \qquad d_{fx} = 45$

$R_1 = 200\ \text{k}\Omega \qquad R_2 = 2\ \text{k}\Omega \qquad R_3 = 100\ \Omega$

$V_i = 0.1$ V p-p dc source = 10 V (dc calculations not needed)

(7) To incorporate the effect of a load resistor on a typical circuit, add a load resistor ($R_L = R_C = 4\ \text{k}\Omega$) to the circuit used in Example 7.10. Repeat the calculations; this time use an R' for the collector load value (instead of R_C). Write a brief description of the effect of adding this resistor to the circuit.

SUMMARY

In this chapter quite a few equations and formulas were developed for the BJT and JFET. Both the dc bias and ac signal flow equations have been presented for a variety of device support circuits. At the end of this summary, a chart is presented that summarizes these dc and ac equations. It should be noted that most of these formulas and equations were developed using common circuit analysis practices. Each circuit was shown to have different dc and ac equivalent circuits. These circuits were independently solved for voltage and current levels around the circuit. A composite solution that included the dc and ac results constituted the final answers for each circuit's voltage and current levels.

It was also demonstrated that the significant factors for the ac model, the one that translates an input into an output, could be summarized into five distinct factors:

- Z_i Input impedance as seen from the source terminals.
- Z_o Output impedance as seen from the output connection (capacitor).
- A_v Voltage gain for the circuit. Even if the device is not rated with a voltage-gain factor, the final circuit can represent a value for voltage gain, since internal components such as resistors are fixed in value.
- A_i Current gain for the circuit. This figure was most significant for the BJT, but even the FET had a current-gain figure that could be computed.
- A_p Power gain, a product of voltage and current gain.

Future chapters will simply presume that these five factors can be defined for any amplifying circuit. Therefore, it will be assumed that any amplifier that can meet the requirements of the application will be usable. Since input and output capacitors are used (in most amplifiers), it is unnecessary to complicate the application of these amplifiers with calculations of their dc bias conditions. These amplifiers are described well enough internally that we can use them as functional amplifier blocks in a block diagram.

KEY EQUATIONS

7.1 $V_{RB} = V_{CC} - (V_{BE} + V_{RE})$

7.2 $V_{RB2} = \dfrac{V_{CC} \times R_{B2}}{R_{B1} + R_{B2}} = V_B$ (base voltage on the BJT)

7.3 $V_{RE} = V_B - 0.7\text{ V}$ (normal forward drop)

7.4 $I_{RE} = \dfrac{V_{RE}}{R_E} = I_E \simeq I_C$

7.5 A_i (current gain) $= \dfrac{R_{B2}}{R_E}$ (for ac inputs)

7.6 $I_B = \dfrac{V_{CC}}{R_f + \beta(R_C + R_E)}$

7.7 (current gain) $A_i = \dfrac{R_F}{(R_C + R_E)}$

7.8 $Z_i \simeq h_{ie} \parallel R_B$

7.9 $Z_o \simeq R_C$

7.10 $A_i \simeq h_{fe}$(beta)

7.11 $A_v \simeq \dfrac{(-1) \times h_{fe} \times R_c}{h_{ie}}$

7.12 $Z_i \simeq R_{B1} \parallel R_{B2} \parallel h_{ie}$

7.13 $Z_i \simeq \dfrac{R_F}{A_v} \parallel (h_{fe} \times R_E)$

7.14 $A_i \simeq \dfrac{R_F}{R_C}$

7.15 g_m (operational value of g_m) $= g_{mo}\left[1 - \left(\dfrac{V_{GSQ}}{V_p}\right)\right]$

7.16 $Z_i \simeq R_G$

7.17 $Z_o \simeq R_D$

7.18 $A_v \simeq (-1) \times g_m \times R_D$

7.19 R_{ac}' – (true ac load on amplifier) $= R_{\text{internal}} \parallel R_L$

PROBLEMS

Section 7.2

7.1. Solve for the voltage drops on each resistor and the three currents noted in Figure 7.36.

7.2. If the resistor R_4 was made variable ($0 - 3\text{ k}\Omega$), find the maximum and minimum values for I_1, I_2, and I_3 in Figure 7.36.

7.3. In the circuit of Figure 7.37, find the value of V_2 and I_1.

7.4. In the circuit of Figure 7.37, find the values of V_2 and I_1 if resistor R_3 is set to the following values.
(a) 0 Ω (b) 500 Ω (c) 15 kΩ

7.5. For the circuit in Figure 7.38, find the values of V_1, V_2, V_3, and I_1, I_2, plus the value of R_3.

7.6. Repeat problem 7.5 for the following I_2 to I_1 relationships.
(a) $I_2 = 20 \times I_1$ (b) $I_2 = 40 \times I_1$ (be cautious)

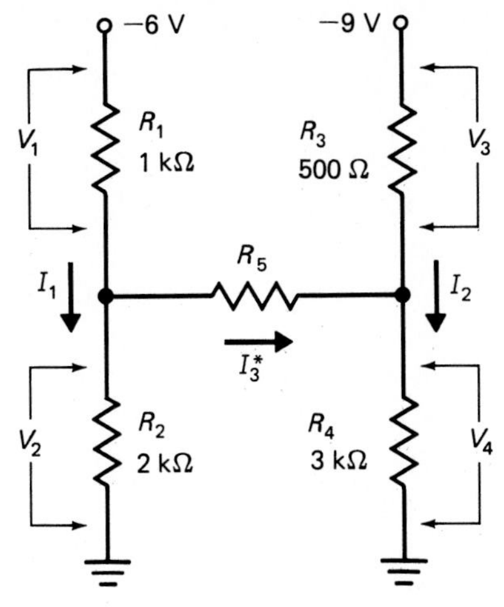

(*Also find the direction for I_3?)

FIGURE 7.36 Complex Circuit Analysis Problem

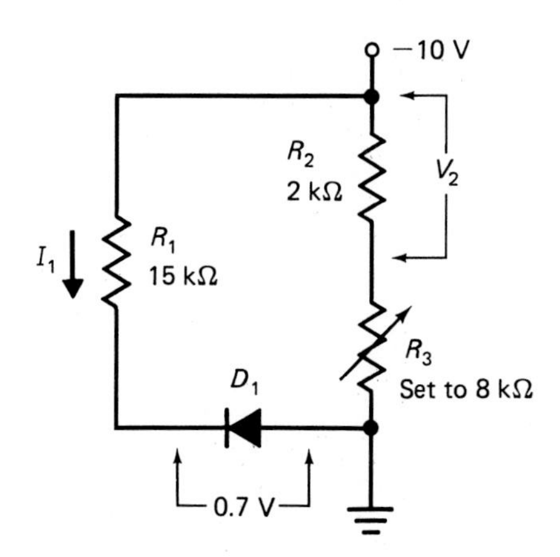

FIGURE 7.37

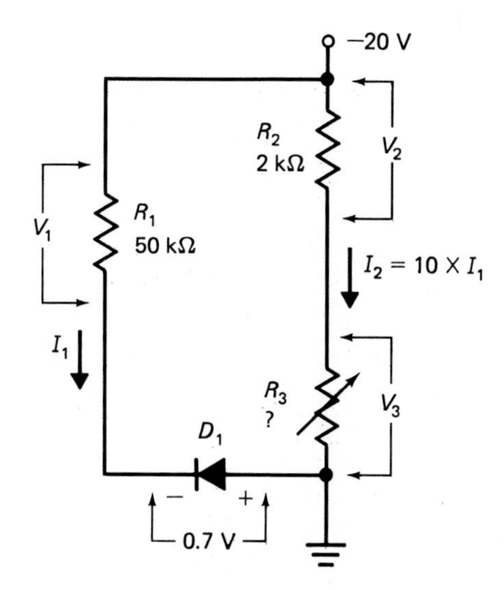

FIGURE 7.38

7.7. For the circuit of Figure 7.39, find V_1 through V_5, I_1, and I_2.

7.8. Sketch the waveforms (ac + dc) for the two resistors shown in Figure 7.40.

7.9. Briefly indicate the function of the collector load resistor on a BJT.

7.10. Briefly indicate the function of the drain load resistor on a JFET.

7.11. Describe what effect the collector load resistor has on the value of collector current from a BJT. (*Hint:* Your answer should make two statements about the effect.)

7.12. What effect do the load resistors (R_C and R_D, respectively) on a BJT and JFET have on the output voltage across that device?

7.13. In the circuit shown in Figure 7.41, calculate the value of I_2, V_1, and V_2 if the collector resistor is set to the following values.
(a) 50 Ω (b) 200 Ω (c) 1 kΩ (be cautious)

7.14. For the circuit and resistances listed in problem 7.13, find the maximum collector (saturation) current that would be possible.

7.15. For the beta-independent circuit shown in Figure 7.42, calculate the following values.
(a) Voltages V_1 through V_5
(b) Collector current I_C

7.16. For the circuit presented in problem 7.15, find the ac voltage-gain value.

7.17. For the voltage feedback circuit shown in Figure 7.43, find the following values.
(a) Base current (I_B)
(b) Collector current (I_C)
(c) Voltage across the device (V_{CE})
(d) Current gain (A_i)

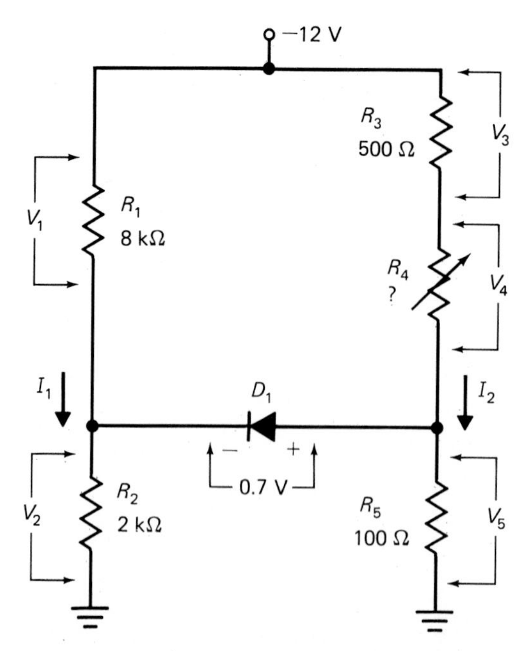

(*Note*: Assume that the current through D_1 is negligible!) **FIGURE 7.39**

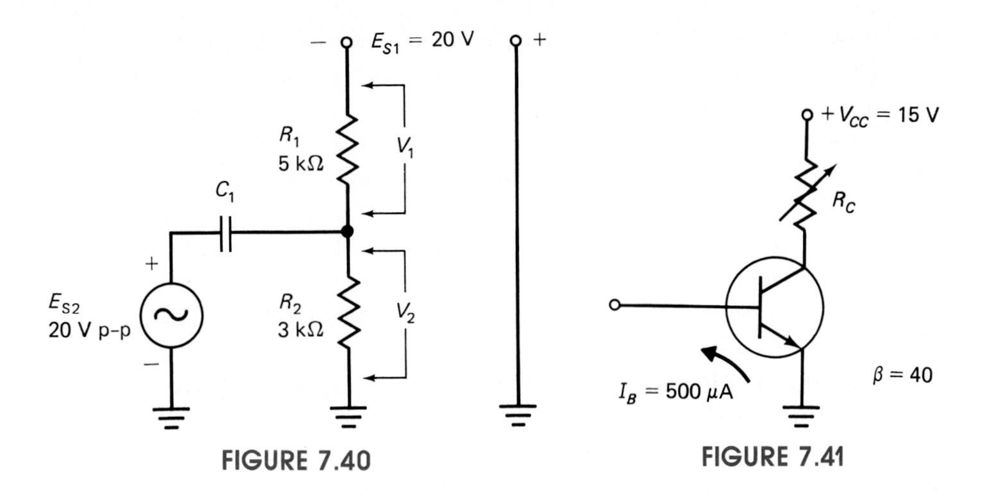

FIGURE 7.40

FIGURE 7.41

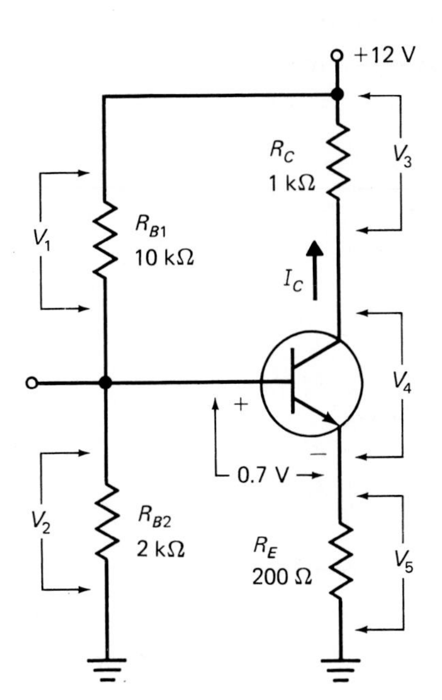

FIGURE 7.42

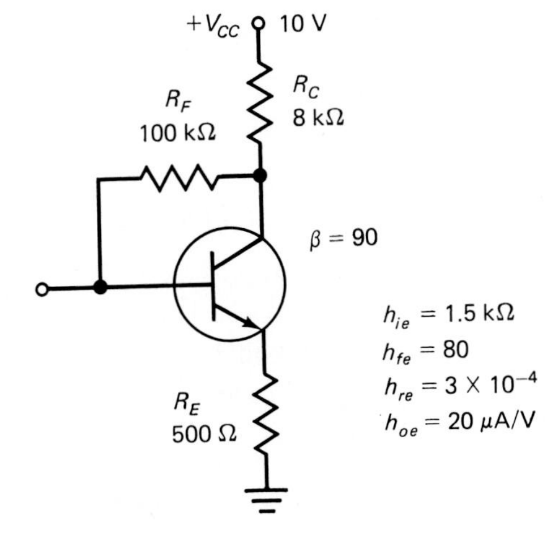

FIGURE 7.43

7.18. Draw the dc and ac equivalent circuits for the circuit shown in Figure 7.44.

7.19. Draw the ac equivalent of the circuit shown in Figure 7.44, assuming that a load has been attached through a coupling capacitor.

7.20. Define (briefly) the following parameter names for a BJT.
(a) h_{fe} (b) h_{ie} (c) h_{re} (d) h_{oe}

7.21. Which of the parameters listed in problem 7.20 closely relates to the dc-quantity beta?

7.22. For the circuit shown in Figure 7.45, find the following values.
(a) Z_{in} (b) Z_{out} (c) A_i (d) A_v (e) A_p

7.23. If the input voltage to the circuit shown in Figure 7.45 was 110 mV p-p, find the following values.
(a) V_{out} (b) I_{in} (c) P_{out}

7.24. Repeat problem 7.22 for the circuit shown in Figure 7.46.

7.25. Repeat problem 7.23 for the circuit shown in Figure 7.46.

7.26. Repeat problem 7.22 for the circuit shown in Figure 7.47.

7.27. Repeat problem 7.23 for the circuit shown in Figure 7.47.

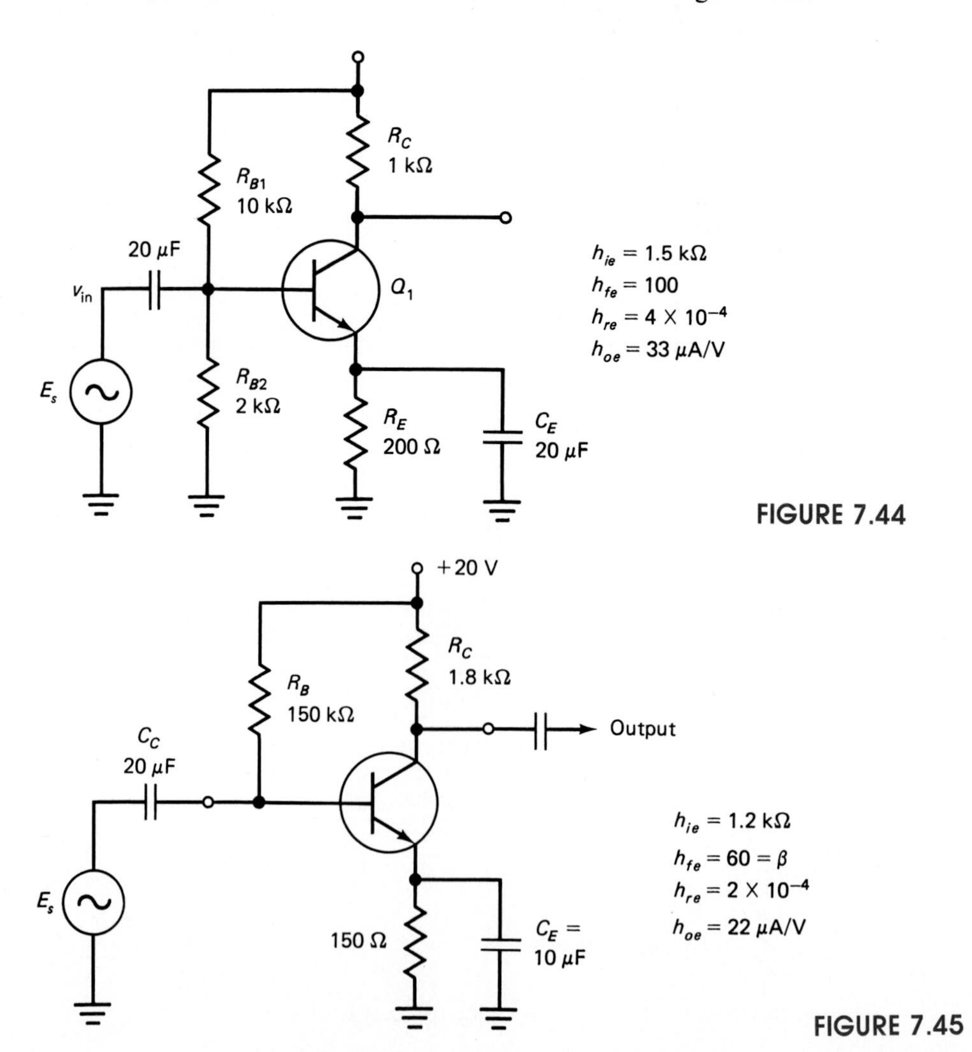

FIGURE 7.44

FIGURE 7.45

7.28. Calculate the dc bias levels for the circuit in Figure 7.46 (all voltages and currents). Redraw this Figure including both the dc levels and ac levels of voltage and current. (See Figure 7.31 as an example.)

7.29. For a JFET, what do the terms Y_{ds} and g_{mo} mean?

7.30. If a JFET is being operated with the following conditions, what is the operational value of transconductance (g_m)?

$$g_{mo} = 4.2 \text{ mS} \quad V_P = -6 \text{ V} \qquad V_{GSQ} = -2.5 \text{ V}$$

7.31. For the circuit shown in Figure 7.48, find the following values.
(a) Z_{in} (b) Z_{out} (c) A_v (d) A_i

7.32. For the circuit of Figure 7.48, calculate the value of the dc bias voltages and currents.

7.33. For the circuit of Figure 7.46, what values would be affected by the addition of a load resistor attached to the output coupling resistor (to ground)?

FIGURE 7.46

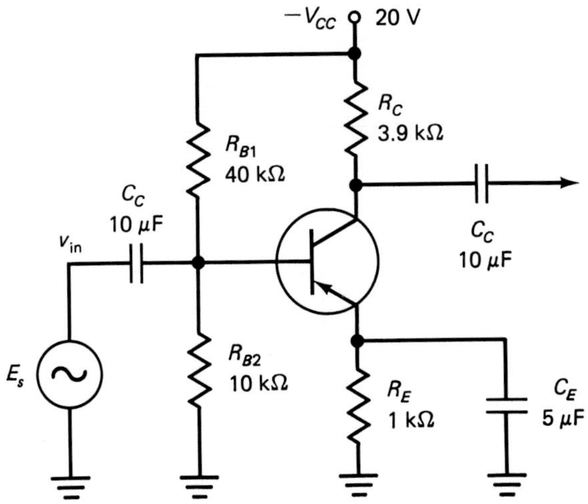

$-V_{CC}$ 20 V
R_C 3.9 kΩ
R_{B1} 40 kΩ
C_C 10 μF
V_{in}
C_C 10 μF
E_s
R_{B2} 10 kΩ
R_E 1 kΩ
C_E 5 μF

FIGURE 7.47

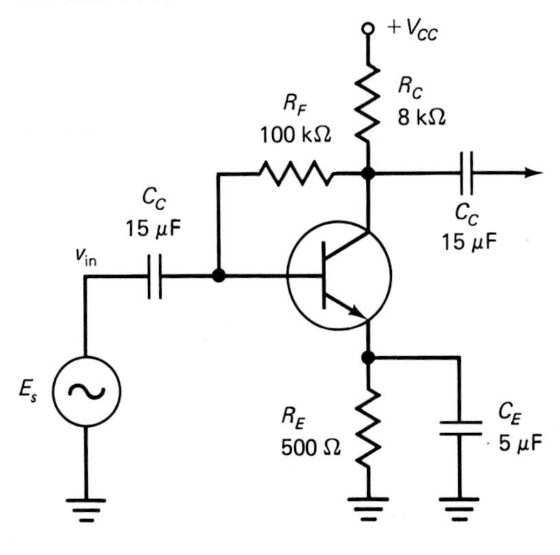

$+V_{CC}$
R_C 8 kΩ
R_F 100 kΩ
C_C 15 μF
C_C 15 μF
V_{in}
E_s
R_E 500 Ω
C_E 5 μF
h_{ie} = 1 kΩ
h_{fe} = 50
h_{re} = 2 X 10^{-4}
h_{oe} = 20 μA/V

FIGURE 7.48

$-V_{DD}$
g_m = 3.6 mS
R_D 1.5 kΩ
C_C
C_C
Load
E_s
R_G 250 kΩ
R_S 400 Ω
C_S 5 μF

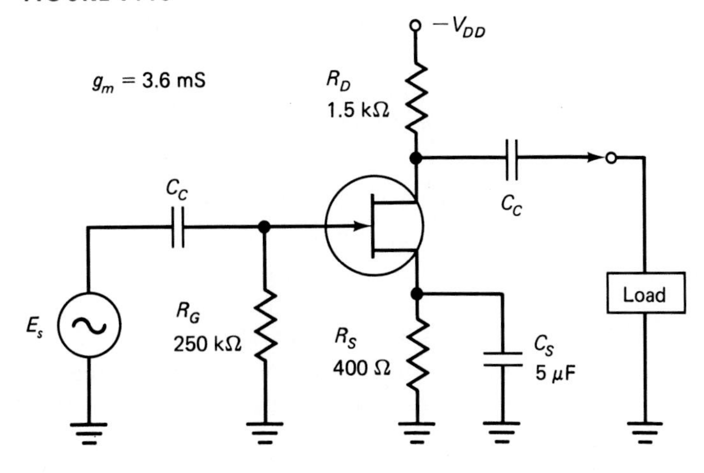

7.34. Calculate the new values of the quantities you noted as changing in problem 7.33 if the load resistor is a 5.6-kΩ resistor.

7.35. Repeat problem 7.33 for the circuit of Figure 7.48.

7.36. Repeat problem 7.34 for the circuit of Figure 7.48.

ANSWERS TO REVIEW PROBLEMS

Section 7.2

1. Answers should be compared to those in the text.

Section 7.3

1. (a) 4.3 mA
   (b) 215 mA
   (c) 219.3 mA
   (d) 0.7 V
   (e) 5 V

2.

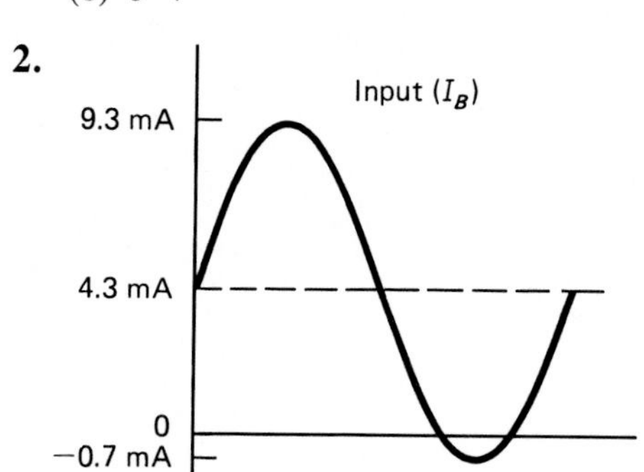

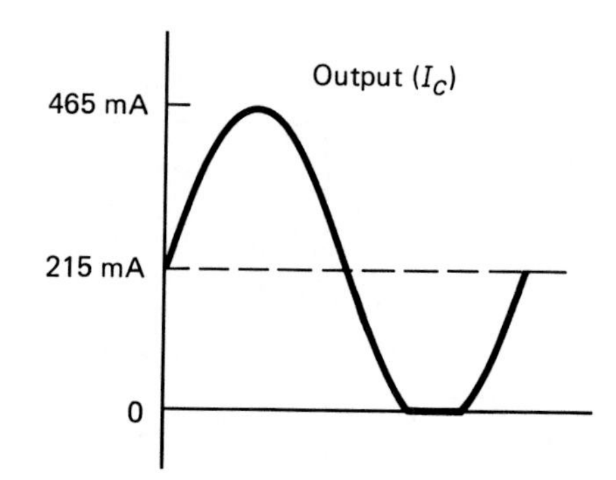

Section 7.4

1. (a) 1 V
   (b) 10.3 V
   (c) 7 V
2. (a) 2.73 V
   (b) 2.03 V
   (c) 1.35 mA
   (d) 6.67
3. A drop in source current would produce a lowering of the voltage drop on R_s. This lower R_s voltage would create a lower V_{GS} for the device. Lower V_{GS} would result in a higher value of source current, thus returning the device to a stable condition.
4. (a) 49.3 μA
   (b) 5.92 mA
   (c) 4.94 V
   (d) 58.8

Section 7.5

1. Chart based on data in the text.
2. All of the energy responsible for the increased output levels from these amplifiers is provided by the dc source for that circuit.

3. Dc power input = 27.4 mW; ac power input = 400 μW; ac power output = 16 mW; efficiency = 58%. The lost energy (power) is being dissipated in the circuit as heat.
4. As per example solutions.
5. $Z_{in} = R_1 \parallel (d_{ix} + d_{fx}R_3)$, $Z_{out} = R_2 \parallel (d_{ox} + R_3)$, $A_v = d_{fx}$, $A_i = R_2/R_3$
6. $Z_{in} \simeq 200\ k\Omega$, $Z_{out} \simeq 200\ k\Omega$, $A_v \simeq 45$, $A_i \simeq 20$
7. $Z_i = 820\ \Omega$, $Z_{out} = 4\ k\Omega$ (not changed due to new load, $A_i = 50$, $A_v = -100$)

8 Specifications and Additional Analysis

Objectives

Upon completing this chapter, the reader should be able to:

- Evaluate devices, specifications, and circuit requirements.
- Interpret device specification sheets.
- Perform graphical analysis of transistor "curves."
- Identify various transistor forms.
- Perform various transistor tests (in and out of circuit).
- Identify causes of transistor failure (and resolutions).

Introduction

In this chapter we introduce the use of specification sheets as a means of determining device function and parameters. How to use these specifications and how to identify operational limits is emphasized.

Additionally, the process of graphical analysis to determine device function within a circuit is presented. Certain variations on the basic BJT and FET devices are presented to round out the overall knowledge of these categories of semiconductor devices.

Testing the devices and basic failure analysis completes this chapter. By having an understanding of device specifications, the quick means of graphical analysis, and troubleshooting methods, a clearer knowledge of the application, limits, and functional operation of these devices will be achieved. ■

8.1 DISCRETE DEVICE SPECIFICATIONS—GENERAL FORMS

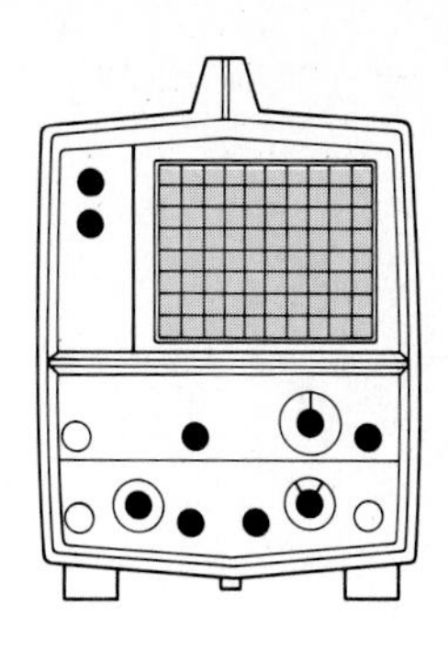

Discrete devices come in a wide variety of sizes, shapes, and forms. There are literally hundreds of BJTs to select from and quite a large number of FETs as well. Each version is given specific part numbers. They each have certain characteristics that make them different from one another.

It may seem quite confusing to the user of these devices just which one to select or investigate when looking for a device. There are some general classifications that should be established first when looking for a device. These are discussed in this section. In later sections, more specific details are added to these to narrow the search to a limited number of devices.

First, it will be important to select the classification of discrete semiconductor that is needed, BJT or FET. In most applications they are not interchangeable. The following is a list of the overall features of both devices along with pro and con discussions of each:

BJT The BJT is the "workhorse" of discrete devices. It is the most popular of the discrete semiconductor devices. The BJT provides a good balance of both voltage and current gain. They are available in a wide variety of current and power-handling capacities. BJTs are available which work at frequencies from the audio range all the way to giga-hertz frequencies. Circuits for BJTs are quite easily developed, and if they are stabilized, will accept a variety of BJT substitutions. The BJT is extremely sensitive to temperature, which is a drawback. It can become unstable and be destroyed if not placed in a stabilizing circuit. It is also somewhat noisy (internally generated electrical interference), although low-noise versions are available. Its input resistance is relatively low and as such requires some current capacity from a input device.

FET The high input impedance of the FET makes it quite useful as an initial amplifier in circuits that are operated by sensitive, low-power capacity inputs. Its inverse temperature coefficient (reduced gain with increased temperatures) makes it more self-stabilizing than most discrete semiconductors. Its output has the current capacity similar to BJT devices. Due to its high input resistance and thermal stability, a special high-power version of the FET, called a V-FET, is available. It is popular as a highly stable and dependable high-power device, especially for high current-switching applications. One draw back of the FET is its (in circuit) voltage gain. Usually quite low, this small gain factor makes it a less effective amplifier than the BJT. The JFET is also less linear (discussed later in this chapter) than the BJT, a factor that could result in higher distortion of an input signal. Finally, the FET can have a relatively high input capacitance (also discussed later in this chapter), which could cause limits to its maximum operating frequency.

Although there are other specifications for these devices, these are the primary distinctions between the FET and BJT. To be able to select either the BJT or FET and then to qualify the selection to a few specific models of these devices, the following factors are usually primary to that decision process.

1. *What will be the polarity of the dc source powering the device?* This factor will help in deciding if the device is NPN or PNP BJT, N-type or P-type FET. If the power source polarity is optional, often the NPN BJT or N-type FET are selected.
2. *What power (or output current) levels will be required?* Devices are usually classified as being either "small-signal" or "power"-type devices. Small-signal devices are usually those dissipating 500 mW or less of power (as a rule of thumb). Typical output currents are less than 500 mA, with inputs as low as microvolts. Power devices, starting with power capacities from 500 mW up to the kilowatt range, are available. They typically require input levels in the high millivolt to tens of volts range.
3. *What is the operating frequency?* Most devices are frequency sensitive. It is necessary to define the range of frequencies that will be applied to the device. Classifications such as general-purpose audio, RF (radio frequencies), VHF, UHF, and microwave will be used to segregate groups of devices.
4. *What are the input and output device characteristics?* What type of input device will be used? (As an example, a sensor that produces little current would most likely be matched with an FET.) What is the impedance of the input device? What is the impedance of the load device? What are the maximum and minimum levels of input that will be expected from the input device?
5. *What are the expected temperature (or environmental) conditions in which the device will be expected to operate?* If the circuit is operated in a controlled environment (such as a clock radio for use indoors only), many device forms may be used. If it may be in an unusually high- or low-temperature situation, or used where temperatures may vary widely (such as the cargo bay of a space vehicle), special devices may need to be used. Some of the environmental conditions that may affect a semiconductor device (beyond temperature) are radiation, rapid changes in temperature, shock, or vibration, chemicals which could damage the transistor packaging, and so on. Even the storage (nonoperational) temperature of the device is a potential problem. If some devices are exposed to extreme heat or cold, even if not in a circuit, damage to the semiconductor material or internal connections could result.

   Special note: Some devices are designed to meet what is called "MIL-SPEC." Primarily, these military specifications require a wide range of allowable operating temperatures, a lower "failure rate," and closer tolerance (greater accuracy) to published specifications. Most manufacturers who supply products to the military, space program, or for use in extreme environments will use MIL-SPEC versions of devices. (Often the same device will be available in either a standard or a MIL-SPEC version.)

6. *Are there any physical requirements for mounting or manufacturing?* In some cases, size, shape, material used in the device's housing, or other physical parameters are important to device selection. Discrete semiconductors are available packaged in a variety of ways. Metal housings ("cans"), molded plastic, flat button-shaped packages, integrated circuits like DIPs (dual-in-line packages), and even unmounted silicon chips are available.

All in all, there are a variety of reasons for selecting one out of the literally thousands of discrete devices currently available. The basic qualifications noted above may assist in narrowing the selection to a few dozen devices. In the next sections of this chapter more details are shown which may help reduce the choice to a relative few. (Often the final selection is based on nontechnical factors. Cost, delivery time, availability of assistance from the manufacturer, and even which companies your company commonly deals with will most often constitute the final decision.)

REVIEW PROBLEMS

(1) For the following basic conditions, indicate whether it would be best to select a BJT or an FET device for use. Write at least one sentence in defense of your selection.
  (a) The input amplifier to a sensitive measurement instrument (the instrument should not draw much current from the system it is measuring)
  (b) An amplifier to be used between other amplifiers where reasonably high voltage and current gain values are expected
  (c) A small amplifier circuit that will be placed in an environment where occasional overheating is common

(2) Write a brief response to each of the six questions used in this section for the following applications for a BJT:
  (a) The BJT used as the final amplification stage of a high-power in-home stereo amplifier
  (b) A BJT used in the FM receiver section of a car radio
  (c) A BJT used inside a miniature in-the-ear hearing aid
  (d) A BJT used in a microwave relay satellite (keep in mind that weight and size are also critical factors)

8.2 BJT SPECIFICATIONS

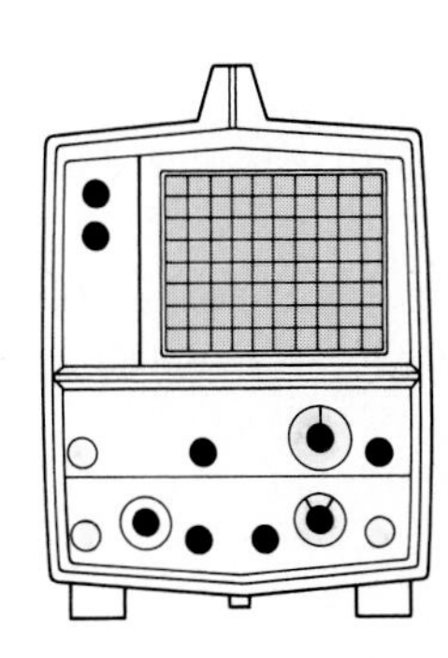

To work with the BJT, it is necessary to know what types of specifications are normally given for that device. Most device specifications (BJT or other devices) are broken down into four major categories.

1. *Primary device features.* These specifications highlight the main features of that specific device, noting what major feature makes it different from others.
2. *Maximum ratings.* These are the electrical and environmental limits for the device. Exceeding any one of these limits may cause damage to or total failure of the device.
3. *Dc operating conditions.* These are the basic bias ratings, often listed as maximum, average, and minimum. In some specifications, these values are listed along with item 2 or 4.
4. *Ac or (small/large)-signal characteristics.* In this section the *h*-type parameters (or similar parameters) are presented. Together with these are other ac characteristics.

(In some cases, the device may be used for digital circuitry. In this case switching characteristics may be added to section 4 or be in a separate section.)

It is also common for the manufacturer to list information about the test conditions used to determine the values listed in the dc and ac characteristic sections. This information is often helpful if the company that buys the device needs to test each one purchased to assure that it is working as specified. Usually, the company that manufactures a component does not test each one but simply checks a random sample. Some companies test all purchased components before using them to help in maintaining quality assurance. Most companies that have military contracts to manufacture a product are required to test all purchased components.

Component data sheets are usually available in three forms. The first type is a very general overview of the component, highlighting its major features and maximum limits. These are usually less than a page in length; often they are simple catalog listings with other similar devices, produced by companies that sell a wide variety of devices from several manufacturers. The second type, which is most commonly used, is a two-page "spec sheet" (spec = specification). These are often included in device handbooks supplied by the manufacturer. They include all of the data above (items 1 to 4) plus suggested applications and drawings of the device to show how it is packaged. The last form is a detailed set of design specifications. These usually include the two-page description, but add several (typically, two to six) more pages of detailed graphs, charts, and suggested circuit applications. The data included in the one- or two-page description is sufficient for most applications.

Some manufacturers will produce a "family" of devices. What this means is that they manufacture a group (e.g., 4, 6, or 10) of transistors that have nearly the same characteristics. They are given different part numbers because one (or only a few) characteristics differ between each of the different types.

> For the following description of devices, refer to the 2N2218 (and associated family) transistor specification sheets included in Appendix A.

The following is an expanded description of the type of data included in each of the four categories listed at the beginning of this section. Statements within brackets refer to the 2N2218.

1. Primary device features
   (a) Device type [NPN silicon (BJT)]
   (b) Typical use [switching and amplification]
   (c) Basic highlight of major facts about device [see first "boxed in" details]
   (d) Drawing of the devices packaging [e.g., case 33-03, TO-5]

(The "selection guide" for this family is there to describe the major differences between each device in the "family.")

In the following characteristics, a three-letter subscript may be used to describe a voltage or current (e.g., V_{CEO}) in this book; so far we have used only two-letter subscripts (e.g., V_{CE}). The third subscript describes the condition of the third lead when measuring the voltage or current related to the two listed leads. Typically, this third letter is O, meaning open. Therefore,

V_{CEO} means the voltage from collector to emitter (*CE*-) with the base open (--*O*), or with zero base current.

Also, the term *BV* or *BR* may be added to the subscript to indicate a breakdown value. Such a value will damage the device or cause it to operate in a nonacceptable manner. The term "sat" may be added to indicate a condition called *saturation*. This condition is where the device is producing maximum conduction and additional base current will not cause more conduction. Finally, the term "co" or "cutoff" or "off" may be used to indicate the condition called cutoff.

2. Maximum ratings
   (a) Usually, the maximum allowable voltage across any two terminals is listed (V_{CEO}, V_{CB}, V_{EB}). In the case of the base–emitter connection, the letters are listed as emitter to base to indicate a reverse breakdown voltage. (A letter *B* ahead of the *V* (e.g., BV_{CEO}) simply means "breakdown.") [V_{CEO}, V_{CB}, V_{EB}]
   (b) The maximum collector current is usually listed. [I_C] (There may be two listings, one for continuous and the other for peak-instantaneous current.)
   (c) The total device power dissipation is noted P_D, where $P_D = V_{CE} \times I_C$. Also listed is what is called a "power derating factor," which reduces this limit if the device is in a high-temperature environment. Use of this factor will be presented later in this chapter.
   (d) Operating temperature range, sometimes called junction temperature. [T_J]
   (e) Storage temperature T_{stg} (often the same as T_J).
   (f) Lead temperature T_L. This is the "soldering" temperature allowed when installing the device. It is usually listed along with "distance from the case" (how far down a lead the iron is) and time (how long the lead is heated). Exceeding this value may cause the device to become internally damaged or actually melt internal wiring.
3. Dc operating conditions (None are listed for the 2N2218; the MC34071 may be used for examples of these data.)
   (a) The saturation voltage (minimum to turn the device on) for both the collector and base-to-emitter potentials may be noted.
   (b) Other dc bias conditions may be listed, often as a min/typical/max charting.
   (c) The value of a dc beta may be noted.
4. Ac or (small/large)-signal characteristics
   (a) Here *h* or hybrid parameters would be noted which relate to special ac equivalent-type circuitry. [h_{ie}, h_{re}, h_{fe}, h_{oe}]
   (b) Certain values of internal capacitance would be noted. [C_{ob}, C_{ib}]
   (c) A maximum frequency for the input signal would be noted. It is usually rated in terms of the frequency where the gain (h_{fe}) would drop to a value of 1 (output = input). Later it will be shown that as frequency increases, the gain of a BJT will begin to drop from its normal rating. [f_T]
   (d) Additional information may be supplied to help guide the user in applying the device to a circuit. [Since the 2N2218 is described as a switching transistor, certain switching functions are noted in the specification sheet.]

By using this information decisions on the applicability of the device will be made. Usually, the user of this device will have a rough idea of what values are desired in each of the categories. If a device meets or exceeds those values, it may be chosen for use in the circuit. In some cases the repair technician may refer to these data charts for one of two reasons:

1. If a certain device is no longer available, it may be necessary to look up that device's specifications and then look to find another one that meets or exceeds those "specs." In that way a substitute device can be found that will function equally as well as the original, without failure.
2. If a certain device continues to fail (or fall short of functioning), it may be necessary to review these specs to see where it may exceed some limit, or to do something it is not designed to do. Once the "limit" is isolated, either a device with higher limits can be looked for, or new restrictions can be placed in the circuit to prevent limits from being exceeded.

EXAMPLE 8.1

If the 2N2218 is to be used in a circuit with a V_{CC} of 32 V, an expected current gain of 40, at a maximum frequency of 20 kHz, can we be assured that it will work? If not, why, and is there a device in the "family" that will work?

Solution:

The 2N2218 cannot be assured of working in the circuit described, because:

1. With a V_{CC} of 32 V, it is possible that the maximum rating for V_{CEO} of 30 V may be exceeded if the base current drops to zero.
2. The minimum ac gain (h_{fe}) for the device is listed as being 30. It is possible that the device may drop to this level of gain and therefore not meet the gain demand of the circuit (40).

(Since the device is rated at 250 MHz for f_T, it should operate at the low 20 kHz desired.) Within this family of devices, the 2N2219A, the 2N2222A, and the 2N5582 seem to meet the required specifications.

Each manufacturer of semiconductor devices will have individual variations on these specifications. Some will supply more data, some less; others may reword the specifications; but usually the same basic information is available on all devices.

REVIEW PROBLEMS

(1) To become more familiar with semiconductor data sheets and books, prepare a list of all data found in the following resources for one of the BJT devices used in your laboratory kit (or one of your instructor's choosing).

Distributor's catalog listing for the device.

Short (one- or two-page) manufacturer's data sheet (found in a data book by that manufacturer).

A detailed data page (three to eight pages, if available)—simply describe the data presented in chart or graph form. If such a page is not available, note at least two ways that such a page may be obtained.

(2) Using any one of the forms noted in problem 1 (device data sheet is suggested) write a brief comparison of the device used in problem 1 to the 2N2218 described in Appendix A. Contrast as many characteristics as possible. (If you used the 2N2218 in problem 1, contrast it to either the MJ6018 device listed in Appendix A or to one of your instructor's choice.)

8.3 JFET SPECIFICATIONS

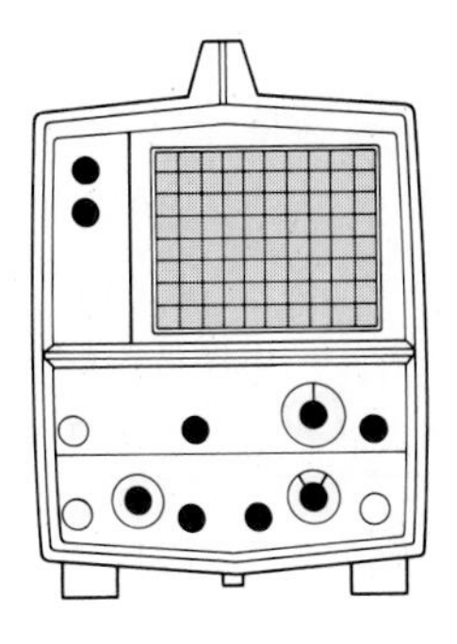

For the JFET, the same characteristics, translated into JFET terms, are important as were for the BJT. Although current gain is not a factor for a JFET, the g_m of the device is. JFET data sheets are arranged much the same as BJT data sheets. In order to discuss the JFET, then, we will skip immediately to the important characteristics normally seen on the data sheet.

For this analysis, refer to the MFQ5460P JFET (quad) data sheet in Appendix A.

Information in brackets in the following description is for the device noted above. The following is an outline of the data found in the four important groups of data, as presented in Section 8.2.

1. Primary features
   (a) The type of JFET being described [P-channel (type)]
   (b) The packaging [This is a quad dual-in-line package, which means that four (quad) devices are packaged in an "integrated circuit type" of package.]
   (c) Major features [The first block indicates basic application data.]
   (d) Maximum ratings (similar to maximum limits for BJT devices) [V_{DS}, V_{DG}, V_{GSR}, I_{GF}, P_D, and T_{stg}]

2. Maximum ratings
   (a) Maximum voltages such as:
      V_{DS} (40V)
      V_{DG} (40V)
      V_{GSR} (40V)
      In this example reverse polarity values are indicated using the letter R (see V_{GSR}).
   (b) Maximum currents are then noted:
      I_D (20mA)
      I_{GF} (10mA)
   (c) Finally, the maximum power dissipation is given as:
      P_D = (0.5w – each JFET in the package)
   with derating for high temperature operation.
3. Dc operating conditions. These are often broken down into "off" and "on" conditions. The off information relates values that cause the drain current to be reduced to nearly zero. (Remember, increased input voltage causes reduced drain current.) The on conditions relate to basic operational conditions.
   (a) "Off" characteristics usually relate to a full reduction of drain current [$V_{(BR)GSS}$, $V_{GS(\text{off})}$, I_{GSS}]
   (b) "On" characteristics [I_{DSS}, V_{GS}]
4. Ac (small/large)-signal characteristics
   (a) Basic gain and output ac conditions [$y_{fs}(g_m)$ and y_{os} (drain resistance)—each stated as an admittance value (ac admittance can be interpreted here as conductance).]
   (b) Capacitance values (These can assist in determining operating frequency.) [C_{iss}, C_{rss}]
   (c) Noise figures (internally generated noise is a common problem with JFETs) [NF, e_n]

These are the primary characteristics for JFETs. Other manufacturers may present the data in a slightly different form, but usually these major parameters are included for all JFETs.

REVIEW PROBLEM

(1) Using the data sheet in Appendix A for the MFQ5460P, copy the connection diagram and around one of the devices shown, illustrate a standard FET amplifier circuit (include R_D, R_S, R_G, C_S, and V_{DD}). Indicate the following values based on information in the data sheet.
   (a) Maximum safe value for V_{DD}
   (b) (Using the V_{DD} above) The minimum value for $R_D + R_S$ (based simply on the maximum value of I_D)
   (c) (Using the maximum I_D) The maximum value of R_S which will keep V_{GS} from going too high
   (d) The maximum value of voltage gain for the device if R_D = 2 kΩ ($A_v = -g_m \times R_D$)

8.4 GRAPHICAL ANALYSIS OF DISCRETE DEVICES

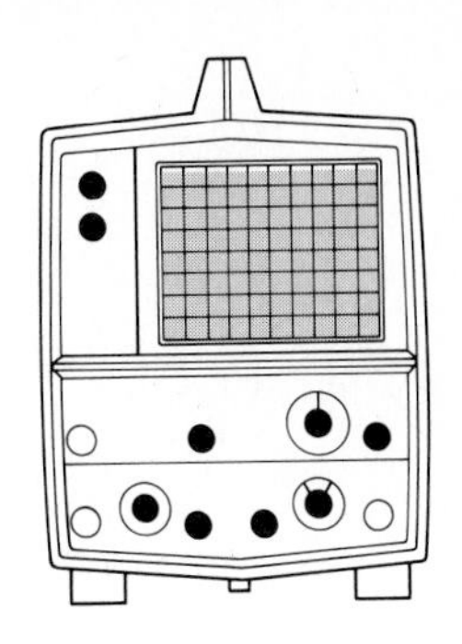

An alternative means of resolving bias and amplification problems is by using graphs that represent the function of the specific device. The two most popular set of graphs are the **collector family of curves** for the BJT and the **transfer curve** for the JFET. Each of these curves is developed by taking data from the circuits by experimentation. The BJT curves display the relationship of output current and voltage, as established by input current. The JFET curves represent the relationship of output current to input voltage. Some manufacturers' data sheets include these curves; otherwise, they can be determined by simple experimentation with the device in question. Most BJTs have individual curves for each device (except those which have identical or nearly identical specifications). FETs, on the other hand, usually fit what is called a **universal JFET transfer curve**; therefore, with simple modification, one curve will describe a majority of FET versions.

The BJT: Collector Family of Curves

Basically, these curves for a BJT are a representation of the cause/effect relationship between base input current and collector output current. Additionally presented on the same graph is information about the voltage across the output side of the device V_{CE}.

The circuit shown in Figure 8.1(a) is the type that would be used to develop a family (set) of curves. The base current would be set to some constant value. Then the potentiometer on the collector side (representing R_C) would be adjusted. By changing that "pot's" value, a variable voltage divider forms between the device and the pot. By adjusting the value of this resistor from zero to a very high resistance, the voltage on the transistor will go from V_{CC} ($R_C = 0$) to nearly zero volts ($R_C \gg$ device resistance). The value of the current I_C is noted for every value of voltage across the device. Once those data are taken, the points (I_C versus V_{CE}) are plotted on a graph such as shown in Figure 8.1(b.) This shows the resulting collector current due to a single value of I_B.

This process would then be repeated for a variety of base currents, usually selected to be in some reasonable steps (such as every 5, 10, or 100 μA, depending on the desired operation of the device). The curves shown in Figure 8.1(c) are the results of taking several sets of data for selected base currents.

> *Note:* These curves show only the response of the device to a select group of base currents. There are as many curves possible as there are possible base currents. Although there may be a distinct line for a base current value of 50 μA, there would be a curve for either 49 or 50.1 μA; they are just not a part of *these* data.

Several important factors can be added to or derived from these curves. The curves shown in Figure 8.2 detail the added features that can be either interpreted from that curve or (due to maximum ratings from the data sheets) have been added to the curves. This complete set of data can be used to evaluate the device's response to being biased and to having an ac input. The following is a description of each of the highlighted points and areas that are noted on the graph (Figure 8.2):

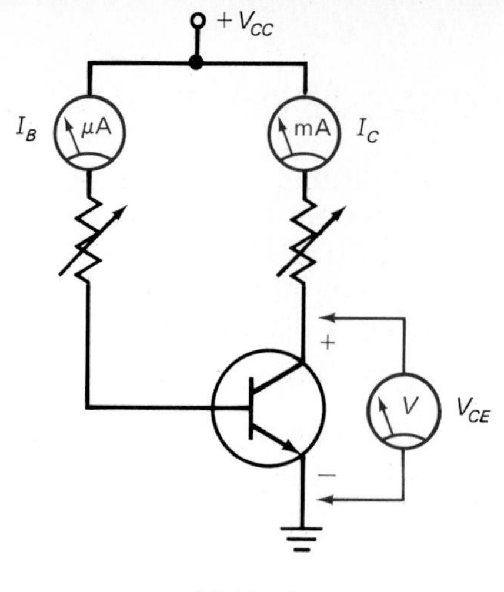

(a) Circuit

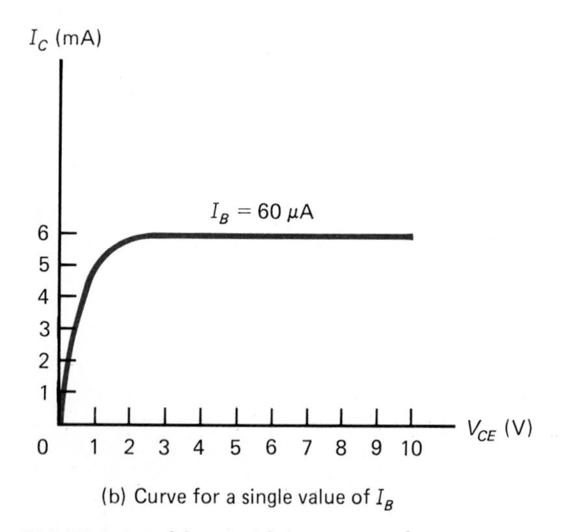

(b) Curve for a single value of I_B

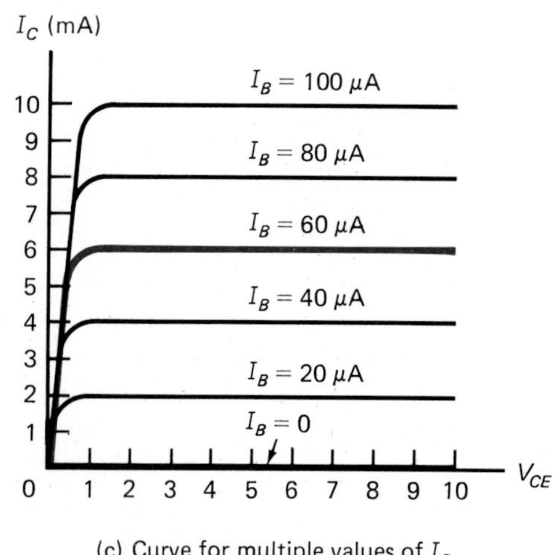

(c) Curve for multiple values of I_B

FIGURE 8.1 V_{CE} and I_C versus I_B

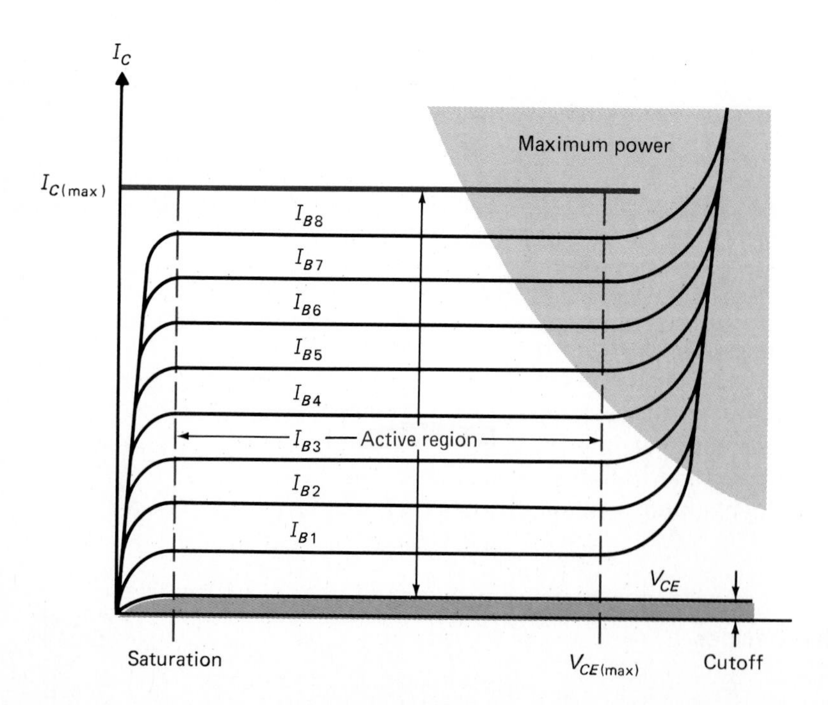

FIGURE 8.2 Boundaries and the Active Region for the BJT Family of Curves

Saturation The saturation area (region), roughly vertical from the low-voltage side of V_{CE}, is a boundary along the edge of the curves, before they become stable (where I_C is relatively constant). This boundary indicates that if the voltage V_{CE} is low, the value of I_C is not clearly predictable. Generally, the minimum V_{CE} that is usable is called $V_{CE(\text{sat})}$. This boundary shows that the transistor cannot be expected to have extremely low voltage and still support a constant value of I_C.

Cutoff Since saturation indicated a minimum acceptable level of device voltage, it is appropriate at this moment to investigate cutoff—a minimum value of collector current. This minimum value occurs when the base current is zero. Even with the base current at zero, some collector (leakage) current will still exist. Another name for this current is I_{CEO}. On most modern BJTs this value is so small that it is often ignored.

I_C Maximum This boundary at the top of the curve indicates the limit of collector current allowed through the device. This quantity appears in the maximum ratings for the device (spec sheets). It cannot usually be measured without damaging the device. Actual collector current levels should stay well below this value.

V_{CE} Maximum This boundary on the right side of the curves is also taken from the data sheets and not experimentally. This limit also should not be exceeded.

Maximum Power Curve This curve is added to the graph by calculation. The maximum power for the device is found in the device specifications. Selected values of V_{CE} are placed into the power formula ($P_D = V_{CE} \times I_C$) and coordinated values of I_C are found. These coordinates are then plotted on the graph. This power curve then sets an upper limit to the operation of the device.

Active Region The device's active region is the area bordered by all of these limiting factors. Within the active region the device can be expected to function without distorting an input and/or causing device failure or breakdown.

Figure 8.2 then shows all the functional relationships between input current, output voltage, and output current. The central area, the active region, is where the BJT can be expected to preform as has been assumed in prior chapters. (The value of β shows up in this graph as a simple ratio of the constant line of I_C to the I_B that labels that line.)

The JFET Transfer Curves

Although a family of drain curves could be developed for the JFET in the same manner as was used to produce the BJT collector curves, a more common practice is to use the JFET's transfer curve. This curve is simply an illustration

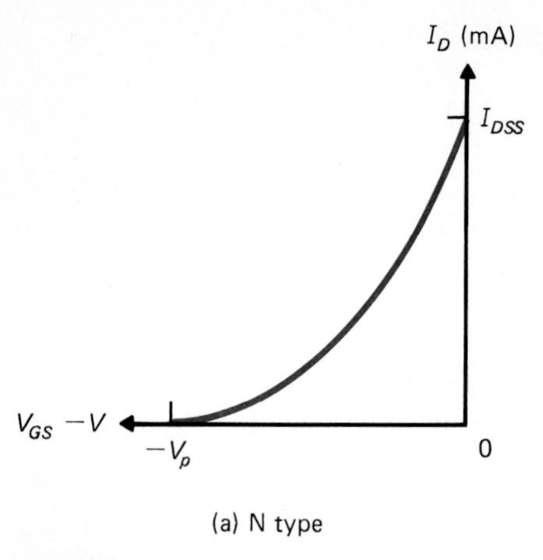

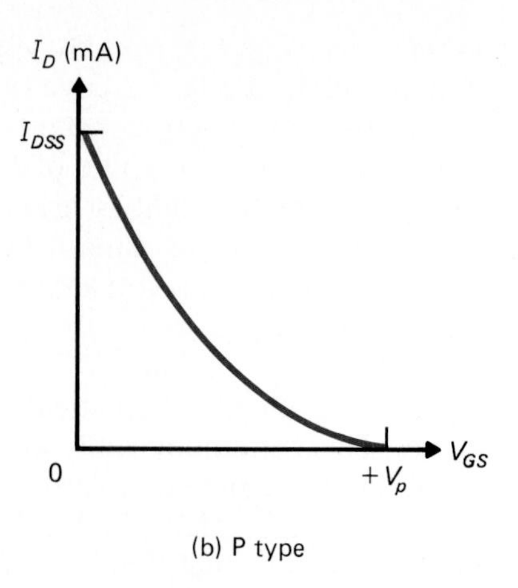

FIGURE 8.3 JFET Transfer Curves

of the relationship of gate voltage to drain current. It does not display the effects due to drain-to-source voltage. Figure 8.3 illustrates a JFET transfer curve. It will be shown later in this section that since this curve is not a linear (straight line) curve, only small segments of the curve will be used to predict the function of the FET. Simply, this is a plot of the gate voltage versus the drain current. It should be noted that this curve illustrates the reverse relationship of gate voltage to drain current. Note that as gate voltage increases, the drain current is reduced. Only when gate voltage is at zero volts is the drain current at its maximum. Following is a description of the two special points on this curve.

The Point I_{DSS} This is the maximum drain current for the device and it occurs when the gate voltage is at zero. (The same value appears in the JFET specifications.)

The point V_P This point is defined as the pinch-off point for gate voltage. Note that at that point, any increase in gate voltage will not produce a response in drain current, since the drain current at V_P is at its lowest value (0 A).

Except for these two points, there are no special borders or boundaries as seen in the transistor. The self-limiting effects for the FET noted in previous chapters show quite well in this illustration.

Applications: Load Lines and Signal Analysis

These characteristic curves can be used to test a device's reaction to circuit components and input signals. In some cases this graphical analysis is simpler than a group of calculations. A load resistor can be artificially applied to the device and the resulting dc currents and voltages can be predicted. Once this is done, an ac input signal can be superimposed on the graph and the resulting output can be estimated.

Graphing the Effect of a Load Resistor

In Figure 8.4 a load line representing the load resistor on a BJT is shown. The endpoints of this line run simply from $V_{CE} = V_{CC}$ (bottom edge) to V_{CC}/R_C on the vertical axis. The intersection of this line and the device's output curves predicts the operating points for that specific circuit. If one of these base current lines is selected for use by the choice of a base bias resistor, the intersection of these two lines predicts the value of I_C and V_{CE} that will occur when the circuit is energized. This intersecting point is often called the *Q-point* (quiescent point).

In Figure 8.5 a load line has been plotted and is shown to intersect three operating curves. The center curve represents the dc bias value of base current ($I_B = 400\ \mu$A). Plotted across these curves is the value of the collector resistor (R_C). The intersection of these two lines represents the Q-point ($V_{CE} = 5$ V, $I_C = 40$ mA). Therefore, the electrical conditions for this circuit (dc bias) are

$$I_B = 400\ \mu\text{A} \qquad I_B = 40\ \text{mA} \qquad V_{CE} = 5\ \text{V}$$

(β could be found to be $I_C/I_B = 100$.)

If an input signal were to cause the base current to rise and fall to the values represented by the remaining two curves (300 and 500 μa), the output conditions could be read from the graph by looking at the intersections of these curves with the load line. Therefore, the electrical conditions with an ac input would be

$$I_B = \text{(from) } 300\ \mu\text{A to } 500\ \mu\text{A (peak-to-peak change)}$$

$$I_C = \text{(from) } 30\ \text{mA to } 50\ \text{mA (peak to peak)}$$

$$V_{CE} = \text{(from) } 7\ \text{V to } 3\ \text{V (peak to peak)}$$

(It should be noted that as predicted before by using a $-A_v$ for the voltage gain of the circuit, the voltage changes on the collector of the device are the inverse of the current change.)

Although this graphical method produces exactly the same results as

FIGURE 8.4 Q-Point Location

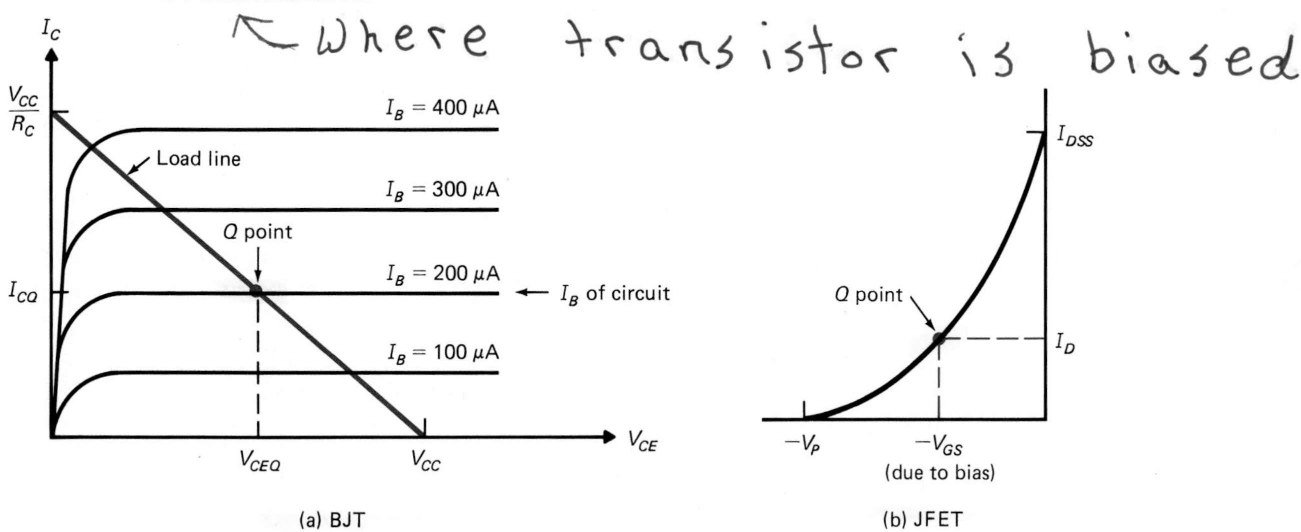

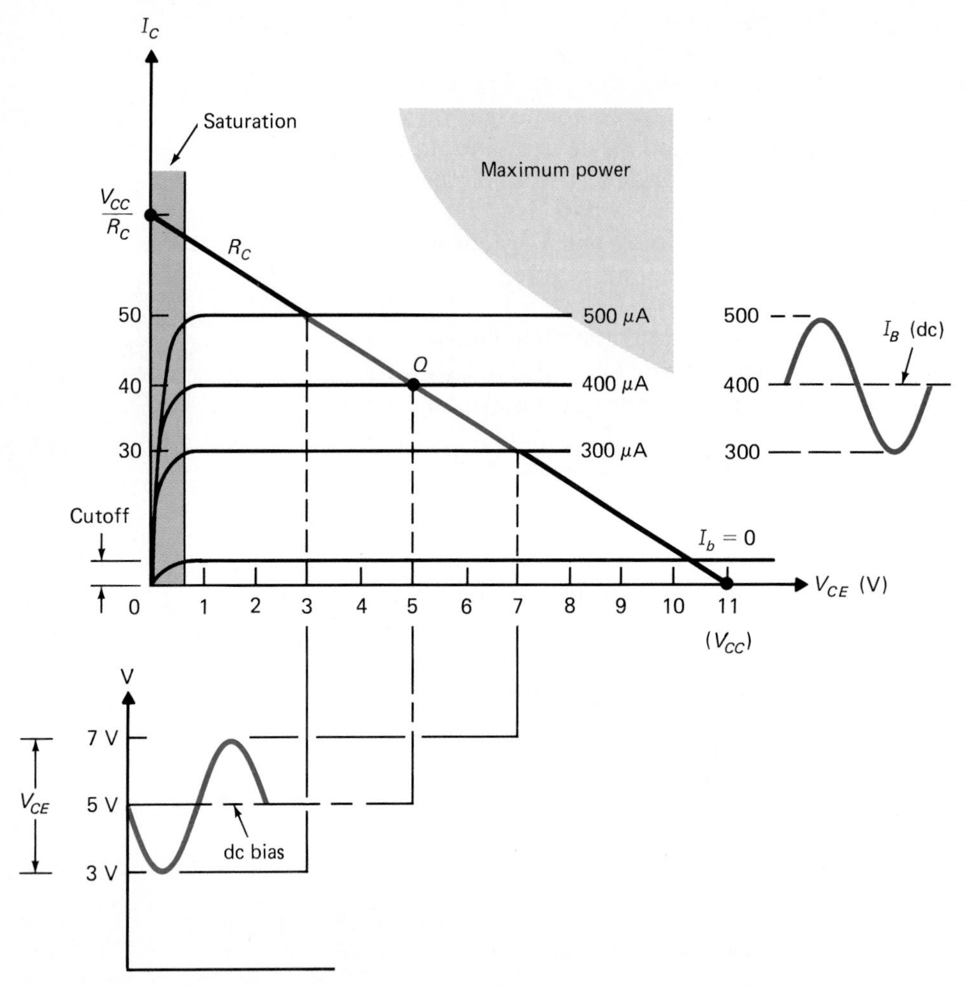

FIGURE 8.5 Effect of AC Input in Relation to the Q-Point (BJT)

would be obtained from calculations, there are some advantages to using the graph instead of calculations:

1. If it would be desirable to investigate other load resistors besides the one shown, their plot could be easily compared to this one.
2. The saturation and cutoff values for the device are easily seen. It is important to avoid these areas when working with the device (in most applications) since the waveform would become distorted if it tried to push the device into these areas. (From the curve shown, it can be seen that if, using the load resistor shown, the base current was pushed beyond 500 μA, the intersection of the resistor line with that value would put the device into saturation. A distorted waveform would result (see Figure 8.6).)
3. The power limit can also be seen, and therefore easily avoided.

In Figure 8.6 the distortion noted in item 2 is shown. Here the swing (ac) of the input current to the base causes the load line to intersect with the upper base current line in the saturation region of the device. The output waveform, either voltage or current, will be distorted as shown in this illustration.

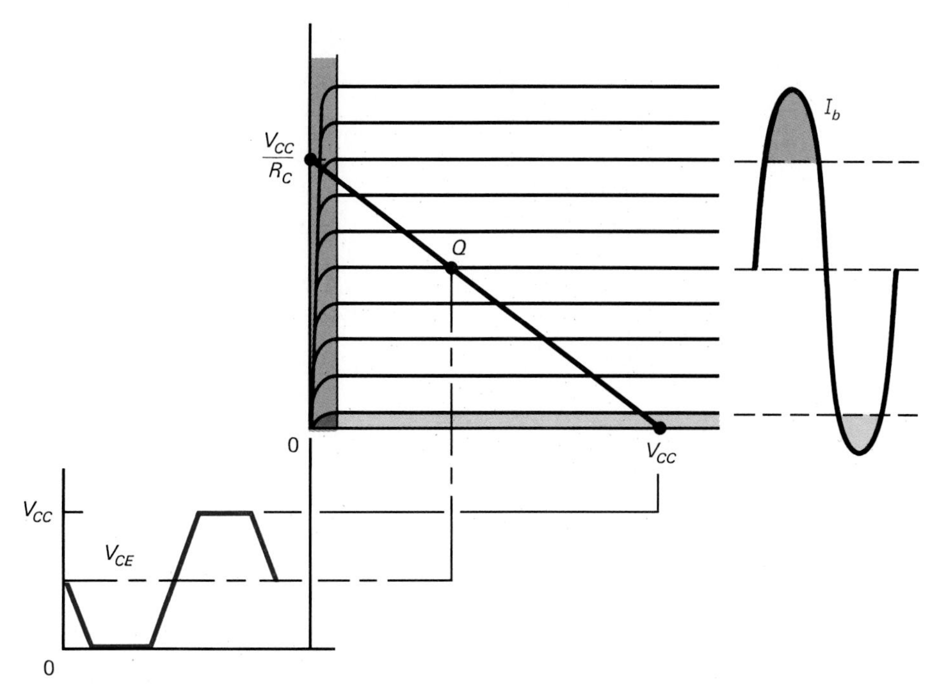

FIGURE 8.6 Distortion Occurring Due to Excessive Peak-to-Peak Input

Another advantage of the graphical method is to allow the user of a BJT to have a mental, if not physical, picture of how the device is acting and reacting. If, in a functional circuit, distortion is occurring, this problem may be related to the picture of the BJT curves more easily than to the formulas. Changes may be made based on this mental picture rather than by a volume of calculations. Once the problem was resolved, actual values could be confirmed by calculation. Changing the Q-point to move the entire graph into the active region of the device (away from saturation for example) is easier to picture than it is to derive by formulas.

The JFET Transfer Curves

Since the transfer curves for the JFET are less detailed than the characteristic curves for the BJT, less information can be derived from their use. Still, these curves can be used to predict the operating conditions of the FET. Before using these curves, it is necessary to know the value of gate voltage, or to have an approximation for gate voltage.

In Figure 8.7 the normal plot of a JFET transfer curve is shown. On that graph the value of gate voltage is selected along the horizontal axis. (If estimating the gate voltage, some center value would be chosen.) The value of I_D could be found simply by referencing the vertical axis. (These are also called Q-points.) Before going further, the value for the source resistor, which determines gate voltage, can be calculated:

$$R_S = \frac{V_{GS}}{I_D} \qquad (\text{since } I_S = I_D) \tag{8.1}$$

(A value for R_D would have already been selected.) Therefore, the basic operating conditions for the circuit are established. Then, just like viewing the

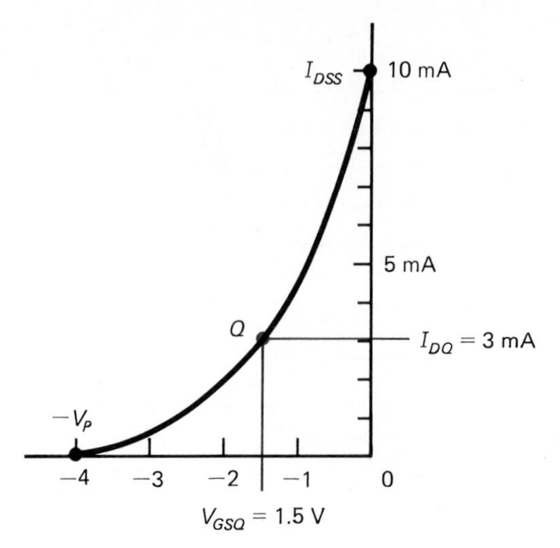

FIGURE 8.7 N-JFET Q-Point Data

changing base current for the BJT, if an ac voltage is applied to this device, these voltage variations can be plotted to see the output changes that would result.

The peak-to-peak voltage would be marked off the horizontal axis (V_G), drawn up to the curve, and "reflected" to the horizontal axis as values of drain current. Figure 8.8 shows just such a plot. Such a plotting is simple and requires little evaluation. It can be seen that there are limits to the input values, just as there were for the BJT. If the gate voltage runs too high (above V_P) or goes down to zero (or even is forced to be the "wrong" polarity), the drain current will not produce a representative output.

There are, though, some significant details that can be derived for the JFET based on the curve shown.

1. As shown before, V_P and I_{DSS} show operational limits.
2. One major problem with the JFET is shown quite clearly in this curve. Since the curve is not linear (a straight line), the gain factor (g_m) at one point on the line is not the same as for another point. What this indicates

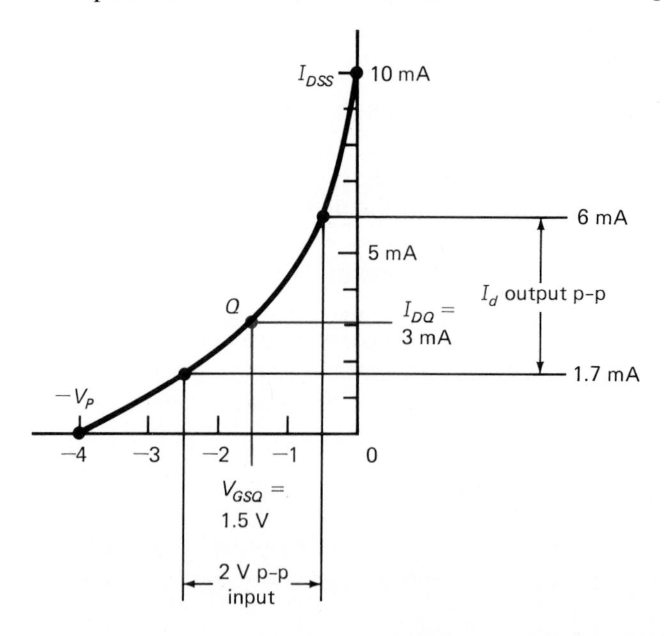

FIGURE 8.8 Effect of AC Input in Relation to the Q-Point (JFET)

is that even if the changes in gate voltage are evenly centered around the bias value of V_G, the output current (I_D) will not be an equally even change. This creates a problem called "nonlinear distortion." The output shape is not the same as the input shape. Such distortion is not easily corrected. The only way to solve the problem is to reduce the amount of change (peak-to-peak ac voltage) at the input, so the output is "reflected" from only a small part of the curve. Taking such a small section of the curve will result in a closer approximation to a straight line. Therefore, as illustrated graphically (and not as a part of the circuit equations): *To avoid nonlinear distortion caused by the JFET, input signal levels should be relatively small (minimal peak-to-peak variations).* This problem therefore creates another restriction on the use of the typical small-signal JFET.

3. It was noted in Chapter 6 that the value of g_m (transconductance) used in the equations was derived from an equation based on the graphical representation of the device. To calculate the "operational" value of g_m, based on data sheet values of g_m, the following equation should be used:

$$g_m = g_{mo}\left(1 - \frac{V_{GS}}{V_P}\right) \tag{8.2}$$

where g_m = operational value of transconductance (ac)
g_{mo} (or y_{fs} from data sheets) = devices listed transconductance
V_{GS} = operational gate-to-source voltage (bias)
V_P = pinch-off or maximum gate voltage on the curve.

EXAMPLE 8.2

The plot of a JFET curve is shown in Figure 8.9. Calculating g_m (operational):

$$g_m = 3000\ \mu\text{s}\left(1 - \frac{4\text{ V}}{6\text{ V}}\right) = 3000\ \mu\text{s}\ (0.333)$$
$$\simeq 1000\ \mu\text{s}$$

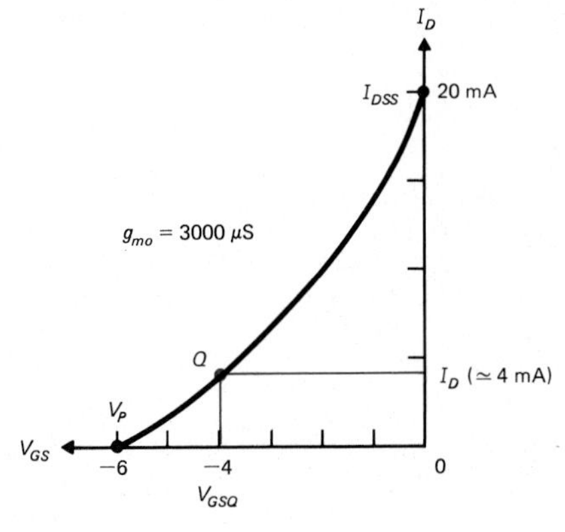

FIGURE 8.9 Higher $V_{GS}(Q)$ Resulting in Lower $I_D(Q)$

(Note that 3000 μs was chosen as an average value.) This will produce (from past equations)

$$I_D = g_m \times V_{GS} \text{ (operational)}$$

$$\simeq 1000 \text{ μs} \times 4 \text{ V} = 4 \text{ mA}$$

This is confirmed on the graph; V_{GS} reflected results in $I_D \simeq 4$ mA.

The graphical analysis of the FET is done occasionally to determine the operating conditions. More often, though, the modified equation for transconductance (g_m) will be used along with typical circuit equations when working with a JFET. Again, the mental picture of the JFET's operating curves can be an aid when working with the device. If distortion of the input is observed, it can be related to either:

1. Driving V_{GS} beyond the acceptable V_P value
2. Attempting to force I_D beyond the maximum of I_{DSS}
3. Having too much of a variation on the input signal and thus running into the normal nonlinear distortion caused by the JFET

The use of a graphical approach to establishing BJT and JFET circuits can have some advantages over calculating these values. Although the solutions are the same, a conceptual picture of what is happening with the device can lead to a more logical analysis of the overall system. Also, keeping a mental picture of the device's operating curves can assist when encountering faults in circuit performance.

REVIEW PROBLEMS

(1) In Figure 8.5, one load line is shown (R_C). Copy this drawing but add two other load lines, one indicating a lower value of R_C and another for a higher value of R_C. [Note that the V_{CE} intercept is the same for all three lines, since V_{CC} is not changed, but the intercept at the I_C axis will change for each resistor ($I_C = V_{CC}/R_C$).] Then write a short evaluation of what values would change due to these two different resistors.

(2) Redraw the curves shown in Figure 8.8, but this time for a P-type (P-channel) FET. (Note that V_{GS} will change polarity.)

(3) The waveforms shown in Figure 8.10 are distorted outputs from both BJT and FET circuits. Based on graphical analysis, what may have caused this distortion in either an NPN BJT or an N-type FET?

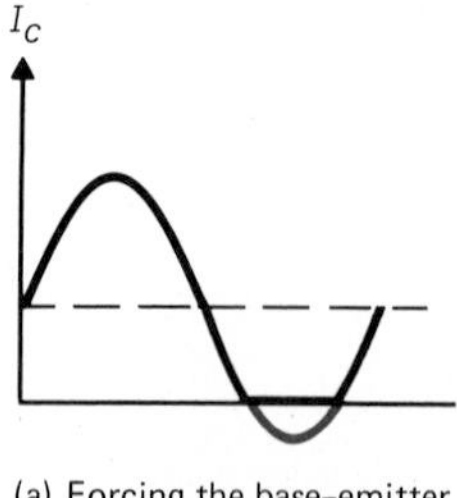

(a) Forcing the base-emitter into reverse bias

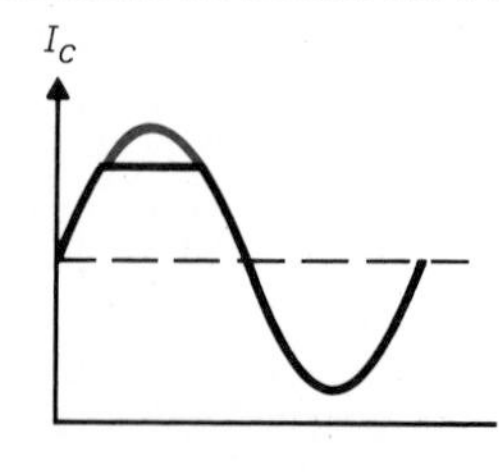

(b) Attempting to exceed I_C maximum set by the collector resistor

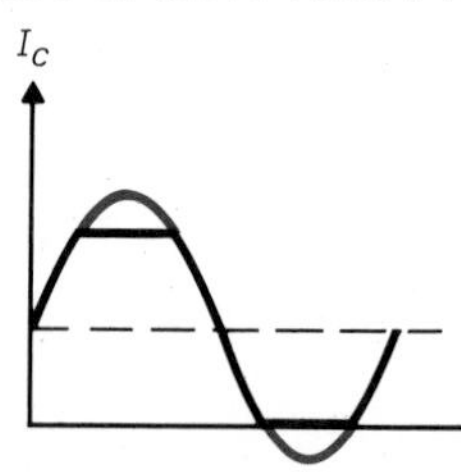

(c) Input forcing both conditions (a) and (b)

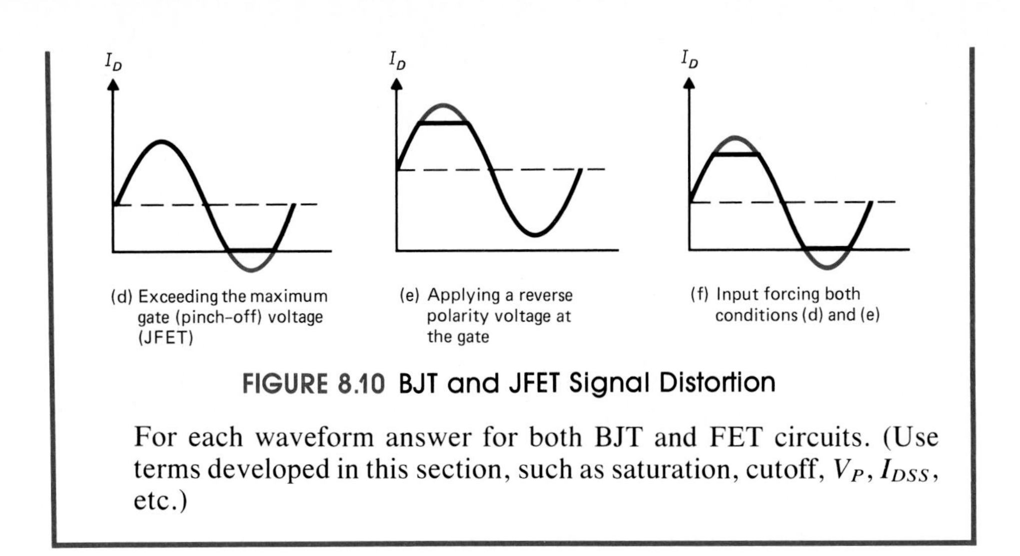

FIGURE 8.10 BJT and JFET Signal Distortion

For each waveform answer for both BJT and FET circuits. (Use terms developed in this section, such as saturation, cutoff, V_P, I_{DSS}, etc.)

8.5 VARIATIONS ON STANDARD BJT AND JFET TYPES

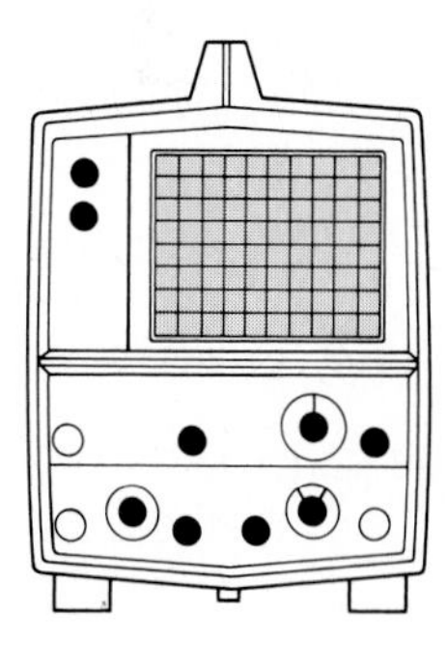

Although we have presented the basic BJT (both NPN and PNP) and the JFET (N and P types), there are other devices that fit into the classifications of discrete BJT and FET types. Some of these have been mentioned before; some are newly presented in this section. Although these new elements function similar to the standard BJT or FET, it is important to note their differences. This section is divided into BJT-like devices and JFET-like devices. For some devices, new schematic symbols may be presented; for others the standard BJT and FET symbols apply. Some of these variations are subtle, others are radical variations on the standard application of the BJT or FET.

Forms of BJT Devices

The following is a basic description of devices that are classified as operating like the fundamental BJT. Their performance differs from the standard small-signal bipolar junction transistor, and therefore need to be described separately. The polarity sensitivity (PNP, NPN) seen in the BJT will also be seen in these devices. Such sensitivity to polarity will not be explained separately.

Germanium Transistors The germanium BJT is identical in operation to the silicon BJT (highlighted in this book). The only major difference is that the forward voltage across the base–emitter junction is assumed to be $\simeq 0.3$ V (not 0.7 V). They are less popular than the silicon type and are often more temperature sensitive.

Power Transistors BJT transistors (silicon usually) are also available as high-power devices. Power transistors usually have the following variations in parameters from small-signal BJTs:

Maximum ratings are higher.

Beta is usually quite low, often <10.

Allowable collector currents are usually quite high (typically 10s to 100s of amperes).

Output impedance is usually quite low.

The packaging for these devices is different from that used for small-signal devices. They are usually large with metal connections used to bolt the device to a heat sink (a metal surface used to draw heat away from the device). Their schematic symbol is the same as for a standard BJT. Usually, they are found as the last output device (or set of devices) in a multiple-stage, high-power amplifier. (See the data sheets for the MJH6018 in Appendix A.)

RF Transistors Some BJTs are manufactured to operate at very high radio frequencies, working in the range of the broadcast radio band (5.4 kHz to 108 MHz) all the way to the microwave (gigahertz) range of frequencies. Most of these devices are manufactured especially for high-frequency applications. They are built into special packaging which both insulates them from picking up stray interference and promotes the amplification of high-frequency signals. They are available in small-signal or power types. Their schematic symbol is the same as for a standard BJT. (See the MRF2628 data sheet in Appendix A.)

Optical Transistors The base junction of all BJTs are light sensitive; therefore, they are built into special "dark" packages. Some are manufactured with clear housings, thus exposing the base to light. They function the same as a regular BJT, except that the base input is a light level instead of a sourced current. The optical transistor has a different schematic symbol (Figure 8.11). It may or may not have an electrical base lead, used to establish a base bias condition.

Darlington Transistors The Darlington transistor (Figure 8.12) is basically two BJT devices connected together in such a way that the output of one directly feeds the input of the next. They are housed in a single package

FIGURE 8.11 Schematic Symbol for an NPN Phototransistor

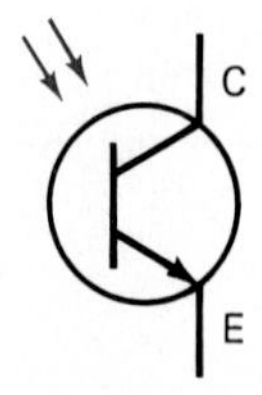

FIGURE 8.12 Schematic Symbol for an NPN Darlington Transistor

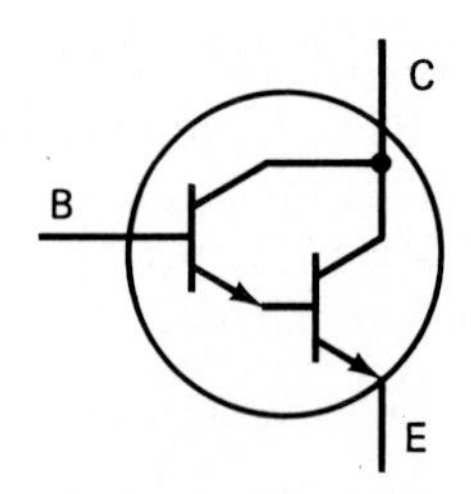

Basically, the Darlington produces a gain value (beta) that is the product of both transistors' beta. Darlington transistors are used where very high gain is needed, often at input points to a circuit. The Darlington can be used in circuitry much like the BJT, as long as its special parameters are taken into account:

Beta: in the tens of thousands typical
Input impedance: very high
Output impedance: often similar to that of a standard BJT

The Darlington transistor is available also as a photo-Darlington and as a power Darlington.

Many of these special-purpose devices will be applied in later chapters. It is important, though, to know that the BJT comes in a variety of forms to fit various applications.

Forms of FET Devices

The FET is also available in a variety of forms. One of the major variations from the JFET is the group of field-effect devices called MOSFETs. These devices are constructed quite differently from the normal JFET and have different operational characteristics, although they still have the basic gate impedance and transconductance function seen in the JFET. Also in the category of field-effect devices, power FETs are available.

MOSFETs: A Special Category of FETs The MOS-FET (metal-oxide semiconductor-FET) is manufactured in a way that is quite different from that used to make a JFET. Many of its characteristics are similar to those of JFETs, with a few exceptions. [MOSFETs are some times called IGFETs (insulated gate-FETs).]

Note: The MOSFET is available in either P-type or N-type forms. In this text the N-type is used extensively. To apply the P-type, simply reverse the bias polarities.

The MOSFET is available in two forms: Enhancement (E-MOS) and depletion (DE-MOS). The DE-MOS has almost the same characteristics as the JFET, except that it can have a gate voltage that is a "reverse" polarity and still produce drain current. Therefore, it is possible to operate the DE-MOS with zero gate bias voltage and still have it reproduce an input that is both positive and negative in polarity. The transfer curve for a DE-MOS device (Figure 8.13) will illustrate this process.

The enhancement MOSFET (E-MOS) is quite different from either the JFET or the DE-MOSFET. The enhancement type actually operates with a

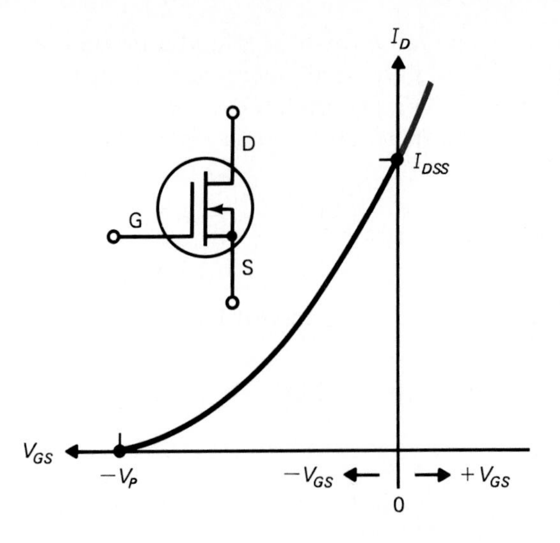

FIGURE 8.13 Schematic Symbol and Transfer Curve of an N-Type DE-MOSFET

gate voltage *opposite in polarity* from that seen in a JFET (or DE-MOSFET). The shape of the transfer curve is the same as that for other FETs, but it is drawn in a different location on the graph from that seen in the other FET devices.

Figure 8.14 illustrates the transfer curve for an N-type E-MOS device. There are some major changes in the operating points observed on this graph compared to a JFET. First, instead of having a maximum gate voltage V_P, it has a minimum value, called $V_{GS(\text{th})}$ (gate-to-source voltage threshold). This is a minimum (not maximum) voltage that must be present on the gate of the device. Also, there is not an I_{DSS} for this device. Actually, up to a maximum rating value, there is no limit to the drain current. In fact, this device works opposite to the JFET.

The gate voltage on a JFET caused a lessening of the drain current out of the device. Increased gate current created a lessening of drain current. In the E-MOS, gate voltage "encourages" drain current. Thus, with increases in gate voltage, there will be coordinated increases in drain current. The E-MOS FET can be placed in a circuit much like the beta-independent BJT

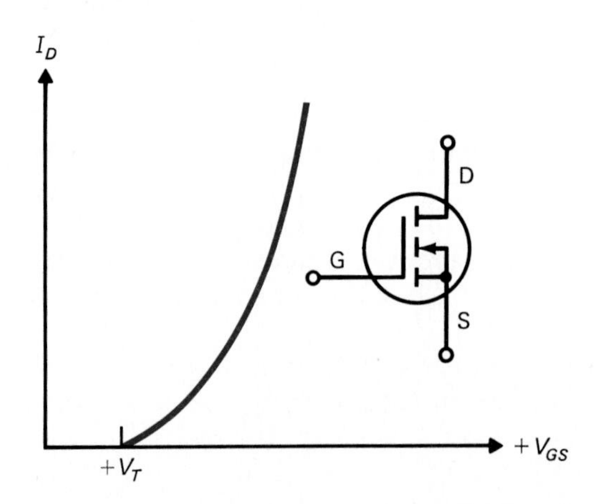

FIGURE 8.14 Schematic Symbol and Transfer Curve for an N-Type E-MOSFET

circuit, where gate voltage is derived from a voltage divider. (Gate polarity can now be the same as drain polarity.)

Both of these devices are available in both small-signal and power-type devices. The MOSFET is also quite popular for use in digital integrated circuits, because more devices can be packaged in an integrated circuit than if using BJTs or standard JFETs. Since the numerical characteristics of MOSFET devices are similar to those for BJTs, they will not be discussed separately.

> *Note:* MOSFETs are, due to their internal structure, quite sensitive to static voltages. If a brief static charge is applied to the gate lead, internal damage can result. Therefore, these devices must be handled quite carefully. Grounded equipment and even ground straps on assembly personnel are used to eliminate static electricity during the assembly of MOS devices.

Individual MOS devices are packaged in such a way as to eliminate static electricity. Some are shipped with a small wire wrapped around all three leads. Others have their leads inserted in conductive foam or plastic. This process prevents a voltage potential form developing between the gate and either source or drain leads. If using a MOS device, the "shorting" material should not be removed until the device is installed in the circuit. If the device must be removed from the shorting material before installation, care should be taken not to touch its leads.

This presentation covered the majority of the BJT and FET discrete devices currently in use. In later chapters, some of these devices are worked with in greater detail as they are applied to functional circuits. Other devices will be introduced which serve quite specific purposes and which do not coordinate with the basic BJT or FET use.

REVIEW PROBLEM

(1) Prepare a reference sheet that includes the schematic symbols for the standard BJTs and FETs. Along with these, include any additional symbols presented in this section. On this composite list, write a short description of the major function of these devices and any special variations specific to them.

8.6 TESTING DEVICES

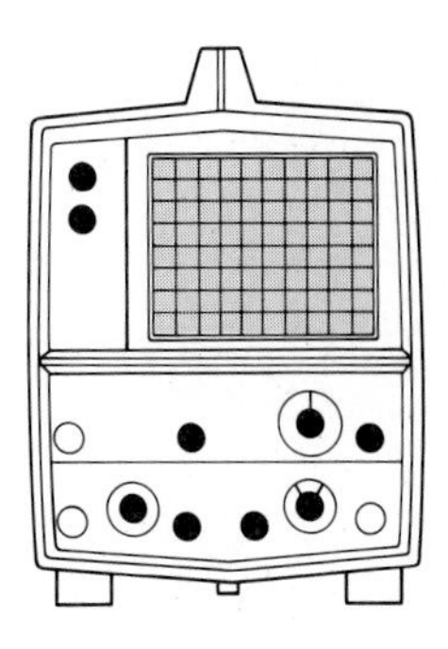

In past sections the normal operating characteristics of BJT and FET devices have been detailed. A clear knowledge of a device's characteristics is necessary whenever working with that device. There are occasions when circuits fail to perform as expected; tests then need to be conducted on the devices used in the circuit to isolate the source of the failure. The following is a basic list of test procedures that can be used to help isolate a defective device and to assist in preventing future failures.

It is assumed that in testing, the fault in a system has been isolated to a few circuits. Now it is necessary to test the devices within these circuits to determine which may be defective. It is first necessary to specify just how semiconductors can fail in a circuit. Although there are several ways in which the devices will fail, there are some common indicators of device failure. Divided by device type, the following are some of the prime factors in device failure.

BJT Devices

There are two primary ways in which the BJT will fail. First, if it is subjected to extreme heating, the characteristics of the device change. When that occurs, the device most often will produce a nearly constant resistance across the collector/emitter leads. No longer does it display the variable "semiconductor" effect, but rather a nearly constant resistance. Usually, the gain factor beta is nonexistent. The second form of failure (and probably the most common) is caused when too much current passes through the device. Usually, this causes the emitter's internal connections to open, thus making the device totally nonconducting. A simple set of tests can be conducted on the device to determine either of these failures. There are two ways the device can be tested, in the circuit or out of the circuit. The procedures are different depending on the form of test to be conducted.

Out-of-Circuit Testing If the device is out of the circuit, the base–emitter "open" condition can be tested. Most modern digital volt/ohm meters have a semiconductor test feature. Basically, the meter is connected to the base and emitter of the device and the test is run. If the device indicates an open circuit, the leads of the meter are reversed (to cross-check the single-polarity sensitivity of the device) and the test is repeated. (Specific test procedures can be found in the manual for that meter.)

A standard VOM (volt-ohm-milliammeter) can also be used for this test. The meter is set on the ohms scale in a range of about 1 kΩ. The polarity (+/−) of the test leads must first be known (see the VOM manual). The leads are then connected to the base–emitter leads of the BJT in such a way as to forward bias the device (observing meter lead polarity). If the meter deflects to indicate a value of resistance (not infinite), forward conduction can be assumed. The test is completed by reversing the test leads to observe reverse bias. In this case the meter should read (near) infinite resistance.

> *Note:* The actual resistance reading of the meter is *not* a valid value and should be ignored. This is a simple "good/bad" test of the BJT. This "ohms scale" test usually cannot be made with a digital meter (DVM or DVOM).

If the device passes this "forward conduction" test, it is necessary to build a simple circuit to check the relationship of I_B to I_C. The circuit shown in Figure 8.15 can be used for this test. In this circuit the base potentiometer is adjusted while monitoring both base and collector current. One ratio of collector to base current (beta) is made, then another is made using a different value of base current. These two values are first compared to each other, then to the specification sheet for that device. If in comparing the two values for beta, they are not similar, or if they fall short of the value in the device specifications, the device can be considered defective.

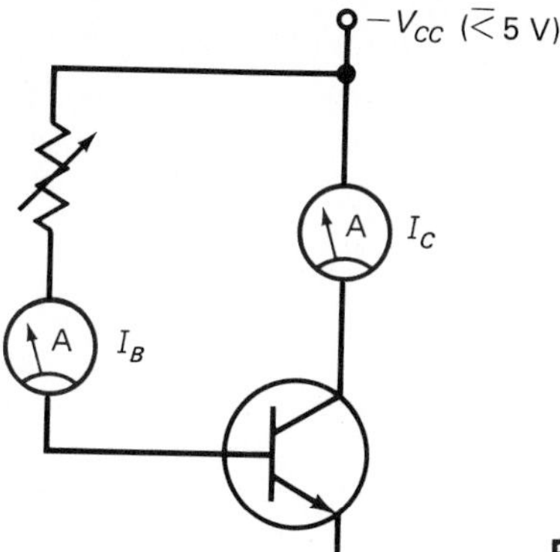

FIGURE 8.15 Test Condition for Determining BJT Gain

To save time, there are several types of test equipment that will test a semiconductor device's function. One such piece of test equipment is called a **semiconductor curve tracer**. It will produce a family of curves for the BJT or for just about any semiconductor device.

In-Circuit Testing It is quite common to have a device fail while being assembled in a circuit. It is usually inconvenient to remove the device from the circuit just to test if the device is operational. However, a BJT can be effectively tested while in a circuit.

To test the BJT in an operational circuit simply requires a voltmeter. (Naturally, proper safety precautions should be observed when testing any "on" or active circuit. The risk of electrical shock presents a true hazard to anyone conducting tests.) First, the operational base-to-emitter dc voltage is checked. Although 0.7 V has been used as a value in this book as a standard drop, in actual circuits this value can range above or below that standard. Primarily, the test should neither produce a zero voltage level nor a voltage above about 1 V (except for power devices; for them the specification sheet should be referenced).

Assuming that the device passes this test, the voltage levels from base to collector and collector to emitter should be tested. In most schematics, these values are listed. (Some schematics indicate voltage levels in reference to the circuit ground; in that case, one lead of the meter should be grounded to the circuit ground and the other used for measurement.) If these levels match those in the schematic, the device should be considered functional. An exception to this is if the problem of the system is distortion. Then an ac test using an oscilloscope should also be conducted.

FET Devices

There is one primary way in which the FET will fail. Usually (although other problems are possible), the FET's gate lead will fail to maintain control over the drain current. Often, this is caused by excessive voltage on the gate (static or dynamic), which causes internal damage to the device. Since the drain current is limited by the device, failure on the output side is not too common. (MOS-type devices are far more sensitive to overvoltage on the gate lead.

They are also sensitive to excessive drain current, since they do not have the traditional current limiting feature of the JFET. Thus they are more susceptible to damage than is the JFET. A special note is added at the end of the testing information especially for the MOS types.)

Out-of-Circuit Testing (These procedures apply only to the JFET; for MOS types, see the end of this test section.) The JFET can be tested out of a circuit, but care must be taken while making these tests. The JFET can be damaged by high gate voltages, either static or dynamic. Although it is not as sensitive as the MOS types, damage can occur if care is not taken. One simple test can be conducted on the JFET generally to test if the output circuit is functional. Simply measuring the resistance between the drain and source leads can give a rough indication if either lead is open. Normal values in the range 1 to 100 kΩ are normal (except for power devices; inspect their data sheets for normal values). This test, though, is not absolute; it will show only a definite open-circuit condition on the drain/source leads.

The only effective way to test a JFET out of the circuit is to build a simple test circuit. Figure 8.16 indicates the basic test circuit.

Note: If the manufacturer's data sheet indicates an approved test circuit, that should be used. This circuit is not to be used for MOS devices or for power FETs.

It can be noted that two power supplies are used for test purposes. The value of gate bias voltage can be carefully controlled by using a separate supply (note polarity!). The 10-kΩ gate resistor is used to reduce the possibility of developing unusually high voltages on the gate lead (some supplies can have high voltage transients if working into a basically open circuit). To reduce the problem of transient voltages, the circuit should be constructed with all voltage supplies set at zero (but turned on); then the supply voltages should be increased from zero to the proper test voltage. To test the device, set the gate voltage to two voltage levels (less than 5 V) and measure the corresponding drain current. Calculate a simple value of g_m by the formula

$$g_m = \frac{I_{D2} - I_{D1}}{V_{G2} - V_{G1}} \tag{8.3}$$

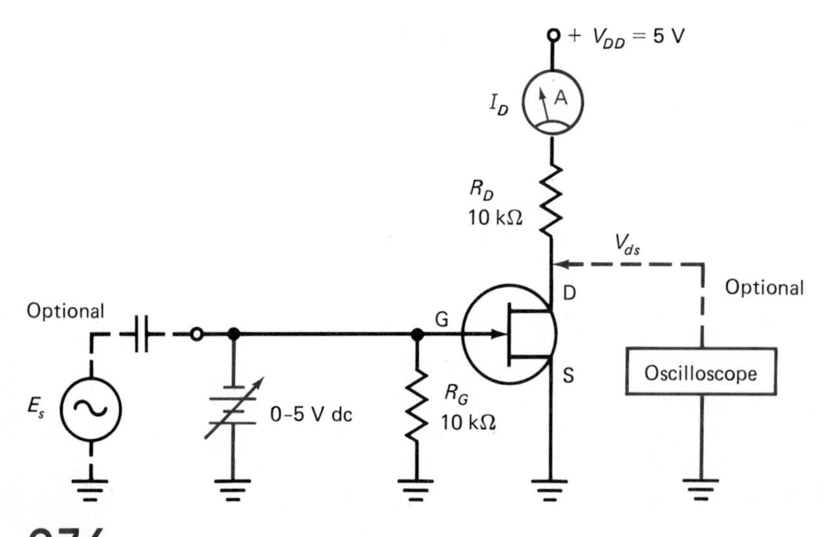

FIGURE 8.16 Test Condition for Determining JFET Gain (and Optional AC Function)

(The numbers 1 and 2 refer to first and second readings.) This value of g_m should be compared to the device's data sheets. If necessary, an input ac voltage can be used, with an oscilloscope to conduct an ac test.

Note: Unless there are specific manufacturer instructions which indicate the exact procedure for testing, MOS devices should not be subjected to an out-of-circuit test. The risk of damage due to static charges and transients usually overrides the advantages of out-of-circuit testing for these devices. Special test equipment handling and circuit construction are required when testing MOS devices.

In-Circuit Testing FETs are most effectively tested while in a circuit. Measurement of voltage levels on the output side of the device should confirm if the device is functioning properly. Because of the high impedance, input of the JFET measurements at the gate lead should not be taken, again due to static voltage problems. (Remember, the gate is usually at virtual ground; thus it would also show no potential difference from ground.) A measurement of voltage on the source resistor will indicate both if drain current is flowing and the level of gate-to-source (V_{GS}) bias potential. If the level of the voltage on the source resistor is above (approximately) 8 V or is below (approximately) 1 V, the device may be defective (unless the schematic indicates that these levels are normal).

For MOS devices, in-circuit testing can cause device failure. First, only test equipment that is approved for MOS testing should be used. Second, it is best not to test the device directly, but rather, to measure levels on attached resistances such as the drain or source (if used) resistors. *Direct measurement of levels at the gate lead of a MOS device can destroy the device.* It is quite common to use only ac measurements with MOS devices. The input level before the coupling capacitor and the output level after the coupling capacitor are the safest measurements to take around a MOS device.

These are basic guidelines to be used when testing these devices. Naturally, certain modifications of these procedures would be necessary to coordinate with specific circuitry that would be used. In later chapters, more detailed test procedures will be explored when dealing with specific application circuitry. Testing for certain specialty devices will also be presented. These procedures are general guidelines for testing just the BJT or FET device within a circuit.

REVIEW PROBLEMS

(*Note:* For problems 1 and 2, the data sheets in Appendix A may be referenced.)

(1) If a 2N2218 transistor was tested in a circuit like that shown in Figure 8.15 and the following values were measured, indicate what may be defective (if anything).

$$V_{CE} = V_{CC} \qquad V_{BE} = V_{CC} \qquad I_C = 0\text{ A} \qquad I_B = 0\text{ A}$$

(2) If it was felt that another 2N2218 transistor was defective in the

circuit shown in Figure 8.17, and the following measurements were made, indicate what may be defective (if anything) (circuit dc source $= 10\text{ V} = V_{CC}$).

$V_{CE} = 5\text{ V}$ $V_{BE} = 0.62\text{ V}$ $V_{RC} = 5\text{ V}$ $V_{RB} = 9.38\text{ V}$

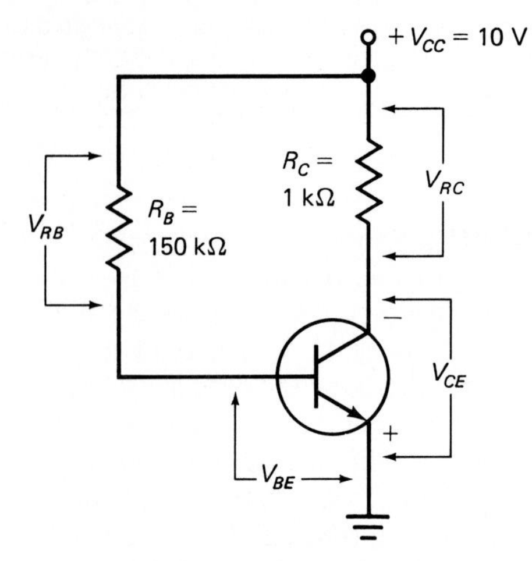

FIGURE 8.17 BJT Test Voltages

(3) Using equation 8.3, calculate the effective g_m of a JFET that was tested to have the following values.

Test 1: $V_{GS} = 2\text{ V}$ $I_D = 3\text{ mA}$
Test 2: $V_{GS} = 4\text{ V}$ $I_D = 1.2\text{ mA}$
y_{fs} from the data sheet: min. $= 1200\ \mu\text{S}$, max. $= 3200\ \mu\text{S}$

Is this device's transconductance close enough to the listed value to consider it functional?

8.7 CAUSES OF DISCRETE DEVICE FAILURE

In Section 8.6 basic test and troubleshooting procedures were outlined for discrete BJT and FET devices. In this section those conditions that can cause these devices to fail in a circuit are outlined. Naturally, one process of troubleshooting is finding a defective device and replacing it. If the cause of that devices failure is not determined, further failures are likely if steps are not taken to correct the problem. The following is a general cataloging of some of the major causes of BJT or FET failure in a (small signal) circuit:

Defective Device In some cases, especially when a circuit is initially constructed, faults may result from a device that was initially defective. Usually, though, if the system has been operational, initial device defects are not the cause of failure.

Solution Preassembly testing of devices should reduce failure.

Device Specifications Do not Meet Circuit Demands In some cases the maximum ratings of the device being used are exceeded by the circuit in which it is operating. Usually, a check of the manufacturer's specifications will reveal such an error.

Solution It is a common rule of thumb to select a device whose maximum ratings exceed all circuit values by a factor of 2. For example, if the dc source voltage for a circuit is to be 9 V, all device breakdown voltages should be 18 V (2×9 V) or greater.

Inappropriate Substitution If a device was defective in the past and a substitution was used (not the same part number), that substitution may not be able to meet the standards used for the initial device selection.

Solution All specifications for substitutions should exceed the ratings of the device they are replacing. (In some situations, substitutions are not permitted without a "design change" approval from the engineering group responsible for the design of the circuit.)

Short or Open Circuits One of the most common defects in any circuit is the open circuit (short circuits are usually not common). A bad solder connection, broken wiring, or cracked "trace" on a circuit board will often make a device appear to be defective.

Solution Once a defective circuit is isolated, the wiring (PC board) should be tested for open circuits. Usually, with the system off, an ohmmeter is used to test for "continuity." Tests should be made from component lead to component lead. Care must be taken when testing MOS circuitry this way!

Overheating Overheating is one of the most common causes of device failure. If a circuit becomes too hot, either due to the environment or to electrically generated heat (power), device damage can occur. In some cases, circuit defects do not show up until the system "warms up." There are products available that can help test for heat-generated defects; these either heat or cool the circuit in order to isolate the heat-sensitive element in the system.

Solution Careful testing should be done to isolate the source of the heating and eliminate the overheating problem. In some cases the circuit or physical arrangement of components will have to be redesigned to reduce the thermal effects.

Turn-On Fatigue and Power Supply Surges Both of these problems are related to the production of occasional surges in circuit current. When a circuit is initially turned on, all capacitors must become charged (such as an emitter bypass capacitor). The rush of current caused when they charge can cause fatigue (excessive wear) within semiconductor devices. The same problem can occur if the power supply voltage should have an occasional surge (spike of higher voltage). Damage due to this problem may be immediate or could be a long-term effect. Turn-on surges can be monitored by observing the current from the power supply when it is initially turned on. If the peak value of this current exceeds the peak current rating (again the factor of 2 should be used) of the device, a device change or power supply change is in order. Power supply surges are more difficult to detect, but one common indicator is that if unrelated devices in the same system are found to be defective, a surge in the main power supply should be suspected.

Solution Corrective action is usually able to be taken at the power supply for both of these problems. There are devices designed to regulate the current allowed to a circuit to some upper limit. These could be attached to the power supply to reduce turn-on surges. Items called *surge suppressors* can also be added to the supply to reduce the occasional surges that originate from the supply itself. (Both of these devices are discussed later in this book.)

Excessive Input Signal In some cases an input signal can be so large that it will cause a device to be destroyed. Usually, the signal has to be sufficiently large to cause the device's reverse polarity limitation to be exceeded (input from either an original input or from a previous amplifier). Usually, this problem is accompanied by distortion of the input signal at the output. Ac measurements are meant to detect such faults.

Solution The source of the excessive input should be identified and eliminated.

Capacitor Leakage With age, many capacitors (especially electrolytic types) can become "leaky." This means that the capacitor is passing an unacceptable amount of dc current (its internal dc resistance is low). Two problems can be created by capacitor leakage.

1. If an input or output capacitor is defective, the bias levels from one stage of the system can affect the bias levels of another. When this happens, the signal becomes distorted since the device is rebiased by this additional voltage, possibly causing saturation (base of a BJT, gate of an FET). The dc level at the input terminal of the device should be tested as well as any dc levels on the circuit's load. An error in these levels from those listed on the schematic may imply capacitor leakage. (The capacitor should be removed from the circuit and tested individually.)
2. Bypass capacitor leakage (such as on a BJT's R_E) will cause the bias level on the element it is bypassing to deviate from the listed value on the schematic. (The effective resistance of these two parallel elements is lowered, causing an imbalance in the bias potential.)

Solution If one capacitor is found to have excessive leakage, it is common practice to replace all capacitors like it in the system (unless they are relatively new). Leakage is a by-product of aging and therefore if one device starts to become defective, it will only be a matter of time before the others also become defective. Many of the newer forms of capacitors (tantalum, polystyrene, etc.) have improved leakage/life statistics, and may thus be preferred as replacements for older electrolytic types.

Overloading In some cases, especially for the final stage of an amplifier, the load tied to the system may create an overload condition on the circuit. The overload can either be due to a load that draws too much current from the circuit (too low in resistance), or surprisingly, an open-circuit load condition. The open-circuit load condition may cause a problem wherein the circuit must consume the full power level of its normal output. Although most circuits will compensate internally for an open-circuit load condition, some will produce power levels high enough to cause internal overheating. In many cases, this overload condition can be traced to a physical action that caused a short-term overload (e.g., the user connected the system to an unapproved load, or unplugged a connector while the system was on).

Special note: In some cases, if the load to a circuit is not attached when the initial power is turned on and then is attached later, a large surge of dc current may be produced when the capacitor coupling the load balances its dc charge.

Solution The load should be checked to assure that it is correct and, as necessary, always in place. A prime indicator of this problem is if the output side of the active device which is connected to the load has been damaged. If the load is normally connected and disconnected, or may be variable, it is common practice to attach an internal load to the system (often a resistor across the output "connector"), which assures that the system has a load even if the "connector" is opened. In other cases, a resistor will be attached in series with the load to assure a minimum load, even if the "connector" terminals are shorted.

Static Charges Although the MOS device is most susceptible to damage to static charge, any devices can be damaged (or fatigued) by excessive static charges. Device damage due to static charge is not easily detected, as such a defect is often identical to one of the other forms of device failures. Generally, static damage is based on environmental conditions and circuit applications. There are no effective tests to detect static charge damage (unless the presence of high static charges is obvious by observation).

Solution To reduce the possibility of static damage, all circuits should be either directly attached to earth-ground connections, or indirectly protected by shielding. (Shielding is achieved by protecting the circuit(s) within a grounded conductive container.) Some semiconductor devices have a fourth lead, which is connected to an internal ground inside the device. This lead should be attached to an external grounding connection. Shielding is never 100% effective, and on some equipment great measures are taken to eliminate static charge.

Internally Generated Oscillations In some circuits an undesirable generation of sine-wave signals, usually at high frequencies, can occur. This process is discussed later in this book, together with means for eliminating the problem.

This basic cataloging of potential circuit defects only scratches the surface of possible defects that can be seen in a system. Often defects are as different as there are different circuits. This listing, though, highlights the most probable initial causes of problems that show up when discrete transistors become defective. In later chapters, more potential defects are explored, together with troubleshooting methods when more sophisticated or different forms of circuits are introduced.

Some manufacturers use sophisticated computer-aided circuit board testers. These units are used to test mass produced system assemblies and subassemblies. Defective circuit boards, returned by field service personnel, are "bulk-tested" to locate the source of malfunction. The computer conducts a battery of tests in a very short time and indicates either the defective section (block diagram coordination) or specific component that is to be replaced. Figure 8.18 illustrates one form of such a test system.

Recording Test and Repair Data

Whenever a system is found to be defective, it is important to repair it and make it operational. It is equally important to keep good records of the repairs made and to evaluate the probable cause of the defect. Only occasionally is a system failure due to an initially defective semiconductor. In most cases, semiconductors can be considered to remain within tolerance for quite a long time. External problems that overload or fatigue the device are usually the cause of device failure.

It is important that a good record of the conditions under which the device failed, the condition of the defective device, and other circuit conditions be maintained. This "history" of the system can be a valuable tool if future defects are to be prevented. A major part of the job of working with prototype circuits (the first circuits built when a system is designed), with sample test circuits, with custom-installed systems, and when conducting troubleshooting and repair is to keep a concise record of good and bad results found. This "history" can be evaluated to determine weaknesses in the system, to eliminate recurring problems, and to improve on the design of similar systems.

Most manufacturers usually require concise reporting ("field report") of all steps taken to resolve a problem, revise a system, or troubleshoot and repair a defective system. This information is evaluated, and probable defects in the system can be detected and corrected. The electronics professional should therefore realize that concise written reports, documented evaluations of systems, and clear communication of ideas and developments are essential to the electronics profession.

REVIEW PROBLEMS

(1) Based on all the above-mentioned potential circuit defects, prepare a basic step-by-step procedure for resolving the cause of a faulty circuit. You may assume that the device in the circuit has been found to be appropriate for the intended use. (Start your list with "shorts and opens.")

(2) Using a semiconductor substitution manual, find a device that is recommended as a substitution for the 2N2218 listed in Appendix A. (Do not select devices within the "family" shown with that de-

vice.) If you do not have access to a substitution guide, prepare a list of parameters for a substitute device based on those listed for the 2N2218 (e.g., maximum $V_{CEO} > 30$ V).

(3) A BJT in a circuit has been found to be defective for the third time in an otherwise functional circuit. Suggest evaluations that should be made before simply replacing the device. The device was found (again) to have an open base–emitter connection. (Consider both device and circuit tests that should be conducted.)

SUMMARY

In this chapter the basic specifications of discrete semiconductor devices has been presented along with ways to use these specifications. Several new forms of BJT- and FET-type devices have been introduced together with a basic description of their function. MOS-type FETs were described in detail. These devices (MOSFETs) differ greatly in operation from the basic JFET. The MOSFET, though, is most commonly found as an internal element in integrated circuits (either linear or digital) rather than as discrete devices. Their sensitivity to static charge, and their similarity to the JFET, make the JFET more popular for use in a discrete application devices.

Test procedures were also outlined which will assist in determining if these discrete devices are functioning as required. A basic list of potential failures in such devices and the probable cause of such failures was presented to allow for a more in-depth understanding of how to work with these circuit elements. In future chapters we introduce more forms of devices and provide continued evaluation of specifications and troubleshooting information.

KEY EQUATIONS

8.1 $R_S = V_{GS}/I_D \qquad \text{(since } I_S = I_D\text{)}$

8.2 $g_m = g_{mo}\,[1 - (V_{GS}/V_P)]$

8.3 $g_m = \dfrac{I_{D2} - I_{D1}}{V_{G2} - V_{G1}}$

PROBLEMS

Section 8.1

8.1. List the major advantages and disadvantages (as noted in this chapter) of FETs and BJTs.

8.2. What are the six primary factors noted in this chapter as being important when selecting an active device?

8.3. Why is operating temperature a factor in selecting an active device?

8.4. What does the term "MIL-SPEC" mean?

Section 8.2

8.5. What are the four major categories into which BJT specifications break down?

8.6. In Appendix A locate a small-signal BJT and a power BJT. Make a table listing the major specifications of these devices for comparison.

8.7. A 2N2218 transistor is to be used in a circuit with the following operational characteristics. Indicate if the device is usable in that application and defend your argument.

$V_{CC} = 40$ V current gain = 30 maximum frequency = 2 MHz

Section 8.3

8.8. Locate the data sheet for a single FET small-signal device (use a library, lab, or classroom resource) and contrast it to the MFQ5460P (quad) FET listed in Appendix A. Compare their significant specifications in chart form.

8.9. Contrast the maximum ratings for the 2N6762 FET to those of the MFQ5460P. (Both are listed in Appendix A.)

Section 8.4

8.10. Sketch a representative set of BJT collector family of curves, illustrating at least three values of base current. Be sure to label all axis.

8.11. Indicate on the curves sketched in problem 8.10 the approximate location of saturation, cutoff, $I_{C(max)}$, $V_{CE(max)}$, maximum power, and the active region.

8.12. Sketch a representative JFET transfer curve (P type). How does the curve differ from the collector curves shown for the BJT?

8.13. For a JFET, define I_{DSS} and V_P.

8.14. On the BJT characteristics shown in Figure 8.18, sketch a load line for a load resistance of 2 kΩ given that $V_{CC} = 12$ V.

8.15. Using the graph and plot prepared for problem 8.14, find the Q-point for the following operational base currents.
(a) 37.5 μA (b) 62.5 μA (c) 87.5 μA
Indicate the values of V_{CEQ} and I_{CQ} from these plots.

8.16. If (using the results of problem 8.15) the dc base current were set to 37.5 μA and then an ac base current was produced that was 25 μA (p-p), what would be the resulting V_{CE} and I_C (p-p) output values.

8.17. Again using the data from problem 8.15, what would be the maximum dc base current that could be used and still produce an undistorted output from the input listed in problem 8.16?

8.18. On the JFET curves shown in Figure 8.19, find the resulting I_{DQ} if the V_{GSQ} is −2 V.

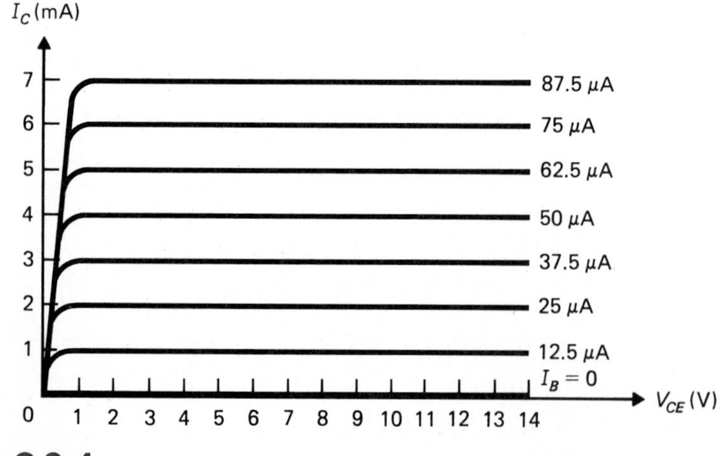

FIGURE 8.18 BJT Common-Emitter Family of Curves

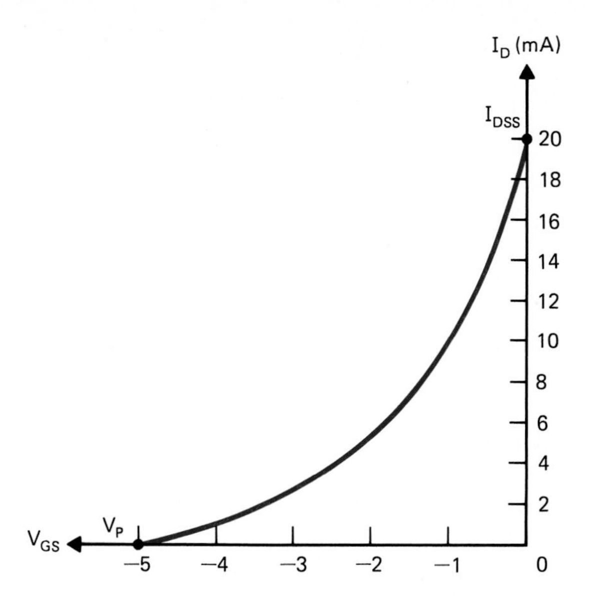

FIGURE 8.19 JFET Forword Transfer Curve

8.19. If the ac gate input voltage were to be 2 V (p-p), what would be the resulting drain current variations. (List maximum and minimum values.)

8.20. What would be the value of source resistor that would produce the JFET *Q*-point values seen in problem 8.18?

8.21. Discuss briefly the distortion seen in the resulting drain current for problem 8.19.

8.22. For the distorted waveforms shown in Figure 8.20, indicate what may have caused this distortion in a BJT circuit.

8.23. What could be done to remedy the problems seen in the waveforms in Figure 8.20 (BJT circuitry)?

Section 8.5

8.24. Define briefly the differences between a small-signal BJT device and a power BJT device.

8.25. Illustrate the symbol for an optical BJT (phototransistor) and write a brief description of its function.

8.26. Repeat problem 8.25 for a Darlington transistor (NPN type).

FIGURE 8.20 Three Forms of Output Distortion

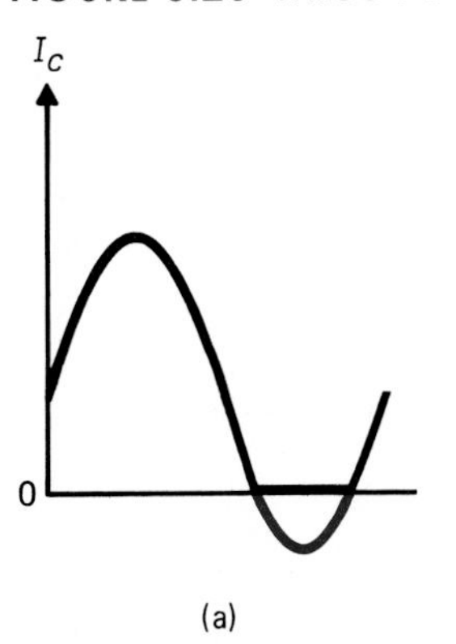

(a)

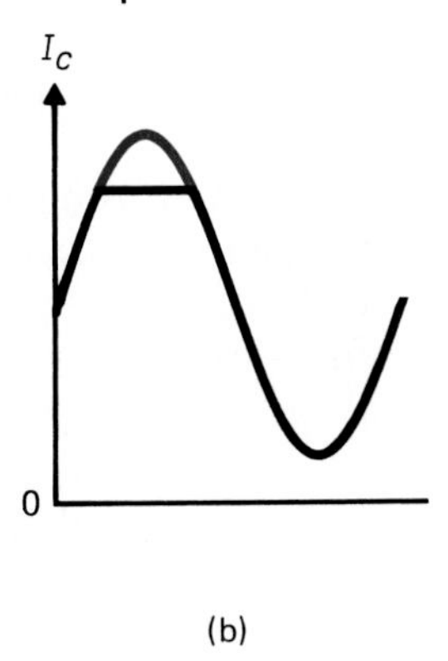

(b)

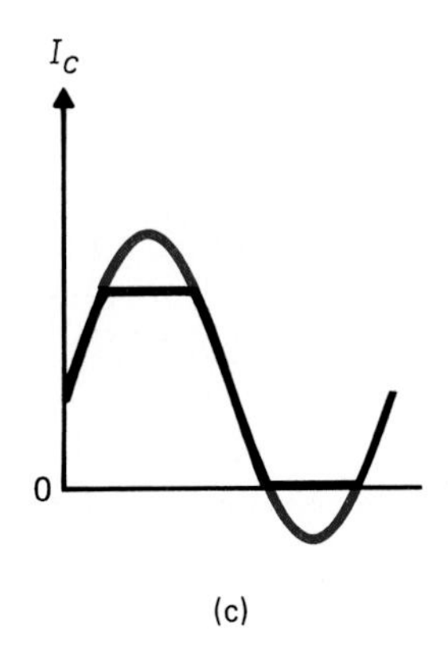

(c)

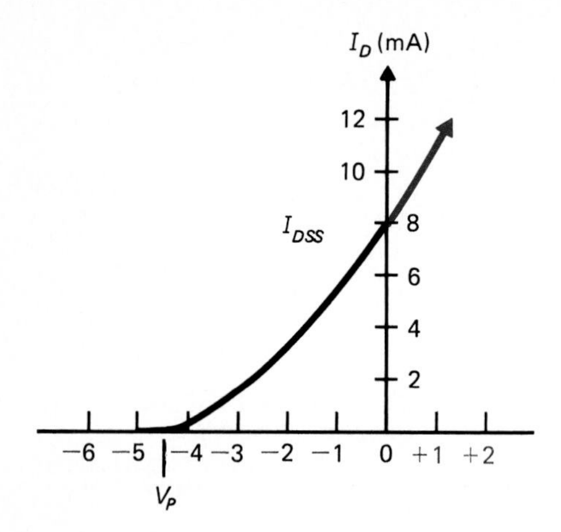

FIGURE 8.21 DE-MOSFET Transfer Curves

8.27. What is the electronic difference between the DE-MOSFET and the JFET?

8.28. What is the electronic difference between the E-MOSFET and the JFET? (You may base this description on an ''N''-type device for each.)

8.29. One application of the DE-MOSFET is to use it in a configuration where there is no dc gate bias voltage. (It is biased at $V_{GSQ} = 0$. Using the DE-MOSFET curves shown in Figure 8.21, indicate what the maximum peak-to-peak input voltage could be to obtain an undistorted output under this bias condition.

Section 8.6

8.30. A test was conducted on an MFQ5460P FET (listed in Appendix A) and the following data were collected. Determine if the g_m of the device is within the listed specifications.

First reading: $V_{GS} = -4$ V $\quad I_D = 3.3$ mA
Second reading: $V_{GS} = -2$ V $\quad I_D = 4.0$ mA

8.31. The FET that is shown in the curves in Figure 8.19 was tested and the following data were collected. Indicate if it is operating properly and if not, what could be defective.

First reading: $V_{GS} = -2$ V $\quad I_D = 20$ mA
Second reading: $V_{GS} = -3$ V $\quad I_D = 20$ mA

ANSWERS TO REVIEW PROBLEMS

Section 8.1

1. (a) An FET should be chosen due to its high input impedance.
(b) A BJT amplifier should be chosen since it can produce higher voltage and current gain (combined) figures than the FET (also since it is implied that input current is to be had).
(c) Either a BJT (stabilized) or a FET circuit should be adequate for this application.

2. (Numbers coordinate to question numbers in the text.)
(a) (1) Select polarity of BJT to match system; (2) high-power device; (3) audio frequencies; (4) must handle high input levels and produce very

high level outputs; (5) normal room-temperature environment; (6) probably will be mounted on a heat sink due to high power used.

(b) (1) BJT polarity to match car "hot" polarity; (2) low power levels; (3) high radio frequencies (88 to 108 MHz, probably); (4) input and output from and to other transistors; (5) wide temperature swing inside car (−30 to +150°F); (6) probably mounted on a printed circuit board.

(c) (1) Dc polarity to match circuitry; (2) very low power; (3) audio-frequency range; (4) input and output undefined; (5) slightly above and below normal body temperature; (6) device must be extremely small, may be best to look for IC rather than discrete devices.

(d) (1) Polarity to match system; (2) potentially high power levels; (3) high microwave frequency range; (4) input and output are yet defined; (5) extreme temperature ranges; (6) device must be very small and lightweight.

Section 8.2

1. Answer depends on resources found and devices selected.
2. Answer depends on resources and device selected.

Section 8.3

1.

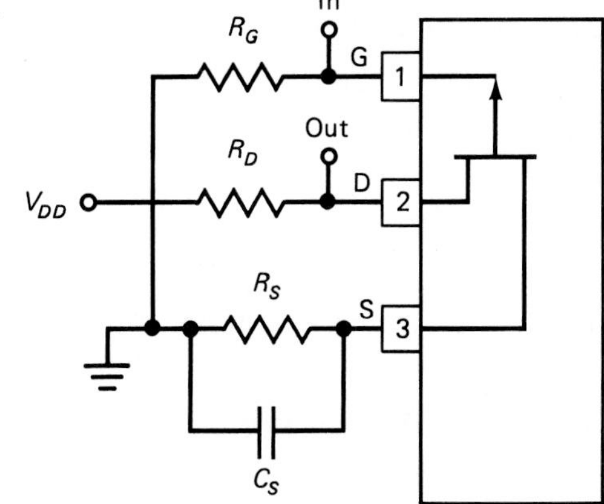

(a) 40 V
(b) 2000 Ω
(c) 200 Ω
(d) $A_v = 8$

Section 8.4

1. I_C

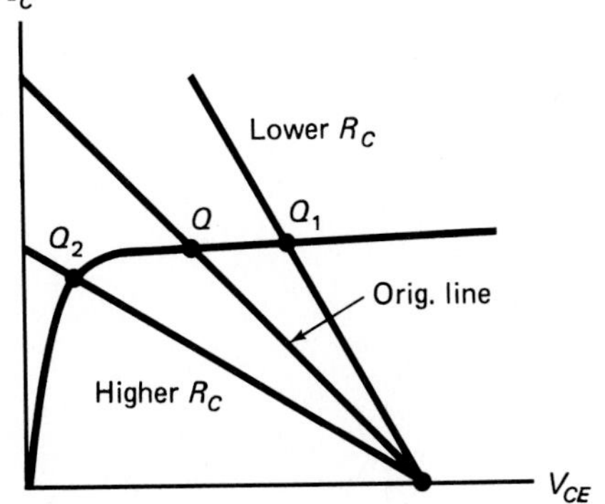

The maximum value of I_C would change along with a change in the location of the Q point.

2. 

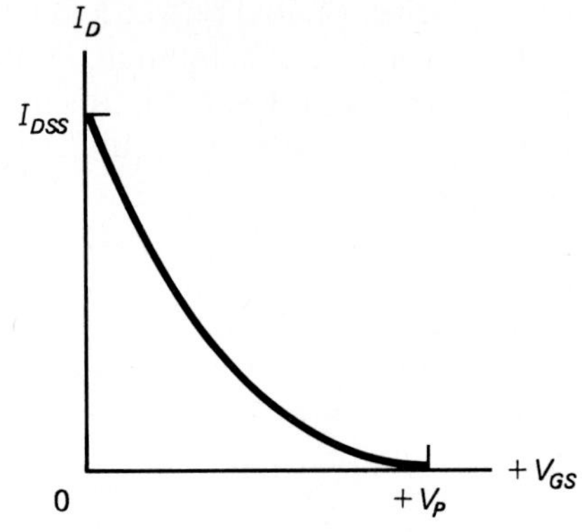

3. (a) BJT, saturated
   (b) BJT, cut off
   (c) BJT, input too high in peak-to-peak value
   (d) FET, V_{gs} too high
   (e) FET, V_{gs} too low
   (f) FET, input too high in peak-to-peak value

Section 8.5

1. Sheet preparation based on text descriptions.

Section 8.6

1. Base to emitter open
2. Device seems to check
3. 900 μs; no, this value is too far below the rated minimum.

Section 8.7

1. Answer is to be a procedure summarizing the information in this section.
2. Answer is based on substitute selected.
3. Was the device an appropriate selection (checking the spec sheet)? Is it possible that there is some way the circuit could develop a short circuit (a mechanical problem)? Is the device exposed to excessive heat or other improper environments? Could there be a problem with "turn-on fatique"? (Check the turn-on currents.) Could the input signal occasionally be excessive? Are there any leaky capacitors? Could an occasional overload exist? Could there be random static charges present? Are there any internally generated oscillations? (Use an oscilloscope.)

2N2218,A/2N2219,A 2N2221,A/2N2222,A 2N5581/82

JAN, JTX, JTXV AVAILABLE

2N2218,A
2N2219,A
CASE 79-02
TO-39 (TO-205AD) STYLE 1

2N2221,A
2N2222,A
CASE 22-03
TO-18 (TO-206AA) STYLE 1

2N5581
2N5582
CASE 26-03
TO-46 (TO-206AB) STYLE 1

GENERAL PURPOSE TRANSISTOR

NPN SILICON

MAXIMUM RATINGS

Rating	Symbol	2N2218 2N2219 2N2221 2N2222	2N2218A 2N2219A 2N2221A 2N2222A	2N5581 2N5582	Unit
Collector-Emitter Voltage	V_{CEO}	30	40	40	Vdc
Collector-Base Voltage	V_{CBO}	60	75	75	Vdc
Emitter-Base Voltage	V_{EBO}	5.0	6.0	6.0	Vdc
Collector Current — Continuous	I_C	800	800	800	mAdc
		2N2218,A 2N2219,A	**2N2221,A 2N2222,A**	**2N5581 2N5582**	
Total Device Dissipation @ $T_A = 25°C$ Derate above 25°C	P_D	0.8 4.57	0.4 2.28	0.6 3.33	Watt mW/°C
Total Device Dissipation @ $T_C = 25°C$ Derate above 25°C	P_D	3.0 17.1	1.2 6.85	2.0 11.43	Watts mW/°C
Operating and Storage Junction Temperature Range	T_J, T_{stg}	−65 to +200			°C

ELECTRICAL CHARACTERISTICS ($T_A = 25°C$ unless otherwise noted.)

Characteristic		Symbol	Min	Max	Unit
OFF CHARACTERISTICS					
Collector-Emitter Breakdown Voltage ($I_C = 10$ mAdc, $I_B = 0$)	Non-A Suffix A-Suffix, 2N5581, 2N5582	$V_{(BR)CEO}$	30 40	— —	Vdc
Collector-Base Breakdown Voltage ($I_C = 10\ \mu$Adc, $I_E = 0$)	Non-A Suffix A-Suffix, 2N5581, 2N5582	$V_{(BR)CBO}$	60 75	— —	Vdc
Emitter-Base Breakdown Voltage ($I_E = 10\ \mu$Adc, $I_C = 0$)	Non-A Suffix A-Suffix, 2N5581, 2N5582	$V_{(BR)EBO}$	5.0 6.0	— —	Vdc
Collector Cutoff Current ($V_{CE} = 60$ Vdc, $V_{EB(off)} = 3.0$ Vdc)	A-Suffix, 2N5581, 2N5582	I_{CEX}	—	10	nAdc
Collector Cutoff Current ($V_{CB} = 50$ Vdc, $I_E = 0$) ($V_{CB} = 60$ Vdc, $I_E = 0$) ($V_{CB} = 50$ Vdc, $I_E = 0$, $T_A = 150°C$) ($V_{CB} = 60$ Vdc, $I_E = 0$, $T_A = 150°C$)	Non-A Suffix A-Suffix, 2N5581, 2N5582 Non-A Suffix A-Suffix, 2N5581, 2N5582	I_{CBO}	— — — —	0.01 0.01 10 10	μAdc
Emitter Cutoff Current ($V_{EB} = 3.0$ Vdc, $I_C = 0$)	A-Suffix, 2N5581, 2N5582	I_{EBO}	—	10	nAdc
Base Cutoff Current ($V_{CE} = 60$ Vdc, $V_{EB(off)} = 3.0$ Vdc)	A-Suffix	I_{BL}	—	20	nAdc
ON CHARACTERISTICS					
DC Current Gain ($I_C = 0.1$ mAdc, $V_{CE} = 10$ Vdc)	2N2218,A, 2N2221,A, 2N5581(1) 2N2219,A, 2N2222,A, 2N5582(1)	h_{FE}	20 35	— —	—
($I_C = 1.0$ mAdc, $V_{CE} = 10$ Vdc)	2N2218,A, 2N2221,A, 2N5581 2N2219,A, 2N2222,A, 2N5582		25 50	— —	
($I_C = 10$ mAdc, $V_{CE} = 10$ Vdc)	2N2218,A, 2N2221,A, 2N5581(1) 2N2219,A, 2N2222,A, 2N5582(1)		35 75	— —	
($I_C = 10$ mAdc, $V_{CE} = 10$ Vdc, $T_A = -55°C$)	2N2218A, 2N2221A, 2N5581 2N2219A, 2N2222A, 2N5582		15 35	— —	
($I_C = 150$ mAdc, $V_{CE} = 10$ Vdc)(1)	2N2218,A, 2N2221,A, 2N5581 2N2219,A, 2N2222,A, 2N5582		40 100	120 300	

2N2218/19/21/22, A SERIES, 2N5581/82

ELECTRICAL CHARACTERISTICS (continued) (T_A = 25°C unless otherwise noted.)

Characteristic		Symbol	Min	Max	Unit
(I_C = 150 mAdc, V_{CE} = 1.0 Vdc)(1)	2N2218,A, 2N2221,A, 2N5581		20	—	
	2N2219,A, 2N2222,A, 2N5582		50	—	
(I_C = 500 mAdc, V_{CE} = 10 Vdc)(1)	2N2218, 2N2221		20	—	
	2N2219, 2N2222		30	—	
	2N2218A, 2N2221A, 2N5581		25	—	
	2N2219A, 2N2222A, 2N5582		40	—	
Collector-Emitter Saturation Voltage(1)		$V_{CE(sat)}$			Vdc
(I_C = 150 mAdc, I_B = 15 mAdc)	Non-A Suffix		—	0.4	
	A-Suffix, 2N5581, 2N5582		—	0.3	
(I_C = 500 mAdc, I_B = 50 mAdc)	Non-A Suffix		—	1.6	
	A-Suffix, 2N5581, 2N5582		—	1.0	
Base-Emitter Saturation Voltage(1)		$V_{BE(sat)}$			Vdc
(I_C = 150 mAdc, I_B = 15 mAdc)	Non-A Suffix		0.6	1.3	
	A-Suffix, 2N5581, 2N5582		0.6	1.2	
(I_C = 500 mAdc, I_B = 50 mAdc)	Non-A Suffix		—	2.6	
	A-Suffix, 2N5581, 2N5582		—	2.0	

SMALL-SIGNAL CHARACTERISTICS

Characteristic		Symbol	Min	Max	Unit
Current-Gain — Bandwidth Product(2)		f_T			MHz
(I_C = 20 mAdc, V_{CE} = 20 Vdc, f = 100 MHz)	All Types, Except		250	—	
	2N2219A, 2N2222A, 2N5582		300	—	
Output Capacitance(3) (V_{CB} = 10 Vdc, I_E = 0, f = 100 kHz)		C_{obo}	—	8.0	pF
Input Capacitance(3)		C_{ibo}			pF
(V_{EB} = 0.5 Vdc, I_C = 0, f = 100 kHz)	Non-A Suffix		—	30	
	A-Suffix, 2N5581, 2N5582		—	25	
Input Impedance		h_{ie}			kohms
(I_C = 1.0 mAdc, V_{CE} = 10 Vdc, f = 1.0 kHz)	2N2218A, 2N2221A		1.0	3.5	
	2N2219A, 2N2222A		2.0	8.0	
(I_C = 10 mAdc, V_{CE} = 10 Vdc, f = 1.0 kHz)	2N2218A, 2N2221A		0.2	1.0	
	2N2219A, 2N2222A		0.25	1.25	
Voltage Feedback Ratio		h_{re}			X 10^{-4}
(I_C = 1.0 mAdc, V_{CE} = 10 Vdc, f = 1.0 kHz)	2N2218A, 2N2221A		—	5.0	
	2N2219A, 2N2222A		—	8.0	
(I_C = 10 mAdc, V_{CE} = 10 Vdc, f = 1.0 kHz)	2N2218A, 2N2221A		—	2.5	
	2N2219A, 2N2222A		—	4.0	
Small-Signal Current Gain		h_{fe}			—
(I_C = 1.0 mAdc, V_{CE} = 10 Vdc, f = 1.0 kHz)	2N2218A, 2N2221A		30	150	
	2N2219A, 2N2222A		50	300	
(I_C = 10 mAdc, V_{CE} = 10 Vdc, f = 1.0 kHz)	2N2218A, 2N2221A		50	300	
	2N2219A, 2N2222A		75	375	
Output Admittance		h_{oe}			μmhos
(I_C = 1.0 mAdc, V_{CE} = 10 Vdc, f = 1.0 kHz)	2N2218A, 2N2221A		3.0	15	
	2N2219A, 2N2222A		5.0	35	
(I_C = 10 mAdc, V_{CE} = 10 Vdc, f = 1.0 kHz)	2N2218A, 2N2221A		10	100	
	2N2219A, 2N2222A		25	200	
Collector Base Time Constant (I_E = 20 mAdc, V_{CB} = 20 Vdc, f = 31.8 MHz)	A-Suffix	$r_b'C_c$	—	150	ps
Noise Figure (I_C = 100 μAdc, V_{CE} = 10 Vdc, R_S = 1.0 kohm, f = 1.0 kHz)	2N2222A	NF	—	4.0	dB
Real Part of Common-Emitter High Frequency Input Impedance (I_C = 20 mAdc, V_{CE} = 20 Vdc, f = 300 MHz)	2N2218A, 2N2219A	$Re(h_{ie})$	—	60	Ohms
	2N2221A, 2N2222A				

(1) Pulse Test: Pulse Width ≤ 300 μs, Duty Cycle ≤ 2.0%.
(2) f_T is defined as the frequency at which $|h_{fe}|$ extrapolates to unity.
(3) 2N5581 and 2N5582 are Listed C_{cb} and C_{eb} for these conditions and values.

2N2218,A/2N2219,A/2N2221,A/2N2222,A/2N5581/82

ELECTRICAL CHARACTERISTICS (continued) (T_A = 25°C unless otherwise noted.)

Characteristic		Symbol	Min	Max	Unit
SWITCHING CHARACTERISTICS					
Delay Time	(V_{CC} = 30 Vdc, $V_{BE(off)}$ = 0.5 Vdc, I_C = 150 mAdc, I_{B1} = 15 mAdc) (Figure 14)	t_d	—	10	ns
Rise Time		t_r	—	25	ns
Storage Time	(V_{CC} = 30 Vdc, I_C = 150 mAdc, I_{B1} = I_{B2} = 15 mAdc) (Figure15)	t_s	—	225	ns
Fall Time		t_f	—	60	ns
Active Region Time Constant (I_C = 150 mAdc, V_{CE} = 30 Vdc) (See Figure 12 for 2N2218A, 2N2219A, 2N2221A, 2N2222A)		T_A	—	2.5	ns

FIGURE 1 – NORMALIZED DC CURRENT GAIN

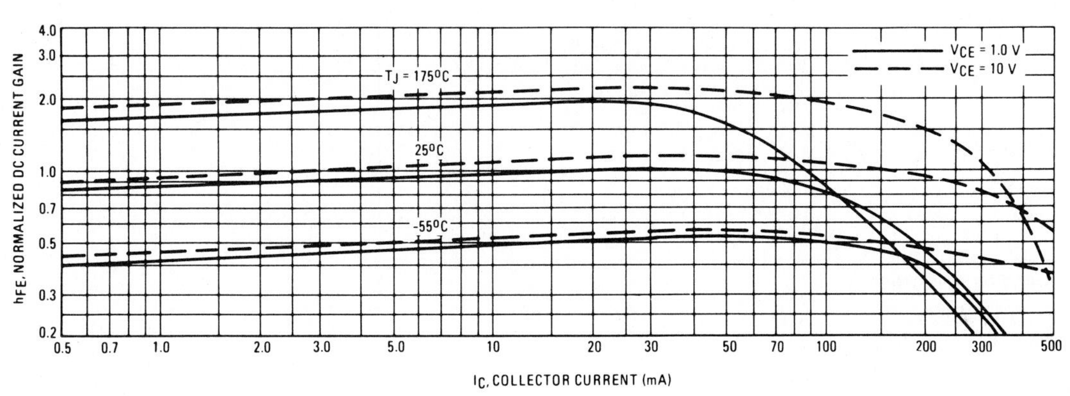

FIGURE 2 – COLLECTOR CHARACTERISTICS IN SATURATION REGION

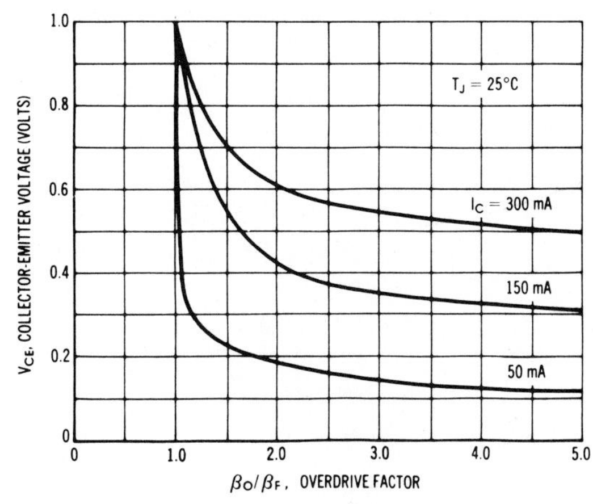

This graph shows the effect of base current on collector current. β_O (current gain at the edge of saturation) is the current gain of the transistor at 1 volt, and β_F (forced gain) is the ratio of I_C/I_{BF} in a circuit.

EXAMPLE: For type 2N2219, estimate a base current (I_{BF}) to insure saturation at a temperature of 25°C and a collector current of 150 mA.

Observe that at I_C = 150 mA an overdrive factor of at least 2.5 is required to drive the transistor well into the saturation region. From Figure 1, it is seen that h_{FE} @ 1 volt is approximately 0.62 of h_{FE} @ 10 volts. Using the guaranteed minimum gain of 100 @ 150 mA and 10 V, β_O = 62 and substituting values in the overdrive equation, we find:

$$\frac{\beta_O}{\beta_F} = \frac{h_{FE} @ 1.0\,V}{I_C/I_{BF}} \qquad 2.5 = \frac{62}{150/I_{BF}} \qquad I_{BF} \approx 6.0 \text{ mA}$$

Designer's Data Sheet

MJ16018
MJH16018

10 AMPERE

NPN SILICON POWER TRANSISTORS

800 VOLTS
150 AND 175 WATTS

1.5 kV SWITCHMODE III SERIES NPN SILICON POWER TRANSISTORS

These transistors are designed for high-voltage, high-speed, power switching in inductive circuits where fall time is critical. They are particularly suited for line-operated switchmode applications.

Typical Applications:
- Switching Regulators
- Inverters
- Solenoids
- Relay Drivers
- Motor Controls
- Deflection Circuits

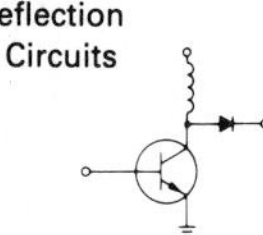

Features:
- Collector-Emitter Voltage — V_{CEX} = 1500 Vdc
- Fast Turn-Off Times
  - 280 ns Inductive Fall Time – 100°C (Typ)
  - 470 ns Inductive Crossover Time – 100°C (Typ)
  - 2.6 μs Inductive Storage Time – 100°C (Typ)
- 100°C Performance Specified for:
  - Reverse-Biased SOA with Inductive Load
  - Switching Times with Inductive Loads
  - Saturation Voltages
  - Leakage Currents

MAXIMUM RATINGS

Rating	Symbol	MJ16018	MJH16018	Unit
Collector-Emitter Voltage	$V_{CEO(sus)}$	800		Vdc
Collector-Emitter Voltage	V_{CEX}	1500		Vdc
Emitter-Base Voltage	V_{EB}	6.0		Vdc
Collector Current — Continuous — Peak (1)	I_C I_{CM}	10 15		Adc
Base Current — Continuous — Peak (1)	I_B I_{BM}	8.0 12		Adc
Total Device Dissipation @ T_C = 25°C @ T_C = 100°C Derate above 25°C	P_D	175 100 1.0	150 50 1.0	Watts W/°C
Operating and Storage Junction Temperature Range	T_J, T_{stg}	−65 to 200	−55 to 150	°C

THERMAL CHARACTERISTICS

Characteristic	Symbol	Max		Unit
Thermal Resistance, Junction to Case	$R_{\theta JC}$	1.0	1.0	°C/W
Lead Temperature for Soldering Purposes, 1/8" from Case for 5 Seconds.	T_L	275		°C

(1) Pulse Test: Pulse Width ≤ 5.0 μs, Duty Cycle ≥ 10%.

Designer's Data for "Worst Case" Conditions

The Designer's Data Sheet permits the design of most circuits entirely from the information presented. Limit Curves — representing boundaries on device characteristics — are given to facilitate "worst case" design.

Designer's and SWITCHMODE are trademarks of Motorola Inc.

MJ16018

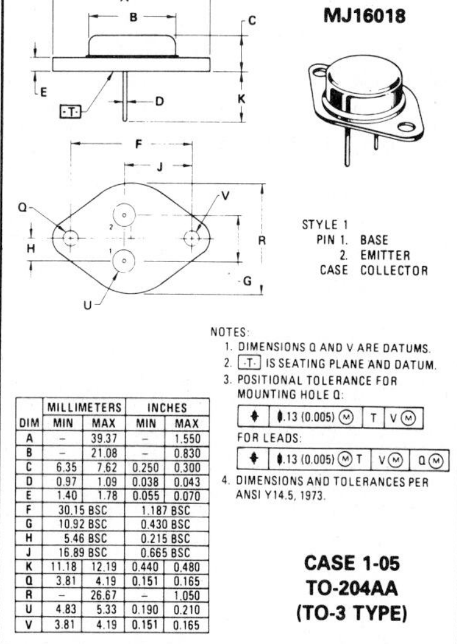

STYLE 1
PIN 1. BASE
2. EMITTER
CASE COLLECTOR

DIM	MILLIMETERS MIN	MILLIMETERS MAX	INCHES MIN	INCHES MAX
A	–	39.37	–	1.550
B	–	21.08	–	0.830
C	6.35	7.62	0.250	0.300
D	0.97	1.09	0.038	0.043
E	1.40	1.78	0.055	0.070
F	30.15 BSC		1.187 BSC	
G	10.92 BSC		0.430 BSC	
H	5.46 BSC		0.215 BSC	
J	16.89 BSC		0.665 BSC	
K	11.18	12.19	0.440	0.480
Q	3.81	4.19	0.151	0.165
R	–	26.67	–	1.050
U	4.83	5.33	0.190	0.210
V	3.81	4.19	0.151	0.165

NOTES:
1. DIMENSIONS Q AND V ARE DATUMS.
2. -T- IS SEATING PLANE AND DATUM.
3. POSITIONAL TOLERANCE FOR MOUNTING HOLE Q:
   ⌖ | Ø.13 (0.005) Ⓜ | T | V Ⓜ
   FOR LEADS:
   ⌖ | Ø.13 (0.005) Ⓜ T | V Ⓜ | Q Ⓜ
4. DIMENSIONS AND TOLERANCES PER ANSI Y14.5, 1973.

CASE 1-05
TO-204AA
(TO-3 TYPE)

MJH16018

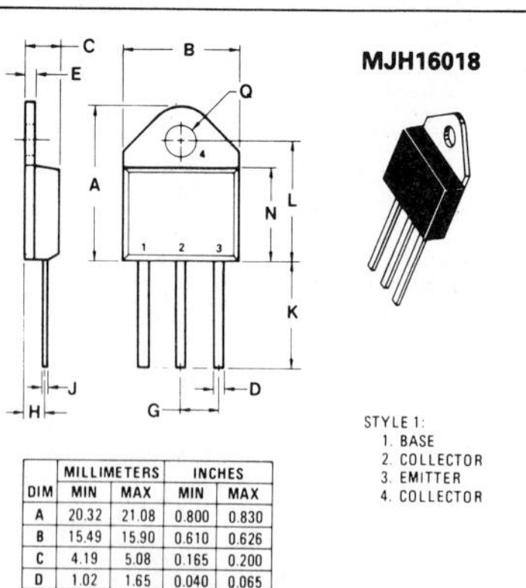

STYLE 1:
1. BASE
2. COLLECTOR
3. EMITTER
4. COLLECTOR

DIM	MILLIMETERS MIN	MILLIMETERS MAX	INCHES MIN	INCHES MAX
A	20.32	21.08	0.800	0.830
B	15.49	15.90	0.610	0.626
C	4.19	5.08	0.165	0.200
D	1.02	1.65	0.040	0.065
E	1.35	1.65	0.053	0.065
G	5.21	5.72	0.205	0.225
H	2.41	3.20	0.095	0.126
J	0.38	0.64	0.015	0.025
K	12.70	15.49	0.500	0.610
L	15.88	16.51	0.625	0.650
N	12.19	12.70	0.480	0.500
Q	4.04	4.22	0.159	0.166

CASE 340-01
TO-218AC

MJ16018
MJH16018

ELECTRICAL CHARACTERISTICS (T_C = 25°C unless otherwise noted)

Characteristic	Symbol	Min	Typ	Max	Unit
OFF CHARACTERISTICS					
Collector-Emitter Sustaining Voltage (Table 2) (I_C = 100 mA, I_B = 0)	$V_{CEO(sus)}$	800	—	—	Vdc
Collector Cutoff Current	I_{CEV}				mAdc
(V_{CEV} = 1500 Vdc, $V_{BE(off)}$ = 1.5 Vdc)		—	—	0.25	
(V_{CEV} = 1500 Vdc, $V_{BE(off)}$ = 1.5 Vdc, T_C = 100°C)		—	—	1.5	
Collector Cutoff Current (V_{CE} = 1500 Vdc, R_{BE} = 50 Ω, T_C = 100°C)	I_{CER}	—	—	2.5	mAdc
Emitter Cutoff Current (V_{EB} = 6.0 Vdc, I_C = 0)	I_{EBO}	—	—	1.0	mAdc
SECOND BREAKDOWN					
Second Breakdown Collector Current with Base Forward Biased	$I_{S/b}$	See Figure 12			
Clamped Inductive SOA with Base Reverse Biased	RBSOA	See Figure 13			
ON CHARACTERISTICS (1)					
Collector-Emitter Saturation Voltage	$V_{CE(sat)}$				Vdc
(I_C = 5.0 Adc, I_B = 1.0 Adc)		—	—	1.5	
(I_C = 10 Adc, I_B = 4.0 Adc)		—	—	1.5	
(I_C = 5.0 Adc, I_B = 1.0 Adc, T_C = 100°C)		—	—	2.0	
Base-Emitter Saturation Voltage	$V_{BE(sat)}$				Vdc
(I_C = 5.0 Adc, I_B = 1.0 Adc)		—	—	1.5	
(I_C = 5.0 Adc, I_B = 1.0 Adc, T_C = 100°C)		—	—	1.5	
DC Current Gain (I_C = 5.0 Adc, V_{CE} = 5.0 Vdc)	h_{FE}	7.0	—	—	—
DYNAMIC CHARACTERISTICS					
Output Capacitance (V_{CB} = 10 Vdc, I_E = 0, f_{test} = 1.0 kHz)	C_{ob}	—	—	400	pF

SWITCHING CHARACTERISTICS

Characteristic	Conditions		Symbol	Min	Typ	Max	Unit
Resistive Load (Table 1)							
Delay Time	(I_C = 5.0 Adc, V_{CC} = 250 Vdc, I_{B1} = 1.0 Adc, PW = 30 μs, Duty Cycle ≤ 2.0%)	(I_{B2} = 2.0 Adc, R_{B2} = 3.0 Ω)	t_d	—	50	100	ns
Rise Time			t_r	—	300	400	
Storage Time			t_s	—	2000	3000	
Fall Time			t_f	—	900	1200	
Storage Time		($V_{BE(off)}$ = 2.0 Vdc)	t_s	—	1600	2400	
Fall Time			t_f	—	500	650	
Inductive Load (Table 2)							
Storage Time	(I_C = 5.0 Adc, I_{B1} = 1.0 Adc, $V_{BE(off)}$ = 2.0 Vdc, $V_{CE(pk)}$ = 400 Vdc)	(T_J = 25°C)	t_{sv}	—	2000	3000	ns
Fall Time			t_{fi}	—	200	400	
Crossover Time			t_c	—	350	500	
Storage Time		(T_J = 100°C)	t_{sv}	—	2600	3600	
Fall Time			t_{fi}	—	280	460	
Crossover Time			t_c	—	470	620	

(1) Pulse Test: PW — 300 μs, Duty Cycle ≤ 2.0%.

MOTOROLA *Semiconductor Products Inc.*

MC34071, MC34072
MC35071, MC35072
MC33071, MC33072

Advance Information

HIGH PERFORMANCE SINGLE SUPPLY OPERATIONAL AMPLIFIERS

SILICON MONOLITHIC INTEGRATED CIRCUIT

HIGH SLEW RATE, WIDE BANDWIDTH, SINGLE SUPPLY OPERATIONAL AMPLIFIERS

A standard low-cost Bipolar technology with innovative design concepts are employed for the MC34071/MC34072 series of monolithic operational amplifiers. These devices offer 4.5 MHz of gain bandwidth product, 13 V/μs slew rate, and fast settling time without the use of JFET device technology. In addition, low input offset voltage can economically be achieved. Although these devices can be operated from split supplies, they are particularly suited for single supply operation, since the common mode input voltage range includes ground potential (V_{EE}). The all NPN output stage, characterized by no deadband crossover distortion and large output voltage swing, also provides high capacitive drive capability, excellent phase and gain margins, low open-loop high frequency output impedance and symmetrical source/sink ac frequency response.

The MC34071/MC34072 series of devices are available in standard or prime performance (A Suffix) grades and specified over commercial, industrial/vehicular or military temperature ranges.

- Wide Bandwidth: 4.5 MHz
- High Slew Rate: 13 V/μs
- Fast Settling Time: 1.1 μs to 0.10%
- Wide Single Supply Operating Range: 3.0 to 44 Volts
- Wide Input Common Mode Range Including Ground (V_{EE})
- Low Input Offset Voltage: 1.5 mV Maximum (A Suffix)
- Large Output Voltage Swing: −14.7 V to +14.0 V for $V_S = \pm 15$ V
- Large Capacitance Drive Capability: 0 to 10,000 pF
- Low T.H.D. Distortion: 0.02%
- Excellent Phase Margins: 60°
- Excellent Gain Margin: 12 dB

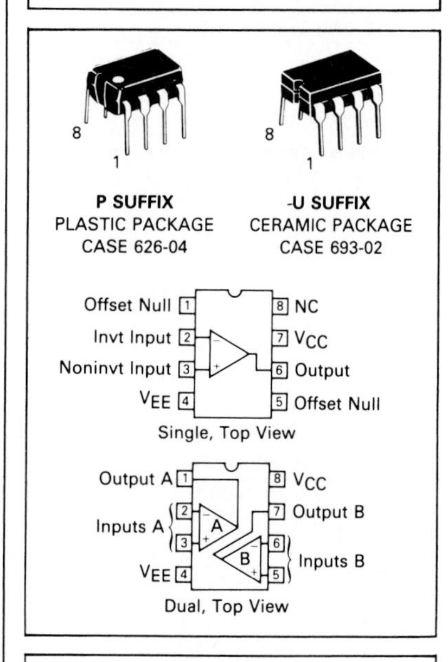

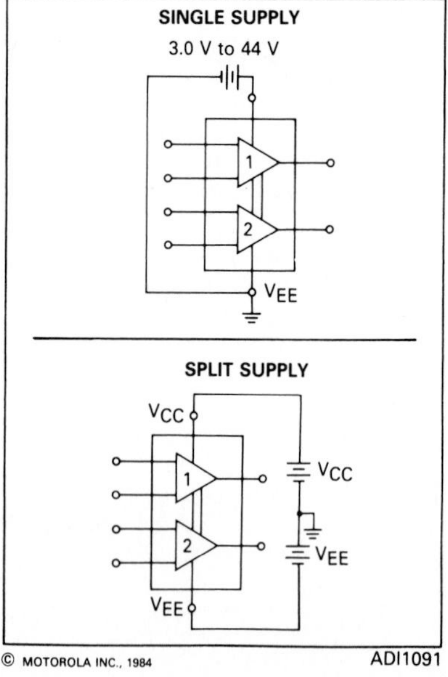

ORDERING INFORMATION			
Op Amp Function	**Device**	**Temperature Range**	**Package**
Single	MC35071U,AU MC33071U,AU MC33071P,AP MC34071U,AU MC34071P,AP	−55 to +125°C −40 to +85°C −40 to +85°C 0 to +70°C 0 to +70°C	Ceramic DIP Ceramic DIP Plastic DIP Ceramic DIP Plastic DIP
Dual	MC35072U,AU MC33072U,AU MC33072P,AP MC34072U,AU MC34072P,AP	−55 to +125°C −40 to +85°C −40 to +85°C 0 to +70°C 0 to +70°C	Ceramic DIP Ceramic DIP Plastic DIP Ceramic DIP Plastic DIP
Quad	MC34074 Series	Refer to MC34074 Data Sheet	

This document contains information on a new product. Specifications and information herein are subject to change without notice.

ADI1091

DC ELECTRICAL CHARACTERISTICS (V_{CC} = +15 V, V_{EE} = −15 V, R_L connected to ground, T_A = T_{low} to T_{high} [Note 3] unless otherwise noted)

Characteristic	Symbol	MC3507_A/MC3407_A/ MC3307_A Min	Typ	Max	MC3507_/MC3407_/ MC3307_ Min	Typ	Max	Unit
Input Offset Voltage (V_{CM} = 0)	V_{IO}							mV
V_{CC} = +15 V, V_{EE} = −15 V, T_A = +25°C		—	0.5	1.5	—	1.0	3.5	
V_{CC} = +5.0 V, V_{EE} = 0 V, T_A = +25°C		—	0.5	2.0	—	1.5	4.0	
V_{CC} = +15 V, V_{EE} = −15 V, T_A = T_{low} to T_{high}		—	—	3.5	—	—	5.5	
Average Temperature Coefficient of Offset Voltage	$\Delta V_{IO}/\Delta T$	—	10	—	—	10	—	μV/°C
Input Bias Current (V_{CM} = 0)	I_{IB}							nA
T_A = +25°C		—	100	500	—	100	500	
T_A = T_{low} to T_{high}		—	—	700	—	—	700	
Input Offset Current (V_{CM} = 0)	I_{IO}							nA
T_A = +25°C		—	6.0	50	—	6.0	75	
T_A = T_{low} to T_{high}		—	—	300	—	—	300	
Large Signal Voltage Gain V_O = ±10 V, R_L = 2.0 k	A_{VOL}	50	100	—	25	100	—	V/mV
Output Voltage Swing	V_{OH}							V
V_{CC} = +5.0 V, V_{EE} = 0 V, R_L = 2.0 k, T_A = +25°C		3.7	4.0	—	3.7	4.0	—	
V_{CC} = +15 V, V_{EE} = −15 V, R_L = 10 k, T_A = +25°C		13.7	14	—	13.7	14	—	
V_{CC} = +15 V, V_{EE} −15 V, R_L = 2.0 k, T_A = T_{low} to T_{high}		13.5	—	—	13.5	—	—	
	V_{OL}							
V_{CC} = +5.0 V, V_{EE} = 0 V, R_L = 2.0 k, T_A = +25°C		—	0.1	0.2	—	0.1	0.2	
V_{CC} = +15 V, V_{EE} + −15 V, R_L = 10 k, T_A = +25°C		—	−14.7	−14.4	—	−14.7	−14.4	
V_{CC} = +15 V, V_{EE} = −15 V, R_L = 2.0 k, T_A = T_{low} to T_{high}		—	—	−13.8	—	—	−13.8	
Output Short-Circuit Current (T_A = +25°C) Input Overdrive = 1.0 V, Output to Ground	I_{SC}							mA
Source		10	30	—	10	30	—	
Sink		20	47	—	20	47	—	
Input Common Mode Voltage Range	V_{ICR}							V
T_A = +25°C		V_{EE} to (V_{CC} − 1.8)			V_{EE} to (V_{CC} − 1.8)			
T_A = T_{low} to T_{high}		V_{EE} to (V_{CC} − 2.2)			V_{EE} to (V_{CC} − 2.2)			
Common Mode Rejection Ratio (R_S ≤ 10 k)	CMRR	80	97	—	70	97	—	dB
Power Supply Rejection Ratio (R_S = 100 Ω)	PSRR	80	97	—	70	97	—	dB
Power Supply Current (Per Amplifier)	I_D							mA
V_{CC} = +5.0 V, V_{EE} = 0 V, T_A = +25°C		—	1.6	2.0	—	1.6	2.0	
V_{CC} = +15 V, V_{EE} = −15 V, T_A = +25°C		—	1.9	2.5	—	1.9	2.5	
V_{CC} = +15 V, V_{EE} = −15 V, T_A = T_{low} to T_{high}		—	—	2.8	—	—	2.8	

NOTES: (continued)

3. T_{low} = −55°C for MC35071,A/MC35072,A
 = −40°C for MC33071,A/MC33072,A
 = 0°C for MC34071,A/MC34072,A

 T_{high} = +125°C for MC35071,A/35072,A
 = +85°C for MC33071,A/33072,A
 = +70°C for MC34071,A/34072,A

MAXIMUM RATINGS

Rating	Symbol	Value	Unit
Supply Voltage (from V_{CC} to V_{EE})	V_S	+44	Volts
Input Differential Voltage Range	V_{IDR}	Note 1	Volts
Input Voltage Range	V_{IR}	Note 1	Volts
Output Short-Circuit Duration (Note 2)	t_S	Indefinite	Seconds
Operating Ambient Temperature Range	T_A		°C
MC35071,A/MC35072,A		−55 to +125	
MC33071,A/MC33072,A		−40 to +85	
MC34071,A/MC34072,A		0 to +70	
Operating Junction Temperature	T_J	+150	°C
Storage Temperature Range	T_{stg}		°C
Ceramic Package		−65 to +150	
Plastic Package		−55 to +125	

NOTES:
1. Either or both input voltages must not exceed the magnitude of V_{CC} or V_{EE}.
2. Power dissipation must be considered to ensure maximum junction temperature (T_J) is not exceeded.

EQUIVALENT CIRCUIT SCHEMATIC (EACH AMPLIFIER)

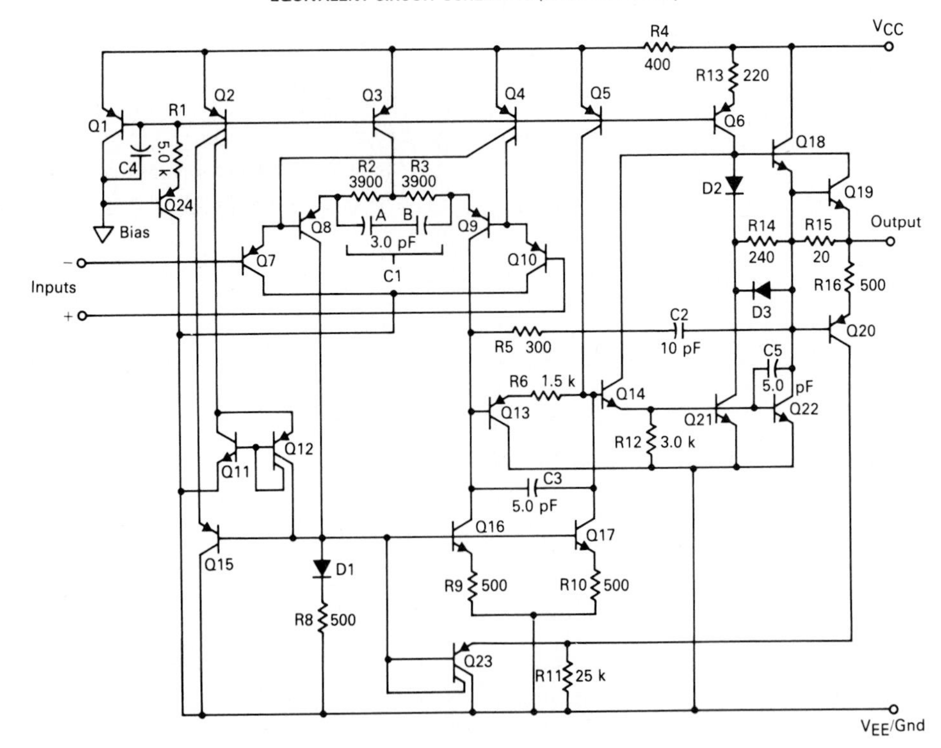

MC34071,A • MC34072,A Series

AC ELECTRICAL CHARACTERISTICS (V_{CC} = +15 V, V_{EE} = −15 V, R_L connected to ground, T_A = +25°C unless otherwise noted)

Characteristic	Symbol	MC3507_A/MC3407_A/ MC3307_A Min	MC3507_A/MC3407_A/ MC3307_A Typ	MC3507_A/MC3407_A/ MC3307_A Max	MC3507_/MC3407_/ MC3307_ Min	MC3507_/MC3407_/ MC3307_ Typ	MC3507_/MC3407_/ MC3307_ Max	Unit
Slew Rate (V_{in} = −10 V to +10 V, R_L = 2.0 k, C_L = 500 pF)	SR							V/μs
A_V + 1.0		8.0	10	—	—	10	—	
A_V − 1.0		—	13	—	—	13	—	
Settling Time (10 V Step, A_V = −1.0)	t_s							μs
To 0.10% (±1/2 LSB of 9-Bits)		—	1.1	—	—	1.1	—	
To 0.01% (±1/2 LSB of 12-Bits)		—	2.2	—	—	2.2	—	
Gain Bandwidth Product (f = 100 kHz)	GBW	3.5	4.5	—		4.5	—	MHz
Power Bandwidth A_V = +1.0, R_L = 2.0 k, V_O = 20 V_{p-p}, THD = 5.0%	BWp	—	200	—	—	200	—	kHz
Phase Margin	ϕm							Degrees
R_L = 2.0 k		—	60	—	—	60	—	
R_L = 2.0 k, C_L = 300 pF		—	40	—	—	40	—	
Gain Margin	A_m							dB
R_L = 2.0 k		—	12	—	—	12	—	
R_L = 2.0 k, C_L = 300 pF		—	4.0	—	—	4.0	—	
Equivalent Input Noise Voltage R_S = 100 Ω, f = 1.0 kHz	e_n	—	32	—	—	32	—	nV/ √Hz
Equivalent Input Noise Current (f = 1.0 kHz)	I_n	—	0.22	—	—	0.22	—	pA/ √Hz
Input Capacitance	C_i	—	0.8	—	—	0.8	—	pF
Total Harmonic Distortion A_V = +10, R_L = 2.0 k, 2.0 ≤ V_O ≤ 20 V_{p-p}, f = 10 kHz	THD	—	0.02	—	—	0.02	—	%
Channel Separation (f = 10 kHz, MC34072,A Only)	—	—	120	—	—	120	—	dB
Open-Loop Output Impedance (f = 1.0 MHz)	z_O	—	30	—	—	30	—	Ω

For typical performance curves and applications information refer to MC34074 series data sheet.

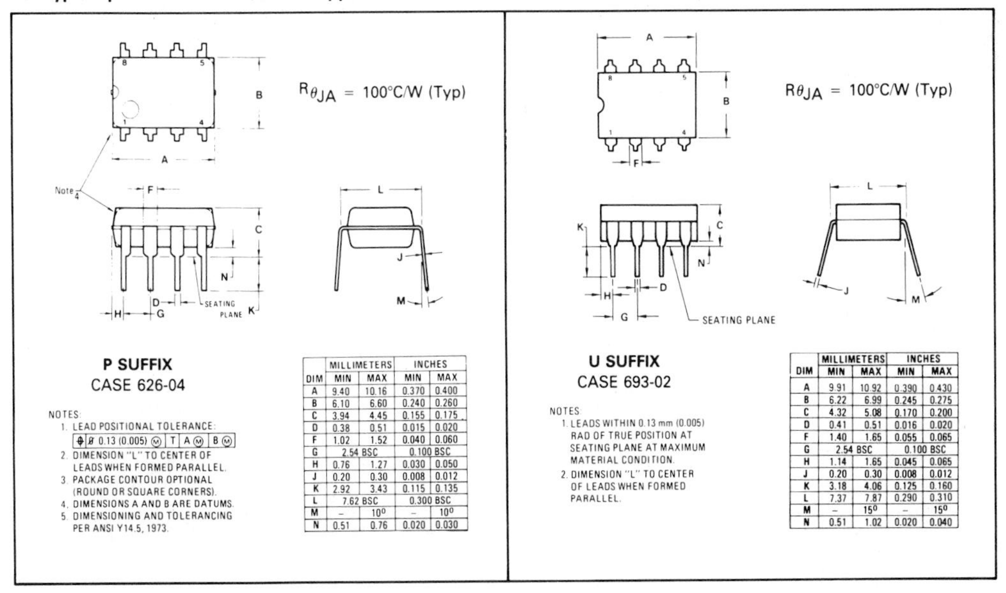

P SUFFIX
CASE 626-04

NOTES:
1. LEAD POSITIONAL TOLERANCE: ⌖ Ø 0.13 (0.005) Ⓜ | T | A Ⓜ | B Ⓜ
2. DIMENSION "L" TO CENTER OF LEADS WHEN FORMED PARALLEL.
3. PACKAGE CONTOUR OPTIONAL (ROUND OR SQUARE CORNERS).
4. DIMENSIONS A AND B ARE DATUMS.
5. DIMENSIONING AND TOLERANCING PER ANSI Y14.5, 1973.

DIM	MILLIMETERS MIN	MILLIMETERS MAX	INCHES MIN	INCHES MAX
A	9.40	10.16	0.370	0.400
B	6.10	6.60	0.240	0.260
C	3.94	4.45	0.155	0.175
D	0.38	0.51	0.015	0.020
F	1.02	1.52	0.040	0.060
G	2.54 BSC		0.100 BSC	
H	0.76	1.27	0.030	0.050
J	0.20	0.30	0.008	0.012
K	2.92	3.43	0.115	0.135
L	7.62 BSC		0.300 BSC	
M	–	10°	–	10°
N	0.51	0.76	0.020	0.030

U SUFFIX
CASE 693-02

NOTES:
1. LEADS WITHIN 0.13 mm (0.005) RAD OF TRUE POSITION AT SEATING PLANE AT MAXIMUM MATERIAL CONDITION.
2. DIMENSION "L" TO CENTER OF LEADS WHEN FORMED PARALLEL.

DIM	MILLIMETERS MIN	MILLIMETERS MAX	INCHES MIN	INCHES MAX
A	9.91	10.92	0.390	0.430
B	6.22	6.99	0.245	0.275
C	4.32	5.08	0.170	0.200
D	0.41	0.51	0.016	0.020
F	1.40	1.65	0.055	0.065
G	2.54 BSC		0.100 BSC	
H	1.14	1.65	0.045	0.065
J	0.20	0.30	0.008	0.012
K	3.18	4.06	0.125	0.160
L	7.37	7.87	0.290	0.310
M	–	15°	–	15°
N	0.51	1.02	0.020	0.040

MOTOROLA *Semiconductor Products Inc.*

BOX 20912 • PHOENIX, ARIZONA 85036 • A SUBSIDIARY OF MOTOROLA INC.

17857 PRINTED IN USA 5-84 IMPERIAL LITHO C22510 12,000

AD11091

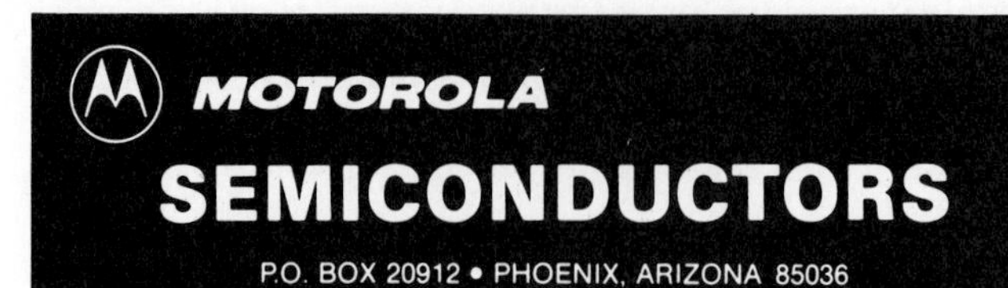

MFQ5460P

QUAD DUAL-IN-LINE P-CHANNEL JUNCTION FIELD-EFFECT TRANSISTORS

. . . depletion mode (Type A) junction field-effect transistors designed for use in general-purpose amplifier applications.

- High Gate-Source Breakdown Voltage — $V_{(BR)GSS}$ = 40 Vdc (Min)
- Low Noise Figure — NF = 1.0 dB (Typ) @ f = 100 Hz
- Low Reverse Transfer Capacitance — C_{rss} = 2.0 pF (Max)
- Refer to 2N5460 Data Sheet for Performance Graphs

P-CHANNEL

QUAD DUAL-IN-LINE JUNCTION FIELD-EFFECT TRANSISTORS

Type A

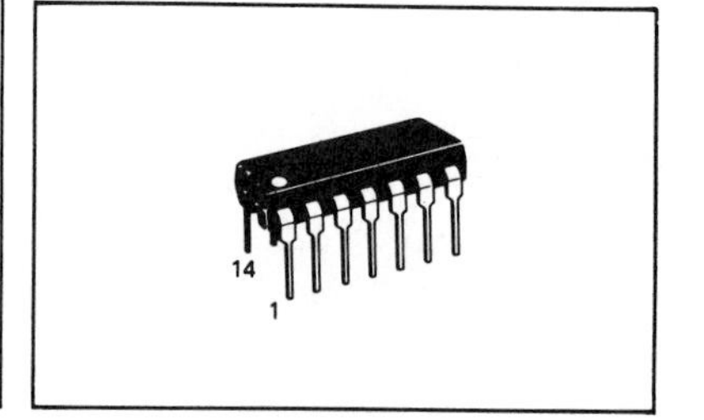

MAXIMUM RATINGS

Rating	Symbol	Value		Unit
Drain-Source Voltage	V_{DS}	40		Vdc
Drain-Gate Voltage	V_{DG}	40		Vdc
Reverse Gate-Source Voltage	V_{GSR}	40		Vdc
Drain Current	I_D	20		mAdc
Forward Gate Current	I_{GF}	10		mAdc
		Each Transistor	Total Device	
Total Device Dissipation @ T_A = 25°C Derate above 25°C	P_D	0.5 2.86	1.5 8.58	Watts mW/°C

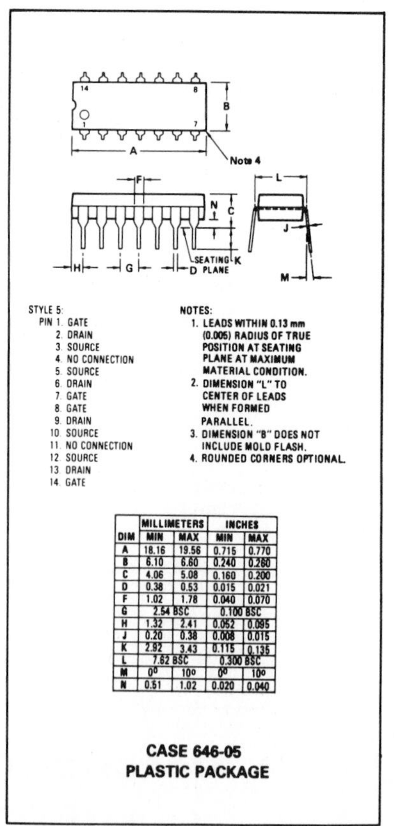

DIM	MILLIMETERS		INCHES	
	MIN	MAX	MIN	MAX
A	18.16	19.56	0.715	0.770
B	6.10	6.60	0.240	0.260
C	4.06	5.08	0.160	0.200
D	0.38	0.53	0.015	0.021
F	1.02	1.78	0.040	0.070
G	2.54 BSC		0.100 BSC	
H	1.32	2.41	0.052	0.095
J	0.20	0.38	0.008	0.015
K	2.92	3.43	0.115	0.135
L	7.62 BSC		0.300 BSC	
M	0°	10°	0°	10°
N	0.51	1.02	0.020	0.040

CASE 646-05
PLASTIC PACKAGE

CONNECTION DIAGRAM

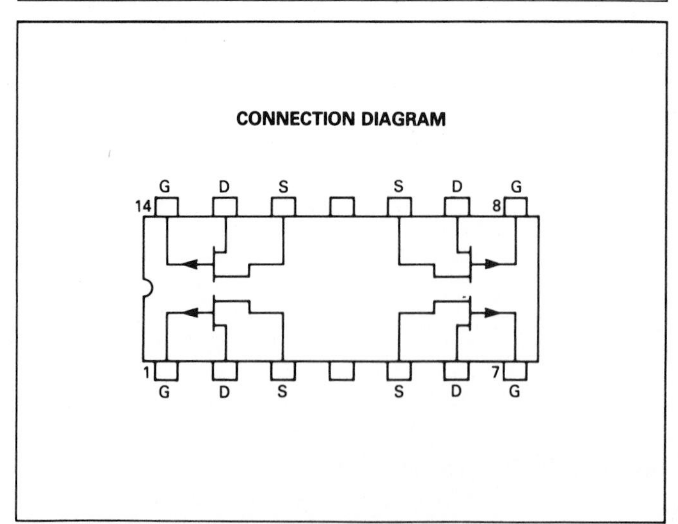

 DS4635

MFQ5460P

ELECTRICAL CHARACTERISTICS (T_A = 25°C unless otherwise noted)

Characteristic	Symbol	Min	Typ	Max	Unit
OFF CHARACTERISTICS					
Gate-Source Breakdown Voltage (I_G = 10 μAdc, V_{DS} = 0)	$V_{(BR)GSS}$	40	—	—	Vdc
Gate-Source Cutoff Voltage (V_{DS} = 15 Vdc, I_D = 1.0 μAdc)	$V_{GS(off)}$	0.75	—	6.0	Vdc
Gate Reverse Current (V_{GS} = 20 Vdc, V_{DS} = 0)	I_{GSS}	—	—	5.0	nAdc
(V_{GS} = 20 Vdc, V_{DS} = 0, T_A = 100°C)		—	—	1.0	μAdc
ON CHARACTERISTICS					
Zero-Gate Voltage Drain Current (V_{DS} = 15 Vdc, V_{GS} = 0)	I_{DSS}	1.0	—	5.0	mAdc
Gate-Source Voltage (V_{DS} = 15 Vdc, I_D = 0.1 mAdc)	V_{GS}	0.5	—	4.0	Vdc
SMALL-SIGNAL CHARACTERISTICS					
Forward Transadmittance (V_{DS} = 15 Vdc, V_{GS} = 0, f = 1.0 kHz)	$\|y_{fs}\|$	1000	—	4000	μmhos
Output Admittance (V_{DS} = 15 Vdc, V_{GS} = 0, f = 1.0 kHz)	$\|y_{os}\|$	—	—	75	μmhos
Input Capacitance (V_{DS} = 15 Vdc, V_{GS} = 0, f = 1.0 MHz)	C_{iss}	—	5.0	7.0	pF
Reverse Transfer Capacitance (V_{DS} = 15 Vdc, V_{GS} = 0, f = 1.0 MHz)	C_{rss}	—	1.0	2.0	pF
Common-Source Noise Figure (V_{DS} = 15 Vdc, V_{GS} = 0, R_G = 1.0 MΩ, f = 100 Hz, BW = 1.0 Hz)	NF	—	1.0	—	dB
Equivalent Short-Circuit Input Noise Voltage (V_{DS} = 15 Vdc, V_{GS} = 0, f = 100 Hz, BW = 1.0 Hz)	e_n	—	60	—	nV/√Hz

MOTOROLA *Semiconductor Products Inc.*

BOX 20912 • PHOENIX, ARIZONA 85036 • A SUBSIDIARY OF MOTOROLA INC.

17443 PRINTED IN USA (3/84) MPS 12M

P.O. BOX 20912 • PHOENIX, ARIZONA 85036

MRF2628

The RF Line

15 W 136–220 MHz

RF POWER TRANSISTOR

NPN SILICON

NPN SILICON POWER TRANSISTOR

Designed for 12.5 volt VHF large-signal power amplifiers in commercial and industrial FM equipment.

- Compact .280 Stud Package
- Specified 12.5 V, 175 MHz Performance
  Output Power = 15 Watts
  Power Gain = 12 dB Min
  Efficiency = 60% Min
- Characterized to 220 MHz
- Load Mismatch Capability at High Line and Overdrive

MAXIMUM RATINGS

Rating	Symbol	Value	Unit
Collector-Emitter Voltage	V_{CEO}	18	Vdc
Collector-Base Voltage	V_{CBO}	36	Vdc
Emitter-Base Voltage	V_{EBO}	4.0	Vdc
Collector-Current — Continuous	I_C	2.5	Adc
Total Device Dissipation @ T_A = 25°C Derate above 25°C	P_D	40 0.23	Watts W/°C
Storage Temperature Range	T_{stg}	−65 to +150	°C
Junction Temperature	T_J	200	°C

THERMAL CHARACTERISTICS

Characteristic	Symbol	Max	Unit
Thermal Resistance, Junction to Case	$R_{\theta JC}$	4.0	°C/W

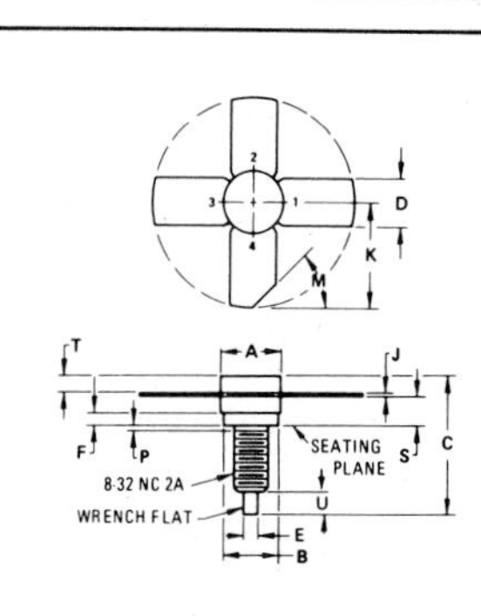

STYLE 1:
PIN 1. EMITTER
2. BASE
3. EMITTER
4. COLLECTOR

DIM	MILLIMETERS MIN	MILLIMETERS MAX	INCHES MIN	INCHES MAX
A	7.06	7.26	0.278	0.286
B	6.20	6.50	0.244	0.256
C	14.99	16.51	0.590	0.650
D	5.46	5.97	0.215	0.235
E	1.40	1.65	0.055	0.065
F	1.52	–	0.060	–
J	0.08	0.18	0.003	0.007
K	11.05	–	0.435	–
M	45^0 NOM		45^0 NOM	
P	–	1.27	–	0.050
S	3.00	3.25	0.118	0.128
T	1.40	1.78	0.055	0.070
U	2.92	3.68	0.115	0.145

CASE 244-04

DS5881

MRF2628

ELECTRICAL CHARACTERISTICS (T_C = 25°C unless otherwise noted.)

Characteristic	Symbol	Min	Typ	Max	Unit
OFF CHARACTERISTICS					
Collector-Emitter Breakdown Voltage (I_C = 25 mAdc, I_B = 0)	$V_{(BR)CEO}$	18	—	—	Vdc
Collector-Emitter Breakdown Voltage (I_C = 25 mAdc, V_{BE} = 0)	$V_{(BR)CES}$	36	—	—	Vdc
Emitter-Base Breakdown Voltage (I_E = 5.0 mAdc, I_C = 0)	$V_{(BR)EBO}$	4.0	—	—	Vdc
Collector Cutoff Current (V_{CB} = 15 Vdc, I_E = 0)	I_{CBO}	—	—	1.0	mAdc
ON CHARACTERISTICS					
DC Current Gain (I_C = 500 mAdc, V_{CE} = 5.0 Vdc)	h_{FE}	10	70	150	—
DYNAMIC CHARACTERISTICS					
Output Capacitance (V_{CB} = 15 Vdc, I_E = 0, f = 1.0 MHz)	C_{ob}	—	33	60	pF
FUNCTIONAL TESTS (Figure 1)					
Common-Emitter Amplifier Power Gain (V_{CC} = 12.5 Vdc, P_{out} = 15 W, f = 175 MHz)	G_{pe}	12	13	—	dB
Collector Efficiency (V_{CC} = 12.5 Vdc, P_{out} = 15 W, f = 175 MHz)	η	60	68	—	%
Load Mismatch (V_{CC} = 15.5 V, P_{in} = 2.0 dB Overdrive, Load VSWR = 30:1)	ψ	No Degradation in Output Power			

FIGURE 1 — BROADBAND CIRCUIT

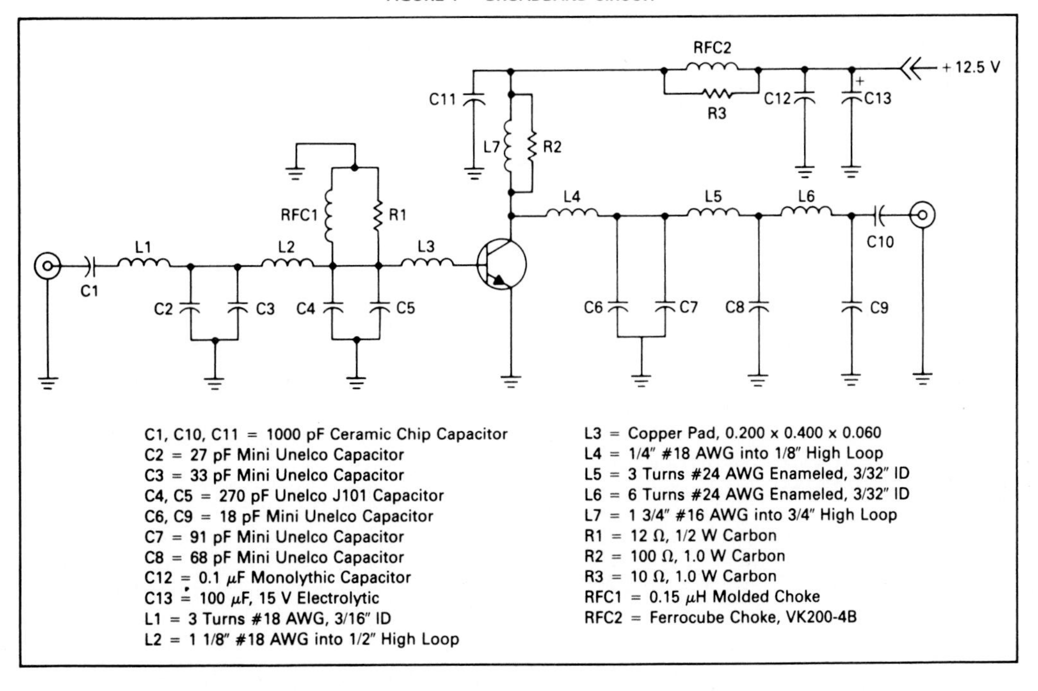

C1, C10, C11 = 1000 pF Ceramic Chip Capacitor
C2 = 27 pF Mini Unelco Capacitor
C3 = 33 pF Mini Unelco Capacitor
C4, C5 = 270 pF Unelco J101 Capacitor
C6, C9 = 18 pF Mini Unelco Capacitor
C7 = 91 pF Mini Unelco Capacitor
C8 = 68 pF Mini Unelco Capacitor
C12 = 0.1 μF Monolythic Capacitor
C13 = 100 μF, 15 V Electrolytic
L1 = 3 Turns #18 AWG, 3/16″ ID
L2 = 1 1/8″ #18 AWG into 1/2″ High Loop
L3 = Copper Pad, 0.200 x 0.400 x 0.060
L4 = 1/4″ #18 AWG into 1/8″ High Loop
L5 = 3 Turns #24 AWG Enameled, 3/32″ ID
L6 = 6 Turns #24 AWG Enameled, 3/32″ ID
L7 = 1 3/4″ #16 AWG into 3/4″ High Loop
R1 = 12 Ω, 1/2 W Carbon
R2 = 100 Ω, 1.0 W Carbon
R3 = 10 Ω, 1.0 W Carbon
RFC1 = 0.15 μH Molded Choke
RFC2 = Ferrocube Choke, VK200-4B

2N6761
2N6762

Designer's Data Sheet

4.0 and 4.5 AMPERE

N-CHANNEL TMOS POWER FET

$r_{DS(on)}$ = 2.0 OHMS
450 VOLTS
$r_{DS(on)}$ = 1.5 OHMS
500 VOLTS

N-CHANNEL ENHANCEMENT MODE SILICON GATE TMOS POWER FIELD EFFECT TRANSISTOR

These TMOS Power FETs are designed for high voltage, high speed power switching applications such as switching regulators, converters, solenoid and relay drivers.

- Silicon Gate for Fast Switching Speeds — Switching Times Specified at 100°C
- Designer's Data — I_{DSS}, $V_{DS(on)}$, $V_{GS(th)}$ and SOA Specified at Elevated Temperature
- Rugged — SOA is Power Dissipation Limited
- Source-to-Drain Diode Characterized for Use With Inductive Loads

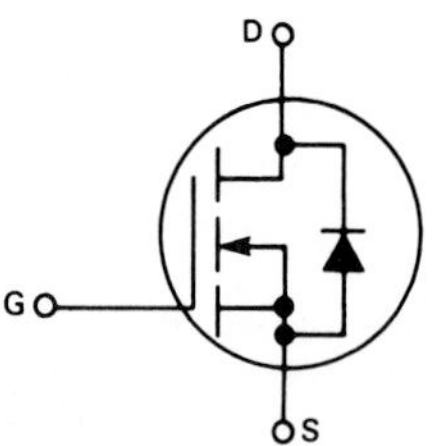

MAXIMUM RATINGS

Rating	Symbol	2N6761	2N6762	Unit
Drain-Source Voltage	V_{DSS}	450*	500*	Vdc
Drain-Gate Voltage (R_{GS} = 1.0 MΩ)	V_{DGR}	450*	500*	Vdc
Gate-Source Voltage	V_{GS}	±20		Vdc
Drain Current Continuous T_C = 25°C T_C = 100°C Pulsed	 I_D I_{DM}	 4.0* 2.5* 6.0	 4.5* 3.0* 7.0	Adc
Gate Current — Pulsed	I_{GM}	1.5		Adc
Total Power Dissipation @ T_C = 25°C Derate above 25°C	P_D	75* 0.6*		Watts W/°C
Operating and Storage Temperature Range	T_J, T_{stg}	−55* to 150*		°C

THERMAL CHARACTERISTICS

Rating	Symbol	Value	Unit
Thermal Resistance Junction to Case Junction to Ambient	 $R_{\theta JC}$ $R_{\theta JA}$	 1.67* 30*	°C/W
Maximum Lead Temp. for Soldering Purposes, 1/16″ from case for seconds	T_L	300*	°C

*JEDEC Registered Values

Designer's Data for "Worst Case" Conditions

The Designer's Data Sheet permits the design of most circuits entirely from the information presented. Limit data — representing device characteristics boundaries — are given to facilitate "worst case" design.

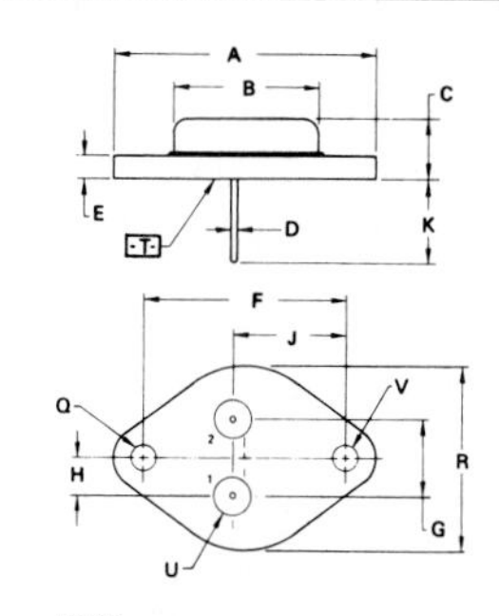

NOTES:
1. DIMENSIONS Q AND V ARE DATUMS.
2. -T- IS SEATING PLANE AND DATUM.
3. POSITIONAL TOLERANCE FOR MOUNTING HOLE Q:
   ⌖ | ⌀.13 (0.005) Ⓜ | T | V Ⓜ
   FOR LEADS:
   ⌖ | ⌀.13 (0.005) Ⓜ T | V Ⓜ | Q Ⓜ
4. DIMENSIONS AND TOLERANCES PER ANSI Y14.5, 1973.

DIM	MILLIMETERS MIN	MILLIMETERS MAX	INCHES MIN	INCHES MAX
A	–	39.37	–	1.550
B	–	21.08	–	0.830
C	6.35	7.62	0.250	0.300
D	0.97	1.09	0.038	0.043
E	1.40	1.78	0.055	0.070
F	30.15 BSC		1.187 BSC	
G	10.92 BSC		0.430 BSC	
H	5.46 BSC		0.215 BSC	
J	16.89 BSC		0.665 BSC	
K	11.18	12.19	0.440	0.480
Q	3.81	4.19	0.151	0.165
R	–	26.67	–	1.050
U	4.83	5.33	0.190	0.210
V	3.81	4.19	0.151	0.165

CASE 1-05
TO-204AA
(TO-3 TYPE)

TMOS and Designer's are trademarks of Motorola Inc.

DS3611

ELECTRICAL CHARACTERISTICS (T_C = 25°C unless otherwise noted)

Characteristic		Symbol	Min	Typ	Max	Unit
OFF CHARACTERISTICS						
Drain-Source Breakdown Voltage (V_{GS} = 0, I_D = 4.0 mA)	2N6761	$V_{BR(DSS)}$	450	—	—	Vdc
	2N6762		500	—	—	
Zero Gate Voltage Drain Current (V_{DSS} = Rated V_{DSS}, I_D = 1.0 mA)		I_{DSS}	—	—	1.0*	mAdc
T_J = 125°C			—	—	4.0*	
Gate-Body Leakage Current, Forward (V_{GSF} = 20 Vdc)		I_{GSSF}	—	—	100*	nAdc
Gate-Body Leakage Current, Reverse (V_{GSR} = 20 Vdc)		I_{GSSR}	—	—	100*	nAdc
ON CHARACTERISTICS						
Gate Threshold Voltage (I_D = 1.0 mA, V_{DS} = V_{GS})		$V_{GS(th)}$	2.0*	2.7	4.0*	Vdc
T_J = 100°C			1.5	2.2	3.5	
Static Drain-Source On-Resistance (1) (V_{GS} = 10 Vdc, I_D = 2.5 Adc)	2N6761	$r_{DS(on)}$	—	—	2.0*	Ohms
T_C = 125°C			—	—	4.4*	
(V_{GS} = 10 Vdc, I_D = 3.0 Adc)	2N6762		—	—	1.5*	
T_C = 125°C			—	—	3.3*	
Drain-Source On-Voltage (V_{GS} = 10 V) (1) (I_D = 4.0 Adc)	2N6761	$V_{DS(on)}$	—	—	8.0*	Vdc
(I_D = 4.5 Adc)	2N6762		—	—	7.7*	
Forward Transconductance (1) (V_{DS} = 15 V, I_D = 3.0 A)		g_{fs}	2.5*	—	7.5*	mhos
CAPACITANCE						
Input Capacitance	(V_{DS} = 25 V, V_{GS} = 0 f = 1.0 MHz)	C_{iss}	350*	—	800*	pF
Output Capacitance		C_{oss}	25*	—	200*	
Reverse Transfer Capacitance		C_{rss}	15*	—	60*	
SWITCHING CHARACTERISTICS						
Turn-On Delay Time	(V_{DS} = 225 V, I_D = 3.0 Adc Z_o = 15 Ω) See Figs. 1 and 2	$t_{d(on)}$	—	—	30*	ns
Rise Time		t_r	—	—	30*	
Turn-Off Delay Time		$t_{d(off)}$	—	—	55*	
Fall Time		t_f	—	—	30*	
SOURCE-DRAIN DIODE CHARACTERISTICS						
Diode Forward Voltage (V_{GS} = 0) I_S = 4.0 A	2N6761	V_{SD}	0.65*	1.10	1.3*	Vdc
I_S = 4.5 A	2N6762		0.70*	1.15	1.4*	
Continuous Source Current, Body Diode	2N6761	I_S	—	—	4.0*	Adc
	2N6762		—	—	4.5*	
Pulsed Source Current, Body Diode	2N6761	I_{SM}	—	—	6.0	A
	2N6762		—	—	7.0	
Forward Turn-On Time	(I_S = Rated I_S, V_{GS} = 0)	t_{on}	—	250	—	ns
Reverse Recovery Time		t_{rr}	—	420	—	

*JEDEC registered values.
(1) Pulse Test: Pulse Width ≤ 300 μs, Duty Cycle ≤ 2%.

RESISTIVE SWITCHING

FIGURE 1 — SWITCHING TEST CIRCUIT

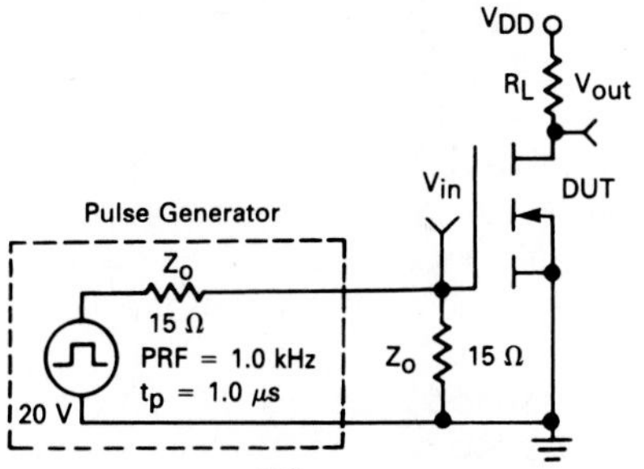

FIGURE 2 — SWITCHING WAVEFORMS

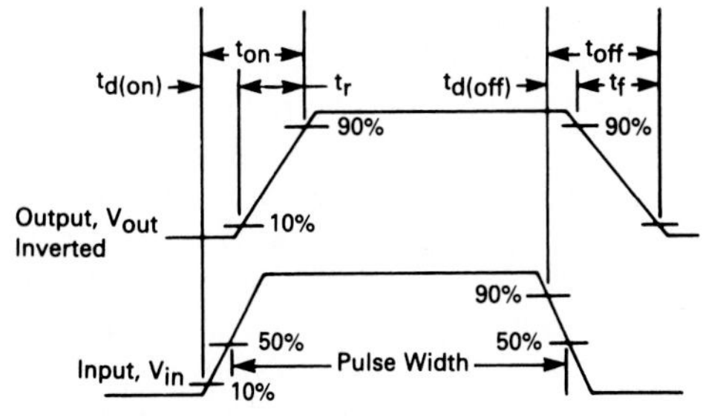

9 The Operational Amplifier

Objectives

Upon completing this chapter, the reader should be able to:

- Interpret op-amp specifications.
- Identify operational amplifier applications.
- Calculate op-amp gain and impedances (open and closed loop).
- Identify special op-amp specifications (and calculate certain specific values).
- Identify special op-amp forms.
- Recognize the universal application of the op-amp.

Introduction

The operational amplifier is an integrated circuit that contains dozens of active (BJT or FET type) and passive devices. This device has nearly ideal voltage amplification characteristics. It has the capacity to produce voltage gains far in excess of those found for the discrete semiconductor device.

The op-amp is internally biased (dc) so there is no necessity to provide more than a dc source to make it fully functional. Since the device presents a complete, unmodifiable amplifier system, it will be treated as a distinct semiconductor device. (It is understood that since it contains multiple internal devices, it is not a single device in the sense that a BJT is, but from an application standpoint, it can be treated as such from the outside.)

In this chapter the op-amp is contrasted with other devices (BJT and FET types). It is also compared to the ideal voltage amplifier. The circuitry necessary to make the op-amp fully functional and to provide specific gain values will be presented. There are some characteristics that are exclusive to the op-amp; these are detailed as well.

Finally, a general model of amplifiers that use feedback (as the op-amp does) to produce gain is established. This model will provide a basis for future evaluations and applications of amplifier systems in general. The op-amp is then shown to be an electronic equivalent of the amplification process described in this model. As such, the op-amp will be used in future chapters (where possible) to represent the amplification circuit that would be used, representing either an actual op-amp system or one that used functional BJT or FET amplifier circuits (fully biased and stabilized). ■

9.1 CONTRASTING THE OPERATIONAL AMPLIFIER TO DISCRETE DEVICES

In the past few chapters a lot of circuitry was developed to support discrete devices (BJTs and FETs). Special elements had to be added to the circuits to protect against the device becoming unstable or to protect it from exceeding functional limits. Even then, the gain of the circuit was on the order of 5 to perhaps 100. The operational amplifier, an integrated circuit containing 20 or more transistors, a dozen or more resistors, and a capacitor or two, can replace several discrete transistor circuits. All of this can be done at a cost that is little more than that for a single transistor, and in a package that is not much larger than a couple of transistors.

The operational amplifier is nearly an ideal voltage amplifier. It should be recalled that the ideal voltage amplifier had as its specifications:

1. A perfect voltage gain value, unaffected by current demands, and the potential for infinite gain
2. An infinite input impedance (no input current drawn)
3. Zero output impedance

The op-amp has characteristics that closely match these specifications, with the following typical values:

1. Voltage gain values as high as 10^5
2. Input impedance approximating an open circuit
3. Output impedance $\simeq$ 10 Ω

Also, it has the following advantages over discrete amplifiers:

1. It does not require stablization elements.
2. Its gain value is variable, simply by establishing a feedback to input ratio.
3. It can amplify ac signals (+/− polarity) without having to be biased.
4. Its output voltage will be (near) zero for zero input; thus isolation capacitors may not be required. (It does not have a dc bias level on the output.)
5. It has both inverting and noninverting input terminals, which means that the output can be either inverted (having a 180° phase shift relative to the output) or will be in phase with the input.

For many applications the op-amp is nearly an ideal amplifier. There are a few drawbacks to the op-amp. First, many op-amps require the use of two supply voltages in reference to a circuit ground. Second, most op-amps cannot be used in high-frequency applications. Op-amp gains are directly tied to frequency. (Some of the newest versions of op-amps have eliminated the dual supply requirement and have increased frequency-handling capacities.) Finally, op-amps are not available as power-amplifying devices.

Figure 9.1(a) shows the standard schematic symbol for the op-amp. As can be seen, the symbol for the op-amp is identical to the standard block diagram symbol for an amplifier. The illustration shown in Figure 9.1(b) is an expanded version of that shown in part (a). This symbol includes the dual power supply connection and terminals used to establish the amplifier's offset (discussed later in the chapter).

Note: Again it must be noted that most op-amps require two power supplies, a positive voltage (+) and a negative voltage (−) in reference to a circuit ground. The $+V_{CC}$ and $-V_{CC}$ terminals do not refer simply to the two terminals of the same supply (such as the two terminals of a battery).

Figure 9.1(c) illustrates a standard integrated-circuit-type package (called a mini-DIP). The pins on the package coordinate to connections on the op-amp symbol. It is a common practice to show the symbol [as in Figure 9.1(b)] and simply use these pin numbers to identify functional connections. The standard orientation of pins is shown in Figure 9.2. Each of the parts identified in Figure 9.2 is defined as follows.

−Input This is the inverting input terminal. Signals input to this terminal will undergo a 180° phase inversion (or for dc inputs, the output will be of opposite polarity to the input).

+Input This is the noninverting input terminal. The same polarity seen at the input will be reproduced at the output.

Output This is the output of the op-amp.

FIGURE 9.1 Op-Amp symbols

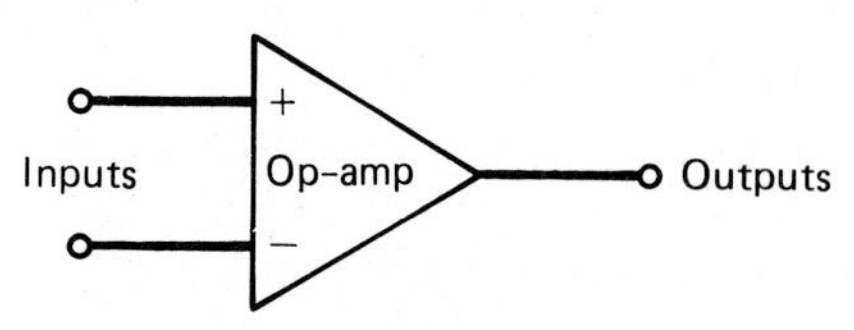

(a) Simplified schematic symbol

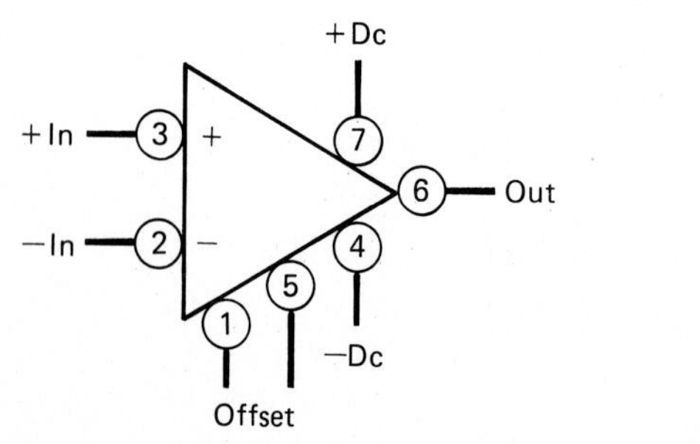

(b) Schematic with pin numbers

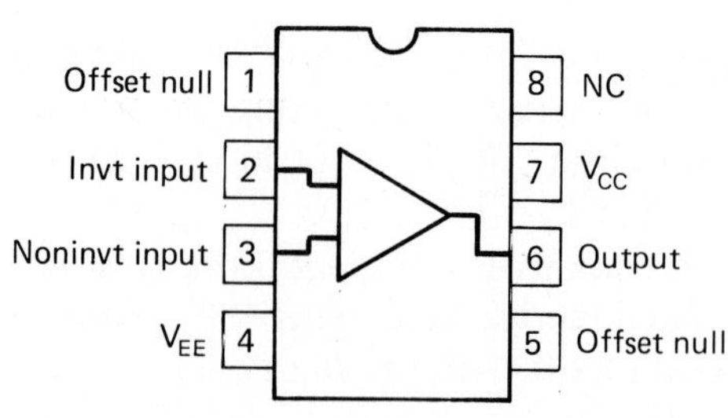

(c) Pin-out

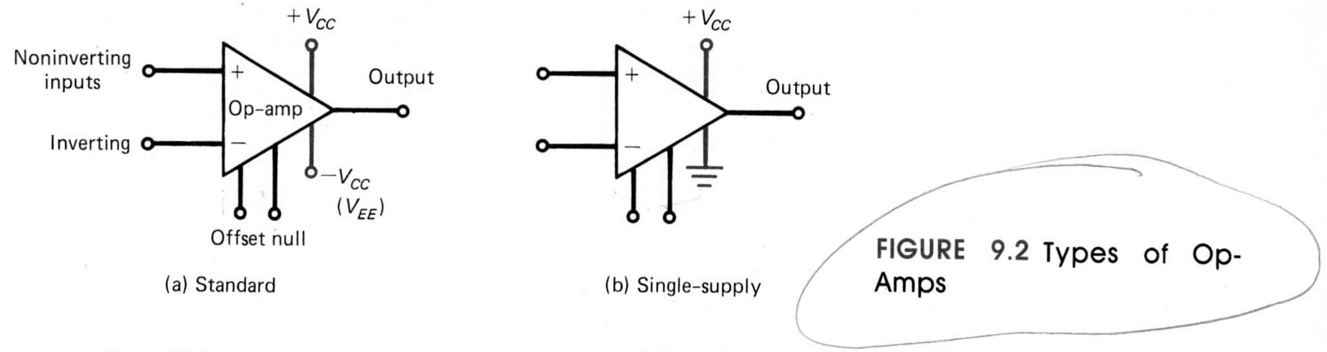

FIGURE 9.2 Types of Op-Amps

$+V_{cc}$ This is the connection for the + side of the first supply (its negative terminal is grounded).

$-V_{cc}$ This is the connection for the − side of the second supply (its positive terminal is grounded).

Offset, Null These terminals are used to balance the output to zero voltage with a zero voltage input.

It may be of interest to note that there is not a direct ground connection for the dual-supply op-amp, even though the supplies are referenced to ground. Input and output connections are referenced to circuit ground, which is also used for supply grounding.

The initial setup of the op-amp is rather simple. The dc terminal(s) are connected to appropiate power supplies set to the correct voltage level(s). In Figure 9.3 several arrangements of power supplies are illustrated which will yield the correct dual polarity in reference to ground.

For the dual-supply op-amp, it is important to have the two supplies set to the same voltage. Voltage levels are determined by the type of op-amp being used (from manufacturer data sheets); the most common range is from about 5 V (+/−) to 15 V (+/−). Again, some op-amps can operate with a single supply, thus eliminating the need for a dual-supply arrangement.

The next basic setup needed (for some op-amps) is to connect the nulling terminals. The objective of nulling is to set the output dc level to zero volts when there is no input to the op-amp. Figure 9.4 illustrates a basic connection

FIGURE 9.3 Methods of Obtaining Dual-Voltage Supply (Common Ground)

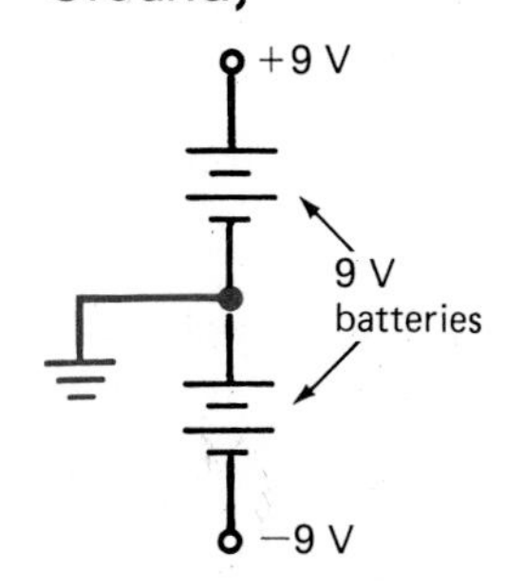

(a) Using two batteries

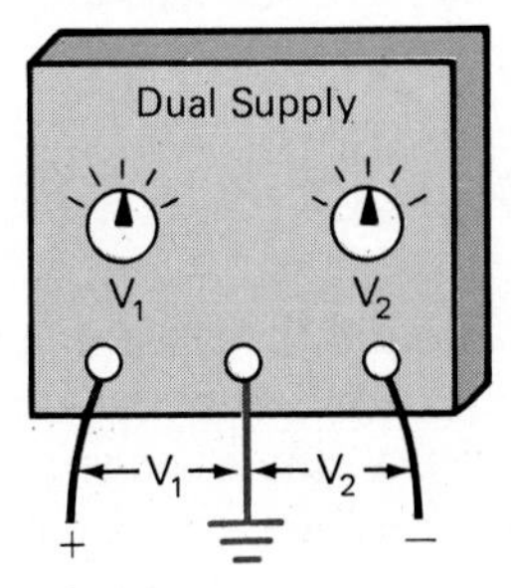

(b) Using dual output power supply with common ground capacity

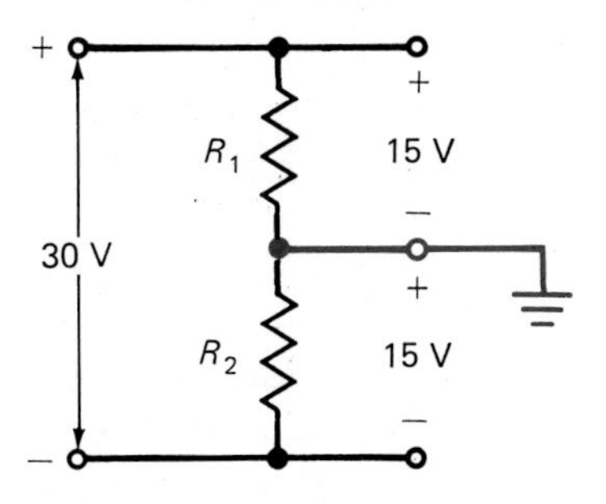

(c) Using a voltage-divider network with center reference ground ($R_1 = R_2$)

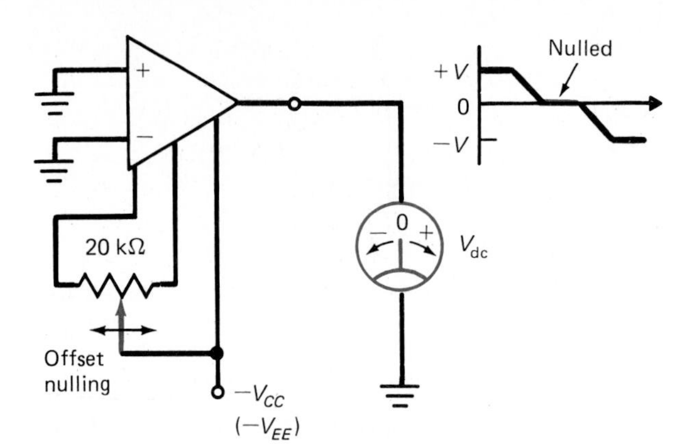

FIGURE 9.4 Op-Amp Connection to Obtain Offset Nulling

for nulling. To null the amplifier, the input terminals are grounded to eliminate noise input. A high-impedance dc voltmeter (such as a DVOM) is connected to the output. The nulling "pot" is adjusted until the output voltage reads zero. (In practical application, the nulling is done as close to zero as possible.) Once these connections and nulling adjustments (if necessary) are made, the op-amp is ready to be used as a fully functional amplifier. In the following sections of this chapter, the applications of the op-amp will be presented.

REVIEW PROBLEMS

(1) Refer to the op-amp data sheet listed in Appendix A. Draw the symbol for the op-amp and label the terminals of the drawing with the coordinated pin numbers found on the mini-DIP package.

(2) Although it may, on the surface, appear that the op-amp could directly replace a BJT or FET amplifier, that is not the case. An op-amp circuit could be designed to replace BJT or FET circuitry, but could not simply be placed in the same circuit instead of a BJT or FET amplifier. Investigate the differences between the op-amp and BJT or FET circuits and note these down as reasons why direct replacement cannot be done.

(3) Assume that an op-amp was set up with a simple voltage gain value of 2. Illustrate how a +3-V dc input would appear if it were first connected to the noninverting and then to the inverting input terminals. Repeat this exercise using a 2.5-V p-p sine wave.

9.2 ESTABLISHING OP-AMP CIRCUITS

In this section the process of setting the op-amp up as an amplifier is detailed. Since it was said that the op-amp could have a very high voltage gain, it would seem logical to try to use it as a single-stage "super amplifier." There are limits and restrictions to how much gain and output the op-amp can produce and still remain a usable amplifier. Feedback will be used to control the op-amp's gain (and other) values.

There are two major factors that cause initial restrictions to the op-amp's gain value. First, since it is operated by set dc voltages, a limit will be set to the amount of output voltage that can be produced. This limit is about 1.3 V less than the supply value. The second limit is that an op-amp's gain is tied to its frequency response. If the desired frequency is very low, a high gain can be had. But if higher frequencies are desired, the potential gain will be reduced. This gain-to-frequency relationship is listed as the **gain–bandwidth product**. The gain of the op-amp is multiplied by the top frequency it will effectively amplify. Therefore, an op-amp with a gain of 500 and an upper frequency limit of 20 kHz will have a GBW (gain–bandwidth product) of 1 million. (The units used for GBW is hertz since gain is unitless and bandwidth is already in hertz.)

The GBW is a rating of the op-amp given on the data sheets. If the desired gain is known, the GBW is divided by this figure to determine a limit to frequency. Conversely, if the operating frequency is known, dividing the GBW by that will indicate the limit to maximum gain.

$$\text{GBW} = \text{gain} \times \text{bandwidth} \tag{9.1}$$

EXAMPLE 9.1

If an op-amp has a GBW of 1.5 MHz, find (a) the frequency limit if the gain is to be 200; (b) the gain at a frequency of 300 kHz.

Solution:

(a) $\text{Frequency limit} = \dfrac{\text{GBW}}{\text{gain}} = \dfrac{1.5\text{ MHz}}{200} = 7500\text{ Hz}$

(b) $\text{Gain} = \dfrac{\text{GBW}}{\text{frequency}} = \dfrac{1.5\text{ MHz}}{300\text{ kHz}} = 5.$

Open-Loop Configuration

If no feedback is used on the op-amp, this is called an **open-loop** application. In this application its maximum gain figure is seen. The problem with this application is that in this mode of operation, it is not difficult to have the output immediately become limited to the maximum set by the dc supplies.

In Figure 9.5 an open-loop op-amp circuit is shown. Input is applied at the noninverting terminal. The inverting input terminal is grounded, and the output is measured from the output terminal to ground. For this op-amp, V_{CC}

FIGURE 9.5 Op-Amp Circuit with Open-Loop Gain (Saturated Output)

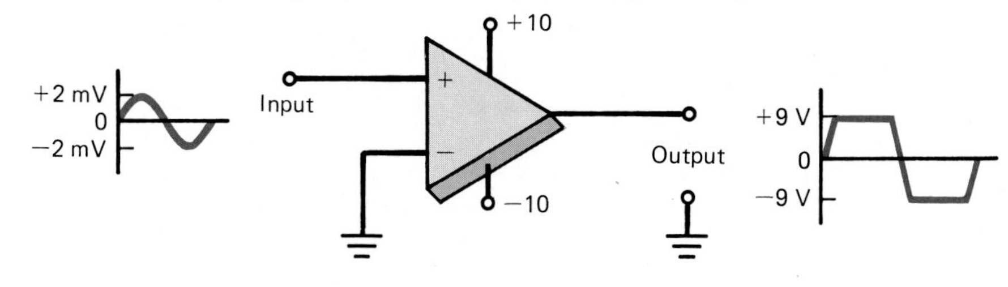

= 10 V (both supplies). Assume that it has been nulled. Maximum gain (manufacturer rating) = 100,000. The maximum output therefore is about ±8.7 V (10 V ± 1.3 V).

Since this amplifier is at such a high gain value, let's calculate what amount of input it would take to drive the output to the maximum limit (8.7 V): Since gain = V_{out}/V_{in}, then $V_{in(max)} = V_{out(max)}/\text{gain}$, which will produce

$$V_{in(max)} = \frac{8.7\text{ V}}{100{,}000} = 87\ \mu\text{V}$$

An input value of 87 μV is almost immeasurable. If any voltage at or above this level is seen at the input, the output will "saturate" or jump immediately to the maximum of 8.7 V (either + or −, following the polarity of the input).

Since the level of voltage necessary to saturate an op-amp in its open-loop configuration is usually so small, there are only limited applications for this circuit arrangement.

Closed-Loop (Feedback) Configuration

Most commonly, the op-amp is used in a configuration that controls the gain to some realistic limit. This control of gain is done by a simple voltage feedback process (shunt/shunt feedback). The output voltage is sampled; then it is returned to the input terminal as a negative feedback value. Such a process causes a reduction in the overall gain of the amplifier system.

One of the more popular circuits for providing closed-loop gain control of the op-amp is shown in Figure 9.6. In this figure, the input is fed to the inverting terminal; therefore, the output is an amplified and inverted value of the input. A portion of this output is fed back to the input terminal, thus providing negative feedback. When the gain of this circuit is calculated, the following equation results:

$$A_v = \frac{-R_F}{R_{in}} \tag{9.2}$$

This equation is developed by as follows: In this mode (inverting with the noninverting terminal at ground potential), the input current is equal to the feedback current:

$$I_{in} = I_{f(\text{feedback})}$$

Since

$$I_{in} = \frac{V_{in}}{R_{in}}$$

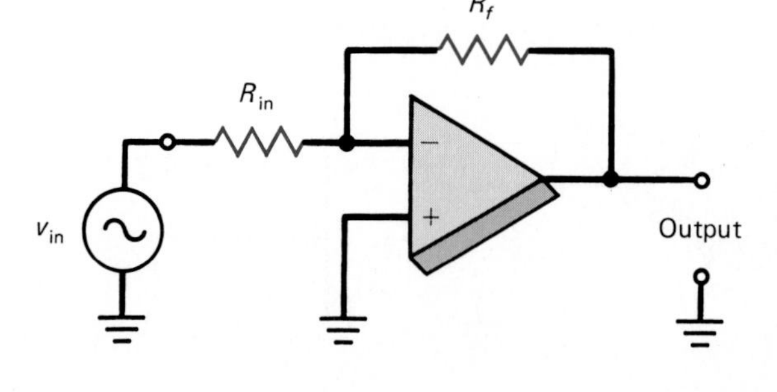

FIGURE 9.6 Standard Inverting Amplifier Configuration (Op-Amp)

and

$$I_f = \frac{-V_{\text{out}}}{R_F}$$

then

$$\frac{V_{\text{in}}}{R_{\text{in}}} = \frac{-V_{\text{out}}}{R_F}$$

By rearranging this, we have

$$\frac{V_{\text{out}}}{V_{\text{in}}} = \frac{-R_F}{R_{\text{in}}} = A_v$$

Thus the equation for gain is a result of the ratio of feedback to input resistance. (The negative sign indicates the use of the inverting terminal for input.)

EXAMPLE 9.2

If an op-amp circuit has the following resistors used, find the circuit's voltage gain:

$$R_F = 20\ \text{k}\Omega \qquad R_{\text{in}} = 500\ \Omega$$

Solution:

$$A_v = \frac{20\ \text{k}\Omega}{500\ \Omega} = 40$$

The effect that feedback has on the upper frequency limit (due to bandwidth) is another reason for using the closed-loop configuration. Now that gain is controlled to within a reasonable range, the frequency limits are expanded.

EXAMPLE 9.3

Using the calculations shown in Example 9.2, find the frequency limit for an op-amp with a GBW = 1.5 MHz.

Solution:

$$\text{BW} = \frac{\text{GBW}}{\text{gain}} = \frac{1.5\ \text{MHz}}{40} = 37.5\ \text{kHz}$$

Since the gain figure for an op-amp is based simply on a ratio of input to feedback resistance, it may seem that any ratio will be suitable. Care should be taken, though, in establishing this ratio. First, if the input (source) imped-

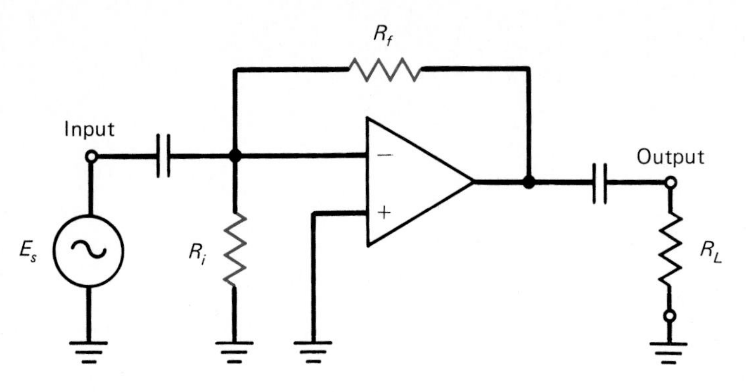

FIGURE 9.7 Inverting Amplifier Using an Internal Input Resistance

ance is known, then given a desired gain, the feedback resistor's value would be specified. If there is freedom in choosing these values (input and feedback resistance), it may be necessary to couple the input to the op-amp through a capacitor. If that is the case, a separate input resistor must be chosen for the op-amp circuit.

In Figure 9.7, an op-amp is shown which is capacitively coupled to the input and output elements. When this is done, there must be a specific input resistor applied to the inverting terminal of the op-amp. This resistance is used when calculating gain.

A final factor in determining the resistor to use with the op-amp is input impedance. Since the op-amp usually has low output impedance, neither resistor plays a significant role in determining output impedance. But they both play a role in setting the input impedance. It is obvious that the input resistor will affect the input impedance. The feedback resistor also plays a role in setting input impedance, since feedback voltage and current will affect current drawn into the input of the circuit. The effect of both of these resistors is shown in

$$Z_{in} \simeq R_{in} + \frac{R_F}{\text{open-loop gain of op-amp}} \tag{9.3}$$

But in most cases, this equation can be reduced to

$$Z_{in} \simeq R_{in} \tag{9.4}$$

REVIEW PROBLEMS

(1) Using the equations developed in this section, specify the appropriate values for R_{in} and R_F to produce an amplifier with a voltage gain value of 75. The op-amp specifications are supply voltages of + and − 12 V, GBW = 100 kHz, and desired input impedance ≃ 400 Ω. (*Hint:* Start with the input impedance and work backward.)

(2) What would be the bandwidth of the amplifier in problem 1?

(3) One application of the open-loop-gain op-amp is to convert a sine wave into a square (like) wave at the same frequency. Assuming a sine wave with an input amplitude of 0.5 V was applied to the open-loop op-amp shown in Figure 9.5, show the resulting output waveform.

In this section the equations used to detail an op-amp's function have been developed. It can be seen that working with op-amps is relatively easier than with discrete devices. In later sections, additional properties of the op-amp will be explored along with additional applications.

9.3 SPECIAL FEATURES OF THE OP-AMP

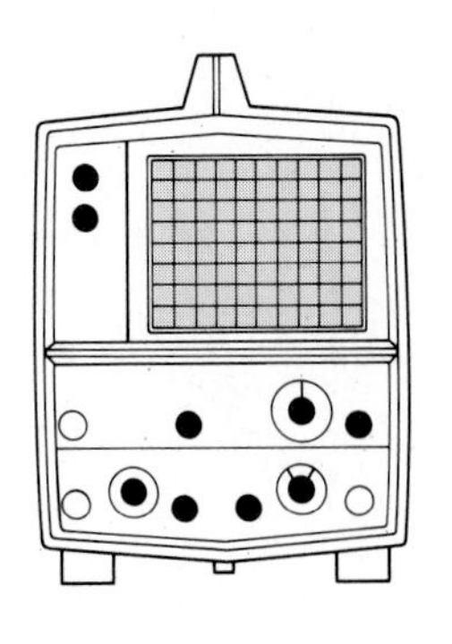

There are features and characteristics of the op-amp that require special explanation. These topics include: differential inputs, noninverting amplifier, and the general form of a gain-with-feedback equation. We discuss each of these topics in the following paragraphs.

Differential Inputs

Let's assume that two inputs are applied to the op-amp, one to the inverting and the other to the noninverting terminal. If this is done, one will be amplified with no phase shift, and the other will be amplified, but with a phase shift. By the time these two reach the output, the inverted signal will have been subtracted from the noninverted signal. The output will represent the difference between the two input values (see Figure 9.8). Figure 9.8 illustrates four possible results of using the differential mode. Each of these is explained below. [For this example, assume that the gain is set to unity (1).]

1. Dc voltages are applied to each terminal in Figure 9.8(a). The output is based only on the difference between the two voltages.
2. In Figure 9.8(b), the stepped input is being compared with the fixed dc level; again their difference is all that becomes amplified. Note that when they are equal, the output is at zero (input 1 − input 2 = 0).
3. The unusual condition of applying identical ac signals to the op-amp's inputs results in their subtracting from each other, point by point, and also producing a zero output value [Figure 9.8(c)].
4. In Figure 9.8(d), sine waves of identical frequency and phase, but of different amplitudes, are applied to the inputs. The result is a sine wave, but only based on the difference in amplitude between these two values.

In the differential mode, only the difference between the two inputs is amplified by the op-amp.

Noninverting Amplifier

The circuit in Figure 9.9 is identical to that shown when the noninverting amplifier was introduced. The process of this amplifier's operation can now be explained in terms of the differential amplification process. First, the output is produced but is in phase with the input (noninverted). A proportion of this signal, based on the simple divider circuit formed by R_F and R_I, is fed back to the inverting input terminal. This proportion is shown in the equation

$$V_{\text{feedback}} = V_f = \frac{V_{\text{out}} \times R_I}{R_I + R_F} \tag{9.5}$$

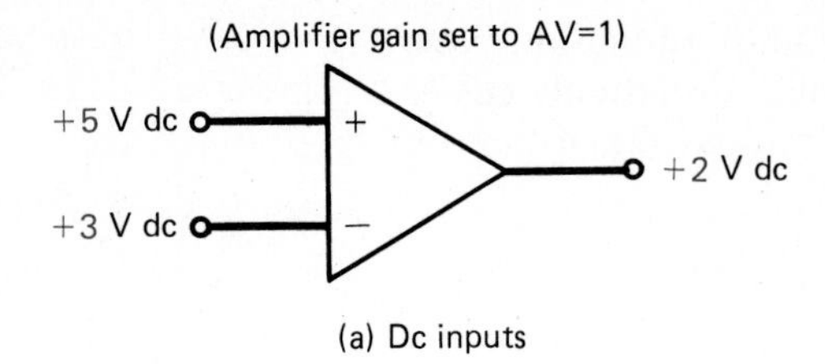

(a) Dc inputs

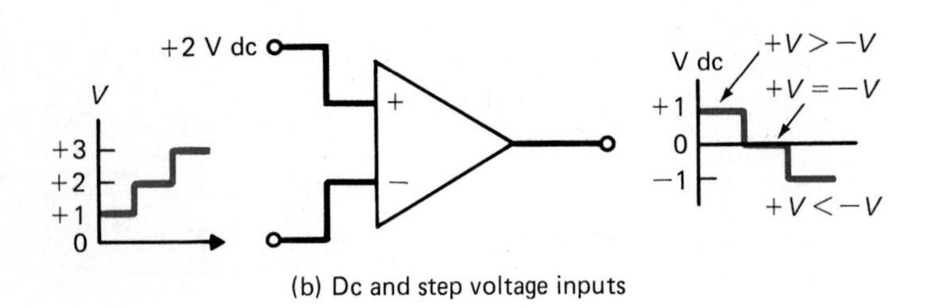

(b) Dc and step voltage inputs

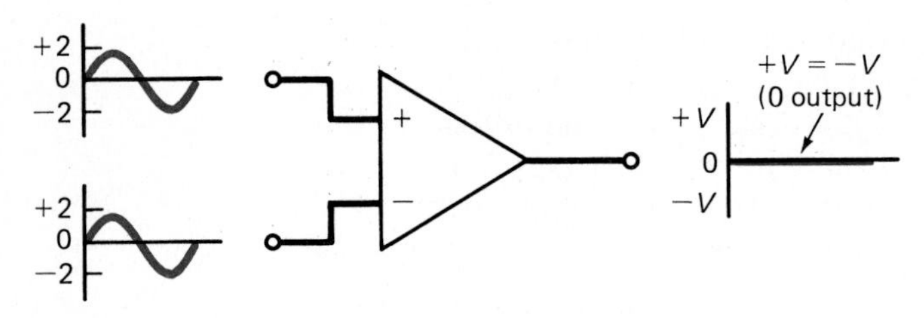

(c) Identical sine-wave inputs

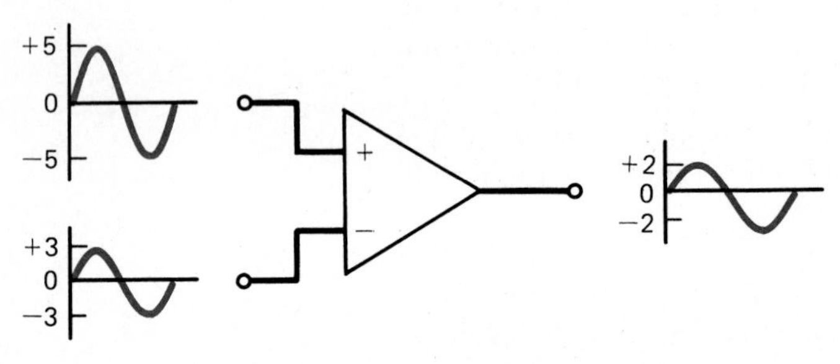

(d) Sine-wave inputs of different amplitudes

FIGURE 9.8 Op-Amp Used as a Differential Amplifier Showing Inputs and Differential Output

This relationship produces what is commonly called a **feedback factor**:

$$B = \frac{R_I}{R_I + R_F} \tag{9.6}$$

Note: The term B is a commonly used term to represent any feedback proportion. This corresponds with the more universal term used for forward gain, A. This terminology will be used in future discussions where a feedback factor is used.

For this amplifier, the output can be represented as

$$V_{\text{out}} = A_{\text{ol}}(V_{\text{in}} - V_f) \tag{9.7}$$

which states that the output is based on the composite (differential) value of input-feedback, multiplied by the open-loop gain.

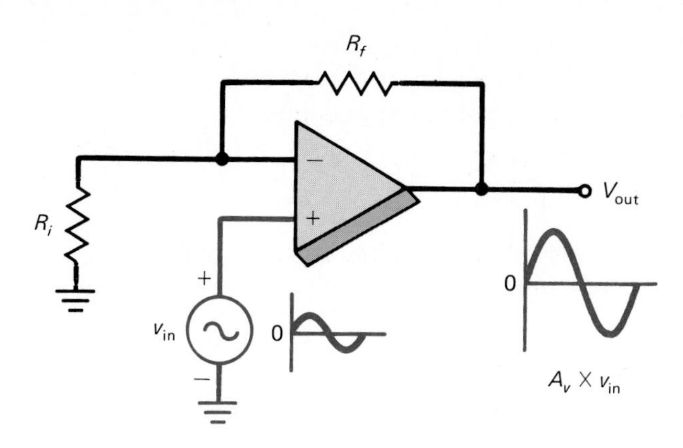

FIGURE 9.9 Noninverting Amplifier Configuration

General Form of the Gain-with-Feedback Equation

When an algebraic solution of equations 9.6 and 9.7 is developed and the result is solved for the true gain of the amplifier, equation results.

$$A_{cl} = \frac{R_I + R_F}{R_I} \tag{9.8}$$

or (9.9)

$$A_{cl} = \frac{1}{B}$$

where A_{cl} is the closed-loop gain value. Equation 9.9 is a simple observation of equation 9.8. This is the general form of the gain-with-feedback equation. This equation will be applied in later analysis processes and is not exclusive to the op-amp.

EXAMPLE 9.4

If in the circuit shown in Figure 9.9, the resistor values were as given below, find the feedback factor (B) and the closed-loop gain (A_{cl}).

$$R_F = 100\ \text{k}\Omega \qquad R_I = 5\ \text{k}\Omega$$

Solution:

$$B = \frac{R_I}{R_F + R_I} = \frac{5\ \text{k}\Omega}{105\ \text{k}\Omega} \simeq 0.0476$$

$$A_{cl} = \frac{1}{B} = \frac{1}{0.0476} = 21$$

Also affected by this configuration is the equation for input and output impedance:

$$Z_{in}' \simeq (1 + A_{ol} \times B)Z_{in} \tag{9.10}$$

where Z_{in}' = actual input impedance
A_{ol} = open-loop gain of the op-amp
Z_{in} = op-amp's input impedance

$$Z_{out}' = \frac{Z_{out}}{1 + A_{ol} \times B}$$

where Z_{out} is the output impedance of the op-amp. In this configuration, a much higher input impedance can be achieved than with the inverting amplifier.

EXAMPLE 9.5

Given the same circuit as that used in Example 9.4, and an op-amp with the following characteristics, find the input and output impedances.

$$A_{ol} = 100{,}000 \qquad Z_{in} = 1.5\ \text{M}\Omega \qquad Z_{out} = 30\ \Omega$$

Solution:

$$Z_{in}' = (1 + 100{,}000 \times 0.0476)(1.5\ \text{M}\Omega) = 4761 \times 1.5\ \text{M}\Omega$$

$$= 7141.5\ \text{M}\Omega\ \text{(gigaohms!)}$$

$$Z_{out}' = \frac{30}{4761} = 0.0063\ \Omega\ \text{(milliohms)}$$

Therefore, it can be seen that using the differential input concept while using the amplifier in its noninverting mode produces an amplifier that even more closely approximates an ideal voltage amplifier. As noted earlier, there are other uses for the differential operation, but these will be seen in later chapters.

The circuits and equations presented in the past two sections of this chapter describe the most common applications of the op-amp. Following is a simple comparison of the inverting and noninverting equations for an op-amp circuit:

Mode	Gain Equation	Z_{in}'	Z_{out}'
Inverting	$\frac{R_F}{R_I}$	$\simeq R_I$	Z_{out}
Noninverting	$\frac{R_F + R_I}{R_I}$	$(1 + A_{ol}B)Z_{in}$	$\frac{Z_{out}}{1 + A_{ol}B}$

$$B = \frac{R_I}{R_I + R_F}$$

REVIEW PROBLEMS

(1) Sketch the outputs that would be produced by the op-amp's differential amplifier function, based on the inputs shown in Figure 9.10.

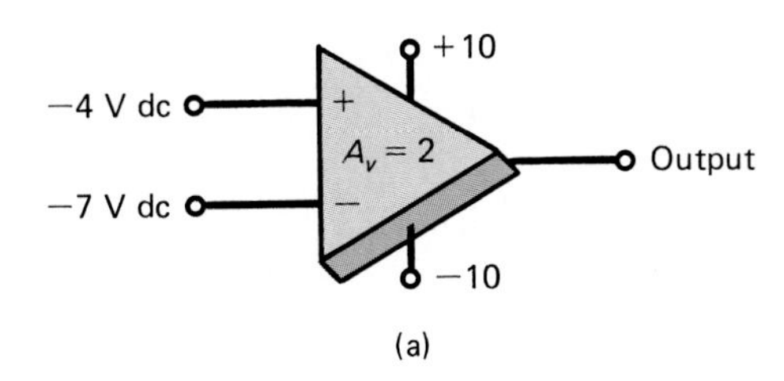

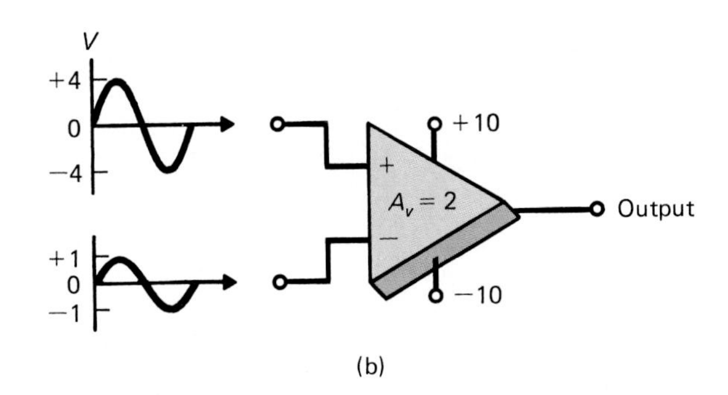

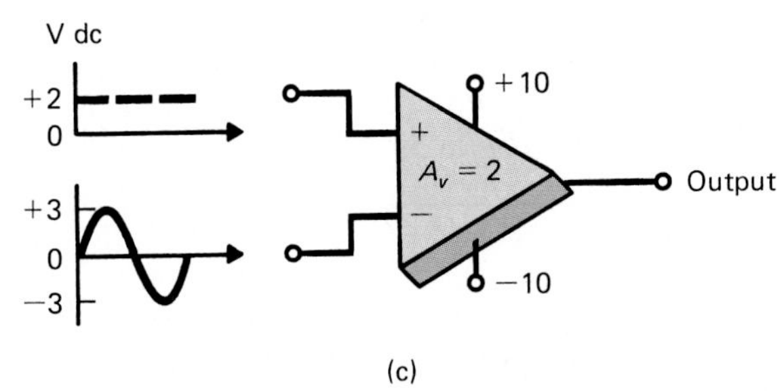

FIGURE 9.10 Three Differential Amplifiers

(2) Given the following op-amp circuit data, find the gain if it is used in both the inverting and noninverting modes:

Op-amp: $A_{ol} = 400{,}000$ $Z_{in} = 1\ \text{M}\Omega$ $Z_{out} = 50\ \Omega$
Circuit: $R_{in} = 8\ \text{k}\Omega$ $R_F = 200\ \text{k}\Omega$

(3) Using the information in problem 1, find the input and output impedances of each of these amplifiers (inverting and noninverting).

9.4 COMMON-MODE REJECTION RATIO AND OTHER SPECIFICATIONS

There are certain special specifications, in addition to those mentioned earlier in this chapter, that need to be described. One of the more common of these is the common-mode rejection ratio (CMRR), which is specifically related to the differential input function of the op-amp. Other relationships that are significant are the common-mode gain, input offset voltage, input offset voltage drift, and slew rate.

CMRR

The **common-mode rejection ratio** is a basic rating of the capacity of the op-amp to maintain an ideal differential input. It will be remembered that the differential input would generate a zero voltage output if the input to the inverting terminal were equal to that at the noninverting terminal. (This is called a common-mode input.)

Common-Mode Gain

In practical applications, the op-amp will not have exactly zero output for a common-mode input. Although this output is usually quite low, it does have a value and could cause problems if the op-amp is expected to produce exactly zero output for common-mode inputs. Basically, the common-mode output level is related to a factor called **common-mode gain** (A_{cm}).

Some op-amps have very good (low) common-mode gain values, whereas others may have less desirable features. If the op-amp is to be used in the common-mode input configuration, the statistic representing common-mode gain is important in order to compare different op-amps to one another.

Instead of listing the common-mode gain, which, for electrical reasons, is related to the "normal" gain, the common-mode gain is expressed in reference to this normal gain figure. This representation, the common-mode rejection ratio (CMRR), is a ratio of common-mode gain to normal gain. It is expressed in equation form as

$$\text{CMRR} = \frac{A_v}{A_{cm}} \tag{9.11a}$$

where A_v is the normal gain and A_{cm} the common-mode gain. It is often the practice to express this ratio in terms of decibels, thus creating the equation form

$$\text{CMRR} = 20 \log \frac{A_v}{A_{cm}} \tag{9.11b}$$

Normally, the CMRR of specific op-amps is a factor that is used for comparison. Op-amps with higher CMRR ratios are more desirable if they are being used for their differential amplifier function (common mode). Otherwise, the CMRR does not produce a major effect when using the device as a basic voltage amplifier.

Input Offset Voltage

Ideally, the op-amp should produce a zero output voltage value if the input voltage is zero. This is not the case with practical op-amps. The rating of input offset voltage (listed as V_{IO} or V_{os}) specifies the amount of input voltage that must be applied to counteract this inability. Usually, this figure is rated in low millivolts. Unless the amplifier is in a very high gain mode, or if having an exact zero output voltage for zero input is a major concern, this can generally be ignored. If it is a concern, a low-level dc voltage (the offset value listed with the device) will have to be added to the input.

Actually, a voltage-divider circuit using a potentiometer for balancing would be placed at the input lead of the op-amp. This potentiometer would be adjusted, as a part of the amplifier's calibration, to offset this nonzero output problem. The rating for the offset voltage listed by the manufacturer would constitute an approximate starting point for the divider's design.

Input Offset Voltage Drift

The offset voltage is not a fixed value but will change due to temperature. The manufacturer's rating for drift will specify how much of a voltage change (usually in microvolts) can be expected for every degree of temperature the device is above (or possibly below) a listed "normal" temperature. Again, if this offset is critical, and if the device's temperature changes, the calibration of the circuit will have to change to match (or use a thermally sensitive resistance in the divider to compensate for temperature variations). Usually, if the offset voltage is a critical issue in applying the op-amp, the offset drift should also be considered.

Slew Rate

The **slew rate** of an op-amp is usually of importance if the amplifier is being used to "trigger" to a high (or low)-voltage output when a change in the input is seen. The slew rate basically tells how fast the device can switch from a zero output to a maximum output level. Some time delay occurs within the device, and this delay is basically displayed by the device's slew rate. This rate is a ratio of the amount of voltage change expected versus the time delay seen. A "V/μS" rating is typically given. Again, these rates can be compared from device to device. (Usually, this rate only shows a subtle time delay in any amplification process and therefore does not create a major problem as long as the input is at a relatively low frequency.)

These are some of the more commonly used op-amp characteristics. Certain other characteristics related to frequency response will be discussed in a later chapter when overall frequency-response analysis is presented.

REVIEW PROBLEMS

(1) For the op-amp listed in Appendix A, look up the following characteristics.
(a) CMRR **(b)** Input offset voltage **(c)** Average temperature coefficient of offset voltage **(d)** Slew rate

(2) If an op-amp has a CMRR of 80 dB (10,000) and an open-loop gain of 50,000, what is the value of the common-mode gain?

(3) For an op-amp, design a simple divider circuit which will provide an offset voltage that could be applied to the noninverting input terminal of this device. Assume that the power supply used is a dual (+ and −) 15 V supply. Assume that the necessary offset voltage should be 10 mV. Since this divider will be attached to the input lead, and thus would affect input impedance, keep any resistance in the divider to no less than 100 kΩ.

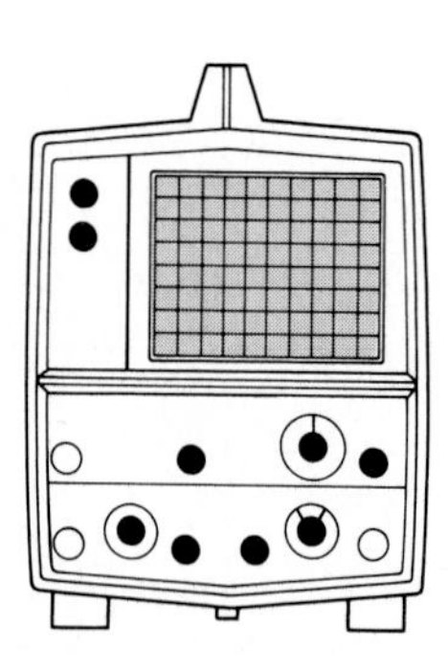

9.5 OTHER OP-AMP TYPES AND SPECIFICATIONS

Because of the wide variety of applications for op-amps a number of versions of the op-amp have been produced. The following is a general listing of some of these types, along with a basic highlighting of some of their different specifications.

General-Purpose Op-Amps

The broad category of general-purpose op-amps includes quite a large selection of op-amps. The following lists the common specifications for the general-purpose op-amp:

Voltage gain: 20,000 to 300,000
Input impedance: 500 kΩ to 20 MΩ
Bandwidth: approximately 1 MHz
CMRR: 70 to 100 dB (≃ 3200 to 10,000)
Offset voltage: 2.0 to 10.0 mV
Slew rate: 0.5 to 5.0 V/μS

There are a few general-purpose op-amps that use an FET input. For these, the input impedance can be in the range of 100 to 2000 MΩ. The other characteristics of these are similar to those listed above.

Single-Supply Op-Amps

Since it is more complicated to operate a circuit using a dual supply, a number of recent op-amps are being produced that use a single power supply (such as the one detailed in the appendix). These usually have the same characteristics as other op-amps which use dual power supplies.

Special Feature Op-Amps

Some op-amps are produced which have improved values for certain specifications. Some of these are listed below.

Low Power These consume less supply power, making them desirable for use where low power consumption is desirable (such as in battery-powered equipment).

High Speed The bandwidth of high-speed op-amps can extend into the 100s of megahertz. Also, they often have higher slew rates and higher voltage gains.

Low Drift When offset drift is of concern, these op-amps can present a very low drift value (decimal range).

Instrumentation These op-amps are used for highly accurate instrumentation. Therefore, they must be extremely stable. Low offset, low drift, and decimal slew rates ($\ll 1$) are critical.

Gain values and bandwidth are often less important for instrumentation purposes; therefore, these values are lower (a by-product of gaining greater stability).

Norton Op-Amp (Current Mirror Amplifier)

The Norton op-amp is one distinct version of the op-amp. (Sometimes this version is called a current mirror op-amp or current-mode op-amp.) This amplifier responds to the level of input current (not voltage as in other op-amps) and produces an output voltage proportional to that value. This corresponds to the transresistance amplifier model.

Like transresistance amplifiers, the Norton amplifier (so named due to its Norton equivalent input circuit) has low input impedance and low output impedance. Also, they typically operate with a single voltage supply, in contrast to the dual supply used for general-purpose op-amps. Since this amplifier differs noticeably from the regular op-amp, it has a slightly different schematic symbol, shown in Figure 9.11. The application of this amplifier is similar to that of other op-amps, and outside of impedance and gain, many of the other parameters are similar to those of the common op-amp. In future chapters, special applications of this form of op-amp will be demonstrated.

Note: Although the Norton op-amp has a low input impedance, it does have a typical forward bias voltage drop on each input terminal, like that of a BJT. The input calculations for this amplifier must include this voltage drop.

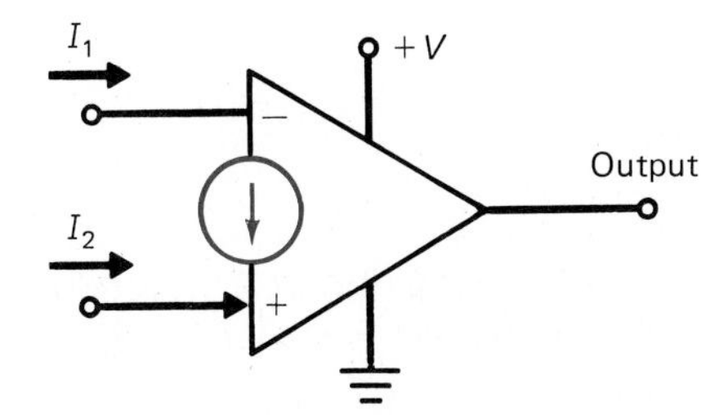

FIGURE 9.11 Norton (or Current-Mirror) Op-Amp

There are two special features of op-amps that have not yet been mentioned. First, op-amps are output-short-circuit-protected, which means that the output terminal can be shorted to ground without damaging the internal circuitry or overheating the device. (BJTs and FETs can both be damaged or destroyed if the output terminal is shorted to ground.) Also, op-amps will not "latch up" at the output. This means that if an input will saturate the device, causing the output voltage to go to the maximum level (+ or −), it will not latch (stay) at that level when the input returns to normal (nonsaturation mode). Some circuits, if driven to a saturated condition and held for any period of time, can "latch up" and stay at the saturated level even after the input is lowered.

REVIEW PROBLEM

(1) Using manufacturers' data books or distributors' catalogs, locate the part number and basic data (as listed under general-purpose types above) for at least one of each of the types of op-amps listed in this section. (*Note:* The op-amp listed in Appendix A can be used as a start.)

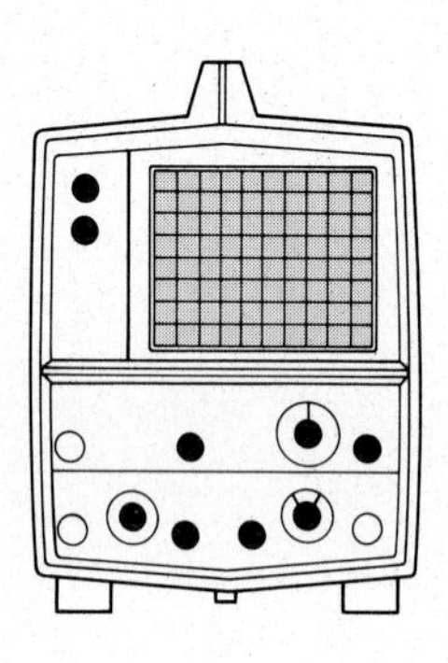

9.6 USING THE OP-AMP TO REPRESENT THE GENERAL AMPLIFIER

In this chapter the functional details of the op-amp have been presented. It was seen that using simple support circuitry, the op-amp can be structured to produce a wide variety of gain and functional characteristics. As such, it can be used to emulate (act like) many of the BJT and FET amplifier circuits presented in earlier chapters.

The op-amp does not require the user to establish dc bias conditions, as do BJT and FET circuits. Therefore, it will be an easier model to use when investigating special ac applications of amplifiers. In future chapters we will use a general-purpose amplifier such as an op-amp to represent the amplification function. Calculations will be simplified by being able to concentrate on the "special-effect" purpose of the circuit and not the internal bias-related function of the specific device used to produce amplification.

To prepare generally for this application, a "generic" model for an amplifier circuit will be used (Figure 9.12). When working with the model shown in Figure 9.12, some basic assumptions are made. First, the amplifier symbol represents a prebiased functional amplifier system. This could be either an

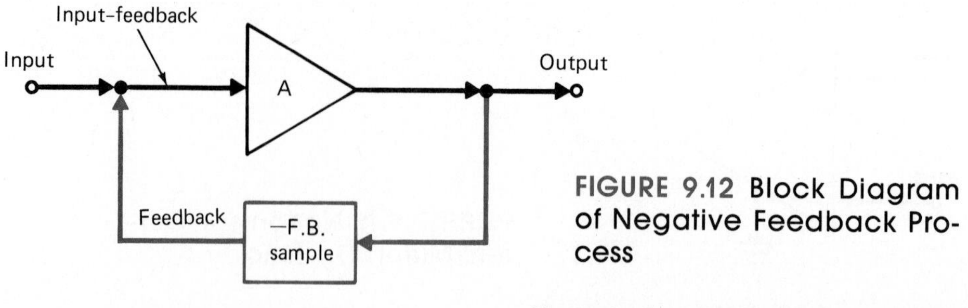

FIGURE 9.12 Block Diagram of Negative Feedback Process

op-amp or a BJT or FET circuit which includes the necessary dc bias components which will produce a functional amplifier. The feedback segment of the system is *specifically* the feedback of only the amplified quantity. It is presumed that the feedback of bias potentials for stability (FET or BJT circuits) is done internally within the amplifier circuit.

Figure 9.13 illustrates the coordination of the "generic" amplifier symbol

FIGURE 9.13 General Amplifier Symbology (Block Diagram) versus Circuitry

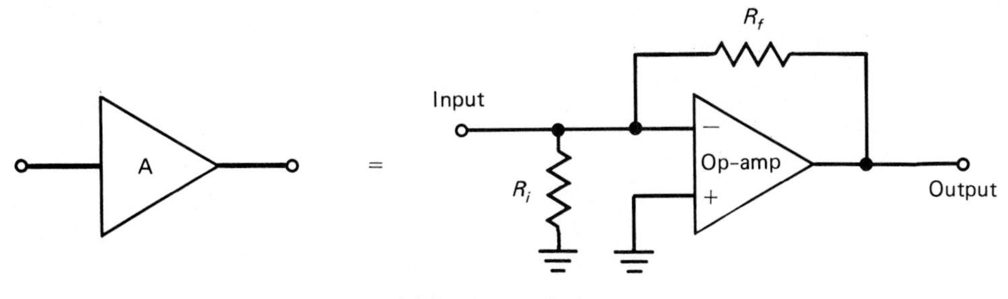

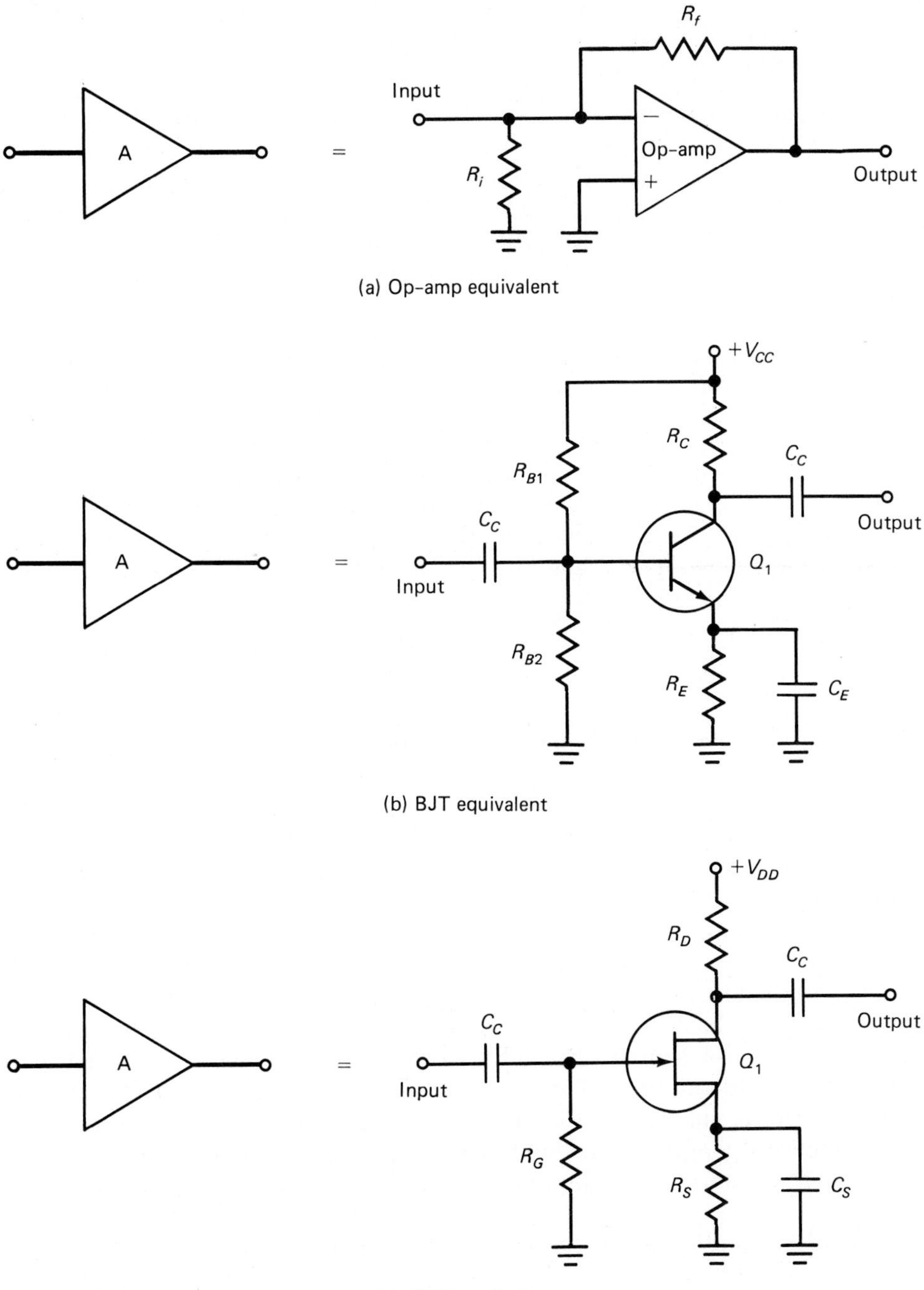

(a) Op-amp equivalent

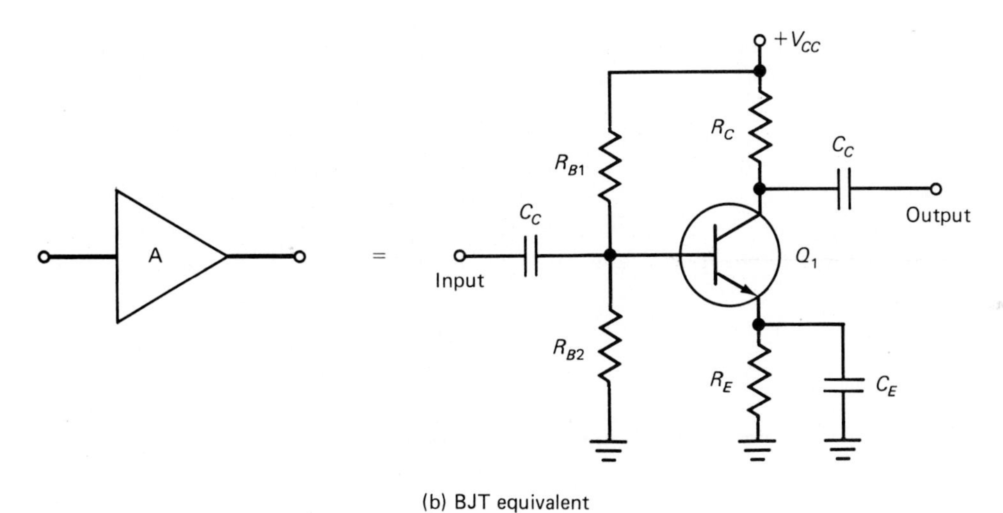

(b) BJT equivalent

(c) JFET equivalent

to op-amp and discrete semiconductor circuitry. It is expected that each of these circuits will present the following characteristics to be used when working with the overall function of the amplifier system (amplifier + feedback):

Gain Each amplifier will produce a voltage, current, or (less used) transconductance or transresistance gain. Naturally, power gain is assumed for these amplifiers (except transconductance types).

Input Impedance (Input resistance) It is necessary to have a value for input resistance for each amplifier; in that way both the power transfer characteristic from previous circuitry and any feedback dividers that correspond to this input terminal can be predicted.

Output Impedance (Output resistance) The value of output resistance is necessary, as was input resistance, in order to compute power transfer to loads and to predict the potential effect it may have on feedback. [Since some feedback quantities are not always obtained from the output terminal of an amplifier, this impedance may not always play a role in the feedback calculations (e.g., using an unbypassed emitter resistor in a BJT amplifier).]

Note: The term "impedance" used here represents only the composite dc form of internal resistances. In later application sections it will be found that certain reactive figures (usually related to capacitance) will have to be incorporated.

The feedback quantity (B) is typically a simple ratio and like amplification, is unitless (such as the ratio of a voltage-divider circuit). When using this generic form of the feedback amplifier several important equations result (many were seen with the op-amp). The gain of a negative-feedback amplifier becomes

$$A_{\text{system}} \simeq \frac{1}{B} \qquad \textbf{(9.12)}$$

where A_{system} is the total gain and B is the feedback ratio. (The only condition is that $A_{\text{ol}} \times B \gg 1$: open-loop gain $\times$ feedback.) At this point it should be noted that the feedback ratio may contain more complex factors than simple resistance ratios. If this is the case, the amplifier system will adopt these characteristics. This process and the use of complex feedback processes are demonstrated later. For now it is important to understand the relationship of feedback to an amplifier system's gain.

From this point forward this simplified model of the amplifier system will be used to demonstrate special amplifier applications. Since op-amps can be described using basically the same form of equations, it will be the natural functioning circuit equivalent used for examples. It should be realized that with some modification, BJT or FET circuits (including bias conditions) can also be used as substitutes for the basic amplifier model. As long as input impedance, output impedance, and gain are known, in most cases any amplifier form can be used to replace the generic amplifier when converting to functional circuits.

One assumption needs to be clearly stated at this point:

It is assumed that when using a "real" (BJT, FET, op-amp) amplifier to replace a generic amplification process, the limits to operation of the "real" amplifier (maximum input, maximum output, etc.) will not be exceeded. It is also assumed that within the boundaries of normal operation, the amplifier will be considered fully stable.

REVIEW PROBLEM

(1) To deal more easily with future circuits, prepare a reference card for the amplifier circuits listed below, detailing the calculations for gain, input impedance, and output impedance. Include a sample schematic and coordinate that with the generic amplifier symbol (identify the input and output terminals).
(a) BJT beta-independent circuit
(b) BJT voltage feedback amplifier
(c) JFET amplifier (include R_G, R_D, R_S)
(d) Op-amp circuits, both inverting and noninverting
For parts (a) to (c) include the influence of a load resistance on any equation.

SUMMARY

The op-amp has been shown to be a rather versatile amplifier system. Since dc bias conditions have been resolved internally, the process of applying the op-amp to functional circuitry is simplified. The gain for the op-amp has been shown to be adjusted by the manipulation of feedback in relation to the input.

It was shown that the op-amp can be used as an inverting or noninverting amplifier. The differential input function was shown to expand the application of the op-amp beyond that of a simple amplifier. Although the op-amp closely approximated the ideal voltage amplifier, limitations on its applications were demonstrated. Since the op-amp is an easily usable, general-purpose amplifier, it can be used in later discussions which concentrate on special applications of amplifier circuitry.

KEY EQUATIONS

9.1 $\text{GBW} = \text{Gain} \times \text{Bandwidth}$
9.2 A_v (voltage gain) $= -R_F/R_{\text{in}}$
9.3 $Z_{\text{in}} = R_{\text{in}} + (R_F/\text{open loop gain of op-amp})$
9.4 $Z_{\text{in}} = R_{\text{in}}$
9.5 $V_{\text{feedback}} = V_f = (V_{\text{out}} \times R_I)/(\text{R}_I + \text{R}_F)$
9.6 $B = R_I/(R_I + R_F)$
9.7 $V_{\text{out}} = \text{A}_{\text{ol}}(V_{\text{in}} - V_f)$
9.8 $A_{\text{cl}} = (R_I + R_F)/R_I$
9.9 $A_{\text{cl}} = 1/B$
9.10 $Z_{\text{in}}' = (1 + A_{\text{ol}} \times B)Z_{\text{in}}$
9.11a $\text{CMRR} = A_v/A_{\text{cm}}$ (A_v = normal gain, A_{cm} = common mode gain)
9.11b $\text{CMRR} = 20 \log (A_v/A_{\text{cm}})$
9.12 $A_{\text{system}} = 1/B$ (A_{system} = Total gain, B = Feeback ratio)

PROBLEMS

Section 9.1

9.1. Draw the schematic symbol for an op-amp:
(a) Using dual power supplies (b) Using a single power supply
(Be sure to include offset null terminals.)

9.2. Briefly describe the difference between the (−) and the (+) input terminals on an op-amp.

9.3. Illustrate a schematic drawing for a dual-supply op-amp connected to a two-battery power supply.

9.4. Briefly describe the process of nulling an op-amp. Why may nulling be needed?

Section 9.2

9.5. If an op-amp has a GBW of 5.8 MHz, determine the maximum operating frequency of that device for the following voltage gain settings.
(a) 20 (b) 100 (c) 1000

9.6. Sketch the schematic for an open-loop op-amp amplifier (inverting). Briefly describe how an op-amp functions when in the open-loop configuration.

9.7. If the supplies attached to an op-amp are both 15 V dc, what is the (approximate) maximum peak-to-peak output of the op-amp?

9.8. What is the gain of the op-amp illustrated in Figure 9.14?

9.9. If the input listed in Figure 9.14 was applied to that op-amp, sketch and label the output waveform.

9.10. What are the input and output impedances of the op-amp in Figure 9.14? What is the maximum frequency it can amplify?

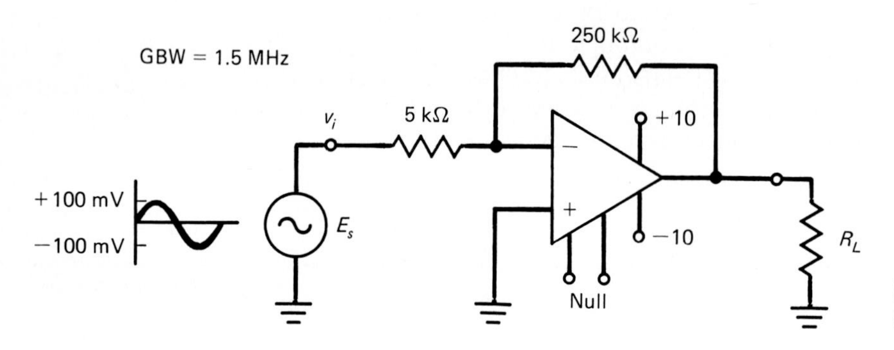

FIGURE 9.14 Typical Op-Amp Circuit

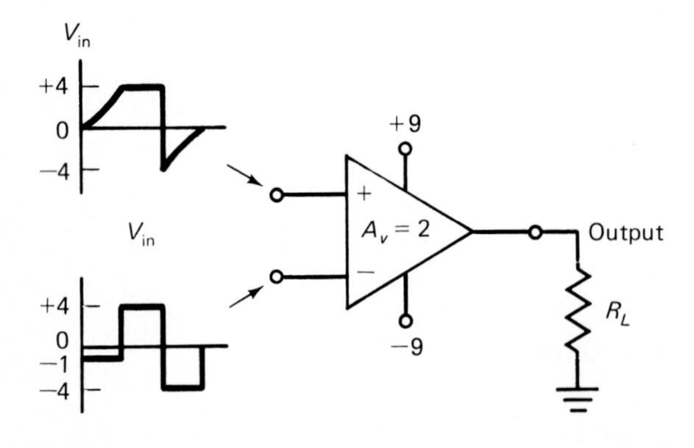

FIGURE 9.15 Differential Inputs to an Op-Amp

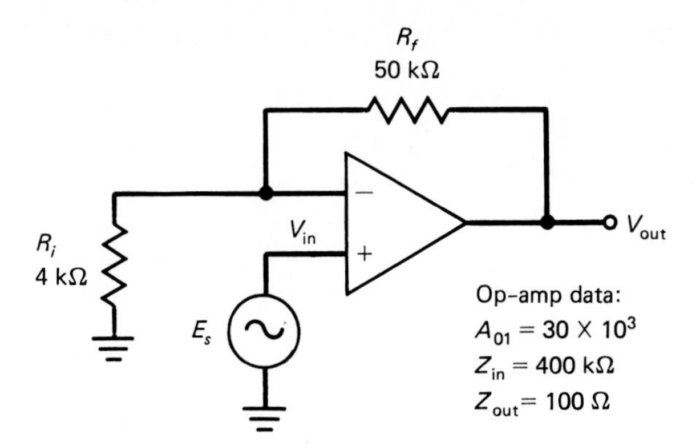

FIGURE 9.16 Noninverting Op-Amp Circuit

Section 9.3

9.11. Illustrate the output waveform and amplitudes for the op-amp shown in Figure 9.15.

9.12. What is the voltage gain of the noninverting op-amp circuit shown in Figure 9.16? What is the feedback factor B?

9.13. What are the input and output impedances of the amplifier shown in Figure 9.16?

Section 9.4

9.14. Briefly define the following terms.
(a) CMRR (b) Input offset voltage (c) Slew rate

9.15. If an op-amp has a CMRR of 90 dB (31,663), what is the common-mode gain (A_{cm}) if the op-amp's open-loop gain is (A_V) = 45,000?

ANSWERS TO REVIEW PROBLEMS

Section 9.1

1.

2. Since the BJT requires additional components and stability, and since the op-amp has the capacity for large gain values, these devices are not one-for-one interchangeable.

3. For a 3 V dc input, noninverting output = +6 V dc and inverting output = −6 V dc. For a 2.5-V p-p ac input, noninverting output = 5 V p-p (in phase) and inverting output = 5 V p-p (180° out of phase).

Section 9.2

1. $R_{in} \simeq Z_{in} = 400\ \Omega$, $R_F = 30$ kΩ

2. 7.5 MHz

3.

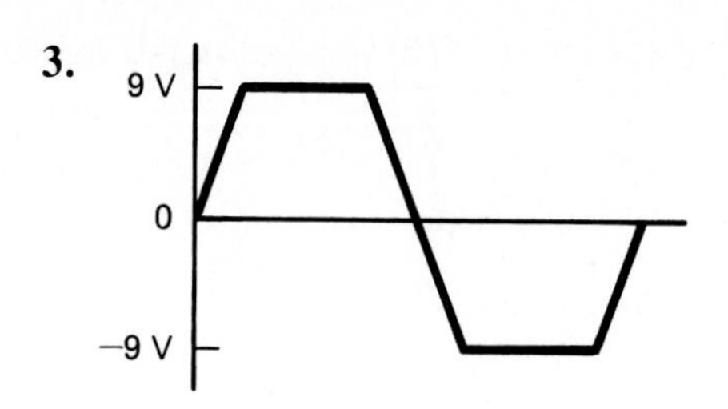

Section 9.3

1.

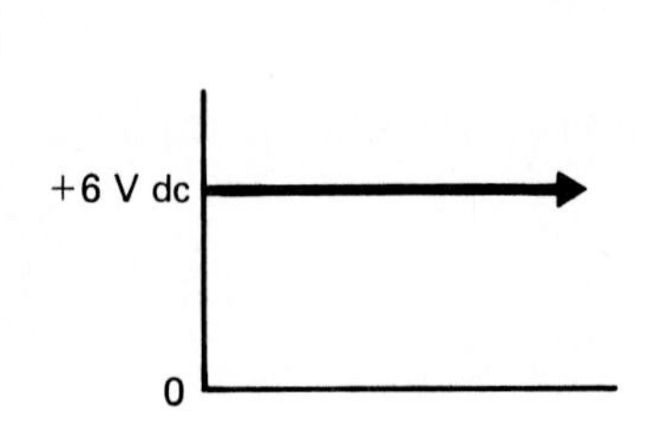

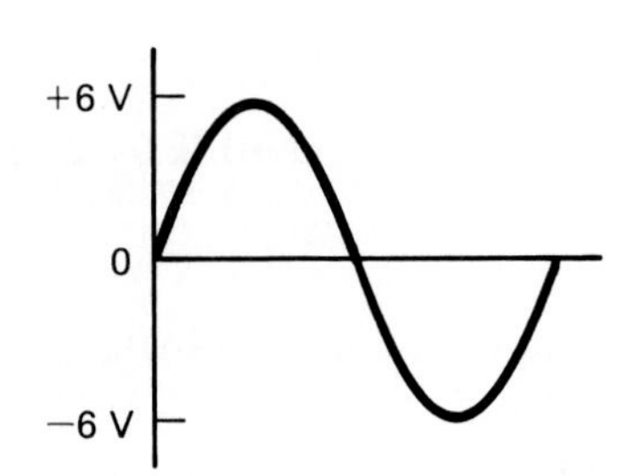

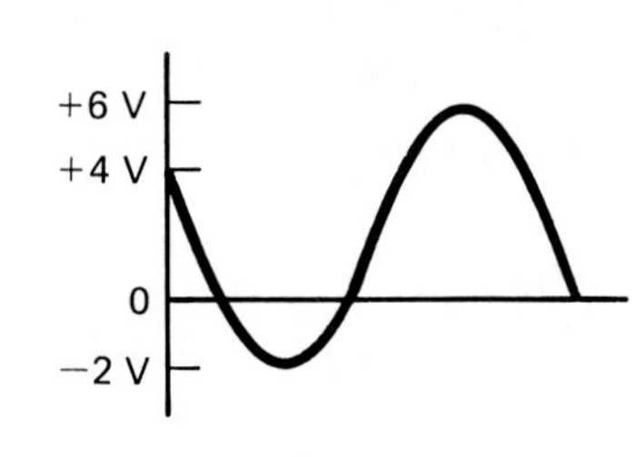

2. Inverting gain = 25, noninverting gain = 26
3. For an inverting amplifier, Z_{in} = 8 kΩ, and Z_{out} = 50 Ω; for a noninverting amplifier, Z_{in} = 1.54×10^{10} and Z_{out} = 3.25 MΩ.

Section 9.4

1. (MC34071)
   (a) 97dB (typical)
   (b) 0.5 mV (typical)
   (c) 10 μV/°C
   (d) 13 V/μS
2. 5
3.

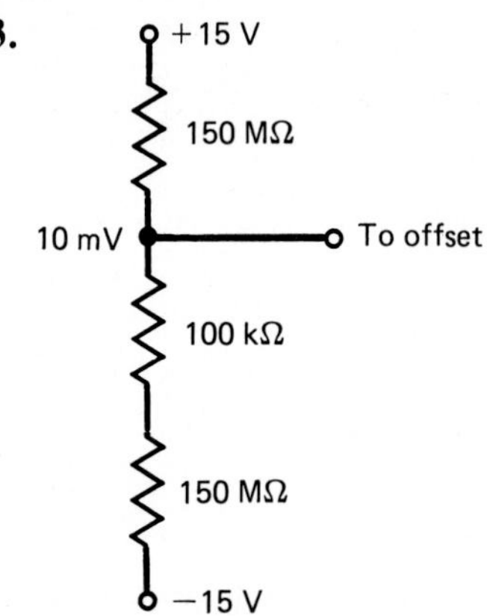

Section 9.5

1. Answer depends on device listings found.

Section 9.6

1. Answer is composite of circuits and values found in this and previous chapters.

10 Detailed Amplifier Analysis

Objectives

Upon completing this chapter, the reader should be able to:

- Recognize amplifier frequency-response limitations.
- Evaluate frequency roll-off.
- Calculate Miller capacitance values.
- Identify phase-shift problems.
- Apply feedback to improve frequency response.
- Do frequency-response testing.

Introduction

Since the basic function of amplifier circuitry is now known, further details will be added to specify amplifier operation more clearly. Dc biasing and gain calculations are assumed to be understood by the reader, based on prior chapter presentations. Special features such as limits to linear amplification, frequency response, and the causes of limits to that response are presented. Ways to improve these limitations by the manipulation of components used in the circuits, along with the application of feedback, are explored. Procedures used to test amplifiers are also presented. ■

10.1 BASICS OF LINEAR AMPLIFICATION

In earlier chapters, the process of amplification has been treated without specific reference to frequency response or any limitations on the input signal. When an amplifier is able to produce even, consistent amplification for all possible input signals, this is referred to as **linear amplification**. As limits or restrictions begin to be imposed on the form of the input signal, this ideal condition becomes less realistic.

It is common to have input and output coupling capacitors used with amplifiers that were designed to block the flow of the dc bias values out of (or into) the circuit. As this is done the capacity of the system to amplify any dc input level is eliminated. This is a first step in reducing the capacity of the circuit to amplify "any" input.

If other capacitors are added to the system, similar problems may begin to occur. The emitter bypass capacitor used in the BJT circuit, for example, may cause certain low frequencies not to be bypassed but to be fed back negatively through the emitter resistor/feedback process. This also lowers the ability of the amplifier to amplify "any" input.

The gain–bandwidth product of the op-amp indicates an upper limit to frequency response caused internally in the amplifier. At some high frequency, the gain of the amplifier will begin to be reduced and as such, a loss of output is created.

Finally, even in discrete devices, there are limits to the frequency at which gain is effective. This is due to the internal structure of the device and to capacitances, which are also a part of their internal makeup.

There are a number of conditions beyond these that cause any amplification circuit to be nonlinear. The following is a listing of additional features that may create the less-than-ideal function of an amplifier:

1. All devices have an upper frequency limit above which they will not produce usable gain.
2. The gain for (especially) BJT and FET devices is not a constant; rather, it can vary based on the bias conditions and level of signal input.
3. Input levels that are extremely low may not produce sufficient variations within the device to produce a detectable output change.
4. All devices (not just semiconductors) can present capacitive and/or inductive characteristics that may prove to interfere with amplification at certain frequencies.
5. All amplification circuits have limits to the maximum output level that can be maintained; thus high-level input may not be linearly amplified.
6. All devices (especially active devices) produce some amount of internally generated noise. If the level of this noise is substantial in comparison with the input, that noise may interfere with the production of the output.
7. Few dc power supplies are totally noise-free. The injection of noise into the circuit from the supply lines will again cause the system not to faithfully reproduce the input.

This is just a summary of problems that will produce a less-than-ideal linear amplification process.

REVIEW PROBLEMS

(1) It was noted that the input and output capacitors used on an amplifier will produce a problem when using that amplifier for dc amplification. This problem is also true for very low frequencies. Actually, this capacitor forms a high-pass filter with the input resistance of the amplifier. Sketch the schematic of a simple high-pass filter and draw the frequency-response curve for that circuit. Briefly relate that circuit and its response to what may happen with the combination of an input capacitor tied to the input resistance of an amplifier.

(2) Contrast the op-amp to the ideal linear amplifier, concentrating mainly on the acceptable range of input frequencies for an op-amp.

(3) From the problems noted above, indicate which may create a problem if a very low level input is applied to an amplifier. (Assume that the frequency of the input is within the amplifier's range of normal operation.)

10.2 SPECIFYING AMPLIFIER RESPONSE

It was noted in Section 10.1 that amplifiers would not respond to all input signals equally. One of these factors is the **frequency response** of the amplifier. When a system fails to amplify signals equally at different frequencies, this is usually due to a limit in the frequency response of the system. Figure 10.1 illustrates the frequency response of a typical amplifier. There are two primary areas on the graph that are of concern. First, notice that at the lower frequencies, the output of the amplifier begins to decrease. As the output begins to fall in level, less and less information at those frequencies would be amplified by the circuit. Finally, at dc, no output is observed.

At the high-frequency end of the graph, output again begins to fall (or "roll off," to use the more exact term). Here, just as with the low frequencies, at some high-frequency point there is virtually no output. It should be recognized that this is not a feature that was designed into the amplifier; it is the result of the components used to produce the amplifier. [If a drastic roll-off

FIGURE 10.1 Low- and High-Frequency Response (Semilog Plot)

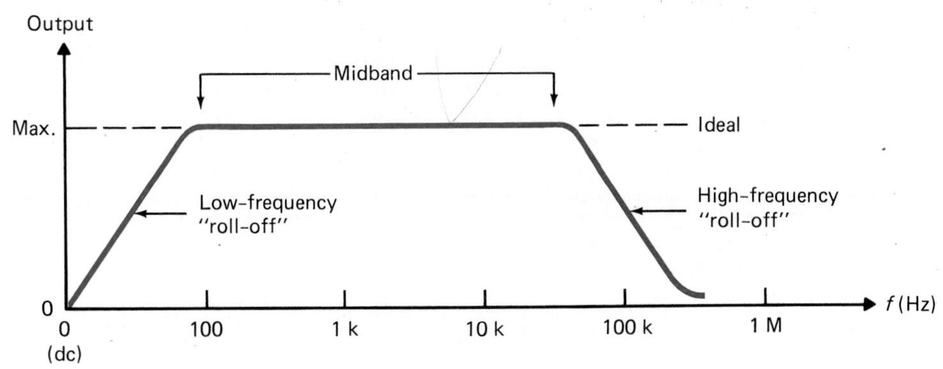

occurred at low and high frequencies for a stereo amplifier, the system would fail to reproduce bass (low) frequencies and treble (high) frequencies. It would be like being able to only use the middle keys on a piano; only the "midrange" frequencies would be reproduced.]

Certain elements can be targeted as being the major causes of frequency-response problems in amplifiers; these are discussed in Section 10.3. To have a clearer understanding of the way in which frequency response is determined, some knowledge of the decibel notation (dB) is necessary.

Decibel System

Earlier the decibel notation was used to represent an amplifier's gain. This is a popular way of rating gain or other losses in a linear system. The decibel system is simply another mathematical way of presenting such gain or loss figures. The decibel system was developed as a means of rating sound levels, such as an increase in volume out of a speaker. Since the human ear responds to sound levels using a logarithmic scale and not a linear scale, it is more logical to represent such levels using logarithms. The decibel rating system has been adopted for use with electrical voltage or power levels. Actually, it is an alternate form of rating voltage or power gain of any amplifier circuit or system.

In terms of power gain, the rating of an algebraic power gain (used in this book so far) in terms of decibels is done using the following equation:

$$A_{P(\mathrm{dB})} = 10 \log A_{P(\mathrm{algebraic})} \tag{10.1}$$

In terms of current gain and voltage gain, the conversion from algebraic gain to gain in decibels (dB) is

$$A_{I(\mathrm{dB})} = 20 \log A_{I(\mathrm{algebraic})} \tag{10.2a}$$

$$A_{V(\mathrm{dB})} = 20 \log A_{V(\mathrm{algebraic})} \tag{10.2b}$$

This, of course, changes the numerical value of the gain figure, but a few more conditions will be modified when using the dB gain representation. First, it is not possible to take the log of a negative number. Therefore, the values of gain used must be absolute values (all positive). Even in the case of an inverting amplifier, where the gain is negative, the gain used must be the absolute value. To signify inversion, a statement must be made outside of the dB value [e.g., $A_{V(\mathrm{dB})} = 3.5$ dB (inverting amplifier)].

For an amplifier (or circuit) that produces a loss instead of a gain (output $<$ input), this is represented in algebraic gain as a decimal value (gain of a voltage divider could be stated as $A_V = 0.5$, meaning that the voltage on the resistor is only 50% of the input voltage). Naturally, this is not a gain but rather, a loss. In dB notation, this is represented by a negative value (for the example above, $A_{V(\mathrm{dB})} = -6$ dB).

EXAMPLE 10.1

Convert the following gains to gains in decibels (dB): (a) $A_p = 100$; (b) $A_v = 1500$; (c) $A_p = 18{,}000$; (d) $A_v = -125$; (e) $A_v = 0.38$ (loss); (f) $A_p = 5$.

Solution:

(a) $A_{P(\text{dB})} = 10 \log(100) = 10(2) = 20$ dB
(b) $A_{V(\text{dB})} = 20 \log(1500) = 20(3.176) = 63.52$ dB
(c) $A_{P(\text{dB})} = 10 \log(18{,}000) \simeq 42.55$ dB
(d) $A_{V(\text{dB})} = 20 \log(125) = 41.94$ dB (inverting amplifier)
(e) $A_{V(\text{dB})} = 20 \log(0.38) = 20(-0.42) = -8.4$ dB
(f) $A_{P(\text{dB})} = 10 \log(5) = 10(0.699) = 6.99$ dB

Using a 0-dB Reference

An additional method of rating output levels is by the use of a 0-dB reference. This involves establishing some output level (or gain value) as the reference point, and then relating other values above or below this level in proportion to this level. For example, if a power gain of 50 were determined to be the reference, all other power gains would be related, as a ratio, to this level. This ratio would then be figured into the standard dB-gain equation. The following example should clarify this point.

EXAMPLE 10.2

A certain amplifier's reference gain is determined to be $A_P = 50$. Using this as a 0-dB reference, compare (via dB) the following other gain values: (a) 20; (b) 45; (c) 70; (d) 50 (at some other point).

Solution:

(a) 0-dB ratio $= 20/50 = 0.4$; $A_{P(0\text{ dB})} = 10 \log 0.4 = -3.98$ dB
(b) 0-dB ratio $= 45/50 = 0.9$; $A_{P(0\text{ dB})} = 10 \log 0.9 = -0.458$ dB
(c) 0-dB ratio $= 70/50 = 1.4$; $A_{P(0\text{ dB})} = 10 \log 1.4 = 1.46$ dB
(d) 0-dB ratio $= 50/50 = 1$; $A_{P(0\text{ dB})} = 10 \log 1 = 0$ dB
(This is where the term 0-dB reference comes from!)

[Audio amplifiers often contain dB meters. These are based on a standard 0-dB reference of 100 mW. They indicate −dB (less than the 0-dB reference), the 0-dB reference point, and +dB (greater than the reference). This 100-mW reference for 0 dB is given the special name of dBm.] This 0-dB reference function will aid in discussing the graph shown in Figure 10.1. That graph represented gains less than ideal at low and high frequencies (see Figure 10.2). Instead of using actual values for the gain, the graph could be scaled in dB, based on a 0-dB reference. This process is also shown in Figure 10.2. In this figure the scaling of the second graph is in dB (versus frequency), not basic gain. This is based on the midband (or midrange) gain being the 0-dB reference. This graph shows the same gain decreases, but now scaled in terms of dB, not basic gain figures.

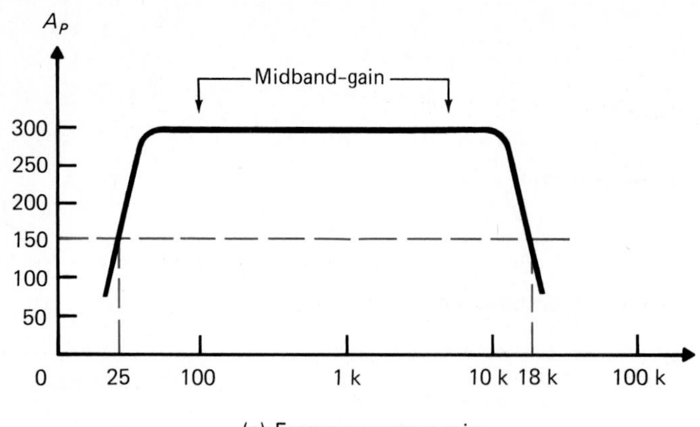

(a) Frequency versus gain

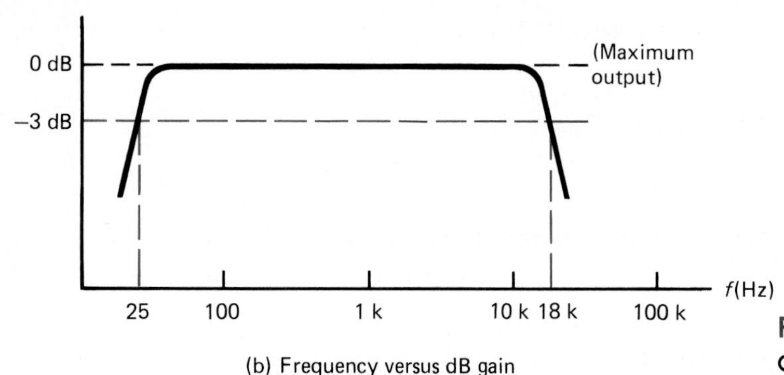

(b) Frequency versus dB gain

FIGURE 10.2 Frequency versus Gain and dB Gain

Using Half-Power Points

One significant point on these graphs occurs when the power out of the amplifier is at 50% (half-power) of the maximum. This has been defined as a major "breakpoint" in the graph. At power levels below this, the system is considered to be relatively ineffective at amplifying. Power above this point is still considered usable. It can be noted that on the second graph, this half-power point occurs at the -3-dB point on this graph [$10 \log(0.5) = -3$ dB]. This point is said to be "3 dB down from the maximum."

The actual definition of the bandwidth of this amplifier is made based on this -3-dB point. Since levels below these high- and low-frequency intercepts are considered ineffective, the actual functional bandwidth of the amplifier is determined as being from the low-frequency intercept at -3 dB to the high-frequency intercept at -3 dB. (For the graph in Figure 10.2, the bandwidth is then said to be from 25 Hz to 18 kHz.) A dB gain plot that is made on semilog graph paper (to compress the representation of a wide range of frequencies) is called a Bode plot. This plot will be used in future sections to represent many of the frequency effects seen in an amplifier.

REVIEW PROBLEMS

(1) Calculate the dB gain values for the following standard gain figures.
(a) $A_v = 20{,}000$ **(b)** $A_p = 5000$ **(c)** $A_v = -3200$ **(d)** $A_p = 0.85$

(2) If a system had a reference voltage gain (0 dB) of 40, find the 0-dB

reference value for the following other gains.
(a) 30 **(b)** 2 **(c)** 65 **(d)** 0.5 **(e)** 40

(3) Sketch a frequency-response curve that has the −3-dB power points (0-dB reference) at 60 Hz and 22 kHz.

10.3 CAUSES OF LIMITS TO FREQUENCY RESPONSE

In Section 10.2 the basic concept of frequency response was presented. It was shown that an amplifier may not respond effectively at low frequencies or at high frequencies. Such response problems are due to capacitive values either within the circuit or inherent in the composition of the active device used. Previously, input, output, and bypass capacitors were illustrated. Also, in many active devices (BJTs or FETs, for example) there are internal values of capacitance. These values are internal capacitive effects that are represented as capacitances between the leads of the devices. (For example, a BJT would have a base-to-emitter capacitance, a base-to-collector capacitance, and a collector-to-emitter capacitance term stated in the specification sheets.) These capacitances are illustrated on Figure 10.3.

In Figure 10.4, all of the capacitors are shown for a complete BJT beta-independent circuit. Several important factors of capacitance should be noted from Figure 10.4. Any capacitance that is in series with the signal path (input to output) will form a high-pass filter configuration. Any capacitance that could conduct the signal to ground (parallel to flow) will form a low-pass filter configuration. In Figure 10.5 a quick review of low-pass and high-pass filter configurations is shown. These capacitances effects can be targeted as the major cause of frequency-response problems in amplifiers.

Low-Frequency Roll-Off

The loss of low-frequency response is caused mainly by the dc blocking and bypassing capacitors used in and between amplifier stages. Each input and output capacitor produces an *R-C* high-pass filter configuration as related to the Thévenin equivalent resistance seen from this capacitor. (Usually, this resistance is the input or output resistance of the amplifier.)

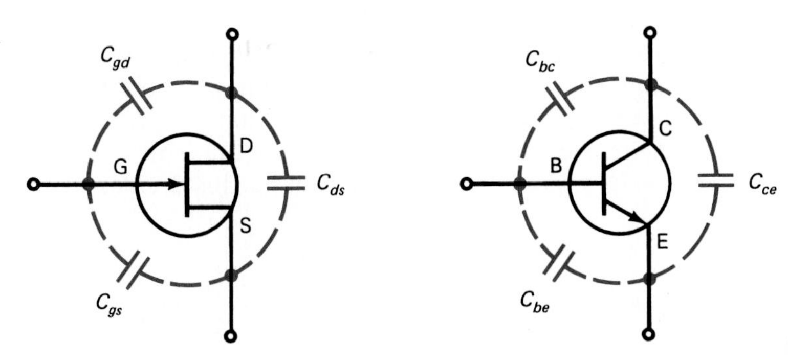

FIGURE 10.3 Capacitances Associated with FET and BJT Devices

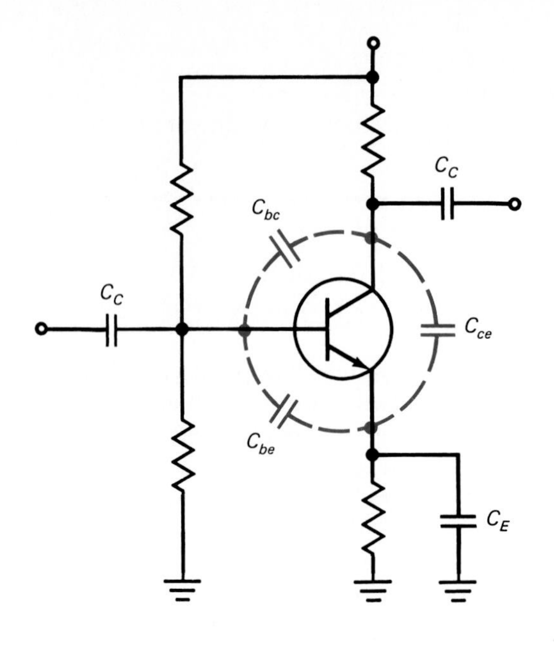

FIGURE 10.4 Beta-Independent BJT Circuit Showing Device Capacitances

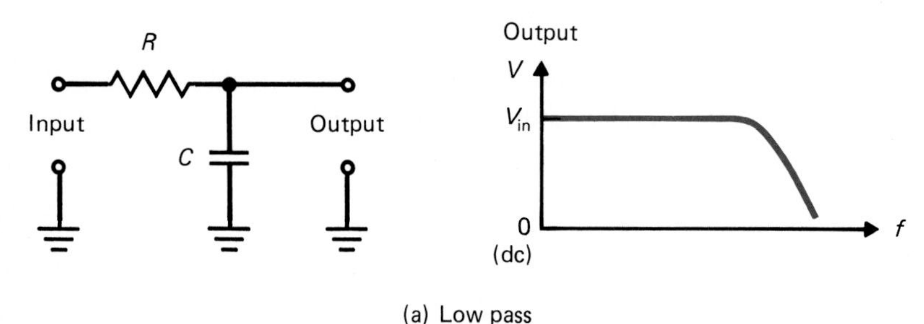

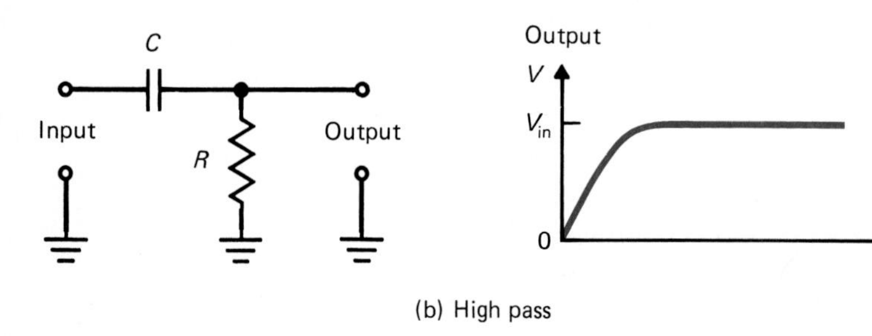

FIGURE 10.5 Filter Circuits and Frequency-Response Curves

In Figure 10.6 it can be seen that the input capacitor and input resistance of the amplifier form a simple divider circuit. As frequency is reduced, the value of capacitive reactance (X_c) will increase. This produces a greater drop in voltage across the capacitor, leaving little voltage to pass on to the amplifier. The critical frequency where a major influence is seen by this divider circuit is shown in

$$f_{\text{low}} = \frac{1}{2\pi R_{\text{in}} C_{\text{in}}} \tag{10.3}$$

At this frequency, the input level to the amplifier has dropped to only about

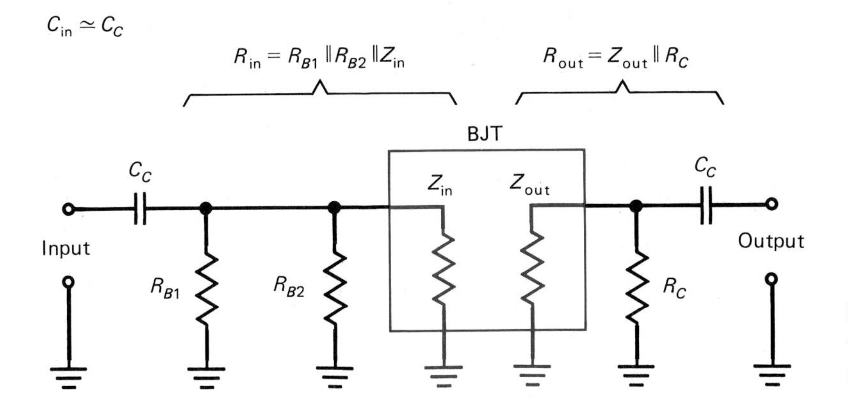

FIGURE 10.6 Filter Equivalents for a Beta-Independent BJT Circuit

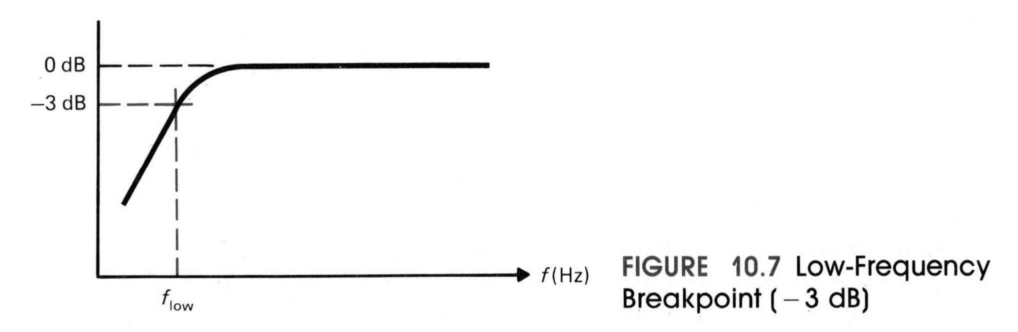

FIGURE 10.7 Low-Frequency Breakpoint (−3 dB)

70% of its potential level. At frequencies below this, the input is even lower. This frequency marks the −3-dB point on a frequency-response curve (Figure 10.7).

Using the same equation, the output frequency limitation (often called the **cutoff frequency**) can also be predicted. If, as demonstrated later in this section, these two frequencies closely line up in value, the problem is magnified. In such a case, only about 70% of the original input reaches the amplifier; then only 70% of the amplifier's output can be passed on to further steps. In that case only about 50% of the signal reaches the actual load attached to the circuit. If within the amplifier circuit bypass capacitors are used to eliminate negative feedback of the amplified signal, these, too, contribute to the frequency-response problem.

The effect of both the input and output capacitors, in conjunction with the input and output resistance (respectively) of the amplifier can magnify the problem of losses at low frequencies. If both roll-off frequencies are the same, then instead of having a −3-dB loss at that frequency (based on a 0-dB curve), there will be a 6-dB loss. (This corresponds to a 50% loss of gain.) What is happening is that both capacitors contribute a loss, so the loss is cumulative. Figure 10.8(a) shows the effect of this dual-loss condition. On the other hand, if the break frequencies are not the same, there are two breakpoints on the curve [Figure 10.8(b)]. The results then show a gradual loss due to the first breakpoint, then a rapid loss after passing the second breakpoint.

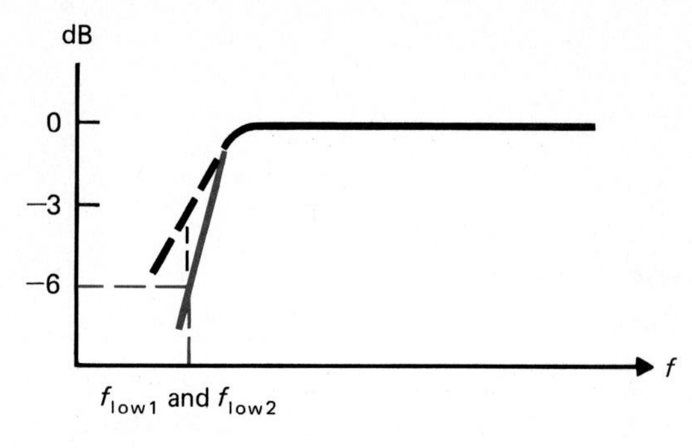

(a) Two identical breaks

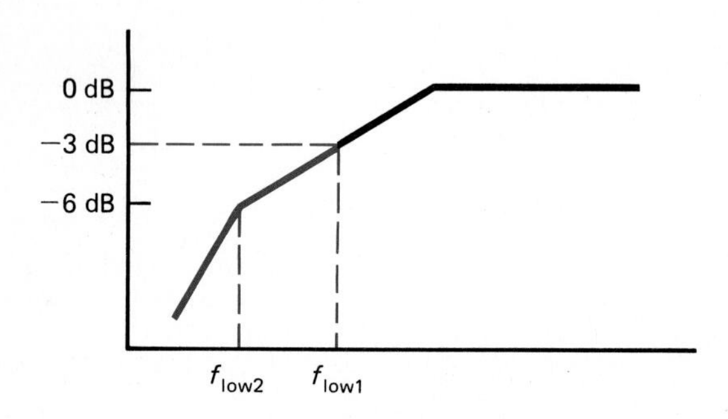

(b) Two different breaks

FIGURE 10.8 Effect of Low-Frequency Breakpoints

EXAMPLE 10.3

If the input and output coupling capacitors used on an amplifier are each rated at 10 μF, the input resistance is 800 Ω, and the output resistance is 1.5 kΩ, find the low-frequency breakpoints and illustrate the low-frequency response.

Solution:

$$f_{\text{low(input)}} = \frac{1}{2 \times \pi \times 800\ \Omega \times 10\ \mu\text{F}}$$

$$\simeq 19.89\ \text{Hz}$$

0 dB
−3 dB
f
10 Hz
19.89 Hz
100 Hz
10.61 Hz

FIGURE 10.9 Location of Two Specific Break Frequencies

$$f_{\text{low(output)}} = \frac{1}{2 \times \pi \times 1.5\text{ k}\Omega \times 10\ \mu\text{F}}$$

$$\simeq 10.61\text{ Hz}$$

See Figure 10.9.

Gain Roll-Off

Since it can be seen that the gain falls as frequency drops in the low-frequency-response curve, there is an additional representation for this *rate* of fall in gain. This rate of drop is termed the (value of) roll-off. By comparing the amount of change of the capacitor's reactance (X_c) to the constant resistance (input and output resistance) over a span of frequencies, it is found that for every factor-of-10 drop in frequency (e.g., from 100 Hz down to 10 Hz), the curve drops by -20 dB (0-dB base). When two reactances play a role (such as input and output capacitance), the roll-off is -40 dB.

Often, the change of a factor of 10 for the frequency is called a **decade**. Therefore, the roll-off is often stated: *For each high-pass filter configuration, the loss in gain is termed as being -20 dB per decade.*

Note: A loss of gain of -20 dB corresponds to a drop in actual (voltage or current) gain by a factor of 0.1. A loss of gain at -40 dB corresponds to a drop by a factor of 0.01. Thus, if the midband gain is 150 and the dB drop is listed as -20 dB/decade, then at a frequency that is 1/10 of the low break frequency (a decade below), the gain would be only 15. If the loss were -40 dB/decade, that gain would be only 1.5!

EXAMPLE 10.4

State the roll-off for the circuit used in Example 10.3.

Solution:

The circuit has a roll-off of -20 dB/decade at frequencies below 19.89 Hz and a roll-off of -40 dB/decade at frequencies below 10.61 Hz.

Basically, the rate of roll-off indicates how much loss can be expected for those frequencies below the calculated breakpoints. This effect will be more dramatic when high-frequency problems are explored in the next section.

High-Frequency Roll-Off (Capacitance Based)

High-frequency losses are usually related to the active device used within the amplifier (BJT, FET, or op-amp). In most cases, the support circuitry for the amplifier does not contain series inductors or capacitors to ground. In most cases the losses are caused by internal capacitance values within the active discrete device. These discrete device capacitances cause a bypassing effect within the discrete device wherein the signal passes through these capacitances and does not react with the control function of the device.

Note: For an op-amp, the high-frequency roll-off is listed in the gain–bandwidth product. Therefore, specific capacitance values are not to be considered in this device. Its high-frequency problems will be investigated later.

These capacitances form two unique relationships. First, any capacitance that could pass the signal flow to ground forms a basic low-pass filter in conjunction with associated resistance values. The other relationship deals with capacitance, which interconnects input circuitry with output circuitry.

In Figure 10.10, the capacitances C_{be} (BJT) and C_{gs} (FET) are the type that will bypass the signal to ground. Capacitances C_{bc} and C_{gd} provide coupling from the output to the input of these devices. The latter capacitors act just like voltage feedback elements. (The reader will recall the effect of voltage feedback bias.) As such, the capacitive effects will be interrelated with the voltage-gain figures. Therefore, this configuration produces a magnified capacitance effect. The "effective" capacitance is

$$C_{\text{in(Miller)}} = C_{bc}(A_v + 1) \tag{10.4}$$

(The term "Miller" relates to the application of Miller's theorem, which proves this magnification of capacitance.)

$$C_{\text{out(Miller)}} = C_{bc}\frac{A_v + 1}{A_v} \tag{10.5}$$

(This usually simplifies to $\simeq C_{bc}$.) Thus the effective voltage gain of the circuit causes a change in the "reflected" capacitance seen in the input circuit for the BJT. The same correction would be true for the C_{gd} of the FET.

FIGURE 10.10 "Bypassing" Effects of Device Capacitance

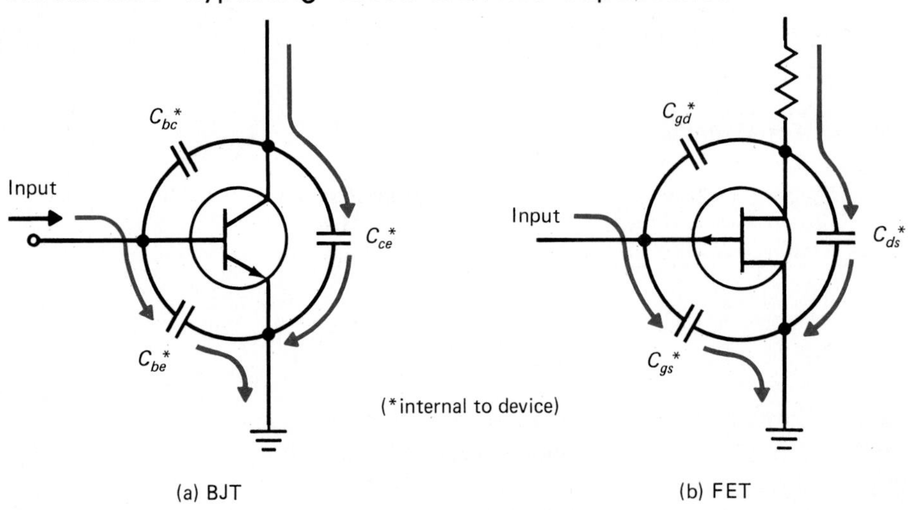

EXAMPLE 10.5

Find the effective (Miller) capacitance for a FET that has a $C_{gd} = 40$ pF and a circuit voltage gain of 21.

Solution:

$$C_{\text{in(Miller)}} = 40 \text{ pF} \times (21 + 1) = 880 \text{ pF}$$

$$C_{\text{out(Miller)}} = 40 \text{ pF} \frac{21 + 1}{21} \simeq 41.9 \text{ pF}$$

or about $C_{gd} = 40$ pF.

Now, to make the solution more complex, these capacitances must be coordinated with the other device capacitances to arrive at a total input and output capacitance. Fortunately, they are considered in parallel with the other capacitances, so the total values are easily calculated as

$$C_{\text{in(total)}} = C_{\text{in(Miller)}} + C_{\text{(input to ground)}} \tag{10.6}$$

$$C_{\text{out(total)}} = C_{\text{out(Miller)}} + C_{\text{(output to ground)}} \tag{10.7}$$

Usually, there is not a functionally high output capacitance to ground, and therefore the output capacitance is usually just the Miller output capacitance.

EXAMPLE 10.6

Given the following BJT and FET specifications and their circuit specifications, find the total input and output capacitance for each:

BJT: $C_{bc} = 6$ pF, $C_{be} = 50$ pF; circuit $A_v = 50$
FET: $C_{gd} = 5$ pF, $C_{gs} = 5$ pF; circuit $A_v = 18$

Solution:

BJT: $C_{\text{in(Miller)}} = 6 \text{ pF} \times (50 + 1) = 306 \text{ pF}$

$$C_{\text{out(Miller)}} \simeq C_{bc} = 6 \text{ pF} = C_{\text{out(total)}}$$

$$C_{\text{in(total)}} = 306 \text{ pF} + 50 \text{ pF} = 356 \text{ pF}$$

FET: $C_{\text{in(Miller)}} = 5 \text{ pF} \times 19 = 95 \text{ pF}$

$$C_{\text{out(Miller)}} \simeq C_{gd} \simeq 5 \text{ pF} = C_{\text{out(total)}}$$

$$C_{\text{in(total)}} = 95 \text{ pF} + C_{gs} = 100 \text{ pF}$$

Thus far, the effective values of input and output capacitance have been computed. Now, these must be applied to an equation that provides details on high-frequency effects.

$$f_{\text{high(input)}} = \frac{1}{2 \times \pi \times C_{\text{in}} \times R_{\text{in}}} \tag{10.8}$$

where R_{in} is the total input resistance seen at the input terminal of the devices. This is computed by placing the input resistance of the circuit in parallel with any source resistance. Thus it is the Thévenin resistance seen from the input terminal of the device.

$$f_{\text{high(output)}} = \frac{1}{2 \times \pi \times C_{\text{out}} \times R_{\text{out}}} \tag{10.9}$$

where R_{out} is the total output ac resistance. This is usually the parallel combination of the internal load resistance (R_C for a BJT or R_D for an FET) in parallel with the load on the amplifier.

EXAMPLE 10.7

If the BJT used in Example 10.6 was in a circuit where the circuit's input resistance was 800 Ω and the output resistance was 1.5 kΩ and the source and load resistances were 1 kΩ each, find the input and output high-frequency breakpoints.

Solution:

$$R_{\text{in(total)}} = 800\ \Omega \parallel 1\ \text{k}\Omega = 444.4\ \Omega$$

$$R_{\text{out(total)}} = 1.5\ \text{k}\Omega \parallel 1\ \text{k}\Omega = 600\ \Omega$$

$$f_{\text{high(in)}} = \frac{1}{2\pi \times 444.4\ \Omega \times 356\ \text{pF}} \simeq 1\ \text{MHz}$$

$$f_{\text{high(out)}} = \frac{1}{2\pi \times 600\ \Omega \times 6\ \text{pF}} \simeq 44.2\ \text{MHz}$$

In a case such as this, the 1-MHz input break frequency would dominate and major frequency losses would occur at that frequency and above. Too much loss would have occurred due to this input capacitance that by the time 44 MHz was reached, the output would be so low that it could be considered nonexistent.

High-Frequency Roll-Off (GBW Based)

When the op-amp was presented, the gain–bandwidth product (GBW) was noted as being related to high-frequency losses in gain. Again, the formula for the GBW is:

$$\text{Voltage-gain elements: GBW} = A_{V(\text{midband})} \times \text{bandwidth} \tag{10.10}$$

where Bandwidth is the frequency at which gain is unity (gain = 1)

$$\text{Current-gain elements: GBW} = A_{I(\text{midband})} \times \text{bandwidth} \quad (10.11)$$

Using the GBW for an op-amp (or any other device for which the GBW is available) makes the calculation of upper-frequency breakpoint much simpler. It is necessary just to divide the GBW by the circuit gain to obtain the upper break frequency:

$$f_{\text{high}} = \frac{\text{GBW}}{A_v} \quad (10.12)$$

EXAMPLE 10.8

An op-amp has a GBW of 4.5 MHz and uses a closed-loop configuration with $R_F = 100\ \text{k}\Omega$ and $R_I = 1\ \text{k}\Omega$. Find the lower and upper frequency limits.

Solution:

$$f_{\text{low}} = \text{dc (0 Hz)} \qquad \text{(standard for op-amps)}$$

$$f_{\text{high}} = \frac{\text{GBW}}{A_V}$$

$$A_V = \frac{R_F}{R_I} = 100$$

$$f_{\text{high}} = \frac{4.5\ \text{MHz}}{100} = 45\ \text{kHz}$$

It can therefore be seen that the frequency sensitivity of an amplifier is based mainly on capacitive terms. Low frequencies are attenuated (loss of gain) due to input and output coupling capacitors. High-frequency losses are due (mainly) to device capacitances which tend to bypass the device by passing the signal to ground. Because of the Miller effect, capacitive values that span (within the device) from the input to output are magnified (in effect) by the voltage-gain figure. Op-amps have no inherent low-frequency problems but do present losses at high frequencies. These are calculated by using the GBW for the device.

Phase-Shift Considerations

If there is a capacitive element that plays a role in the loss of gain at low and high frequencies, that capacitance will also contribute to a phase shift in the amplifier. In some cases phase shift in the amplifier may not be of any real concern, but at other times it could be a critical feature. Phase shift is calculated using standard ac circuit analysis techniques.

Since there is a capacitance (input or output) that is connected to a resistance there will be a composite value of total impedance.

$$\text{impedance} = R - jX_C \qquad (10.13)$$

To obtain the phase shift due to this value, the following equation is used:

$$\text{Low-frequency responses: phase shift} = \theta = \arctan \frac{X_C}{R} \qquad (10.14)$$

$$\text{High-frequency responses: phase shift} = \theta = -90^\circ + \arctan \frac{X_C}{R} \qquad (10.15)$$

At the break frequencies, this phase shift is 45°. For other frequencies the equation above needs to be used to predict the value of phase shift. This phase shift will occur at either the upper or lower frequencies (except for an op-amp, where the shift occurs only at upper frequencies).

To make the analysis a little less complicated, the following assumptions can be made when making a general analysis:

For low-frequency roll-off:

At the break frequency, the phase shift is 45°.

At frequencies a decade or more above (× 10) the break frequency, the shift is assumed to be zero.

At frequencies a decade or more below (÷ 10) the break frequency, the shift is assumed to be 90°.

For high-frequency roll-off:

At the break frequency, the phase shift is −45°.

At frequencies a decade or more below (÷ 10) the break frequency, the shift is assumed to be zero.

At frequencies a decade or more above (× 10) the break frequency, the shift is assumed to be −90°.

EXAMPLE 10.9 (COMPOSITE EXAMPLE, GAIN AND PHASE ESTIMATIONS)

If an amplifier has a midband gain of 150, estimate the dB gain (0-dB base) and phase relationship for the frequencies listed below if the break frequencies are $f_{low} = 120$ Hz and $f_{high} = 45$ kHz. Evaluate for frequencies: (a) the break frequencies; (b) 12 Hz; (c) 1200 Hz; (d) 450 kHz; (e) 4.5 kHz.

Solution:

(a) Gain: at 120 Hz, −3 dB; at 45 kHz, −3 dB
Phase: at 120 Hz, +45°; at 45 kHz, −45°

(b) Gain at 12 Hz, 1 decade below $f_{low} = -20$ dB; phase $= +90°$

(c) Gain at 1200 Hz = 1 decade above f_{low} = 0 dB; phase = 0°
(d) Gain at 450 kHz = 1 decade above f_{high} = −20 dB; phase = −90°
(e) Gain at 4.5 kHz = 1 decade below f_{high} = 0 dB; phase = 0°

Multiple Roll-Off Effects

It has been shown that there are usually two elements that affect low-frequency roll-off and two that affect high-frequency roll-off. The following general rule can be applied when observing both low- and high-frequency roll-off conditions: For every break frequency condition in an amplifier circuit, the rate of fall in dB gain (roll-off) is at a rate of −20 dB/decade. In a typical circuit where there are two low and two high break frequencies, the rate of fall from the first break frequency to the second is −20 dB/decade. The rate then accelerates another −20 dB/decade (total of −40 dB/decade) for frequencies beyond that point. (Typically, the rate of phase shift is assumed to be as noted in the discussion on phase shift, even for multiple break elements.)

EXAMPLE 10.10

Given the following low and high break frequencies, illustrate the frequency-response curve for the circuit.

$$f_{low(1)} = 100\text{ Hz} \qquad f_{low(2)} = 30\text{ Hz}$$

$$f_{high(1)} = 30\text{ kHz} \qquad f_{high(2)} = 100\text{ kHz}$$

FIGURE 10.11 Low- and High-Frequency Response Showing "Breaks" and "Roll-Offs"

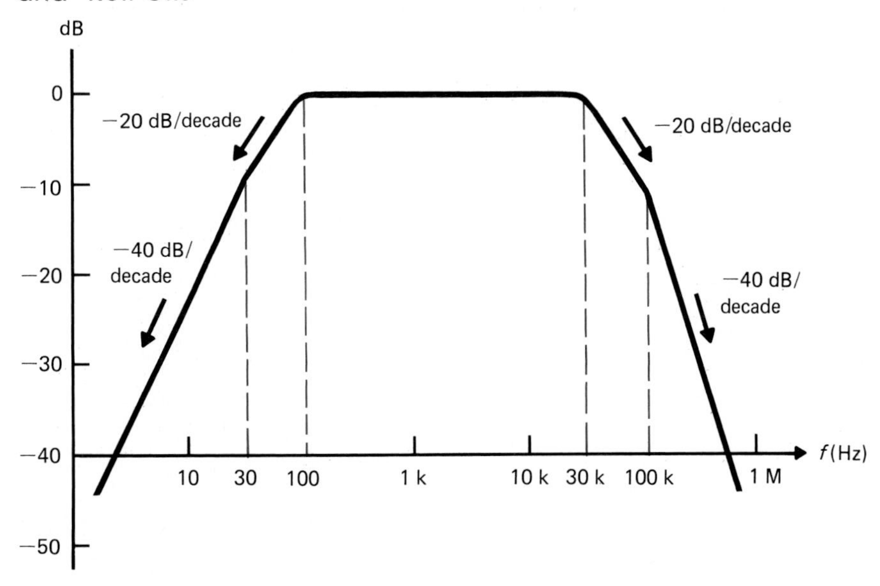

Solution:

At low frequencies, the roll-off from 100 to 30 Hz will be −20 dB/decade. At frequencies below 30 Hz, the roll-off will be −40 dB/decade. At high frequencies, the roll-off from 30 to 100 kHz will be −20 dB/decade. At frequencies above 100 kHz, the roll-off will be −40 dB/decade. This is illustrated in Figure 10.11.

REVIEW PROBLEMS

(1) If an FET amplifier had the following circuit parameters, find the low-frequency breakpoints and sketch a low-frequency response curve.

input resistance = 10 MΩ0.002 μF output resistance = 500 Ω

input capacitor = output capacitor = 20 μF

(2) Find the upper break frequencies for the 2N2218 listed in Appendix A. (*Note:* Use $C_{ib} = C_{be}$ and $C_{ob} = C_{bc}$.) If the following are the remaining parameters of the circuit in which it is being used:

$A_v = 45$ input resistance = 1.4 kΩ $R_C = 1$ kΩ

R_S(source resistance) = 1 kΩ $R_L = 1$ kΩ

(3) Sketch the plots of frequency response for the circuits in problems 1 and 2.

(4) If an op-amp had a GBW = 1.6 MHz and was being operated with a voltage gain of 150, what would be the upper-frequency roll-off point?

(5) If the total input capacitance of an amplifier was 350 pF and the total input resistance was 1.5 kΩ, find the upper roll-off frequency and the amount of phase shift that would be seen at a frequency of 400 kHz.

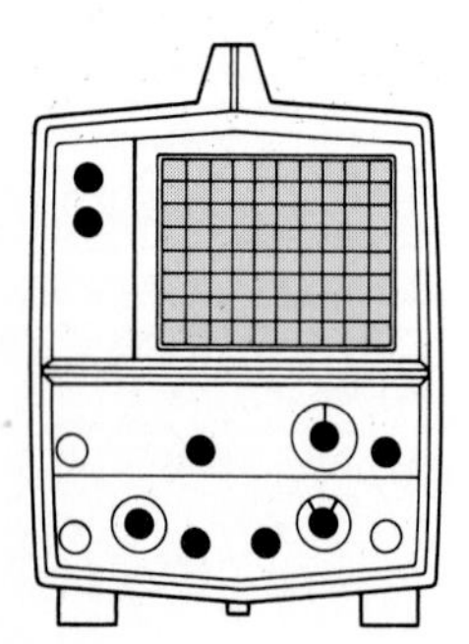

10.4 THE USE OF FEEDBACK TO IMPROVE AMPLIFIER RESPONSE

In past chapters, negative feedback was shown as a way to provide stability in the dc model of active devices. This process worked in the following way. A sample of the dc output is fed back to the input of the circuit. This sample is subtracted from an "original" level of bias at the input. If there were changes in the output due to fluctuations in device gain, these changes, via the feedback, readjusted the bias level to compensate for the changes. Within some boundaries, this adjustment of bias level created a stable output even if the gain of the device failed to remain constant. Even though the gain of the system was below the potential gain of the device, the stability produced was an advantage worth the trade-off. A similar process can be applied to an amplifier to improve the loss in gain due to the limits of frequency response, using negative feedback of the amplified signal.

The Problem Caused by Limits to Frequency Response

So far, frequency response has been treated as a natural effect seen in an amplifier. Ideally, though, an amplifier should *not* be frequency sensitive. The objective of any amplifier is to produce gain, and this gain should exist regardless of the frequency of the input. In the past sections, though, it was shown that the ability of the amplifier to function ideally was affected by frequency. For many applications, limits to frequency response are undesirable. An ideal linear amplifier *should* respond equally for any frequency applied to its input (see Figure 10.12).

An Objective for Extended Response

Since it is often desirable to produce a wider band of amplified frequencies, some objectives can be set for an amplifier system based on this desirable response. First, unless the amplifier can be designed without input or output capacitors (e.g., use of an op-amp or other special configuration), amplification cannot be expected to extend down to the dc level. But an attempt to extend the gain to very low frequencies is realistic. In many cases, though, it is also desirable to extend the high-frequency response as high as possible. Poor high-frequency response not only affects the overall frequency capacity of the amplifier but may also contribute to distortion of complex waveforms which may be a part of the input.

General Solution to Frequency Response

The obvious solution to frequency problems is to attack those elements that created the limits in the first place. This is usually the first step in improving the response of the amplifier.

1. Change the value of the input and output capacitors to improve the low-frequency response. Increased capacitor values will reduce the low-frequency limit. The only restrictions to this change may be either the cost of the capacitors (higher-value capacitors cost more), or restrictions due to size (higher-value capacitors are bigger).

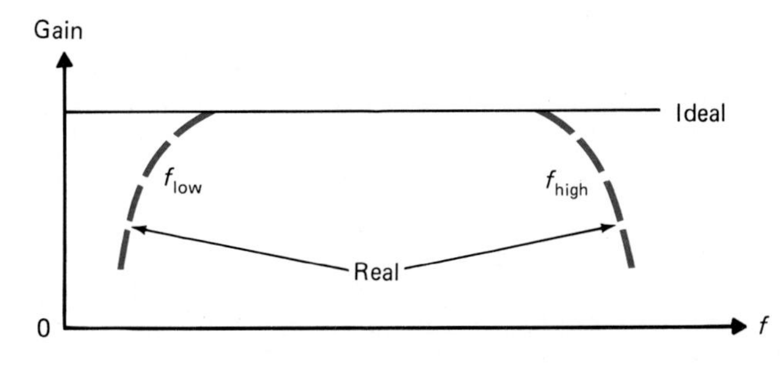

FIGURE 10.12 Comparison of Real to Ideal Frequency Response

2. Change the active device to one that will provide better high-frequency response. For discrete devices, locating a device with lower internal capacitance will extend the high-frequency limit. For op-amps (and discrete devices with a GBW rating) locate a device with a higher GBW rating. The only drawback to this change may be either cost or availability.
3. Lower the voltage gain of the circuit, thus reducing the Miller capacitance figure. This would also improve the high-frequency response. But this would require adding more amplifier stages to make up for the lost gain, which could place the high-frequency limitation back to where it had been before the change.

In some cases it is not possible to make these simple circuit modifications. The replacement components may not be available or may be too costly or bulky to solve the problem.

Signal Feedback Solution

To comprehend how negative signal feedback can extend the frequency response of an amplifier, the following basic observation about frequency response will be used as a reference: *The effect seen in an amplifier due to its frequency response limitations is a reduction of the output level at certain low and high frequencies.* This reduction of output is not unlike the type of reduction that would occur in an unstabilized amplifier if the active device's gain were to fall below the normal level. Since there is this similarity between these two effects, the use of negative feedback to improve losses due to frequency can be related to a use of negative feedback to provide gain stability.

The negative feedback of dc bias levels achieved the following:

1. The circuit gain was set to match the minimum capacity of the device.
2. The lower output levels created less feedback, thus boosting the input bias level, which caused a reproduction of the initial output level (before the drop in output).

The negative feedback of output signals (ac usually) will achieve the following:

1. The circuit gain at midfrequencies will be set lower than the potential of the circuit. (This places the 0-dB reference to a lower value.)
2. At high (or low) frequencies, where the output signal will tend to drop, there will be a reduction in the level of negative feedback, thus increasing the effective input to the amplifier. This increase in effective input will force the output to become larger in magnitude, thus tending to stabilize the gain. This will occur until the level of negative feedback is reduced to a point where it no longer affects the input enough to make up for the reduced capacity of the amplifier. At that point the output level will fall, following the normal −20-dB/decade loss.

This process is illustrated in Figure 10.13. First, the normal gain and roll-off curve for an amplifier are shown. On top of this is the curve for the same

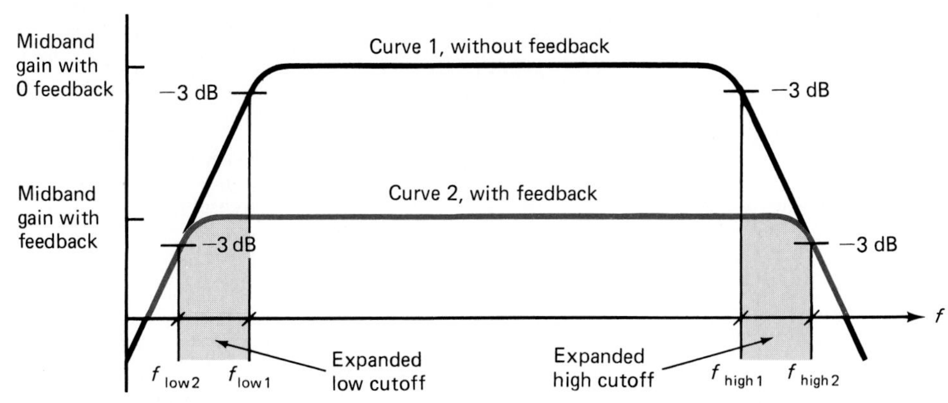

FIGURE 10.13 Effect on Frequency Response by the Application of Negative Feedback

amplifier when negative feedback is applied. The following is a step-by-step description of how the negative feedback is affecting this curve.

1. At midband levels, the feedback curve shows a reduced gain due to the fact that true input = input − feedback.
2. As the input frequency hits the −3-dB point on the original curve, the gain is expected to decrease. If the gain tends to drop, the level of feedback will also drop. Thus true input = input − less feedback. Therefore, the true input will increase, compensating for the drop in output.
3. This process will occur until the gain of the amplifier (actual gain − losses due to capacitor bypassing) is not high enough to support this higher true input level.
4. From that point (frequency) on, the output will simply follow the normal curve, −20 dB/decade drop (without feedback), since the amplifier is not capable of producing a usable output even with the increase in the true input.

What this process achieves is an extension, on both ends of the frequency spectrum (span), of the actual point where the −3-dB loss of output occurs, beyond which the losses continue at a rate of −20 dB/decade.

EXAMPLE 10.11

An amplifier has a frequency-response curve like the one shown in Figure 10.14. If negative feedback is applied, the midband circuit gain is found to be −10 dB below that for the original circuit. On this graph, sketch the new response curve based on feedback.

Solution:

To sketch the new curve:

1. Draw in a midband gain line that is −10 dB below the original curve.
2. Extend this line until it intersects the original low- and high-frequency

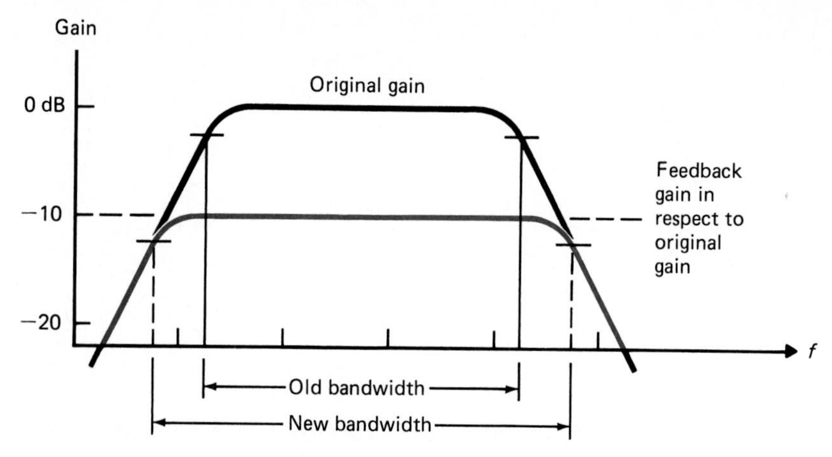

FIGURE 10.14 Detailed Comparison of Response Curves With and Without Negative Feedback

roll-off curves. The new amplifier response will follow this new line and then follow the roll-off from the point of intersection.

3. At the intersection points (roughly) are the new low and high break frequencies. Locate their values down on the horizontal frequency axis.

Producing the Negative Signal Feedback

The production of negative signal feedback can be done either by the addition of a specific signal feedback circuit, or by using means identical to those used for bias feedback. Figures 10.15 through 10.19 circuits will illustrate several possible ways of achieving signal feedback.

In Figure 10.15, the standard op-amp circuit is designed to produce signal (not bias) feedback. This basic configuration, though, a feedback to input divider circuit, is commonly used in other amplifier circuits to provide signal feedback (series/shunt feedback).

A partially bypassed feedback resistor in the voltage feedback circuit shown in Figure 10.16 produces a negative feedback of the output signal. This produces a similar effect to that of the op-amp feedback circuit. [The inversion (−) of the feedback is achieved since the voltage on the collector is the inverse of the input (180° phase shift); series/shunt feedback.]

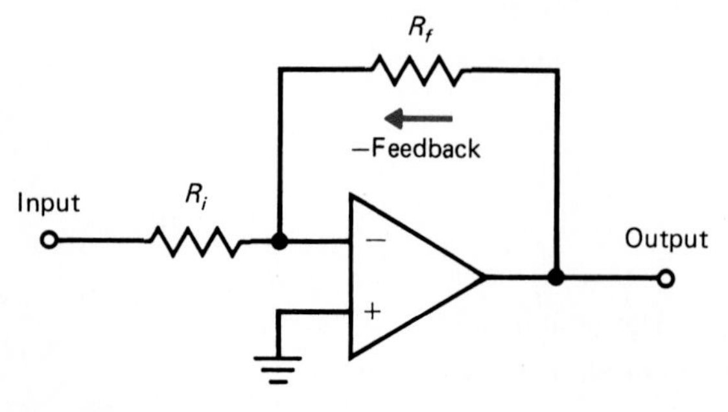

FIGURE 10.15 Op-Amp Using Negative Feedback (Standard Noninverting Circuit)

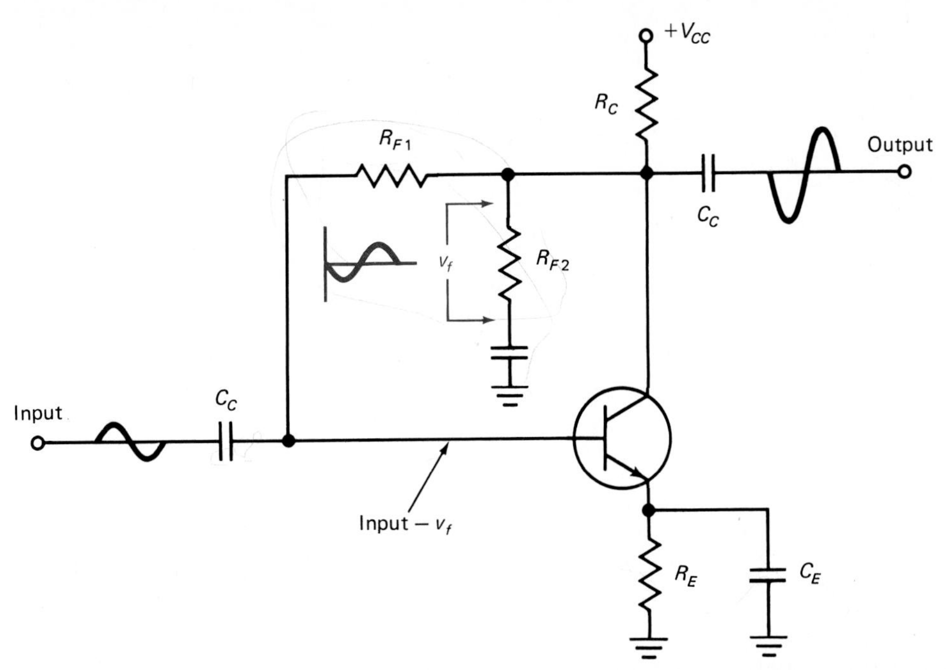

FIGURE 10.16 Obtaining AC Feedback in the Voltage Feedback Bias BJT Circuit

The un-bypassed section of the emitter feedback resistor in Figure 10.17 will produce both dc and signal feedback to the base of the circuit. (Again inversion is achieved due to the 180° phase shift of the output; series/series feedback.) The same form of feedback is achieved for the FET circuit shown in Figure 10.18.

In fact, each of the four forms of feedback, related to the four forms of amplifiers, will achieve the effect of producing extended frequency response.

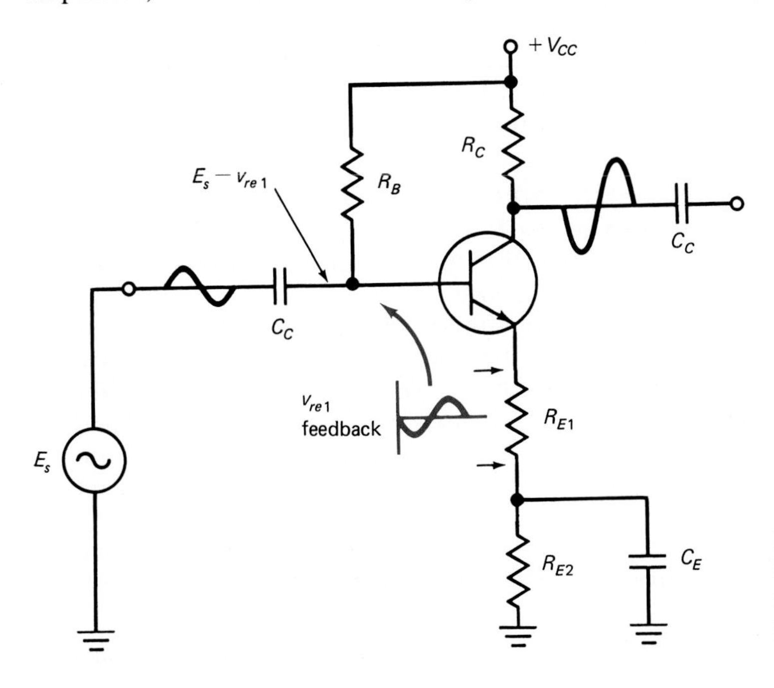

FIGURE 10.17 Using a Partially Bypassed Emitter Resistor to Obtain Negative Signal (AC) Feedback (BJT Circuit)

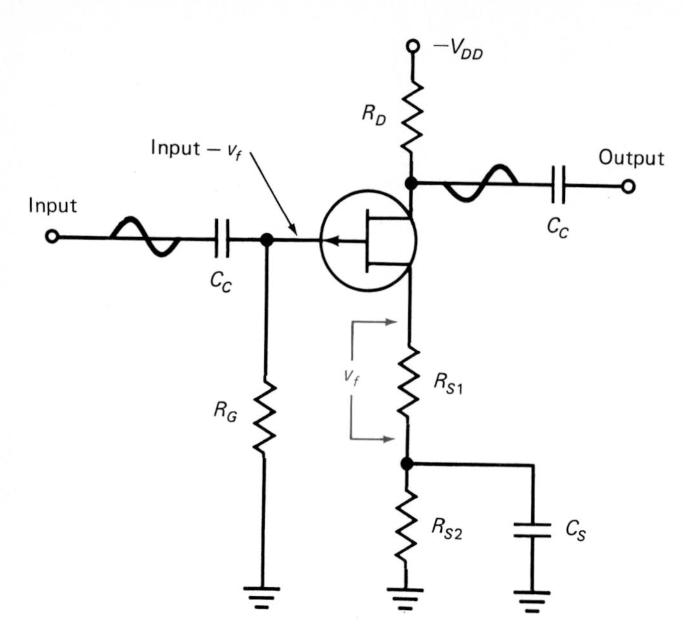

FIGURE 10.18 Using a Partially Bypassed Source Resistor to Obtain Negative Feedback Signal (AC) Feedback (JFET Circuit)

In Figure 10.19 these four forms are presented (again). For each of these forms of amplifiers, the following is the equation for gain:

$$A_{\text{feedback}} = \frac{A}{1 + BA} \qquad (10.16)$$

where A_{feedback} = final gain of circuit with feedback
A = initial gain of amplifier circuit
B = feedback ratio (this is usually some voltage- or current-divider ratio specific to the type of circuit being used for the feedback path)

These forms of feedback (related to types of amplifiers) and the resulting equations can be applied to produce the extended frequency response desired from the amplifier. In later chapters, more extensive applications of these feedback processes will be investigated.

REVIEW PROBLEMS

(1) Prepare a basic list of the changes that may be made to an amplifier to improve the frequency response.
(2) Sketch a basic amplifier's low-frequency response curve if it has only one break frequency at 100 Hz. (Show a slope of −20 dB/decade.) Then, on this same curve, sketch the response curve for the same amplifier if feedback is used which produces a midband gain that is −10 dB below the original gain. Now indicate the approximate frequency at which the new low-frequency breakpoint occurs.
(3) Propose at least two ways in which a basic FET amplifier can have feedback added in order to produce an improved frequency response.

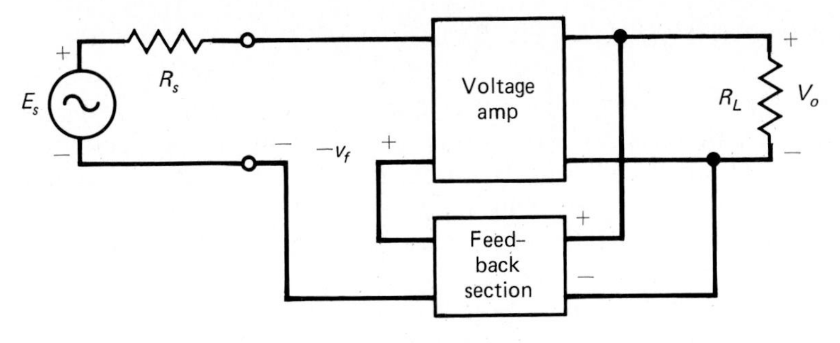

(a) Series–shunt feedback

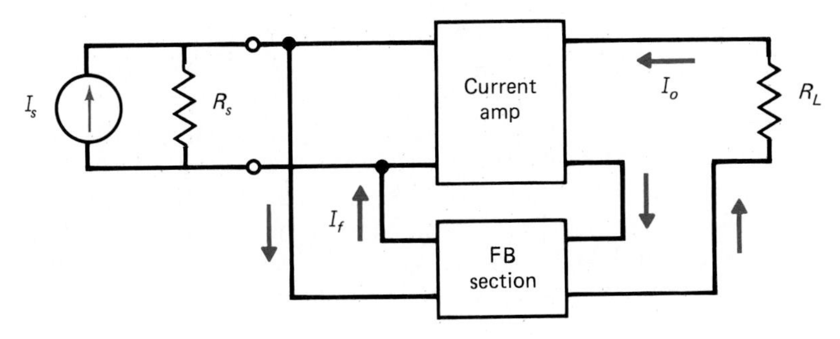

(b) Shunt–series feedback

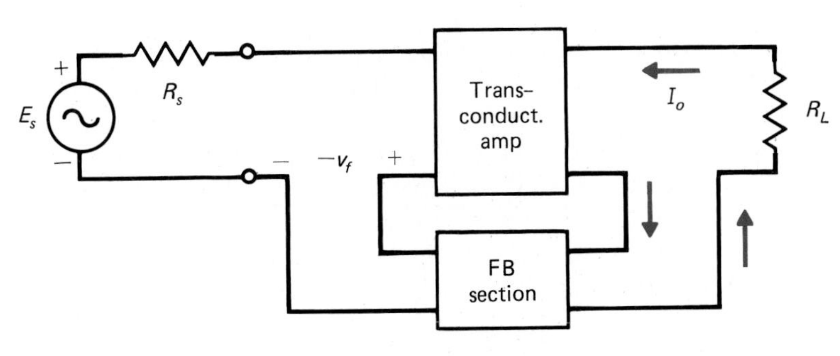

(c) Series–series feedback

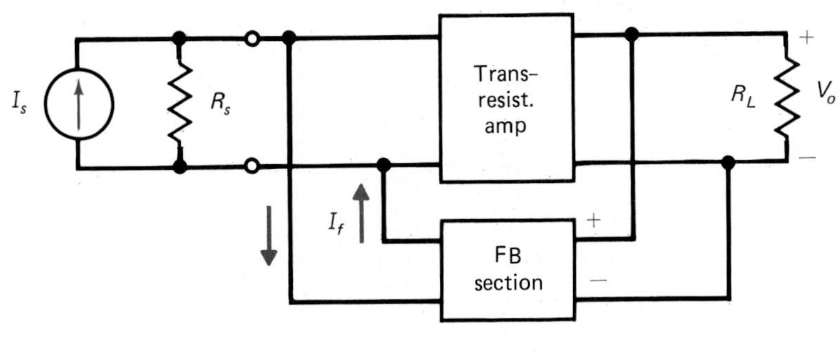

(d) Shunt–shunt feedback

FIGURE 10.19 Four Forms of Feedback Related to the Four Forms of Amplifiers

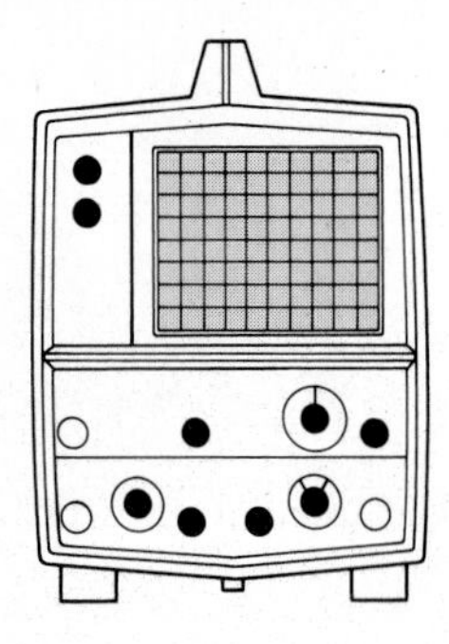

10.5 TESTING AMPLIFIER RESPONSE

In determining amplifier performance, it is quite common to test for its frequency-response characteristics. To do this, a variable-frequency generator is attached to the amplifier. Then an accurate frequency counter (or meter) and voltmeter are attached to the same input terminals as the generator. At the output an ac voltmeter (or dB meter, if available) is attached (see Figure 10.20).

The generator frequency is initially set to some midband level appropriate for the type of amplifier being used, using the frequency counter as an accurate measuring device. (For audio amplifiers 1000 Hz is commonly used as a reference frequency.) The amplitude of the input generator is adjusted so that the voltage level on the output is an easily read "normal" output level (e.g., if the amplifier typically will have an output voltage in the range 0.5 to 2.0 V, select 1.0 V). If a dB meter is used, it would be set to 0 dB or some other appropriate reference value. The amplitude of the input signal is noted.

Then, maintaining the same level of input voltage, the frequency of the generator is turned down. As the low break frequency is reached, the output level will begin to drop. The frequency should be decreased until the output voltage is 70.7% of the output seen at the midband frequencies (or to 3 dB lower than what was seen on the dB meter at midfrequencies). The generator frequency is then observed using the frequency counter. This, then, is the lower break frequency.

The same process is repeated to find the upper break frequency (although, of course, by increasing instead of decreasing the generator frequency). The bandwidth is then illustrated as being between the lower and upper frequency range.

An alternative method of determining this response is by using an audio sweep generator. This generator will automatically change (sweep) its output frequency through a range of frequencies. An oscilloscope, synchronized with the sweep generator, is used to observe the output of the amplifier. A graph of the frequency-response curve will be shown on the oscilloscope. (The actual test conditions are dependent on the type of sweep generator and oscilloscope used, and therefore are not presented here. The operator's manuals for this equipment should be checked to determine the proper test configuration.)

FIGURE 10.20 Measuring Amplifier Frequency Response

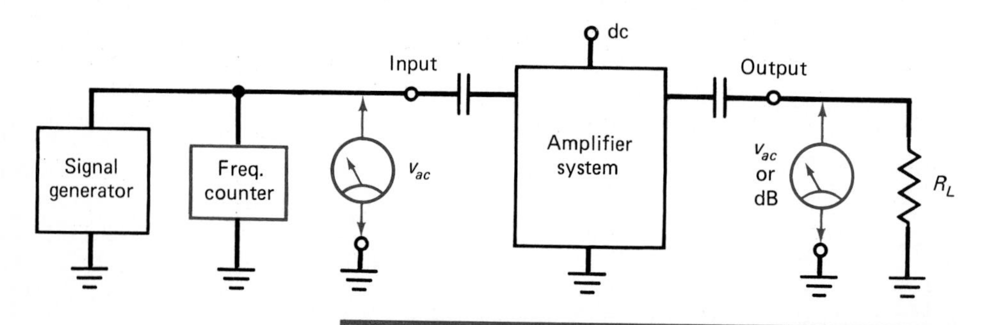

REVIEW PROBLEMS

(1) Write out a clear step-by-step test procedure that can be used to

collect enough data to illustrate a complete plot of the frequency-response curve for an amplifier. Include a basic procedure for determining the two break frequencies (as illustrated in this section), then instructions for completing an accurate set of measurements that span a decade below and above each break frequency.

(2) Assume that a dB meter is used to measure the output of an amplifier while running a frequency-response test. Assume that the amplifier has two separate high-frequency breakpoints. Write out a test procedure that will determine each of these break frequencies approximately. (Use the fact that after the first break frequency, the output will decrease by a factor of −20 dB/decade and that after the second break, the decrease is −40 dB/decade.)

SUMMARY

In this chapter we have explored the topic of frequency response of an amplifier circuit. It was found that certain capacitive elements in the circuit contributed to limiting the frequency response. Input, output, and bypass capacitors created high-pass filter configurations which limited the lower-frequency response of the circuit. Internal capacitances within the active device or (if listed) the GBW of the device coordinated with limits to high-frequency response.

For each reactive (capacitive) term, there was a "break" frequency that identified a −3-dB drop in output level (70.7% voltage loss; 50% power loss). The loss continued to drop at a rate of −20 dB/decade (multiple of 10) in frequency change. If more than one reactive loss was observed, the roll-off was accelerated to −40 dB/decade. Also, a phase shift in the output was seen to coordinate with these breakpoints, contributing up to 90° (or −90°) of phase shift between the output and input.

Corrective action could be taken to extend the response of the amplifier by either selecting different components within the circuit or by applying negative signal feedback. When negative feedback was applied, the total gain of the system was reduced, but the effective bandwidth (low to high frequency range) was extended. The process of adding negative signal feedback could be done within the amplifier circuit (for discrete devices) or could be applied externally as done for op-amp circuitry.

The frequency response of an amplifier could be measured by using a rather simple circuit configuration. In many cases the response of an amplifier is acceptable and no rearrangement of components or application of feedback would be required. For some applications the frequency response of the amplifier may be critical. In these cases, corrective action may have to be taken to produce a more usable response.

In future chapters, the frequency response of an amplifier will continue to be an important characteristic of that amplifier. In some cases, that response will be as significant as the gain of the amplifier.

Special note regarding frequency response: In this chapter specific equations were used to assist in predicting the break frequencies of an amplifier. It is important to note here that these equations mainly provide a basis for predicting the response of the amplifier. In practical applications, the actual break frequencies will often not match those predicted by these calculations.

It is not possible to predict totally every form of capacitance (or inductance) that may exist in a functioning circuit. Therefore, the effect on frequency response that is produced by these capacitances is also quite difficult to predict. The most common error is found when predicting high-frequency breakpoints.

Thus it must be realized that if the low- or (especially) high-frequency breakpoints are considered critical, accurate tests must be conducted on the amplifier to confirm these breakpoints. In fact, conversion of a circuit from a "breadboard" prototype (initial test of a design) to a functioning printed circuit where "stray" capacitance figures may differ could change these breakpoints. It is thus not uncommon to include frequency-response testing in every phase of the development, production, and installation of a system.

It is important to comprehend what factors influence the frequency response of a system. As a general guideline:

1. Any capacitance that is placed (or which occurs) in series with the flow of the information signal will produce a high-pass filter in conjunction with associated resistances. This filter will produce a low-frequency breakpoint in the frequency response of the system.
2. Any capacitance that occurs (or is placed) in a circuit which can bypass the information signal to ground (a parallel configuration) will produce a low-pass filter in conjunction with associated resistances. This filter will produce a low-frequency breakpoint in the frequency response of the system.

KEY EQUATIONS

Decibel gain conversions:

10.1 $A_{P(\text{dB})} = 10 \log A_{P(\text{algebraic})}$

10.2(a) $A_{I(\text{dB})} = 20 \log A_{I(\text{algebraic})}$

10.2(b) $A_{V(\text{dB})} = 20 \log A_{V(\text{algebraic})}$

Amplifier frequency effects:

10.3 $f_{\text{low}} = 1/(2\pi R_{\text{in}}C_{\text{in}})$

10.4 $C_{\text{in(Miller)}} = C_{bc}(A_v + 1)$

10.5 $C_{\text{out(Miller)}} = C_{bc}\,[(A_v + 1)/A_v)]$

10.6 $C_{\text{in(total)}} = C_{\text{in(Miller)}} + C_{\text{(input to ground)}}$

10.7 $C_{\text{out(total)}} = C_{\text{out(Miller)}} + C_{\text{(output to ground)}}$

10.8 $f_{\text{high(input)}} = 1/(2 \times \pi \times C_{\text{in}} \times R_{\text{in}})$

10.9 $f_{\text{high(output)}} = 1/(2 \times \pi \times C_{\text{out}} \times R_{\text{out}})$

Op-amp response:

10.10 $\text{GBW} = A_{V(\text{midband})} \times \text{Bandwidth}$

10.11 $\text{GBW}\ A_{I(\text{midband})} \times \text{Bandwidth}$

10.12 $F_{\text{high}} = \text{GBW}/A_v$

General impedance calculation:

10.13 $\text{Impedance} = (R - jX_C)$

Low frequency phase shift:

10.14 Low-Frequency PHASE SHIFT

$$= \theta = \arctan \frac{X_C}{R}$$

High frequency phase shift:

10.15 High-Frequency PHASE SHIFT

$$= \theta = -90° + \arctan \frac{X_C}{R}$$

General gain-with-feedback equation:

10.16 $A_{\text{feedback}} = \dfrac{A}{1 + BA}$

PROBLEMS

Section 10.1

10.1. Prepare a brief statement about the following limitations noted for linear amplifiers.

(a) High-frequency response
(b) Low input levels
(c) Constant-gain figures
(d) Capacitive effects
(e) Maximum output
(f) Noise (amplifier and power supply)

Section 10.2

10.2. Briefly define the terms "frequency response" and "roll-off."

10.3. For the following algebraic gain values, find the gain in decibels.
(a) $A_p = 6000$ (b) $A_v = 15{,}500$ (c) $A_i = 40$

10.4. How would the dB gain of the amplifiers in problem 10.3 be specified if they were inverting amplifiers (negative gain statistic)?

10.5. Represent the following algebraic gains in decibels.
(a) $A_v = -450$ (b) $A_v = 0.85$ (loss)

10.6. An amplifier's normal voltage gain is $A_v = 480$. Using this as a 0-dB reference, express the following gains for the amplifier under other operating conditions (in decibels).
(a) $A_{v2} = 240$ (b) $A_{v3} = 120$ (c) $A_{v4} = 520$ (d) $A_{v5} = 50$

10.7. Redraw the graph shown in Figure 10.21 using a 0-dB scaling and indicate the half-power points.

10.8. On the graph prepared for problem 10.7, sketch a graph for another amplifier which has half-power points (−3 dB) at 100 Hz and 100 kHz.

Section 10.3

10.9. Illustrate the schematic symbols for a PNP transistor and a P-type JFET. On these symbols illustrate and label the capacitive terms associated with the devices leads.

10.10. If the input resistance of an amplifier is 1.5 kΩ and the input coupling capacitor is 5 μF, at what low frequency will the amplifier begin to lose its response?

10.11. If the output resistance and coupling capacitance of the amplifier in problem 10.10 are identical to those at the input, sketch and label the low-frequency response curve. Use a 0-dB-based curve and label the graph.

10.12. Repeat problem 10.11 for an output resistance of 3 kΩ and a capacitance of 5 μF.

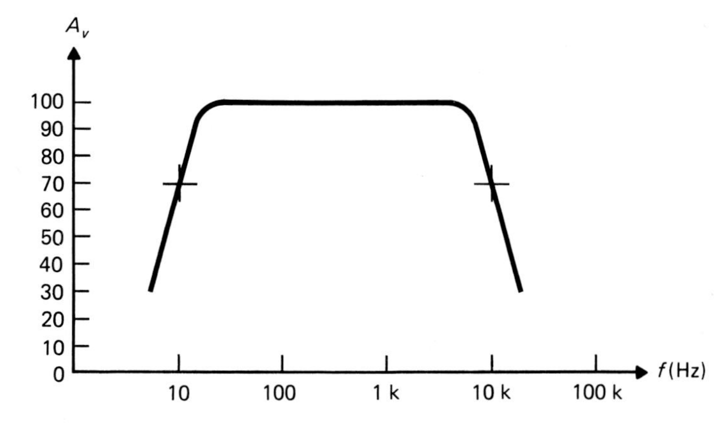

FIGURE 10.21 Typical Plot of Amplifier Frequency Response

10.13. Briefly define what a roll-off of -20 dB/decade means.

10.14. If an amplifier has a voltage gain of $A_v = -35$ and an input base-to-collector capacitance of $C_{bc} = 100$ pF, find the Miller effect input and output capacitance. ($C_{in(Miller)}$ and $C_{out(Miller)}$).

10.15. If the amplifier used in problem 10.14 also has a base-to-emitter (input-to-ground) capacitance and a collector-to-emitter (output-to-ground) capacitance listed below, find the total input and output capacitance of the circuit. (Find $C_{in(total)}$ and $C_{out(total)}$ to ground.)

$$C_{be} = 200 \text{ pF} \qquad C_{ce} = 40 \text{ pF}$$

10.16. If an amplifier has the following calculated capacitance and resistance terms, find the two high-frequency breakpoints due to these terms.

$$C_{in(total)} = 3000 \text{ pF} \qquad R_{in(total)} = 2.7 \text{ k}\Omega$$

$$C_{out(total)} = 2100 \text{ pF} \qquad R_{out(total)} = 1.5 \text{ k}\Omega$$

10.17. If an op-amp with a GBW of 1.6 MHz were used in a circuit with a gain of 240, what would be the maximum frequency that could be expected to the amplifier by the device?

10.18. If the op-amp used in problem 10.17 were required to have a frequency response up to 25 kHz, what would be the highest gain it could be used at while still meeting this response demand?

10.19. Briefly describe the phase-shift problem that occurs when dealing with low-frequency and high-frequency response of an active device.

10.20. Consider the circuit shown in Figure 10.22.
(a) Calculate the Miller input and output capacitance.
(b) Calculate the total input and output capacitance (to ground).
(c) Find the high-frequency breakpoints for the amplifier.
(d) Find the low-frequency breakpoints (due to C_C).

Section 10.4

10.21. For the circuit of Figure 10.22, draw a plot of its frequency response. (Include multiple roll-off effects. Use a 0-dB-based curve.)

10.22. Illustrate the schematic for the following feedback amplifiers.

FIGURE 10.22

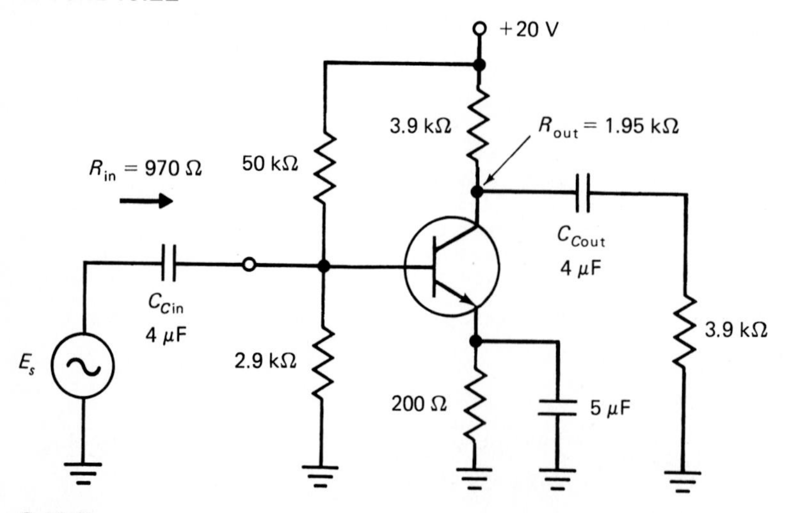

$C_{bc} = 10$ pF
$C_{be} = 45$ pF
$A_v = 104$
$h_{fe} = 80$

(a) Partially bypassed voltage feedback BJT amplifier
(b) Partially bypassed emitter resistor on a beta-independent BJT amplifier
(c) Partially bypassed source resistor in an FET amplifier

10.23. The circuit in Figure 10.22 has the emitter resistor bypassed. If the 5-μF capacitor across the emitter resistor is totally removed, negative signal feedback is produced. The results of this produce the following basic changes:

$$A_V = 19.5 \qquad R_{in} = 2.37 \text{ k}\Omega$$

(Other values remain relatively unchanged.) Calculate the new upper break frequencies for this feedback circuit and compare them to those found in problem 10.20.

10.24. Graph the response of the modified amplifier in problem 10.23 on the graph produced for problem 10.21. The gain of this amplifier is approximately −14.5 dB below the other. Does the intersection of this curve with the original curve occur at approximately the frequencies predicted in problem 10.23?

Section 10.5

10.25. Write a basic test procedure to find the frequency response of an amplifier first without signal feedback, then with signal feedback. You are free to select any amplifier illustrated in this chapter as a test amplifier. Also prepare a chart on which to enter the results of the test.

ANSWERS TO REVIEW PROBLEMS

Section 10.1

1. High-pass filter

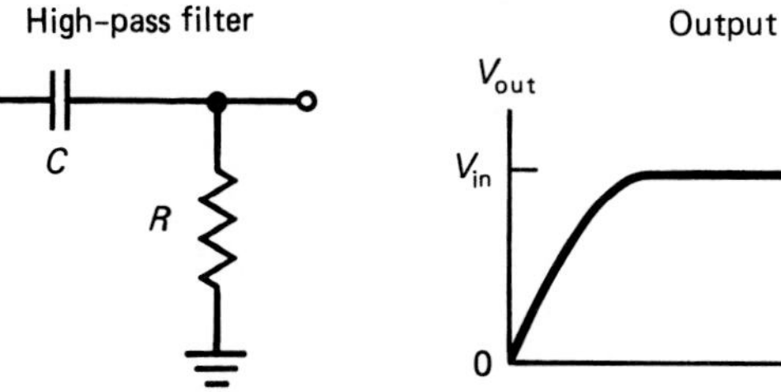

An input capacitor tied to the input resistance of an amplifier will form a high-pass filter. As such, certain low frequencies will not be passed on to the amplifier for amplification.

2. The op-amp closely duplicates the ideal linear amplifier in that it will amplify inputs ranging from dc to potentially several megahertz. Although the ideal does not have an upper-frequency limit, the op-amp most closely (of the more common amplifier forms) duplicates the ideal.
3. Gain may not be sufficient at such a low level; the input level may be so low as not to produce an output; internal noise may interfere with the output; power supply noise may also interfere with the output.

Section 10.2

1. (a) 86.02 dB
   (b) 37 dB
   (c) 70.1 dB (inverted output)
   (d) −0.706 dB (loss)

2. (a) −2.5 dB
   (b) −26 dB
   (c) 4.22 dB
   (d) −38.06 dB
   (e) 0 dB
3.

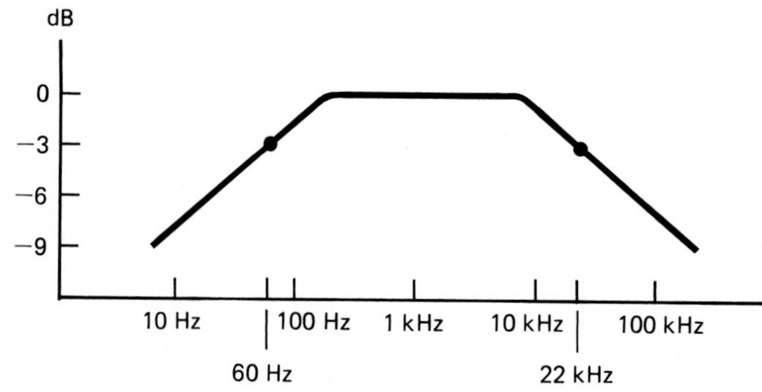

Section 10.3

1. Breakpoints at 7.96 Hz and 15.9 Hz
2. $f_{\text{high(input)}}$ = 686 kHz, $f_{\text{high(output)}}$ = 39.8 MHz
3.

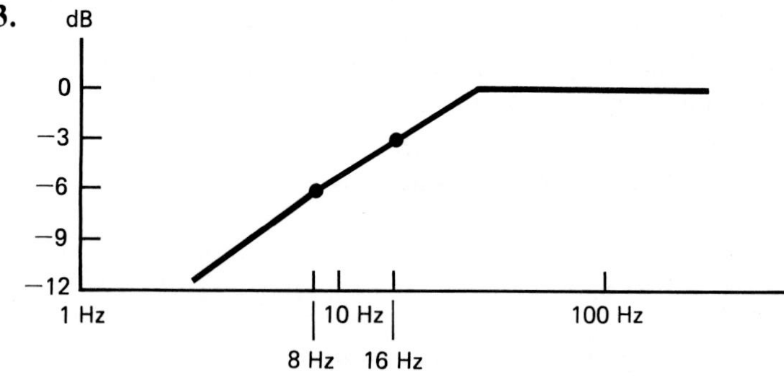

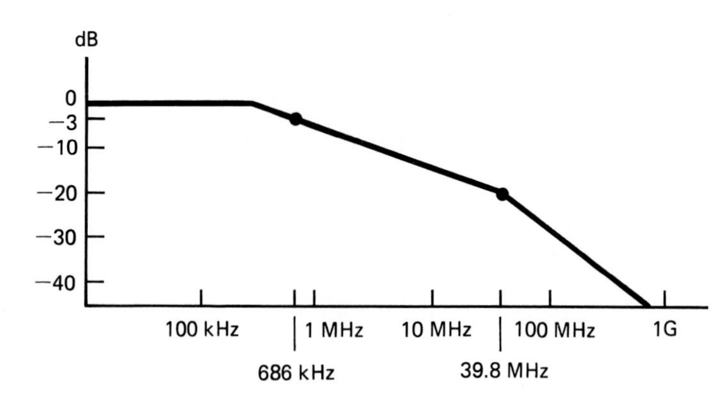

4. 10.7 kHz
5. 303.2 kHz, 37.2° phase shift

Section 10.4

1. Change the physical capacitors used, modify the input and output impedance values, change the device, and/or use negative feedback.

2.

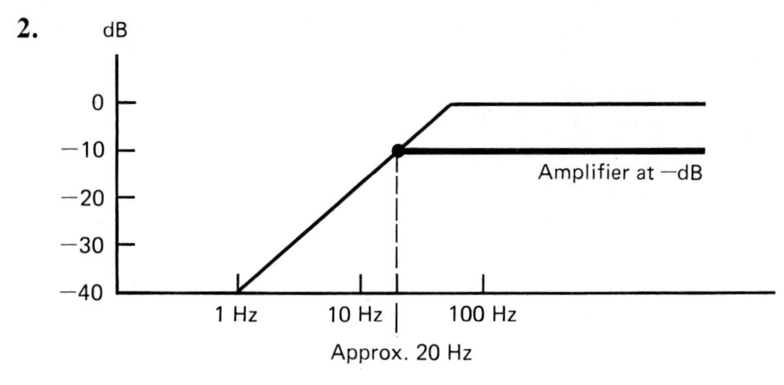

3. By using a partially bypassed source resistor; by adding an output-to-input feedback (resistor) circuit.

Section 10.5

1. The answer is a summary of procedures noted in this section.
2. (1) Establish a 0-dB reference on the dB meter that corresponds to the midband gain. (2) Apply an input at the midband frequency. (3) Increase this frequency until a drop is noted in the dB output. (4) Continue to increase the frequency until −20 dB is noted at the output. This frequency is the first decade above the first break frequency. (Divide it by 10 to determine the first breakpoint.) (5) Continue to increase the input frequency. The output should decrease by −20 dB/decade (or −6 dB/octave) until the second break frequency is encountered. (6) When a roll-off that exceeds the rate in step 5 is observed, the second break has occurred. The generator frequency can be carefully adjusted around this other breakpoint until the −40 dB/decade change point can be estimated. (7) To confirm (or more accurately, find) this breakpoint, note the actual dB loss at this point (example: −32 dB) and increase the generator frequency by a factor of 10. The new dB roll-off should be −40 dB greater than that observed (example: −72 dB). Repeat this estimate/check process to more accurately find this second breakpoint.

11 Frequency-Dependent Amplifiers

Objectives

Upon completing this chapter, the reader should be able to:

- Calculate low- and high-pass filter effects on amplifiers.
- Apply bandpass and band-reject filters to amplifiers.
- Calculate bandpass filter frequency.
- Apply signal feedback.
- Identify and use various oscillator circuits.
- Troubleshoot oscillator circuits.

Introduction

In the previous chapters, simple linear amplification was investigated. It was seen that outside of the limitations placed by the amplifier circuit on frequency response, it could be assumed that all input signals would be treated on an equal basis. If the gain of the system were 100, all inputs within the response of the amplifier would be amplified by a factor of 100.

In this chapter, amplifiers that do not have a linear response to input signals are investigated. Also, oscillators, which use positive feedback and filters to produce an output signal, are discussed. In addition, circuits that perform special tasks are considered. It will be shown that most of these systems use the basic linear amplifier building block. Circuit additions will be made that will place special restrictions on the normal response of the linear amplifier or which utilize that response in a special way. ■

11.1 THE NEED FOR FREQUENCY-DEPENDENT AMPLIFIERS

In earlier chapters the basic processes of amplification were investigated. Circuits were produced that provided gain for any input signal. In Chapter 10 the first form of special conditions related to frequency response were investigated. In this chapter, frequency-dependent amplifiers are investigated. These circuits deal with the input signal in a highly restricted manner to produce a specialized form of output.

In many cases the basic amplifier will remain unchanged. In fact, in many of the applications the op-amp will be used to represent "general" voltage amplification. The major emphasis will be on special circuitry (usually passive elements) designed to manipulate the signal before, during, or after amplification.

It will be found that if a specific circuit that exhibits a special response to frequency is added to an amplifier, the new "system" produced will exhibit amplification *only* with respect to this circuit's function. As a basic primer to this concept, consider the following example:

EXAMPLE 11.1

A basic amplifier is used that has a voltage gain of 10 as shown in Figure 11.1. If a frequency-selective circuit, such as a bandpass filter* which resonates at 10 kHz, were attached to the input of this amplifier, as in Figure 11.2, the circuit (as a system) would have a voltage gain of 10 only at the 10-kHz resonant frequency of the filter. At other frequencies, the output of the filter would be very low in amplitude, and thus the output of the amplifier would also be low.

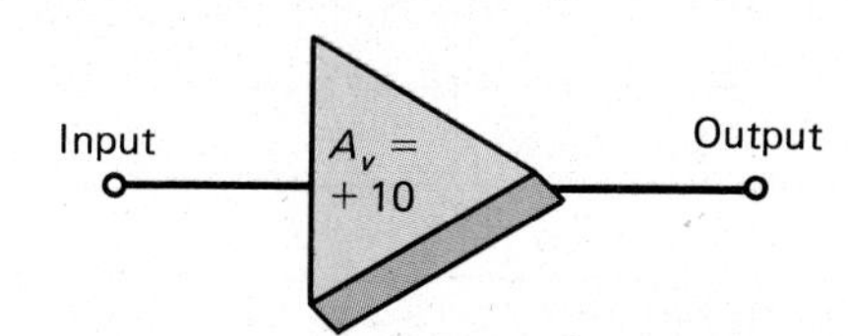

FIGURE 11.1 Typical Voltage Amplifier

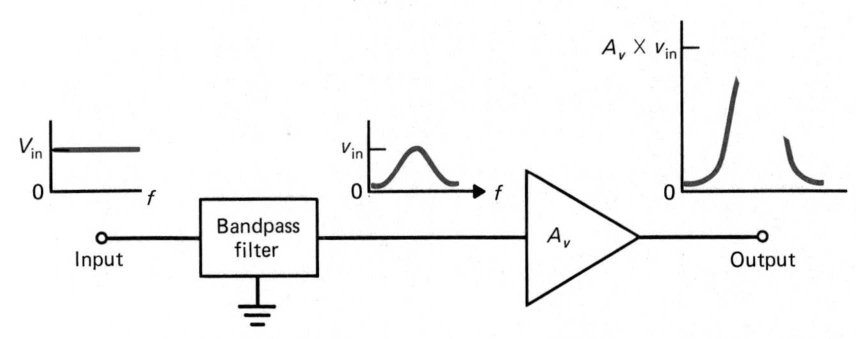

FIGURE 11.2 Bandpass Filter Added to a Typical Voltage Amplifier

* A bandpass filter is a resonant *L/C* circuit which permits only one specific frequency and a limited bandwidth around that frequency to pass through to its output terminals. For a more specific description, refer to a circuit analysis textbook under the general classification "resonant circuits."

In future sections of this chapter, the addition of frequency-dependent circuitry to basic amplifiers will be shown to produce a system that provides amplification only in coordination with the function of the frequency-dependent circuit.

REVIEW PROBLEM

(1) To enhance the evaluation shown in Example 11.1, sketch a possible frequency versus amplitude curve for the bandpass filter used in this example (assuming a 1-V input to the filter). Taking select points on this curve, illustrate the effect of the amplifier on these points by drawing an output curve for the amplifier.

11.2 CATEGORIES OF FREQUENCY-DEPENDENT AMPLIFIERS

This chapter will deal with four categories of frequency-dependent amplifiers. These categories are (1) frequency-selective amplifiers (2) resonant circuit amplifiers (3) signal feedback amplifiers, and (4) oscillators (positive-feedback amplifiers). Figure 11.3 should be studied to see the basic differences between these forms of amplifier systems. Notice that each system has a specific function and a unique means for achieving that function. In many cases, the same form of amplifier could be used to achieve any of these tasks. It should also

FIGURE 11.3 Special-Purpose Amplifiers

Name	Function	Block Diagram	Typical Usage
Frequency-selective amplifier	To amplify only a band of frequencies	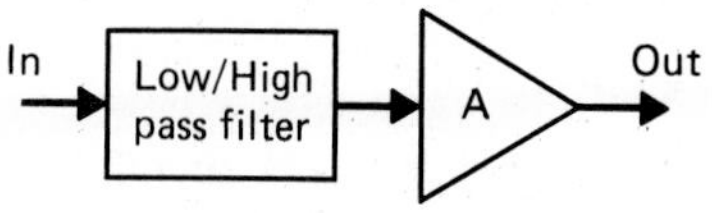 In → Low/High pass filter → A → Out	Audio-frequency amplifier
Resonant circuit amplifier	To amplify a very limited band of frequencies	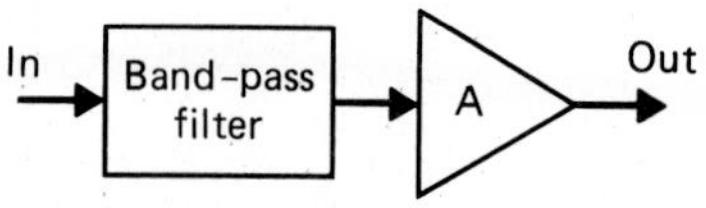 In → Band-pass filter → A → Out	Audio graphic equalizer
Signal feedback amplifier	Refined method of other "tuned" circuit amplifiers	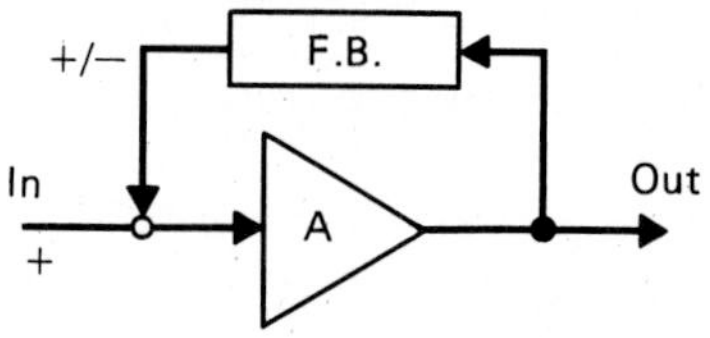 In, +; +/− F.B.; A; Out	
Oscillator	Self generating signal	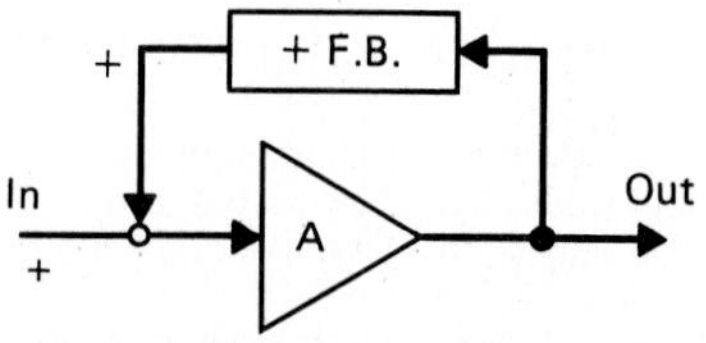 In, +; + + F.B.; A; Out	Used to generate frequencies—signal generator

be noted that the added circuitry that achieves the special function is in addition to the amplifier circuit.

> **REVIEW PROBLEMS**
>
> (1) From memory, list the four forms of frequency-dependent amplifiers and a brief description of their output.
>
> (2) To be prepared for the following sections, look up the schematic and important equations for the following circuits (refer to a circuit analysis text).
>
> **(a)** Low-pass filter
>
> **(b)** High-pass filter
>
> **(c)** Bandpass and bandstop (band-reject) filters

11.3 FREQUENCY-SELECTIVE AMPLIFIERS

In Chapter 10 it was shown that all amplifiers display some limit to frequency response. Low-frequency limits are due to input and output capacitors. High-frequency limits are caused by device limitations (usually). Therefore, it is not possible to assume that an amplifier will be capable of responding to frequencies from dc to hundreds of gigahertz. In this section we deal with the deliberate addition of circuit elements that are chosen to further restrict an amplifier's frequency response. It will be assumed that the amplifiers used in the circuits in the balance of this chapter will initially be capable of responding to a wider band of frequencies than those which are selected by the special circuits added to them. Thus the capacity of the amplifier exceeds the capacity of the special circuit attached to it.

The following terminology will be used in this chapter as special-purpose amplifiers are presented:

Audio frequencies: frequencies (approximately) below 30 kHz
Radio frequencies: frequencies above 30 kHz
Frequency spectrum: a range of frequencies (as in "audio spectrum")

$f_{co} = \frac{1}{2\pi RC}$

$f_{co} = \frac{R}{2\pi L}$

Within the radio-frequency spectrum (often called simply RF), which is quite extensive, there are specific divisions related to specific applications. These will be discussed later in future courses dealing with communication electronics. In this book we simply generalize when referring to this range of frequencies.

Active devices are often classified as to their ability to be used in audio- or radio-frequency circuits. Most active devices can amplify low (down to dc) frequencies. Their ability to maintain gain at high frequencies, though, is often a special rating for the device. This rating relates to their internal capacitance (see Chapter 10) and/or GBW characteristics. Thus the first special restriction to frequency may simply be based on the correct selection of the device to be used in the amplifier. Assuming that the correct device is selected and is properly biased (or connected to a correct power supply), additional circuitry may be added to modify the frequency response of this amplifier.

The use of high-pass or low-pass filters at the input terminals of the amplifier will affect its frequency-handling capacity. It is often desirable to limit the frequency response of an amplifier to a range that coordinates with its intended application. Usually, the reason for applying such restrictions is to prevent unwanted signals from being amplified along with the desirable signal.

For example, in an audio amplifier, it may be necessary to apply a low-pass filter to prevent the amplification of high, nonaudio frequencies that could interfere with the desired signal. Another common application uses a high-pass filter, set approximately to 100 Hz, to eliminate the 60-Hz noise produced by the ordinary electrical power lines.

Either high-pass or low-pass filters can be applied to any form of amplifier. Usually, these filters are made up of an inductor or a capacitor connected either in series or parallel (depending on the desired filter configuration) with the input resistance of the amplifier. The following is a list of the basic configurations of these filters, their response curves, and the equations used to initially determine the break frequency.

1. High-pass amplifier using a capacitive filter (Figure 11.4)

$$f_{\text{low}} = \frac{1}{2\pi R \times C} \tag{11.1}$$

2. High-pass amplifier using an inductive filter (Figure 11.5)

$$f_{\text{low}} = \frac{\text{R}}{2\pi L} \tag{11.2}$$

3. Low-pass amplifier using a capacitive filter (Figure 11.6)

$$f_{\text{high}} = \frac{1}{2\pi R \times C} \tag{11.3}$$

4. Low-pass amplifier using an inductive filter (Figure 11.7)

$$f_{\text{high}} = \frac{R}{2\pi L} \tag{11.4}$$

Therefore, there are four basic configurations that will allow for the production of a low-pass or high-pass amplifier system. (It is possible to place these filters at the output of the circuit and see exactly the same response. It is customary, though, to place these at the input; thus they handle a lower-level signal prior to amplification.)

EXAMPLE 11.2

An audio amplifier uses an op-amp with a low-frequency response down to dc. It is found that too much "line noise" at 60 Hz is being amplified and is interfering with the normal input signal, causing distortion of the output. Establish a filter at the input to this circuit to reduce this problem. The amplifier specifications are

$$A_v = 50 \qquad \text{input } R = 5\ \text{k}\Omega \qquad \text{output } R = 10\ \Omega$$

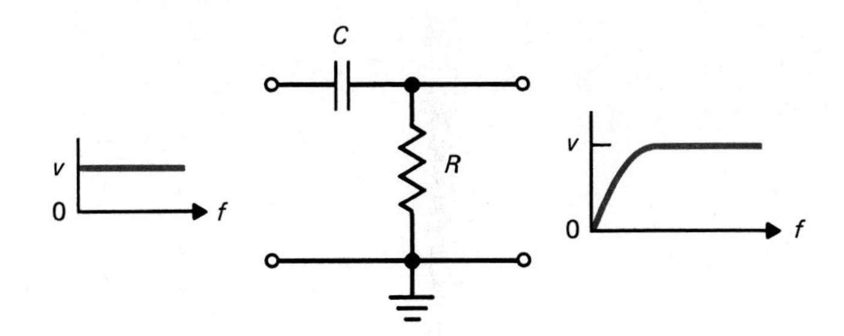

FIGURE 11.4 *R-C* High-Pass Filter and Frequency-Response Curve

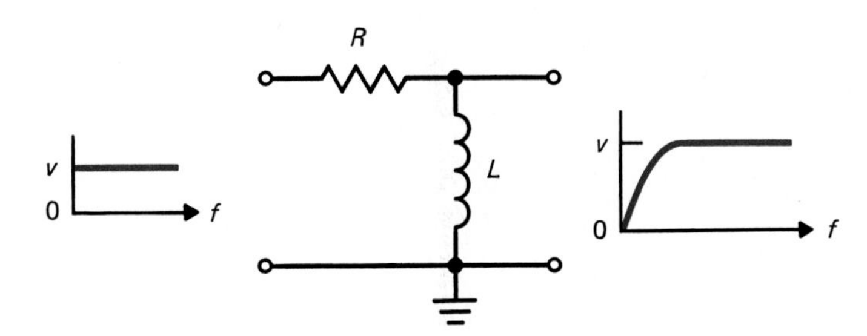

FIGURE 11.5 *R-L* High-Pass Filter and Frequency-Response Curve

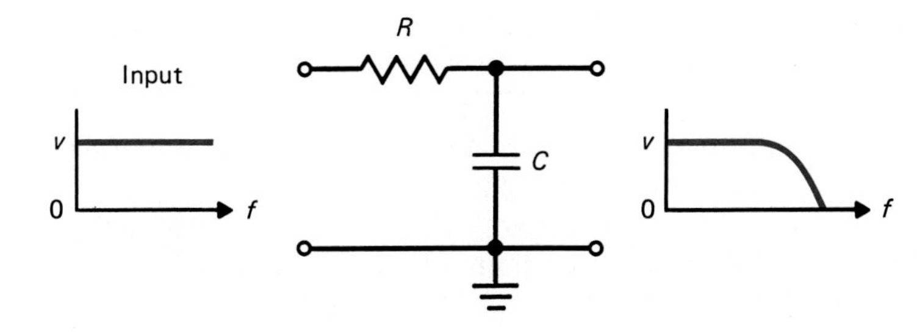

FIGURE 11.6 *R-C* Low-Pass Filter and Frequency-Response Curve

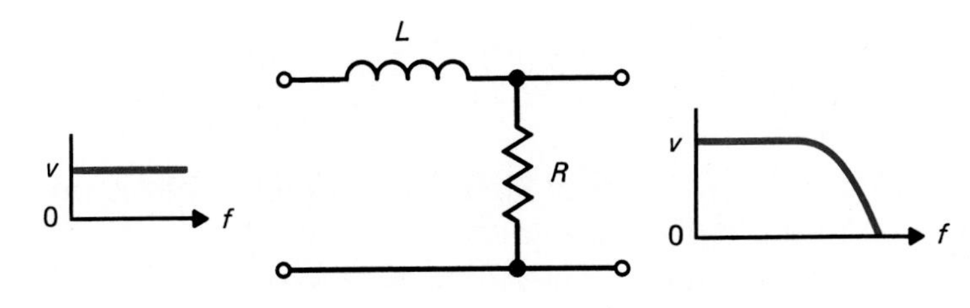

FIGURE 11.7 *R-L* Low-Pass Filter and Frequency-Response Curve

Solution:

A high-pass filter must be added to the input with a pass frequency of 100 Hz (thus assuming a blocking of 60 Hz).

Capacitive filter. Rearranging equation 11.1,

$$C = \frac{1}{R_{\text{in}} \times 2\pi f}$$

gives a value of

$$C = \frac{1}{5\ \text{k}\Omega \times 2\pi \times 100} \simeq 0.637\ \mu\text{F}$$

Inductive filter. Rearranging equation 11.2

$$L = \frac{R}{2\pi \times f}$$

gives a value of

$$L = \frac{5\ \text{k}\Omega}{2\pi \times 100} \simeq 7.96\ \text{H}$$

(Because of the physical size of a 7.96-H inductor, the capacitor version would probably be chosen.)

Argument against the use of inductive filtering: Since the inductor's reactance is zero at dc, it is often undesirable to use an inductor in a simple input or output filtering configuration. The potential of either grounding a dc bias level or of passing one circuit's dc bias on to another amplifier is usually undesirable.

A common application of low-pass or high-pass filters in audio amplifiers is seen in the bass and treble controls of an amplifier. The bass (low-frequency) control is, in many amplifiers, a potentiometer which is a part of a simple high-pass filter circuit. The treble (high-frequency) control is also a potentiometer which is a part of a low-pass filter circuit. By adjusting either of these controls the break frequency of the filter is adjusted either higher or lower. As such, the output of the attached amplifier will follow the low- or high-pass characteristics of these filters, thus producing an output that has a modified frequency response (see Figure 11.8).

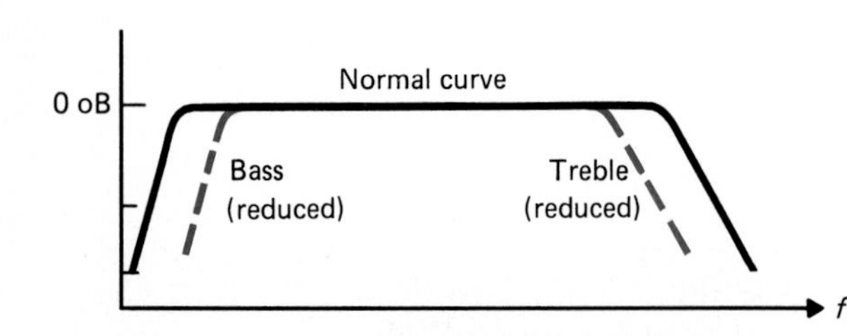

FIGURE 11.8 Effect of Bass and Treble Controls on an Amplifier's Frequency Response

Special Low-Pass Filter Application

When dealing with high-frequency (RF) signals, it is often necessary to prevent these signals from being present on the power supply (dc supply) lines connected to amplifiers. Therefore, it is a common practice to apply special filters to these supply lines, from each amplifier to the supply line. The most common practice is to use an inductive low-pass filter in series with the dc power supplied to the individual circuits. This inductive filter, often called an RF choke (RFC in schematics), is selected to block high-frequency RF signals. Due to its position in the circuit (see Figure 11.9), it does not form an identifiable filter (*L*/*R* combination) since the resistive term is hidden within the amplifier and power supply. It does provide a high-resistance term, although at the high RF frequencies. Such a filter is not usually required in audio-frequency circuits for two reasons. First, the special propagation (flow) effects of RF signals are not the same as for audio signals. Second, the dc power supply is usually capable (due to its structure) of producing efficient low-pass filtering for audio-frequency signals.

Typical RF choke (inductor) values are in the milli- to micro-henry range of inductance. They are often selected based on the RF frequency used in the circuit rather than by specific equations. It is not uncommon to add RF choke (RFC) elements to a circuit once it has been constructed and tested, as often RF "noise" problems surface after the original design has been prepared.

Low-pass and high-pass filtering can be achieved through the use of simple *R*/*C* or *R*/*L* circuitry, typically using the input (or output) resistance of the amplifier as the resistive term in equations. The capacitive form of filters is usually most popular, except for the special application of RFC power supply filtering. In Chapter 10, basic frequency restrictions (filtering effects) were discussed; in this section additional filtering is seen as being usable to specially tailor the response of an amplifier.

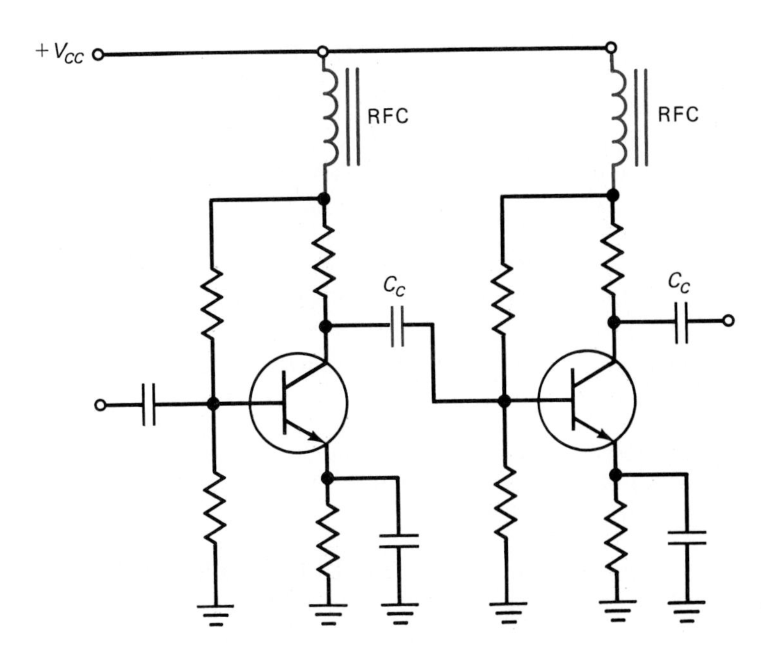

FIGURE 11.9 Use of an RFC [Radio-Frequency Choke (Inductor)]

REVIEW PROBLEMS

(1) State the general equations for filter response for the four forms of filters used in this section.

(2) Illustrate a standard op-amp amplifier which has an added high-pass capacitive filter at the input and uses RFC filters on the power supply lines.

(3) A typical application of a low-pass circuit is in a simple audio tone control circuit. For the tone control shown in Figure 11.10, write a brief summary of how this circuit will affect the output frequency response of the amplifier. Note that the potentiometer forms a simple divider circuit with the input resistance of the second amplifier.

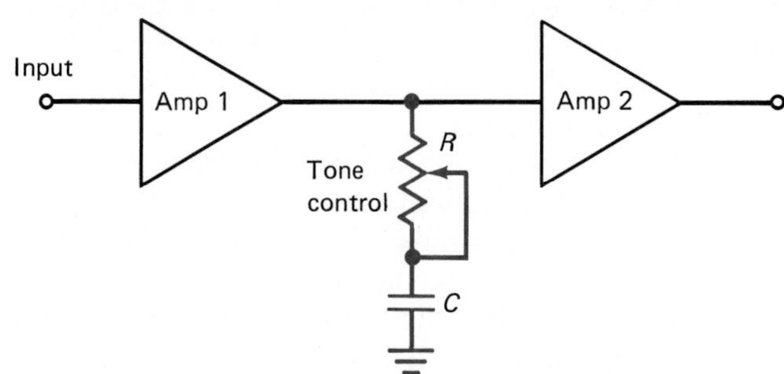

FIGURE 11.10 Tone Control Placed between Amplifier Stages

For comparison, consider the potentiometer at zero ohms, at a value equal to R_{in} of the amplifier, and then at a value that is $10 \times R_{in}$. Sketches of the frequency response of the system under these various conditions will probably assist in presenting the summary.

11.4 RESONANT CIRCUIT AMPLIFIERS

In much the same way that low-pass and high-pass filters are applied to an amplifier circuit, bandpass and bandstop (band-reject) filters can also be used. The purpose of using these filters is to restrict the amplifier's frequency response to a range limited by these filters. For example, a radio receiver uses a collection of basic amplifiers. Without the addition of resonant circuits it would receive and (potentially) amplify the signal of all available radio stations at the same time. The mixture of music programs, news, and so on, would be totally unusable. To select a specific station from all the potential stations, resonant circuits are used.

Note: In the following presentations it is assumed that the amplifier being used has a full-spectrum frequency response. Thus the amplifier will not present any restrictions to the frequencies that it will amplify.

Either series or parallel resonant circuits can be applied to the resonant circuit amplifier. The following is a summary of the formulas used for resonant circuit calculations.

Bandpass Filters

Series Resonant Circuit For the series resonant circuit shown in Figure 11.11, the following equations apply:

$$f_r = \frac{1}{2\pi\sqrt{L \times C}} \tag{11.5}$$

$$\text{BW (bandwidth)} = \frac{f_r}{Q} \tag{11.6}$$

$$Q = \frac{X_L}{R' + R_l} \tag{11.7}$$

where R' is the total resistance seen by the filter [typically, this is $R_{source} + R_{in(amplifier)}$] and R_l is the inductor's dc resistance.

Parallel Resonant Circuit For the parallel resonant circuit shown in Figure 11.12, the following equations apply:

$$f_r = \frac{1}{2\pi\sqrt{L \times C}} \tag{11.5}$$

$$\text{BW} = \frac{f_r}{\text{Q}} \tag{11.6}$$

$$Q = \frac{X_L}{R_l + (X_L^2/R_P')} \tag{11.8}$$

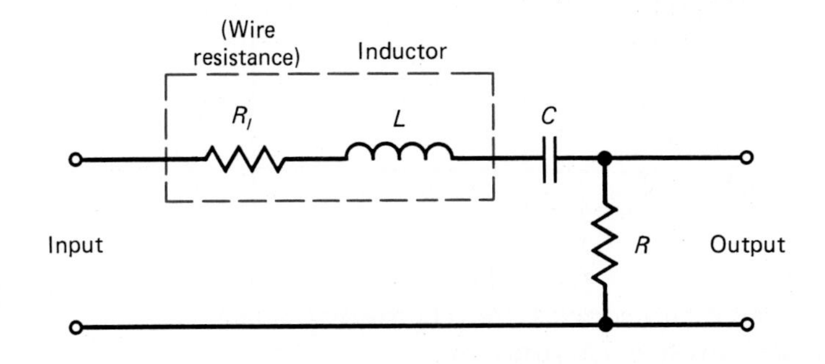

FIGURE 11.11 Series Resonant Circuit (LRC)

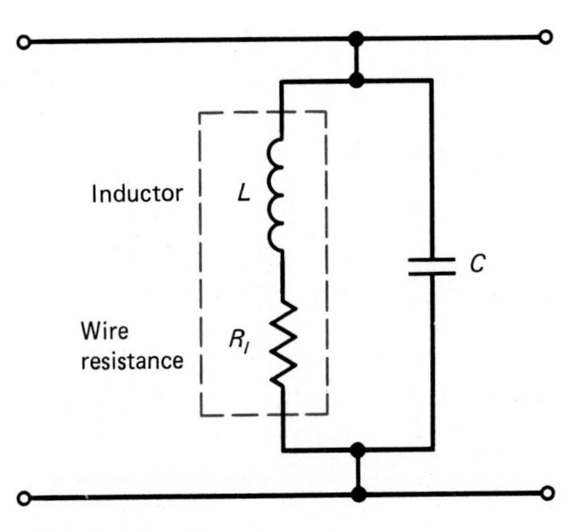

FIGURE 11.12 Parallel Resonant Circuit (LRC)

where R'_P is the parallel resistance seen by the filter [typically, this is $R_{\text{source}} \parallel R_{\text{in(amplifier)}}$] and R_l is the inductor's dc resistance.

Note: Equations 11.5 and 11.8 are valid only when R_P and the quality factor Q of the inductor are large, which is typical for these applications.

It should be noted that it is important to include the amplifier's input resistances and the resistance of any sources supplying the filter, as they will affect the resonant circuit conditions. (These are the brief form of the equations commonly used to estimate the resonant frequency of each of the circuit types.) The input to the amplifier and therefore the output of the amplifier will take on the characteristics of the resonant circuit. Thus the amplifier can be called a "resonant (or tuned circuit) amplifier."

EXAMPLE 11.3

For the "tuned circuit amplifier" shown in Figure 11.13, calculate the resonant frequency, the bandwidth, and Q. Sketch a frequency versus voltage graph for the output, assuming a full-spectrum input signal.

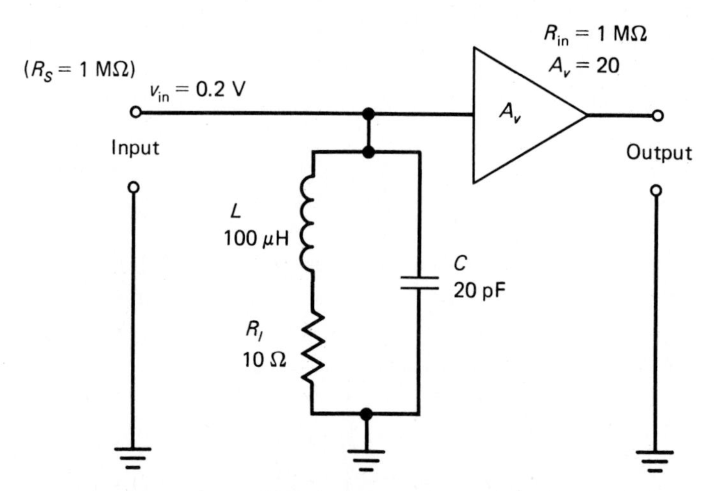

FIGURE 11.13 Use of a Resonant Circuit (Bandpass Filter) in Coordination with an Amplifier

Solution:

$$R' = R_S \parallel R_{\text{in(amplifier)}} = 500\ \text{k}\Omega$$

$$f_r \simeq 3.56\ \text{MHz}$$

$$X_L = 2\pi f_r \times L \simeq 2.24\ \text{k}\Omega$$

$$Q \simeq 111.8$$

$$\text{BW} = \frac{f_r}{Q} \simeq 31.83\ \text{kHz}$$

With an amplifier gain of 20 the maximum output will be

$$0.2 \text{ V (in)} \times 20 = 4 \text{ V}$$

which will occur only at f_r, as shown in Figure 11.14(a).

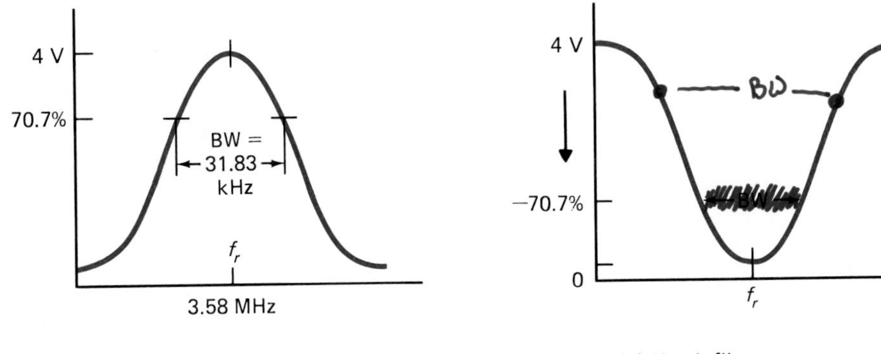

(a) Bandpass filter (b) Notch filter response

FIGURE 11.14 Response Curves

From Example 11.3 it can be seen that by the addition of a parallel resonant circuit to a basic amplifier, the overall system response will follow that of the resonant circuit. Similar to low-pass and high-pass circuits, the active portion of the circuit (amplifier) has an effect on the system's response. Generally, the following considerations are important:

1. The amplifier's input resistance will influence the resonant circuit. If this resistance is low, the Q of the resonant circuit may be lowered substantially. A general rule to be used when setting up such a circuit is to have

$$R_P > X_L^2 \tag{11.9}$$

2. The high-frequency response of the active circuit must exceed the resonant frequency of the filter. Usually, it is common to have the upper break frequency of the amplifier be twice (or greater) the resonant frequency of the resonant circuit.
3. The input capacitance of the amplifier circuit must also be taken into account. Since it is in parallel with the parallel resonant capacitor, and in series with that of the series resonant circuit, it must be a value quite different from the other capacitors (either lower or higher) so as not to affect the resonant circuit's resonant frequency.

Other Resonant Elements It will be found in later communication electronics courses that there are other resonant elements that may be used instead of the common $L/R/C$ resonant circuit. Crystals, ceramic filters, and mechanical filters may be used to produce a resonant-circuit-like effect. Whatever the means, though, the output of the resonant amplifier system will take on the characteristics of the resonant circuit attached.

Although bandstop filters were not specifically detailed in this section, the same form of equations apply to such a circuit. If a bandstop (notch) filter [Figure 11.14(b)] is used, the output of the amplifier will also reflect the characteristics of that resonant filter.

Testing Resonant Circuit Amplifiers

Bandpass and bandstop resonant circuit amplifiers are tested in much the same way as testing is done on frequency-selective amplifiers. (See the information on testing for frequency response in Chapter 10.) Care must be taken, though, so that the test equipment does not introduce stray capacitance into the resonant circuit, thus causing an error in the measurement for resonant frequency. The amplifier section's dc bias levels can simply be checked, ignoring the resonant circuit, to determine if that portion of the system is functional.

The application of bandpass and bandstop filters to amplifier circuitry results in a frequency-selective amplifier system. The filter serves as a block to undesirable frequencies, passing only a select range on to the actual amplifier circuitry. Therefore, the total system (amplifier + filter) has a usable gain at only those specific frequencies.

REVIEW PROBLEMS

(1) Sketch a BJT (beta-independent) circuit, a simple FET amplifier, and a noninverting op-amp amplifier. To these add a sketch of a typical parallel (L/C) resonant circuit. (Values need not be shown.)

(2) Sketch an inverting op-amp circuit that is fed by a parallel circuit bandstop filter. (See a circuit analysis textbook for assistance in the configuration of the filter.)

(3) For the MC34071 op-amp listed in Appendix A, look up all of the circuit specifications that may have an effect on using it with an attached parallel resonant circuit.

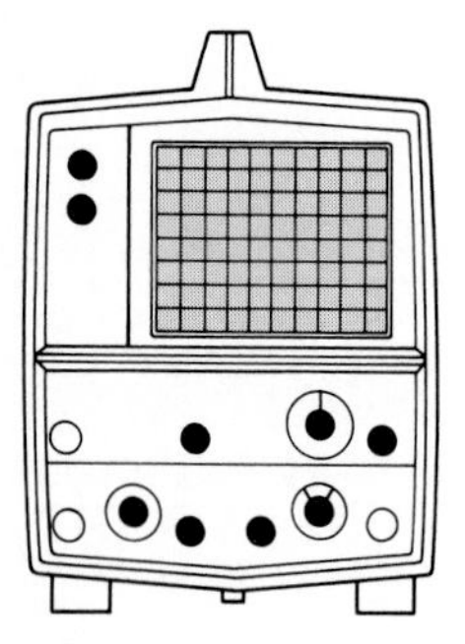

11.5 SIGNAL FEEDBACK AMPLIFIERS (AN EXPANDED COVERAGE)

In previous chapters, the basic process of applying negative feedback to an amplifier system has been detailed. The use of dc feedback to produce stability and ac feedback to improve frequency response have been illustrated. There are further advantages to using negative feedback (sometimes called *degenerative feedback*). Following is a basic list of these advantages.

1. Using dc feedback, greater stability is achieved.
2. Using signal feedback, the bandwidth of the amplifier is extended beyond that of a nonfeedback system.
3. Factors within the system that produce nonlinear distortion can be overcome by using feedback.

4. Noise that is produced within the amplifier circuit can be dramatically reduced by the use of negative feedback.

Items 1 and 2 were discussed previously; following is an expanded discussion of items 3 and 4.

Improvement in Nonlinear Distortion Figures

In many amplifier circuits, the active device does not produce a constant gain for all input/output levels. This was most dramatically noted when the FET was presented; the nonlinear transfer curve illustrated that gain was not constant for all values of gate voltage. Such a problem (in any amplifier) will produce a distortion of the input signal at the output terminals.

Figure 11.15 illustrates the effect of the nonlinear distortion produced by an amplifier. It can be seen that the shape of the output curve is not identical to that of the input curve. This is due to the amplifier not having a constant gain for each value of input. Negative feedback can be used to help correct for this form of distortion. In previous chapters, negative feedback was shown to provide a leveling effect for "long-term" drift in the amplifier's gain. That is, for example, if the gain were to change gradually due to heating, the feedback would produce a corrective action (adjustment of the "true input") to compensate for the change. This compensation can occur on an instantaneous basis as well as a long-term basis.

The negative-feedback correction of the true input will happen on a point-by-point basis along the time axis of the input signal. What this means is that if the output signal is not an exact point-by-point duplication of the input, this "error" is reflected in the feedback signal. Since the feedback is subtracted from the input to produce the true input, these errors are corrected by an adjustment of the true input on a point-by-point basis.

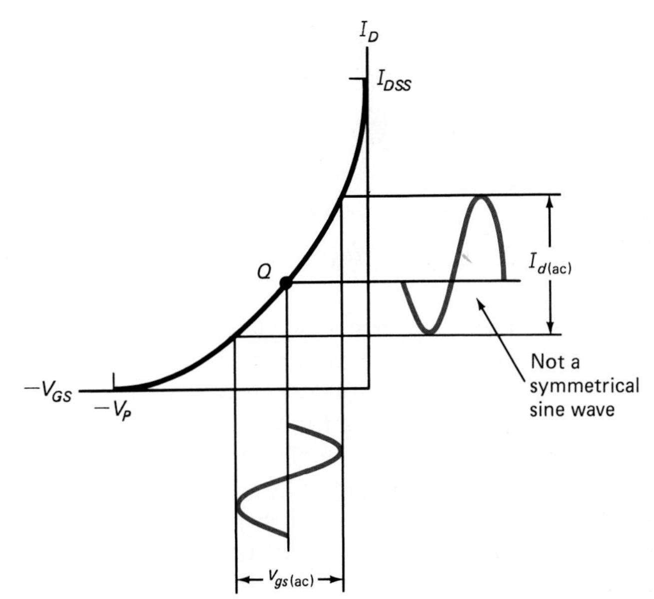

FIGURE 11.15 Nonlinear Distortion Demonstrated on a JFET Transfer Curve

Thus an amplifier system that is producing a nonlinear distortion of the input can be corrected by the application of feedback. The greater the level of feedback, the higher the compensation, and thus the gain of the system will be more linear.

Noise Reduction

All devices (especially active devices) will produce some form of noise. Noise is generally defined as the production of any signal within a system that is not present in the input to the system. This noise is usually a by-product of the flow of current through the elements in the circuit.

Figure 11.16 illustrates the production of noise within an amplifier system. The input is a pure sine-wave signal, but the output contains an additional signal (noise) which is superimposed on the original signal. Since all devices contribute some amount of noise, it is possible that the noise level is so low that it is acceptable. In such a case, no corrective action is taken. [For example, if the noise level in an audio amplifier is -20 dB in relationship to the actual signal (often called a signal-to-noise ratio), the level of noise at the output may be so low that it cannot be detected by the human ear under normal listening conditions.]

If the noise level is significant, feedback can be used to correct this

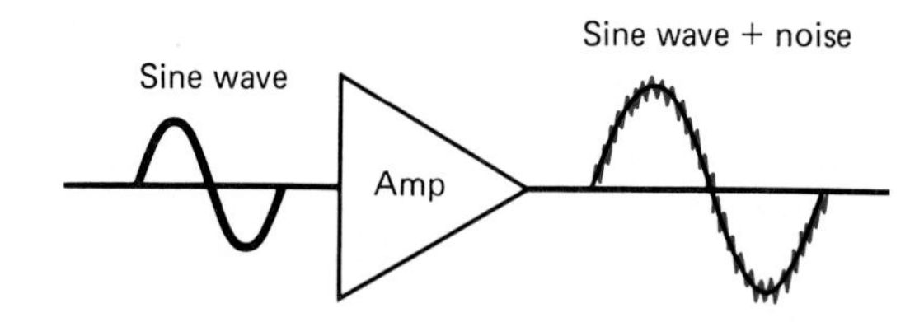

FIGURE 11.16 Noise Occurring at the Output of an Amplifier

FIGURE 11.16b The distortion analyzer uses a highly accurate acitve notch (band-reject) filter to assist in measuring the level of distortion (including noise) produced in an amplifier system. Essentially it uses the active filter to eliminate one frequency from the output of the amplifier (identical to the input frequency). It then measures the level of "other" output signals, thus measuring the noise and distortion level produced within the amplifier system. Photo courtesy of Hewlett-Packard Company.

problem. Since the noise is produced within the amplifying system, it is a signal that is present at the output that was not present at the input. If this is fed back to the input (negative feedback, of course) it is applied directly to the input of the system as a subtractive value. This value then produces a quick cancellation of (most of) the noise produced.

It should be noted that it is not possible to cancel 100% of the noise produced in any amplifier system. Application of enough feedback initially to cancel the noise would produce an amplifier with an extremely low gain. Also, because of the complexity of most noise "signals," the feedback circuit may not be able to reproduce 100% of the noise "signal" at the input. Therefore, total cancellation cannot be achieved. Finally, it is possible that the application of feedback itself may contribute an additional noise factor that could not be canceled.

Special note: Most active devices have included in their specifications a value for noise production. This is often listed as a dB figure or as some other form of ratio. If noise is critical for a certain system, the selection of "low noise" active devices may be desirable, thus reducing the initial production of noise.

At this point it is important to note that the reduction of noise or distortion by the use of negative feedback is effective *only* when the distortion or noise occurs *within* the amplifier system. If noise or distortion is present at the input of the system, it will be amplified as any other form of input would be amplified. In every case, the source of noise or distortion must be isolated and cancellation (if possible) must be done at this source, as it is nearly impossible to eliminate it later without (potentially) affecting some portion of the desirable input as well.

Using Reactive Devices in the Negative-Feedback Path

So far, feedback has been applied by the use of resistors. A constant proportion of feedback to input is thus maintained, regardless of the frequency of the signal. If a frequency-dependent element is used in the feedback path, the feedback effect is frequency dependent. Using capacitors and inductors in the feedback path will produce amplifiers which have gain that are dependent on frequency. As an illustration of this, consider the following example.

EXAMPLE 11.4

The amplifier shown in Figure 11.17 uses a capacitor instead of a resistor in the feedback path. Calculate the system gain for various frequencies.

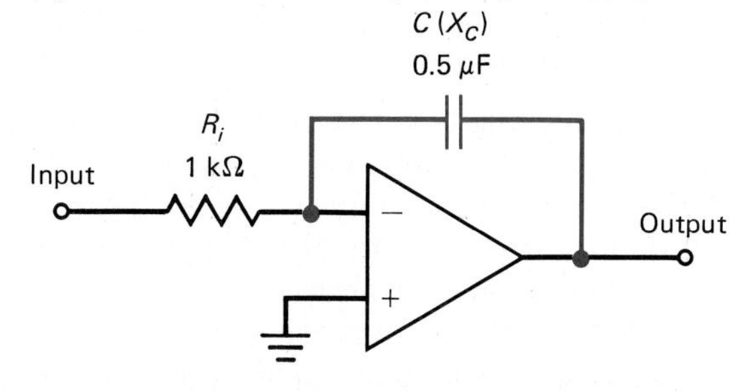

FIGURE 11.17 Op-Amp Integrator

Solution:

The normal gain equation is

$$A_V = \frac{-R_F}{R_I}$$

but this is modified as

$$A_V = \frac{-X_C}{R_I}$$

Therefore, as X_C changes due to frequency, the following values for gain will be produced:

Frequency (Hz)	X_C	A_V
10	31.83 kΩ	−31.83
50	6.37 kΩ	−6.37
100	3.18 kΩ	−3.18
500	637 Ω	−0.637 (loss)

There are two major points that can be shown to contrast this circuit with a simple low-pass filter circuit attached to the input of the device. First, the amount of change in the output due to the reactive element is dramatically increased when using the filter section to change the gain of the amplifier. The roll-off for the circuit is more rapid. The second feature is even more dramatic. By using this capacitor in the feedback path, the actual circuit response is identical to that of an inductive, not a capacitive, filter. This surprising result is due to two functions.

1. Since the amplifier is used in its inverting mode, a −180° phase shift is seen through the amplifier. When this −180° shift is added to the −90° shift produced by the capacitor, a total phase shift of −270° is seen. Since a −270° shift is equal to a +90° shift, the phase relationship is that of an inductor, not a capacitor. (If review of this phase-shift principle is needed, the reader is advised to refer to a textbook dealing with ac circuit analysis.)
2. In the basic configuration used (ignoring the amplifier inversion) the capacitor appears to be in a high-pass filter configuration. But due to the manipulation of the feedback term and its effect on gain, the circuit actually produces a low-pass form of output. Again, with this configuration, only an inductor would produce the low-pass conditions if the circuit was a simple filter (not using the amplifier).

This process, where a capacitor is used in the feedback path to simulate an inductor, is commonly used in electronics. Since real inductors are often quite bulky and are nonideal (due mainly to wire resistance), the simulation of inductance using this capacitor-feedback circuit is quite popular. This inductance-simulating circuit is called a **girator**.

In a similar manner, other filters can be used in the feedback path of an inverting amplifier to produce these "amplified" filter responses. Figure 11.18 summarizes the effect of such an application. It should be noted that since

Type	Impedance	Response Curve
Low-pass filter	For low frequencies, Z = low For high frequencies, Z = high	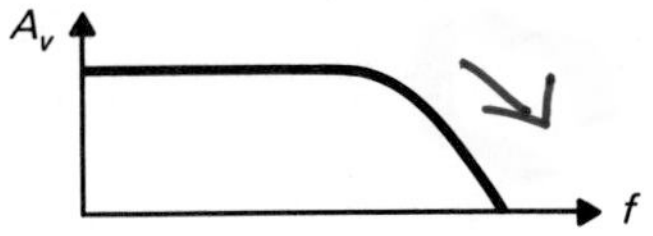
High-pass filter	For low frequencies, Z = high For high frequencies, Z = low	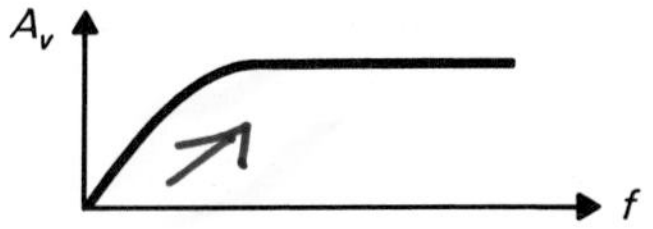
Bandpass filter	At f_r, Z = minimum At other frequencies, Z = high	
Band-reject filter	At f_r, Z = maximum At other frequencies, Z = low	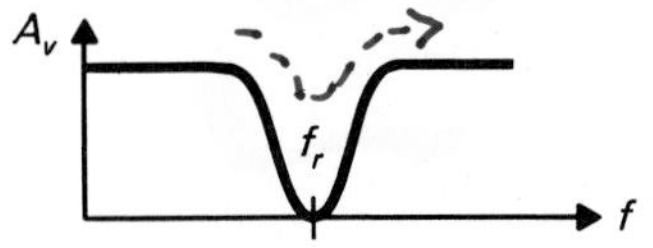

FIGURE 11.18 Filter Effects When Used in Feedback Circuits

these filters are used with an inverting amplifier in a negative-feedback path, the actual output response of the amplifier is a mirror image of the filter's effect. This is the case since the filter's impedance is used in the gain formula. It should also be noted that with bandpass and band-reject filters, an additional series resistor is added in the feedback path. This resistor constitutes the gain limit when the impedance of the resonant circuit is virtually zero ($A_V = -R/R_I$).

There are quite a few other variations on active filters (filters associated with amplifying circuits). One of the more popular forms is called a **Butterworth filter**. This filter incorporates both a feedback filter network and an input filtering network. Such a system would produce a multiple roll-off at a resonant frequency of −40 dB/decade. An attractive point of this filter is that its output prior to the break frequency is quite constant and the roll-off is even above (or below) the break frequency. Low-pass and high-pass Butterworth filters are shown in Figure 11.19. The filter types shown are often called two-pole Butterworth filters. The term "two-pole" relates to the fact that there are two resonant circuits at the input of this circuit.

Following is the equation used to estimate the break frequency for the Butterworth filter:

$$f_c = \frac{1}{2\pi\sqrt{R_1 \cdot R_2 \cdot C_1 \cdot C_2}} \tag{11.10}$$

As can be seen in equation 11.10, the break frequency (or critical frequency) is predicted by incorporating both R/C filters in the same equation.

An additional feature of active filters is that the gain of each amplifier can be adjusted to produce the desired output level of the filter. (Varying the break or resonant frequency is still done using variable capacitors in the filter circuit.) This variable output can be produced by adding a feedback resistor (R_F) and making the R_I resistance variable.

A popular application of the variable active filter is seen in graphic equalizers used in audio circuits. These equalizers are usually a collection of active

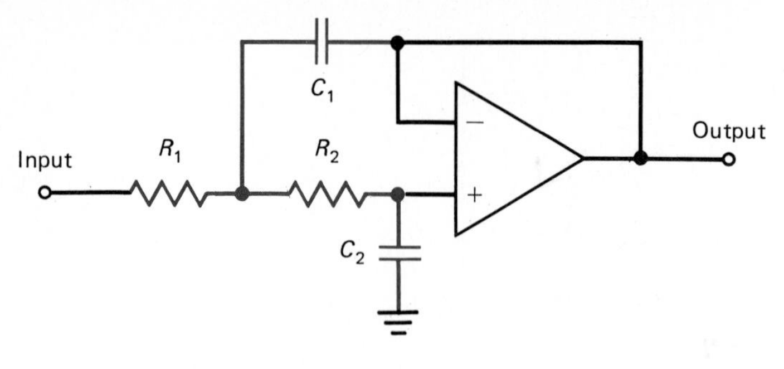

(a) Low pass

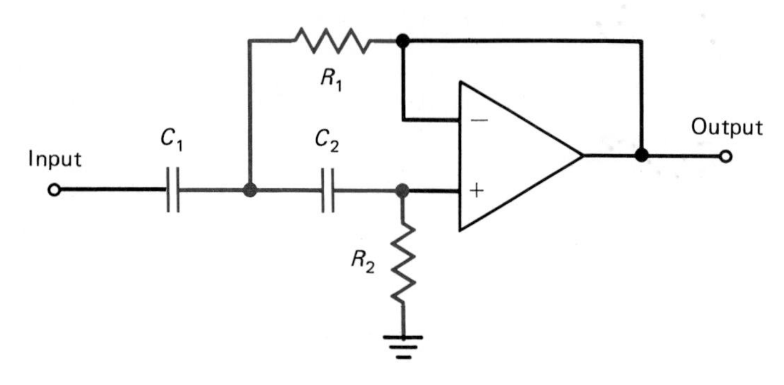

(b) High pass

FIGURE 11.19 Butterworth Filters

filters that use resonant circuits in the feedback loop. The gain of each active filter is made variable so that signals at the resonant frequency can be either "boosted" (dB gain > 1) or "cut" (dB gain < 1). The resonant frequencies of each filter overlap in such a way that the complete audio-frequency range is covered by the group of filters (see Figure 11.20).

Another means of producing a bandpass or bandstop filter is to produce a circuit that incorporates both low- and high-pass filters, each set to the same break frequency. One major advantage of using this circuit is that it eliminates the need for the inductor typically used in a resonant circuit. Two filter forms are integrated into one feedback arrangement on the same amplifier in order to achieve this process. In Figure 11.21, capacitor C_1 and resistor R_1 form the low-pass filter arrangement, while C_2 and R_2 produce the high-pass segment. (These filters, though, cannot be analyzed independently but must be treated as a composite effect!)

The equations for the resonant frequency for either of these configurations is the same as for the Butterworth filter, except that the correct values of resistance need to be incorporated into the formula.

$$\text{Bandpass configuration: } f_r = \frac{1}{2\pi\sqrt{(R_1 \parallel R_3)R_2 \cdot C_1 \cdot C_2}} \tag{11.11a}$$

$$\text{Bandstop configuration: } f_r = \frac{1}{2\pi\sqrt{R_1 \cdot R_2 \cdot C_1 \cdot C_2}} \tag{11.11b}$$

Using these configurations is desirable since bulk and associated problems of using an inductor can be eliminated.

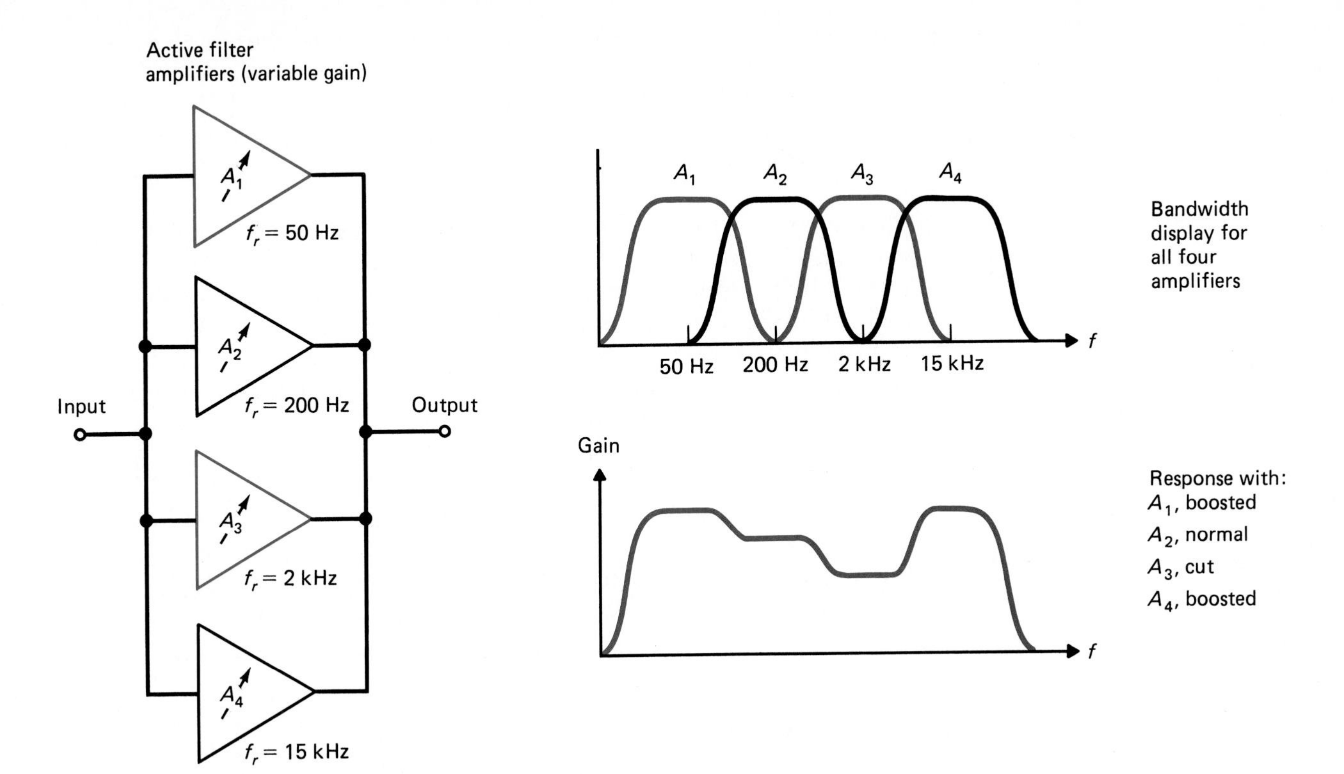

FIGURE 11.20 Graphic Equalizer (Simple Schematic and Filter Effects)

There are quite a few variations of both the filtering application and of other special applications of negative feedback used in amplifiers. In future chapters other examples will show different applications for manipulating amplifier feedback to create many special electronic effects. For example, this section did not explore the use of positive feedback; that topic is presented in the next section.

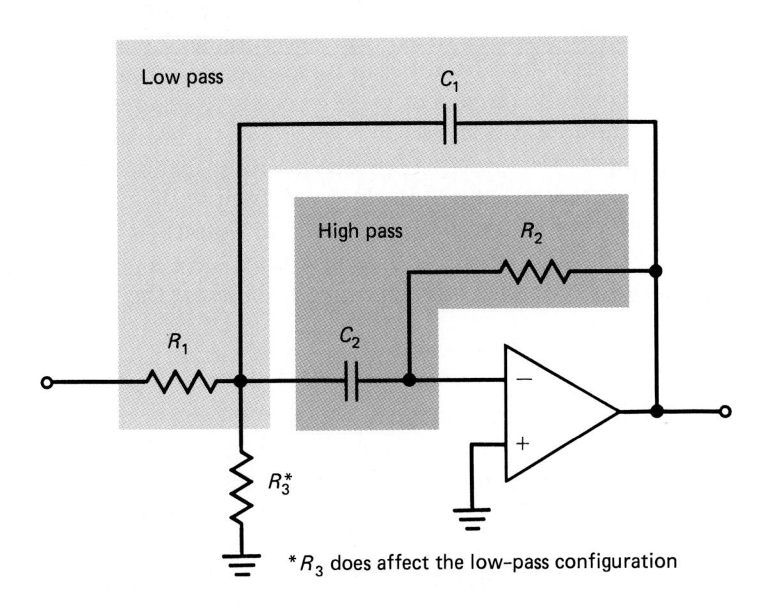

FIGURE 11.21 Op-Amp Bandpass Active Filter (Filter Occurring in Feedback Path)

REVIEW PROBLEMS

(1) List the basic advantages of both dc and ac negative feedback in amplifier circuits. (If necessary, refer to previous chapters for details.)

(2) Illustrate two forms of amplifier systems that will produce an output which has a bandpass-filter-shaped response. Show both a circuit using a filter input and one that uses a filter in a negative-feedback path. (For simplicity, use an op-amp as the active device.)

(3) If a low-pass Butterworth filter uses the components listed below, calculate the critical (break) frequency of the amplifier and illustrate the amplifier, showing component values.

$R_1 = 16\ \text{k}\Omega \qquad R_2 = 16\ \text{k}\Omega \qquad C_1 = 0.02\ \mu\text{F} \qquad C_2 = 0.01\ \mu\text{F}$

(4) Generally describe how resonant filter effects are duplicated in the circuits shown in Figure 11.21. To aid in your description, illustrate the overlapping effect of a low-pass and a high-pass filter, each with the same break frequency.

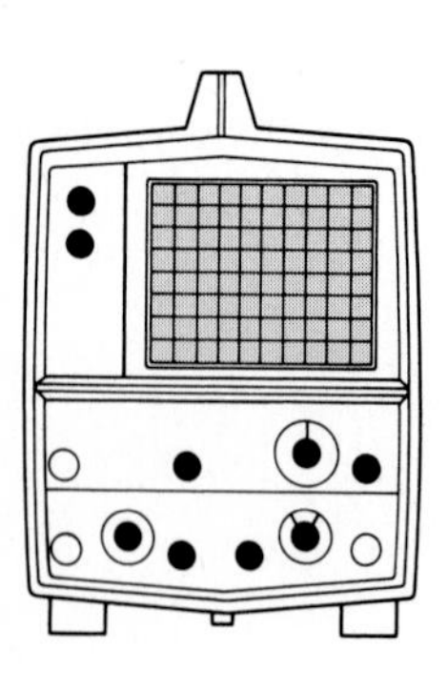

11.6 OSCILLATORS

Earlier, degenerative (negative) feedback was investigated. In this section the topic of regenerative (positive) feedback is explored. In negative feedback, the feedback quantity was subtracted from the input to produce an effective input that was less than that of the original. In regenerative feedback, the feedback quantity is added to the input, thus producing a higher effective input than the original. When negative feedback was used, the upper limit to the actual input was the value of the original input. In positive feedback circuits, the actual input is always in excess of the original.

If positive feedback were to be used with a standard amplifier configuration, the output level would immediately rise to excessively high levels. This condition would saturate the amplifier, driving it to its maximum output. It is possible that the amplifier would be damaged in the process, as excessively high voltage and currents would be produced.

It may be surmised that it is impractical to use positive feedback in a standard amplifier. This is a correct assumption. Not only would the positive feedback be potentially damaging to the circuitry, but the output (in nearly every application) would be unusable. Positive feedback does have a practical purpose, though, if the amplification circuit is modified to support the effects of positive feedback.

Circuit Modifications to Support Positive Feedback

To apply positive feedback, the following basic modifications of the amplification circuit need to be made:

Gain The gain of the circuit cannot be above 1. A gain that exceeds unity (1) will produce successively higher output levels, which will lead to saturation of the circuit. Basically, this means that the product of the amplifier's gain multiplied by the feedback factor must be (approximately) 1. Actual figures for these quantities will be shown later.

Feedback The actual amount of feedback level produced is not significant. Even if minute values of positive feedback are produced, eventually the amplifier's output will increase. Again, the product of gain and feedback should be around a value of 1.

Input Once the system has begun to produce the positive feedback quantity, an actual input term is not necessary. This is true since the positive feedback will produce enough level at the input that a final regeneration of the input signal will occur, even if the input is removed. In the applications shown later in this section, a physical input connection is often unnecessary.

In the schematic shown in Figure 11.22, a unity-gain amplifier system is shown. First, the basic amplifier is established with a simple gain (using negative feedback) of 4. The positive feedback loop (at the bottom) produces a feedback that is one-fourth of the output. Therefore, the total gain of the system will be $4 \times \frac{1}{4} = 1$. If an input is initially applied to the amplifier, the following steps will occur:

1. The circuit will produce an output that is four times as large as the input. Let us then assume that as soon as this output is produced, the input signal is removed.
2. This output will then pass to the feedback circuit, where one-fourth of it will be reapplied to the input. This value is exactly equal to the original input (input $\times$ gain $\times$ feedback = input $\times 4 \times \frac{1}{4}$ = input).
3. Since the feedback is now the same amplitude as the input had been, the process will continue, without interruption and without the need for the original input.

Note: This presentation is for the sake of illustrating the use of the positive-feedback process. Since it would be impossible immediately to remove the input signal, the actual output would tend to rise quickly and would, in all probability, saturate the amplifier before the input could be removed!

The term **regenerative feedback** is appropriate since the feedback process regenerates the original input value.

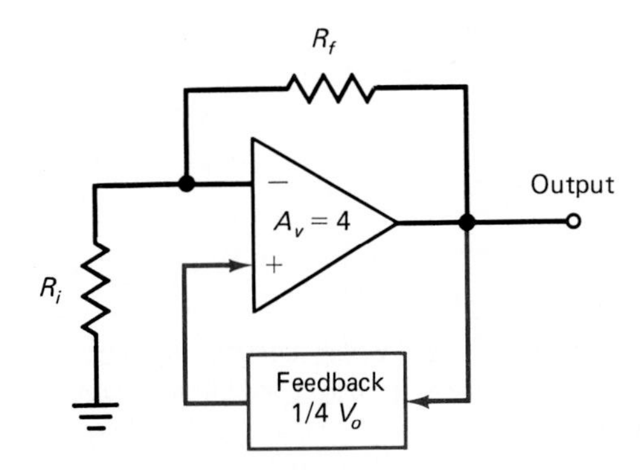

FIGURE 11.22 Positive Feedback Used to Sustain Oscillations

Oscillation

Positive-feedback amplifiers are used to produce oscillators. Oscillators are circuits that will constantly produce an output frequency without the need for an input. To produce oscillation, the regenerative effects of the positive-feedback amplifier will provide the basis for sustaining an output. It will be necessary, though, to add special circuit elements to the positive-feedback amplifier to produce this output quantity. Oscillation is produced at only one frequency, as determined by these added circuit elements.

Basic Oscillator Configuration

The circuit shown in Figure 11.23 will produce an output frequency determined by the resonant frequency of the resonant circuit in the feedback loop. This circuit does not have an input terminal, but an initial starting input is produced within the amplifier. This startup signal is created either by the generation of noise within the amplifier or by the nonlinear effect seen in the amplifier circuit.

Special discussion of noise: When noise is produced within an amplifier, the noise signal is quite random. As such, it produces a wide variety of signals at a wide spectrum of frequencies. Mathematically, this noise signal can be broken down into a nearly infinite combination of individual sine waves, all at different frequencies.

Following is a general analysis of the steps that occur within the oscillator circuit to cause the production of an output signal.

1. When the circuit is first turned on, there will be an automatic generation of some level of noise.
2. This noise signal is present at the output terminal of the amplifier.
3. The positive feedback path returns a portion of this output back to the input, to be regenerated.
4. But in the feedback path there is a bandpass filter. This filter attenuates all of the noise signals except the one at which it resonates. That signal, at the one frequency selected by the filter, is the only one presented at the input terminal of the device (via this feedback path).

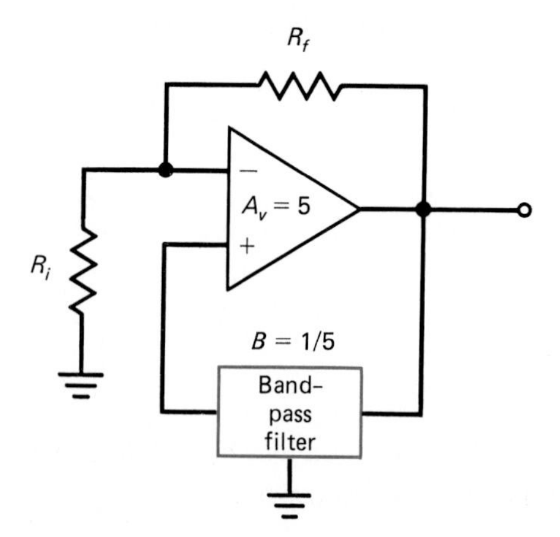

FIGURE 11.23 Frequency-Selective Circuit Used in the Oscillator Feedback Path

5. This one frequency is then amplified by the amplifier and becomes the signal that is again fed back to the input terminal.
6. The process of producing an output at one frequency, feeding that back to be regenerated, and again reproducing that signal continues until the dc power supplied to operate the (gain of) amplifier is removed.

Further cycles of the system continue to eliminate all nonresonant filter frequencies (due to the loss of the resonant circuit) and to pass through the one frequency acceptable to the resonant circuit. Further signals continue to be reprocessed by the amplifier/feedback combination. This, then, is the basic process of oscillation.

The usual waveform to be produced by an oscillator is the sine shape. Mathematically, it is the most fundamental wave shape. Actually, to produce a nonsinusoid, special configurations of the amplifier must be determined. Even in these cases, the signal handled by the resonant circuit is still a sine shape, just the output of the amplifier is modified to produce the different waveform.

Special note: It should be noted that if there is not a resonant circuit in the feedback path (or associated with the circuit in some way), the circuit will not generate any usable output. Initially, it may be thought that the circuit would reproduce the noise signal. In reality, some one frequency (usually quite randomly generated) would be filtered out and amplified more than others. This is caused by random reactance terms which are a part of the elements used to construct the circuit. This frequency is called the **natural resonant frequency**.

Occasionally, an ordinary amplifier will have positive feedback applied in a test condition in order to determine this natural resonant frequency. The objective of doing this is later to add filters, either low-pass, high-pass, or notch type, to the circuit (not in a positive-feedback configuration) to assure that this resonant condition will not occur inadvertently.

Oscillation will occur if the following set of conditions are made within any amplifier circuit:

1. Overall system gain is unity (1).
2. Positive feedback is used for signal regeneration.
3. Frequency-selective elements are applied (usually in the feedback path) so that only one frequency is maintained in the system.

Special note: Items 1 and 2 are called the *Barkenhausen criterion* for oscillation.

It is necessary that the total system gain be exactly at unity. A gain higher than this will cause the output to saturate; a gain below this will cause the output to die out. Careful use of feedback elements (usually in the gain-controlling negative-feedback section) is necessary to support the oscillation. Any instability in the amplifier will create problems in maintaining stable oscillation. It is not uncommon to make one of the elements in the gain-controlling feedback loop variable (potentiometer) so that minor adjustments can be made to correct for errors in the overall gain. (This would usually be a one-time calibration adjustment, set when the circuit is first tested.)

Sample Oscillator Circuitry

Before presenting actual oscillator circuits, three basic points need to be reviewed:

1. Positive feedback must be used. Since many amplifiers have an output that is inverted (180° phase shift), further inversion of that signal is necessary to produce an in-phase signal.
2. The overall gain must be maintained at unity.
3. Stability is an absolute necessity. It will be assumed that any amplifier circuit (BJT, FET, or op-amp) must first be highly stable.

The following circuits demonstrate several popular methods of producing oscillation.

The Phase-Shift Oscillator Figure 11.24 illustrates a phase-shift oscillator. In this oscillator, basic gain is established by the R_F and R_I ratio. The feedback path is through the resistor/capacitor combination attached to the input. First, it can be noted that an inverting amplifier configuration is used. From this inverting format, a 180° shift in the phase of the input is seen at the output terminal. (This will not, of course, sustain oscillation.)

Each of the resistor/capacitor combinations will produce an additional phase shift. If they were ideal, only two capacitors would be needed, as each will contribute a 90° phase shift. Thus the total phase shift would be 180° + 90° + 90°, a total of 360°, or back in phase. Since the R/C circuits cannot produce a pure 90° shift, an additional R/C combination is added. These are then selected so that the resulting phase shift of each R/C combination is 60°, which is easily obtained.

Thus there is a proper in-phase feedback signal. Due to voltage dividers between each resistor in the shifting network and the input resistance of the amplifier (including R_I), a percentage of the output, not all of it, is feedback. (The feedback factor B is less than 1.) The product of this factor and the gain of the amplifier must be unity. Therefore, the ratio of R_F to R_I must be such that it corrects for the loss in the feedback path.

All three resistors and all three capacitors in the feedback phase shifter/frequency selection circuit are chosen to be identical. The actual frequency

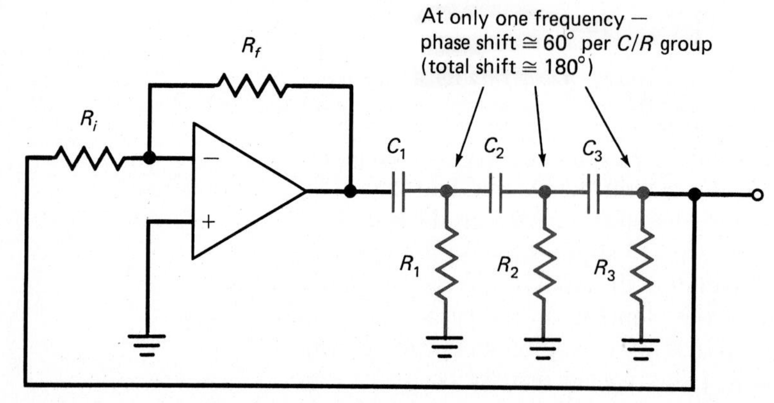

FIGURE 11.24 Op-Amp Phase-Shift Oscillator

at which this circuit oscillates is determined by

$$f_r = \frac{1}{2\pi \sqrt{6} \times R \times C} \tag{11.12}$$

EXAMPLE 11.5

If the phase-shift oscillator shown in Figure 11.24, the values are as follows:

$$R = 10 \text{ k}\Omega \qquad C = 0.001 \ \mu\text{F} \qquad \text{feedback ratio} = 0.025$$

Find the frequency at which this circuit oscillates and find the necessary gain to be set by the R_F-to-R_I ratio.

Solution:

$$f_r = \frac{1}{2\pi \sqrt{6} \times 10 \text{ k}\Omega \times 0.001 \ \mu\text{F}} = 6.5 \text{ kHz}$$

Computing the gain, we have

$$\text{gain} \times B = 1$$

$$\text{gain} \times 0.025 = 1$$

$$\text{gain} = \frac{1}{0.025} = 40$$

The Colpitts Oscillator Figure 11.25 shows another form of oscillator, called the **Colpitts oscillator**. In this oscillator, the traditional L/C bandpass filter combination is seen. The center connection between the two capacitors is established in order to produce the necessary phase shift that will produce, along with the amplifier's 180° shift (inverting), a total phase shift of 360°. The resonant frequency of this combination fits a standard parallel resonant equation, except that the capacitance used in that equation is based on

$$C_{\text{res}} = \frac{C_1 C_2}{C_1 + C_2} \tag{11.13}$$

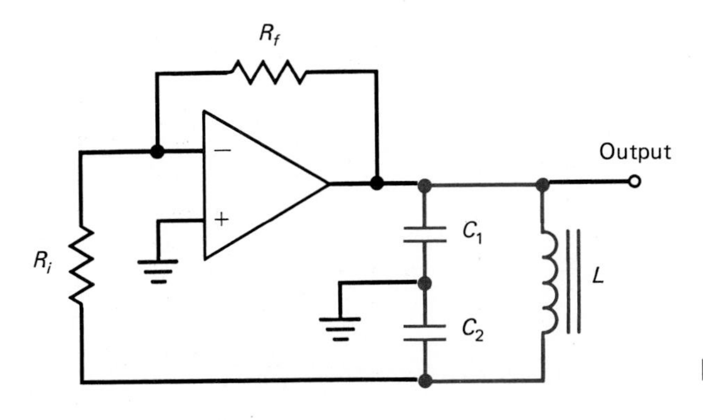

split capacitors on output

FIGURE 11.25 Colpitts Oscillator

The Hartley Oscillator The **Hartley oscillator** shown in Figure 11.26 is functionally the same as the Colpitts oscillator except that the inductive leg contains the phase correction (grounding of the center terminal). The equation used for its resonant frequency is identical to the standard resonant-frequency equation. (Both inductor values are summed up to produce the total inductance for the resonant-frequency equation.)

Other Oscillator Forms A wide variety of oscillator circuits are available. Some are crystal controlled to provide greater stability. Others are special forms of the basic configuration which use standard resonant circuits, but apply special feedback controls which are used to stabilize the output level. Certain other forms use additional components selected to modify the shape of the output waveform.

Oscillators can be made which have variable-frequency outputs. A typical example of this is the signal generator commonly used for laboratory experimentation. A common method of making the frequency variable is to use a variable capacitor (or a switch that selects various capacitance values). By adjusting the variable capacitor the resonant frequency of the circuit is changed and thus is the output frequency changed.

Delicate Balances in Oscillators

The unity-gain figure of the oscillator has been emphasized throughout this discussion. Therefore, it is necessary that we handle this system with great care so as not to disturb the delicate balance. In fact, even the output of the oscillator must be treated carefully. Should a load be applied to this circuit that would demand even modest current flow, it is quite possible that this demand could drive the overall gain below unity, which would cause oscillation to stop. The output of the oscillator must therefore be treated carefully.

Often the output is fed through a buffer circuit. The objective of the buffering element is to reduce the loading effects of attached devices (see Figure 11.27). An FET circuit could be used as a buffer since it draws no appreciable current from the input. Thus it would isolate the oscillator from

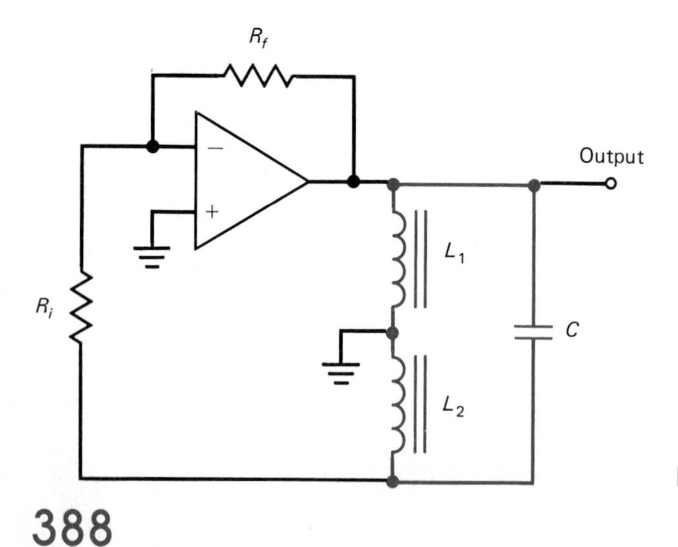

FIGURE 11.26 Hartley Oscillator

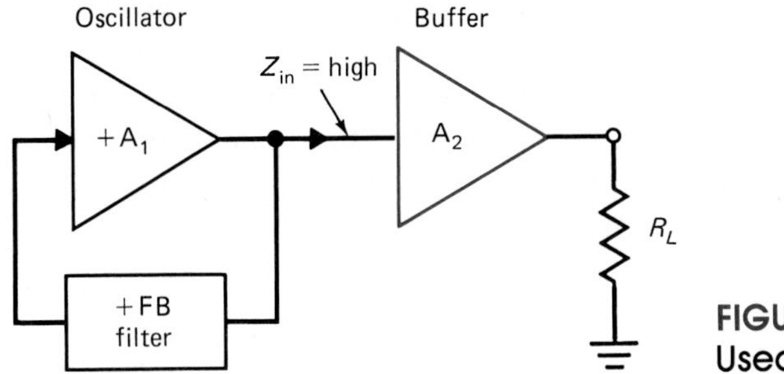

FIGURE 11.27 Buffer Amplifier Used with Oscillator Circuit

the load element. It is also common to use transformers to provide isolation. The transformer may be used both to couple the load and to provide the inductive segment of the resonant circuit. Finally, the effects of load capacitance may cause the resonant circuit to function at a value that is off from the predicted value. Again, a buffer could be used to serve to reduce this effect.

Troubleshooting Oscillators

Because of their extremely delicate balance, oscillators are prone to problems. Also due to this balance, oscillators are not simple to troubleshoot. If the direct output of an oscillator is measured, it is possible to load the output enough to actually cancel the oscillation or possibly throw it off frequency due to the capacitance of the measuring device (or even the test leads). Therefore, an oscillator is typically tested by an indirect method.

The most obvious place to measure the output of the oscillator is after it has passed through the buffering element. If a signal is present there, it can be assumed that the oscillator is functioning properly. If this does not adequately confirm that the oscillator is functioning, a second method is usually employed.

Often the dc bias conditions around the oscillator circuit are measured, again avoiding direct measurement of the ac oscillations. If the bias conditions around the amplifier match those listed on the circuit's schematic, it can be assumed that the oscillator is functioning correctly. If the circuit is not oscillating, the circuit will show up as either being saturated, cut off, or highly out of balance in respect to these bias conditions.

The most common failure in an oscillator is an error in the gain–stability circuitry. Usually, this circuit fails to maintain the ideal unity gain. Often this is resolved by replacing one of the gain-determining resistors with a potentiometer. This potentiometer can then be adjusted, while the circuit is functioning, until the correct output is obtained. It is not too common to have the active device fail, since the circuit is running with unity gain and thus is not "overdriving" the active device.

One other common failure in an oscillator is that it may "drift" or change the output frequency. This problem is often resolved by selecting more accurate resonant elements (e.g., a crystal oscillator circuit). Occasionally, the drift is due to the effect the amplifier's resistance, which can possibly change, has on a resonant circuit. In this case the circuit needs to be redesigned so that these resistance values will have a minimal effect on the circuit's function.

REVIEW PROBLEMS

(1) List, from memory, the basic requirements for an oscillator circuit.

(2) Given the following data for a Colpitts oscillator, calculate the frequency at which it will oscillate:

$$C_1 = 300 \text{ pF} \qquad C_2 = 300 \text{ pF} \qquad L = 100 \text{ mH}$$

(3) If a variable-frequency Hartley oscillator were to be constructed that was to range from 1 to 50 kHz, using two 10-mH inductors, find the range of capacitance for the variable capacitor that would be needed to produce this range of output frequencies. Also sketch the circuit.

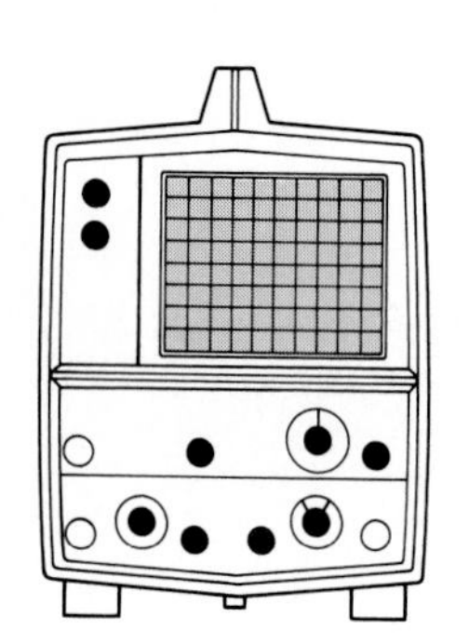

11.7 ADDITIONAL TROUBLESHOOTING COMMENTS

This chapter has concentrated on the specific application of resonant circuits to amplifiers to produce either frequency-selective systems or oscillators. An extension of this concept is the production of a system that has unplanned resonant effects.

Any circuit component arrangement that places a capacitive element in series or parallel with an inductive element may produce one of the effects described in this chapter. It is therefore possible to have a circuit that is either frequency selective or oscillates, but was not intended to do so. In fact, one of the most serious problems when working with amplifier faults is encountering a system that is "self-resonant."

An amplifier system can be destroyed by the production of unwanted oscillator effects. Thus in troubleshooting, it is necessary to consider the possibility that an oscillator configuration may exist in any amplifier system. Unusual combinations such as wiring capacitance coupled with a "girated" form of inductance (phase-inverted capacitance) can produce oscillations or a frequency-selective system.

There are three basic procedures for the elimination of such undesirable circuit conditions:

1. Eliminate the resonant condition that produced the undesirable effect. Often, the elimination of inductive elements or unusual capacitive effects can reduce or eliminate the resonant problem.
2. Eliminate the resonant frequency effect by *adding* an opposite form of resonant circuit to the system (e.g., add a notch filter set to the same frequency or add negative feedback through a resonant circuit).
3. Change the resonant components or add series/parallel components to shift the resonant frequency "out of the range" of the amplifier's intended application.

For example, if an audio amplifier had a resonant condition that produced an oscillation at 10 kHz, locate the source of the resonance first. Assuming the components could not be removed without creating a problem, and a notch filter at 10 kHz would interfere with the amplifier's expected frequency response, this third approach may have to be used. If the resonant elements could be changed in value so that the resonance would occur at (for example) 50 kHz, this may solve the problem since this audio amplifier's natural response (bandwidth) may cut off at 20 kHz, well below the new resonant frequency. Thus, although the circuit resonated, the effect would not be supported by the balance of the circuit to which it was attached. Thus in troubleshooting even the most basic of amplifier circuits, the potential for resonant circuit problems must be considered.

SUMMARY

In this chapter we have presented several amplifier circuits, all of which, in one way or another, are frequency selective. It was demonstrated that by attaching resonant filter circuits to amplifiers, the amplifier output became frequency dependent. Therefore, with the application of resonant circuits, amplifiers can be produced which provide output values at only specific frequencies or bands of frequencies.

Resonant circuits could be applied to amplifiers either as input signal filters or as a composite part of the signal feedback process of the amplifier. Active filters were shown that use resonant circuits in their negative signal feedback paths to produce a system that is highly frequency selective.

The use of resonant circuits in an amplifier that applies positive feedback produced an oscillator. Positive feedback is necessary to sustain the generation of an output. In order neither to lose the output signal nor saturate the amplifier, the gain of the amplifier must be maintained at unity (1). This is achieved by balancing positive and negative feedback levels against one another.

KEY EQUATIONS

11.1 $f_{low} = \dfrac{1}{2\pi R \times C}$ (capacitive filter)

11.2 $f_{low} = \dfrac{R}{(2\pi L)}$ (inductive filter)

Low pass capacitive filter:

11.3 $f_{high} = \dfrac{1}{2\pi R \times C}$ (capacitive filter)

Low pass inductive filter:

11.4 $f_{high} = \dfrac{R}{(2\pi L)}$ (inductive filter)

Series resonant circuit:

11.5 $f_r = \dfrac{1}{2\pi\sqrt{L \times C}}$

11.6 B_w (Bandwidth) $= \dfrac{f_r}{Q}$

11.7 $Q = \dfrac{X_L}{(R' + R_1)}$ (series)

Parallel resonant circuit:
(11.5 and 11.6 plus-)

11.8 $Q = \dfrac{X_L}{R_1 + (X_L^2)/R_p'}$ (parallel)

General rule regarding shunt resistance for parallel resonance:

11.9 $R_P > X_L^2$ (parallel)

Butterworth filter break frequency:

11.10 $f_c = \dfrac{1}{2\pi\sqrt{R_1 \cdot R_2 \cdot C_1 \cdot C_2}}$

Modified versions of equation 11.5, for Butterworth filters:

11.11a $f_r = \dfrac{1}{2\pi\sqrt{(R_1 \parallel R_3)R_2 \cdot C_1 \cdot C_2}}$

11.11b $f_r = \dfrac{1}{2\pi\sqrt{R_1 \cdot R_2 \cdot C_1 \cdot C_2}}$

11.12 $f_r = \dfrac{1}{2\pi\sqrt{6} \times R \times C}$

Total capacitance for a Colpitts oscillator:

11.13 $C_{\text{res}} = \dfrac{C_1 C_2}{C_1 + C_2}$

PROBLEMS

Section 11.3

11.1. Briefly define a frequency-selective amplifier.

11.2. Find the value of first a capacitor, then an inductor, which will coordinate with a 5-kΩ resistor to produce both of these filter configurations. Sketch and label all four circuits.
(a) Low-pass filter with $f_{\text{low}} = 80$ Hz
(b) High-pass filter with $f_{\text{high}} = 40$ kHz

11.3. Define an RFC and state what it is used for.

Section 11.4

11.4. What are the resonant frequency, bandwidth, and Q for the following components, first if they are in series, then if they are in parallel?

$C = 150$ pF $\quad L = 4.5$ mH

$R_1 = 40\ \Omega$ (inductor wire resistance)

resistive load on the circuit $= 150\ \Omega$

11.5. Sketch and label a graph of the response of the two circuits calculated in problem 11.4.

11.6. Sketch the schematic of an op-amp (inverting) with a parallel resonant filter at the input.

Section 11.5

11.7. Briefly describe nonlinear distortion and what can cause it in an amplifier.

11.8. Briefly define noise as it relates to an amplifier.

11.9. For the 2N2218 BJT and the MFQ5460P JFET, look up the specification for noise (data sheets in Appendix A). Contrast these two noise figures.

11.10. If a standard inverting op-amp has an input resistor of 2 kΩ and uses a 0.22-μF capacitor in the feedback position, note what effect this will have on the amplifier's response.

11.11. For the circuit described in problem 11.10, calculate the amplifier's gain for the following frequencies.
(a) 20 Hz (b) 400 Hz (c) 5 kHz

11.12. Sketch a basic low-pass Butterworth filter using an op-amp.

11.13. If the filtering capacitors in a Butterworth high-pass filter are each 1000 pF and $R_1 = R_2 = 50\ k\Omega$, find the break frequency of the filter.

11.14. Illustrate an op-amp bandpass filter that uses a low-pass and a high-pass filter combination.

Section 11.6

11.15. Define an oscillator and describe why positive feedback is used in an oscillator circuit.

11.16. Why is the use of a filter necessary when producing an oscillator?

11.17. Define the Barkenhausen criterion for oscillation.

11.18. Illustrate the circuit for an op-amp phase-shift oscillator.

11.19. Find the oscillation frequency for a phase-shift oscillator (as illustrated in problem 11.18) if the resistors used were all 15 kΩ and the capacitors were 500 pF.

11.20. Illustrate both a Hartley and a Colpitts oscillator (using op-amps).

11.21. Describe why it is necessary to "buffer" the output of an oscillator.

11.22. If a variable Hartley oscillator were to be constructed which had a variable output frequency of from 88 to 108 MHz and used a 100-nH inductor, what would the value of the variable capacitor have to be to produce this range of frequencies?

ANSWERS TO REVIEW PROBLEMS

Section 11.1

1.

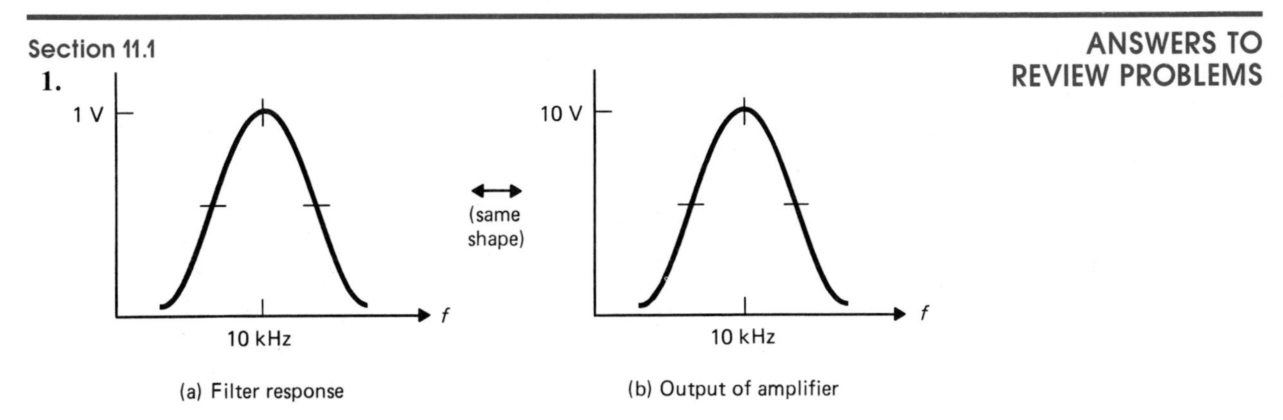

(a) Filter response (b) Output of amplifier

Section 11.2

1. See Section 11.2 for details.
2.

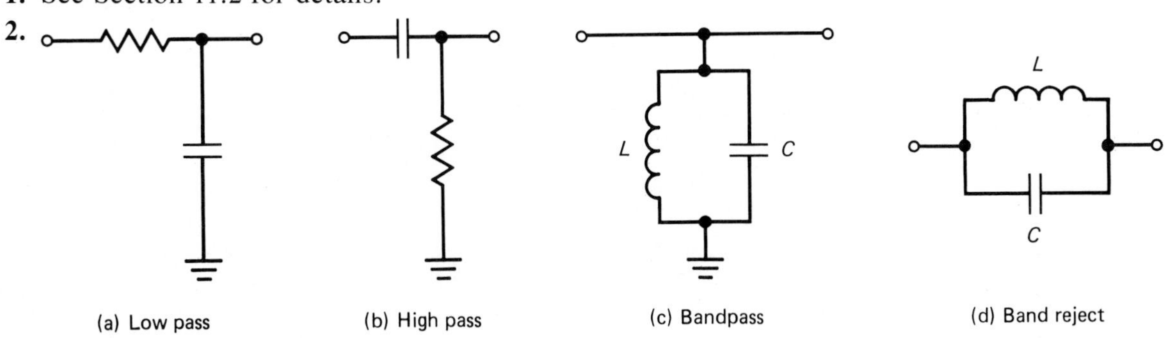

(a) Low pass (b) High pass (c) Bandpass (d) Band reject

(a) Break frequency $= 1/(2\pi R \times C)$

(b) Break frequency same as part (a)
(c) $f_r = 1/(2\pi\sqrt{L \times C})$

Section 11.3

1. For low- and high-pass capacitive, $f = 1/(2\pi R \times C)$; for low- and high-pass inductive, $f = R/(2\pi L)$
2. 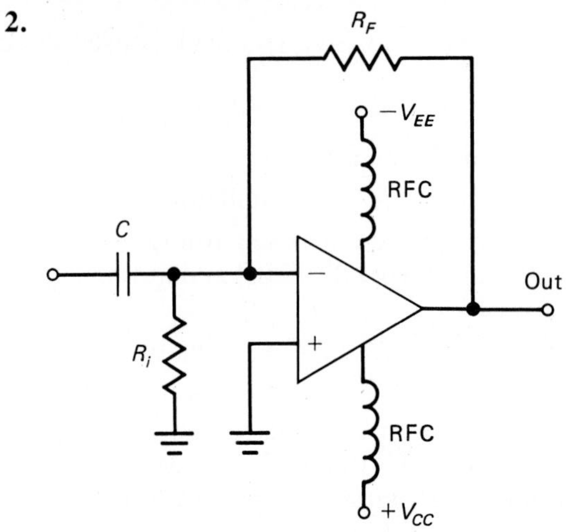

3. When the control potentiometer is set at zero ohms. the effect of the capacitor, in coordination with the amplifiers' resistance, forms a simple low-pass filter. As frequency increases, increasing levels of the signal are shunted to ground, thus reducing the output of amp 2. When the potentiometer is set so that its value equals the amplifier's resistance, when the signal frequency is such that the reactance of the capacitor is near zero, nearly 50% of the signal is passed on. When the potentiometer is set to a very high resistance, the capacitor's impedance has little, if any, effect on the signal passed on. This circuit has the effect of a simple treble (high-frequency) tone control.

Section 11.4

1. 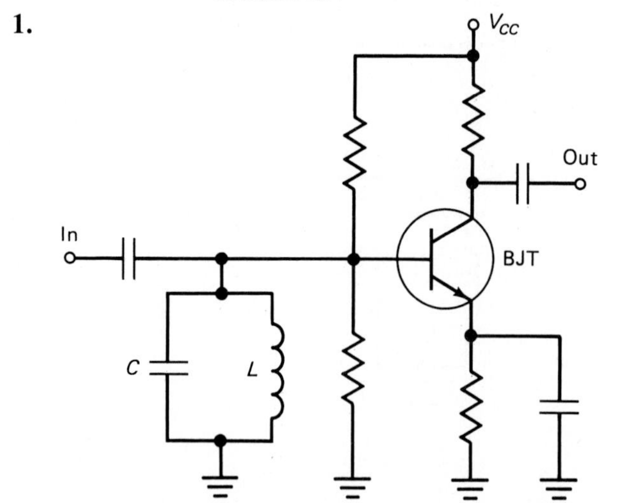

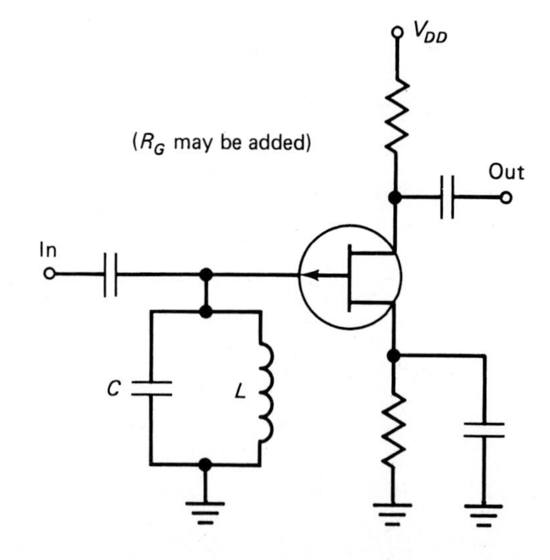

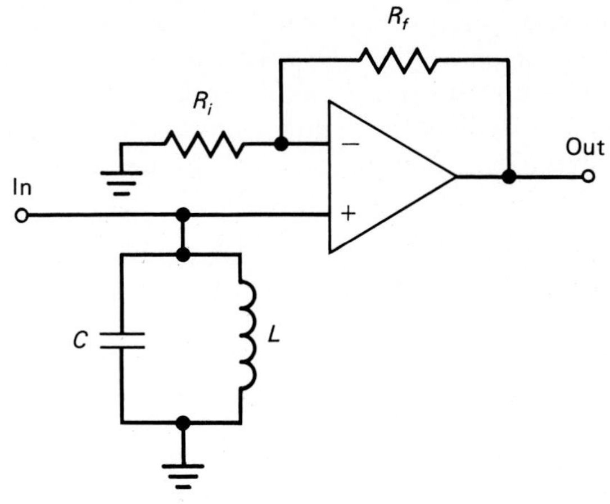

2.

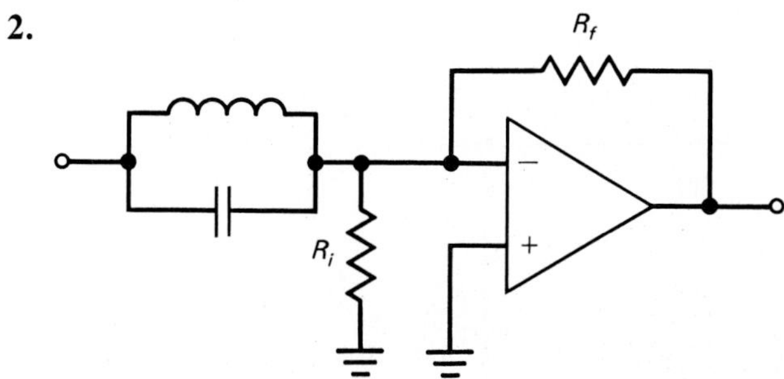

3. Bandwidth = 4.5 MHz, input capacitance = 0.8 pF (phase margin 60°, if important to application); since input resistance is adjustable via application, that is not a fixed figure of the op-amp.

Section 11.5

1. Greater stability, wider frequency response, lower distortion

2. (alternative forms are acceptable)

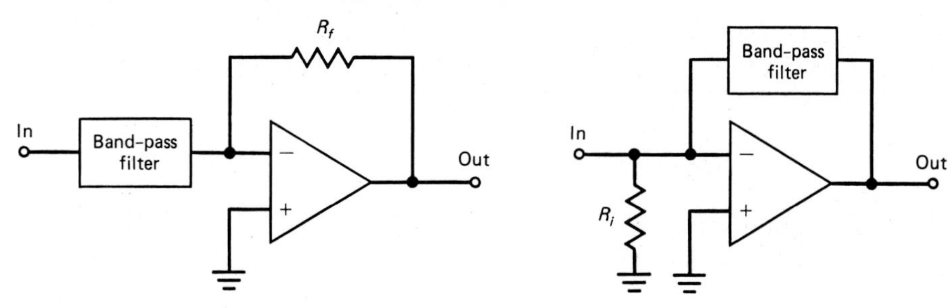

3. f = 703.4 Hz.

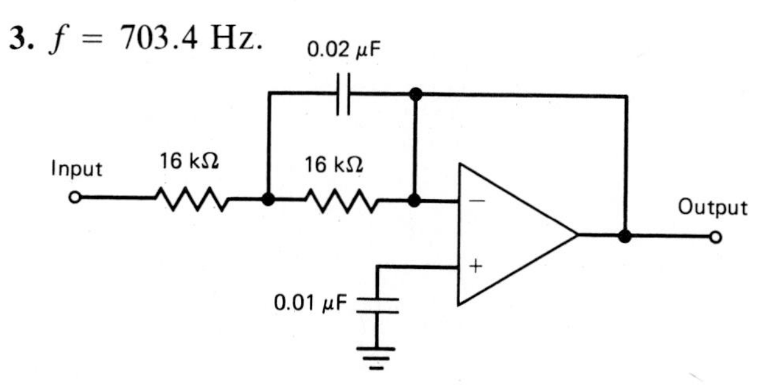

4. The high-pass filter will produce a reduced output for frequencies above its "break frequency." The low-pass filter will produce reduced output for frequencies below its "break frequency." Since these break frequencies match, the overall system will produce low output levels for frequencies both below and above the coordinated break frequency.

Section 11.6

1. The overall total gain must be unity; positive feedback must be employed; a frequency-selective circuit must be used to isolate the oscillator's output frequency; "noise" must be present to start oscillation.
2. 41.1 kHz.
3. Capacitor to range from 507 pF to 1.27 μF

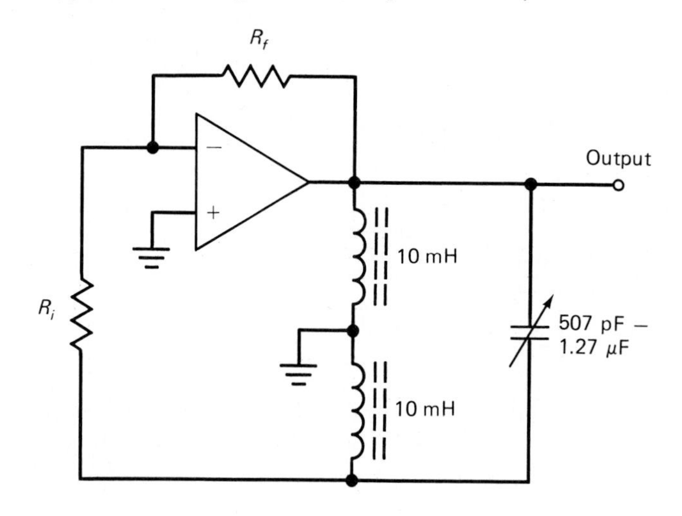

12 Power and Other Special-Purpose Amplifiers

Objectives

Upon completing this chapter, the reader should be able to:

- Identify power amplifier requirements.
- List typical power device specifications.
- Identify classes of operation.
- Troubleshoot power amplifiers.
- Identify the function of a variety of special-purpose solid-state circuits.
- Analyze potential new device applications.

Introduction

In this chapter we cover amplifiers designed to produce high power (over 1 W) outputs and amplifiers which have specialized functions not covered in previous chapters. The manipulation of high levels of electrical power is not done using the more conventional "small-signal" semiconductors. In addition, power circuits often require greater levels of efficiency than is seen in the circuits covered to this point. Therefore, special devices, circuits, and equations are required when dealing with the manipulation of high-power output levels of an amplifier. Finally, some circuits are unique to their function and are presented in this chapter, as they do not coordinate with the other forms presented so far in this book. ■

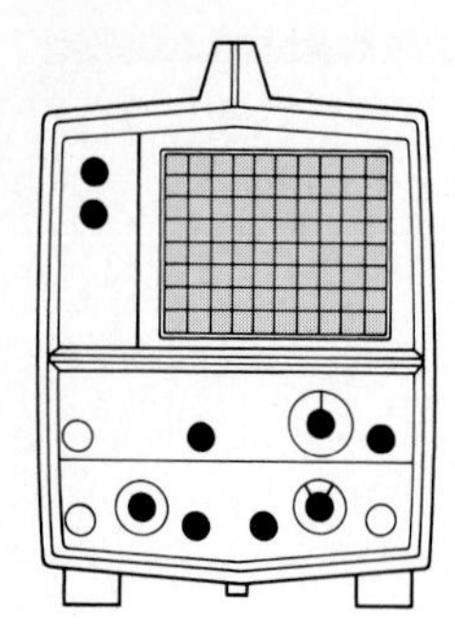

12.1 POWER AMPLIFIERS

So far the amplifiers presented have been small-signal types. They have amplified signals in the millivolt or low volts range. Their current and voltage gains have been such that the power gain and output power have been relatively low. Typical output powers have been in the hundreds of milliwatts. Often, the actual value of output power has not been investigated since it has been relatively low.

Many amplifier systems, though, are used to provide relatively high levels of output power. It is common for an audio amplifier to produce 20, 40, or more than 100 W of output power. This level of output power requires a combination of voltage and current applied to the load. Voltage levels in the 10-, 20-, or 30-V range, along with currents of potentially 50 or 100 A, are possible output levels for high-power amplifiers. Such levels of both voltage and current require the following special circuitry conditions:

1. High-power circuit components are mandatory, including high-wattage resistors and semiconductors capable of supporting these high voltage and current levels.
2. Special power amplification circuits are needed. These circuits are modifications of standard amplifier configurations.
3. Since these circuits deal with high power levels, heat (the by-product of electrical power consumption) becomes a critical factor. It is necessary to use special temperature-control elements in order not to destroy the circuit elements.

The following discussion of power amplification circuitry will concentrate on these three major factors in power amplification.

Preamplifiers

The types of amplifiers discussed so far fit into the category of **preamplifiers** (preamps). The objective of these amplifiers is to increase the average voltage and current levels of the signal sufficiently so that, if desirable, a power amplifier circuit can be used to produce maximum output power. In most cases, the voltage gain for preamps is the most important feature. Typical input levels to an amplifier system are milli- or microvolts. In order to supply energy to a power amplifier circuit, these levels need to be in the volts to tens of volts range. Power amplifiers (as seen later) produce primarily current gain rather than high voltage gain. Therefore, it is necessary to provide a signal that is relatively high in voltage. Also, any special signal conditioning (such as volume control, tone control, etc.) is done in the preamplifier stages. Basically, the amplified signal must be exactly in the form desired at the output prior to entering the power amplifier.

Maximum Power Transfer

It may be remembered from circuit analysis courses that when power delivery to a load is a factor in a circuit, it is desirable to provide maximum power transfer. From the maximum power transfer principles, though, remember that when the ideal power level is delivered to a load, the driving circuit (in this case the amplifier) must also dissipate an equal amount of power. Thus if 100 W is being delivered to a load, the internal circuitry that produced this power must also be dissipating 100 W. This internal power dissipation is converted into thermal power (heat) and is considered a loss to the system.

High-current and high-power devices are necessary when producing power amplification. Power-amplifying devices, BJTs, FETs, or integrated power amplifiers, form a separate category of active devices. Following is a summary of the basic characteristics of these types of devices:

1. High-power devices must be capable of supporting high output current flow.
2. High-power devices must be capable of handling high peak voltage levels.
3. High-power devices must have a means of dissipating (removing or eliminating) the high temperatures that are a by-product of the high power levels they receive.

The last category, the dissipation of thermal energy, places special restrictions and requirements on the form the device may take. Therefore, it is necessary that the device used have a means by which this energy is removed; otherwise, it will be destroyed due to heat.

Power transistors (BJT and FET types) are usually physically larger than the small-signal types. By having a larger surface area, the device can dissipate (spread) the thermal energy over this larger area. Figure 12.1 shows a comparison of the physical size of a small-signal transistor and a power transistor.

(a) Power transistor (b) Small-signal transistor

FIGURE 12.1 Comparison of Power and Small-Signal Device Packages

The power transistor is often attached to a **heat sink**. A heat sink is a metal surface that is able to dissipate the thermal energy further by spreading it out over an even greater surface area (see Figure 12.2). The transistor is attached mechanically to the heat sink. When thermal energy is produced by the transistor, that heat is conducted by surface contact to the heat sink, which then spreads it out over a larger surface area. Therefore, instead of all the energy being concentrated in the device and potentially raising its internal

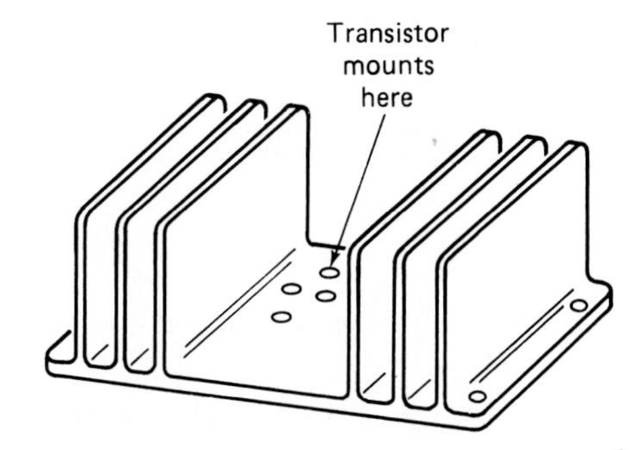

FIGURE 12.2 Transistor Heat Sink

temperature to hundreds of degrees, the energy is spread over the area of the heat sink. Thus the transistor temperature is maintained at a "safe" level.

Because of these heating problems, it is quite difficult to produce integrated-circuit high-power (above 100 W) amplifiers. When the power-amplification devices are concentrated in the small area of an integrated circuit, it is almost impossible to transfer large amounts of thermal energy to the exterior of a device fast enough to keep it from overheating.

Power-amplification integrated circuits are available in low- to medium-wattage forms (often less than 20 W). These have special heat-sink tabs attached to the circuit's package and must be attached mechanically to a larger heat-sink external to the device (see Figure 12.3).

Another form of high-power amplifier circuit is a type that is a blend of discrete devices and integrated circuitry. These amplifiers, sometimes called integrated (package) amplifiers, contain discrete semiconductor chips which are connected internally to bias resistances. These packages, which are larger than standard integrated circuits, contain the complete circuitry for power amplification. The components are not on a single "chip," so they do not fit the classification of integrated circuits. But like an integrated circuit, they are simply connected to a power supply, input signal, and output load to become fully functional (see Figure 12.3).

Active Power Device Specifications

Following is a basic investigation of two power amplification devices' specifications contrasted with small-signal devices.

The MC1306P-½-W Audio Amplifier (see specifications in Appendix A) As can be seen by this integrated circuit's specifications, a complete amplifier system is built into one complete package. The specifications concentrate mainly on figures such as power capacity, input signal values, output impedance (for matching), and other signal- and power-related capacities.

The MJ15022-025 Complementary Silicon Power Transistors (see specifications in Appendix A) These 16-A (output current) power transistors are available as NPN and PNP types with identical characteristics. In a later section the application of these matched types will be explored. Following are the highlighted specifications for these devices:

16-A current capacity

High breakdown voltages (200 to 400 V)

5-A base current capacity

Total power dissipation of 250 W

These obviously exceed the specifications of a standard small-signal device. The following are other specifications that are specialized for power-type devices:

Thermal resistance. This specification indicates how well thermal energy can be transferred out of the device to a heat sink.

Power derating curves. These curves indicate a decrease in the power-

FIGURE 12.3 TDA2002A Integrated Power Amplifier

TDA2002
TDA2002A

8 WATT

AUDIO POWER AMPLIFIER

SILICON MONOLITHIC INTEGRATED CIRCUIT

8 WATT AUDIO POWER AMPLIFIER

The TDA2002 and TDA2002A are Class B power amplifiers designed for automotive and general-purpose audio applications. High output current capability (3.5 A) enables these devices to drive low-impedance loads (down to 1.6 Ω) with low harmonic and crossover distortion. High-voltage protection is available (TDA2002) which enables the amplifier to withstand 40 V transients. These devices provide an output power of 8 watts (typ) with $R_L = 2\ \Omega$ and 4.8 watts (min) with $R_L = 4\ \Omega$ at 14.4 volts.

- Internal Thermal Overload Protection
- Internal Short-Circuit Current Limiting
- Supply Over Voltage Protection
- Wide Supply Voltage Range (8 – 18 Volts)
- Low External Component Count

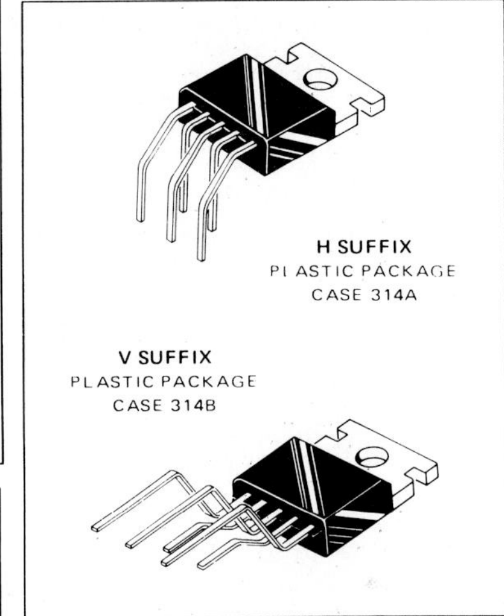

BLOCK DIAGRAM

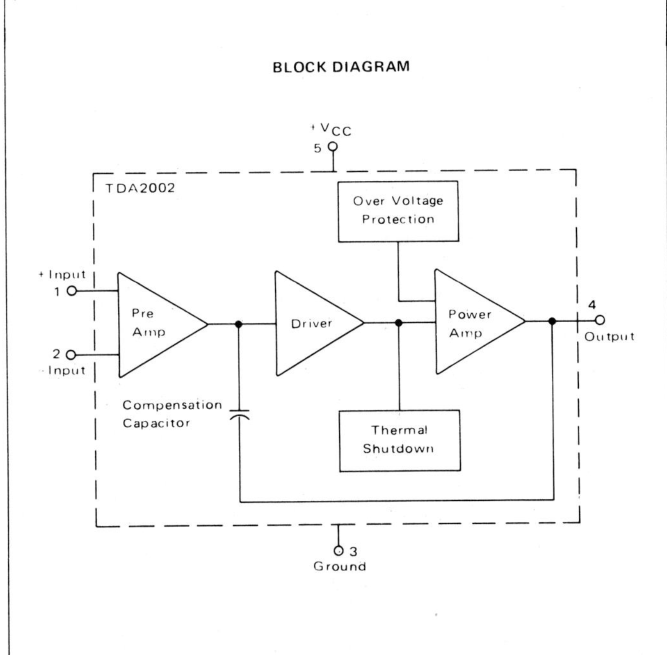

PIN CONNECTIONS

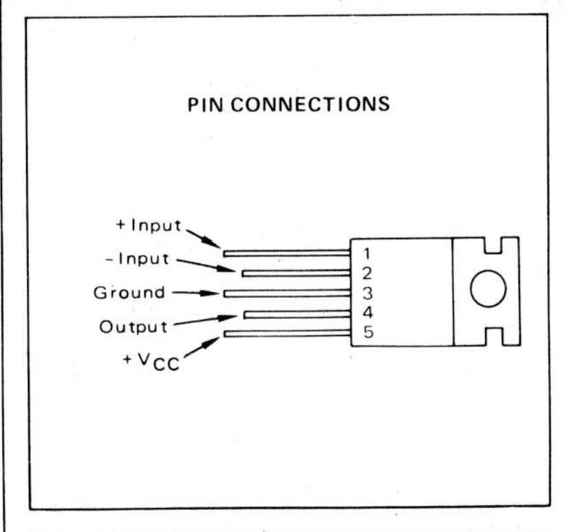

ORDERING INFORMATION

Device	Temperature Range	Plastic Package
TDA2002H *	-40 to +85°C	Case 314A
TDA2002V *	-40 to +85°C	Case 314B
TDA2002AH	-40 to +85°C	Case 314A
TDA2002AV	-40 to +85°C	Case 314B

*High Voltage

DS9445 R1

(Reprinted by permission of Motorola Semiconductor Products Inc.)

handling capacity of the device as it is heated above room temperature. By observing these characteristics, it can be seen how necessary it is to keep the temperature around the device as close to room temperature as possible.

Dc current gain. It can be seen that in contrast to small-signal devices which have typical current gain values in excess of 100, the power device can provide much less gain. This is the trade-off for having high power capacity.

Current–gain–bandwidth product. Since frequency-handling capacity is important, this characteristic is typically given with power devices.

Overall, the specifications for these types of devices include much higher current and voltage capacities, a special observation of temperature as well as other specifications critical to power handling.

It is quite common for the main output terminal of a power device to be electrically connected to the metal case or heat-sink tab attached to the device. This is done in order to obtain the maximum flow of thermal energy from this output terminal to the exterior of the package. It must be remembered that this case or tab is "electrically hot" to the output levels. The device can be damaged if this terminal is accidentally grounded. Also, a shock hazard may exist when working on such a device, as it is possible to touch this terminal when conducting tests on the system.

Power Amplification Circuitry

For relatively low power amplification (approximately 5 W or less) standard amplifier circuitry (except for op-amps) can often be used. The only modification necessary is to select high-power devices, but the balance of the bias and support circuitry can be applied directly.

When higher power levels are desired, it often becomes desirable to use special circuitry to provide power amplification. When high power levels are being manipulated, the efficiency of the system becomes critical. A system's efficiency is related to the amount of power consumed versus the amount of power delivered to the load. For small-signal amplifiers, such low power levels were seen that efficiency was not a problem. When potentially tens or hundreds of watts are being handled by a power amplification circuit, efficiency translates into losses that could prove both costly and potentially damaging to the system.

Often, the simple bias levels common to a small-signal amplifier can produce monumental power losses in a power amplifier. A power amplifier that used a BJT device, having a collector bias current (dc only) of 5 A producing a collector-to-emitter drop of 10 V, is "burning up" 50 W of electrical power—before any output is produced from an input. This is demonstrated in Figure 12.4. In this circuit the power supply is delivering over 50 W of electrical power even when no output is being produced. The efficiency of such a system is very poor. This power simply overheats the device, often reducing its capacity to produce a usable output power.

It is desirable to obtain the maximum output from the device, thus improving the efficiency of the system. This produces a second problem. The base bias level must be such that the changes in input signal can push the

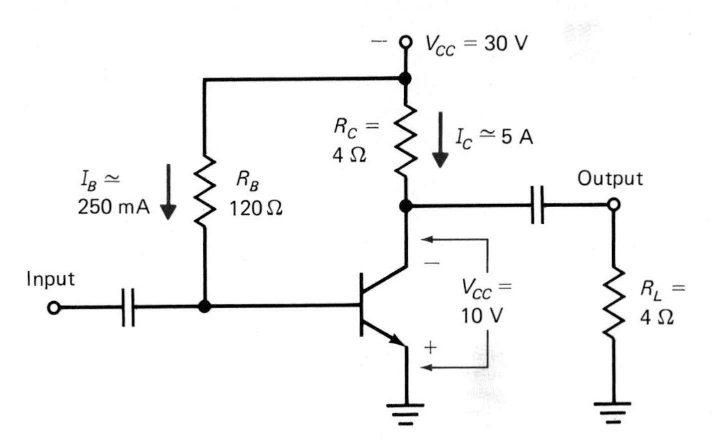

FIGURE 12.4 BJT Class A Amplifier

output to its maximum and minimum output (before saturating or cutting off the device). This requires that the device be biased so that the dc level rides "dead center" on the input signal. If this bias level fails to be centered and the input level is maximum, the output signal will be distorted (see Figure 12.5). For small-signal amplifiers, it was seldom necessary to extend the output from saturation to cutoff; close-to-"center" bias conditions were acceptable.

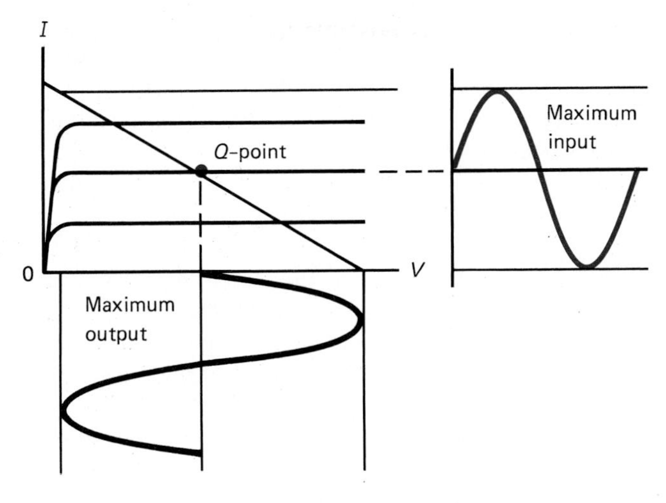

(a) Maximum output swing

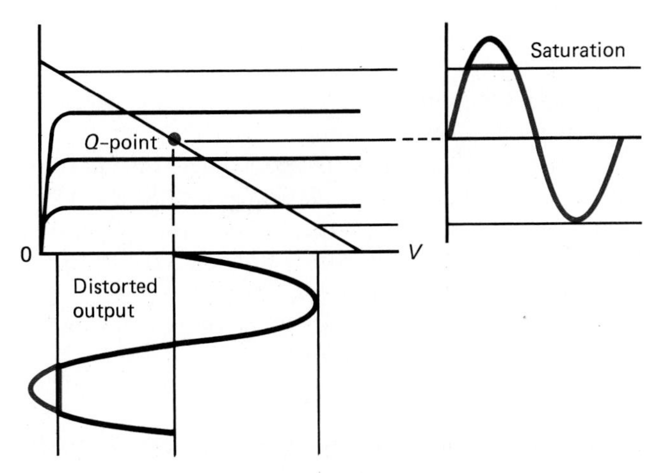

(b) Distortion due to saturation

FIGURE 12.5 Proper and Improper Biasing in a Class A Amplifier

In Figure 12.5 a properly biased signal that pushes the output to the maximum limit is shown. The effect of a subtle shift in the bias, producing a distorted waveform, is also shown. To reduce both the lost power due to the bias conditions and to help avoid distortion due to critical bias conditions, an alternative configuration of the circuit is used.

Classes of Operation Before proceeding with the analysis of power circuits, it is necessary to demonstrate alternative methods of handling the signal. These methods are called **classes of operation**:

Class A operation. An amplifier that produces a full sine-wave (both + and − cycles) output from a sine-wave input is termed to be operating as a class A amplifier.

Class B operation. An amplifier that reproduces only the positive (+) or only the negative (−) half of an input wave is called a class B amplifier.

Class C operation. An amplifier that reproduces less than half of the input waveform is called class C. The use of class C amplifiers is quite similar to that of class B amplifiers. The following descriptions of class B amplifiers can be interpreted for class C functions; thus a specific discussion of class C amplifiers has been omitted.

Class A, B, and C outputs from an amplifier are illustrated in Figure 12.6. As can be seen, the class B amplifier reproduces only half of the input signal. There are other classes of operation, related to the percentage of the signal reproduced at the output, but these have rather specialized applications and are not discussed in this book.

It may seem that the use of a class B amplifier is not very acceptable when attempting to produce linear amplification of the whole sine-wave input. But the use of two specially coordinated class B amplifiers can allow for an undistorted reproduction of the input signal. Figure 12.7 illustrates the basic process used to apply two class B amplifiers, basically working in parallel with each other, to reproduce a complete sine wave. The positive half of the input signal is amplified by one class B amplifier, set up to amplify only positive-going inputs. The negative half is amplified by the other amplifier, which amplifies only negative-going inputs. The outputs of these are blended together in the output circuitry, recomposing the original sine wave.

The advantage of this form of amplification is seen when the dc bias conditions needed to produce a class B amplifier are investigated. Since BJTs and FETs are polarity sensitive, without bias at the input, they will automatically amplify using a class B mode. Therefore, it is unnecessary to use a bias circuit with the device, thus eliminating the losses of power associated with these bias conditions. An additional benefit is that when there is no signal at the input, there will be no current flow at the output.

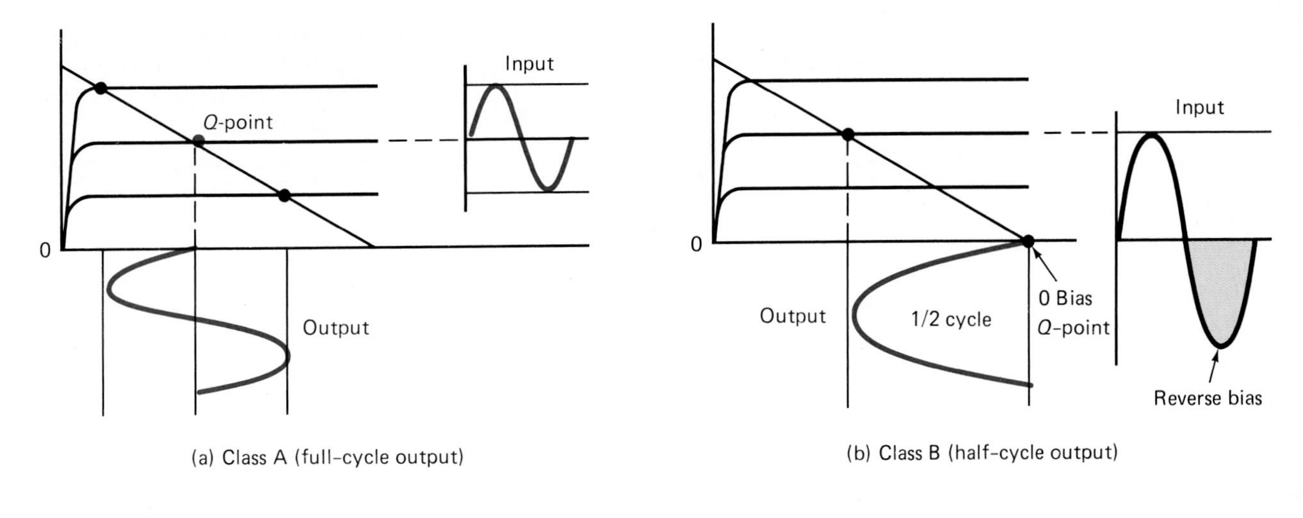

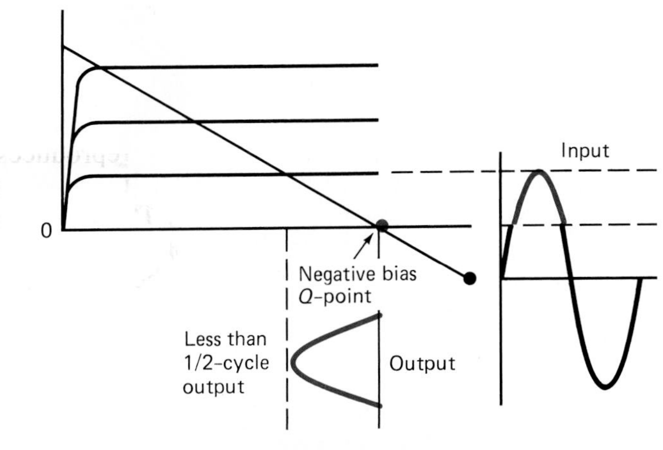

FIGURE 12.6 Classes of Operation

FIGURE 12.7 Two Class B Amplifiers Used to Reproduce a Full Cycle

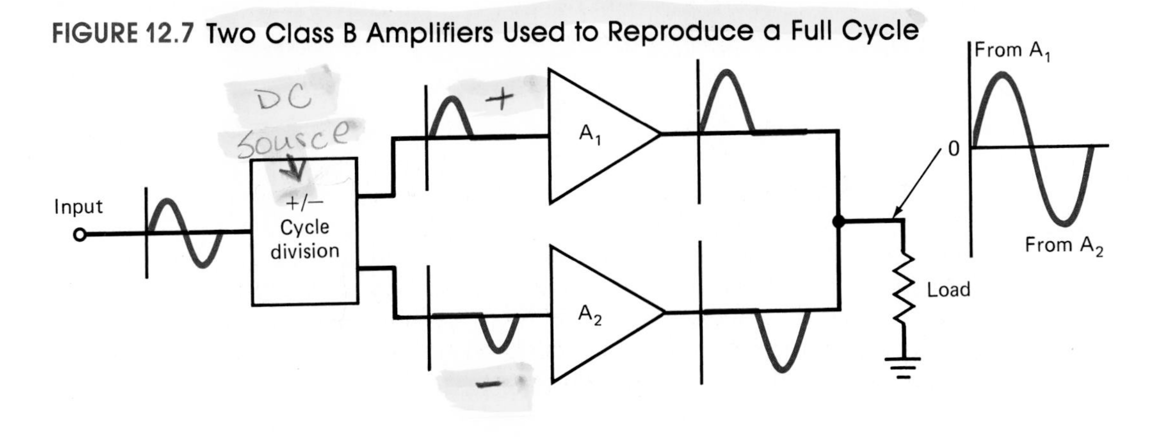

Note: Should class C amplification be used, enough individual amplifier stages would be used in parallel to totally recompose the sine-wave output. Since class C operation deals with from 0 to 180° of the input wave, the number of amplifiers needed is not fixed. But the overall process of amplification is the same as for class B operation.

EXAMPLE 12.1

The circuits shown in Figure 12.8 illustrate a PNP and an NPN transistor being used without base bias levels. It can be seen that the NPN transistor will amplify the positive portion of the input wave, while the PNP device amplifies the negative portion of the wave. When there is no input signal, the output current is zero; thus the only power that is dissipated is the power used to amplify the input signal.

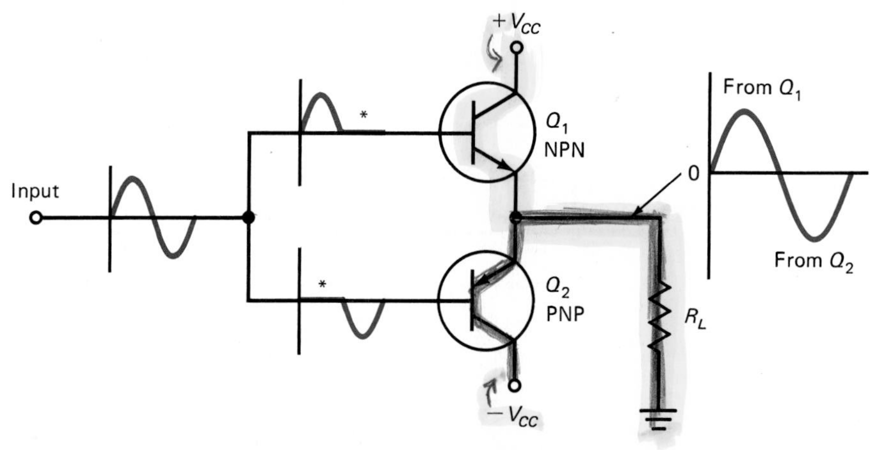

FIGURE 12.8 Simple BJT Complementary Amplifier (two Class B Amplifiers)

There are various means of connecting these devices or producing the negative versus positive signal splitting needed to produce class B operation. (Some additional circuits are demonstrated in the problems at the end of the chapter.) Power amplifiers that use a combination of two class B amplifiers to reproduce the input signal are often called either *push-pull amplifiers* or *complementary-symmetry amplifiers*. This is an application of the MJ15022-025 complementary power transistors noted previously.

Troubleshooting Power Amplifiers

Since power amplifiers work with large magnitudes of current and voltage, they are more prone to becoming damaged by excessive power levels than are small-signal devices. Also, since they are often connected to external loads (such as speakers) they are subject to being damaged by connection of inappropriate loads or by accidental short circuits. Therefore, they are often the most common device to fail in an amplifier system.

To troubleshoot a power amplifier, it is first necessary to determine which form of amplifier is being used, class A or class B (or another classification). If the amplifier is class A, standard measurements of bias and signal levels will provide an adequate test of the device.

If class B amplification is used (usually, the appearance of two or more transistors connected in parallel is an indicator of this class of operation), it is possible that only one of the devices is defective. An inspection of the output signal will often reveal if only one device is defective. If only one device fails, the output will characteristically be only a half-wave (distorted) signal. An indirect means of isolating the defective device is to monitor the temperature of the two devices; the one that displays heating is the one that is functioning, while the cool one is the device that is failing to conduct.

Occasionally, bias conditions on power amplifiers may change in such a manner that the output waveform becomes distorted. Therefore, it is necessary to monitor carefully the dc bias levels around the device to establish if it or its related circuitry is the source of distortion.

Special Troubleshooting Information A power amplifier should always be tested with an appropriate load attached. Failure to attach a load (often a "dummy load," which is a power resistor with the same resistance as the typical load) may cause the amplifier to overheat and become damaged.

Unless approved in the design of the amplifier, a square-wave signal should not be applied to a power amplifier for a prolonged period. The switching on and off that occurs within the device can damage it.

Some power amplifier circuits use either thermal resistors or temperature-sensitive diodes to assist in controlling the temperature of the active device. These are often tied mechanically to the heat sink of the device. They are used to sense temperature variations and adjust bias levels to compensate for these variations. These devices must remain attached to the heat sink and should be replaced with exact duplicate replacements if found defective. Power devices should never be left unmounted from the heat sink, as excessive temperatures can develop which would destroy the device within seconds of turning on the main power.

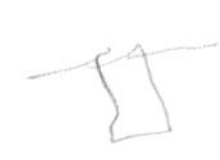

Finally, standard transistor testers may not be usable to check power devices. They may not have enough power to energize the device sufficiently and run an adequate test. The manual for the tester should be used to determine the correct way to test power devices.

REVIEW PROBLEMS

(1) Prepare a comparison of the specifications of the power transistor listed in Appendix A to a similar (NPN or PNP) small-signal device. Include a discussion of the maximum and minimum ratings for each.

(2) Write a short evaluation of the effect of using a class A power amplifier to produce a 100-W amplifier. Pay special attention to the need for maintaining bias levels. Assume, for statistics, that the 100 W is to be delivered to an 8-Ω load. (For this evaluation, calculate the actual current and voltage that would be delivered to the 8-Ω load under normal operation. The dc bias levels that would have to be maintained on the device are basically one-half of these values.)

(3) Write a short comparison of what may have caused the output of a power amplifier to be distorted, using first a class A amplifier, then a class B amplifier. Assume that the output is just the positive half of the input wave. There will be two obvious causes for the distortion, based on the types of amplifiers used.

12.2 SIGNAL-MIXING AMPLIFIERS

So far only single inputs have been applied to the amplifier system. There are cases where multiple inputs are used to produce a common output. A typical application of a signal mixing amplifier is in an audio microphone mixer. In the microphone mixer, two, three, or more microphone inputs are blended to produce a single output, which is then amplified (see Figure 12.9). The microphone mixer produces an output that is a linear blend of the two microphone signals. These signals are simply added together in a blending process. Another form of mixing is seen in a circuit called a **comparator**. In the comparator, although two inputs are used, the output is not a blend of the two inputs but represents only the difference between the two signals. The comparator has a wide application in electronic instrumentation and special control circuitry. The comparator actually subtracts one input's levels from those of the other and then amplifies this difference. A typical configuration for the comparator is illustrated in Figure 12.10.

Basically, the op-amp comparator takes advantage of the differential input function of the op-amp. This differential input function automatically causes an inverted versus a noninverted addition of the two input signals, which results in their being subtracted [signal 1 + (– signal 2)]. This difference is what is amplified through the op-amp, using the standard voltage-gain function of the op-amp.

Two Basic Forms of Mixing

In a mixing amplifier, there are two basic forms of mixing the signals. One is a linear mixing, which produces an even blend (either addition or subtraction) of the input signals. The other form is nonlinear mixing, which does not produce an even addition of the signals, but rather a unique blend of the signals.

Linear Mixing As noted in the two forms of mixers presented at the beginning of this section, linear mixing creates an even blend of the signals input

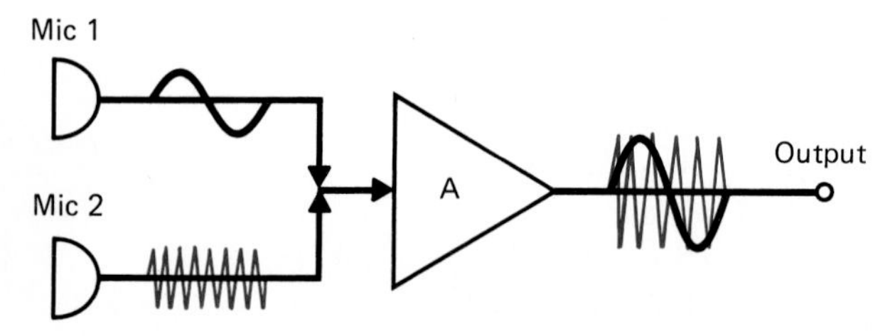

FIGURE 12.9 Mixing Amplifier

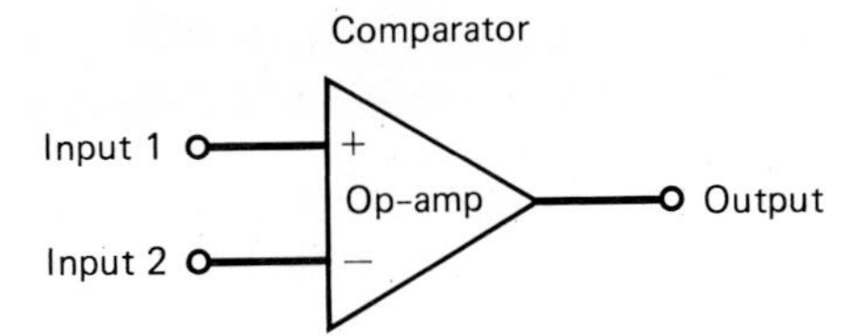

FIGURE 12.10 Comparator (Differential-Mode Op-Amp)

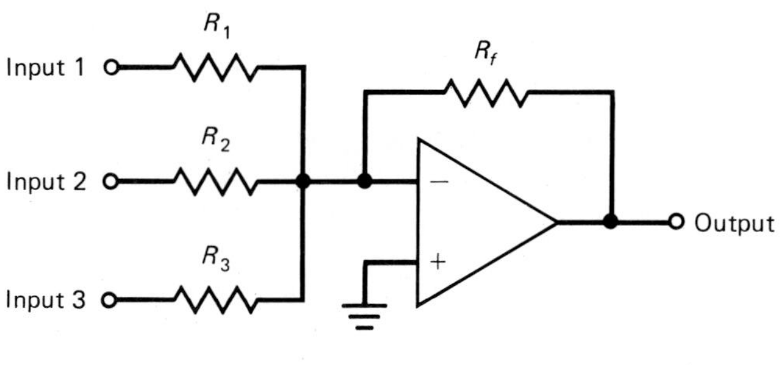

FIGURE 12.11 Three-Input Summing Amplifier

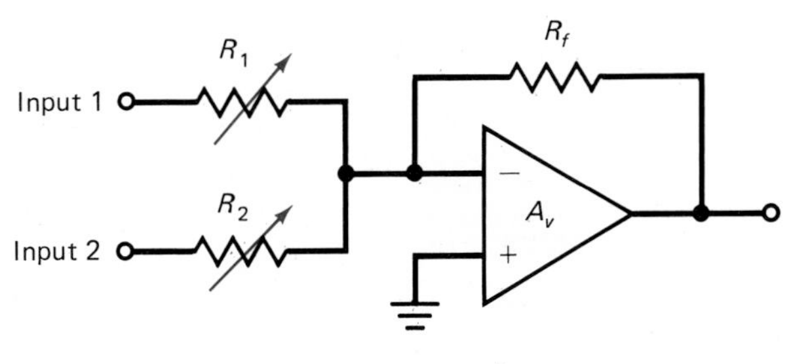

$$A_v \text{ (input 1)} = -\frac{R_f}{R_1}$$

$$A_v \text{ (input 2)} = -\frac{R_f}{R_2}$$

FIGURE 12.12 Variable-Ratio Summing Amplifier

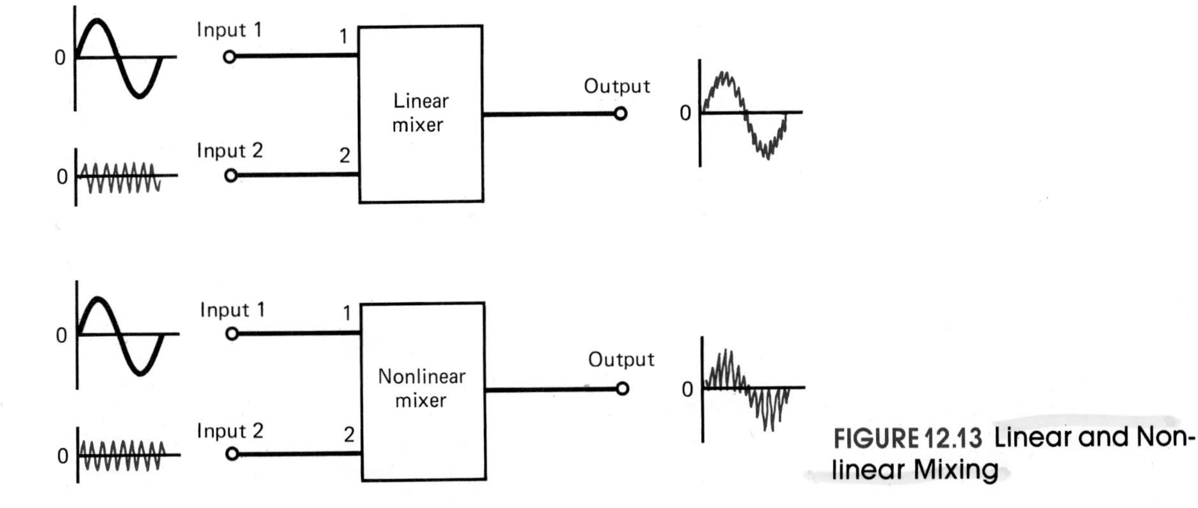

FIGURE 12.13 Linear and Nonlinear Mixing

to the mixing amplifier (system). This mixing process actually produces a new signal, one that is a composite blend of the other inputs. Another name for this form of amplifier is **summing amplifier**.

In the amplifier shown in Figure 12.11, all three inputs are presented to the input of the op-amp. These are blended at the input terminal and this blended signal is amplified by the op-amp. (*Note:* This is not a differential amplification process.) For each of the signals the gain is determined by the R_F-to-R_I ratio, which is the same since all R_I values are the same. If the individual input resistors are made to be either variable or are of fixed but different resistances, the amplification of each signal will depend on the different R_F-to-R_I ratios.

Figure 12.12 illustrates what is called a **scaling amplifier**. In this amplifier the ratio of each input resistance to the common feedback resistor produces various gain values for the amplifier and thus various levels of each of the inputs would be present in the composite output.

In each of these mixing amplifiers some level of isolation of one signal from the other is achieved. This can be seen by observing the total resistance between any two of the inputs. This form of mixing is superior to connecting the inputs directly to each other, as problems may occur—one input may affect the other when they are directly connected.

Nonlinear Mixing Nonlinear mixing involves a unique blending of two input signals. One of the signals is applied in the normal manner to the input of an amplifier. The other signal is applied in such a manner as not to blend with the first signal, but to affect the gain of the amplifier itself. Thus, as the first signal is being amplified, the second is controlling the proportion of gain for this signal.

In this configuration, the two signals do not simply blend, but one signal's rises and falls are superimposed on the other signal. Figure 12.13 more clearly illustrates a comparison of linear and nonlinear mixing. Nonlinear mixing is commonly used in communication electronics. Due to this special application, we will not concentrate on the process of nonlinear mixing. Since linear mixing amplifiers are nearly identical to single-input amplifiers, most components and equations are the same as for single-input amplifiers.

REVIEW PROBLEMS

(1) Briefly describe the difference between linear and nonlinear mixing.

(2) Illustrate the output from first, a summing amplifier and then a differential (mode) amplifier, using the signals shown in Figure 12.14, assuming a gain of 2 for the amplifier being used. (Pulsed waveforms were chosen for this analysis because they are easier to compare on a point-by-point basis.)

(3) If a 50-kΩ resistor were used for a feedback resistance on a scaling amplifier used as a microphone mixer, specify the value to which input potentiometers must be set to meet the following considera-

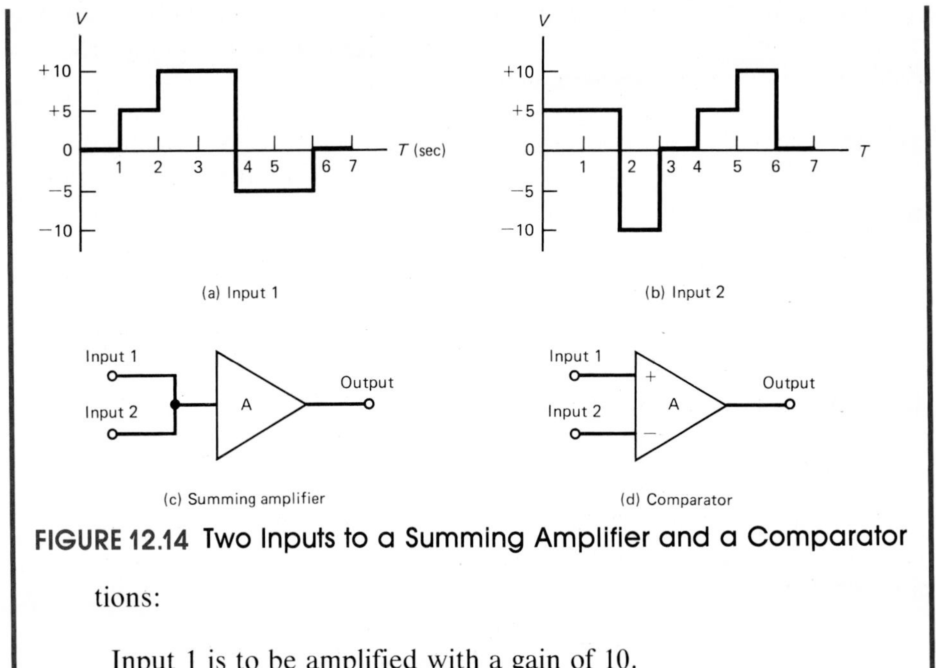

FIGURE 12.14 Two Inputs to a Summing Amplifier and a Comparator

tions:

Input 1 is to be amplified with a gain of 10.
Input 2 is to be amplified with a gain of 5.
Input 3 is to be amplified with a gain of 10.
Input 4 is to be amplified with a gain of 2.

Also illustrate the circuit, indicating the resistance values.

12.3 OTHER SPECIAL-PURPOSE AMPLIFIERS

There are quite a large number of amplifier applications. There are also quite a wide variety of other configurations of amplifiers. It would be impossible to list all the possible combinations of devices, bias arrangements, and signal-handling functions seen in these amplifiers. There are, though, a number of somewhat common amplifier configurations that need to be presented. Many of these configurations are exclusive to particular devices (BJT, FET, etc.). Following is a discussion of several of these configurations.

The BJT Common-Base Amplifier

A BJT can be used in three popular configurations. The common-emitter configuration is the most popular and hence was the one highlighted in earlier chapters. [It will be remembered that the term "common" emitter indicated that the emitter current was a composite of input (base) current and output (collector) current.] It is possible to orient the BJT so that either the base lead or the collector lead is common to both the input and output. This presentation concentrates on the common-base configuration.

Figure 12.15 illustrates the common-base configuration of the BJT. The principal use of this configuration is as a buffer amplifier, providing primarily impedance matching and some voltage gain. It functions as follows:

1. Input current is supplied to the emitter terminal. A bias current flows at this terminal, as set by the emitter resistor and emitter power supply.
2. This input and bias current determines the current flow both through the base terminal to ground and to the collector lead. (It will be remembered that emitter current is the sum of base and collector currents in the common-emitter configuration. The same is true in the common-base mode.) Therefore, the sum of base current and collector current is equal to emitter current.
3. The collector current is the output current of the circuit, supplied through the collector resistor (which is the load resistor) via the collector power supply.

As can be imagined, the current gain of this circuit is less than 1 since the emitter current determines both base and collector (output) current. The current gain of this amplifier is given a new term (not beta). It is called alpha (α) and is represented by the formula

$$\alpha = \frac{I_C}{I_E} \qquad (12.1)$$

In some specification sheets, the alpha value is given; otherwise, it can be calculated in terms of beta:

$$\alpha = \frac{\beta}{1 + \beta} \qquad (12.2)$$

The voltage gain of this circuit is given in the equation

$$A_V = \frac{R_C}{r_e} \qquad (12.3)$$

where the term r_e is the ac emitter resistance, usually in tens of ohms, and is either a listed specification or is found by

$$r_e = \frac{26 \text{ mV}}{I_E} \qquad (12.4)$$

Thus, by a manipulation of the collector resistance value, the voltage gain can be large (20, 30, etc.) while the current gain is less than unity. The input

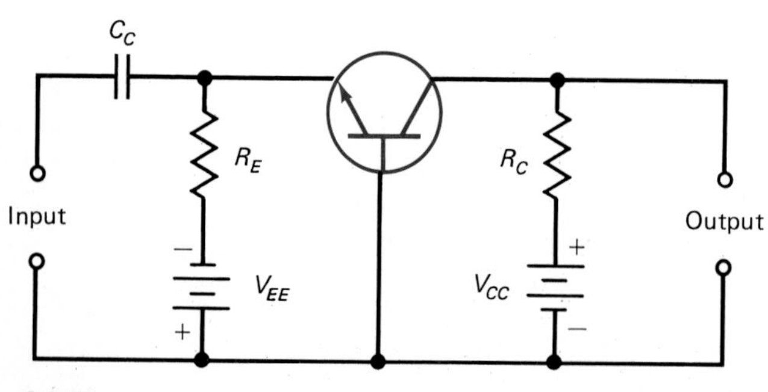

Impedance Matcher or Buffer

FIGURE 12.15 Common-Base Amplifier (BJT)

resistance (as shown below) is usually quite low, while the output impedance can be set to nearly any usable level (again see below).

$$R_{\text{in}} = R_E \parallel r_e \tag{12.5a}$$

$$R_{\text{out}} = R_C \tag{12.5b}$$

Since this configuration requires the use of two dc sources, one at the emitter and one at the collector, it has a limited use in electronic circuitry. Only when the critical match of a low to a high impedance is necessary is this circuit used.

The BJT Common-Collector Configuration

As mentioned in the past coverage, the common-collector configuration of the BJT is another form of application. This configuration is more popular than the common base and like the common-emitter form, uses a single power supply.

The common-collector configuration (Figure 12.16) is almost identical to the popular common-emitter form. In fact, the design of the bias elements is identical to that of a similar common-emitter configuration. It will be noted that the output is taken not on the collector lead but at the emitter lead. (This configuration is also called an emitter-follower configuration.) By taking the output at this lead, several statistics about the circuit change dramatically from using a competitive common-emitter circuit.

Since the voltage at the emitter is nearly identical to the voltage at the base, the voltage gain is assumed to be

$$A_V \simeq 1 \tag{12.6a}$$

$$A_i \simeq \beta \tag{12.6b}$$

The input impedance of the amplifier is strongly affected by the value of the emitter resistor and load resistor attached at the emitter terminal:

$$R_{\text{in}} = R_{B1} \parallel R_{B2} \parallel \beta(R_E' + r_e) \tag{12.7}$$

where $R_E' = R_E \parallel R_L$.

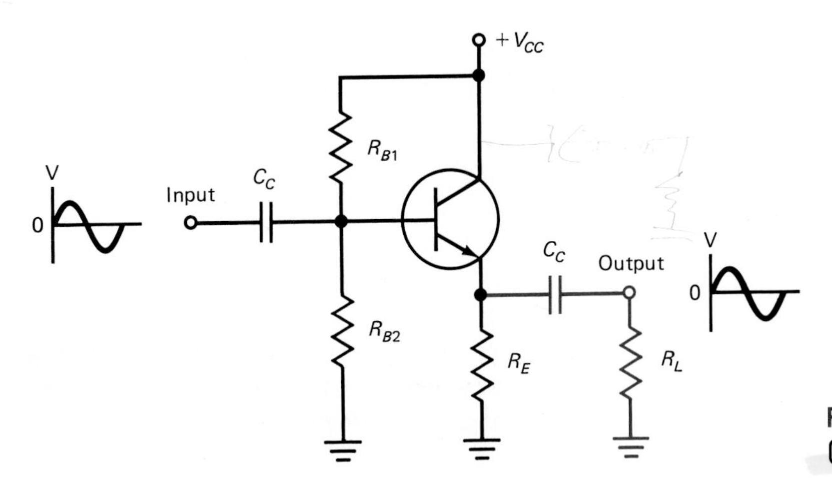

FIGURE 12.16 Common-Collector (Emitter-Follower) Amplifier (BJT)

The output resistance is

$$R_{out} = R_E' \parallel \left(\frac{R_S'}{\beta} + r_e\right) \tag{12.8}$$

where $R_S' = R_S \parallel R_{B1} \parallel R_{B2}$. This output impedance is usually quite low, often about the value of r_e.

The current gain of this circuit is computed in the same way as for a common-emitter circuit. The advantage of this circuit is its potentially low output impedance and high input impedance, determined mainly by the R_{B1}–R_{B2} parallel combination. Such a circuit is often used in power amplification, especially in the complementary-symmetry mode of operation, where one transistor's output is taken from its collector while the other's output is from its emitter. A final advantage of this configuration (and why it is usable in the complementary-symmetry mode) is that the output of the amplifier is not an inverted version of the input but is totally in phase with the input. If the phase inversion of a common-emitter amplifier produces a problem, the common-collector mode may be selected.

It is possible on a single BJT amplifier to take outputs simultaneously from the emitter and collector terminals (Figure 12.17). This is done to obtain identical signals but with a 180° phase difference between them. This amplifier is sometimes termed a "phase-splitting amplifier." In this case the voltage gain for both the collector and emitter outputs is set to the same value, thus producing identical amplitude signals.

The FET Source-Follower Configuration

In a like manner, the output can be taken from the source lead on the typical FET circuit. The effects are similar to those of the emitter-follower amplifier (except the input impedance remains nearly infinite, as for any FET amplifier). A similar phase-splitting, two-output configuration can also be used.

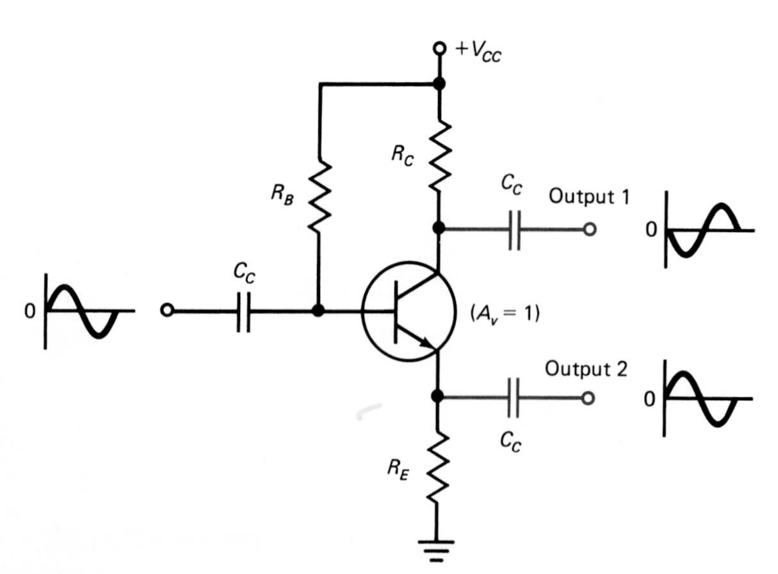

FIGURE 12.17 BJT Amplifier Used in Both Common-Emitter and Common-Collector Modes

MOSFET Biasing

In the chapter that described MOSFETs it was noted that there are two types of MOSFETs, the depletion and enhancement types. It was noted that these forms functioned quite similar to the standard JFET, but had different parameters.

The depletion MOSFET (DE-MOS) is made such that it will accept an input of either polarity, up to some limits. What this means is that it will produce conduction levels that are more or less in magnitude than the zero-bias conduction value. Thus for certain applications, it is unnecessary to provide a dc gate bias level and still have the device amplify both the positive and negative portions of an ac input. In Figure 12.18 both P- and N-type depletion MOSFET circuits are shown. It will be noted that a source resistor, used with a JFET for gate bias, is not used. Even though there is no bias level at the gate, the input signal is still amplified without regard for polarity. For each type, P and N, the amount of "reverse" potential that can be fully amplified is less than the typical "forward" amount. If too high an amplitude is input, the output will become clipped, much like the signal from a saturated amplifier. Such a circuit can, however, be used for relatively low input signals.

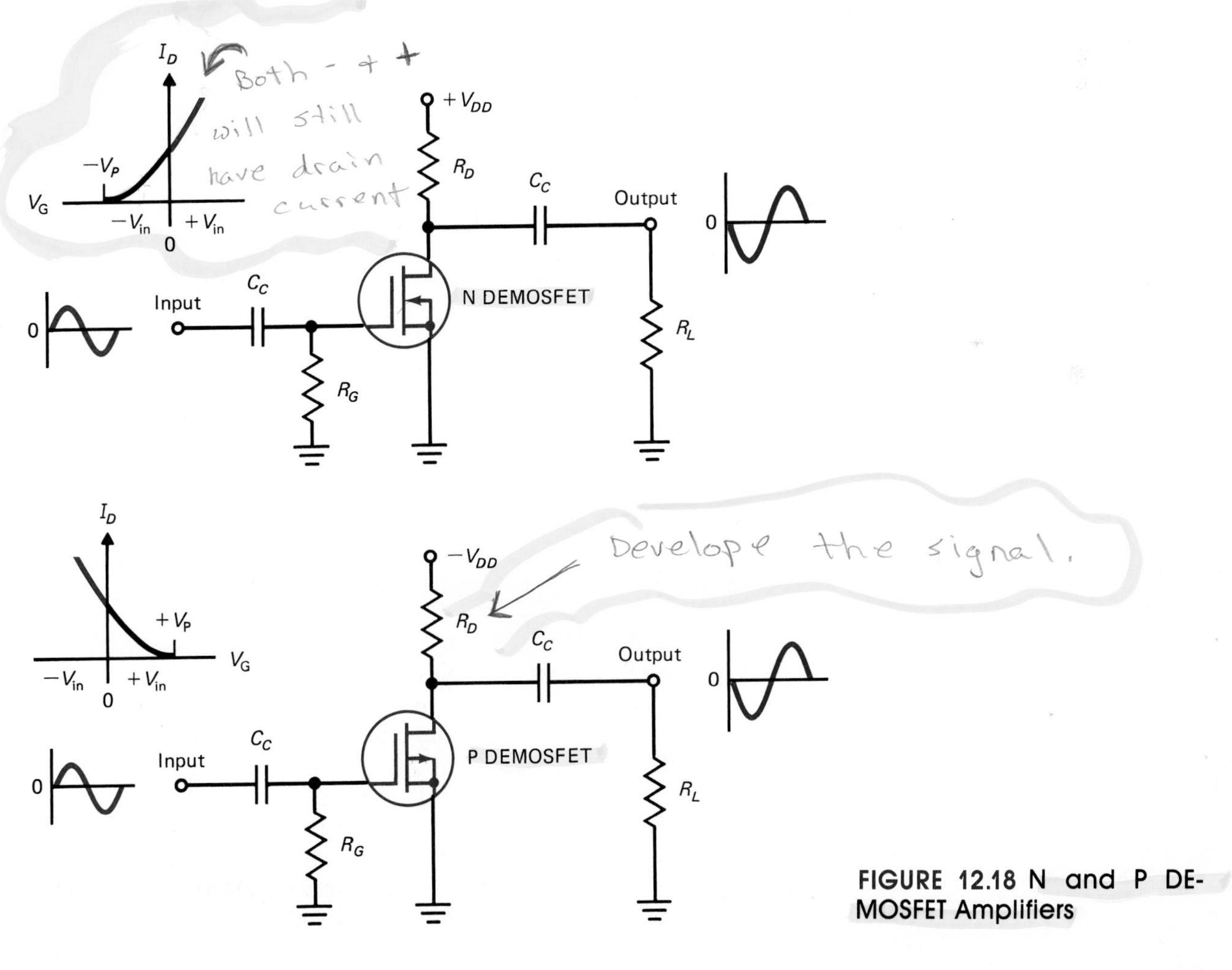

FIGURE 12.18 N and P DE-MOSFET Amplifiers

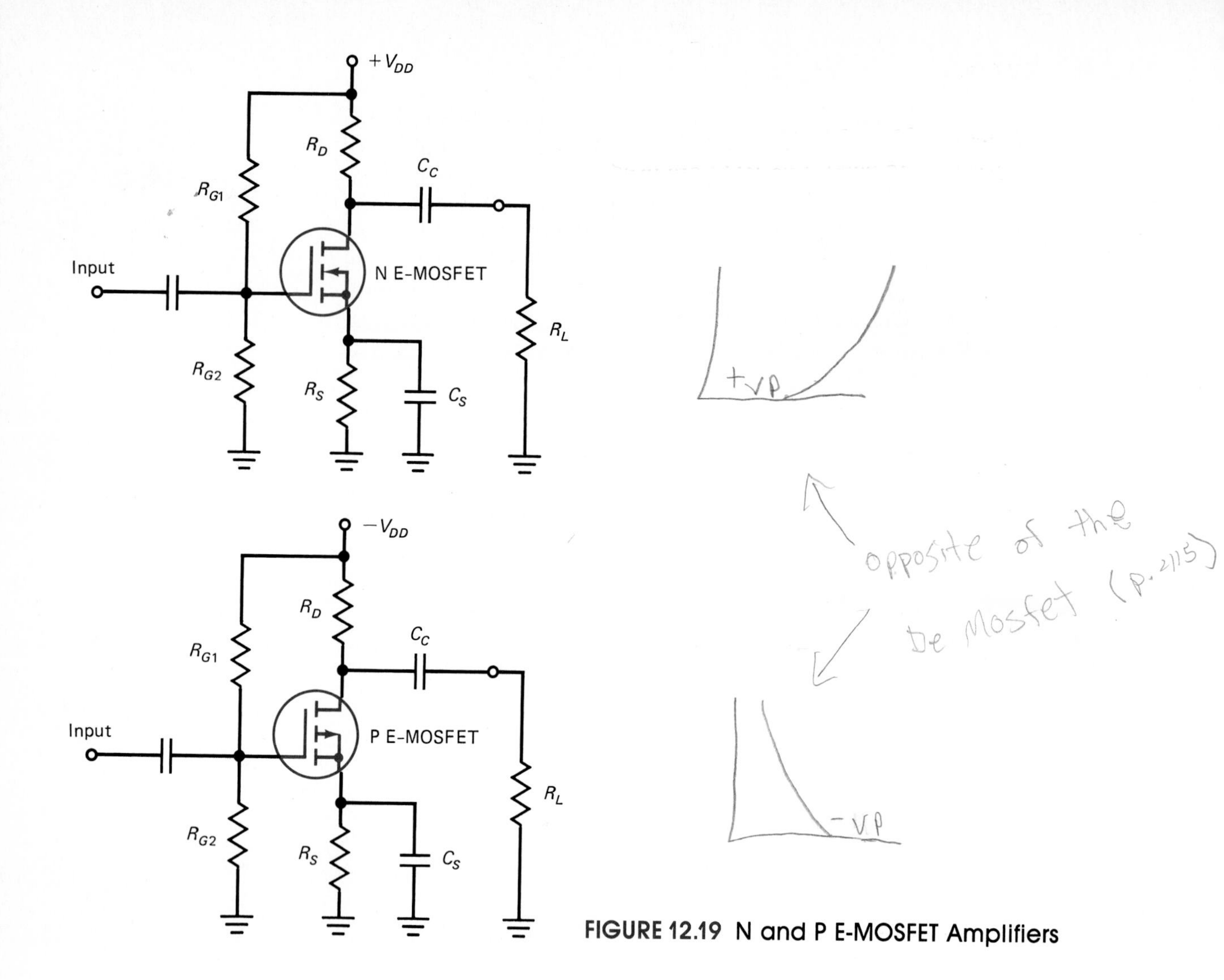

FIGURE 12.19 N and P E-MOSFET Amplifiers

The enhancement MOSFET (E-MOS; Figure 12.19) is constructed in a manner quite different from the JFET or the DE-MOS FET. This device uses gate bias to encourage the flow of drain current. Thus there is no output current flow if there is not an input voltage level. Although still showing a very high input impedance, the E-MOS device can use a bias configuration much like the beta-independent BJT circuit. The two resistors at the gate of the device form a simple voltage divider. This divider produces a bias voltage that creates a constant drain current at the output. Input signals raise or lower this gate voltage and thus produce a rise and fall in the responding drain current. Generally, the same form of operational calculations can be used for this device as is used for other FET types. The only special function is that there is a minimum gate voltage that must be maintained to obtain an output. This is usually termed V_T, or threshold voltage. If the gate level drops below this value, no output is seen. (This can be somewhat related to the minimum base–emitter drop on a BJT.) The voltage is typically around 2 V.

Discrete Differential Amplifier

In the discussion of the op-amp, the differential input function of that amplifier was presented. The same function can be duplicated using discrete devices. In the circuit of Figure 12.20(a), two BJTs are used to produce a differential output. The two transistors can each accept an input [see Figure 12.20(b)]. If only one input is applied, the output of one transistor is an inverted representation of that input, the output of the other is a noninverted representation.

FIGURE 12.20 Discrete (BJT) Differential Amplifier

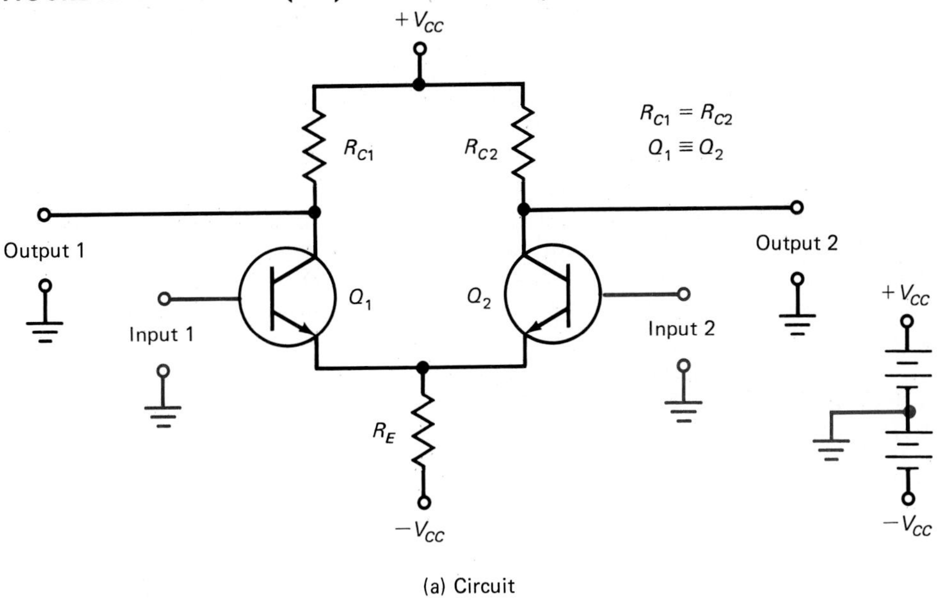

(a) Circuit

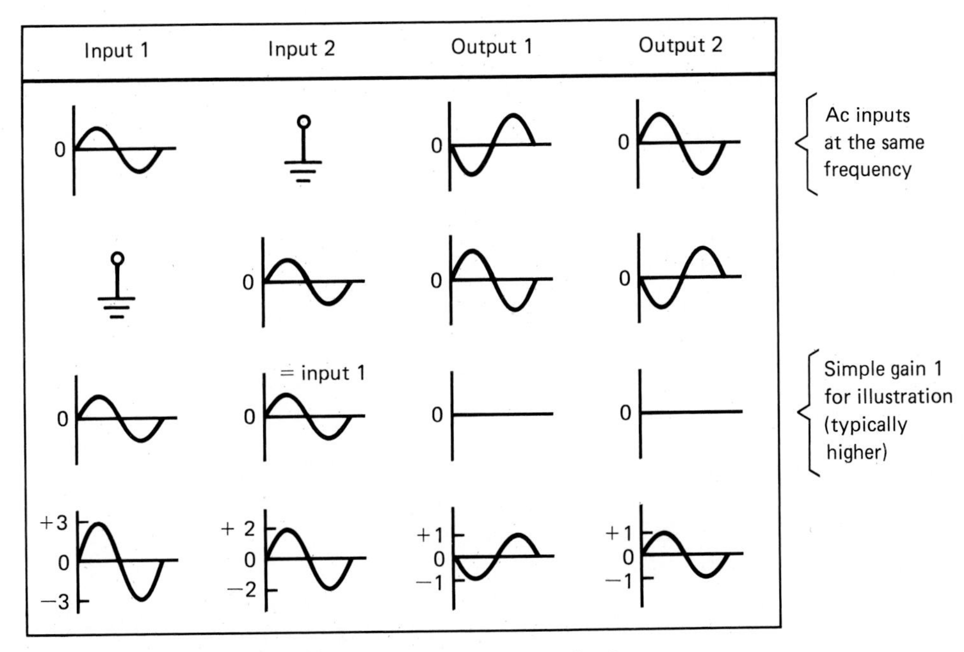

(b) Input and output evaluation

If both transistors receive the same input signal, they affect each other in such a manner that the output from either is zero. If there is a slight variation in the input to one in respect to that of the other, the output will be an inverted and a noninverted representation of only the variation. This is due to the fact that the conduction of one device will, through the shared emitter resistor, have a countereffect on the current through the other (and vice versa).

The application of such a circuit is similar to the application of an op-amp used for its differential input function. [It should be noted that a nearly perfect match of one transistor to the other is mandatory, since any imbalance will cause an error in this "countereffect" process. It is for this reason that op-amps are usually used when differential amplifier functions are needed.]

Op-Amp Differentiators and Integrators

The mathematical process of integration and differentiation is a primary topic for calculus courses. Electronic circuitry can be made that duplicates this mathematical process when dealing with electrical signals. We will not attempt to develop the mathematical background to understand these complex processes, but the electronic circuitry that produces differentiation and integration is fairly simple.

So far in this book we have dealt with sine waves and have viewed these waveforms passing through amplifiers without change (except for inversion) in their shape, aside from distortion, which is not desirable. Other waveforms, such as square and triangular waves (to name a few), can be changed in shape. The mathematical process of integration and differentiation can prove this shape-change function.

An integrator can be used to change the shape of a square wave into a triangular wave shape of the same frequency and amplitude. A differentiator

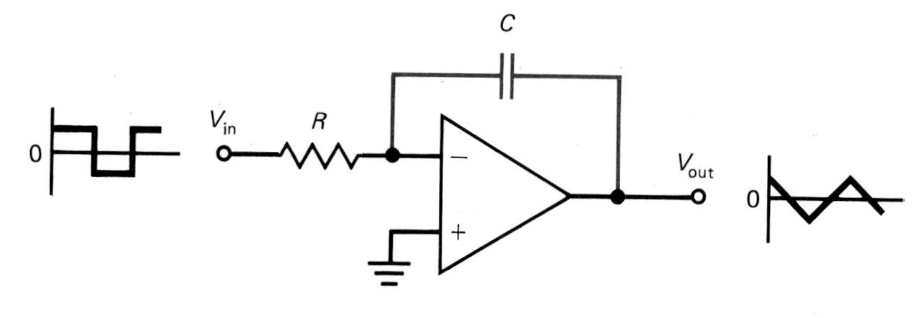

(a) Integrator (inverting)

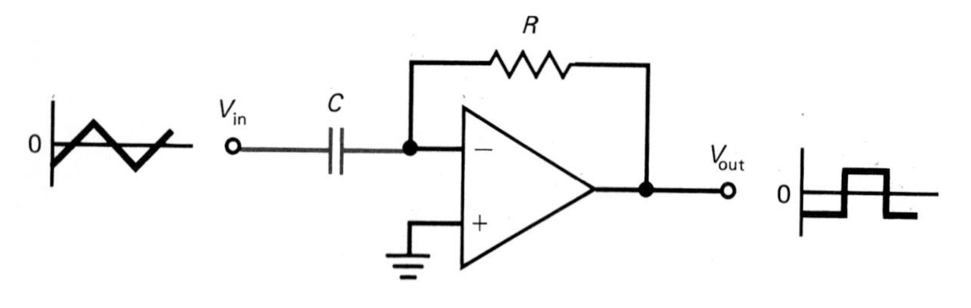

(b) Differentiator (inverting)

FIGURE 12.21 Capacitor-to-Resistor Relationships

(which performs the opposite mathematical function of an integrator) can change a triangular wave shape into a square wave shape. These two circuits use a capacitor-to-resistor relationship to achieve this function (see Figure 12.21). In these two circuits, the process of integration and differentiation is produced by the relationship of the resistor to the capacitor's charge/discharge process. The specific position of the capacitor details how the circuit will respond to the input (integrate it or differentiate it). The major application of these circuits is when a change in the form of a wave is desirable. (By the way, neither the integrator nor the differentiator will affect the shape of a pure sine wave.) Naturally, to understand all of the applications and internal functions of these circuits, a knowledge of calculus is required. At this point it is sufficient to know that such functions can be achieved in relatively uncomplicated circuits.

Although there are quite a few more configurations for active amplification circuits, in this section some of the more popular variations have been presented. The reader is encouraged to research more circuit variations, especially using the guidelines presented in the next section.

REVIEW PROBLEMS

(1) Following is a list of possible circuit applications. Match the type of circuit presented in this section to these applications.
  - **(a)** To couple (join) a very low impedance sensor to a high-input-impedance circuit
  - **(b)** To amplify a low-level input signal (voltage), providing both a high input impedance and a minimum number of bias components
  - **(c)** To provide both an inverted and a noninverted output signal simultaneously from the same input
  - **(d)** To convert the square-wave output of a signal generator into a triangular wave shape

(2) A disadvantage of the common-collector amplifier is that it can not usually be used with a variable-resistance load. Discuss why this is a problem.

(3) Discuss briefly why it is a problem if the transistors used in a differential amplifier are not perfectly matched. (To help in your description, consider if one had a current gain of 50 and the other had a gain of 90.)

(4) Identify at least one possible malfunction in each of the circuits presented in this section. Discuss what effect that malfunction could have on the circuit's operation.

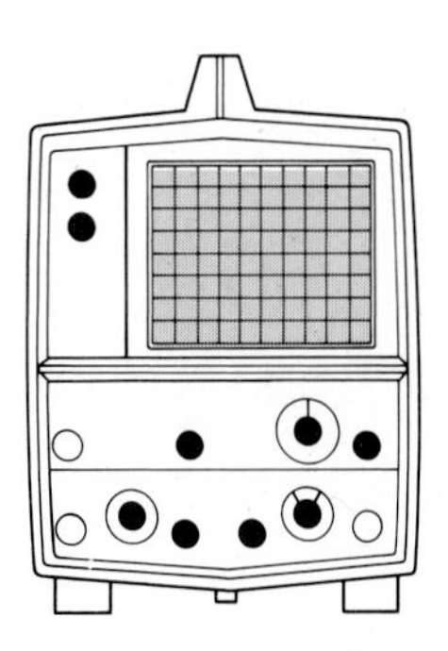

12.4 INVESTIGATING NEW APPLICATIONS

In this chapter has presented a wide variety of circuits and applications. By no means have we been able to cover all the possible variations or circuit combinations that are currently in use. Also, there are always potential new uses for semiconductor devices and certainly the potential for the development of new discrete devices. The production of new linear integrated circuits has become almost a daily event. Manufacturers are designing special-task integrated circuits at a pace that is almost impossible to keep up with.

To prepare for these developments it will be necessary to present guidelines and observations that can be followed when approaching these new developments. The following is a generalized breakdown of the primary techniques that can be used when making such an investigation. They are written on a generalized ("generic") basis and should be modified according to the specifics encountered when actually applying them to circuitry.

New Circuitry Applications

Taking a known device and placing it in a different configuration or unique bias/signal flow arrangement can produce a circuit that does not exactly fit the traditional applications. When observing this new application, the following general techniques can be used to evaluate this application:

1. Whenever possible, a block diagram for the circuit should be used to determine the "goal" of the circuit. This diagram provides an overview of the entire system's processing of signal information. By identifying the role of the specific circuit in question as it relates to other circuitry sections, one can easily identify the purpose of that circuit. Knowing the role played will often define the role of each component in the circuit itself.
2. Second, the bias arrangement associated with the circuit will often describe how that circuit functions in respect to the rest of the system. If the bias arrangement appears to follow a traditional form (such as shown in this text), the circuit can be roughly identified (e.g., ". . . a common-emitter BJT amplifier with beta-independent biasing . . ."). If the arrangement does not fit a more traditional form, a simple cause/effect relationship should be developed, relating input bias to output bias (these bias levels are traditionally included with the schematic of the circuit for just such a purpose).

   To assist in this analysis, it is common to refer to the manufacturer's data sheets which describe the specific device used. By both investigating the normal dc limits and "average" values, one can form a clearer picture of the circuit's function. The data sheet will also help to qualify the device's application. For example, if the device were described as a general-purpose audio-amplification device, one could begin to narrow down the circuit's function to fit into that category.
3. The next step would involve the flow of signal information within the circuit. Such questions as "Where is the input signal applied?", "Where is the output taken?", and "Is there feedback of the signal, and how is it achieved?" should be answered. Such information is obtained by looking at the circuit's positioning within the overall system. In some cases, sample signals (waveforms) will be illustrated at key input, feedback, and output points, either within the circuit or at key positions in the entire system. Determining how the signal "gets from point *a* to point *b* and what happens to it in the process" can assist in identifying the actual role played by any one circuit.

   At this point, signal conditioning elements such as filters, resonant circuits, and dividers can be segregated from the amplification function of the circuit. Since the dc bias elements are identified and the input and output connections are located, additional elements can be highlighted as fitting (usually) the role of signal conditioning. Included in this category would be elements (often capacitors) used to contain the bias levels within the specific amplifier.

By the time these three specific investigations are conducted, the role of the circuit and how it will achieve that role becomes fairly well known. There are a few cases where a process of elimination will have to be done to identify the role of an individual component. By that time, though, so much is known about the rest of the circuit that it is a simple challenge to ''jigsaw'' the final pieces into the entire circuit's picture.

EXAMPLE 12.2

For the circuit and block diagram pictured in Figure 12.22, identify the role of the separate amplifiers and components used in them, applying the three steps noted above.

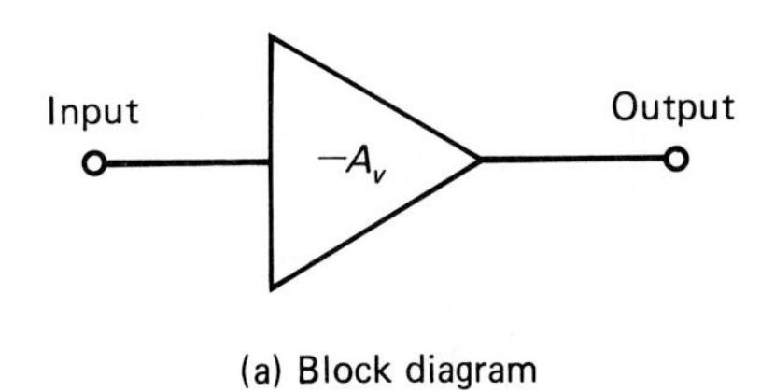

(a) Block diagram

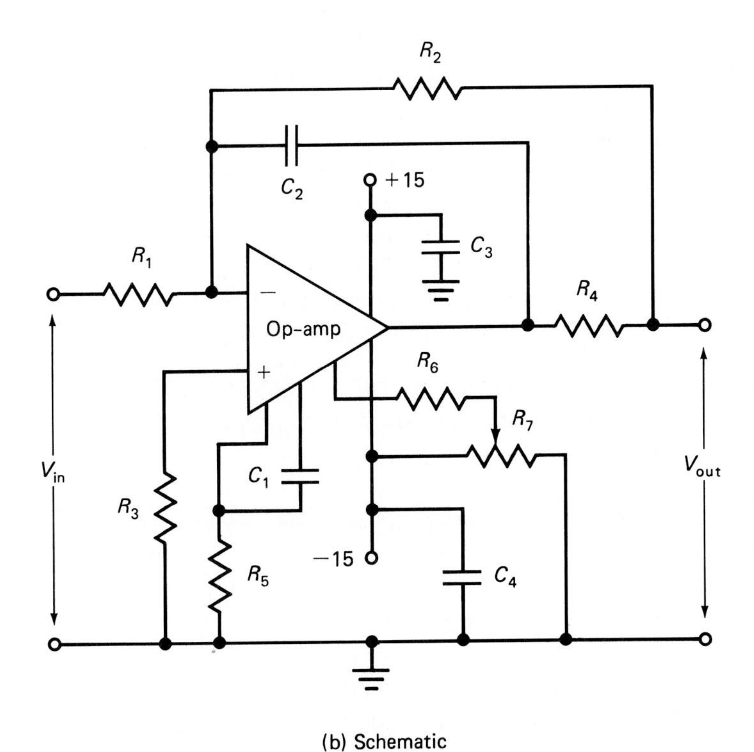

(b) Schematic

FIGURE 12.22 Simple Amplifier

Solution:

1. The block diagram that corresponds with this amplifier identifies it as a simple inverting amplifier. Although there are more components shown in this amplifier than have been shown previously, a basic objective will be to isolate the primary "inverting amplifier" components from those that have been added to the circuit.
2. Since this is an op-amp circuit, most of the bias conditions are met internally. The positive and negative dc bias voltages are applied directly. In addition to this, the potentiometer R_7 and resistor R_6 provide output offset nulling (setting output exactly to zero with zero input).

   In addition to this information, the manufacturer's data sheet is checked (not presented here). It was found that the $R_5 + C_1$ combination were recommended to provide frequency compensation (an inprovement in the op-amp's frequency-handling capacity). This would be used to extend the gain–bandwidth product.
3. The signal-handling functions of the circuit are investigated next. A simple feedback to input gain control (R_2 to R_1) can be seen (gain of -5). Also, R_1 sets (approximately) the input resistance of the amplifier. The output resistor, R_4, through a process of elimination must be necessary for coupling loads to the amplifier, since the actual output resistance will be much less than this value.

   The capacitors C_3 and C_4, since they span the power supply leads to ground, must be there to "bypass" any undesirable signals that may enter the system through these lines.

   The only elements remaining are the capacitor C_2 and resistor R_3. By a general inspection, C_2 seems to form a low-pass filter when viewed in connection with R_2. By reviewing the circuit's design notes (note listed here), it was found that this capacitor was used to reduce the circuit's phase shift and to improve stability at high frequencies. Otherwise, it would be speculated that this capacitor may be used to form an active low-pass filter (breaking at approximately 16 kHz).

   Resistor R_3 seems to be misplaced. Traditionally, the positive terminal of an inverting amplifier would be grounded. By again inspecting the manufacturer's recommended applications, it was found that by using a resistance in this place, related in value to R_1 and R_2 [$R_3 = R_1 \parallel (R_2 + R_4)$], the error in offset current could be reduced. This is a common practice for inverting-op-amp design. Thus each element's role has been described in relationship to either bias, manufacturer's specifications, or to the signal handled by the circuit.

New Device Applications

New discrete (or op-amp) devices are being produced on nearly a daily basis. To understand the application of these devices, some basic steps can be taken.

1. First, the manufacturer suggests a typical application for the device. In

these "Headlines" of the device specification sheets, the manufacturer will indicate the classification of the device (BJT, FET, . . . , new form, etc.). Then special features of the device will be highlighted; those functions that make it different from other similar devices may be detailed.

2. Next, attention must be paid to the maximum ratings of the device, both for dc and signal handling (ac). Usually, a device can be qualified as to typical circuitry applications by using these ratings as a guideline. (For example, if a circuit would use a 40-V power supply, any device with a breakdown voltage below 40 V would not be usable.)
3. The next figures to consider about the device would be the specific signal-handling features of the device. The following are the most commonly listed and most commonly investigated specifications:
   (a) Is the device mainly a voltage-, current-, or power-amplifying device?
   (b) What are the input and output resistance (impedance) specifications of the device?

   By qualifying items (a) and (b), the device can be related to one of the four ideal amplifier models (voltage, current, transconductance, transresistance).
   (c) What are the frequency-handling limitations of the device? Does it have a gain–bandwidth product? Does it have high- or low-frequency limitations listed?
   (d) Are there values of input and output capacitance that may affect frequency response?
4. What are the manufacturer's recommended applications? Often the manufacturer will suggest typical circuitry uses and may even illustrate a recommended circuit for that device.
5. Outside of the maximum ratings of the device, the dc bias figures for the device are not used much in qualifying the device but are applied later when establishing the support circuitry (bias) for its function.

Other information about the device, such as write-ups in professional journals or special publications from the manufacturer, can aid in further investigation of the device's potential applications.

EXAMPLE 12.3

From the specification sheet reprinted in the Appendix A, identify the application of the MRF2628 transistor, using the five steps listed above.

Solution:

1. The manufacturer (Motorola Inc.) has identified this as a 15-W power transistor for use at RF frequencies (136 to 220 MHz). It is an NPN silicon-type device (BJT) designed for use in systems with a 12.5-V dc power supply. (Other numerical specifications are listed in the top "box" on the specification sheet.)
2. The maximum ratings can be used to identify the dc limits for operation. Maximum bias voltage and current will help qualify in which form of circuit it can be used (e.g., due to the 36-V limit in collector-

to-base voltage, it would not be safe to use it in a system with a power supply of 50 V). Maximum power and other data indicate limits to the typical application of the device. [Due to these power levels, the thermal data, and the way it is packaged (heat-sink fins on top), it seems obvious that it would be connected to a larger heat sink in order to dissipate thermal energy during operation.]

3. Turning to the signal-handling capacities, it is seen that it is a current and power amplifier, signified by the dc current gain and common-emitter amplifier power gain specifications. Capacitance and other specifications round out the description of the device.
4. The manufacturer has illustrated a circuit that would utilize this component. By doing this, the user could either directly copy the circuit or use it as a basis for developing a similar circuit.
5. If another circuit were built around the device or it was to be substituted into another circuit, the dc characteristics would be used as a guideline to preparing the design.

Thus, by careful inspection of the manufacturers' data sheets, much can be learned about the applications and function of new discrete devices.

New Linear-Integrated-Circuit Applications

Investigating new linear integrated circuits is the simplest form of investigation. Typically, the new large-scale (and other forms) linear integrated circuits (except for op-amps, which are classified more by using the process described above) are manufactured to achieve one or a relatively limited number of tasks. Therefore, investigation of these devices involves looking up the manufacturer's intended application for the device and simply following the manufacturer's instructions for use.

EXAMPLE 12.4

Investigate the function of the TCA5550 integrated circuit listed in Appendix A.

Solution:

By definition of the manufacturer, this integrated circuit provides control over the sound in a stereo system. It controls the bass (low) and treble (high) frequency response, and provides the left-to-right channel balancing and total volume control for a stereo system. Additional specifications detail the actual parameters of its operation and indicate the type of circuit elements to be used in conjunction with it. Finally, a complete application circuit is given.

Thus for integrated circuits, the bulk of all information on the appropriate application of that device are specified by the manufacturer.

REVIEW PROBLEMS

(1) Indicate the types of resources that could be used to identify the actual role played by a single circuit within a complete system. (Do not overlook "people" resources.)
(2) Prepare a basic checklist that could be used to identify the application of a new device. Test this checklist by looking up a new (to you) device in the appendix of this book. Then write a short survey of that device's applications.
(3) Manufacturers are always eager to encourage the use of their devices and often publish applications manuals (books) for specific categories of devices. In your library (or with the help of your instructor) locate such an applications manual and investigate the application of one device. Prepare a brief list of the information available on that device (specifications, what types of circuits are shown for its application, what design information is given, etc.).
(4) Locate at least one electronics trade (professional) journal or magazine. Scan this publication and list the number of new or unusual devices or device applications tht are listed either in articles or advertisements (either discrete devices or integrated circuits).

SUMMARY

In this chapter we have presented various forms of specialized amplifier circuits. The power amplifier was shown to have special design and functional characteristics. Classes of operation (A, B, C) were shown to produce various levels of power-amplifier efficiency. Several forms of power-amplifier circuits were demonstrated. Other miscellaneous circuits, often with special applications, were illustrated. It was shown that the BJT, FET, or op-amp could be used in nontraditional (amplification) applications.

KEY EQUATIONS

Common Base configuration:

12.1 $\alpha = \dfrac{I_C}{I_E}$

12.2 $\alpha = \dfrac{\beta}{1 + \beta}$

12.3 $A_V = \dfrac{R_c}{r_e}$

(a.c. emitter resistance-)

12.4 $r_e = \dfrac{26 \text{ mV}}{I_E}$

12.5a $R_{\text{in}} = R_E \parallel r_e$

12.5b $R_{\text{out}} = R_C$

Common Collector configuration:

12.6a $A_V \simeq 1$

12.6b $A_i \simeq \beta$

12.7 $R_{in} = R_{B1} \parallel R_{B2} \parallel \beta(R'_E + r_e)$

12.8 $R_{out} = R'_E \parallel \left(\frac{R'_S}{\beta} + r_e\right)$

PROBLEMS

Section 12.1

12.1. List some of the physical differences between small-signal amplifiers and power amplifiers.

12.2. What electrical specifications are usually different for a power transistor when compared to a small-signal transistor?

12.3. Define the use of both thermal resistance information and power derating curves.

12.4. What is the difference between class A and class B output for an amplifier?

12.5. Illustrate the basic block diagram for using two class B amplifiers to produce a complete sine-wave output. Illustrate waveforms on this sketch.

12.6. Convert the block diagram of problem 12.5 into a simple PNP/NPN BJT amplifier circuit.

Section 12.2

12.7. Illustrate a four-input summing amplifier (op-amp), with the voltage gain for each input to be −20. (Include resistance values. Use a 15-kΩ feedback resistor.)

12.8. Illustrate a four-input scaling amplifier where the gain for each stage is to be as listed below. (Use a 15-kΩ feedback resistor.)

Input 1: voltge gain of −10
Input 2: voltage gain of −20
Inputs 3 and 4: voltage gain of −30 for each

12.9. Describe the basic differences between linear and nonlinear mixing.

Section 12.3

12.10. Briefly describe the application and typical forward gain of a common-base BJT amplifier.

12.11. For the common-base amplifier illustrated in Figure 12.23, find the following values.
(a) Alpha
(b) I_C
(c) Voltage gain of the amplifier
(d) Input and output resistance of the amplifier

12.12. For the common-collector circuit shown in Figure 12.24, find the following values.
(a) Voltage gain
(b) Input resistance

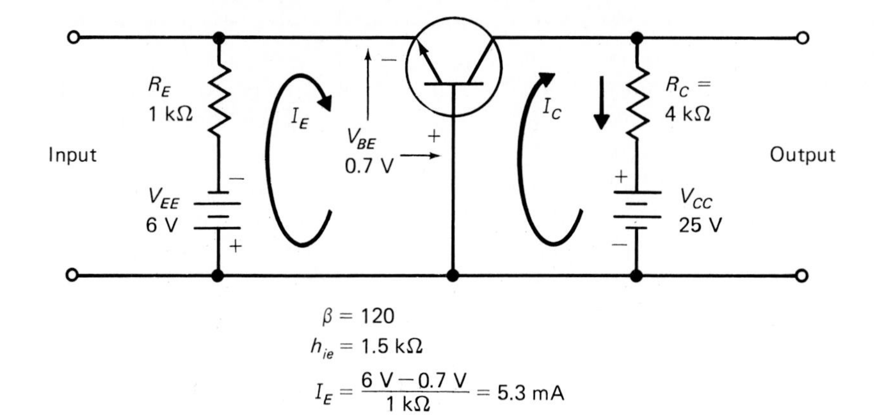

FIGURE 12.23 Common-Base Amplifier

(c) Output resistance
(d) Current gain

12.13. (An exercise related to previous chapters) Calculate the dc bias conditions for the circuit in Figure 12.24, finding the following values.
(a) V_{RB1} (b) V_{RB2} (c) V_{RE} (d) V_{CE} (e) I_E

12.14. Illustrate a BJT used as a "phase splitter" where there are two out-of-phase signals available at the output from a single input.

12.15. Illustrate the circuit for a P-DE-MOSFET circuit using no gate bias voltage (V_{GS} = 0 V).

12.16. Illustrate the circuit for a discrete (BJT) differential amplifier using NPN devices.

12.17. What form of amplifier would be used to convert a triangular waveform into a square wave? Sketch the schematic of this amplifier.

Section 12.4

12.18. Briefly summarize the three major techniques used to evaluate a new circuit application.

12.19. Assume that a new electronic device has been placed on the market. List at least five important questions or resources that you could use to determine the function and applications for that device.

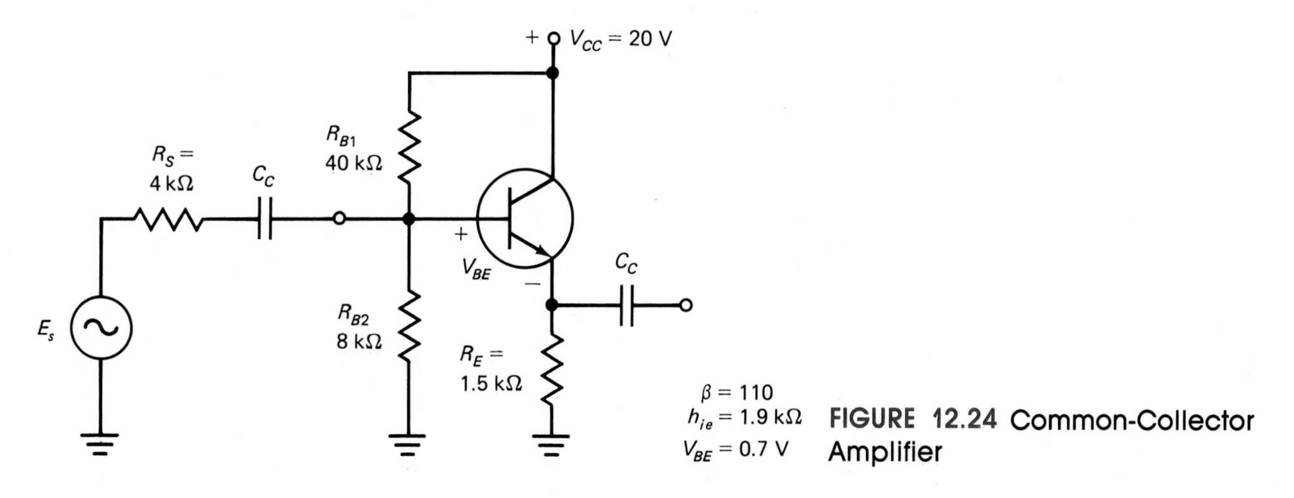

FIGURE 12.24 Common-Collector Amplifier

ANSWERS TO REVIEW PROBLEMS

Section 12.1

1. Across the board, the maximum ratings for the power transistor are greater in magnitude than those for a small-signal type (e.g., 2N2218). Most minimum ratings are also higher.
2. Using this class A amplifier would require that the device used would have to have a minimum bias current of 3.54 A and a minimum bias voltage of 28.3 V. This would require the device to dissipate at least 100 V of thermal energy, even if it was not producing an output. (Totally, the device would have to handle a minimum of 200 W of electrical power while operating.)
3. The cause for the distortion in a class A amplifier could be that the amplifier may be biased at saturation or cutoff, thus producing the half-wave output. The causes for the distortion in the class B amplifier could be that one of the amplifiers in the pair needed for operation may be defective, or the signal-splitting circuit used to drive these amplifier may be defective.

Section 12.2

1. Linear mixing produces an output that is an even blend of two or more input signals, basically a point-by-point mixing of the signals. Nonlinear mixing produces an output in which (typically) the gain of the amplifier is controlled by one signal and the other is amplified by this variable gain, thus producing a product of the two signals not an addition.
2. 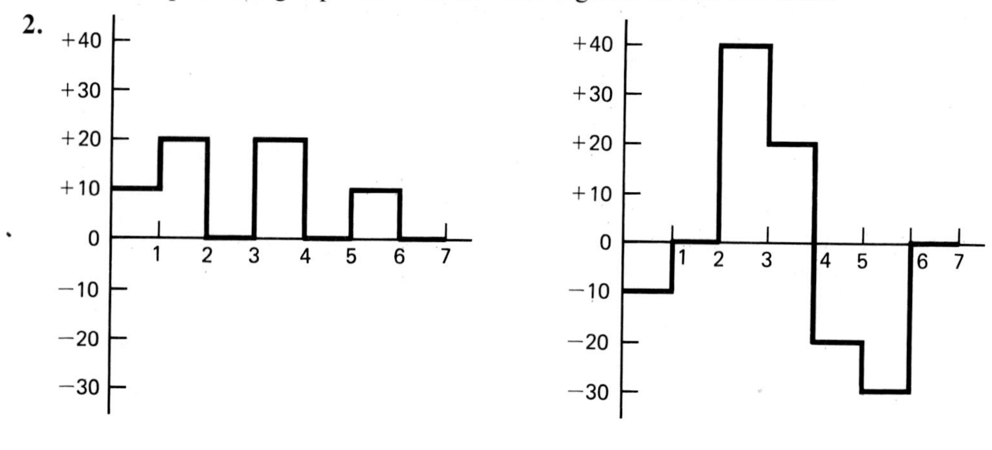

(a) Summing output (b) Differential output

3. Input 1, 5 kΩ; input 2, 10 kΩ; input 3, 5 kΩ; input 4 = 25 kΩ

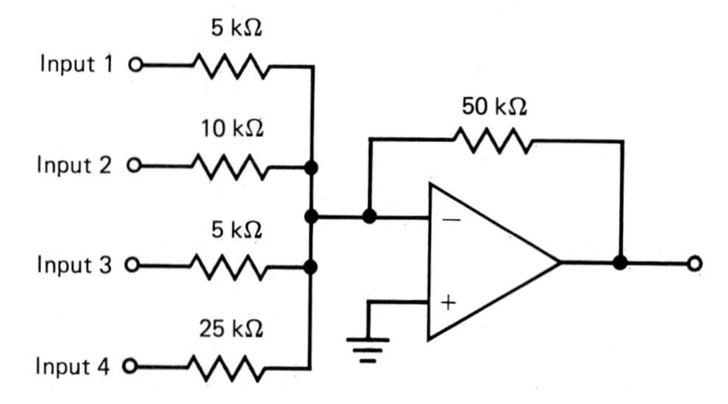

Section 12.3

1. (a) A BJT common-base amplifier
   (b) A DE-MOSFET amplifier
   (c) Either a combined common-emitter, common-collector BJT circuit or a common-drain, common-source FET circuit
   (d) An integrator
2. The common-collector circuit reflects the value of the load resistance back to set the resistance at the input. Problems may occur if a variable load was used with such an amplifier, as such variability may change the input signal delivered to the amplifier.
3. If the transistors used in a discrete differential amplifier were not matched, the output would not reflect the exact differential between the two inputs. Both devices would not treat their inputs identically; thus the output would be both a differential of the inputs and a relative differential of the transistor gains.
4. Answers to this question are based on the problems selected. (For assistance, students should exchange problems and solutions with one another as a cross-check.)

Section 12.4

1. (to include) Manufacturers' schematics, block diagrams, and written descriptions; past experience with similar circuitry; information gained from design and repair personnel familiar with the circuit in question
2. The solution is based on the component selected and the list prepared.
3. The answer is based on the component selected and the resources available.
4. The answer is based on the information source and device selected.

MC1306P

1/2-WATT AUDIO AMPLIFIER

The MC1306P is a monolithic complementary power amplifier and preamplifier designed to deliver 1/2-Watt into a loudspeaker with a 3.0 mV(rms) typical input. Gain and bandwidth are externally adjustable. Typical applications include portable AM-FM radios, tape recorder, phonographs, and intercoms.

- 1/2-Watt Power Output (12 Vdc Supply, 8-Ohm Load)
- High Overall Gain – 3.0 mV(rms) Sensitivity for 1/2-Watt Output
- Low Zero-Signal Current Drain – 4.0 mAdc @ 9.0 V typ
- Low Distortion – 0.5% at 250 mW typ

1/2-WATT AUDIO AMPLIFIER

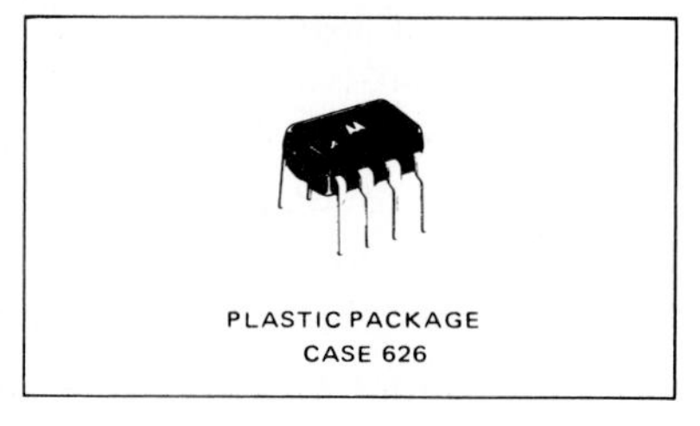

PLASTIC PACKAGE
CASE 626

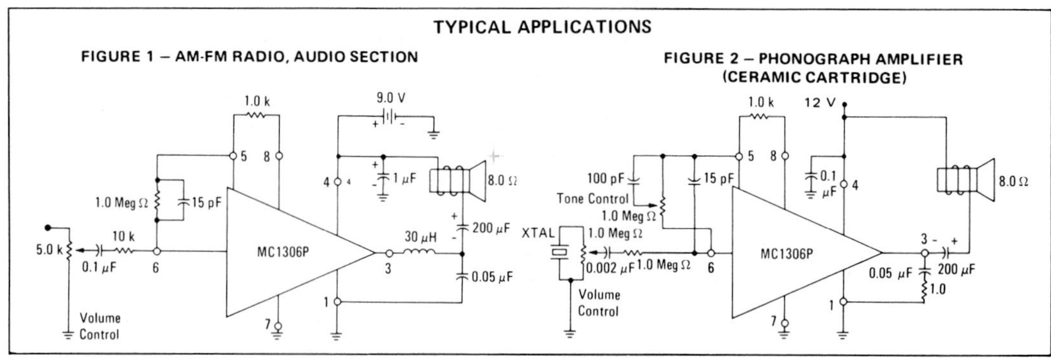

TYPICAL APPLICATIONS

FIGURE 1 – AM-FM RADIO, AUDIO SECTION

FIGURE 2 – PHONOGRAPH AMPLIFIER (CERAMIC CARTRIDGE)

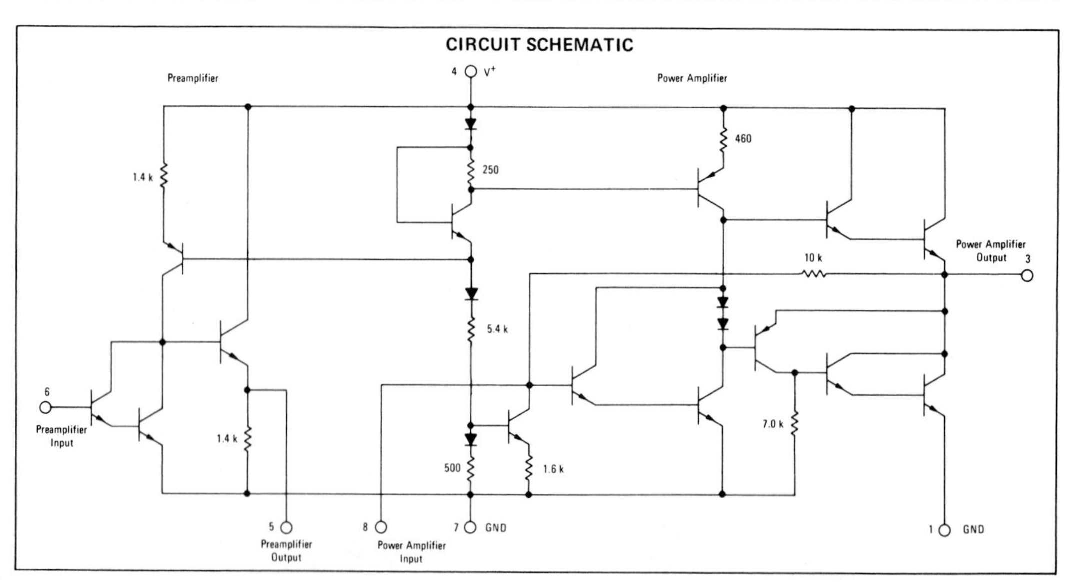

CIRCUIT SCHEMATIC

DS 9137 R2

MC1306P

MAXIMUM RATINGS (T_A = +25°C unless otherwise noted)

Rating	Symbol	Value	Unit
Power Supply Voltage	V^+	15	Vdc
Load Current	I_L	400	mAdc
Power Dissipation (Package Limitation)	P_D		
T_A = +25°C		625	mW
Derate above T_A = +25°C	$1/\theta_{JA}$	5.0	mW/°C
Operating Temperature Range	T_A	0 to +75	°C
Storage Temperature Range	T_{stg}	-65 to +150	°C

Maximum Ratings as defined in MIL-S-19500, Appendix A.

ELECTRICAL CHARACTERISTICS (V^+ = 9.0 V, R_L = 8.0 ohms, f = 1.0 kHz, (using test circuit of Figure 3), T_A = +25°C unless otherwise noted.)

Characteristic	Symbol	Min	Typ	Max	Unit
Open Loop Voltage Gain	A_{VOL}				V/V
Pre-amplifier R_L = 1.0 k ohm		–	270	–	
Power-amplifier R_L = 16 ohms		–	360	–	
Sensitivity (P_o = 500 mW)	S	–	3.0	–	mV(rms)
Output Impedance (Power-amplifier)	Z_o	–	0.5	–	Ohm
Signal to Noise Ratio (P_o = 150 mW, f = 300 Hz to 10 kHz)	S/N	–	55	–	dB
Total Harmonic Distortion (P_o = 250 mW)	THD	–	0.5	–	%
Quiescent Output Voltage	V_o	–	$V^+/2$	–	Vdc
Output Power (THD ≤10%, V^+ = 12 V)	P_o	500	570	–	mW
Current Drain (zero signal)	I_D	–	4.0	–	mA
Power Dissipation (zero signal)	P_D	–	36	–	mW

FIGURE 3 – TEST CIRCUIT

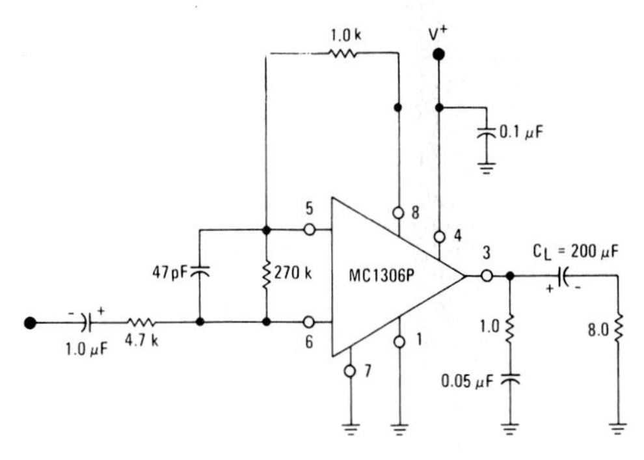

FIGURE 4 – ZERO SIGNAL BIAS CURRENT

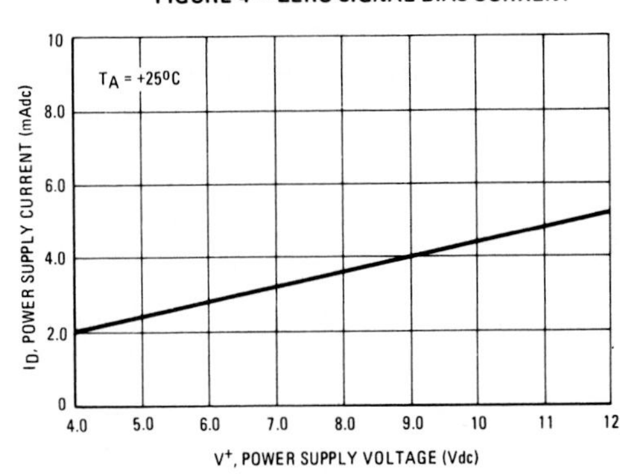

FIGURE 10 – TYPICAL CIRCUIT CONNECTION

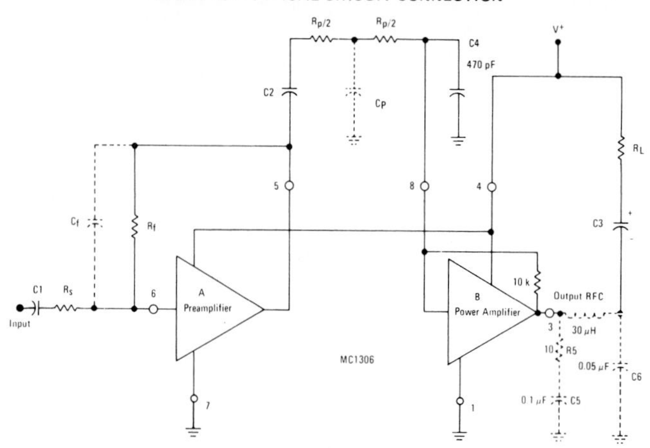

DESIGN CONSIDERATIONS

The MC1306P provides the designer with a means to control preamplifier gain, power amplifier gain, input impedance, and frequency response. The following relationships will serve as guides.

1. Gain

The Preamplifier Stage Voltage Gain is:

$$A_{V_A} \approx \frac{R_f}{R_s}$$

and is limited only by the open-loop gain (270 V/V). For good preamplifier dc stability R_f should be no larger than 1.0-megohm.

The Power Amplifier Voltage Gain is controlled in a similar manner where:

$$A_{V_B} \approx \frac{10\text{ k}}{R_p}$$

The 10-k ohm feedback resistor is provided in the integrated circuit.

Recommended values of R_p range from 500-ohms to 3.3-k ohms. The low end is limited primarily by low-level distortion and the upper end is limited due to the voltage drive capabilities of the pre-amplifier. (A resistor can be added in the dc feedback loop, from pin 6 to ground, to increase this drive). The Overall Voltage Gain, then, is:

$$A_{VT} = \frac{R_f\ 10\text{ k}}{R_s\ R_p}$$

2. Input Impedance

The Preamplifier Input Impedance is:

$$Z_{in\,A} \approx R_s$$

and the Power Amplifier Input Impedance is:

$$Z_{in\,B} \approx R_p$$

3. Frequency Response

The low frequency response is controlled by the cumulative effect of the series coupling capacitors C1, C2, and C3. High-frequency response can be determined by the feedback capacitor, C_f, and the -3.0 dB point occurs when

$$X_{C_f} = R_f$$

Additional high frequency roll-off and noise reduction can be achieved by placing a capacitor from the center point of R_p to ground as shown in Figure 10.

Capacitor C4 and the RC network shown in dotted lines may be needed to prevent high frequency parasitic oscillations. The RF choke, shown in series with the output, and capacitor C6 are used to prevent the high-frequency components in a large-signal clipped audio output waveform from radiating into the RF or IF sections of a radio (Figure 10).

4. Battery Operation

The increase of battery resistance with age has two undesirable effects on circuit performance. One effect is the increasing of amplifier distortion at low signal levels. This is readily corrected by increasing the size of the filter capacitor placed across the battery (as shown in Figure 8; a 300-μF filter capacitor gives distortions at low-tonal levels that are comparable to the "stiff" supply). The second effect of supply impedance is a lowering of power output capability for steady signals. This condition is not correctable, but is of questionable importance for music and voice signals.

5. Application Examples: (1) The audio section of the AM-FM radio (Figure 1) is adjusted for a preamplifier gain of 100 with an input impedance of 10-k ohms. The power amplifier gain is set at 10, which gives an overall voltage gain of 1000. The bandwidth has been set at 10-kHz. (2) The phono amplifier (Figure 2) is designed for a preamplifier gain of unity and a power amplifier gain of 10. The input impedance is 1.0-megohm. An adjustable treble control is provided within the feedback loop.

NPN	PNP
MJ15022	**MJ15023**
MJ15024	**MJ15025**

HIGH-CURRENT COMPLEMENTARY SILICON POWER TRANSISTORS

. . . designed for use in high-power amplifier and switching circuit applications, audio amplifiers, disk head positioners and motor controls.

- High Gain Complementary Silicon Power Transistors
- High Safe Operating Area 100% Tested
  50 V, 5.0 A, 0.5 Sec.
  80 V, 2.0 A, 0.5 Sec.
- Excellent Frequency Response — f_T = 4.0 MHz min.
- Low Thermal Resistance — 0.70°C/W max.
- High Power Capability — 250 Watts

16 AMPERES

COMPLEMENTARY SILICON POWER TRANSISTORS

200 AND 250 VOLTS
250 WATTS

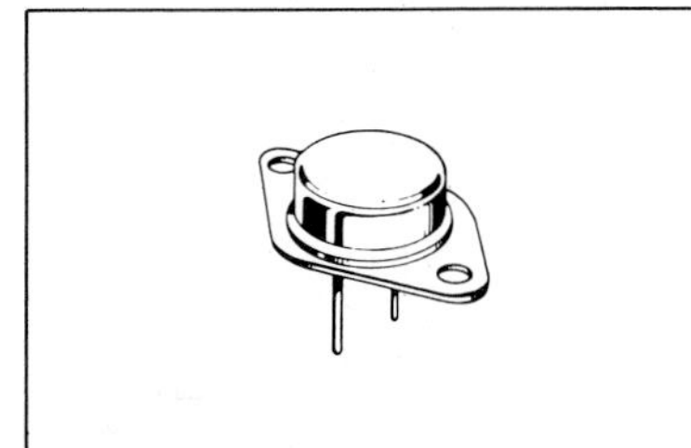

MAXIMUM RATINGS

Rating	Symbol	MJ15022 MJ15023	MJ15024 MJ15025	Unit
Collector-Emitter Voltage	V_{CEO}	200	250	Vdc
Collector-Base Voltage	V_{CBO}	350	400	Vdc
Emitter-Base Voltage	V_{EBO}	7.0		Vdc
Collector Current — Continuous	I_C	16		Adc
Base Current — Continuous	I_B	5.0		Adc
Emitter Current — Continuous	I_E	21		Adc
Total Power Dissipation @ T_C = 25°C Derate above 25°C	P_D	250 1.43		Watts W/°C
Operating and Storage Junction Temperature Range	T_J, T_{stg}	-65 to +200		°C

THERMAL CHARACTERISTICS

Characteristic	Symbol	Max	Unit
Thermal Resistance, Junction to Case	$R_{\theta JC}$	0.70	°C/W

FIGURE 1 — POWER DERATING

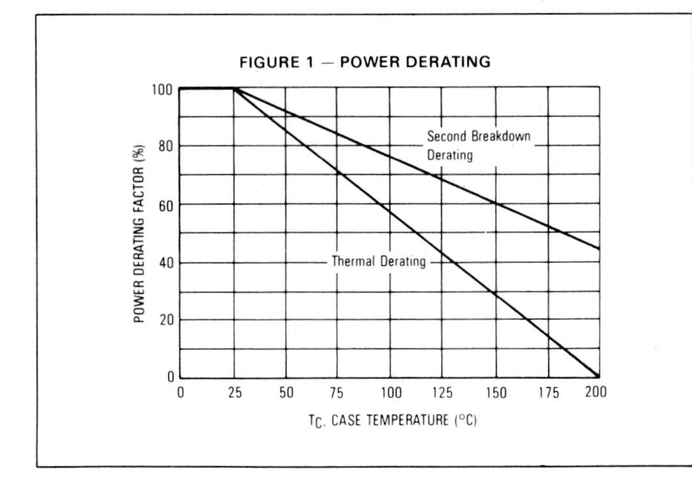

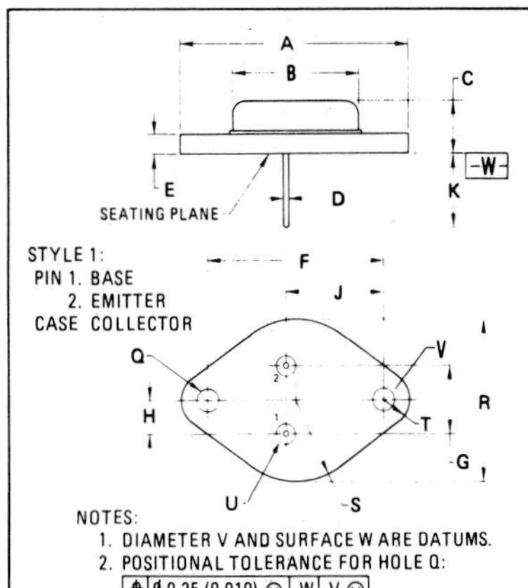

NOTES:
1. DIAMETER V AND SURFACE W ARE DATUMS.
2. POSITIONAL TOLERANCE FOR HOLE Q:
⌖ ∅ 0.25 (0.010) Ⓜ | W | V Ⓜ
3. POSITIONAL TOLERANCE FOR LEADS:
⌖ ∅ 0.30 (0.012) Ⓜ | W | V Ⓜ | Q Ⓜ

DIM	MILLIMETERS MIN	MILLIMETERS MAX	INCHES MIN	INCHES MAX
A	–	39.37	–	1.550
B	–	21.08	–	0.830
C	6.35	7.62	0.250	0.300
D	0.97	1.09	0.038	0.043
E	1.40	1.78	0.055	0.070
F	30.15 BSC		1.187 BSC	
G	10.92 BSC		0.430 BSC	
H	5.46 BSC		0.215 BSC	
J	16.89 BSC		0.665 BSC	
K	11.18	12.19	0.440	0.480
Q	3.81	4.19	0.151	0.165
R	–	26.67	–	1.050
U	2.54	3.05	0.100	0.120
V	3.81	4.19	0.151	0.165

CASE 1-04
TO-204AA
(TO-3)

DS3356R2

MJ15022 ● MJ15024 NPN
MJ15023 ● MJ15025 PNP

ELECTRICAL CHARACTERISTICS (T_C = 25°C unless otherwise noted)

Characteristic		Symbol	Min	Max	Unit
OFF CHARACTERISTICS					
Collector-Emitter Sustaining Voltage (1)		$V_{CEO(sus)}$			Vdc
(I_C = 100 mAdc, I_B = 0)	MJ15022, MJ15023		200	—	
	MJ15024, MJ15025		250	—	
Collector Cutoff Current		I_{CEO}			μAdc
(V_{CE} = 150 Vdc, I_B = 0)	MJ15022, MJ15023		—	500	
(V_{CE} = 200 Vdc, I_B = 0)	MJ15024, MJ15025		—	500	
Collector Cutoff Current		I_{CEX}			μAdc
(V_{CE} = 200 Vdc, $V_{BE(off)}$ = 1.5 V)	MJ15022, MJ15023		—	250	
(V_{CE} = 250 Vdc, $V_{BE(off)}$ = 1.5 V)	MJ15024, MJ15025		—	250	
Emitter Cutoff Current (V_{EB} = 7.0 Vdc, I_C = 0)		I_{EBO}	—	500	μAdc
SECOND BREAKDOWN					
Second Breakdown Collector Current with Base Forward-Biased		$I_{S/b}$			Adc
(V_{CE} = 50 Vdc, t = 0.5 s (non-repetitive)			5.0	—	
(V_{CE} = 80 Vdc, t = 0.5 s (non-repetitive)			2.0	—	
ON CHARACTERISTICS (1)					
DC Current Gain		h_{FE}			—
(I_C = 8.0 Adc, V_{CE} = 4.0 V)			15	60	
(I_C = 16 Adc, V_{CE} = 4.0 V)			5.0	—	
Collector-Emitter Saturation Voltage		$V_{CE(sat)}$			Vdc
(I_C = 8.0 Adc, I_B = 0.8 Adc)			—	1.4	
(I_C = 16 Adc, I_B = 3.2 Adc)			—	4.0	
Base-Emitter on Voltage (I_C = 8.0 Adc, V_{CE} = 4.0 Vdc)		$V_{BE(on)}$	—	2.5	Vdc
DYNAMIC CHARACTERISTICS					
Current-Gain — Bandwidth Product (I_C = 1.0 Adc, V_{CE} = 10 Vdc, f_{test} = 1.0 MHz)		f_T	4.0	—	MHz
Output Capacitance (V_{CB} = 10 Vdc, I_E = 0, f_{test} = 1.0 MHz)		C_{ob}	—	500	pF

(1) Pulse Test: Pulse Width ≤300 μs, Duty Cycle ≤2%

TYPICAL DYNAMIC CHARACTERISTICS

FIGURE 2 — CURRENT GAIN BANDWIDTH PRODUCT

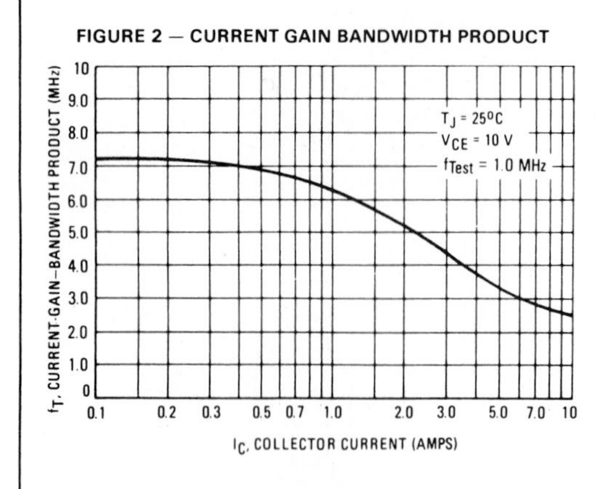

FIGURE 3 — CAPACITANCE

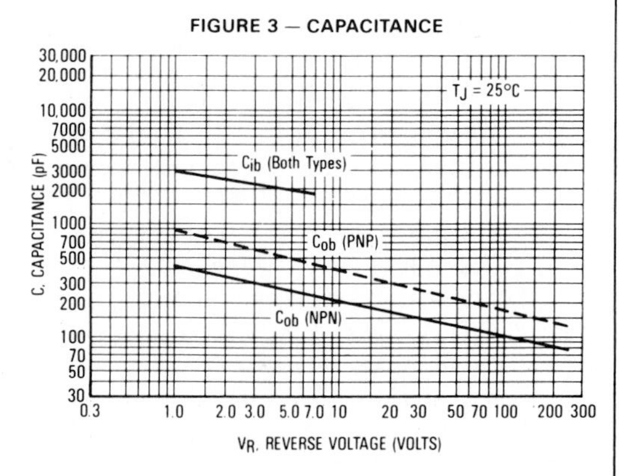

13 Multistage Amplification

Objectives

Upon completing this chapter, the reader should be able to:

- Identify and calculate overall function of multistage amplifiers.
- Identify special coupling forms.
- Identify and calculate special conditions for multistage amplifiers.
- Apply multistage feedback.
- Troubleshoot multistage amplifiers.

Introduction

In earlier chapters the block diagram function of multistage amplifiers was presented. Now specific details about the use and manipulation of amplifiers in multiple stages are presented. It is important to pay close attention to the means by which multistage amplifiers are interconnected. This process of interconnection, called coupling, is quite important when attempting to maintain efficiency and other important characteristics.

The accumulated effects of frequency response and power transfer are investigated as they relate to multiple-stage applications. Additional features and drawbacks involved in the process of establishing multiple-stage amplifiers are shown. Finally, test procedures that relate to both the multistage amplifier's block diagram and the circuitry used in such amplifiers are explored. ■

13.1 THE OVERALL FUNCTION OF MULTISTAGE AMPLIFIERS

Early in this book it was noted that amplification is commonly done in steps called stages. It is not possible for a signal in the millivolt range to be amplified to the high-voltage range typical of a power amplifier in a single amplification circuit. The process of amplification requires multiple stages of amplification. The total gain for the system is the product (multiplication) of each stage's gain, as shown by

$$A_{I(\text{total})} = A_{I1} \times A_{I2} \times A_{I3} \times \cdots \tag{13.1a}$$

$$A_{V(\text{total})} = A_{V1} \times A_{V2} \times A_{V3} \times \cdots \tag{13.1b}$$

$$A_{P(\text{total})} = A_{P1} \times A_{P2} \times A_{P3} \times \cdots \tag{13.1c}$$

Therefore, in such a system, the overall gain is computed as a product of all the individual gains within the system. (The only exception to this is if amplifiers are placed in parallel with each other, such as in a push-pull power amplifier. In that case the gains are simply added before being included in the multiplication process.)

This multiplication of gains helps to justify why, in prior chapters, it was not considered a problem to add stability and feedback, which drastically reduced the potential gain for a circuit. It seemed wasteful to take a transistor capable of producing a gain of 120 and placing it in a beta-independent circuit where the gain was reduced to 40. This gain loss can be made up simply with the addition of an amplifier with a gain of *just* 3 ($3 \times 40 = 120$)! In fact, three amplifiers in a row that suffered this reduction in gain would require only one more amplifier with a gain of 27 to make up the difference.

EXAMPLE 13.1

If the five amplifiers with the following specifications were placed in a multistage configuration, what would be the total gain for all stages? Also, if an input voltage of 0.3 μV were applied with an immeasurable current, what would be the resulting output levels (voltage, current, power)?

Stage	A_V	A_I	A_P	Other Data
1	50	30	1500	
2	50	30	1500	
3	40	22	880	
4	40	22	880	
5	20	6	120	$R_{load} = 10\ \Omega$

Solution:

Totals: $A_V = 80 \times 10^6$, $A_I = 2.6 \times 10^6$, and $A_P = 2.1 \times 10^{14}$.
Input value: $V_{in} = 0.3 \times 10^{-6}$.
Output values: $V_{out} = 24$ V, $I_{out} = 2.4$ A, and $P_{out} = 57.6$ W.

Several factors become important when dealing with multiple-stage amplifiers:

1. Any noise signal or distortion that is produced will be amplified. If the noise or distortion occurs in the initial stages of amplification, it will have the greatest effect on the final noise level or distortion at the output, since it will pass through the other gain elements (amplifiers).
2. The efficiency with which signal voltages and currents are passed from stage to stage is critical. If a loss of signal occurs, due say to some voltage or current divider, this loss may have to be made up by having to add more stages to the amplifier.
3. Certain limiting factors, such as frequency response, will also have a cumulative effect when occurring in multiple stages.

In later sections of this chapter, many of the special effects seen in multiple stages will be explored.

REVIEW PROBLEMS

(1) If four amplifiers in a multistage circuit each have a voltage gain of 22 and a current gain of 15, find the total amount of voltage and current gain for the system.
(2) In the circuit in problem 1, if a noise signal were introduced at the input that had a level of 20 μV, calculate the level of that noise at the final output. Repeat the analysis, but investigate the noise output if the noise were introduced at the input to the second, third, and fourth amplifiers.
(3) If, again for the amplifiers in problem 1, there was a circuit between each stage which produced a 50% loss in voltage, calculate the resulting gain for the system. [*Hint:* Each time the gain is calculated at each stage, multiply this by the loss factor (50%) before multiplying by the next gain.]

13.2 BLOCK DIAGRAM COORDINATION

When investigating a complete system, it is necessary to form a correspondence between the schematic and block diagram describing that circuit's function. In many cases, each block of the diagram coordinates with an amplifier in the schematic. In many cases, a single transistor, op-amp, or integrated circuit can be identified as being the central element in a circuit that coordinates with a block in the diagram. For example, the block diagram may specify a preamplifier and on the schematic there exists one transistor, with all of its bias and other related elements, which can be identified as being that preamplifier.

The block diagram can be used quite effectively to assist in isolating a defective circuit or the one that needs modification (due to an error in the output). The block diagram serves as a guide to exploring just what form of input/output paths are to be taken through the more complex schematic. When a correction to the circuit is required, spot checks at each step indicated by the block diagram will often lead to the circuit that is responsible for the error. Often as an aid to troubleshooting and modification, the specific names used on the block diagram will be repeated alongside the transistor (or other active device) that performs that specific task in the circuit. Figure 13.1 illustrates this type of coordination.

Following is a brief narration of the coordination of the block diagram to the circuit illustration shown in Figure 13.1. (The reader is not expected to comprehend all of the functions within this circuit, but simply to see the overview as presented.)

Block 1. This block simply indicates an input tuning circuit. It does not coordinate with a transistor, but with a tunable resonant circuit.

Block 2. This block calls for a local (internal) oscillator. The oscillator is seen to be associated with transistor Q_1.

Block 3. The mixer (which combines the outputs of the first two blocks) coordinates with transistor Q_2.

FIGURE 13.1 Complete AM Radio Schematic and Block Diagram

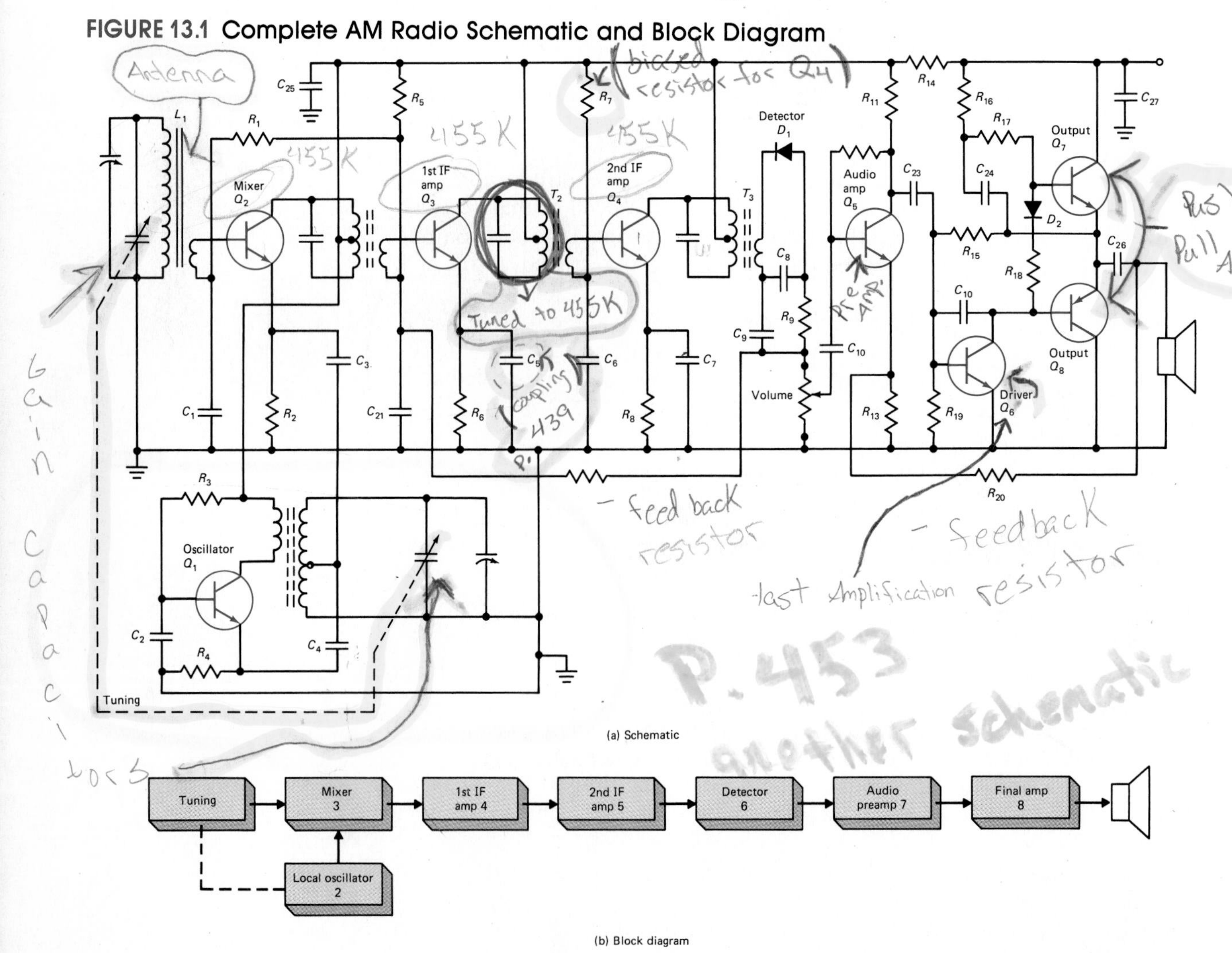

(a) Schematic

(b) Block diagram

Block 4. The first IF (intermediate frequency) amplifier, which is a resonant amplifier, is seen at transistor Q_3.

Block 5. The second IF amplifier is a duplicate of the first and is based at Q_4.

Block 6. This block coordinates with the diode D_1 and the circuitry associated with that diode.

Block 7. The audio preamplifier appears to be a simple voltage feedback (see R_{12}) amplifier centered around Q_5.

Block 8. This block coordinates with the two output transistors Q_7 and Q_8. These two transistors form a basic push-pull (class B) power amplifier set.

The only transistor not identified by a block is Q_6, the driver transistor. This device is in place to provide the out-of-phase signal needed for the second transistor (Q_8) in the push-pull power-amplifier circuit.

Thus, from this example and the previous discussions, the need to understand and work with a circuit's block diagram is essential if the function of any part in the circuit is to be clearly understood.

REVIEW PROBLEMS

(1) With which of the following elements would a block in a block diagram most likely coordinate when matching it to a schematic?
(a) Capacitor **(b)** Resistor **(c)** BJT **(d)** FET
(e) Op-amp **(f)** Potentiometer **(g)** Integrated circuit

(2) If in the block diagram shown in Figure 13.1 it was desirable to illustrate the function of volume control, where would it be placed (between which blocks)? Use the schematic in that figure to assist in placing that element.

(3) If an integrated circuit was available that could replace the complete circuit functions performed by Q_1, Q_2, Q_3, and Q_4, what changes, if any, would be made to the block diagram? In the latter part of this section, most blocks in the block diagram were shown to coordinate with a transistor or group of transistors in the schematic. How would this description change if the IC were to be used?

13.3 COUPLING

The process of interconnecting the various stages of an amplifier is called **coupling**. It is seldom possible simply to wire the output of one circuit to the input terminal of another. Certain special considerations usually need to be made. The following is a basic list of those factors that must be considered:

1. It is usually mandatory to keep the bias conditions of one amplifier isolated within that amplifier. Since the output of one stage is fed to the sensitive input of another stage, the bias levels at that output would often saturate the second amplifier. Thus one of the effects of coupling is to isolate the dc bias levels while passing the ac signal to the next stage.

Block

2. It is usually desirable to obtain maximum power transfer from one stage to another. This is sometimes not mandatory if the only functional level that is passed from one stage to another is just a voltage level or only a current flow. Usually, though, if there is an inefficiency in transferring the power from one stage to another, the inefficiency results in a loss of potential output.
3. It is possible to use the coupling element to change the form of the signal. The waveform is usually not changed, but it is possible to cause a step up in voltage or a step up in current. A transformer is the most popular element for this process. (Often this change in form aids in the maximum power transfer.)
4. Some additional effects can be seen by specially manipulating the coupling elements. It is possible to blend two roles into one function, coupling plus another special effect. (A common use of this blended process is to use resonant circuits to provide coupling as well; thus a frequency-selective amplifier is produced in the process of coupling it to another stage.)

Therefore, the process of coupling one stage to another can be a bit more challenging than simply running a wire between them. In the next sections, specific methods of providing coupling are detailed. It should be noted at this point that there are circuits that are directly connected (direct-coupled amplifiers) without the use of a special coupling circuit.

REVIEW PROBLEM

(1) Based on this brief description of coupling and the type of elements used (or implied in the descriptions), go back to Figure 13.1 and highlight as many elements used for coupling as possible. (To start this off, transformer T-1 couples both the oscillator and mixer and then couples that signal to Q_5! Also, it should be noted that the output of transistor Q_6 is coupled directly to transistor Q_8.)

13.4 SPECIFIC COUPLING ELEMENTS

To achieve the various tasks involved in coupling, a variety of elements are used to provide coupling. The most popular element is the capacitor. Transformers are also relatively common coupling elements. By contrasting the properties of various components to the effects desired in coupling, the basic process of joining individual amplifiers together into a composite system can be explored. The following presentations highlight the more popular coupling techniques.

Capacitive Coupling

In previous chapters, capacitors were shown at the input and output terminals of amplifiers. At that time their application as coupling elements was defined. When charged, a capacitor will block the flow of dc current. When ac is

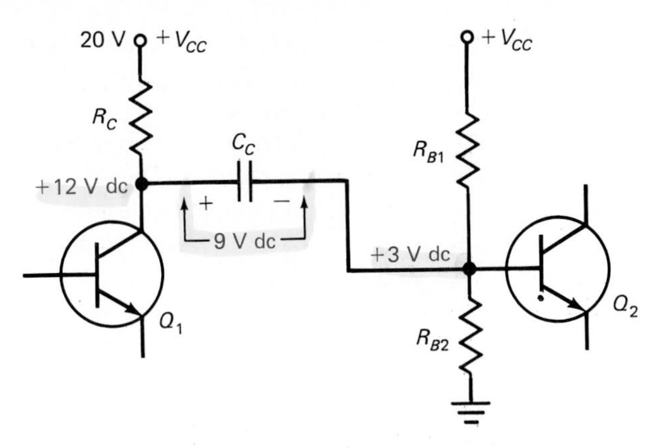

FIGURE 13.2 DC Blocking Effect of Coupling Capacitor

applied, the reactance of the capacitor (X_c) can be sufficiently low to allow a free flow of the ac current through it. Thus if a capacitor is used to couple amplifier stages, the dc bias levels will be kept from passing from one amplifier to another, while the ac signal will flow freely from stage to stage.

Capacitive coupling is not ideal, as it does effect the low-frequency response of the amplifier to which it is attached. But since capacitors are relatively inexpensive and are easy to work with, they are the most popular coupling elements. Figure 13.2 illustrates one application of a capacitor used to couple one amplifier to another.

Transformer-Coupled Amplifiers

The process of isolation of dc levels while passing on ac signals can also be achieved by the use of a transformer between amplifier stages (often called an interstage transformer). As illustrated in Figure 13.3, the interstage transformer couples the signal from one amplifier stage to a following stage using the magnetic coupling between coils within the transformer.

There are two advantages to using transformer coupling over capacitive coupling. The transformer can be selected so that it presents an ideal match between output resistance of the first stage and input resistance of the second stage. Second, gain can be achieved with transformer coupling in that the transformer can be used in a step-up or step-down voltage mode. As such,

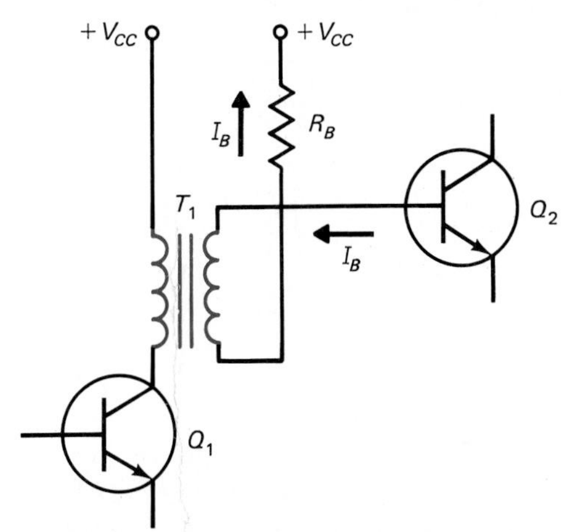

FIGURE 13.3 Passage of DC Base Bias Current through Transformer Secondary

more (or less) voltage gain can be had by the action of the transformer. This is done at the cost of lower current gain, but if voltage gain is the primary consideration, it may be advantageous to use transformer coupling.

There are two serious drawbacks to using transformer coupling. First the transformer is bulkier and much more costly than a capacitor. The second problem stems from the transformer's increased reactance (X_l) at high frequencies. The use of a conventional transformer will usually place an undesirable limit on the high-frequency response of an amplifier.

Resonant Circuit Coupling

A special resonant circuit, composed of a transformer connected to a primary and secondary capacitor, will form a double-tuned resonant circuit (see Figure 13.4). On the primary side of this transformer, the inductance of the primary forms an L/C resonant circuit. On the secondary side, the inductance forms another resonant circuit, with the capacitor connected in parallel with it. Magnetic coupling (typical transformer coupling) will cause the signal produced on the primary to be transferred to the secondary.

Such a circuit will not only produce an effective dc block/ac pass circuit needed for coupling, but can exhibit very high efficiency at the resonant frequency. (It is assumed that each side is tuned to the same resonant frequency.) By using this resonant system, the process of frequency selectivity and coupling are achieved simultaneously.

This coupling method is commonly used in radio-frequency circuits, where frequency selectivity is essential. In fact, there are preassembled transformer/capacitor packages commonly available for such use. (They are often called RF or IF "cans.")

Due to the transformer's impedance-matching characteristics, it is also popular to use a transformer to couple the output of an oscillator to a following amplifier stage. In such a configuration one side of the transformer provides the inductive element of the L/C resonant circuit used in the oscillator circuit, while the other side couples the signal to a load. The transformer is selected so as to provide very high reflected impedance from the load back to the oscillator circuit. Oscillator circuits are sensitive to current drawn by the load. If too much current is drawn, oscillation may cease. The high impedance reflected by the transformer will reduce the loading effect to the circuit.

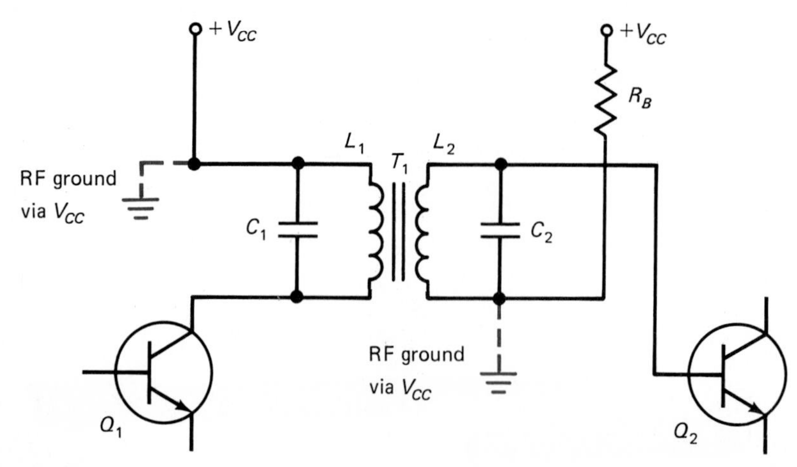

FIGURE 13.4 Establishing Primary and Secondary RF (AC) Grounds

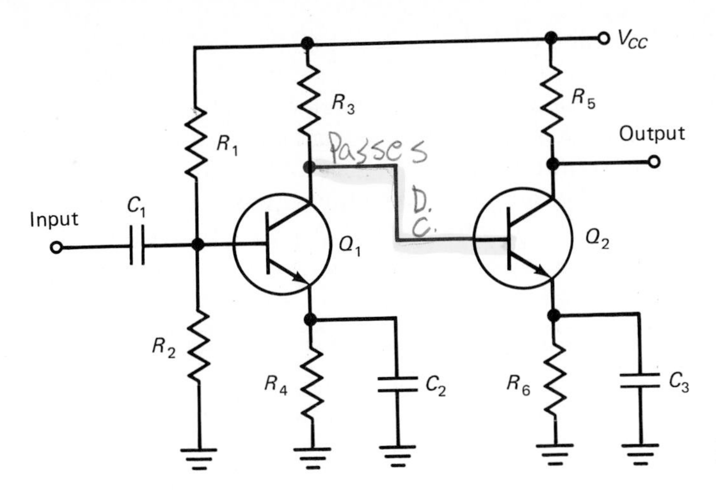

FIGURE 13.5 Direct-Coupled Amplifiers

Direct Coupling

Most devices cannot be directly coupled without revising the bias equations of each stage. Since the bias of one stage affects the bias of the other, bias values must be solved for using simultaneous equations. In such a case, a delicate balance of bias values is necessary in order not to cause circuitry malfunction or device damage.

Figure 13.5 illustrates direct-coupled amplifier stages. The first stage (related to Q_1) is a basic beta-independent amplifier. But R_3 plays a special role between this circuit and the second amplifier (Q_2). R_3 provides both the collector load for Q_1 and the base bias for Q_2. The value of collector voltage (V_C) for Q_1 becomes the value of base voltage (V_B) for Q_2.

The advantages of such a configuration over a capacitor-coupled amplifier are:

1. Reduction in cost and bulk of the circuit by the elimination of both a coupling capacitor and a base bias resistor for Q_2
2. Improved lower-frequency response due to the elimination of the X_c of the coupling capacitor (fewer capacitors results in improved response)

But there are also some drawbacks to this process:

1. Such circuits are more complex to design. More circuit design
2. A fault in either stage will affect both. This complicates troubleshooting, as it may be harder to find the single defect. Also, it is possible that a single fault could damage both transistors (Q_1 and Q_2).

EXAMPLE 13.2

For the circuit shown in Figure 13.5, describe what would happen if the base–emitter junction of Q_1 were to open.

Solution:

1. If the base–emitter junction of Q_1 were to open, Q_1 would cease to conduct collector–emitter current.

2. This would cause the collector voltage (V_C) of Q_1 to rise to the value of V_{CC}.
3. The voltage (V_C) would then be the base voltage for Q_2.
4. Q_2 could be damaged by its base voltage being at V_{CC} by:
   (a) Too much base current due to this voltage.
   (b) Overheating by having too much collector current:

$$I_C = \frac{V_{CC} - V_{BE}}{R_E}$$

A popular application of direct coupling is seen in the complementary power amplifier, described in Chapter 12. Both power transistors are direct-coupled to each other. It is common to provide direct coupling of the input to this form of amplifier as well.

REVIEW PROBLEM

(1) Return again to the circuit shown in Figure 13.1 and identify all the coupling elements and name them as to type, as noted in this section.

13.5 SPECIAL EFFECTS SEEN IN MULTIPLE-STAGE AMPLIFIERS

Multiple-stage amplifiers are used to serve two purposes. First, as presented earlier, it is often necessary to use multiple stages to provide the gain necessary to amplify low-level signals. Second, multiple effects are seen in the multistage amplifier. Several processes may need to be done to manipulate the input signal properly before an appropriate output can exist. One stage cannot usually produce more than one or possibly two effects at one time; thus several stages may be needed to produce the appropriate manipulation of the signal.

In the process of coupling the stages together, several undesirable effects can occur. First the losses due to maximum power transfer and reflected resistance (detailed later) will tend to reduce the potential gain of the system. Because of this, more stages of amplification may have to be added. Another problem is an accumulated frequency-response problem. As more stages are added, the roll-off (drop in gain on a dB/decade basis) is increased. This may reduce the effective frequency response of the system.

Effects of Power Transfer and Reflected Resistance

As the output resistance of one amplifier is attached to the input resistance of a following amplifier, a voltage- and/or current-divider circuit is produced. This divider will tend to reduce the capacity of the output amplifier to transfer power to the next stage. A basic maximum power transfer problem is therefore introduced.

From circuit analysis course work it will be remembered that maximum

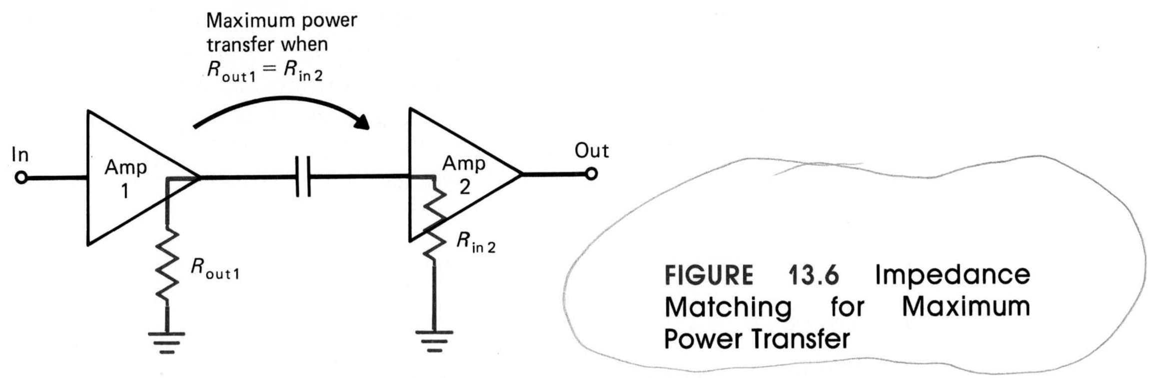

FIGURE 13.6 Impedance Matching for Maximum Power Transfer

power transfer is achieved only when the load resistance matches the source resistance. In the case of amplifier circuits, this means that to have maximum efficiency, the input resistance of one amplifier should match the output resistance of the previous stage (see Figure 13.6). Maximum power transfer is achieved only when the two resistances (input and output) are matched. Even when this occurs, the maximum power delivered to the input of the second stage is

$$P_{in} = \frac{P_{out}}{2} \tag{13.2}$$

(This can be verified by referring to a circuit analysis textbook.)

When dealing with power amplifiers, this factor can be critical. The amplifiers that drive (supply) the power amplifier should be matched, as to resistance, to the power stage. The power stage must also match the resistance of the load in order to transfer the maximum possible output power to that load.

Another major reason for wanting to work carefully with these resistances is the effect a load resistance (input of another stage) has on the gain of an amplifier. In the ac analysis of multiple-stage amplifiers, the input resistance of a second stage is in parallel with the output resistance of the first. Figure 13.7 illustrates the load seen by a first stage when a second stage is attached.

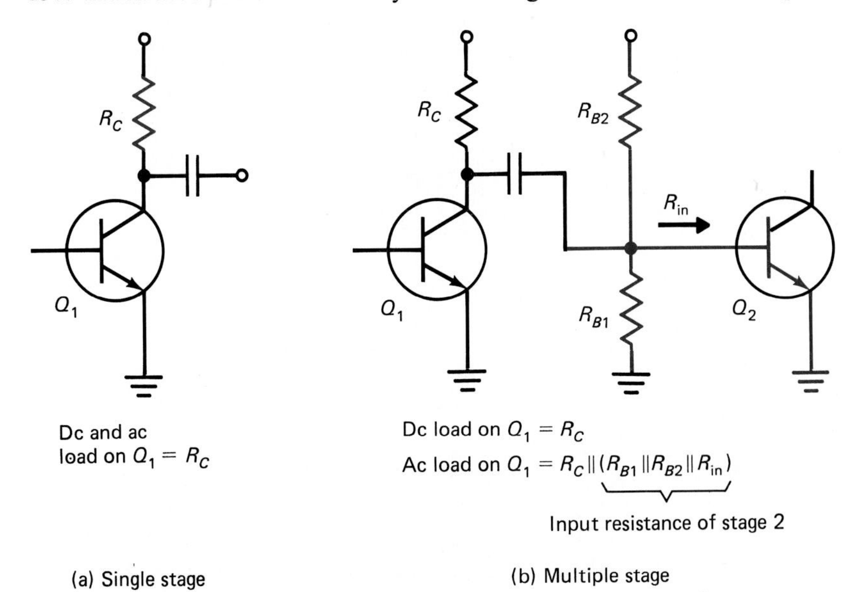

FIGURE 13.7 Amplifier Output Loads

Therefore, the attachment of a second stage will cause a reduction in the effective load on the amplifier (output load resistance). The same effect was seen in previous chapters when a simple load resistor was attached to an amplifier. For some amplifiers the second stage may cause a reduction in the voltage gain for the first stage. Such a loading effect, seen throughout all of the stages of an amplifier, may require the addition of more stages of amplification in order to compensate for the losses.

EXAMPLE 13.3

For the BJT amplifier circuit shown in Figure 13.8, indicate the effect seen by coupling that stage to an amplifier with an input resistance of 1 kΩ.

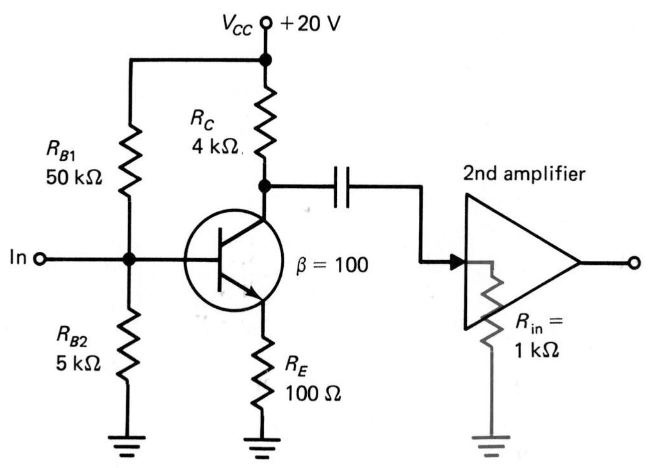

FIGURE 13.8 Total Loading Effect on a Single Stage (in Multistage System)

Solution:

The voltage gain of this circuit without the load factor of the next stage is

$$A_v = -\frac{R_C}{R_E} = -\frac{4000}{100} = -40$$

With the load of the second stage, the load on this circuit is

$$R_L' = R_C \parallel R_{in(stage\ 2)} = 800\ \Omega$$

Substituting this figure for the collector resistance gives us

$$A_v' = -\frac{800}{100} = -8$$

Thus the effect of the load of the second stage has caused the gain of this circuit to drop from −40 down to −8. If this were to occur at every stage of a multiple-stage system, the number of additional stages that would have to be added would almost double the size of the overall circuit. In such a case it would be desirable to change either the value of the input resistance of the second stage or to revise the values in this stage (especially R_E) to help compensate for the potential loss in gain.

Even in a case where maximum power transfer is attempted, matching the load (input of stage 2) with the output resistance of stage 1, the voltage gain of the circuit, as shown in Example 13.3, would be cut in half. Therefore, the effect of the input resistance of one stage must be considered when dealing with the gain of an initial stage. Two circuits with independent gains, when connected together and then to a load, may not produce a composite gain which is the simple product of their independent gains.

Generally, it can be stated that:

1. If voltage gain is a goal of a stage, the value of the input resistance of any following stage should be as high as possible so as to have a minimal effect on the voltage gain.
2. If current gain is a goal of a stage, the value of the input resistance of any following stage should be as low as possible so as to have a minimal effect on the current gain. (The proof of this follows the same lines as that for a voltage amplifier.)
3. If maximum power transfer is a goal between two stages, the input resistance of any stage should be equal to the output resistance of the prior stage. Although voltage and current gain will be affected by this, the maximum transfer of power between stages will be achieved.

As can be seen, the effect of one stage on another can be significant. To determine the best combination of input and output resistance, the objective of the amplifier circuit must be well known.

Frequency Response in Multiple Stages

The accumulated effects of each amplifier's (and the coupling element's) frequency-response problems can create problems in multistage systems. In Chapter 10 the concept of roll-off, a dB/decade loss, was presented for individual amplifiers. When amplifiers are placed in a multistage system, the frequency-response problems are cumulative.

As in single stages where each frequency-dependent element produces an increase in the roll-off (loss of gain at low and high frequencies) for the circuit, in multiple stages each *circuit's* roll-off increases the roll-off for the entire system. The following basic statements can be made about these losses:

1. The low-frequency breakpoint (where the first roll-off occurs) is determined by the circuit with the highest low-frequency breakpoint.
2. The high-frequency breakpoint is determined by the circuit with the lowest high-frequency breakpoint.
3. For each additional breakpoint there is an increased roll-off of 20 dB/decade added to any previous roll-off. The slope of the losses continues to increase for every additional breakpoint encountered.

OVER →

EXAMPLE 13.4

If each stage of a three-stage amplifier has the following break frequencies, find the composite (Bode) frequency-response plot.

> *Amplifier 1:* low-frequency break at 30 Hz, high at 45 kHz
> *Amplifier 2:* low-frequency break at 10 Hz, high at 20 kHz
> *Amplifier 3:* low-frequency break at 40 Hz, high at 30 kHz

Solution:

For this circuit, the low-frequency would begin to break at 40 Hz and roll off at a rate of −20 dB/decade until 30 Hz was reached; past that value the roll-off would be −40 dB/decade. When 10 Hz was reached, the roll-off would become −60 dB/decade. The high-frequency curve would break first at 20 kHz, fall at a rate of −20 dB/decade until hitting 30 kHz, when the fall would increase to −40 dB/decade, and finally drop to −60 dB/decade as the curve encountered 45 kHz. This is illustrated in Figure 13.9.

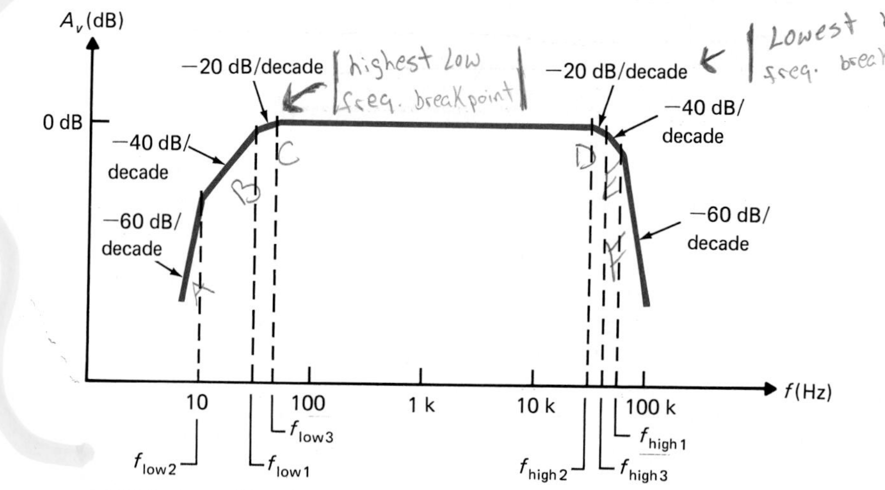

FIGURE 13.9 Multiple-Stage Frequency-Response Effects

EXAMPLE 13.5

If four identical amplifiers were placed into a multiple-stage configuration, and they had the following response conditions, indicate the frequency response of this system. All four amplifiers have a roll-off of −40 dB/decade at 20 Hz (due to dual breakpoints) and two high-frequency roll-offs, one at 18 kHz and another at 40 kHz.

Solution:

For the composite response, at 20 Hz, the system's frequency response would break and begin a fall of −160 dB/decade. At the high-frequency end, the roll-off would begin at 18 kHz and fall at a rate of −80 dB/decade until it reached 40 kHz, where it would begin to fall at a rate of −160 dB/decade.

As can be seen by these examples, the frequency response of the multiple-stage system can suffer drastically. What is not apparent through these calculations is that the actual frequency response becomes narrower when the roll-off increases. Remember that these roll-off functions are caused (mainly) by the effects of capacitive reactance. This function (X_C) does not simply "switch on" when a critical frequency is reached; it is a gradually increasing (or decreasing) effect. Thus the gradual changes in each stage become additive and the entire system's response begins to be seriously affected.

For example, for the circuit described in Example 13.5, the actual low-frequency −3-dB breakpoint would not be 20 Hz but would move up to almost 46 Hz. The high-frequency breakpoint would be almost at 10 kHz. Thus, using individual circuits with responses fairly acceptable for audio amplification, when these are tied together their composite response may become undesirable.

This problem can be solved in one of two ways. First, the response of each amplifier should be made as good as possible. That way, when they are tied together, the composite response, although less desirable, may still be in a reasonable range. The second method is to use feedback, across the multiple stages, to increase the frequency response.

Multiple-Stage Feedback

Feedback need not be applied to a single stage of an amplifier system. It is possible to treat a complete system as a composite amplifier block (Figure 13.10), and based on the type of amplification performed (voltage, current, transconductance, or transresistance), feedback can be applied.

The feedback calculations would be the same as used for single-stage amplifiers. The feedback method can be a simple voltage divider of the output fed back to the input stage (as long as the feedback is negative). The advantage would be an increase in the frequency response. Naturally, care would be

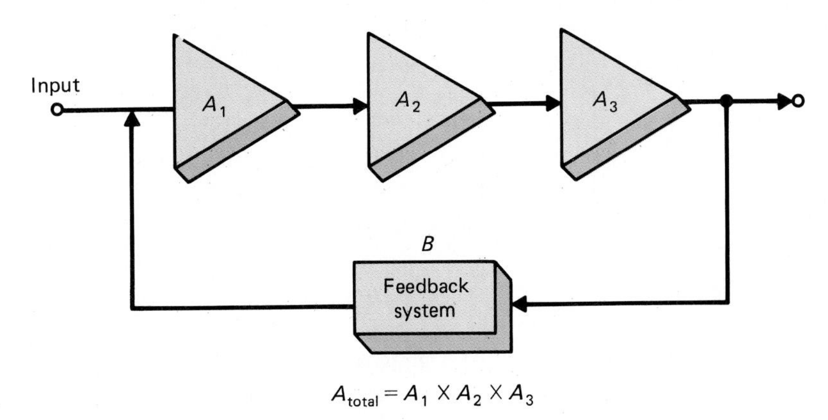

FIGURE 13.10 Multiple-Stage Feedback Process

taken to isolate any bias levels from the output so that they would not affect the bias of the input stage. It would, of course, be necessary to add another stage (or two) to make up for the loss in system gain, but the overall improvement in frequency response would be worth the effort.

Figure 13.11 shows a multistage amplifier using voltage-series feedback spanning two stages. The combination of R_1 and R_2 provides a divider circuit that samples the output. This level is fed back to the input. To achieve negative feedback, this feedback signal is injected at the emitter side of the first amplifier. (Actually, the output is in phase with the input, due to the double inversion of the two stages, so it could not be returned directly to the base of Q_1.) Since this feedback is at the emitter, it becomes subtractive from the input, just as the voltage on an unbypassed emitter resistor would be in a single stage. And as occurs when feedback is used in single stages, the overall frequency response of the system is improved by the addition of this feedback process. (An added benefit of this feedback is to raise the effective input resistance and to lower the effective output resistance of the system. This improves this voltage amplifier's characteristics, making it more like an ideal voltage amplifier.)

The sample shown has been but one of a wide variety of possible multistage feedback processes. Many special circuit effects can be achieved simply by adding special-function feedback. Most of the feedback applications described in this book for use with single-stage circuits can also be applied across multiple stages.

Thus it can be seen that quite a few special effects are associated with multistage amplifiers. Some of these effects are undesirable and have to be compensated for by the addition of more stages in the system. It is not uncommon to see one or more stages added to an amplifier system just to com-

FIGURE 13.11 Two-Stage BJT Amplifier Using Multistage Feedback

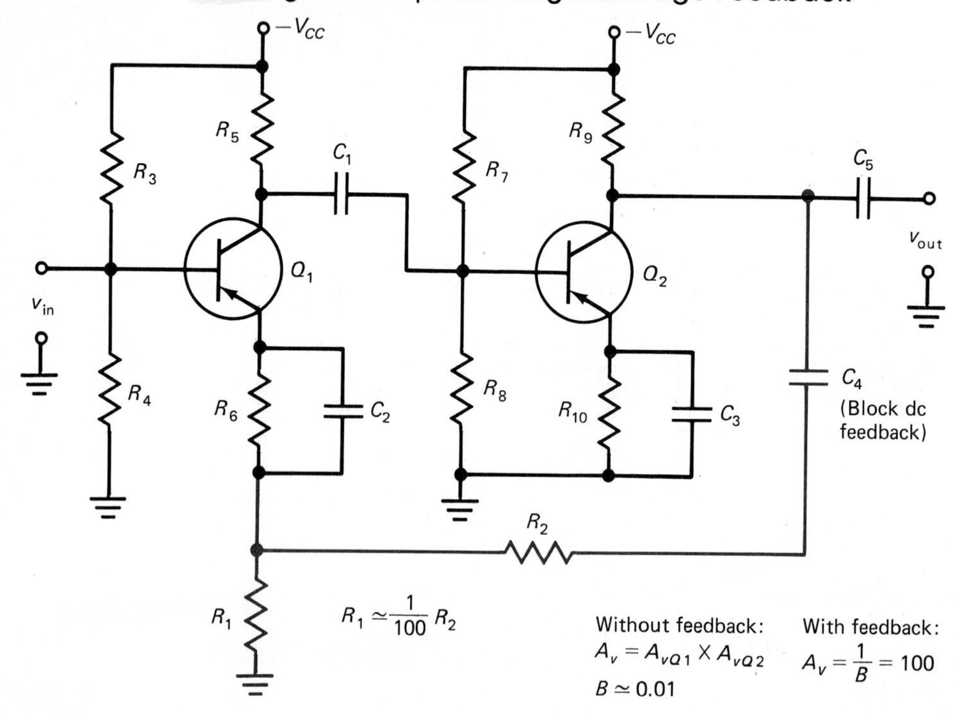

pensate for the losses seen and for feedback used to overcome frequency problems.

REVIEW PROBLEMS

(1) Prepare a composite list of the detrimental effects seen when amplifiers are placed into a multistage configuration.
(2) For the circuit in Figure 13.8, repeat the calculations for circuit gain if the load were to be matched to the collector resistor to achieve maximum power gain.
(3) Return to the circuit shown in Figure 13.1 and prepare a brief description of the use of resistor R_{20} in that circuit. (*Hint:* It works in conjunction with R_{13}.) (It should be noted here that resistor R_{10} does not perform the same task in the circuit. It actually produces a gain control for transistor Q_3. How this process is achieved is beyond the scope of this book but is explained in electronics communication courses.)

13.6 TROUBLESHOOTING MULTISTAGE AMPLIFIERS

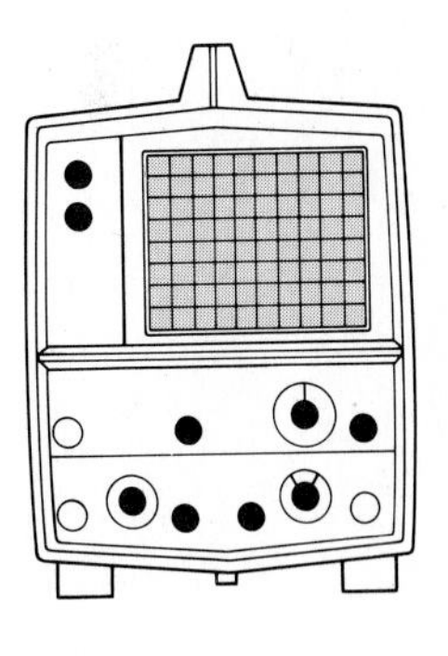

The primary objective in troubleshooting any system is to be able to detect the source of the fault, eliminate that source, replace defective or damaged components, and place the system back into operation. All of this is to be done in the minimum amount of time. When working with multiple-stage systems, it is unrealistic to begin with the first component in the system, test it, and then proceed to the next until the fault is "discovered." This would be like beginning to disassemble an automobile engine by searching for one defective part whenever the engine failed to start.

To detect the source of the problem in any system, a certain pattern of investigation needs to be formed. For an electronic circuit (with some variations) the following is the typical pattern used to detect a fault:

1. Investigate the block diagram and compare the detected fault to the blocks in the diagram that could be the source of the fault.
2. Locate the specific circuits which relate to the block(s) that were assumed to be related to the fault.
3. Test the input and output signals associated with each circuit, again coordinating with the block diagram, working from the output segment (where the fault was isolated) back to the input (the beginning of the block pattern suspected to be at fault).
4. Once a circuit is found to have an acceptable input but produces an unacceptable output, the potential source of the defect has been isolated (in most cases).
5. Then the signal and bias conditions associated with the active device in the circuit are measured. If these levels are found to be unacceptable, that active device, or nearby support components, are tested further to determine where the fault lies.

6. Once a defect has been isolated to one or a small group of components, the actual source of the fault must be detected. A cause/effect relationship should be made to determine just what condition could have created the defect or component failure.
7. Once the source of the fault is found, both it and any damaged components are changed. Thus not only are the "broken" parts changed, but the cause of the defect is corrected.
8. Next, an investigation of both the block diagram and the circuit is made to determine if that fault could have created another fault or "fatigued" some other portion of the system. If necessary, these parts, too, are replaced.
9. Finally, the system is tested and placed back into service. A complete write-up of the defect and solution is done so that further problems can be related to the one defect seen at this time, hopefully to prevent further faults from occurring.

By using this basic process, or one modified for a specific system, an effective use of the schematic and block diagram can be seen. The following example will help illustrate these points.

EXAMPLE 13.6

For the circuit shown in Figure 13.12 (a duplication of Figure 13.1), there is no output to the speaker (no sound). Determine the cause of this fault.

Solution:

1. The block diagram does not exactly isolate this problem since many errors could produce the problem. But since a preamplifier and power amplifier group could first be suspected, they will be inspected first.
2. Comparing the block diagram to the circuit resulted in determining that the circuitry related to the audio amplification section was that associated with Q_5, Q_6, Q_7, and Q_8.
3. An audio signal (from a generator) was applied to the wiper of the volume control potentiometer. That way a known signal could be applied to the circuit, which would be easier to trace.

   Then a test for signal was made at the speaker terminals. When the audio generator was set to a typically low level, a barely detectable signal was found at the terminals. If the generator output was increased, a fairly good output was seen. But it was found that the generator level was much higher than would generally be found at that potentiometer during normal operation.

   The signals on Q_7 and Q_8 were tested. Both outputs test as a sharp sine wave. Since Q_8 had a proper input, it could be assumed that Q_6 was operational. (Q_6, as noted before, is used for simple phase inversion; thus no substantial gain is expected.)

   Then the signal on the collector of Q_5 was tested. It was a bit larger than the signal at the bases of the two power transistors, but that was to be expected, since that voltage is run through several

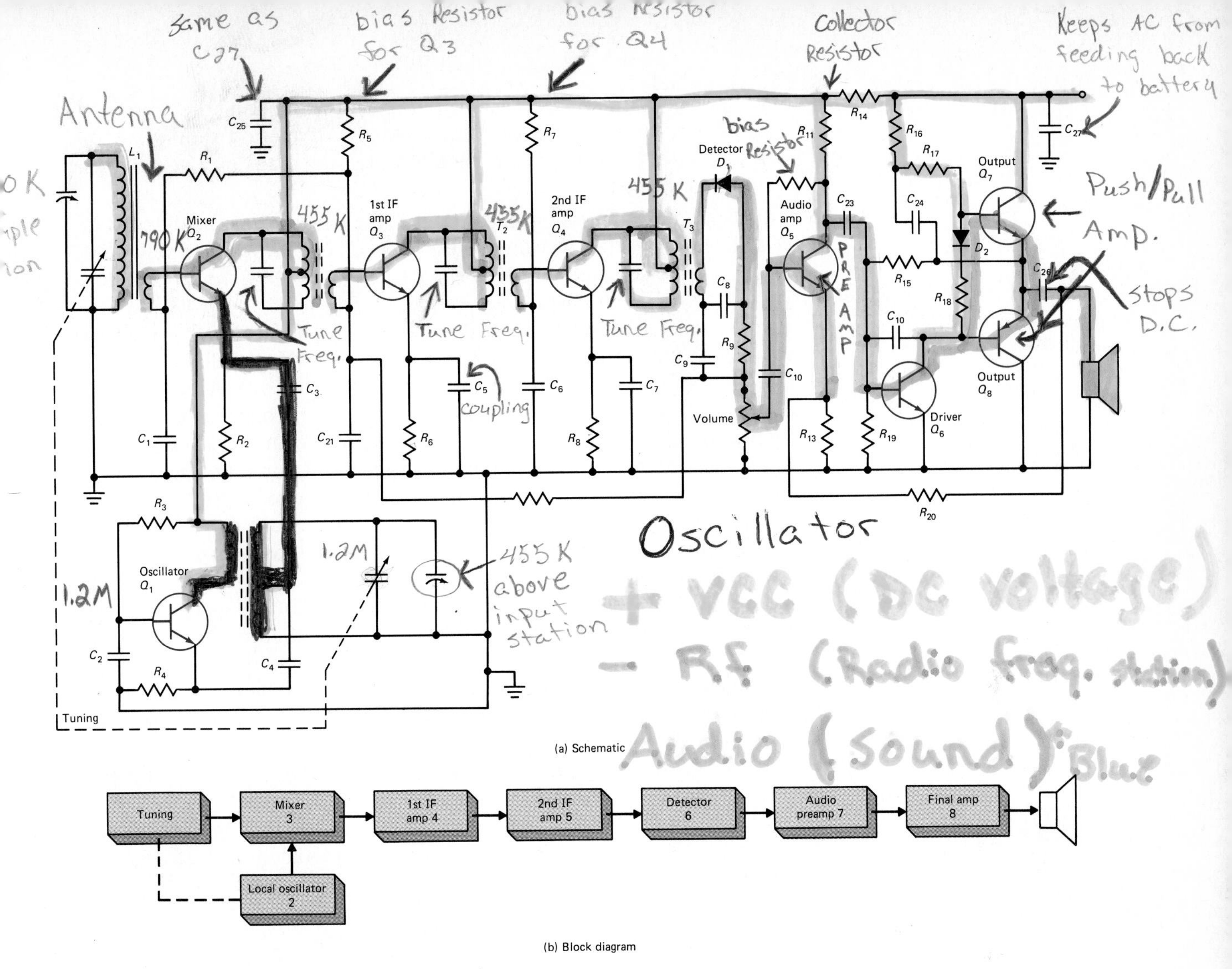

(b) Block diagram

FIGURE 13.12 Complete AM Radio Schematic and Block Diagram

dividers. But it was found that that voltage was equal to the voltage out of the generator. Q_5 was immediately suspected of being defective (no gain!).

4. This basically indicates that the preamplifier section must be defective. Further tests will be conducted to determine the source of the defect.
5. The following measurements were made around Q_5. The base and collector signal levels were identical. There was a feedback potential on the emitter resistor but no value for the input signal. (Since the emitter resistor is un-bypassed, it would be assumed that some of the input signal should be on it.) Because of this, it was speculated that the transistor itself was defective, since no signal was passing from the collector to the emitter.

   Dc values were measured next. The collector voltage was found to be equal to the supply voltage and the emitter voltage was zero. Now it was certain that the device was not conducting. (If it was conducting, there should be a drop on R_{11}, thus making the collector voltage less than the source voltage. Also, conduction should create a dc drop on R_{13}.) Next, the base voltage was measured and it was found also to be at the supply voltage.

6. This last dc measurement certified that the device must be defective. If the base–emitter was open, no base or collector current would flow. Therefore, there would be no drop on R_{11}, R_{12}, or R_{13}. Since Q_5 was not in a high-power circuit, and by looking up its specifications it was found that its maximum ratings were fairly high in respect to the circuit it was in, it was a puzzle why the device failed.

   Upon inspecting the circuit it was determined that if either R_{11} or R_{12} were shorted or open or if R_{13} was open, no bias current would be produced. (Even if R_{13} shorted, bias current would still flow. Also, since there was some signal feedback on R_{13}, it could not be shorted to ground.) R_{11} could not be open, or there would not be supply voltage on the collector of Q_5. Similarly, R_{12} could not be open or there would not be voltage to the base, as well as signal passed to the collector of Q_5.

   That left either R_{11} or R_{12} to be shorted. If R_{11} were shorted, the full dc supply voltage would be on the collector of Q_5 during operation, leaving it saturated. But checking the amount of base current passed through R_{12} and multiplying that by the maximum beta for the device (200 from the spec sheets for Q_5) resulted in less than a 4-mA collector current, certainly not enough to destroy the device. Also, if R_{11} were shorted, the signal that was found on the collector would have been shorted to ground by either C_{27} or the battery itself.

   Thus there only remained R_{12} to be shorted to have caused the destruction of the device. The battery was disconnected and an ohmmeter check of R_{12} proved it to be shorted. With R_{12} shorted a high flow of current would pass through the base–emitter junction. This could cause the base connection (internal) to open. Upon close inspection of the circuit it was found that the printed circuit board on which it was mounted had a stray piece of solder bridging the terminals of R_{12}. (Since resistors do not usually just short out, it was necessary to find the source of this short-circuit condition. Some time could have been saved by running an ohmmeter check of the three suspect resistors. Being analytical proved to be more challenging, though!)

7. The short across R_{12} was removed and Q_5 was replaced.

8. The block diagram did not reveal any special problems that may have been caused by these defects. The circuit was inspected and it was found that since the problem was basically a dc fault and since C_{22} and C_{23} would contain the short-circuit current within this stage, no further damage or fatigue could be presumed. Even with R_{12} shorted, the signal coming from the volume control could not "overload" (draw too much current) from the previous circuits since the only direct paths to ground were through R_{11} (via C_{27}) or through R_{19}, both of which are quite large in value.

9. With a new Q_5 in place, the circuit was turned on and tested and was found to function exactly as expected. A quick check of the dc levels

around Q_5 found them to be well within expected ranges (base voltage at 0.68 V, collector voltage at 3.2 V). A brief notation of the defect was made, and a suggestion for more careful inspection of the soldering practice was made for reference by the manufacturer.

Although these steps have been taken to describe the details involved in evaluating this relatively simple circuit, the procedure worked and is one that will also be applicable to other, much larger circuitry. To summarize these troubleshooting procedures:

1. Relate the fault to the block diagram.
2. Identify the circuit that corresponds to the defective block.
3. Test the inputs/outputs of the suspect circuit(s).
4. Isolate the single defective circuit.
5. Conduct dc and ac voltage tests on that circuit.
6. Isolate the defective part(s).
7. Repair the defect (replace the defective part).
8. Investigate for coordinated faults.
9. Test the system and return it to service.

REVIEW PROBLEM

(1) Working with a partner, prepare a possible defect in the audio-amplifier stage of this circuit. (Avoid the earlier stages, as a description of cause/effect relationships may be a problem.) Analyze the symptoms that would occur due to this defect. Exchange your lists of symptoms and, using the steps outlined above, see if you can troubleshoot each other's circuits. Write brief notes as to the responses detected at each step of the evaluation. You may ask each other about results of tests made (e.g., "What is the dc voltage on R_{19}?"). Your notes plus the number of tests run will help judge how efficiently you have conducted the troubleshooting.

SUMMARY

The process of using multistage amplifiers to accumulate the high gain levels needed for the amplification of low-level input signals has been illustrated. Many of the components and methods used to couple these amplifiers together and the need for maximum power transfer between stages were illustrated.

The accumulated effects of gain and of frequency-response problems were shown to be of major importance when dealing with these amplifier systems. It was shown that even though circuitry may function adequately on an independent basis, modification of circuit elements may be needed when attempting to place a circuit into a multistage application.

Finally, it was shown that the block diagram of a multistage amplifier can be quite useful when attempting to troubleshoot problems in such a system.

The block diagram was shown to be a time- and trouble-saving tool when determining the faults in multistage, multiple-effect amplifiers.

KEY EQUATIONS

13.1a $A_{I(\text{total})} = A_{I1} \times A_{I2} \times A_{I3} \ldots$

13.1b $A_{V(\text{total})} = A_{V1} \times A_{V2} \times A_{V3} \ldots$

13.1c $A_{P(\text{total})} = A_{P1} \times A_{P2} \times A_{P3} \ldots$

13.2 $P_{\text{in}} = \dfrac{P_{\text{out}}}{2}$

PROBLEMS

Section 13.1

13.1. If the amplifiers in a multistage amplifier had the following individual specifications, what are the total specifications for the amplifier (A_v, A_i, and A_p)?

Stage 1: $A_v = -30$, $A_i = 20$
Stage 2: $A_v = -50$, $A_i = 40$
Stage 3: $A_v = -50$, $A_i = 40$
Stage 4: $A_v = -1$, $A_i = 2$ (buffer amplifier)
Stage 5: $A_v = 20$, $A_i = 15$

Section 13.2

13.2. The blocks in a block diagram often coordinate with what element in a schematic diagram?

Section 13.3

13.3. List three electronic functions that can be achieved in the process of coupling amplifiers in a multistage circuit.

Section 13.4

13.4. Between the two circuits shown in Figure 13.13, illustrate the following coupling methods.
(a) Simple capacitive coupling
(b) Transformer coupling

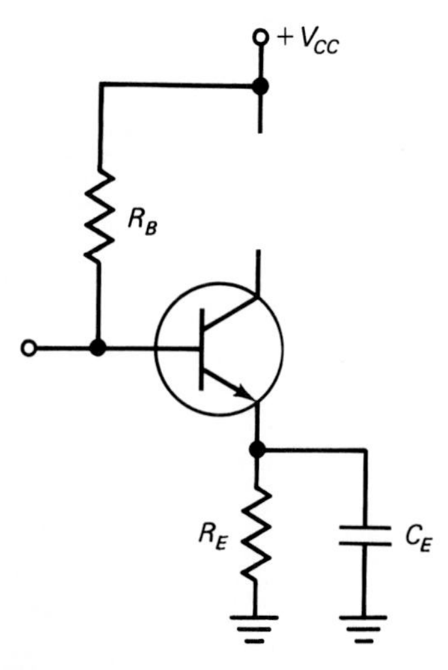

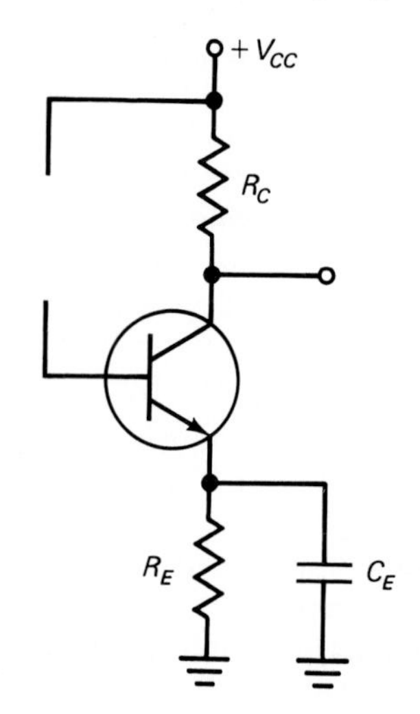

FIGURE 13.13

(c) Resonant circuit coupling (double tuned)
(Be sure to complete the dc bias for the output of the first stage and the input for the second stage!)

Section 13.5

13.5. If a power amplifier is to be attached to an 8-Ω load, what value should the output resistance of that amplifier be? If the amplifier is delivering 60 W (rms) to that load resistor, assuming that the resistance specified in the first part of this question is used, how much power does the amplifier actually have to produce?

13.6. The voltage gain of the amplifier shown in Figure 13.14 is

$$A_V = -\frac{R_C'}{R_E}$$

(where R_C' is the parallel combination of R_C and the load resistor). Calculate the voltage gain of this amplifier first without the load resistor attached ($R_C' = R_C$), then with it attached. Compare these results.

13.7. Sketch the frequency-response curve of a five-stage amplifier if all the stages are identical, each having an $f_{\text{low}} = 20$ Hz and an $f_{\text{high}} = 30$ kHz. (Use a 0-dB-based plot.)

13.8. Since the roll-off of the amplifier illustrated in problem 13.7 is at a rate of −100 dB/decade, the actual break frequencies will be moved closer to the midband range. The following chart indicates the correction factor for the multistage effect:

Number of Stages	Correction Factor (Approx.)
1	1
2	0.64
3	0.51
4	0.44
5	0.39
6	0.35

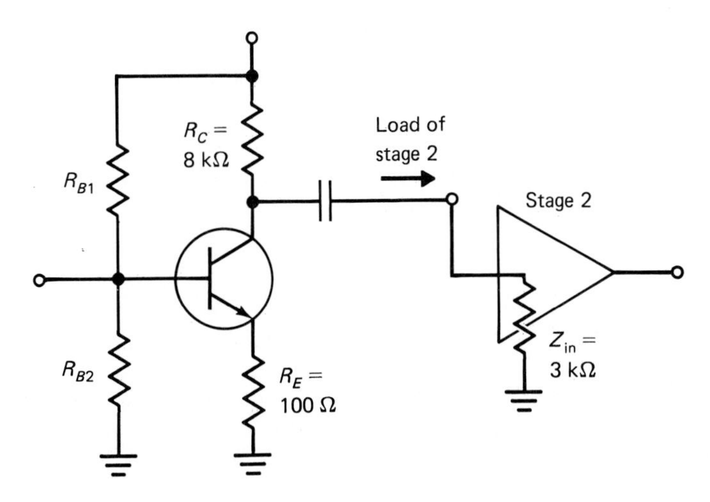

FIGURE 13.14

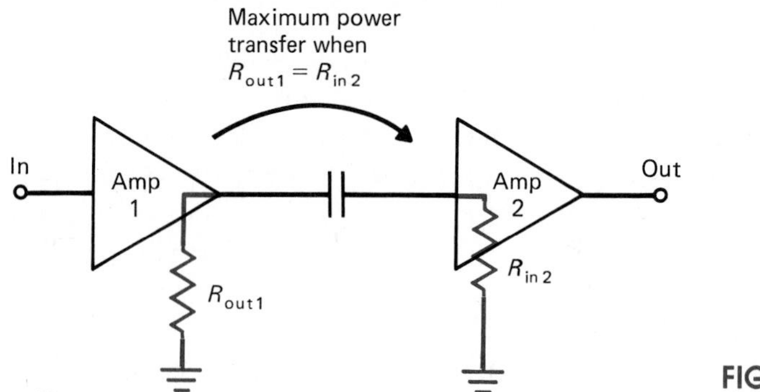

FIGURE 13.15

The way this factor is used is to multiply upper break frequencies and divide lower break frequencies by it. Using this information, predict the actual break frequencies for the multistage amplifier noted in problem 13.7. Also illustrate these points on the graph produced for that problem.

13.9. A three-stage amplifier has a voltage gain of 25 for each stage. The frequency response was found to be too limited, so feedback was applied. The feedback factor is $B = 0.0016$. Find the gain both before and after feedback.

13.10. In problem 13.9 there was obviously a loss of gain for the system when feedback was used. Determine the voltage gain of an amplifier that could be added after this multistage circuit to compensate for this loss. (This will be a fourth stage outside the feedback loop.)

Section 13.6

13.11. For the circuit shown in Figure 13.15, the following parts defects (individually) occurred. Indicate first what effect they may have on the output, then indicate one, two, or three voltage levels that may be off value in the nearby circuit, thus indicating the approximate source of the defect. [*Example:* If Q_7's base–emitter were to open, the output would be distorted since only Q_8 would produce an output. Indications: The collector–emitter voltage would be too high. The base–emitter voltage would not be 0.7 V. The ac base–emitter voltage would be a complete sine wave (not class B).]

(a) Base–emitter open on Q_6.
(b) Diode D_2 would open.
(c) R_{13} would open.
(d) The volume control wiper would open.
(e) The left side of the transformer (T_4) became open (concentrate only on the circuitry right around this transformer).
(f) Signal feedback resistor R_{20} would open.

ANSWERS TO REVIEW PROBLEMS

Section 13.1

1. $A_v = 234{,}256$, $A_i = 50{,}625$
2. Noise input at stage 1, output = 4.69 V; at stage 2, output = 0.213 V; at stage 3, output = 9.68 mV; at stage 4, output = 440 μV
3. 14,641

Section 13.2

1. c, d, e, and g
2. Between the detector and audio preamp blocks 6 and 7
3. Since the block diagram illustrates how the process is achieved, there would be little, if any, change. The diagram-to-circuit coordination would be different, in that blocks 1 through 4 would all coordinate to a single IC. (*Note:* They may have a dashed line around them with an indicator that they all were a part of the single IC.)

Section 13.3

1. Coupling is achieved through L_1 (input to Q_2); T_4 (output of Q_1); T_1 (as noted); T_2 (Q_3 to Q_4); T_3 (Q_4 to detector); C_{10} (detector to Q_5); C_{23} (Q_5 to Q_6); C_{24} and C_{10} (Q_5 to Q_7 and Q_8); C_{26} (Q_7 and Q_8 to speaker); Q_6 is directly coupled to Q_8, as is the output of Q_7.

Section 13.4

1. L_1, T_1, T_2, and T_3: transformer coupling; C_{10}, C_{23}, C_{24}, C_{10}, and C_{26}: capacitive coupling; Q_6 to Q_8 and Q_7 to Q_8: direct coupling

Section 13.5

1. Interstage losses due to voltage or current dividers, loading effects on previous stage gains, reduction in frequency response
2. -20
3. Resistor R_{20} provides signal feedback to the audio preamp stage (Q_5). It forms a basic divider between R_{13}. The signal fed back is reduced (by the divider) to a low level; this level is then supplied to the emitter of Q_5 as either more or less signal feedback than that already provided in that stage by the unbypassed resistor R_{13}. Since the composite ac voltage on this emitter is fed back as a negative signal, greater stability and less distortion of the system's output will exist.

Section 13.6

1. The solution depends on the defect selected. A "partner" cross-check should help resolve a correct solution.

14 Power Conversion (Power Supplies)

Objectives

Upon completing this chapter, the reader should be able to:

- Identify the steps necessary for ac-to-ac, ac-to-dc, and dc-to-dc conversion.
- Identify the specific components used in these conversions.
- Illustrate and explain three forms of rectifier circuits.
- Define regulation and illustrate typical regulator circuits.
- List typical power supply component specifications.
- Analyze and troubleshoot power supply circuits.

Introduction

In all of the circuits shown in this book so far, the need for dc voltages and sources of energy has been demonstrated. Such energy can be provided by batteries. This is not the most desirable source of energy, as batteries are bulky and need to be replaced. Instead of using batteries, dc energy sources can be made using the ac supply lines present in homes and industry. The process of this ac-to-dc conversion, along with ac-to-ac conversion and dc-to-dc conversion, is illustrated in this chapter. A variety of new circuits and devices are introduced which relate to the process of power conversion. ■

14.1 THE NEED FOR POWER CONVERSION

Throughout this book the source of energy in the amplifier circuits has been described as a dc power supply. One common source of that power is through the use of batteries. It is inconvenient and often impractical to power all circuits using bulky batteries which will eventually wear out. Therefore, it is desirable to produce dc energy sources that operate off common ac electrical lines. In other cases there is the need to produce multiple levels of dc for use in various circuits within a system. To avoid having several "battery packs" for one piece of equipment, it may also be desirable to have a power supply (or supplies) which produce several levels of dc voltage.

From the standpoint of electronic circuitry applications, it would be more practical to have dc outlets in homes and industry. Since so many circuits require dc for operation, the need to convert the conventional ac supplied through power lines requires that most equipment have an ac-to-dc converter built in. Amplifiers cannot operate using ac sources, since the sinusoidal variations in this source voltage would either damage the circuitry or would produce an "extra" signal that would interfere with the signal that was to be amplified.

Basically, the conversion of ac supply voltages to dc is done in four primary steps (Figure 14.1):

1. The incoming ac voltage levels are adjusted to provide a level compatible with the demands of the attached circuitry. Typically, transformers are used to achieve this process. (For example, a 120-V ac input would be transformed to a 24-V ac level.)
2. This ac level would be converted to a pulsed dc level which has a varying amplitude but is of one polarity/current direction. This is commonly done using diodes.
3. The pulsed dc voltage would be smoothed out to a constant dc level. This is commonly done using low-pass filters.
4. This dc level would (often) be fixed using what is called a regulator circuit. Thus a stable dc level would be produced.

There are other forms of power conversion which we will discuss briefly. These fall into the category of ac-to-ac conversion (such as in transformers) and dc-to-dc conversion (a dc level is raised or lowered efficiently).

REVIEW PROBLEMS

(1) Investigate your home, apartment, or dormitory and prepare a list of all of the electronic circuits that are plugged into the ac outlets. Almost every one of these circuits requires some form of ac-to-dc conversion.

(2) To understand more clearly why dc is not supplied to homes and industry rather than ac, refer to a circuit analysis or ac power distribution textbook and prepare a short description of the reasons.

(3) In the following sections of this chapter, it will be assumed that the reader is familiar with the actions and operation of transformers. If necessary, refer to a circuit analysis textbook to review these principles and the related equations.

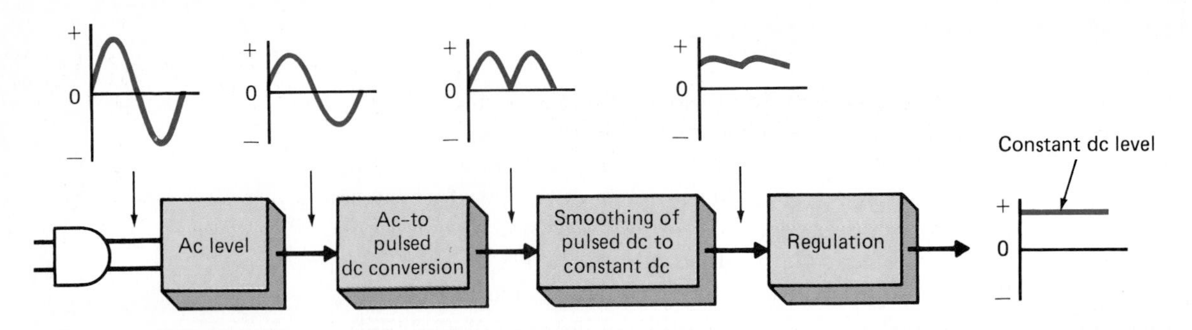

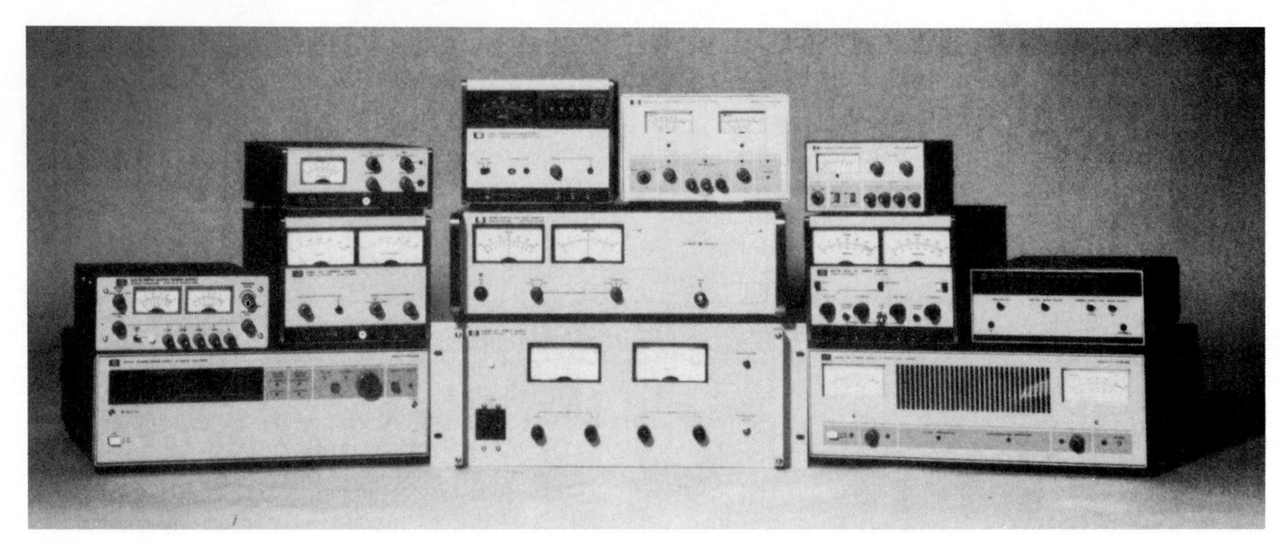

FIGURE 14.1 Block Diagram of AC-to-DC Conversion (DC Power Supply) (b) Photo Courtesy of Hewlett Packard Company; (c) Courtesy of Lambda Electronics, a Division of Veeco Instruments Inc.

14.2 FORMS OF POWER CONVERSION

In this section various forms of power conversion are explored. In Section 14.1 several forms of conversion were introduced. In this section each of these forms is analyzed on a block diagram basis and basic principles of the conversion introduced.

AC-to-AC Conversion

The most common form of ac-to-ac conversion (Figure 14.2) is by use of a transformer. In most cases, such conversion is done to produce an output voltage or current level that is different from the input level. As noted, the most popular method of achieving this conversion is by using either step-up or step-down transformers. When such conversion is done, the output voltage (or current) differs from the input, but the frequency of the ac signal is identical to that of the input. (In some cases the output is identical in both amplitude and frequency as the input. Such a process is done to provide **isolation**. The output of the isolation circuit is not related to the ground connection of the original ac input. This is often usable for test and measurement purposes.)

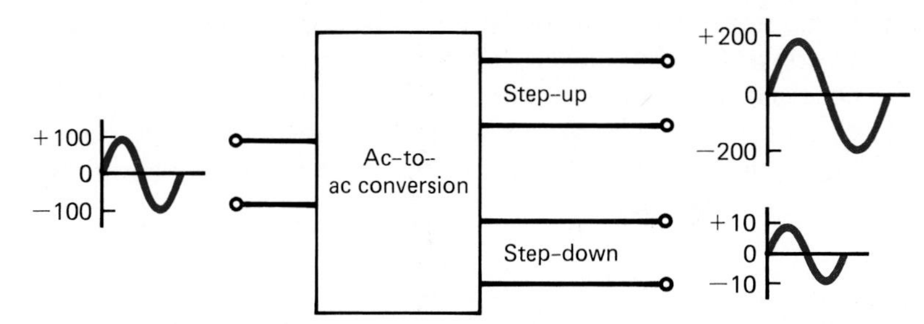

FIGURE 14.2 Detail of function of AC-to-DC Level Conversion

An ac-to-ac conversion can produce several special effects. First, the conversion device (again usually a transformer) can produce more than one voltage level. Therefore, the output could be several ac voltages, each a by-product of the single input level. Also, special phase relationships can be produced at the output, since there usually is not an established ground at the output of the device.

In Figure 14.3 a transformer has been illustrated which demonstrates this capacity for multiple output levels. Also shown is what is called a **center-**

FIGURE 14.3 Transformers

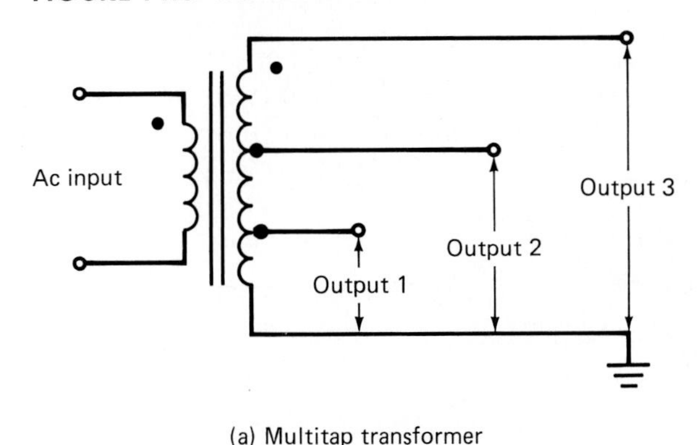

tapped transformer connection, which can be used to produce waveforms that are 180° out of phase with each other. The way the center-tapped output can be manipulated is as follows:

1. Since there is no specific ground associated with the output of this center-tapped section, ground (in this case, a common connection) can be specified at any of the three leads.
2. If the center lead is identified as the common lead for other circuits, the top and bottom voltages can be measured in reference to this ground.
3. The phase of the top lead, in reference to this center terminal, will be considered to be the "in-phase" lead (0° phase shift in respect to the input line).
4. In respect to this phase definition, the phase of the bottom lead will be shifted from that top lead by 180°. Thus as the top lead begins its positive cycle, the bottom is beginning a 180° out-of-phase negative cycle.

Thus, through a simple transformer an ac-to-ac conversion can be developed. This conversion can be used to adjust the voltage level of the ac signal and to (if necessary) produce in-phase or out-of-phase signals. Other forms of ac-to-ac converters (not transformers) can be found which will change the frequency of the input wave. These use sophisticated electronic circuitry.

Ac-to-Dc Conversion

The basic steps used to provide ac-to-dc conversion were presented in the preceding section. Each of these blocks is examined here in greater depth to explore the basic forms of conversion. Figure 14.4 is a reproduction of the earlier diagram.

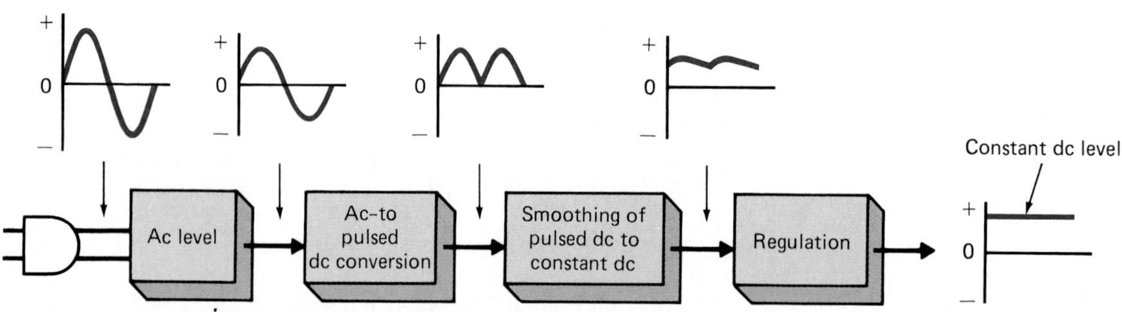

FIGURE 14.4 Block Diagram of DC Power Supply

Ac Conversion Block

The first block is not mandatory but is used in nearly every conversion circuit. The objective of this block is to produce an ac level that is compatible with the dc level that is being produced as the output of the system. In most cases, circuitry does not use a dc level that is a direct product of the common 120- or 240-V ac voltage levels available from the power lines.

The method by which the ac-to-pulsed dc conversion circuit works may also require dual-phase ac. To produce this form of output, the ac-to-ac converter must perform as described in the earlier discussion about center-tapped transformers. As can be surmised, transformers are usually used to produce the ac-to-ac conversions steps.

Ac-to-Pulsed Dc Conversion Block

The second block is needed to produce a signal that is of a single polarity. Since dc voltages are of one constant polarity and the ac is of an alternating polarity (+ and −), the next step in the conversion requires that the ac be converted to a single polarity (pulsed dc).

This process is done in one of two ways. The first way involves the elimination of the undesirable polarity from the ac signal. In this process only ac signals of the correct polarity in respect to a circuit ground are passed on to the balance of the circuit. The other section of the wave is dropped on an internal device with high impedance.

Figure 14.5 illustrates what is called a half-wave output from such a converter. A complete sine wave enters the converter, but only one polarity

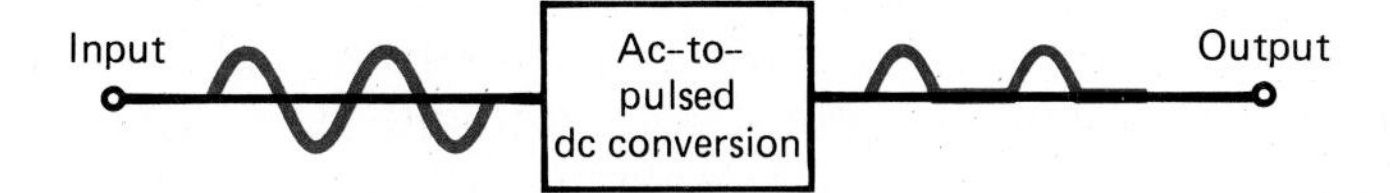

FIGURE 14.5 AC-to-Pulsed DC Waveform Conversion

of that wave exits the block and is passed on to further circuit elements. The other half of the wave is dropped across a high-resistance element inside the block. The efficiency of such a conversion is low. First, since part of the wave was dropped internally, that loss reduces the energy capacity of the waveform entering the block. Second, since a part of the output is zero, further conversion processes must work harder to maintain a constant level during this "off period" of the wave. A common name applied to this form of converter is **half-wave rectifier**.

Another more desirable conversion process uses the special effect seen from the ac converter, which can produce two signals, each 180° out of phase with the other. In this process, both of those waves are run through the half-wave rectification process. Then the two outputs (each a half wave) are blended as the output of the converter (see Figure 14.6).

The results of this process produce a waveform which does not have the "off" period characterized by the half-wave rectifier. Since the two input sine waves have their positive peaks occurring when the other wave is in a negative cycle, there is a positive cycle available at every instant in time. Thus these cycles can be isolated and combined to produce what is called a full-wave output. The circuit used to achieve this function (in conjunction with the split-phase input) is called a **full-wave rectifier**.

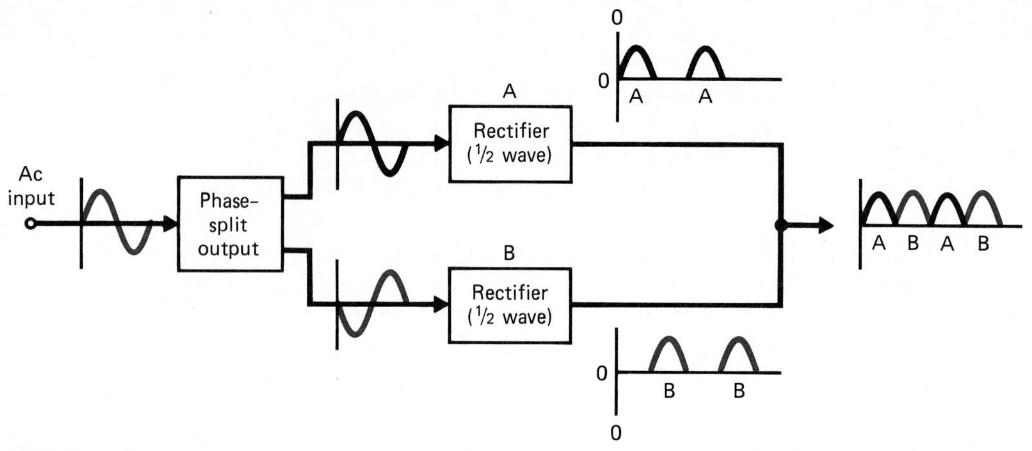

FIGURE 14.6 Use of Two Half-Wave Rectifiers to Compose Full-Wave Output

Pulsed Dc-to-Constant Dc Level Conversion

Although the output of the ac-to-pulsed dc converter is of a single polarity, the changes in level are still undesirable. To operate electronic equipment, the dc output must be of a constant level. (Imagine what would happen to an amplifier if the dc source began to vary in level. Consider the bias problems, problems in saturation or cutoff of the amplifier, and so on. Thus there must be little if any variation in the dc supply voltage level.)

The process of producing a constant output level (fully dc) from a pulsating input is called **filtering**. In past chapters the use of low-pass filters was presented. Special forms of low-pass filters are used in this final step in the conversion process. The most typical form of filter is a capacitive type of low-pass filter. The charge and discharge of the capacitor in the filter is designed so that it is capable of maintaining a relatively constant level as the input to the filter varies. Some of the basic timing effects necessary in this circuit are demonstrated in Section 3.4.

Output-Level Regulation

Once the basic dc level is produced from the filter an optional step to take is to provide **regulation**. Regulation involves electronic circuitry designed to maintain a constant output level under various load conditions. It will be found that the output of an unregulated power supply can change under various load conditions.* Some circuits are quite sensitive to such variations, and therefore a regulation circuit is used to maintain a constant output level.

Regulation of the output level from the power supply can be achieved by the use of a variety of electronic circuit elements. A special form of diode called a **zener diode** can be used for simple regulation. (Transistor and IC regulator circuits are also used.) The schematics and applications of each of these circuits is presented in Section 14.4.

* For power supplies, the load is considered to be the circuit to which it is supplying energy. A heavy load is one that draws large amounts of current; a light load draws less current.

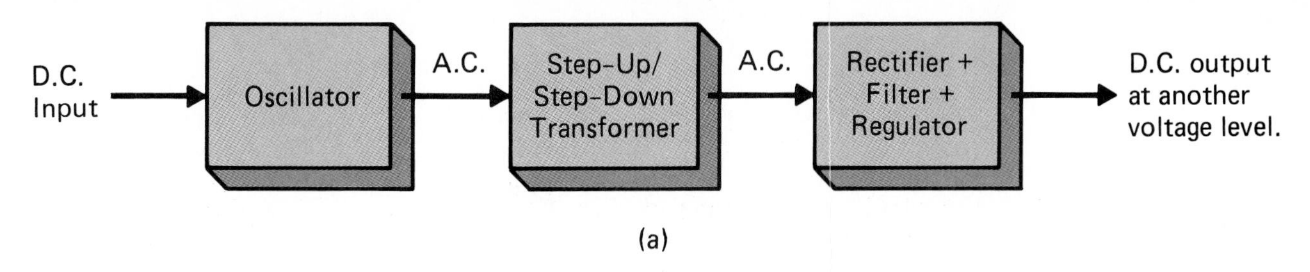

(a)

(b)

(c)

FIGURE 14.7 DC-to-DC Conversion (Also Called an Inverter) (b) & (c) DC to DC Converter Module (Courtesy of Lambda Electronics, A Division of Veeco Instruments, Inc.)

Dc-to-Dc Conversion

Once dc levels have been produced (usually by a power supply) it may be necessary to change the voltage level for some special application. Since transformers cannot be used for dc voltages, special circuits must be used to shift the level of a dc voltage.

There are simple circuits that are used to produce dc voltages at higher levels than the typical output seen from the transformer of a standard ac-to-dc converter. These circuits are not true dc-to-dc converters, but are special forms of the ac-to-dc converters. Such converters are typically called **voltage doublers/triplers** or **quadruplers**. In effect, what they do is produce a dc voltage which is substantially higher than the level that would be produced by a standard converter connected to the same ac source (transformer). These are presented in Chapter 16 as a special-purpose application.

The basic process of dc-to-dc conversion is as follows:

1. A dc level is the input to the converter.
2. This dc level is supplied to an oscillator, which converts the dc voltage to ac voltage.
3. This ac voltage is fed through a transformer, where its level is increased or decreased.
4. The ac output of this transformer is then run through a rectifier circuit, a filter, and often a regulator. Thus it is converted to another dc level.

Figure 14.7 illustrates the basic steps used for dc-to-dc conversion. It can be seen that there are certain steps taken to produce these various conversions. In the following sections, the function of each of these steps and conversions is presented.

REVIEW PROBLEMS

(1) Draw the basic block diagrams used to represent ac-to-dc conversion and for dc-to-dc conversion.

(2) Sketch a transformer schematic symbol and illustrate a center tap on it. Using the center tap as a reference (ground) point, illustrate the input and two output (top and bottom) waveforms.

(3) Based on the principles of electromagnetic circuits, describe why a transformer cannot be used to provide dc-to-dc conversion. (Refer to a presentation of transformers and or electromagnetic circuits in a circuit analysis textbook if assistance is needed.)

14.3 CIRCUITRY USED IN POWER CONVERSION

Sections 14.1 and 14.2 introduced the block diagram process for power conversions. In this section actual circuitry will be developed which does the conversion.

Ac-to-Ac Conversion

The conversion of ac line voltage amplitudes (120 or 240 V rms) to the level needed by the circuit being operated is done using transformers. A step-down transformer is used when the dc voltage is to be lower than the peak value of the ac input. For dc levels above the line voltage values, a step-up transformer is used. Since, as in any circuit, there are some voltage losses in the power supply circuit, the peak output voltage of the transformer is usually chosen to be slightly greater than the desired dc output level of the power supply.

EXAMPLE 14.1

A power supply is to be operated from the 120-V ac line and is to deliver 15 V dc to the circuit to which it is connected. Generally, specify the transformer that should be used.

Solution:

Since the peak voltage of the line is approximately 169.7 V, a step-down transformer will be needed when making the 15-V dc power supply. Also, to overcome losses in the power supply circuit, the transformer may need to produce about 17 V peak. Thus a 120-V rms input/12-V rms output step-down transformer should be sufficient.

Ac-to-Pulsed Dc Conversion (Rectification)

The process of ac-to-pulsed dc voltage conversion (usually called rectification) is typically done with diodes. It will be remembered that diodes were polarity-sensitive devices. The diode will conduct current only when the correct, forward-bias voltage polarity is applied. When an opposite (reverse bias) polarity is used, the diode fails to conduct.

Half-Wave Rectifiers If the diode is connected to an ac source and a simple load resistance, as shown in Figure 14.8, electron current flow to the load occurs only when the correct bias polarity is on the diode. When the polarity of the ac source is such that the diode is reverse biased, the diode does not

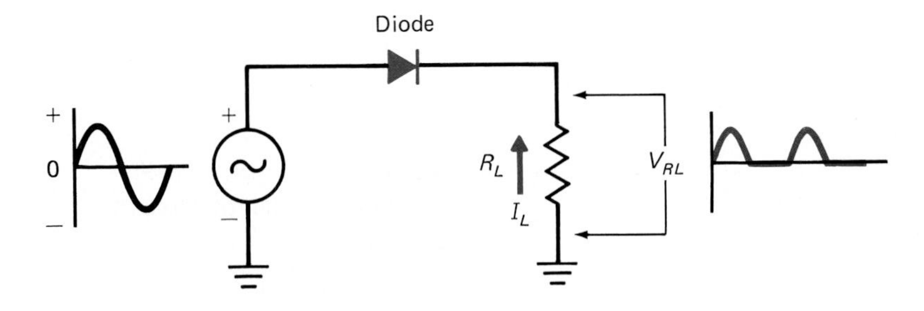

FIGURE 14.8 Semiconductor Diode Used as a Rectifier

conduct. What results is that the load resistance will have a voltage drop on it only during the portion of the ac cycle that forward biases the diode. During the other half of the cycle, the voltage is dropped on the reverse-biased diode—thus the waveforms seen in Figure 14.8 would be produced in the locations shown. The result of this process is that only one-half of the ac cycle (a single polarity, although of variable magnitude) will be present on the load resistance. Such a circuit is called a simple half-wave rectifier. (If the diode were reversed, as shown in Figure 14.9, the opposite polarities would be manipulated by the diode.)

There are two drawbacks to the use of this circuit and the resulting waveform. First, the average value of the waveform shown on the load (in either Figure 14.8 or 14.9) is quite low. This means that this wave is low in energy in respect to the potential energy (rms value) of the initial ac waveform. Also, this wave will be much harder to filter than the full-wave output of other forms or rectifiers. During the long "off" period in the wave, it will be hard to maintain a constant output level from the attached filter. Such rectifiers are not commonly used in power supplies.

Full-Wave Rectifiers To realize a higher energy level at the output of the rectifier circuit and to produce a more easily filtered output, the full-wave rectifier is used. The basic full-wave rectifier uses a center-tapped transformer and two diodes (fundamentally forming two half-wave rectifiers together).

First it must be remembered that the center-tapped transformer produces two waves, each out of phase with the other when referenced to the shared center tap, as shown in Figure 14.10. Each of the diodes in the circuit forms a simple half-wave rectifier, passing only the positive cycle of the ac wave to the load. The blend of these two factors, the transformer action and the rectifiers, will produce a full wave output (as shown in Figure 14.10).

The process that occurs to produce this full wave is as follows:

1. Assuming that the system starts with the top wave of the transformer passing through its positive cycle in respect to the center tap. Thus the bottom wave will be passing through its negative cycle.
2. During the period that this polarity is present (the first half of the cycle) the top diode (D_1) is conducting current to the load. At the same time, the bottom diode (D_2) is blocking current (it is reverse biased).
3. As the ac cycle passes into the next half-cycle, the roles are reversed. With the top polarity negative, diode D_1 is now reverse biased. Since the bottom polarity is now positive, diode D_2 now conducts current to the load.

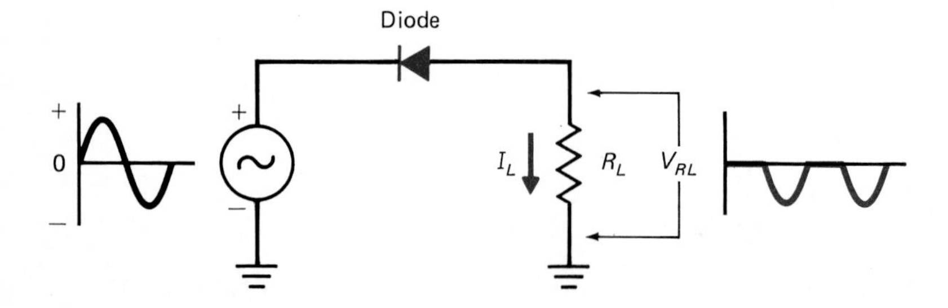

FIGURE 14.9 Rectifier with Polarity Opposite That Shown in Figure 14.8

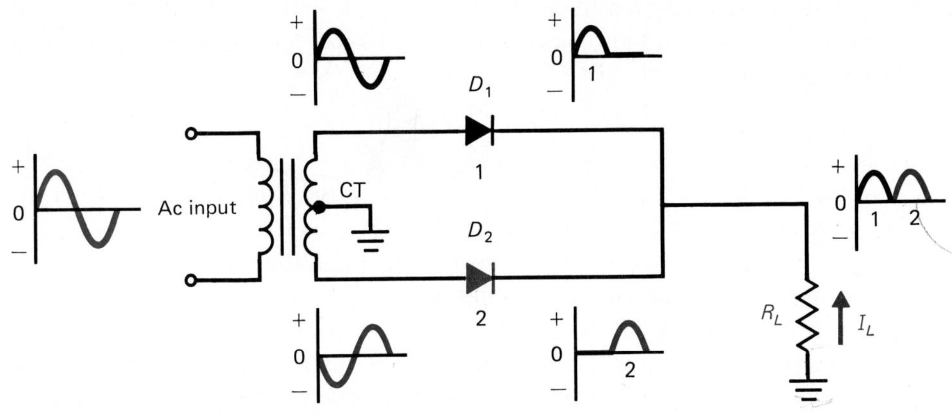

FIGURE 14.10 Full-Wave Rectifier Circuit

4. It must be noted that both diodes conducted current to the load in the same direction, thus producing the same polarity of voltage on the load during each half of the ac cycling.

Thus, a full (uninterrupted) wave is produced on the load. This produces twice the electrical energy that was seen from the half-wave rectifier. As well, this wave does not have a "dead spot" or time when the voltage remains zero. This will allow the filter to maintain a constant output level.

The only drawback in using the standard full-wave rectifier is that the transformer (because of the center tap) may be a bit more expensive than one without a center tap. Another drawback is that in order to produce the same amount of peak output voltage as a half-wave circuit, the transformer voltage (top to bottom) must be double that of the type used in the half-wave. This is due to the need to supply peak voltages to both of the diodes during operation.

To eliminate the need for the center tap on the transformer and to use a transformer with a lower ac voltage, the **bridge rectifier** configuration was established. The bridge rectifier circuit is shown in Figure 14.11. This rectifier circuit can be thought of as more of a current-steering (or directing) circuit than a current-blocking/passing type as in the other rectifiers. Basically, the objective of the diodes is to steer (pass) current always in the same direction through the circuit regardless of the polarity of the source. (For ease of illustration, the load is placed in the center of the "bridge"; later, it will be shown that the load can be placed in a more practical location.)

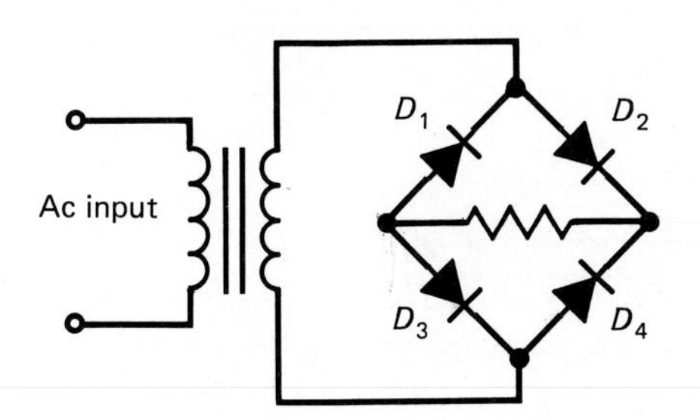

FIGURE 14.11 Full-Wave Bridge Recifier Circuit

Figure 14.12 is shown to clarify the steering process that occurs in the bridge rectifier. The following actions occur for each half-cycle of the ac wave. (*Note:* Current directions are related to electron flow.)

1. When the ac cycle is positive at the top of the transformer and negative at the bottom, this creates a polarity on diodes D_2 and D_3 which forward biases them (diodes D_1 and D_4 are "off").
2. These two diodes then conduct current through the load in a left-to-right direction. Thus the voltage on that resistor is positive (+) on the right, negative (−) on the left.
3. In the second half of the ac wave the polarity on the transformer reverses. This causes diodes D_1 and D_4 to be forward biased (D_2 and D_3 are "off").
4. These diodes then conduct current through the resistor, again in a left-

FIGURE 14.12 Bridge "Steering" Process

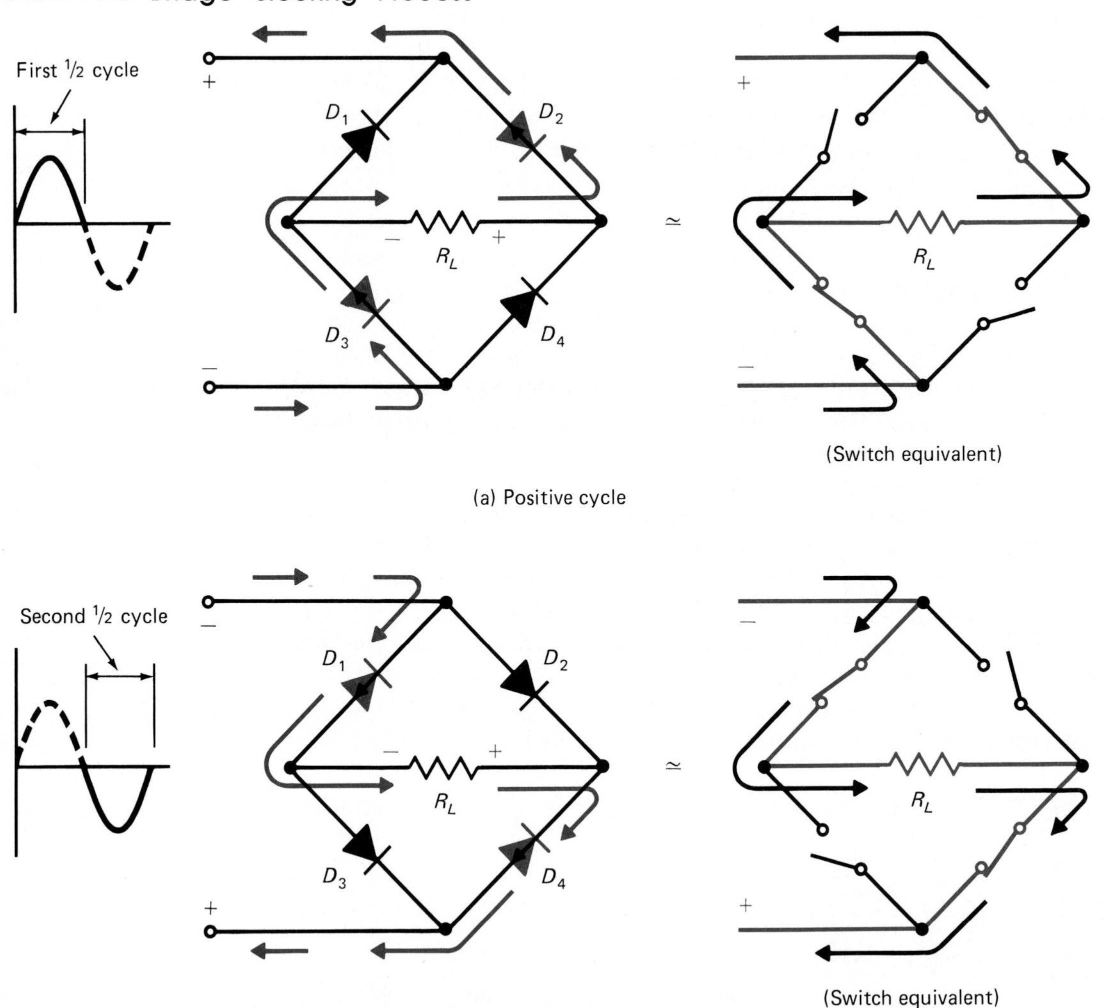

to-right direction. Again the drop on the load resistor is positive on the right and negative on the left.

In effect, then, the diodes have assured that the current from the supply has always passed through the load resistor in a left-to-right direction, always producing the same polarity of voltage on that resistance. Thus none of the energy from the transformer is wasted as a drop on a reverse-biased diode.

It may not be obvious why a pair of diodes (D_1/D_4 and D_2/D_3) is needed. In order that current is not misdirected around the load resistor, neither of these diodes could be a simple conductor. To verify this, if diode D_1 were simply a wire, when the polarity of the source went positive on the top, current would flow through this wire, to D_2, and back to the transformer, thus bypassing the load resistance.

It was noted earlier that this circuit configuration was a bit impractical since the load (which could be the schematic for a complete television) is placed inside the bridge. The more common way of illustrating this circuit is shown in Figure 14.13. If a bridge circuit is needed to supply a positive polarity to the load (referenced to ground), simply reverse the ground connection and the output connection from the bridge to the load, or leave these symbols intact and reverse the direction of each of the diodes (which is the more acceptable method of illustrating this polarity change).

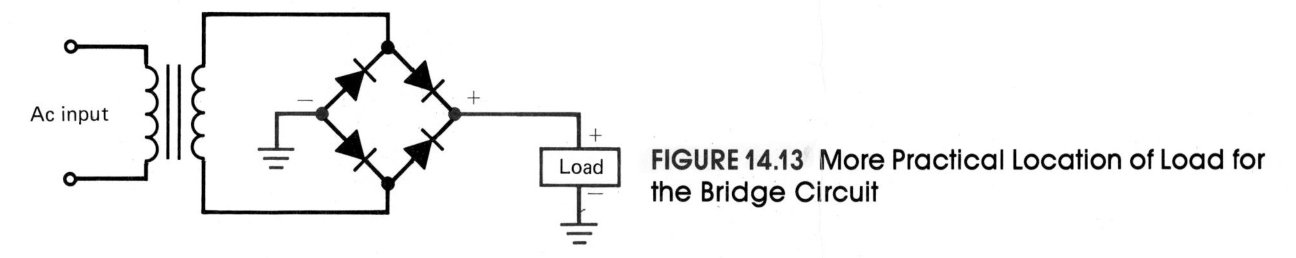

FIGURE 14.13 More Practical Location of Load for the Bridge Circuit

Special note: It should be noted that the actual polarity of the voltage delivered to the load in any of these rectifier forms is determined by the orientation of the diode(s). None of the connections to the transformer have any relationship to this polarity. (All transformer leads, or the transformer's input-to-output "dot notation," could be changed without affecting the polarity of the voltage delivered to the load.) In every case, to reverse the polarity of the voltage delivered to the load, reverse the diode's anode and cathode positioning.

Filter Circuits

The objective of the power supply filter is to accept the pulsed dc input and produce a smooth, continuous dc output. The most popular element used for filtering is a capacitor. The capacitor can store voltages over a period of time. The objective in using a capacitor is to store a dc level during the cycling of the output of (for example) a full-wave rectifier.

In Figure 14.14 this process is illustrated. The first peak of the input signal charges the capacitor to that peak level. Then the voltage out of the rectifier drops below that level. Since the diodes in the rectifier prevent current

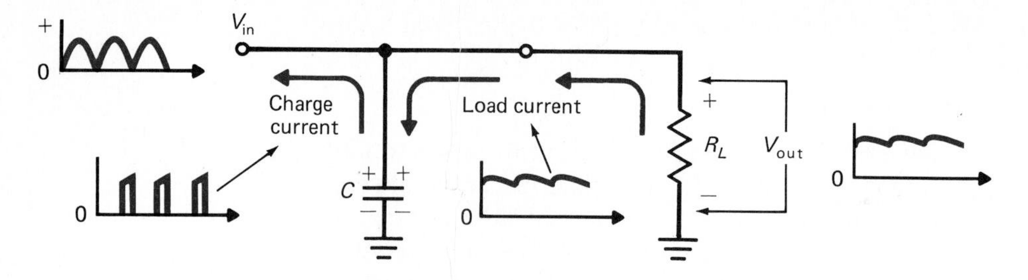

(a) Simple capacitive filter (*R-C* circuit)

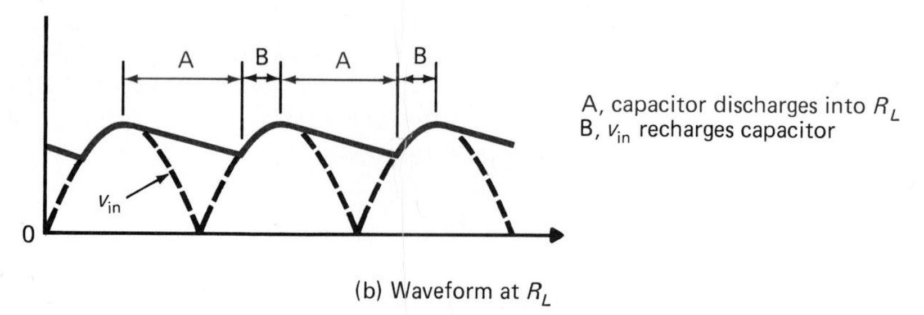

(b) Waveform at R_L

FIGURE 14.14 Simple Capacitor Filter and Output Waveform

from flowing from the capacitor (backward to the rectifier), the capacitor discharges into the load resistance. A slow drop in capacitor voltage occurs as this discharge occurs. This discharge continues until the level of voltage from the rectifier begins to exceed the capacitor voltage again; then the capacitor voltage is brought back up to the peak output level by being recharged by the rectifier voltage.

As long as the capacitor's rate of discharge into the load is much longer than it takes for the rectifier's voltage peak again, the current passed to the load from the capacitor remains relatively constant. Thus, as long as the *R/C* time constant for this capacitor and load resistance is much longer than the time it takes for the rectifier to peak again, the voltage on the load remains basically constant. The capacitor is then being used as a storage device for the rectifier's peak output voltage.

This process should help explain why it is more desirable to have a full-wave output from the rectifier than a half-wave. With a half-wave output, the capacitor must maintain the voltage level during the complete "off" period of the wave. This delay in recharge voltage (in comparison to full-wave voltages) could cause the capacitor's voltage to drop rather low, thus not maintaining a constant output level. This function is shown in Figure 14.15. As can be seen in Figure 14.15, there is a difference in the amount of variation in the output voltage from the two circuits shown. This variation in output voltage is called the **ripple voltage**. An ideal power supply produces no ripple (no loss in capacitor charge level). Power supplies can be rated in terms of how much **ripple** is present in the output. This rating is called the **ripple factor**. The equation for ripple factor is

$$r = \frac{V_r}{V_{dc}} \tag{14.1}$$

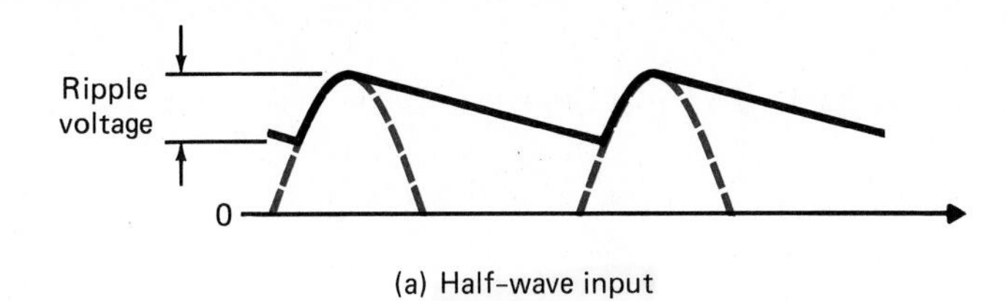

(a) Half-wave input

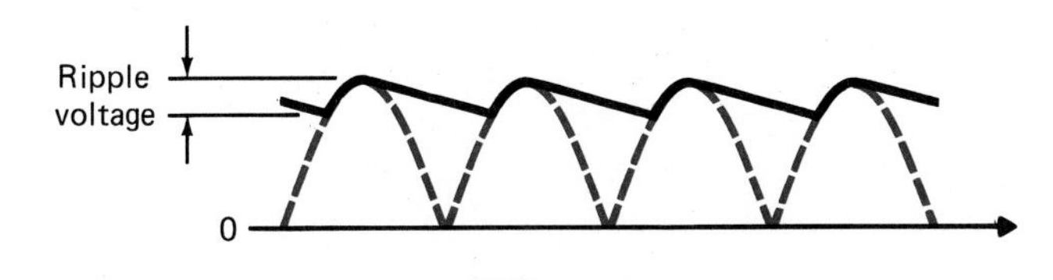

(b) Full-wave input

FIGURE 14.15 Comparison of Filter Outputs

where V_r is the amount of ripple voltage peak to peak and V_{dc} is the average dc output of the filter.

Also, it can be seen that ripple voltage levels are related to the frequency of the "recharge" voltage applied to the capacitor (the pulsed dc input) and to the time constant of the filter capacitor and load combined. Since most power supply circuits operate with a standard 60-Hz (U.S. frequency standard) input to the rectifier circuit and since most usable rectifier circuits are of the full-wave type, the following formulas describe the expected ripple and dc output from a capacitor filter:

$$\text{Ripple voltage: } V_r \simeq V_P \frac{2.4 \times 10^{-3}}{R_L C} \tag{14.2}$$

$$\text{Average dc output voltage: } V_{dc} \simeq V_P \left(1 - \frac{4.17 \times 10^{-3}}{R_L C}\right) \tag{14.3}$$

where V_P = peak voltage out of the rectifier (usually assumed to be the peak value of the ac voltage into the rectifier)
R_L = total resistance of the load applied to the filter
C = total capacitance of the filter

It must be noted that these formulas are valid *only* for a 60-Hz input to the rectifier and *only* when the filter is used with a full-wave rectifier.

EXAMPLE 14.2

Given the following statistics about a power supply circuit, find the expected dc voltage output and the ripple voltage. Also find the ripple factor of this supply.

Ac input to the rectifier: 48 V p-p at 60 Hz
Rectifier type: full-wave bridge
Capacitor value: 100 μF
Load equivalent resistance: 2 kΩ

Solution:

$$V_P = 24 \text{ V } (\tfrac{1}{2} \text{ of p-p input})$$

$$V_{dc} = 24 \text{ V} \left(1 - \frac{4.17 \times 10^{-3}}{(2 \text{ k}\Omega)(100 \ \mu\text{F})}\right) \simeq 23.5 \text{ V dc}$$

$$V_{r(\text{p-p})} = 24 \text{ V} \frac{2.4 \times 10^{-3}}{(2 \text{ k}\Omega)(100 \ \mu\text{F})} \simeq 0.288 \text{ V p-p}$$

$$r = \frac{0.288 \text{ V p-p}}{23.5 \text{ V dc}} \simeq 0.0122$$

(Ripple factor is often expressed as a percent; thus for this circuit the ripple factor in percent is 1.22%.)

There are several other forms of filters used in power supplies; these are illustrated in Figure 14.16. Primarily, these take advantage of time constants (R/C or R/L) to achieve the filtering necessary to produce a constant dc output. Each form has a set of equations that differs from the simple capacitive type. Only the equations for the simple capacitive form are presented here, as this filter represents both the most common form of filter used and is representative of the type of filter that would be added to a functioning power supply circuit to improve filtering (as illustrated in Section 14.6).

Regulation Circuits

If there are either variations in the ac input to the rectifier circuit of a power supply or variations in the load conditions, the actual value of output voltage

FIGURE 14.16 Filters

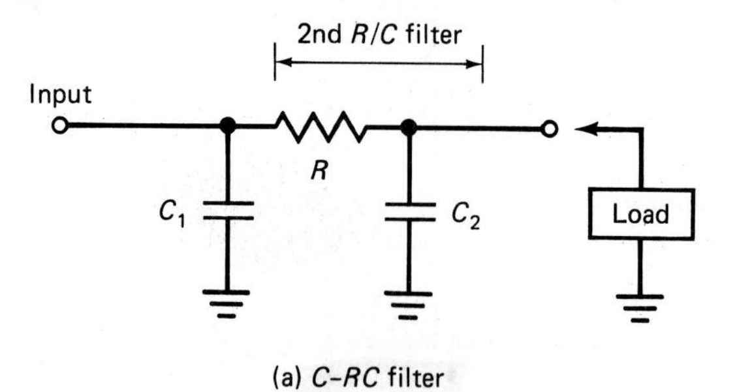

(a) C-RC filter

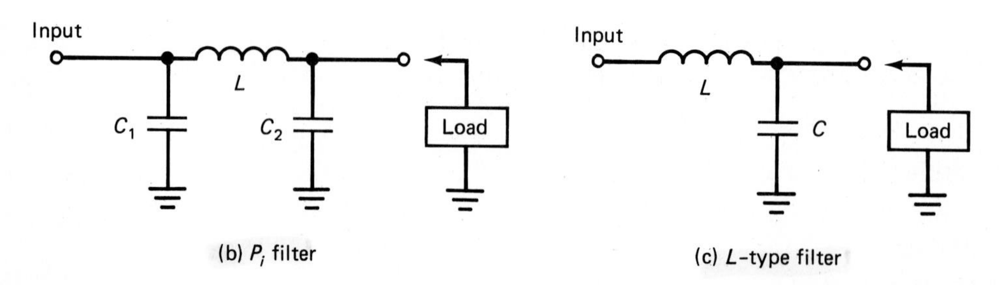

(b) P_i filter

(c) L-type filter

(dc) and ripple would also change. For some circuits these fluctuations in the supply values would be acceptable. Some circuits, though, are extremely sensitive to even minor variations in these supply values. For these circuits, it is necessary to use output **regulation**, a process of maintaining a constant dc level regardless of minor variations in the ac input or load.

The process of regulation typically uses special semiconductor devices or integrated circuits to produce a constant output. Functionally, regulators work by supplying them with a dc input that is higher than the expected output of the regulator. Then the regulator produces a constant output, using the range that exists between the input and lower output as a "buffer zone." As long as the input level stays within the range of the normal value and (basically) the minimum acceptable level, the output will be held to a constant value. This is illustrated in Figure 14.17.

Regulators fit into three major categories:

1. Constant-output (analog function) regulators
2. Variable-output (analog function) regulators
3. Switching regulators

The first two classifications of regulators depend on the use of either special discrete devices (zener regulator diodes), special integrated-circuit devices (IC regulators), or specially designed circuits using transistors, op-amps, and specialized devices. The degree of complexity of the circuit used depends on the amount of accuracy needed from the regulation process. Many use the process illustrated in Figure 14.17 to maintain a constant output. The function of each circuit and specific equations related to each are illustrated in Section 14.4.

Switching regulators use a different process to achieve regulation. The basic block diagram for switching regulators is presented in Figure 14.18.

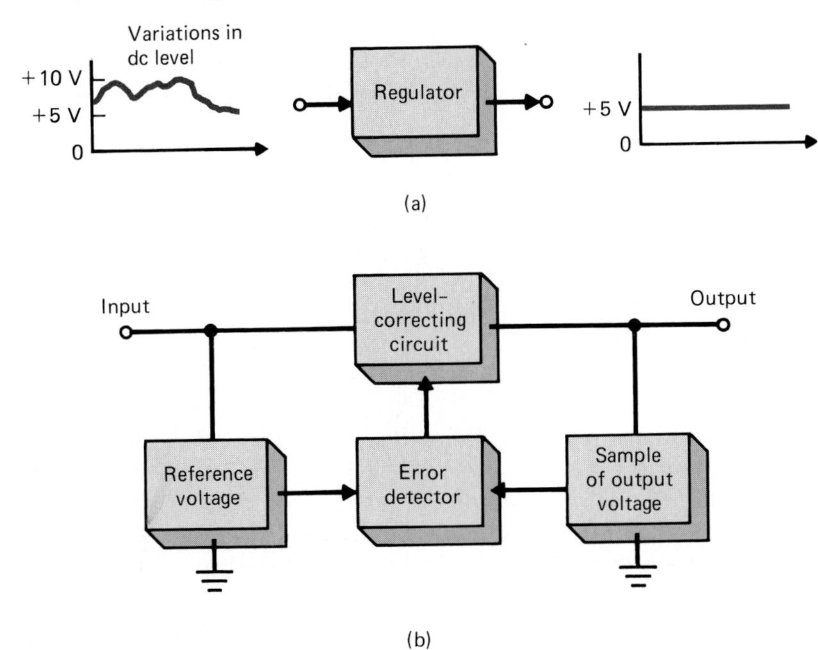

FIGURE 14.17 Regulator Function and Block Diagram

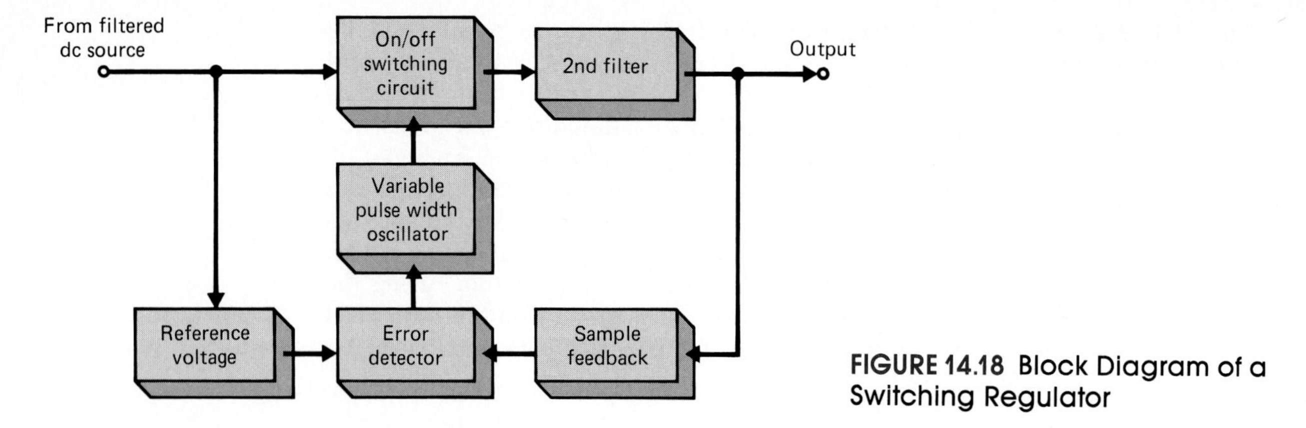

FIGURE 14.18 Block Diagram of a Switching Regulator

Fundamentally, the switching regulator works in the following manner (coordinated with Figure 14.18) (the description is made from the output backward, since special feedback is used to control the dc input to the regulator).

1. At the output terminals of the regulator, the amount of current and/or voltage supplied to the load is constantly sensed.
2. Variations in the dc value supplied to the load, as feedback in the circuit, are compared to a standard level.
3. The amount of variation from that standard controls the output of a pulse (square-wave) oscillator. If the variation is minimal, the output of the oscillator is a set of short-duration pulses. If the variation is large, the output of the oscillator changes to long-duration pulses.
4. These pulses are then used to control the amount of time (duration) that a switching circuit turns on the dc input to the regulator. (Short pulses turn it on for only brief durations; long pulses turn it on for longer durations.)
5. This dc output from the switching circuit is then applied to an additional filtering circuit, thus smoothing out the variations caused by the switching action.

Thus the output is maintained at a constant level by using feedback to produce a constant *average* level at the output terminals, via the switching function within the regulator circuit. This may seem like a rather awkward means for producing a regulated output, but there are some specific advantages of the switching regulator:

1. The switching regulator is more efficient than a conventional analog regulator, since a portion of the input voltage is not discarded (internally dissipated) as the regulator maintains a constant output. The switching regulator draws current from the input supply only when that current is needed to supply energy to the load. Analog regulators draw a continuous current from the input.
2. The switching regulator can be made easily variable in output level by

manipulating the average pulse width of the oscillator. Once the average level is set, the regulator will maintain that level through the feedback process.

3. Switching regulators can be had in either step-down or step-up configurations. Thus the output voltage can be less or more than the input dc voltage level. Analog regulators can produce only a step-down function.

Because of their complexity and thus expense, switching regulators are usually used only in equipment where load conditions would demand an accurate control of the dc level over a wide variation in load conditions. Circuitry (such as computers) that produces widely varying current demands on the power supply can run with improved efficiency and accuracy by using switching regulators. In contrast, an analog regulator would have to be used that was capable of handling the maximum peak demand of the circuit, and when that level of demand was not present, the analog regulator would simply dissipate the excess energy (in the form of heat).

Thus it can be seen that regulators, where needed, are available in several forms. Greater details on regulator circuits will be presented when an investigation of specific circuit elements is made (Section 14.4).

Dc-to-Dc Conversion

The process of dc-to-dc conversion (also called inversion), where one dc level is increased or (occasionally) decreased, involves a wide variety of circuit configurations. Most of these circuits, though, involve some of the basic principles covered in this chapter, plus the function of oscillators. Figure 14.19 is an illustration of the basic block process used to achieve this conversion. This circuit operates in the following manner:

1. The dc input voltage is applied to a conventional oscillator circuit. This circuit often uses power transistors and is commonly a Hartley-type oscillator, due to its association with the transformer.
2. The oscillator voltage is applied to a transformer where the ac voltage can be increased (or decreased) in magnitude (step-up or step-down transformer).

FIGURE 14.19 Block Diagram of a DC-to-DC Converter (Also Called Inverter)

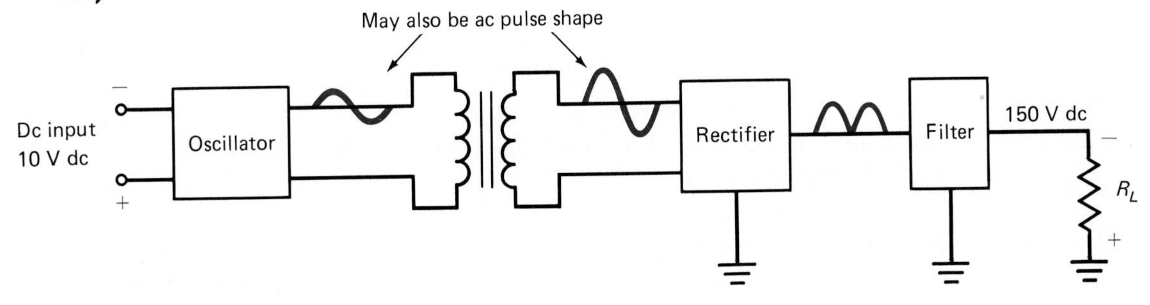

3. The output of the transformer (ac) is applied to a conventional full-wave rectifier circuit.
4. The pulsed dc output of the rectifier is then filtered and (if necessary) regulated to produce a new dc output level.

The oscillations used in this conversion process may be either sine wave or square wave in shape, depending on the design of the converter. Since the construction of such a converter is often specialized to the application for which it is intended, no single specific circuit will be presented in this book.

REVIEW PROBLEMS

(1) Sketch the basic block diagram for a power supply (ac-to-dc conversion). On that block diagram illustrate the approximate waveforms that would be produced at each stage of the conversion.

(2) Sketch the schematic for a basic full-wave rectifier and one for a full-wave bridge rectifier. Connect these to a load resistance where the polarity on that resistance is negative in respect to ground [top of R_L (−), bottom (+)].

(3) Prepare a composite schematic for a power supply (ac-to-dc converter). Use the following elements:

Input transformer
Full-wave bridge rectifier
Capacitive filter
(Simple block symbol) regulator

(It should be noted that only one load resistance is attached to the total supply circuit. In the explanations used in this section, the load resistance for each stage is assumed to be simply the input resistance of the next stage.)

(4) Sketch the block diagram for the switching regulator. Write a short explanation of what would happen in this regulator if the average dc output level were to drop below a normal standard. (A simple step 1-to-step 2 description can be used.)

(5) A common use for a dc-to-dc converter (inverter) is in the electronic strobe used for photography. The (typically) 3 V from two batteries is boosted in the strobe circuit to the 500 to 2000 V dc needed to "fire" a xenon flash tube. Using the description of the dc-to-dc converter made in this section, describe how such an electronic strobe circuit functions.

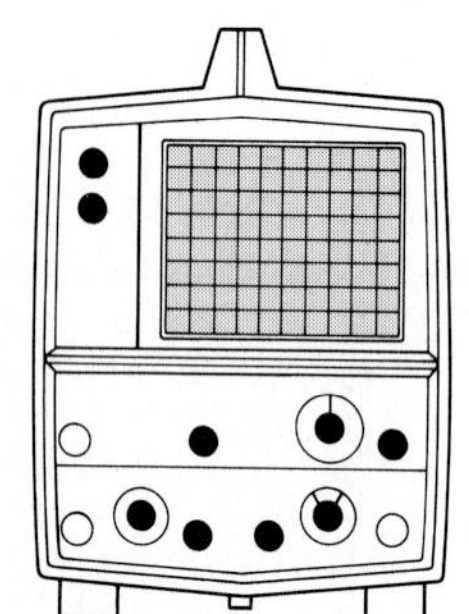

14.4 POWER SUPPLY DEVICE DATA AND SPECIAL CIRCUITRY

In Sections 14.1 to 14.3, the functional circuitry and method of operation of power supplies has been detailed. In this section the specifications for the devices used in power supplies, plus an elaboration on special circuitry, are presented.

Transformer Ratings

The transformers used in power supplies must be capable of delivering enough voltage and current to support the demands of the load on the power supply circuitry. Therefore, the following data must be observed when applying a transformer in a power supply circuit.

Input Voltage and Frequency The input voltage and operating frequency of the transformer must match the voltage and frequency specifications of the ac supply to which it is to be attached (e.g., 120 V ac, 60 Hz).

Output Voltage and Current Naturally, the output voltage of the transformer must be sufficient to produce the dc output of the circuit. Also, the current capacity of the transformer must match the expected current demands of the load and circuit. One major rating for a transformer is its current-handling capacity (e.g., output voltage = 28 V ac at a maximum of 3 A).

Other Features Since multiple-tap and center-tapped transformers may be used in power supplies, such specifications should also be considered [e.g., a center-tapped transformer may be specified either as −28-V ac output CT (center tap) or as −14–0–14 V ac (indicating a 0 V reference between two 14-V terminals)].

It should be noted that transformer ratings are specified in rms voltages and currents. Since power supply circuits use the peak voltage of the ac input, conversion is necessary.

Rectifier Diode Ratings

There are two major ratings that must be considered when applying diodes in power supply applications. They are:

Maximum Forward Current The maximum forward current drawn from the diode by the load and filter circuit must be considered when using a diode in a rectifier circuit. First, it must be noted that the peak value of current drawn from the diode is not just the current draw of the load. As the filter capacitors are recharged briefly during the cycles of the pulsating dc output of the rectifier, they can draw large values of current in these brief periods. Therefore, it is common to specify diode current ratings well in excess of the load current.

Peak Inverse Voltage (PIV) Rating Since rectifier diodes conduct on only one-half of the ac input cycle, they must be able to withstand a reverse bias voltage equal to at least the peak voltage of the opposite cycle of the ac input. The PIV rating of the diode therefore specifies a limit to the reverse diode voltage that can be applied to the diode without it breaking down and conducting.

There are two means of finding the *minimum* PIV rating, based on the type of rectifier circuit being used:

1. If the diode is used in a half-wave or full-wave *bridge* rectifier, the PIV rating must be equal to at least the peak voltage out of the transformer.
2. If the diode is used in a full-wave circuit (nonbridge), it must have a minimum PIV rating equal to twice the peak voltage output from one terminal of the transformer to the center tap.

For both maximum current and PIV, the "rule of thumb" that is followed is that devices are chosen with limits that are no less than twice those found through the methods described above (e.g., if it was determined that for a diode the PIV = 14 V and the maximum current should be I = 0.5 A, a diode with a PIV of at least 28 V and a maximum current capacity of at least 1 A would be chosen for use).

EXAMPLE 14.3

From the specifications found in Appendix A, determine the maximum PIV and forward current for a 1N4002 diode.

Solution:

The specification sheet illustrates seven diodes (1N4001 through 1N4007). Looking up the specifications for the 1N4002, we find that PIV = 100 V (listed on the sheet as V_R):

$$I_O = 1.0 \text{ A}$$

Therefore, this diode could be applied to a circuit with an ac peak voltage of 50 V and maximum current of 0.5 A, using the basic selection guidelines.

It should be noted here that in the discussions of power supply applications, the normal forward drop on the diode has not been figured into any equations. Typically, there are several losses within a power supply circuit which could cause a lower output voltage than predicted through equations. Often, these losses (including forward drops) are not substantial enough to have a major effect on the application of the power supply. If an accurate calculation of output potential is necessary, then of course all of these figures must be incorporated in the initial design of the supply.

Filter Circuit Elements

Outside of the basic design of the value of filter capacitance (or inductance, if used) there are few specifications that are required for the filter elements. For a capacitive filter, the capacitor's breakdown voltage must be in excess of the peak voltage from the rectifier. If an inductive filter is used, the maximum current rating of the inductor must be in excess of the load current (usually, it is chosen to be the same as diode current specifications).

Since a limited variety of capacitors are available, when choosing a filter capacitance the actual value used is always greater than or equal to the value specified by design. Outside of cost and physical space limitations it is a common practice to select the largest possible value of capacitance for use in the filter, thus improving the ripple factor. The durability of the capacitor (long-term life) is often as much a concern as is its initial value. Although capacitors meeting these needs may be more costly, their ability to handle the large current surges associated with the brief recharging process without breaking down or becoming "leaky" may make them quite desirable in precision equipment or computers.

Regulator Devices (and Circuits)

The process of regulation uses both specialized circuit elements and special circuitry. The following is a basic listing of these special devices and a summary of applicable circuitry.

Zener Diodes The **zener diode** is a specially produced variation of the standard diode. In the initial discussion of diodes it was found that when a diode is reverse biased, it does not conduct as it does in the forward direction (outside of leakage, there is no reverse current). This lack of conduction was true up to a point, where the reverse voltage was high enough to break down the diode. At that point, current would flow and (in most cases) destroy the diode. (This reverse breakdown voltage is the PIV rating of the diode.)

The zener diode, though, is designed to be operated in the reverse breakdown mode! The zener diode is manufactured with both an accurate breakdown voltage rating and with a relatively high allowable reverse current rating. Thus the zener can have a voltage applied at or even slightly above this breakdown level; it will break down and conduct, maintaining a constant drop across it (the breakdown voltage rating) and will not be destroyed by this reverse current.

Figure 14.20 illustrates the function of the zener diode. Part (a) illustrates the conduction curves for a 5-V zener diode. In the forward direction, it acts like an ordinary diode. In the reverse direction, when the applied voltage exceeds the zener voltage (reverse breakdown) the zener becomes conductive and allows current to flow. Part (c) shows how this 5-V zener diode will react to a rising input voltage. As voltage is below the 5-V zener voltage, it is simply dropped across the zener. When the voltage reaches 5 V, the zener begins to conduct. At voltages above 5 V, the zener maintains the zener voltage of 5

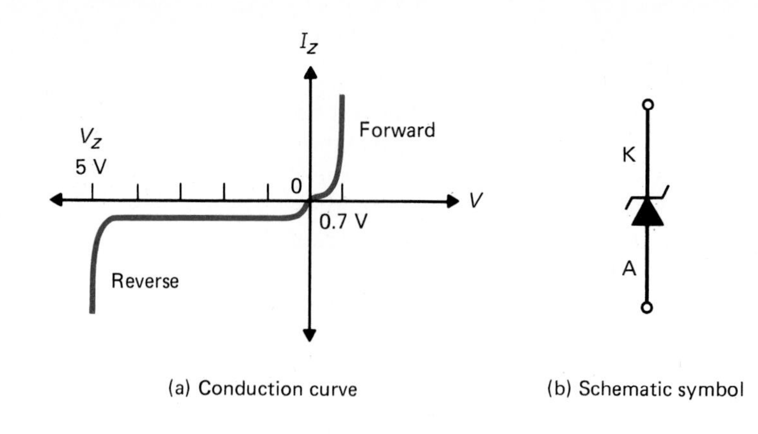

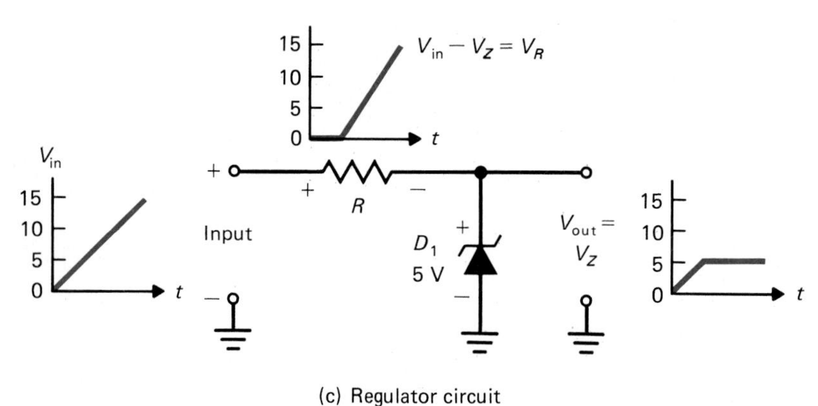

FIGURE 14.20 Zener Diode Function

V across its terminals. (The excess voltage is dropped on the resistor connected in series with the zener diode.)

Zener Diode Ratings and Support Circuitry The prime rating for the zener diode is its reverse breakdown voltage (zener voltage rating). Zener diodes are available in a variety of voltage ratings. There is a maximum limit to the current that can pass through the zener diode. This limit is used to establish the proper support circuitry and to interface the zener with a specific circuit application. Since the zener is not used in its forward direction, forward bias values are not critical.

In the circuit shown in Figure 14.21, the zener is used to provide a regulated output voltage to a load. The resistor R_Z is used as a place to drop the supply voltage, which is in excess of the zener voltage.

$$V_{RZ} = E_S - V_Z \tag{14.4}$$

Thus from Figure 14.21 and equation 14.4 it can be seen that the zener resistor must be selected so that will drop the supply voltage that is in excess of the zener voltage, based on the current flow to the load and the current necessary to operate the Zener diode (labeled I_{ZK} in the figure, typically 1 mA). Since a zener diode is selected with a fixed zener voltage, it is not possible to have a variable-voltage output from the zener regulator. Usually, zener regulation is used when a regulated voltage is to be applied to a fixed load (a load that draws a constant current).

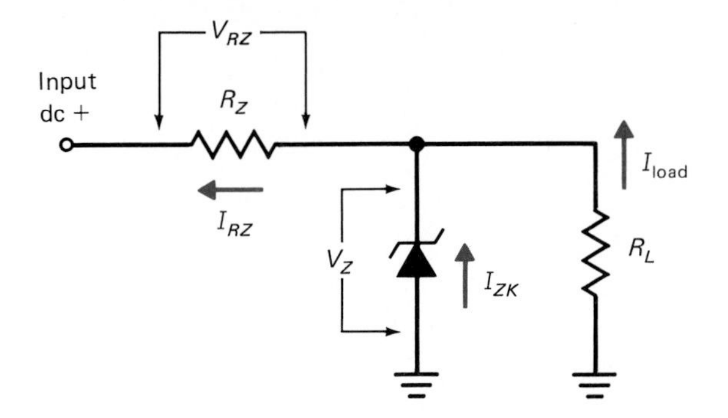

FIGURE 14.21 Identification of Voltage and Currents in the Zener Regulator

EXAMPLE 14.4

From the specification sheets for the 1N4728 through 1N4764 category of zener diodes found in Appendix A, find the specifications of the 1N4740 zener diode.

Solution:

This is a category of 1-W zener diodes. The specifications for the 1N4740 are found on the second page to be:

Zener voltage: 10 V
Zener current: 25 mA (labeled "test current")
Maximum current: 454 mA (labeled "surge current")

Other specifications (impedances and leakage current) would be used when doing detailed design of zener circuits.

Integrated-Circuit Voltage Regulators One of the most popular forms of regulator elements is the integrated-circuit regulator, sometimes called three-terminal regulators. These devices simply produce a regulated output from an unregulated input. They are placed in a power supply connected between the output of the filter and the input to the load. An additional terminal is connected to the circuit ground. These devices produce a specified output voltage and are usually current limiting. If a greater demand for current is made, they will automatically limit the supplied current to the value specified for the device.

IC Regulator Specifications IC regulators are available as either positive (+) or negative (−) or dual-tracking (+ and −) voltage regulators. They are purchased using both the desired output voltage level and the current-limiting capacity desired (e.g., a 1-A 5-V positive voltage regulator). (*Note:* See the specifications for the MPC-100 regulator in Appendix A.)

Some regulators are available as variable-voltage output regulators. As such, they are connected to the power supply in the conventional manner, but a potentiometer (usually on the ground lead) is used to make the output

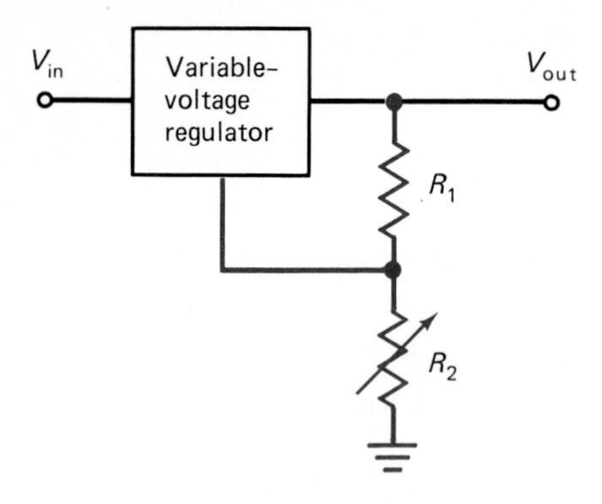

R_1 and R_2 provide a simple divider which produces a control voltage level for the variable-voltage regulator. Adjusting R_2 will vary the output voltage level.

FIGURE 14.22 Connection of an IC Variable-Voltage Regulator

voltage variable. Once the output is set to a desired level, the regulator maintains that level (see Figure 14.22).

All of the IC regulators require a minimum input voltage which is in excess of the regulated output level. Also, they have a limit to the input voltage level (a V_{max}). Variable regulators have both a maximum and minimum (usually nonzero) output level.

One major advantage of the IC regulator is that they are (usually) output short-circuit protected. That means that even if the output terminals are shorted, they will limit current and not be destroyed. In this process they also will draw only a maximum current from the initial power supply; thus the short circuit is not reflected back to the supply, where severe damage could be done. Another advantage of this device is that since it provides continuous regulation of the output, it will even improve the ripple factor of the power supply. (It is not uncommon to have the ripple output of a regulator be only 0.6% of the ripple input.) For this reason it is possible to reduce the value of a filter capacitor when applying an IC regulator. As such, it is possible to recover the cost of the regulator almost entirely in a reduced cost of the balance of the power supply circuit.

EXAMPLE 14.5

In Appendix A, locate the specifications for the MPC100 voltage regulator. Specify the type of power supply it can be attached to and indicate its output specifications.

Solution:

The input voltage from the power supply should be from +6.5 to +20 V (found in the output specifications, page 2) to a maximum of 25 V (maximum rating page 1). Maximum short-circuit current drawn from the supply is 20 A I_{SC}. The output specifications are:

Voltage: +5 V dc (4.75- to 5.25-V limits)
Current: 10 A maximum
Ripple: −45 dB better than the input

A wide variety of IC regulators are available. Many of them have special applications or features. If further information on such devices is desired, it is suggested that these manufacturer references be used.

Transistor Regulators Regulation can be achieved by the use of transistor circuitry. Certain older circuits used discrete transistor regulators (prior to IC regulators); these types may be usable in certain modern-day applications. There are two basic forms of transistor regulator circuits: series regulators and shunt regulators. Figure 14.23 illustrates both the shunt and series forms of regulators.

The basic function of these regulators is similar up to the way in which they set the value of output voltage. Therefore, the following discussion of the regulation process describes both forms up to the point of output control, which is separately described for each:

Error Detector (Comparator) The op-amp in these circuits is used as a differential amplifier which provides an error signal (differential output) based on the input voltages received from:

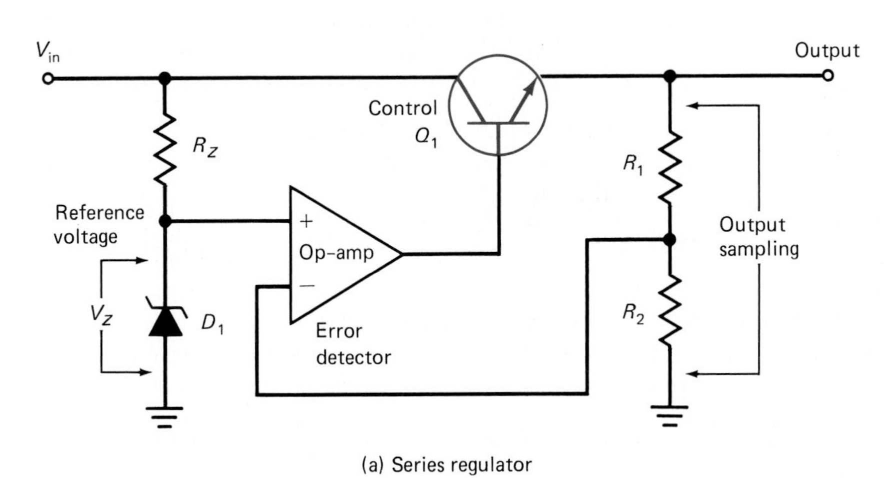

(a) Series regulator

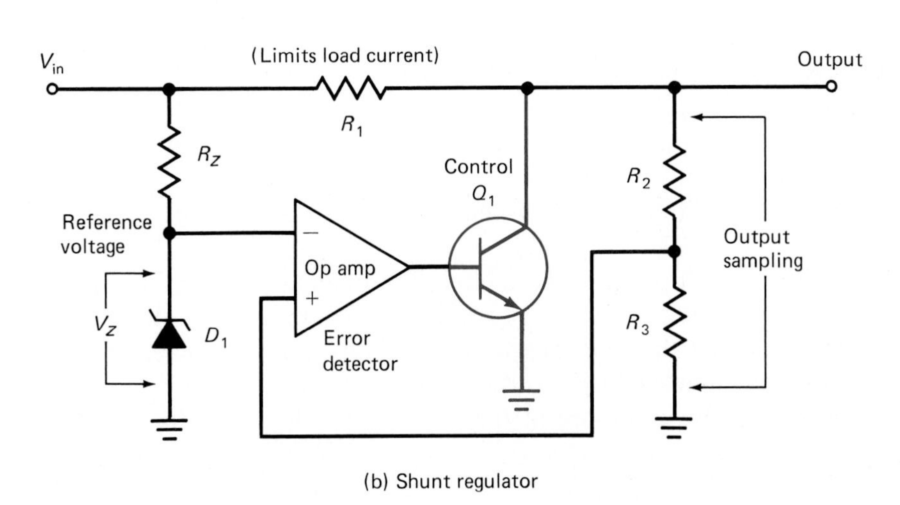

(b) Shunt regulator

FIGURE 14.23 Basic Voltage Regulators (Discrete)

Reference Voltage Circuit The zener diode produces a constant reference voltage which is supplied to the op-amp.

Output Voltage Sampling Circuit A simple divider circuit is used to provide a sample of the output. (The divider is designed to produce a voltage exactly equal to the zener voltage when the output level is exactly as desired from the regulator.)

The amount of output from the op-amp error detector is directly proportional to the difference between the reference voltage and the sample voltage. How this error voltage is used to control the output differs between the two forms of regulators:

Series Regulator (Figure 14.23(a) This error signal is used to control the conduction of the series control transistor. Basically, the error voltage controls the collector–emitter voltage drop on the control transistor. By adjusting this voltage, more or less voltage is passed on to the load (and sample circuit). The value of V_{CE} on the transistor varies to produce a constant voltage on the load.

Shunt Regulator (Figure 14.23(b) The control transistor in the shunt regulator is placed in parallel with the load (and sample circuit). As such, the load voltage is equal to the V_{CE} voltage on the transistor. Excess voltage (source-V_{CE} is dropped on the resistor R. The value of the output voltage therefore matches the value of the transistor voltage.

In either arrangement, the input dc voltage must be greater than the voltage output desired. In that way, the control transistor can be made to increase or decrease its conduction to compensate for output variations.

Switching Regulators Switching regulators operate in a manner similar to the discrete regulator circuits. As presented in the preceding section, rather than present a constant control over the voltage supplied to the load, the switching regulator uses a pulsed recharging circuit to maintain the output level. Figure 14.24 illustrates a simple step-down switching regulator system.

Basically, the process of producing an error detection signal is the same as seen in the analog (nonswitching) regulator. How this signal is used to control the output level is where the switching regulator differs from analog types.

1. The voltage from the control transistor is not on continuously as it is in an analog regulator.
2. The current flow from the input voltage is turned on and off by the control transistor.
3. The current from this transistor is filtered through an inductive filter and is fed to a final filter capacitor (C) which stores the output voltage.
4. The duration that the control transistor is on—pulsing current to the capacitor—is controlled by a variable-pulse-width oscillator.
5. The width (or time duration) of the pulses from the oscillator is determined by the amount of error voltage present from the error detector.

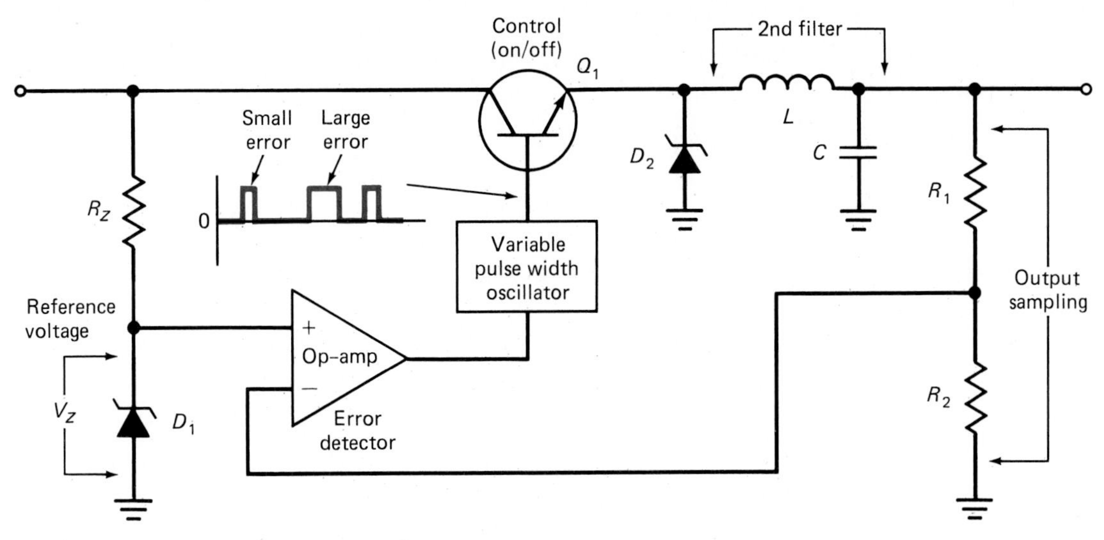

FIGURE 14.24 Basic Diagram of a Switching Voltage Regulator

More error voltage produces longer pulse widths; lower error voltage produces shorter pulse widths.

Thus the average voltage stored in the capacitor is controlled by the pulsing action of the control transistor. If it were desirable to make the output voltage level adjustable, the average pulse rate of the oscillator is all that needs to be adjusted. (The use of diode D_2 in the circuit is to reduce the adverse effects of the inductor's response to the pulsed input.)

This circuit's efficiency is improved since the control transistor need only be turned on when load conditions demand a correction in the output voltage level. In analog control regulators, this transistor would be kept on continuously and thus would dissipate power even if the load were drawing only slight amounts of current. Also, since the control transistor is run from saturation to cutoff, the power on it is also minimal in both cases. (On power is low since the drop on the device is minimal; off power is low since the current is near zero.)

Additional Power Supply Elements

There are two additional elements that may typically be added to a normal power supply circuit. One is a resistor, sometimes called a surge resistor, which is used to limit the current drawn from the rectifier by the filter capacitors. Previously, it was noted that when these capacitors charge, they may draw large amounts of current, this resistor (usually low in value) is used to place an upper limit on this current. The second element used is called a **bleeder resistor**. This resistor is placed across the output of the filter. The objective of the bleeder resistor is to provide a path for the filter capacitor(s) to discharge if the load is disconnected from the power supply and the supply is then shut off. Without this resistor in place, the filter capacitors could hold a dangerously high charge voltage even after the circuit is turned off. This resistor (much

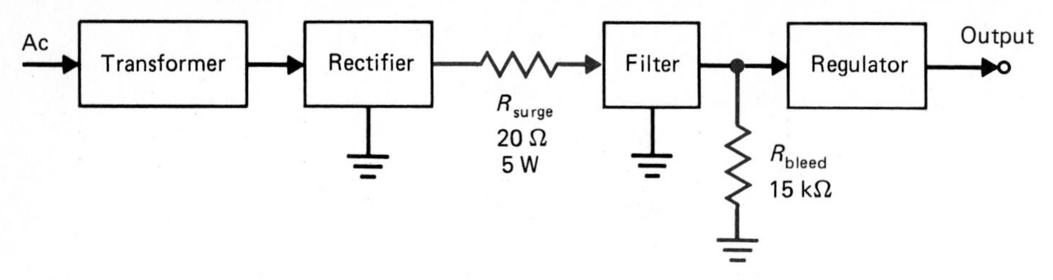

FIGURE 14.25 Location of Surge and Bleeder Resistors

higher in value than load resistance) will slowly discharge these filter capacitors even after the system is turned off and the load is disconnected. The locations of the surge and bleeder resistors are shown in Figure 14.25.

Additional Filtering and Remote Regulation

In order to improve the performance of a power supply, it is not uncommon to encounter additional filtering and remote (away from the power supply circuit) regulator elements. Often, if the power supply is providing dc voltage to a variety of circuits, especially several separate circuit boards, filtering capacitors and remote regulators may be used with these circuits.

In Figure 14.26 the basic application of remote filtering and regulation (VR_1-VR_4) is shown. The additional filter capacitors (C_1-C_4) are used to im-

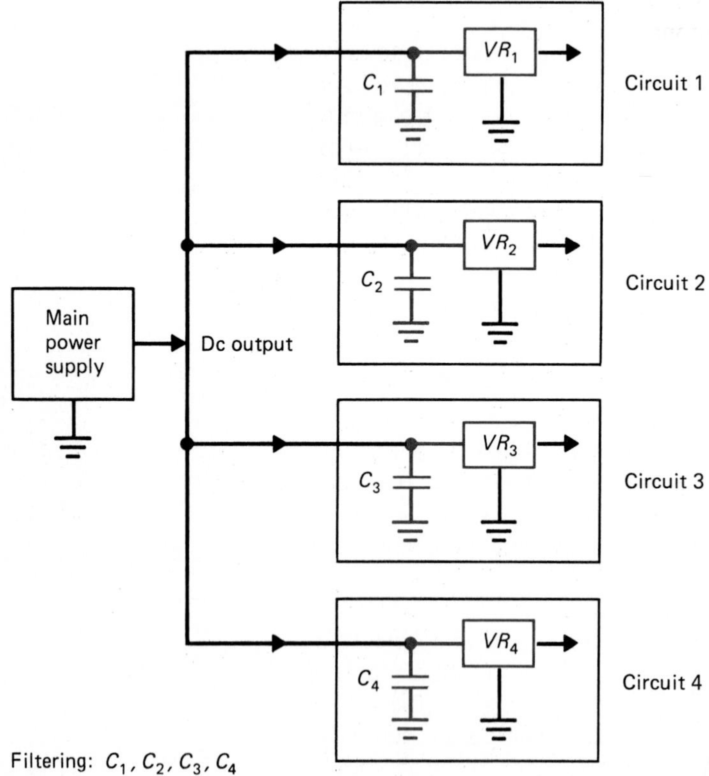

FIGURE 14.26 Remote Filtering and Regulation

prove the filtering and to eliminate possible noise signals that may be introduced into the lines from the power supply to that specific circuit.

The use of the remote regulators instead of one central regulator is done so that if one circuit should fail and draw excessive current (a short circuit on one circuit board, for instance), the short would not be reflected back into the main power supply lines. If such a short were reflected into the supply lines, damage to all the circuitry could result. The remote regulator will reflect only a relatively small surge in current back through the lines, not a major short-circuit condition. Second, the filtering effects seen in IC regulators can also help to eliminate the noise that may be introduced into the supply lines. Thus a more pure dc signal is applied to the circuit attached to the regulator.

REVIEW PROBLEMS

(1) From the discussions above, prepare a list of basic specifications for the components that would be used in a power supply with the following input and output specifications:

Ac input: 120 V ac, 60 Hz
Transformer output: 25 V rms
Load to draw up to 2 A dc at a regulated voltage of 30 V. (Specify for either a zener or an IC regulator.)

(2) Sketch the circuitry for the four popular forms of output voltage regulation (zener, IC, discrete, and switching).

(3) List the purpose of the following additional power supply elements.
- **(a)** Surge resistor
- **(b)** Bleeder resistor
- **(c)** Remotely located filter capacitors
- **(d)** Remotely located regulators

14.5 FAILURE ANALYSIS AND TROUBLESHOOTING OF POWER SUPPLIES

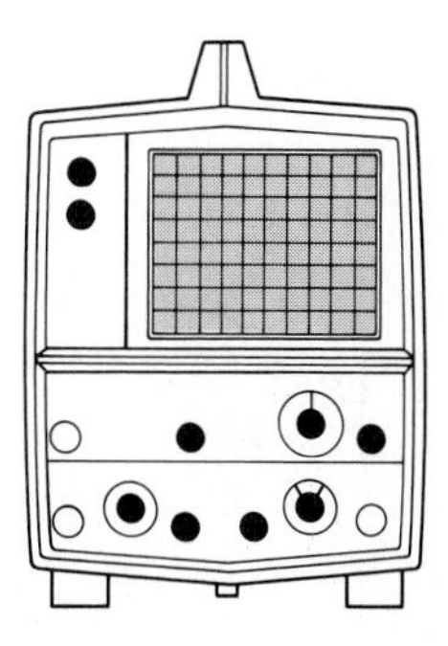

Any failure within a power supply will produce a malfunction and potential damage in the circuits it supplies. Therefore, it is necessary to track down the source of power supply error and correct the problem quickly and effectively. The following are output problems that could be encountered from a normal dc power supply, including the potential cause(s) of the fault.

Total Loss of Dc Output (see Figure 14.27)

1. The ac source may be defective or fuses may be blown which are in line with the ac input.
2. The transformer may be open on either the secondary or primary side.
3. The rectifier diodes may be open.

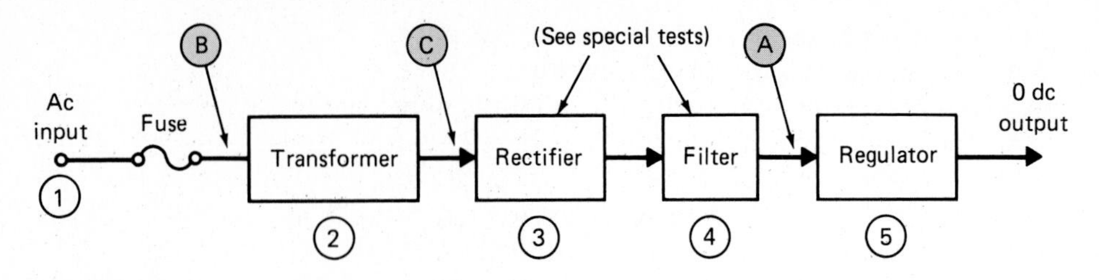

FIGURE 14.27 Block Diagram Coordination for Power Supply Testing

4. There may be a short (mechanical) across the filter circuit.
5. If a regulator is used, it may be defective.

Testing for These Problems: If a regulator is used, the input to it (dc) should be checked; if dc is present, the regulator is defective. Next, the ac input to the transformer should be checked; if that voltage is zero, the defect is in the ac input line. Next, the secondary side of the transformer should be checked. Normally, the waveform on that side is not a perfect sine wave, but if no wave is present, the transformer may be defective. The defect may be either in the diodes in the rectifier circuit or in the filter. See the following presentation on troubleshooting to test these circuits.

Dc Voltage Is Too Low or There Is Excessive Ripple on the Output (see Figure 14.28):

1. If the transformer used is center tapped, one side of the secondary has become open.
2. One of the diodes in the rectifier has opened.
3. The capacitor in the filter has developed excessive leakage.
4. The regulator circuit is defective.
5. The load on the circuit is too heavy (drawing too much current).

Testing for These Problems: Initially, visually inspect the power supply components. A common cause for power supply failure is the occurrence of an "overload" condition. Drawing excessive current from the supply will cause elements (wiring, diodes, resistors, the transformer) to overheat and open. Often signs of overheating, even burned components, wires, or printed-

FIGURE 14.28 Additional Testing of the Power Supply

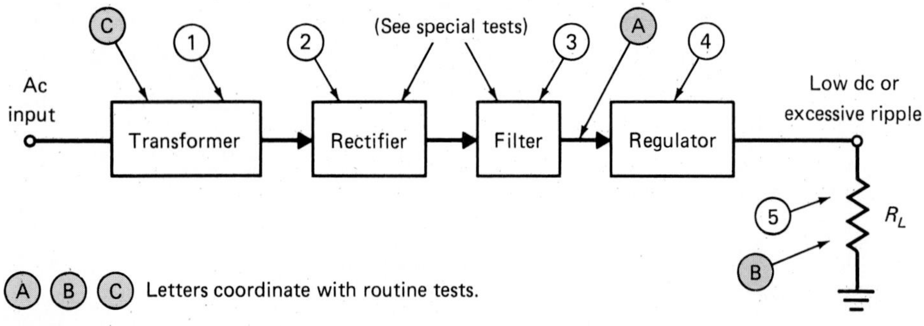

circuit connections, will point to such an overload. Damaged components should be replaced then tests must be made to isolate the source of the overload (e.g., a short circuit in some other circuit). If such obvious faults do not appear, the following tests should be conducted.

If a regulator is used, check the input to the regulator to see if it is at a proper level; if this value is correct, the regulator is defective or improperly adjusted (if it is variable). The specifications for the supply should be checked to see if the load being applied is correct. If necessary, disconnect the load and connect a substitute resistor (high wattage) to represent the proper load condition. If the supply returns to normal, the load was too heavy (drawing too much current) and needs to be redesigned. It is possible that the secondary of the transformer is defective, a reading of the ac voltage on either side of the center tap will reveal this defect. If these steps prove to be correct, the rectifier diodes or filter circuit must be checked. The following procedure should be used to test them.

Filter/Rectifier Troubleshooting

Since the power supply filter is in parallel with the output of the rectifier (Figure 14.29), there is no convenient way to measure the pulsed dc output independent of the rectifier. The voltage on the diodes can be measured though, and this should reveal if one is defective. The waveform across each diode in a rectifier should be a half-wave ac, since during one-half of the cycle the diode will conduct (no drop) and on the other half it will be like an open. Each diode can therefore be tested by measuring the voltage drop on each. If a diode shows a full wave (or distorted signal for both cycles of the supply), it is open. A shorted diode is not common, as excessive current would tend to destroy the shorted condition.

Should all diodes be considered good, the filter should be suspected. A simple test of a filter is to attach a known-good capacitor across the filter capacitor. If this action dramatically improves the ripple and dc output, the

FIGURE 14.29 Testing the Full-Wave Filtered Power Supply

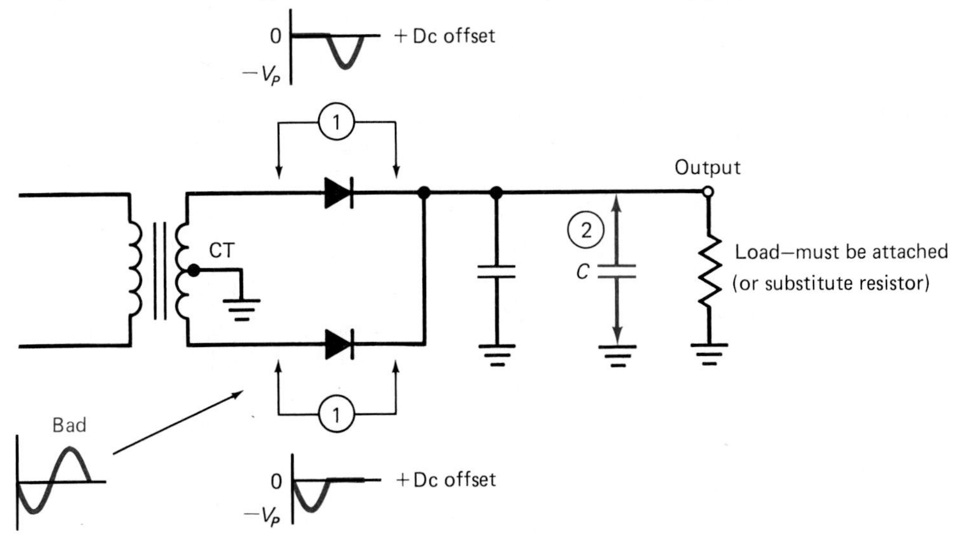

filter capacitor may be defective. Otherwise, the circuit will have to be disassembled to determine the single defective component.

Note: The parallel capacitor test should not be done on high-current power supplies, as component damage may result.

Special Notes on Testing Rectifiers and Filters All tests of rectifiers and filters should be done with some form of load attached to the power supply. It is common to disconnect the supply from the circuit to which it is attached to prevent further damage to that circuit if the supply is defective. In such a case, some characteristic load (often called a "dummy load") should be reattached to the supply. A power supply that does not have a load attached and therefore is not supplying current could show the following conditions.

1. *No-load filter.* If there is not a load on the filter, the capacitor (if used) will charge to the peak supply voltage and hold that level, never discharging. In such a case, a test of that filter will show that a full dc voltage is available. This would imply that the circuit is working properly. Only if the capacitor has excessive leakage (uncommon) or if *all* diodes in the rectifier were opened (again uncommon) would any defect appear.
2. *No-load rectifier.* If there is no load on a rectifier diode, testing it may indicate that it is open when it actually is working properly. A measurement of the ac wave across the diode may show a full sine wave, an indication that the diode is open. This may not necessarily be true. If the current drawn from the diode during its forward conduction is near the leakage current value in the reverse direction, the diode may appear to be conducting in both directions. When this happens, a typical drop across the diode's internal resistance would show a full sine wave (see Figure 14.30). When a load is attached to the rectifier which draws significant current, the ratio of forward to reverse (leakage) current is substantial enough to illustrate a difference between the two conditions.

Thus from these procedures it can be seen that care must be taken to troubleshoot a power supply properly and to resolve the source of the problem.

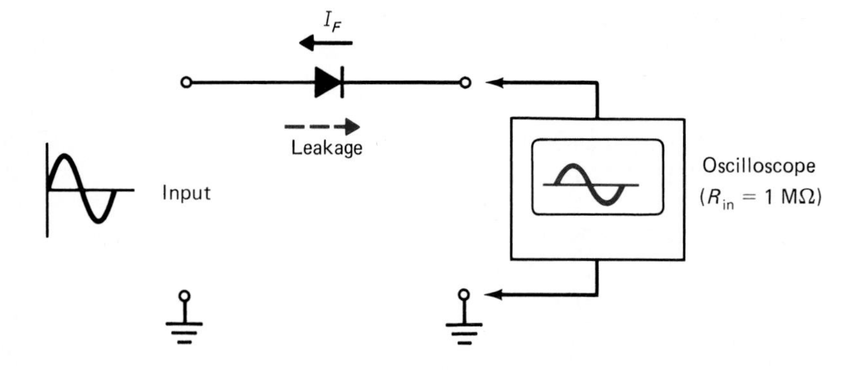

FIGURE 14.30 Improper Testing of Rectifier Diodes

REVIEW PROBLEMS

(1) Prepare a checklist type of troubleshooting procedure for a power

supply (ac-to-dc conversion) which is a composite of the general procedures described above.

(2) Select one component in a power supply schematic (identify the type of supply) to be defective. First, describe the defect, then indicate what the output voltage would be due to that effect. Then indicate any other circuit voltage values.

Example: (Select a different problem!) Full-wave rectifier, capacitor filter, no regulator.

Defect: one-half of secondary (to center tap) open
Output symptom: higher ripple, lower average dc
Other voltages:
Output of rectifier same as circuit output
Voltage across one diode = half-wave
Voltage across other diode = zero
Voltage across one side of transformer = distorted sine wave
Voltage across other side = zero
Voltage on primary of transformer = sine wave

(3) Explain the symptoms that would exist if a fully functioning power supply designed to operate at 10 V dc at a load current of 100 mA were connected to a load which attempts to draw 500 mA.

14.6 REVISIONS/MODIFICATIONS AND SPECIAL EFFECTS

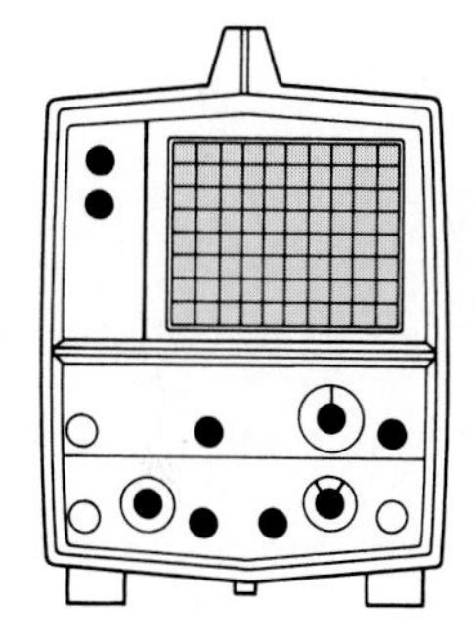

Since power supplies are common in most electronic circuitry, it is often possible to use one design of a supply to operate a variety of equipment. Thus it may only be necessary to modify the internal components of a "basic" supply to fit the new application. Another typical process is to add more complex circuitry to the same electronic system. These additions may be able to operate off the same supply voltages as the initial circuitry. Modification of the supply circuit may be necessary, though, to support the added circuitry. Following is a set of guidelines that can be used when revising or modifying a power supply (ac-to-dc conversion).

Power Supply Voltages

If a different level of supply voltage is needed, it will usually be necessary to change the ac source (transformer) in the circuit. With these new levels it is possible that diodes and capacitors with new voltage ratings be used. In a few cases where regulators are used, it may be possible to add a new or variable regulator to the same circuit without changing the other components.

If multiple supply voltages are desired, they can be obtained by using either a multitap transformer or separate rectifier/filter/regulator circuits for each voltage desired. If the voltages needed are close in value, it is possible to use multiple voltage regulator circuits. In such an arrangement, one regulator feeds input to a second (see Figure 14.31). In such a configuration it is

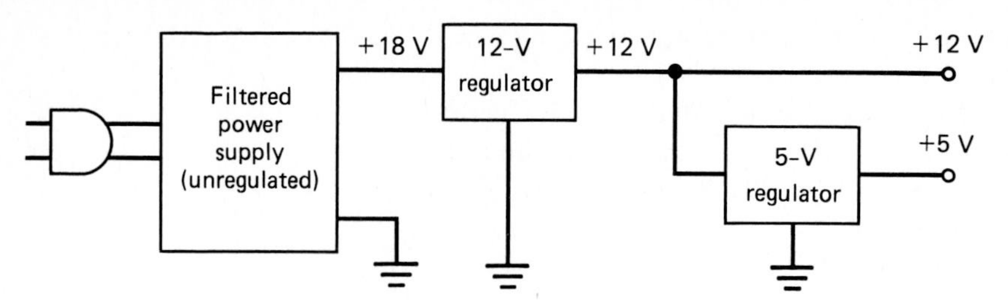

FIGURE 14.31 Multiple Regulator Circuit

necessary to be certain that the supply and first regulator can handle the total current demands of the circuit.

Power Supply Current

If the demands for current from a power supply are changed, modification of the power supply circuitry may be needed. If the current demands are lowered, usually no changes are necessary. If current demands are raised, all elements of the supply usually need to be changed. The current rating of the transformer and diodes need to be increased. The amount of filter capacitance or inductance must be increased (greater current flow would create greater ripple if using the same elements). The current ratings of the regulator or regulator components should also be changed.

Battery Backup for Power Supplies

In some cases it is desirable to provide an uninterrupted supply of dc to a circuit. Thus if the ac lines are interrupted, the circuit will still continue to function, at least for some time. Typically, a battery backup system is added to a power supply between the filter circuit and the regulator (if used) and output (see Figure 14.32). Depending on the type of battery system used, the battery backup circuit may be left permanently connected to the power supply (thus always recharging the batteries) or may be switched on automatically if the main supply is interrupted. If the backup is switched (see Figure 14.32) on when the main supply is disconnected, it is often common to use a second filter capacitor after the batteries in order to hold voltage during the switching period.

FIGURE 14.32 Battery Backup for a Power Supply (Automatic Turn-On)

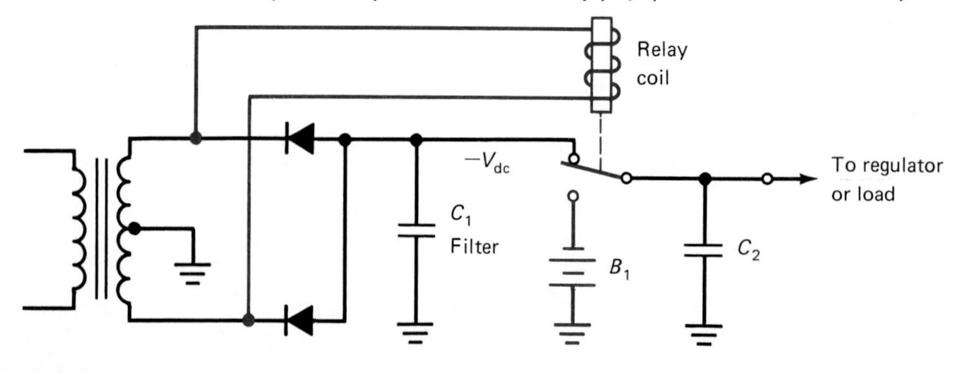

Delayed Turn-On of Circuitry

When a power supply initially is turned on, there is some delay in charging the capacitors used in the filters. When this happens, the output voltage is not stable. For many circuits, this initial turn-on error in voltage output is not critical. For delicate instrumentation and other sensitive circuits (such as computers) these voltage variations may cause other circuit faults.

In some systems there is a time delay built into the circuitry which allows the power supply to become fully functional, and then it energizes the attached circuitry. This delay circuit is placed at the output of the supply circuitry and is (usually) set to a preset time delay of several seconds. Often, this circuit applies a "dummy load" to the power supply and then switches the normal circuitry in place of this load when sufficient time has passed for the supply to reach a proper operating level (see Figure 14.33).

Power Supply Fusing

To protect the wiring and potentially, the electronic circuitry attached to the power supply, fuses are used to interrupt the supply circuitry. The most common location for fuses is:

1. At the ac input. These fuses protect against a major short circuit on the primary or secondary side of the transformer. Normally, they do not protect against short circuits that occur past the rectifier circuit.
2. Between the rectifier or filter and the load. These fuses protect against short circuits or major surges in load current. Normally, though, they are not fast enough in reacting to protect the rectifier diodes or other sensitive electronic circuitry.
3. After regulators or on individual circuit boards. Usually, fast-acting low-amperage (less than 1 A) fuses may be used to prevent severe damage. Usually again, these fuses do not act fast enough to protect all components.

Generally, an open fuse signifies potentially damaged components in a circuit. Since most fuses do not react fast enough to protect sensitive electronic components, an open fuse should be interpreted as a sign of potential other

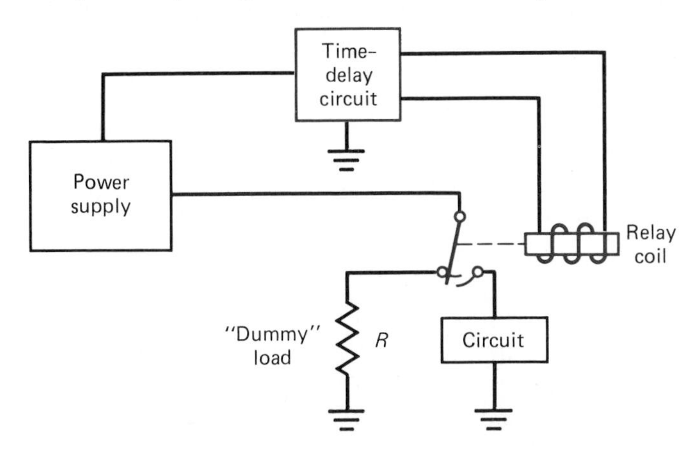

FIGURE 14.33 Delayed-Turn-On Power Supply

damage. Nearby components should be tested prior to simply replacing a fuse and turning on a power supply.

Some circuits use circuit breakers (slower than fuses), while others may use fast-acting electronic circuit interrupters. Usually, these forms of interrupters must be reset to reactivate the power supply. Again, an open breaker or interrupter is a sign of potential damage elsewhere and should not be reset without first testing the associated circuitry.

REVIEW PROBLEMS

(1) Indicate the modifications that would be necessary for converting a 12-V unregulated power supply (1-A maximum current output) to a 20-V unregulated power supply (1-A output).

(2) Indicate what modifications would have to be made to convert a 15-V/0.5-A power supply to supply 15 V at the following current demands.
(a) 250 mA **(b)** 2A

(3) Redraw the block diagram of a power supply (Figure 14.34) and show where the following elements would be added: battery backup, sequential startup, and fuses (every potential position).

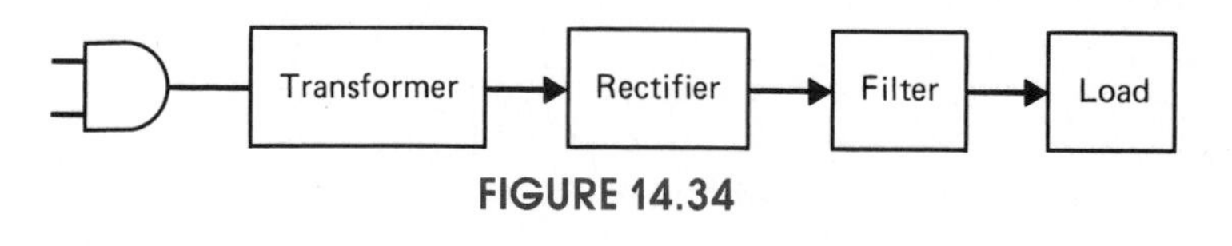

FIGURE 14.34

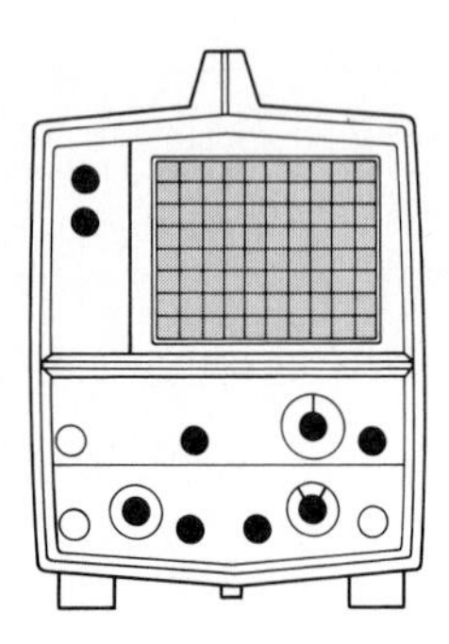

14.7 POWER SUPPLIES: A COMMON SOURCE OF ERROR

Since power supplies are connected to all circuitry within electronic systems, they are the most vulnerable to damage if there is a fault elsewhere. Also, since power supply circuitry carries the highest level of current flow within a system, they are more susceptible to heating and fatigue problems than is most other circuitry.

A large percentage of problems within an electronic system can be either directly traced to power supply faults or if occurring elsewhere, will cause power supply damage. It is then necessary that one of the first circuits within a system that is checked be the power supply. Naturally, if the supply is defective, that defect will affect all circuits that it supplies.

A common procedure when testing any equipment is first to test the power supply output. If that supply is defective, it needs to be repaired before further tests on the rest of the circuit can be considered valid. Of course, the cause of the power supply defect should be determined before returning the circuit to operation.

Safety Precautions

Not only is the power supply a most likely source for circuit malfunctions, but it is often the most dangerous section of the electronic system. It is quite common to have dangerous supply-line voltages within the power supply circuit. As well, the voltages produced in the supply may be an even greater danger. Some supplies step up voltages to the hundreds or thousands of volts. One must always identify these voltage lines within the supply and treat them with extra caution.

Some supplies, although their voltages are low and relatively safe, have current capacities (unfused) that can prove dangerous. It is possible to encounter a 5-V/100-A(or more) supply line. The use of metal tools and other metallic objects around these lines may prove dangerous if the lines are ever accidentally shorted by these objects. (They can become "spot welded" to the lines and then immediately overheat.)

Finally, if a power supply does not have a bleeder resistor or that resistor is not adequate to discharge the system in a brief time, high voltages (and potentially high current capacities) can still be present within the supply circuitry and lines even after the ac supply has been turned off. Therefore, a circuit should be considered electrically "hot" even if the main ac source has been disconnected (or turned off). It is a common practice to test the dc levels on a power supply *before* working on the balance of a circuit to determine that the system is truly safe.

REVIEW PROBLEMS

(1) Note the possible effects that a power supply which has excessive ripple may have on a simple audio-amplifier system.

(2) Write out a basic checkpoint safety procedure to use when servicing a system that uses 350 V-at-1.5 A power supply.

SUMMARY

In this chapter the process of ac-to-ac conversion, ac-to-dc conversion, and dc-to-dc conversion was presented. Most of the presentation concentrated on the major circuits used for ac-to-dc conversion (the power supply). It was shown that basic ac-to-ac conversion is commonly achieved using transformers either to increase or decrease voltage (or potential current) levels.

Ac-to-dc converson involved several steps. Rectification produced a pulsed dc (single polarity) wave from an ac (dual polarity) input. The pulsations were smoothed out to a constant level using filters. Ripple and the ripple factor were shown to indicate the effectiveness of the filtering process. This level could then be regulated to produce a constant level at the output. Several versions of regulators were illustrated, with information on their general function being provided.

Dc-to-dc conversion (inversion) was seen to involve the conversion of the input dc to an ac signal, using a transformer to increase the ac voltage, then converting the ac voltage back to dc with conventional rectifiers and filters.

KEY EQUATIONS

14.1 Ripple Factor: $r = \dfrac{V_r}{V_{dc}}$

14.2 $V_r = V_P \left(\dfrac{2.4 \times 10^{-3}}{R_L C}\right)$

14.3 $V_{dc} = V_P \left[1 - \left(\frac{4.17 \times 10^{-3}}{R_L C}\right)\right]$

For a Zener diode:

14.4 $V_{RZ} = E_S - V_Z$

PROBLEMS

Section 14.1

14.1. Illustrate the basic block diagram for an ac-to-dc power converter (power supply).

Section 14.2

14.2. Define the basic process involved in ac-to-ac conversion and the common electronic component used to make this conversion.

14.3. Sketch the schematic symbol for the following types of transformers.
(a) Multitap (b) Center-tapped

14.4. Define the basic function of the ac-to-pulsed dc conversion block in a power supply block diagram.

14.5. Define the difference between a half-wave and a full-wave output from a pulsed dc conversion.

14.6. Define the term "filtering" as it relates to a power supply circuit.

14.7. What is achieved by a power supply regulator?

14.8. Describe the four basic steps in dc-to-dc conversion.

Section 14.3

14.9. Sketch the schematic for a simple one-diode rectifier. Illustrate the ac input and pulsed dc output. Assume that the voltage polarity on the load resistor is to be negative in reference to ground.

14.10. Sketch the schematic for a full-wave rectifier using a center-tapped transformer. Illustrate the ac input and pulsed dc waveforms in the circuit. Assume that the voltage polarity on the load resistor is to be negative in reference to ground.

14.11. Illustrate the schematic for a full-wave bridge rectifier. Connect it to a load resistor and indicate the polarity on that resistor. Also illustrate the current paths taken for each cycle of the ac input from the transformer.

14.12. On the schematic prepared for problem 14.11, illustrate a capacitive filter. (Orient the leads of the rectifier so that the resistor voltage is negative in respect to ground.)

14.13. Define the term "ripple."

14.14. If a filter circuit has 0.2 V p-p ripple and supplies 18 V dc, calculate the ripple factor. (Also convert this factor to a percentage.)

14.15. For the circuit shown in Figure 14.35, find the expected dc output voltage and amount of ripple voltage. (Assume a 60-Hz-based input from a full-wave rectifier.)

14.16. If the rectifier in the circuit used for problem 14.15 were changed to a full-wave bridge, what effect would this have on the output calculations made in this problem? What if a half-wave rectifier were used?

14.17. Illustrate the following filter circuits.
(a) A *C-RC* filter (b) A pi filter (c) An L-type filter

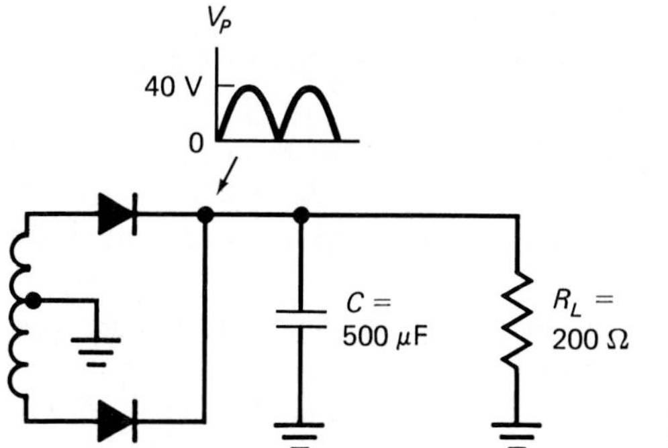

FIGURE 14.35

14.18. Sketch the basic block diagram for a voltage regulator system (the general form).

14.19. Sketch the block diagram for a switching regulator.

14.20. Sketch the block diagram for a dc-to-dc converter (basic form using a transformer).

Section 14.4

14.21. What are the two primary ratings for a rectification diode?

14.22. What must be the PIV rating for any of the diodes in the following types of rectifier circuits?
- (a) Half-wave rectifier, transformer peak voltage of 40 V
- (b) Full-wave rectifier, transformer peak voltage from center-tap to either side of 28 V
- (c) Full-wave bridge rectifier, transformer peak voltage of 18.5 V

14.23. If the load on a power supply draws 5 A of current, is it safe to use diodes with a maximum forward current rating of 5 A? If not, indicate why.

14.24. Illustrate the schematic symbol for a zener diode.

14.25. Illustrate a zener diode regulator and label voltage values for the circuit if the dc power supply provides 28 V dc and uses a 1N4748 zener (listed in Appendix A).

14.26. Illustrate the circuit for a discrete-device shunt regulator.

14.27. Describe the basic function of the shunt regulator illustrated in problem 14.26.

14.28. Illustrate the basic circuit for a discrete-device switching regulator (you need not detail the variable-pulse-width oscillator).

14.29. Briefly describe the function of the switching regulator illustrated in problem 14.28.

14.30. Define the function of a surge resistor in a power supply circuit. Do the same for a bleeder resistor.

14.31. Illustrate a complete power supply schematic incorporating each of the following elements. (You may wish to use an oversized sheet of paper.)
- (a) Transformer
- (b) Full-wave bridge rectifier
- (c) Surge resistor
- (d) Capacitor filter
- (e) Bleeder resistor

(f) Switching regulator
(g) The connection to three remote circuits which each have "on-board" filters and regulators

14.32. On the circuit drawn in problem 14.31, illustrate important ac and pulsed dc waveforms along with potential dc voltages (select your own voltage levels). Also indicate significant test points to use for troubleshooting.

14.33. Locate the schematic for a piece of electronic equipment (through your library, laboratory, or other classroom resource) and isolate the power supply section of the schematic. Redraw that schematic as it is shown and identify the function of the parts related to this power supply. Illustrate potential voltages and test points on that schematic.

14.34. On the power supply schematic drawn for problem 14.31, add the following features.
(a) A temporary battery backup circuit
(b) Delayed turn-on (after the switching regulator)
(c) Fusing

14.35. Indicate the changes that would have to be made to the parts used in the circuit illustrated in problem 14.31 if it was necessary to add six more circuit boards to the output.

ANSWERS TO REVIEW PROBLEMS

Section 14.1

1. The answer is based on investigations.
2. Ac can be stepped down in voltage level efficiently through transformers where dc cannot. By distributing ac at high voltage levels (up to 17,000 V), the percentage of loss due to transmission-line wire resistance (based on $I \times R$) is dramatically reduced. These high levels can be efficiently reduced to safe levels by using transformers at the point of use (home or industry). Were dc to be used, it would have to be originally produced at the safer low levels (120 to 240 V) and due to the wire resistance dramatic losses in energy would occur, rendering it nearly impossible to distribute it over large areas (e.g., cities, states).
3. The reader should be familiar with basic transformer actions, with basic terminology including step-up and step-down, and with center-tapped and multitap transformers.

Section 14.2

1. See Figures 14.4 and 14.7.
2.

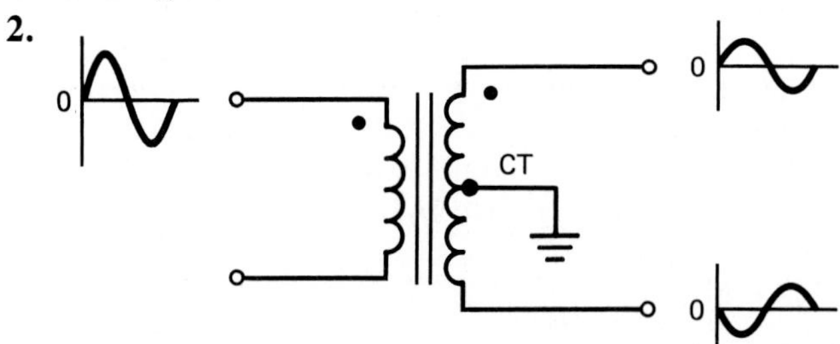

3. For the coil on the secondary side of the transformer to produce an output

current, the magnetic field around it (produced by the primary) must be changing. Since a dc input will produce a constant unchanging field, there would be no output.

Section 14.3

1. See Figure 14.4.
2.

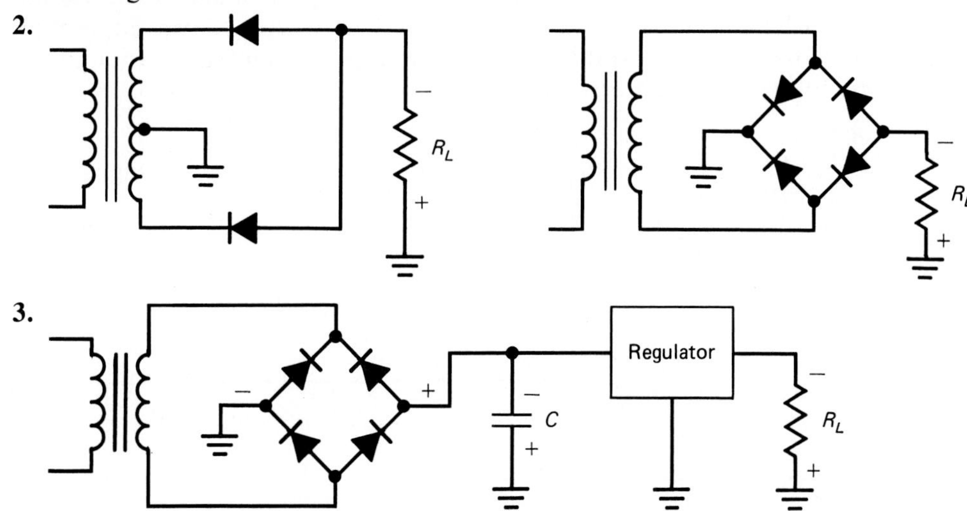

4. See Figure 14.8. If the output went below the standard value, the feedback sample would drop below the reference voltage level. The error detector would indicate this variation to the variable-pulse-width oscillator. The width of the output pulsed from this block would become larger, thus turning on the switching circuit for a longer duration. This would pass a greater average voltage on to the second filter, which would return the output to the normal level.
5. The battery dc voltage is applied to an oscillator which changes it to an ac value. This ac voltage is then applied to a step-up transformer which boosts the voltage level to 500 to 2000 V. This ac voltage is then rectified and filtered to become the 500- to 2000-V dc level needed to fire the xenon flash tube.

Section 14.4

1. Transformer: step-down; 120 V to 25 V (rms); secondary current capacity in excess of 2 A continuous current. Diode: maximum continuous current in excess of 2 A (*Note:* peak currents of 10 A are not unrealistic), PIV $>$ 70.8 V (2 $\times$ 35.4 V); capacitor's dc voltage rating $>$ 35.4 V; zener voltage $=$ 30 V, maximum current specified by the circuit design or regulator voltage $=$ 30 V dc with a normal output current of 2 A continuous.
2. See Figures 14.21, 14.22, 14.23(a) or (b), and 14.24 for details.
3. (a) Used to limit the current drawn from the rectifier circuit
   (b) Used to discharge filter capacitors when the supply is turned off
   (c) Used to provide additional filtering of dc voltages, especially filtering of noise picked up in supply lines
   (d) Used for circuit-by-circuit (system-by-system) regulation, especially used in high-current systems

Section 14.5

1. The chart is based on the summary of descriptions in this section.
2. The answer is based on the defect selected. It is suggested that readers exchange problem/solution notes to guide each other in their evaluation.
3. Drawing this excessive value of load current would produce the following problems: lower average dc output, more ripple voltage at the output, possible damage or imbalance to regulator circuit, damaged rectifier diode(s), overheated transformer or creating an open secondary (burned-out wiring), blown fuse.

Section 14.6

1. Higher transformer output voltage, higher diode PIV rating, higher filter capacitor dc voltage rating
2. (a) None
   (b) Higher diode and transformer current rating; four or more times the filter capacitor value; if a regulator is used, its ratings must be modified to coordinate.
3.

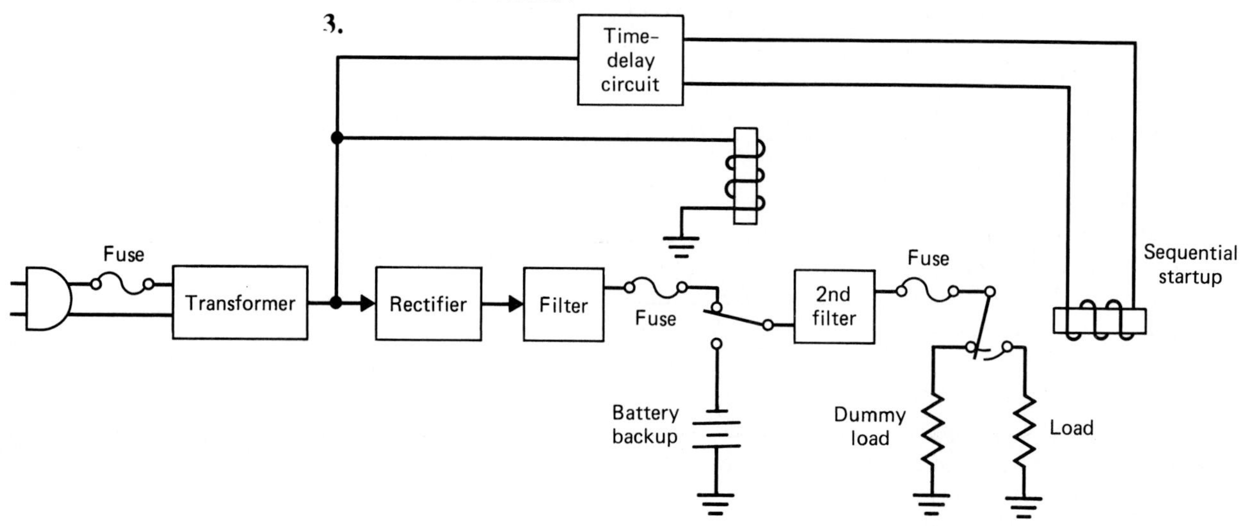

Section 14.7

1. It would create noise on the output of the amplifier. It may damage the amplifier circuitry.
2. Bleed the supply voltage to zero before repairing it; identify any high-voltage (ac or dc) lines before working on the system; conduct "safe test" procedures (with system off and all voltages bleed to zero, connect measuring instruments, then keeping hands out of the system, turn it on and read the value(s) tested for; turn system off and repeat these steps before taking new measurements); observe personal safety procedures (only place one hand inside the equipment at any time; use protective gear (e.g., insulated tools, gloves, and eye protectors); and on a system as potentially hazardous as this (525 W), work with a partner nearby.

Designers▲Data Sheet

1N4001 thru 1N4007

LEAD MOUNTED
SILICON RECTIFIERS

50-1000 VOLTS
DIFFUSED JUNCTION

"SURMETIC"▲ RECTIFIERS

. . . subminiature size, axial lead mounted rectifiers for general-purpose low-power applications.

Designers Data for "Worst Case" Conditions

The Designers▲ Data Sheets permit the design of most circuits entirely from the information presented. Limit curves – representing boundaries on device characteristics – are given to facilitate "worst case" design.

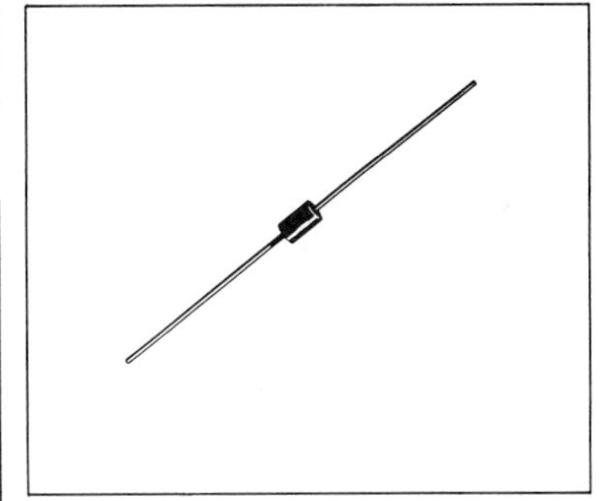

*MAXIMUM RATINGS

Rating	Symbol	1N4001	1N4002	1N4003	1N4004	1N4005	1N4006	1N4007	Unit
Peak Repetitive Reverse Voltage Working Peak Reverse Voltage DC Blocking Voltage	V_{RRM} V_{RWM} V_R	50	100	200	400	600	800	1000	Volts
Non-Repetitive Peak Reverse Voltage (halfwave, single phase, 60 Hz)	V_{RSM}	60	120	240	480	720	1000	1200	Volts
RMS Reverse Voltage	$V_{R(RMS)}$	35	70	140	280	420	560	700	Volts
Average Rectified Forward Current (single phase, resistive load, 60 Hz, see Figure 8, $T_A = 75^oC$)	I_O	1.0							Amp
Non-Repetitive Peak Surge Current (surge applied at rated load conditions, see Figure 2)	I_{FSM}	30 (for 1 cycle)							Amp
Operating and Storage Junction Temperature Range	T_J, T_{stg}	-65 to +175							^oC

*ELECTRICAL CHARACTERISTICS

Characteristic and Conditions	Symbol	Typ	Max	Unit
Maximum Instantaneous Forward Voltage Drop (i_F = 1.0 Amp, $T_J = 25^oC$) Figure 1	v_F	0.93	1.1	Volts
Maximum Full-Cycle Average Forward Voltage Drop (I_O = 1.0 Amp, $T_L = 75^oC$, 1 inch leads)	$V_{F(AV)}$	–	0.8	Volts
Maximum Reverse Current (rated dc voltage) $T_J = 25^oC$	I_R	0.05	10	μA
$T_J = 100^oC$		1.0	50	
Maximum Full-Cycle Average Reverse Current (I_O = 1.0 Amp, $T_L = 75^oC$, 1 inch leads	$I_{R(AV)}$	–	30	μA

*Indicates JEDEC Registered Data.

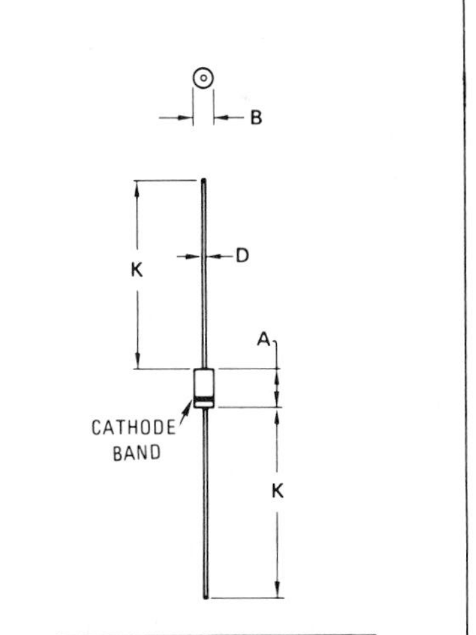

DIM	MILLIMETERS MIN	MILLIMETERS MAX	INCHES MIN	INCHES MAX
A	5.97	6.60	0.235	0.260
B	2.79	3.05	0.110	0.120
D	0.76	0.86	0.030	0.034
K	27.94	–	1.100	–

CASE 59-04
Does Not Conform to DO-41 Outline.

MECHANICAL CHARACTERISTICS

CASE: Void free, Transfer Molded
MAXIMUM LEAD TEMPERATURE FOR SOLDERING PURPOSES: 350°C, 3/8" from case for 10 seconds at 5 lbs. tension
FINISH: All external surfaces are corrosion-resistant, leads are readily solderable
POLARITY: Cathode indicated by color band
WEIGHT: 0.40 Grams (approximately)

▲Trademark of Motorola Inc.

DS 6015 R3

MOTOROLA

1N4728, A thru 1N4764, A

Designers▲ Data Sheet

1.0 WATT

ZENER REGULATOR DIODES

3.3–100 VOLTS

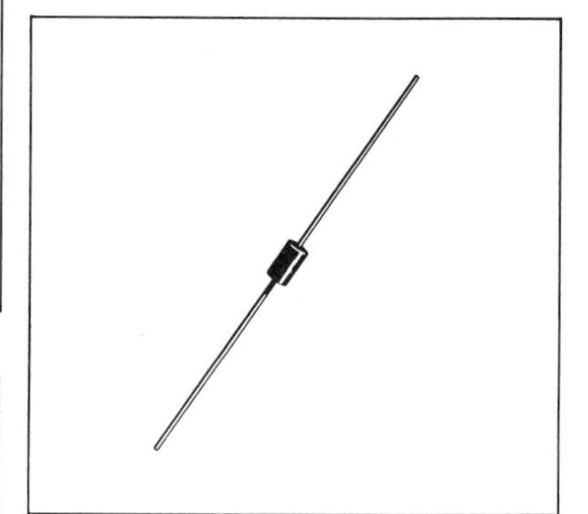

ONE WATT HERMETICALLY SEALED GLASS SILICON ZENER DIODES

- Complete Voltage Range – 2.4 to 100 Volts
- DO-41 Package – Smaller than Conventional DO-7 Package
- Double Slug Type Construction
- Metallurgically Bonded Construction
- Nitride Passivated Die

Designer's Data for "Worst Case" Conditions

The Designers▲ Data sheets permit the design of most circuits entirely from the information presented. Limit curves – representing boundaries on device characteristics – are given to facilitate "worst case" design.

*MAXIMUM RATINGS

Rating	Symbol	Value	Unit
DC Power Dissipation @ $T_A = 50^oC$ Derate above 50^oC	P_D	1.0 6.67	Watt $mW/^oC$
Operating and Storage Junction Temperature Range	T_J, T_{stg}	-65 to +200	^oC

MECHANICAL CHARACTERISTICS

CASE: Double slug type, hermetically sealed glass

MAXIMUM LEAD TEMPERATURE FOR SOLDERING PURPOSES: 230°C, 1/16" from case for 10 seconds

FINISH: All external surfaces are corrosion resistant with readily solderable leads.

POLARITY: Cathode indicated by color band. When operated in zener mode, cathode will be positive with respect to anode.

MOUNTING POSITION: Any

FIGURE 1 – POWER TEMPERATURE DERATING CURVE

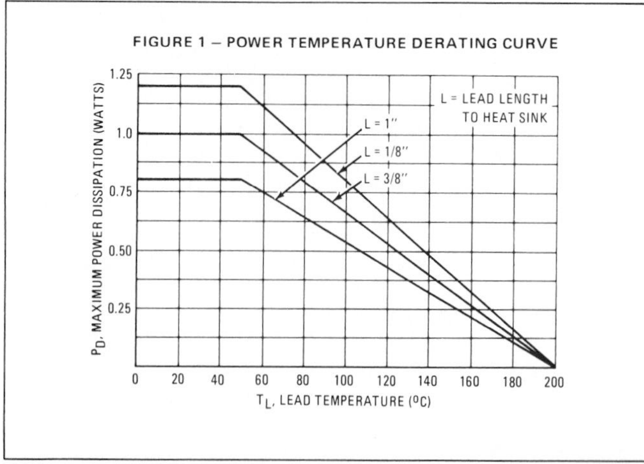

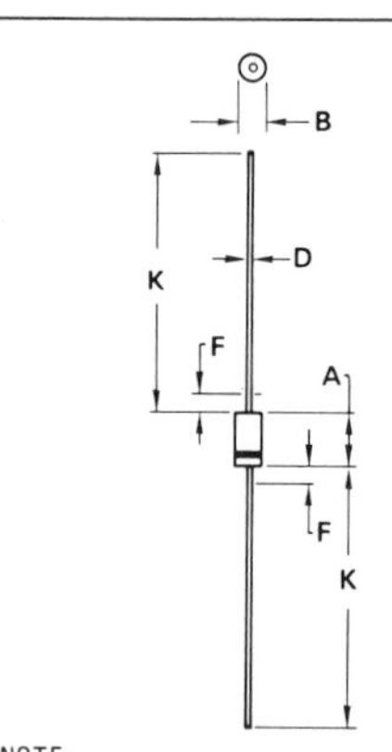

NOTE:
1. POLARITY DENOTED BY CATHODE BAND
2. LEAD DIAMETER NOT CONTROLLED WITHIN "F" DIMENSION.

DIM	MILLIMETERS		INCHES	
	MIN	MAX	MIN	MAX
A	4.07	5.20	0.160	0.205
B	2.04	2.71	0.080	0.107
D	0.71	0.86	0.028	0.034
F	–	1.27	–	0.050
K	27.94	–	1.100	–

All JEDEC dimensions and notes apply.

CASE 59-03
(DO-41)

*Indicates JEDEC Registered Data
▲Trademark of Motorola Inc.

DS 7039 R1

***ELECTRICAL CHARACTERISTICS** (T_A = 25°C unless otherwise noted) V_F = 1.2 V max, I_F = 200 mA for all types.

JEDEC Type No. (Note 1)	Nominal Zener Voltage V_Z @ I_{ZT} Volts (Notes 2 and 3)	Test Current I_{ZT} mA	Maximum Zener Impedance (Note 4)			Leakage Current		Surge Current @ T_A = 25°C i_r – mA (Note 5)
			Z_{ZT} @ I_{ZT} Ohms	Z_{ZK} @ I_{ZK} Ohms	I_{ZK} mA	I_R μA Max	V_R Volts	
1N4728	3.3	76	10	400	1.0	100	1.0	1380
1N4729	3.6	69	10	400	1.0	100	1.0	1260
1N4730	3.9	64	9.0	400	1.0	50	1.0	1190
1N4731	4.3	58	9.0	400	1.0	10	1.0	1070
1N4732	4.7	53	8.0	500	1.0	10	1.0	970
1N4733	5.1	49	7.0	550	1.0	10	1.0	890
1N4734	5.6	45	5.0	600	1.0	10	2.0	810
1N4735	6.2	41	2.0	700	1.0	10	3.0	730
1N4736	6.8	37	3.5	700	1.0	10	4.0	660
1N4737	7.5	34	4.0	700	0.5	10	5.0	605
1N4738	8.2	31	4.5	700	0.5	10	6.0	550
1N4739	9.1	28	5.0	700	0.5	10	7.0	500
1N4740	10	25	7.0	700	0.25	10	7.6	454
1N4741	11	23	8.0	700	0.25	5.0	8.4	414
1N4742	12	21	9.0	700	0.25	5.0	9.1	380
1N4743	13	19	10	700	0.25	5.0	9.9	344
1N4744	15	17	14	700	0.25	5.0	11.4	304
1N4745	16	15.5	16	700	0.25	5.0	12.2	285
1N4746	18	14	20	750	0.25	5.0	13.7	250
1N4747	20	12.5	22	750	0.25	5.0	15.2	225
1N4748	22	11.5	23	750	0.25	5.0	16.7	205
1N4749	24	10.5	25	750	0.25	5.0	18.2	190
1N4750	27	9.5	35	750	0.25	5.0	20.6	170
1N4751	30	8.5	40	1000	0.25	5.0	22.8	150
1N4752	33	7.5	45	1000	0.25	5.0	25.1	135
1N4753	36	7.0	50	1000	0.25	5.0	27.4	125
1N4754	39	6.5	60	1000	0.25	5.0	29.7	115
1N4755	43	6.0	70	1500	0.25	5.0	32.7	110
1N4756	47	5.5	80	1500	0.25	5.0	35.8	95
1N4757	51	5.0	95	1500	0.25	5.0	38.8	90
1N4758	56	4.5	110	2000	0.25	5.0	42.6	80
1N4759	62	4.0	125	2000	0.25	5.0	47.1	70
1N4760	68	3.7	150	2000	0.25	5.0	51.7	65
1N4761	75	3.3	175	2000	0.25	5.0	56.0	60
1N4762	82	3.0	200	3000	0.25	5.0	62.2	55
1N4763	91	2.8	250	3000	0.25	5.0	69.2	50
1N4764	100	2.5	350	3000	0.25	5.0	76.0	45

*Indicates JEDEC Registered Data.

NOTE 1 – Tolerance and Type Number Designation. The JEDEC type numbers listed have a standard tolerance on the nominal zener voltage of ±10%. A standard tolerance of ±5% on individual units is also available and is indicated by suffixing "A" to the standard type number.

NOTE 2 – Specials Available Include:

A. Nominal zener voltages between the voltages shown and tighter voltage tolerances,

B. Matched sets.

For detailed information on price, availability, and delivery, contact your nearest Motorola representative.

NOTE 3 – Zener Voltage (V_Z) Measurement. Motorola guarantees the zener voltage when measured at 90 seconds while maintaining the lead temperature (T_L) at 30°C ± 1°C, 3/8" from the diode body.

NOTE 4 – Zener Impedance (Z_Z) Derivation. The zener impedance is derived from the 60 cycle ac voltage, which results when an ac current having an rms value equal to 10% of the dc zener current (I_{ZT} or I_{ZK}) is superimposed on I_{ZT} or I_{ZK}.

NOTE 5 – Surge Current (i_r) Non-Repetitive. The rating listed in the electrical characteristics table is maximum peak, non-repetitive, reverse surge current of 1/2 square wave or equivalent sine wave pulse of 1/120 second duration superimposed on the test current, I_{ZT}, per JEDEC registration; however, actual device capability is as described in Figures 4 and 5.

APPLICATION NOTE

Since the actual voltage available from a given zener diode is temperature dependent, it is necessary to determine junction temperature under any set of operating conditions in order to calculate its value. The following procedure is recommended:

Lead Temperature, T_L, should be determined from

$$T_L = \theta_{LA} P_D + T_A$$

θ_{LA} is the lead-to-ambient thermal resistance (°C/W) and P_D is the power dissipation. The value for θ_{LA} will vary and depends on the device mounting method. θ_{LA} is generally 30 to 40°C/W for the various clips and tie points in common use and for printed circuit board wiring.

The temperature of the lead can also be measured using a thermocouple placed on the lead as close as possible to the tie point. The thermal mass connected to the tie point is normally large enough so that it will not significantly respond to heat surges generated in the diode as a result of pulsed operation once steady-state conditions are achieved. Using the measured value of T_L, the junction temperature may be determined by:

$$T_J = T_L + \Delta T_{JL}.$$

ΔT_{JL} is the increase in junction temperature above the lead temperature and may be found as follows:

$$\Delta T_{JL} = \theta_{JL} P_D$$

θ_{JL} may be determined from Figure 3 for dc power conditions. For worst-case design, using expected limits of I_Z, limits of P_D and the extremes of $T_J(\Delta T_J)$ may be estimated. Changes in voltage, V_Z, can then be found from:

$$\Delta V = \theta_{VZ} \Delta T_J$$

θ_{VZ}, the zener voltage temperature coefficient, is found from Figure 2.

Under high power-pulse operation, the zener voltage will vary with time and may also be affected significantly by the zener resistance. For best regulation, keep current excursions as low as possible.

Surge limitations are given in Figure 5. They are lower than would be expected by considering only junction temperature, as current crowding effects cause temperatures to be extremely high in small spots resulting in device degradation should the limits of Figure 5 be exceeded.

MOTOROLA SEMICONDUCTORS

PO BOX 20912 • PHOENIX, ARIZONA 85036

MPC100

Advance Information

POSITIVE 5.0 VOLT FIXED VOLTAGE REGULATOR

10 AMPERES
80 WATTS

SMARTpower SERIES
10 AMPERES POSITIVE VOLTAGE REGULATOR

This fixed voltage regulator is a series pass monolithic integrated circuit capable of supplying current up to 10 amperes. SMARTpower technology, utilizing a combination of a high-power bipolar output transistor in conjunction with small-signal CMOS control circuitry offers a unique monolithic chip with the following features:

- Internal Thermal Protection
- Internal Short Circuit Protection
- Low Differential Voltage — Typ 1.5 V @ 10 A

MAXIMUM RATINGS

Rating	Symbol	Value	Unit
Input Voltage	V_{in}	25	Vdc
Output Current	I_o	10	Adc
Total Power Dissipation @ T_C = 25°C Derate above T_C = 25°C	P_D	 80 0.8	 Watts W/°C
Storage Temperature Range	T_{stg}	0 to 150	°C
Operating Junction Temperature Range	T_J	0 to 125	°C

THERMAL CHARACTERISTICS

Characteristic	Symbol	Value	Unit
Thermal Resistance, Junction-to-Case	$R_{\theta JC}$	1.25	°C/W
Maximum Lead Temperature for Soldering Purposes 1/8" from Case for 5.0 sec	T_L	275	°C

STANDARD APPLICATION

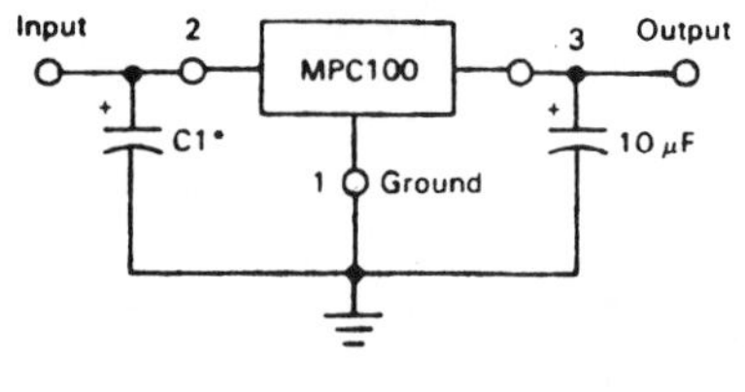

*C1 is required if the regulator is located an appreciable distance from the power supply main filter capacitor.

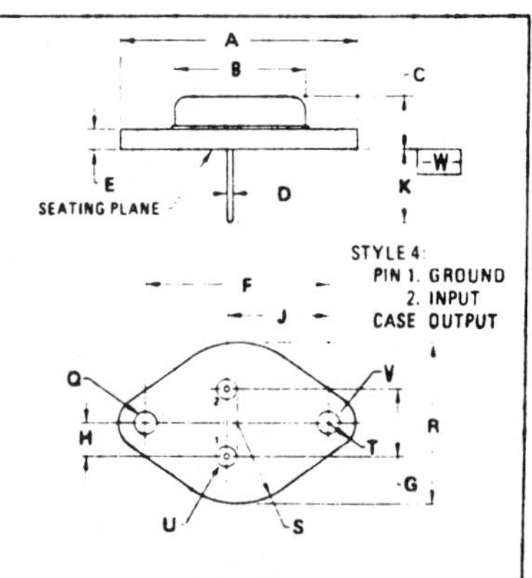

NOTES:
1. DIAMETER V AND SURFACE W ARE DATUMS.
2. POSITIONAL TOLERANCE FOR HOLE Q:
   ⌖ | ∅ 0.25 (0.010) Ⓜ | W | V Ⓜ
3. POSITIONAL TOLERANCE FOR LEADS:
   ⌖ | ∅ 0.30 (0.012) Ⓜ | W | V Ⓜ | Q Ⓜ

DIM	MILLIMETERS MIN	MILLIMETERS MAX	INCHES MIN	INCHES MAX
A	–	39.37	–	1.550
B	–	21.08	–	0.830
C	6.35	7.62	0.250	0.300
D	0.97	1.09	0.038	0.043
E	1.40	1.78	0.055	0.070
F	30.15 BSC		1.187 BSC	
G	10.92 BSC		0.430 BSC	
H	5.46 BSC		0.215 BSC	
J	16.89 BSC		0.665 BSC	
K	11.18	12.19	0.440	0.480
Q	3.81	4.19	0.150	0.165
R	–	26.67	–	1.050
U	2.54	3.05	0.100	0.120
V	3.81	4.19	0.150	0.165

CASE 1-04
TO-204AA (TYPE)

ADI-711

ELECTRICAL CHARACTERISTICS (1)

Characteristic	Symbol	Min	Typ	Max	Unit
Output Voltage $(10\ mA \leq I_O \leq 10\ A)$ $(10\ mA \leq I_O \leq 5.0\ A,\ 6.5\ V \leq V_{in} \leq 20\ V)$	V_O	4.75	—	5.25	Vdc
Line Regulation (2) $(6.5\ V \leq V_{in} \leq 20\ V)$ $(6.5\ V \leq V_{in} \leq 10\ V)$	Reg_{line}	— —	— —	100 50	mV
Load Regulation (2) $(10\ mA \leq I_O \leq 10\ A,\ T_C = 0\ to\ 125°C)$	Reg_{load}	—	—	50	mV
Quiescent Current $(10\ mA \leq I_O \leq 10\ A)$	I_B	—	—	25	mA
Ripple Rejection $(V_{in} = 8.0\ V,\ f = 120\ Hz)$	RR	—	45	—	db
Dropout Voltage $(I_O = 10\ A,\ T_C = 0\ to\ 125°C)$	V_{in}-V_O	—	1.5	2.0	Vdc
Short Circuit Current (3) $(V_{in} = 10\ Vdc)$	I_{SC}	—	—	20	Adc
Averge Temperature Coefficient of Output Voltage $(I_O = 10\ A)$	TCV_O	—	1.6	—	mV/°C
Output Noise Voltage $(10\ Hz \leq f \leq 100\ kHz)$	V_N	—	5.0	—	mV
Output Resistance $(f = 120\ Hz)$	R_O	—	2.0	—	mΩ

(1) Unless otherwise specified, test conditions are: $T_J = 25°C$, $V_{in} = 7.0$ Vdc, $I_O = 5.0$ Adc, $P_O \leq P_{max}$ and $C_O = 10\ \mu F$.
(2) Load and line regulation are specified at constant junction temperature. Changes in V_O due to heating effects must be taken into account separately. Pulse testing is used with pulse width ≤ 3.0 ms and duty cycle $\leq 1.0\%$. Kelvin contacts must be used for these tests.
(3) Depending on heat sinking and power dissipation, thermal shutdown may occur.

SMARTpower VERTICAL PROFILE

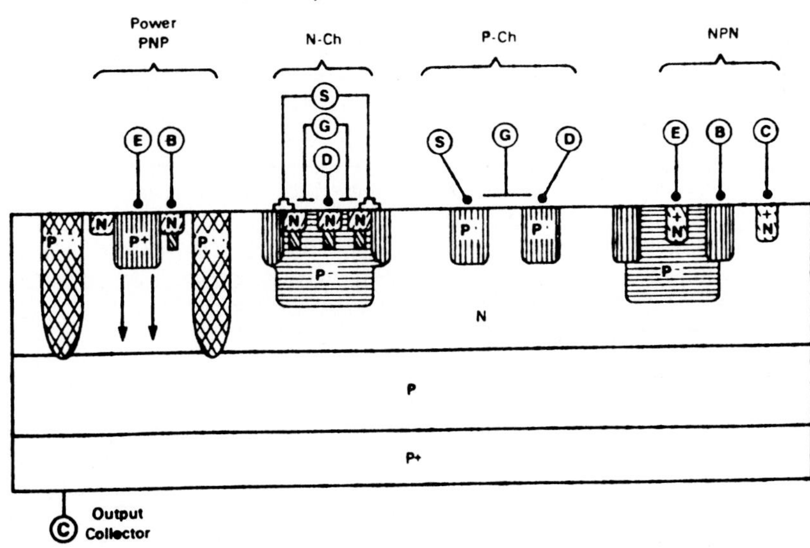

The PNP power transistor uses standard fabrication technology which assures a low saturation voltage and results in a low input-output differential voltage. The die bond is the output collector contact which results in superior load regulation.

15 Control Circuitry

Objectives

Upon completing this chapter, the reader should be able to:

- Contrast control circuitry to amplifier circuits.
- Describe the control process.
- Identify typical control input signals and identify how control circuits handle such signals.
- Identify various control devices (form and function).

Introduction

In earlier chapters the operation of circuits as basic amplifiers or power supplies for amplifiers was the primary goal. The use of semiconductor devices in a control function differs greatly from this process. In basic amplification, the input was a signal and the output was some representation of that signal, often an amplified version of that input.

In control circuits, the output is not a direct representation of the input. The input to a control circuit may be a waveform, a pulse or a dc level. The output may be simply an on/off control of the current to a load. Often control devices are operated in a saturated or cutoff mode rather than being biased in a linear amplification mode. In this chapter we present both the control function and certain special devices used in control applications. ■

15.1 THE NEED FOR CONTROL CIRCUITS

One of the many functions of electronic systems is to provide control over other electrical, mechanical, or other functional circuits or machines. The process of providing this control does not involve a standard amplification process. The input to a control system is typically not a varying ac signal but is an on/off type of signal. The objective of the control circuit is not to amplify the input (in the classical sense) but to control the function of a higher power electrical force as it acts on a load device. Typical control functions may be: a dimmer for a light; control of a motor on an assembly line; operating an alarm when a door is opened.

To help understand the function of control circuits, the following examples of input-to-output relations for several control systems may be helpful:

Input	Output
Potentiometer setting	Average voltage to a light
Resistance of a thermistor	On/off control to a furnace
Resistance of a photocell	Drive voltage to a motor
Time (delay) of an *R/C* circuit	Voltage to a bell

One of the primary functions of a control circuit is to provide either a single on/off control of a voltage (or current) supply, or to provide a continuous control of the average voltage (or current) from such a supply. Normally, the circuit that provides the actual control function to the source is rather simple. Such control can involve a simple transistor amplifier which is either biased at cutoff (off) or is saturated (full-on).

Control functions deal primarily with two important circuit functions. One of the functions is how the actual control of supply voltage and current are to be metered out to a load element (motor, light bulb, etc.). The other important function is how the control signal, that which drives the controlling device, is to be provided.

Finally, just as in amplification circuits, control functions can be done either with or without feedback. Control circuits which operate without feedback (called open-loop control) simply have the control signal feed the control device without sensing the actual effect that circuit has on the system. Closed-loop control, that which uses feedback, has additional sensors which detect the function of the controlled element and use that detected quantity to maintain the control signal's level(s) (see Figure 15.1).

A form of control circuitry has already been presented in Chapter 14. The regulator circuits used to maintain the output level are control circuits. They have a predesigned (or preset) level for the output. Feedback of the output is used to operate a control circuit that maintains the output level at a constant voltage value.

Control circuitry uses either standard solid-state devices as presented previously, or can apply special-purpose devices, designed to perform control functions. There are a wide variety of devices which are specially dedicated to the control function.

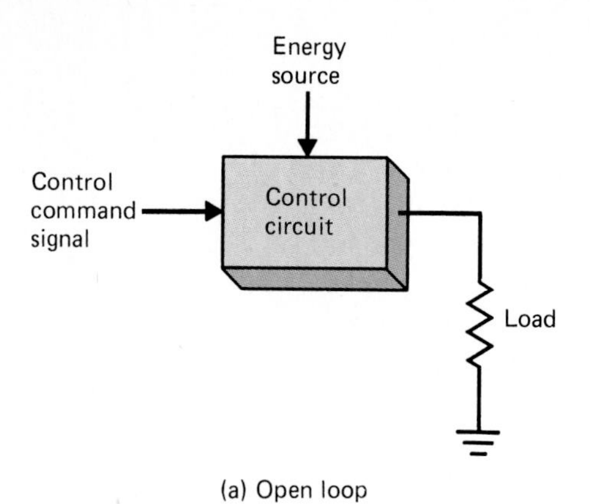

(a) Open loop

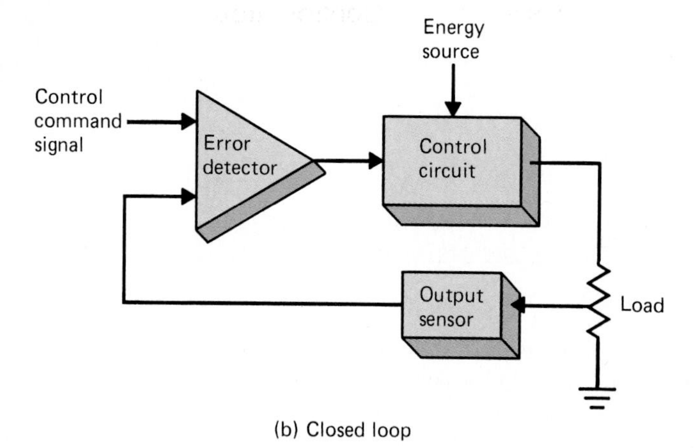

(b) Closed loop

FIGURE 15.1 Basic Open- and Closed-Loop Control

> **REVIEW PROBLEMS**
>
> (1) Illustrate the block diagrams for control circuits with and without feedback.
> (2) Refer back to Chapter 14 and duplicate the block diagram for a switching regulator circuit. Relate this diagram to the basic functions of control circuits noted in this section.

15.2 THE FUNCTION OF CONTROL CIRCUITS

In this section the basic block diagrams used to develop control circuit functions are presented. The development will flow for simplistic control systems to slightly more sophisticated forms of control.

Open-Loop Control Diagram

The open-loop control diagram (Figure 15.2) includes only a few simple functional blocks. The basic control signal block defines that there is available some voltage or current level which is used to produce the control function.

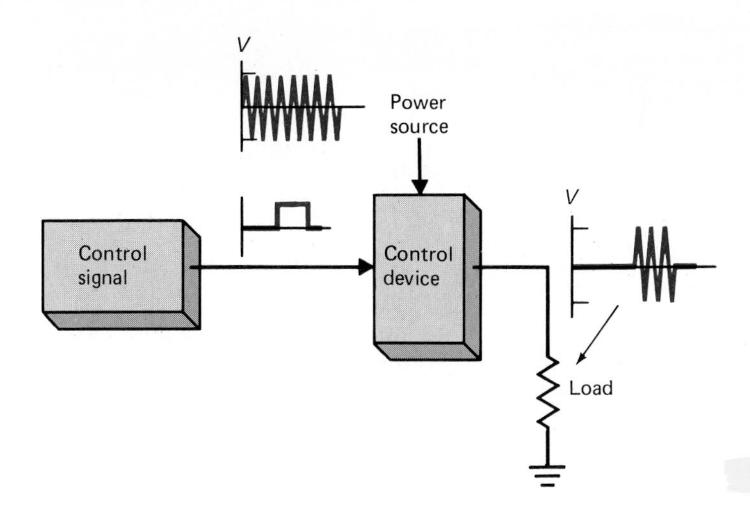

FIGURE 15.2 Function of Control Signal in a Control Circuit

The output of such a block could be either a dc level or an ac (usually pulsed or square-wave forms) signal representative of the appropriate control information.

The control device block could represent either a single control device (transistor) or a circuit that produces the control function. The control signal provides the electrical quantities necessary for this block to control the flow of energy from the main power source.

The power source block is used to indicate the source of controlled energy. This source could be dc voltage(s) or could be an ac voltage source. Naturally, the load is the device to which the output of the control circuit is passed.

EXAMPLE 15.1

Figure 15.3 shows a schematic for a simple motor (fan) speed control circuit. The speed of the motor is controlled by setting a potentiometer.

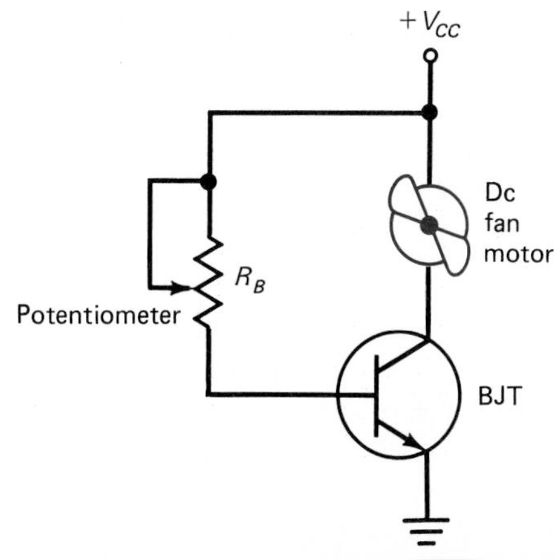

FIGURE 15.3 Simple DC Fan Control Circuit

As the resistance of the "pot" changes, the dc base current to the transistor changes. As this current changes, the collector current changes. Increased collector current makes the fan run faster; less collector current slows the fan down.

Closed-Loop Control Diagram

The closed-loop diagram (Figure 15.4) is more complex than the simple open-loop diagram. In a closed-loop function, the control signal to the control device is derived from both a preset value and from the output of a sensor associated with the load device. The error detector supplies a signal that indicates the amount of error between the preset value and the sensor's reading. Once the sensor's output indicates that the load is functioning identical to the preset value, the error detector either produces a constant output or zero output (depending on the desired function of the load).

The function of this feedback process is identical to the function of feedback used in basic amplifiers. The feedback provides information to the control device which is directly proportional to the amount of error between the desired value and the actual value.

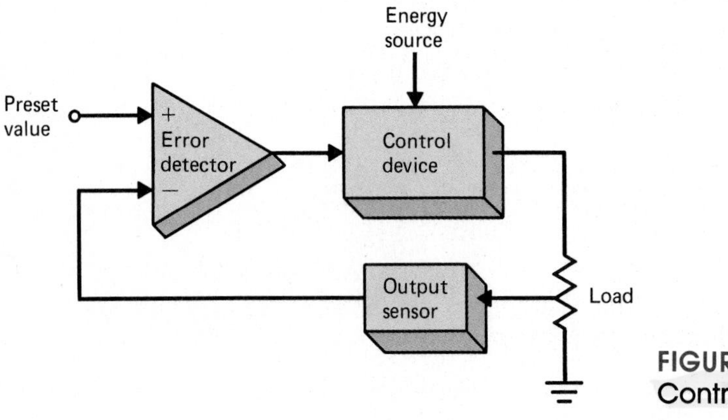

FIGURE 15.4 Closed-Loop Control Block Diagram

EXAMPLE 15.2

The circuit illustrated in Figure 15.5 is similar in function to that in Example 15.1. In this case, though, feedback of the actual temperature is compared to the value of a preset temperature. The output of the error sensor (op-amp in this case) is the differential output derived from each input voltage. As the error between these values is greater, the op-amp's output is higher and thus causes the fan to run faster. (Since the control transistor cannot respond to a reverse polarity input from the op-amp, if the actual temperature is lower than the temperature setting, the fan will not run, but no special heating effect is seen either.)

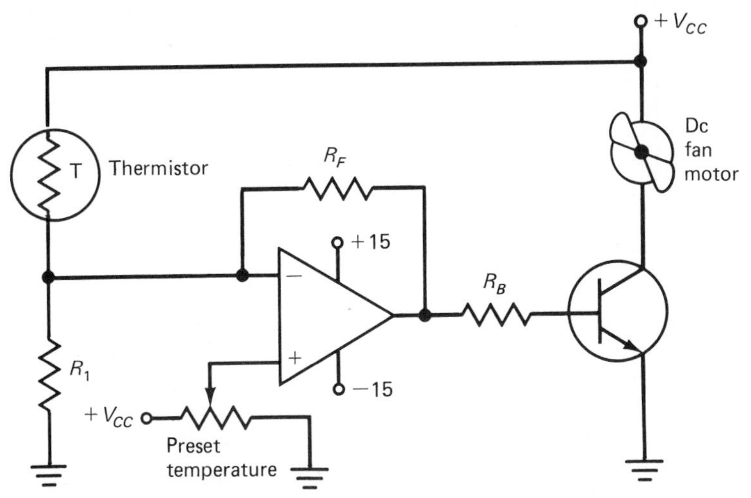

FIGURE 15.5 Closed-Loop Control Circuit—Temperature-Activated Fan

These basic block diagrams have been presented to more fully comprehend the function of control circuitry. There are some unique ways in which the control signal can be supplied to a specific control device. The following descriptions are presented to demonstrate more clearly the use of these signals.

Forms of Control Signals

The signal that is used to operate the control device can take one of three primary forms (there are other forms, but they will not be presented here). These forms are dc amplitude signals, pulsed control signals, and ac phase control signals.

DC Amplitude Signals A dc amplitude signal functions with the control device much like the input signals functioned in amplifiers. These dc levels represent the level of control (or error) dictated by a sensor or preset control. As more current is to flow through the controlled load (via the control device), this amplitude increases. To decrease the flow, the level of this signal decreases. Figure 15.6 illustrates such a coordination. In this figure, as the level of the control signal changes, the amount of output current (or voltage) to the load varies. This form of control is often called linear control.

Pulsed Control Signals In some control circuits, it is not desirable to produce a linear control function. In these circuits the objective is to produce either full-on or full-off control to the load element. In some cases this is because the load is not designed to be operated with other than a full value of supply voltage and current. In such applications, the control signal pulses the control. These pulses turn the device fully on (such as saturating a transistor) or keep it fully off (Figure 15.7). In such a function, the load is either on or it is off. Such a function is seen when remote control of a motor or lamp is desirable.

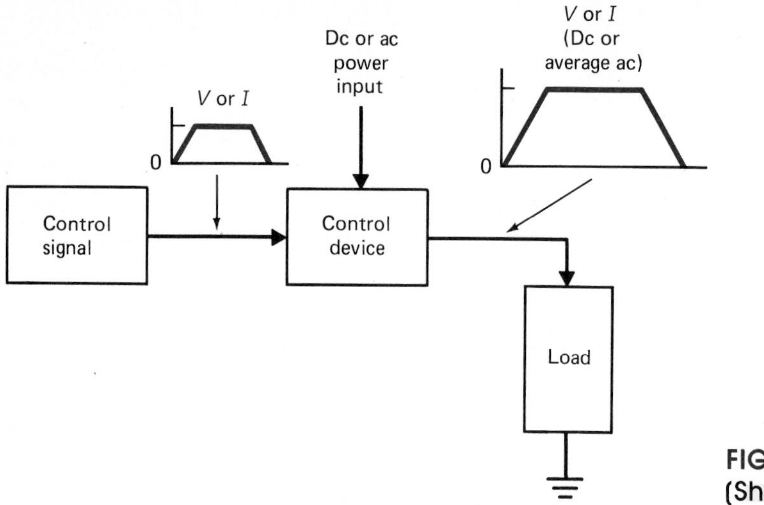

FIGURE 15.6 Linear Control (Shown with DC Power Input)

It is possible to obtain a variable control over the load (if the load can tolerate the changes) by varying either the rate with which the control device is pulsed, or by varying the amount of "on-time" for the pulse. Figure 15.8 illustrates this form of control. When the rate of the pulse is changed, this is called *pulse rate control*; when the on-time duration of the pulse is changed, this is called *pulse-width control*.

The actual averaging of this control value, thus producing a variable output level, is often done by the load device itself. One advantage of the pulse control method is that the control device is not subject to the large values of power consumption typically seen when using linear control. Pulse-width control was illustrated in the section of this text that investigated the switching regulator. Since the average value of the energy delivered to the load is proportional either to the pulse rate or duration of pulse supplied by the control circuit, the accumulated power delivered to the load is thus made variable.

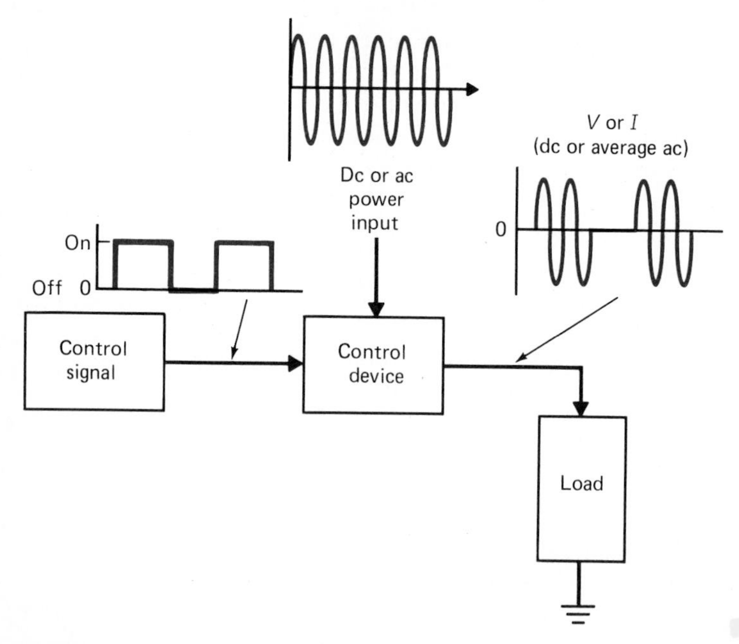

FIGURE 15.7 Pulsed Control (Shown with AC Power Input)

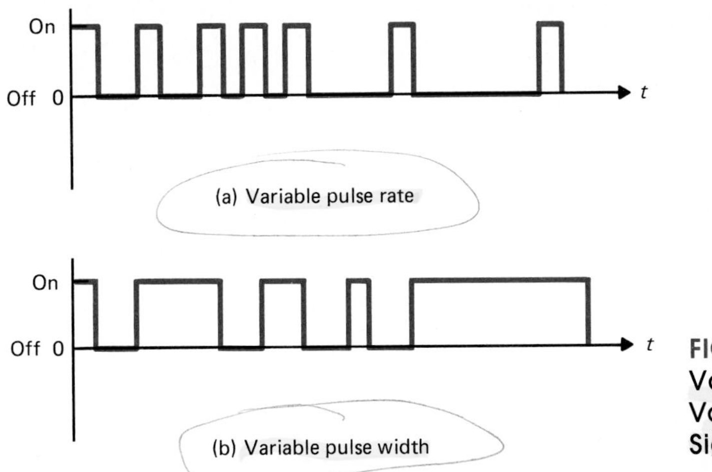

FIGURE 15.8 Comparison of Variable Pulse Rate and Variable Pulse Width Control Signals

AC Phase Control Signals When a system is to control an ac power source, the circuitry to convert this ac to a dc or pulsed control signal may be awkward or involve unnecessary circuitry. When an ac power source is used, control information may be obtained by tapping that ac source. Since a sample of the controlled ac signal is in phase with that signal and is of the same frequency, there would be synchronization between the main source and this tapped signal.

Basically, the process of phase control uses a control circuit which is a pulse-like control circuit. When the control input voltage reaches a minimum level, the control circuit turns on the controlled device. Thus, once a "threshold" level at the input is reached, the control circuit turns "on". This is illustrated in Figure 15.9.

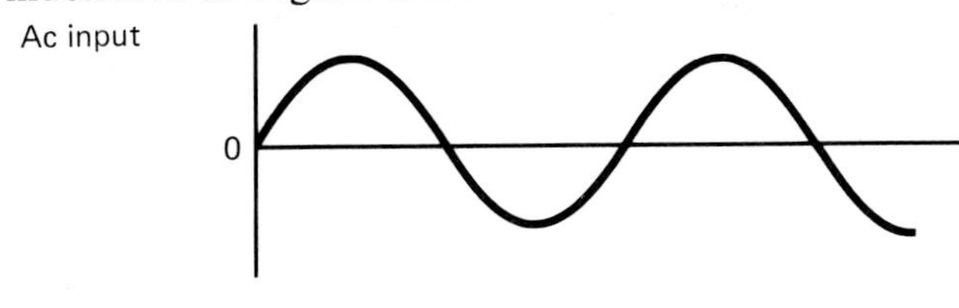

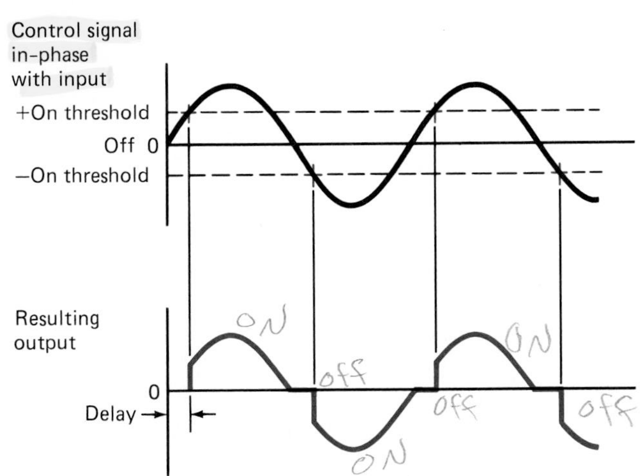

Note: Control device used requires +on and −on signals; off occurs as input (ac) drops to zero (0)

FIGURE 15.9 Phase Control with Control Signal In-Phase with AC Input

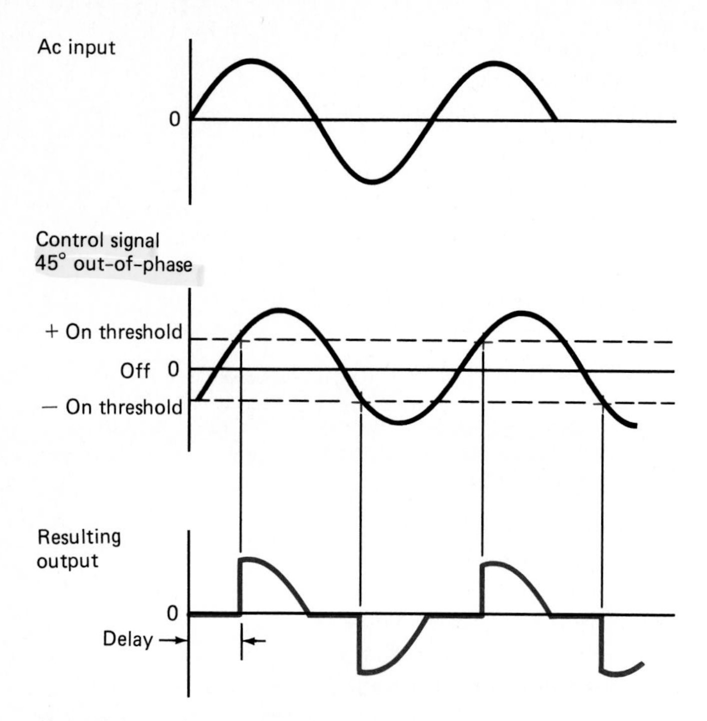

Note: Same control device requirements as in Figure 15.9.

FIGURE 15.10 Phase Control with Control Signal Out-of-Phase with AC Input

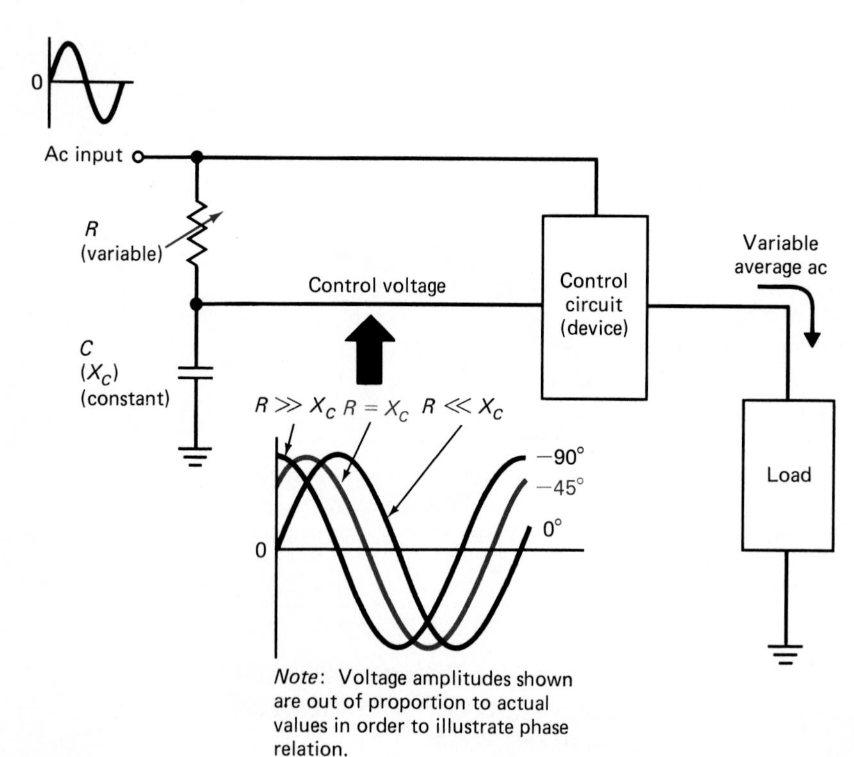

Note: Voltage amplitudes shown are out of proportion to actual values in order to illustrate phase relation.

FIGURE 15.11 Multiple Phase Effects Using Variable *R*/*C* Circuit

If the control signal is a sample of the ac input to the system, then when this level reaches the "threshold" level, the circuit comes on. Depending on the type of control device, the load will continue to operate until the control signal drops below the threshold level or when the actual controlled (main power) source drops to zero. So far, though, the control signal is in synchronization with the input sine wave. Every time the control signal reaches the threshold, the original sine wave is at the same point in its cycle.

When phase control is used, the phase of the signal which is used to trigger the control device is shifted in phase from the original sine wave. When this happens, the control signal no longer reaches the threshold level at the same instant as the original sine wave reaches the same point in its cycle. Figure 15.10 will help to illustrate this point. By shifting the phase of the control signal in respect to the main power signal, the threshold level on the control device becomes "out of sync" with the initial input. Thus a time delay is produced which causes the control device to turn on later in the cycle than it would without the phase-shift process.

Therefore, to make the control process variable, it is simply a problem of controlling the phase of the control signal. To do this (typically) a basic *R/C* circuit with a variable resistor is used. The *R/C* circuit will cause a phase shift of from near 0° to nearly 90°. Using this control process results in an output of the control circuit which is a proportion of the sine wave, related to the amount of phase shift. Figure 15.11 should help clarify this process.

REVIEW PROBLEMS

(1) Sketch the basic block diagrams of the open- and closed-loop control processes.
(2) Briefly describe the difference between the open- and closed-loop control processes. Discuss both the block diagram and the basic effects seen in each process.
(3) List the three main forms of control signal used in control circuits. Briefly describe their differences.

15.3 SPECIALIZED DEVICES USED IN CONTROL CIRCUITS

Although it is possible to use transistors, op-amps, or FETs in control circuits, there are specially designed control devices. These devices are designed to work with and manipulate the control signals used in these circuits and to handle the potentially higher power levels seen when controlling power. Following is a list of the devices that will be covered in this section. Note the abbreviated notation for each type.

Shockley diode
Diac
Silicon bilateral switch (SBS)
Silicon-controlled rectifier (SCR0
Triac
Silicon-controlled switch (SCS)

Unijunction transistor (UJT)
Programmable unijunction transistor (PUT)

All of these devices, except the UJT and PUT, fall under a general category of semiconductor devices called **thyristors**. These devices, each of a four-layer (PNPN) semiconductor construction, shares similar characteristics. They are typically nonconductive devices, capable of withstanding relatively high voltages until they are triggered into a conducting mode. They therefore work quite effectively as solid-state switches which are electronically activated. Once triggered on, they are capable of maintaining conduction until either triggered off or until certain other electrical conditions are met. These devices can be used in the triggered (pulsed) form of control circuitry. (They do not act as linear controllers; BJTs and FETs are typically used in these applications.)

Primarily the Shockley diode, the diac, and the SBS are used to aid in triggering the function of SCRs and triacs, which are the most commonly used control devices. The SCS, UJT, and PUT can also be used with the SCR and triac circuits, but are commonly used in other control applications as well. In order to establish the basic ground work for these control devices and their functions, the Shockley diode, diac, and SBS will be introduced first.

Shockley Diodes, Diacs, and SBSs

The schematic symbol and *I*/*V* curves for the Shockley diode are illustrated in Figure 15.12. By observing the curve presented in this illustration, one can see the basic function of the Shockley diode (or four-layer diode, PNPN). Like an ordinary diode, the Shockley diode conducts only in the forward direction. But there is a special effect seen in this forward curve. Until a minimal voltage is reached, called the **forward breakover voltage** (V_{BR}), there is minimal conduction through this diode. Then, when this voltage is exceeded, the device immediately begins to conduct. Unlike a zener diode which continues to drop the zener voltage level, the forward drop on the Shockley diode falls back to a relatively low value.

The Shockley diode continues to conduct, dropping its minimal forward voltage as long as there is sufficient current to maintain this conduction. This

FIGURE 15.12 Shockley Diode Curve and Schematic Symbols

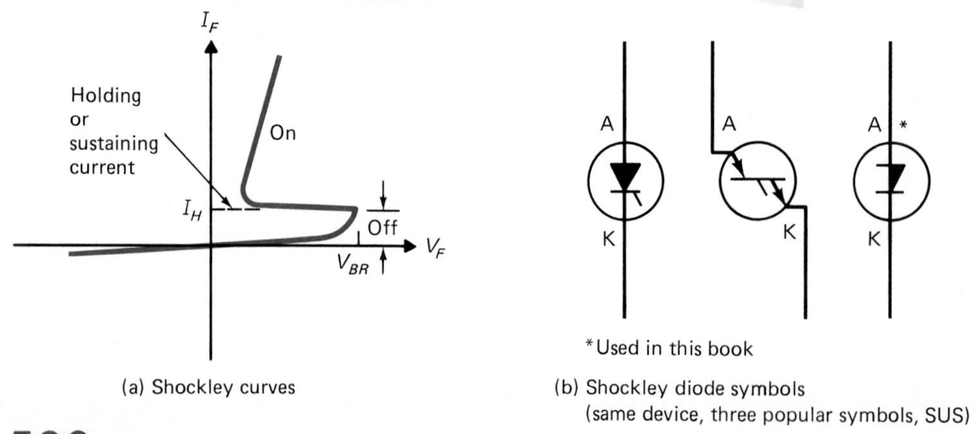

(a) Shockley curves

(b) Shockley diode symbols
(same device, three popular symbols, SUS)

minimal current (a specification of each variety of Shockley diode) is called the holding current or sustaining current. Therefore, once the Shockley diode is triggered with the appropriate breakover voltage it will continue to conduct as long as the circuit to which it is attached will maintain the minimal current flow. Such a function could be used to activate a circuit once an appropriate voltage level is reached. Other unique applications are discussed in Chapter 16, which lists certain special circuit applications.

EXAMPLE 15.3

The Shockley diode could be used in a simple "overvoltage" warning circuit. Such a circuit may be used in delicate instruments to indicate a potentially damaging surge of voltage was present. This function is illustrated in Figure 15.13. As long as the supply voltage remained below the 15-V breakover level of the diode, no significant conduction is seen through the diode. When the voltage surges above 15 V, the diode "switches on" and lights the indicator lamp. This lamp stays on even if the voltage supply drops below 15 V, thus indicating that a surge had occurred. (Naturally, the supply will have to remain on to supply sustaining current through the Shockley diode.)

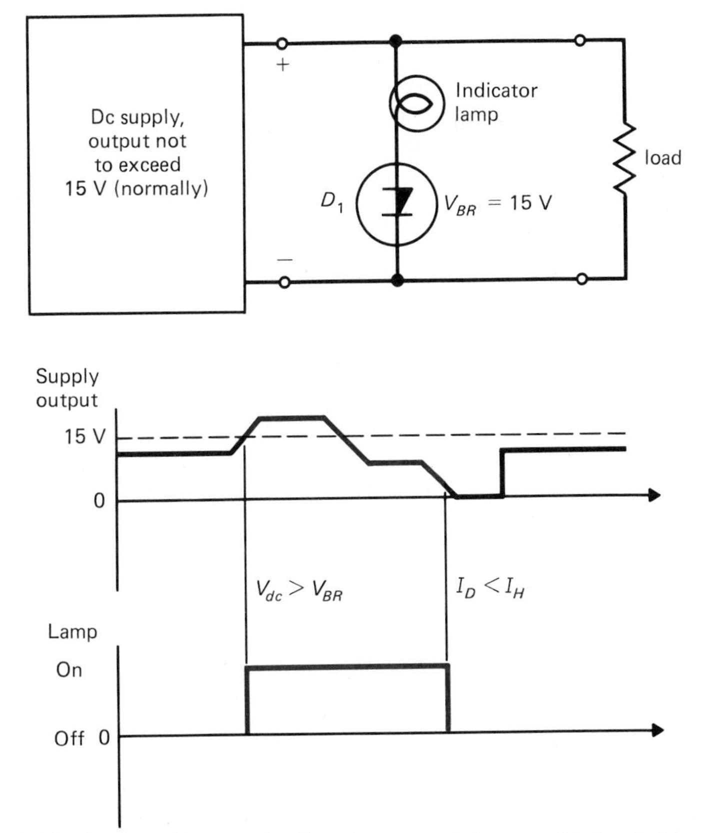

FIGURE 15.13 Overvoltage Indicator and Lamp Output Coordination

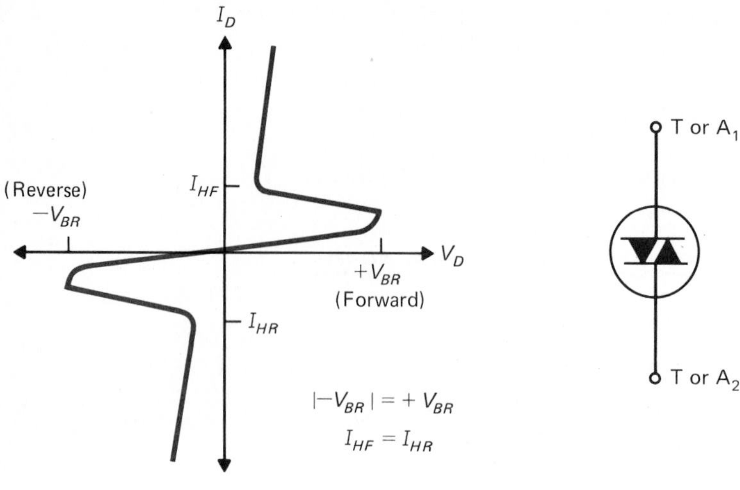

(a) Diac curve

(b) Diac symbol (SBS)

FIGURE 15.14 Diac Curve and Schematic Symbol

Diacs, as illustrated in Figure 15.14, operate much like Shockley diodes, except that they have the capacity to reach threshold voltages in either a "forward" or a "reverse" direction. (Since this device is a bidirectional conductor, the terms "forward" and "reverse" do not actually apply.) The basic functional characteristics of breakover voltage and holding current are the same s for the Shockley diode, but they apply to either polarity. [An application of the diac could involve its use as an overvoltage indicator (as in the past example) for an ac voltage supply.]

Both the Shockley diode and the diac are available in a variety of breakover and holding current ratings. The SBS is simply another version of the diac but typically operates with lower voltages and displays a few slightly different operational characteristics than the diac. The SBS has the same schematic symbol as the diac. [A slight variation of the Shockley diode is the silicon unilateral switch (SUS), but it, too, is functionally the same, sharing the same schematic symbol with the Shockley diode.]

EXAMPLE 15.4

From Appendix A investigate the significant characteristics of the MKP9V120 "Sidac." ("Sidac" is the manufacturer's name for this version of a diac.)

Page 2 of the specifications reveals that this device will "trigger" or break over at a voltage of from 110 to 125 V. It has a forward voltage drop (forward "on" voltage) of 1.3 V typical and a holding current of 100 mA. Naturally, other specifications, such as maximum ratings, are significant as well.

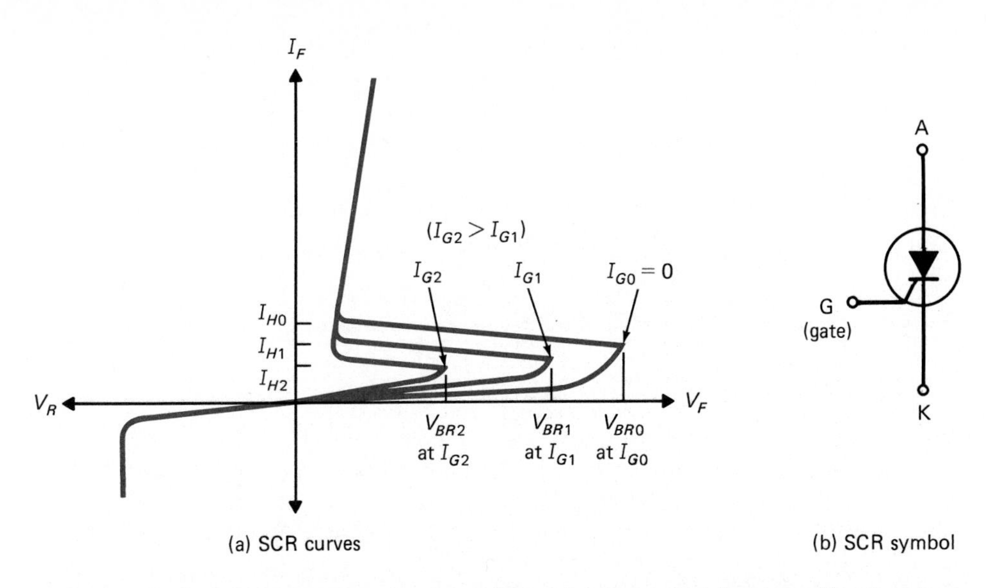

FIGURE 15.15 Silicon-Controlled Rectifier Curves and Schematic Symbol

Silicon-Controlled Rectifier

The SCR (silicon-controlled rectifier) is similar in function to the Shockley diode, but the triggering "on"' of the device is controlled by operating the GATE lead of the device. (*Note:* This gate is not at all related to the function of the GATE lead of an FET.) A selection of operating curves for the SCR are illustrated in Figure 15.15. Several curves are illustrated in this figure since

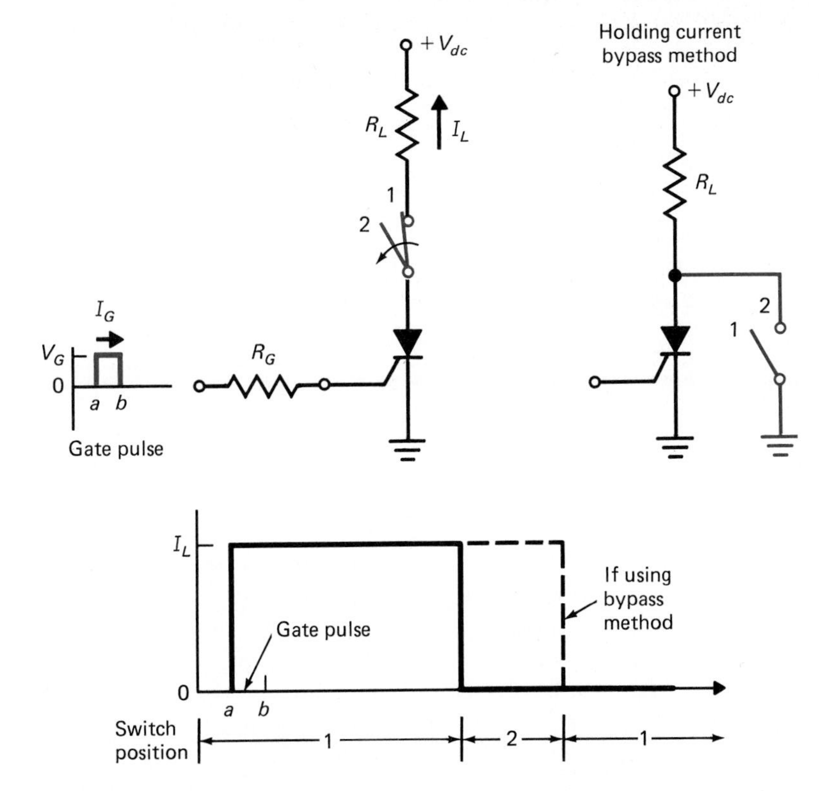

FIGURE 15.16 Pulsing "On" the SCR

a variety of gate voltages will produce a corresponding variety of breakover voltages.

When operated in a dc mode, the SCR will maintain an "on" condition once it is triggered on by a gate potential. This conduction will remain functional as long as holding current is present through the anode-to-cathode path. This is illustrated in Figure 15.16. The only way to turn "off" the current through the SCR is to reduce the anode current to a level below the holding current point.

EXAMPLE 15.5

A simple alarm circuit using an SCR, a trigger switch, a horn, and a reset switch is illustrated in Figure 15.17. The moment the trigger switch is thrown, the gate of the SCR triggers the device into its "on" state. This then supplies current to the horn. If the trigger lead is interrupted, the SCR still maintains conduction to the horn. Only by throwing the reset switch and breaking the holding current through the SCR can the system be reset.

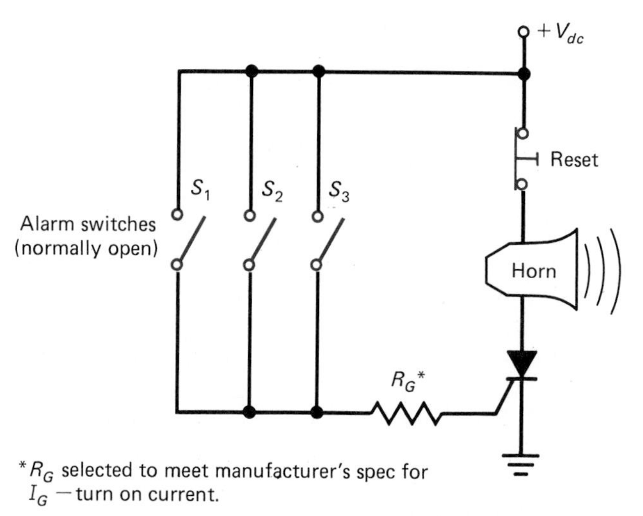

FIGURE 15.17 SCR "Latching" Alarm Circuit

Triac

Functionally, the triac is to the SCR as the diac was to the Shockley diode. The triac has the capacity to conduct either in the "forward" or "reverse" direction when trigger voltage is applied to the gate lead. This makes the triac highly usable for ac control operations. (The SCR will not conduct in the reverse direction, as indicated by the diode indication on its schematic symbol.) Figure 15.18 illustrates the symbol for the triac and also illustrates its *I*/*V* curves.

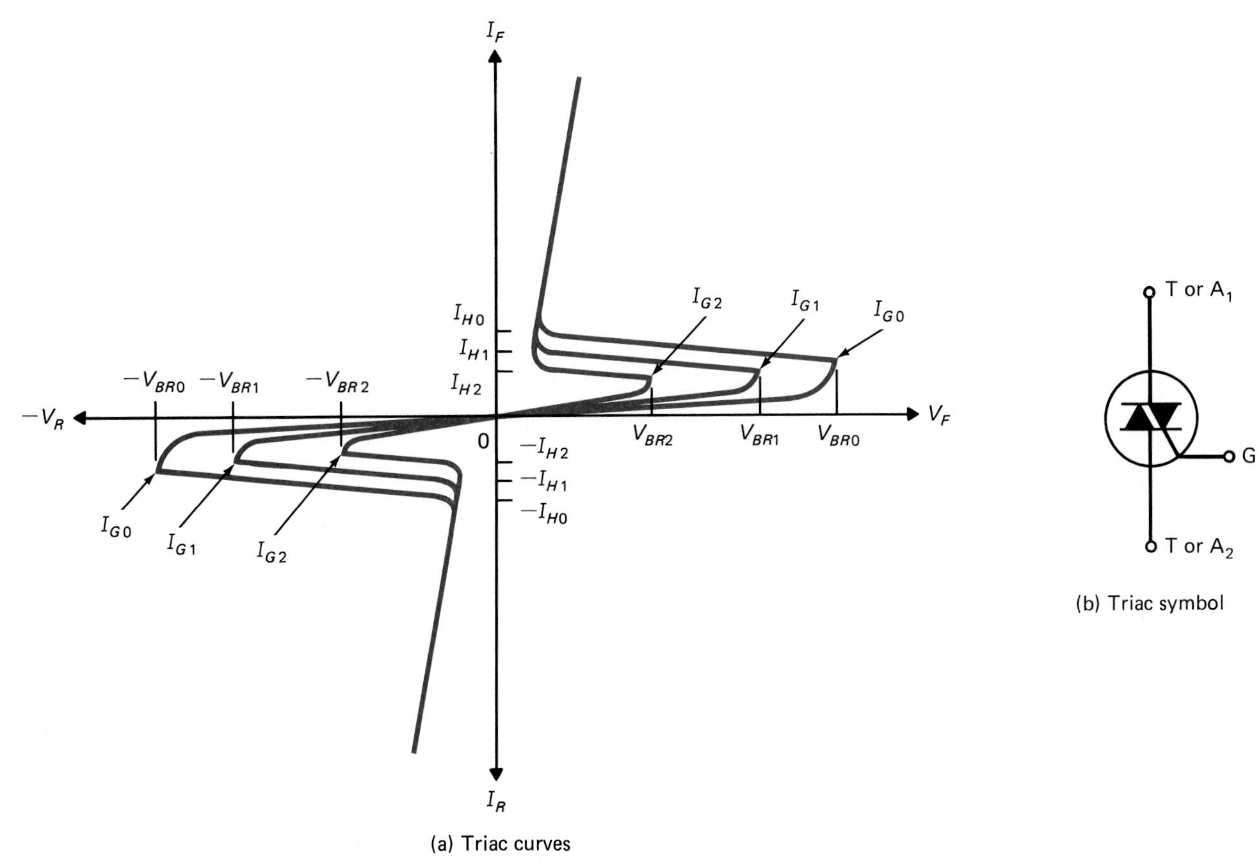

(a) Triac curves

(b) Triac symbol

FIGURE 15.18 Triac Curves and Schematic Symbol

Just like the SCR, the triac must be triggered on by voltage on the gate lead. It will maintain conduction (in either direction) as long as the conducting current does not drop below the minimum holding current specification. The SCR is to be triggered with positive gate voltages, unless otherwise specified by the manufacturer. But the triac can be triggered with either a positive or a negative gate voltage.

Use of SCRs and Triacs with AC Voltages

Before proceeding to the description of other control devices, the special effect of using SCRs and triacs with ac voltage sources needs to be presented. The process of triggering relative to an ac supply involves, in many cases, phase-angle triggering, as discussed in the preceding section.

The unique effect seen with these devices when ac is the controlled voltage results from the basic ccycling of the ac voltage. If ac voltage is being controlled, there will always be a point in the ac cycle where the current flow through the device (SCR or triac) will drop below the minimum holding current for that device (see Figure 15.19). From this illustration it can be seen that if there is no gate voltage on the SCR or triac, when the ac voltage drops below the level that would maintain the proper holding current, the device would turn off. This drop in potential current occurs twice during each cycle of the ac (at each point where it crosses the zero voltage axis).

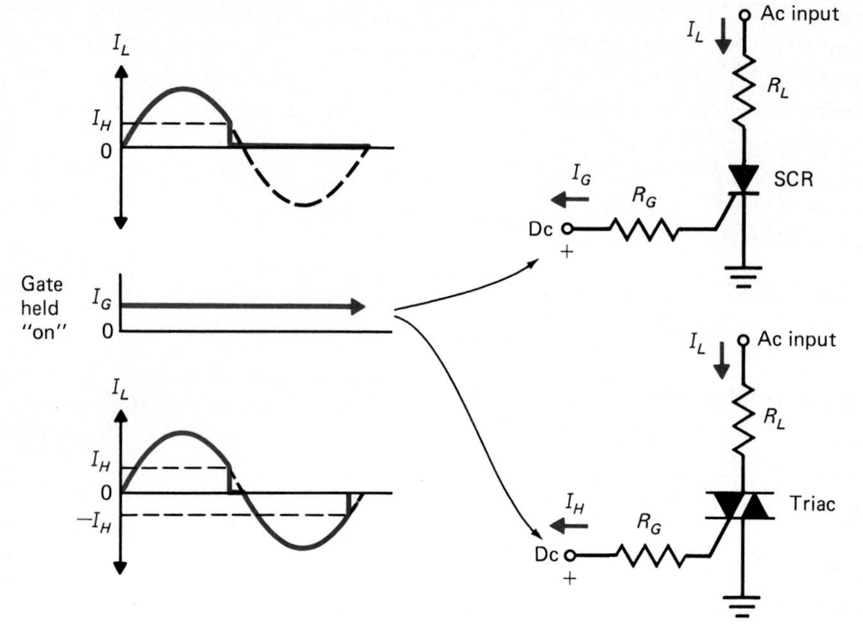

FIGURE 15.19 Effect of Load Current Dropping below the Necessary Holding Current Specification

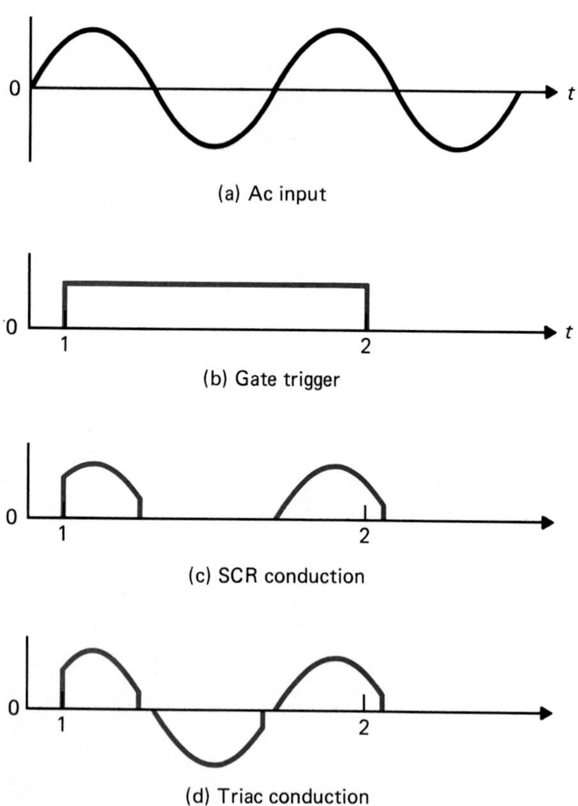

FIGURE 15.20 Comparison of SCR and Triac Control of a Sine-Wave Input

In Figure 15.20 a timing diagram illustrates the coordination between the gate input and sine-wave input and the resulting output for an SCR and a triac. From this diagram it can be seen that:

1. An SCR will produce a half-cycle conduction of the ac wave as long as the trigger voltage is maintained.
2. The triac will maintain a full-wave conduction of the ac wave as long as the trigger voltage is maintained.
3. When the trigger voltage is turned off to either device, the conduction will cease only when the ac cycle reaches a point in time when the holding current through the device drops too low to maintain conduction. (For either device, this is within one-half cycle of the moment the gate voltage is disconnected.)

If the gate control voltage is pulsed in synchronization with the ac supply, the devices will re-trigger at every cycle (or half-cycle for the triac) and thus maintain conduction. Even if a low-level sine wave is obtained from a simple voltage divider of the main ac input and is applied to the gate lead, as soon as this voltage reaches the proper trigger voltage for the SCR or triac, the device will be turned on. Since this wave would be in sync with the main ac wave, re-triggering at every cycle of the input would be had (see Figure 15.21).

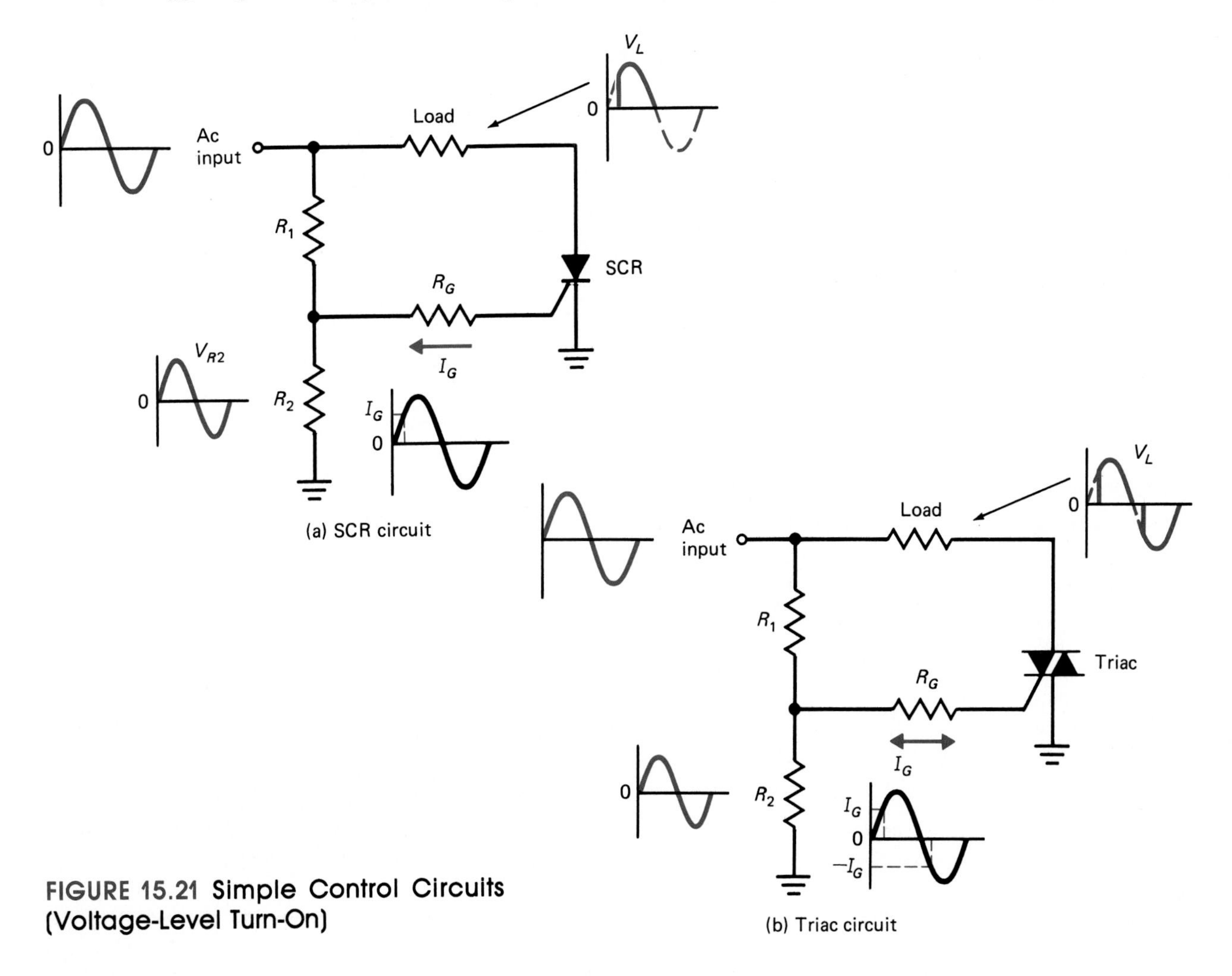

FIGURE 15.21 Simple Control Circuits (Voltage-Level Turn-On)

(Due to their similarity of operation, the following discussion will concentrate on use of the triac, but it is assumed that an SCR could be used as effectively, as long as the half-wave effect of the SCR is kept in mind.)

A more sophisticated control of these circuits can be had by using a phase-shifting network in conjunction with the gate lead of the control device (triac). If the ac waveform at the gate lead of the triac is supplied through a circuit that produces a variable phase shift of this waveform relative to the ac input, flexible control of the conduction of the triac is seen. In Figure 15.22

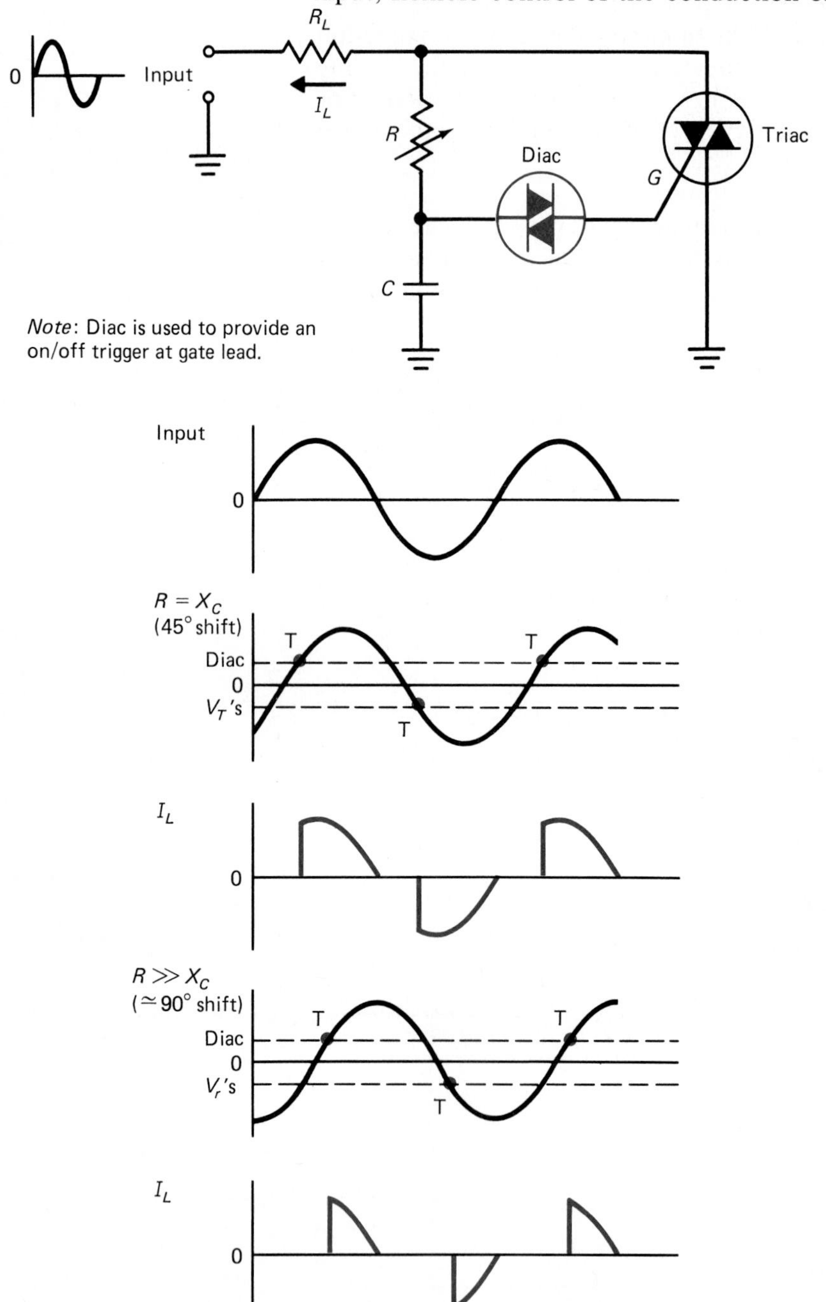

FIGURE 15.22 Use of Phase-Shift Circuit and Triggering Diac with a Triac Controller (Key Waveforms Shown)

such a phase control circuit is shown. In this circuit, the variable resistor *R* and capacitor *C* produce a simple phase-shifting network. The phase of the voltage presented to the triac is therefore variable from about 90 to 180° (due to the layout of the circuit). The voltage necessary to trigger the diac and thus fire the gate lead of the device is delayed in time by this shift. The result of this shifting is a variable control on the "firing" of the triac and thus on the voltage delivered to the load. It should be remembered that the sine wave itself will reset the triac's conduction as it crosses the zero-voltage axis. This form of triggering is termed "conduction angle" triggering, referring to the phase angle of the trigger voltage relative to the initial input voltage.

There are quite a few variations on this form of triggering, both simple and sophisticated. More details about the triggering process will be presented in courses that deal specifically with control (or industrial control) circuitry.

Silicon-Controlled Switch

The silicon-controlled switch (SCS) pictured in Figure 15.23 functions quite like an SCR. There are a few major differences between these two devices, though. first, it will be noted that the SCS has four leads, not three like the SCR. Each of these terminals (the anode gate and cathode gate) can be used to trigger the device on. Access to these two parts of the device will allow it to be turned on from either terminal. A positive voltage pulse on the cathode gate lead will cause the device to conduct, much as it will an SCR. But a negative voltage on the anode gate will also cause the device to turn on. Thus the device can be activated by either a positive or a negative voltage.

Unlike the SCR, the SCS can also be triggered off at the gate lead. An application of a negative pulse to the cathode gate lead or a positive voltage at the anode gate lead will shut off the conduction of the device (cathode-to-anode current). This is the first of the control devices that can be turned off by manipulation of the gate leads. Finally, the SCS is designed for use in low power control circuits. As yet there are no high-power (high-current and/or high-voltage) SCS devices.

Unijunction Transistor

The function of the unijunction transistor (shown in Figure 15.24) is quite similar to that of the SCR. The major difference between the UJT and the SCR is that the SCR has a preset value of gate trigger voltage (and current)

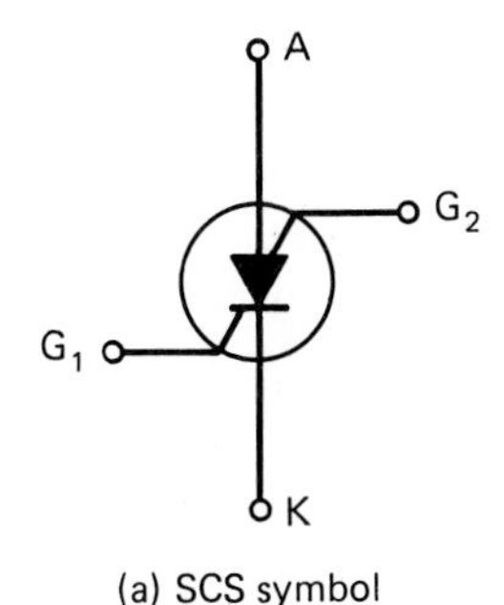

(a) SCS symbol

Lead	Pulse Polarity	Conduction
G_1	+	On
	−	Off
G_2	+	Off
	−	On

(b) Turn-on and turn-off functions

FIGURE 15.23 Silicon-Controlled Switch (SCS) Schematic Symbol and Functions

where the value of UJT triggering voltage is proportional to the voltage across the device's output terminals. (IIt should be noted that the UJT terminals are called E, emitter; B_1, base 1; and B_2, base 2. Although the terms "emitter" and "base" are used, the UJT differs greatly from the makeup of a standard BJT. And though the symbol for the UJT is similar to that for an FET (note that the polarity arrow is at an angle), it also differs dramatically from the FET.)

It was noted that the UJT's trigger voltage, applied between the emitter and base 1 leads, is proportional to the supply (or B_1-to-B_2 voltage). This voltage, called the **peak-point voltage**, is defined by

$$V_P = \eta V_{BB} + V_D \tag{15.1}$$

where V_P = peak (or "firing") voltage of the UJT
η = intrinsic standoff ratio (a UJT specification)
V_{BB} = B_1-to-B_2 voltage (or source voltage)
V_D = normal forward drop (usually 0.7 V)

(Often the value of V_D is so low that it is dropped from the equation.)

FIGURE 15.24 N- and P-Type Unijunction Transistors (UJT)

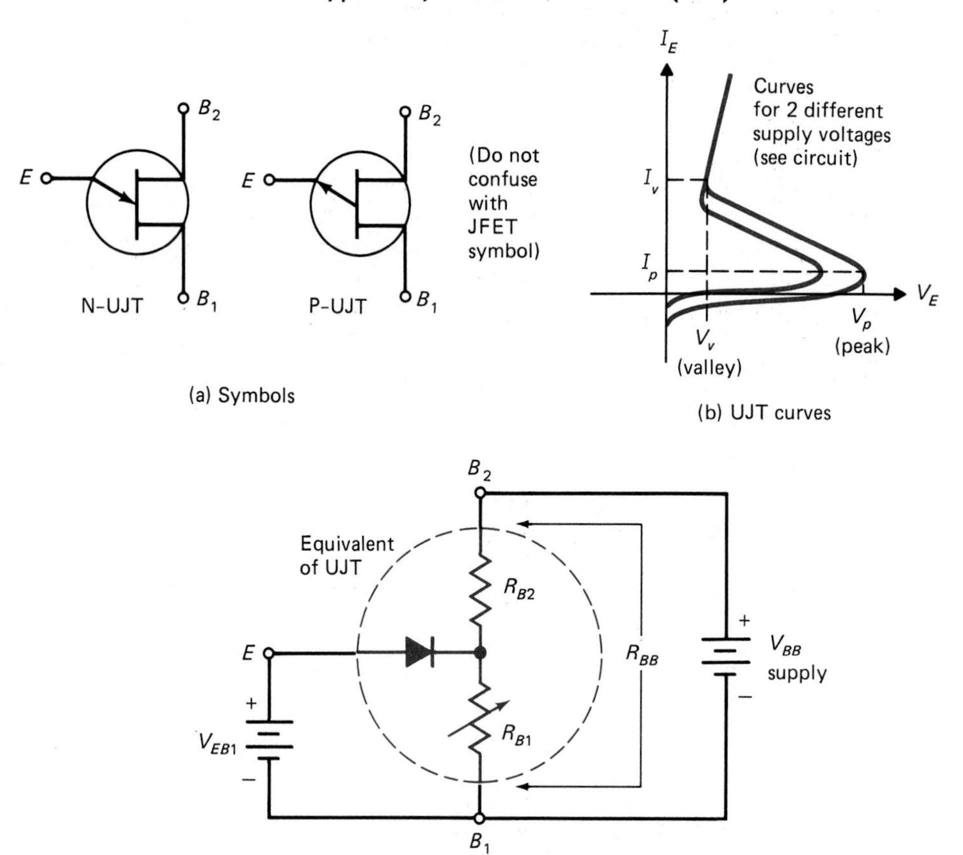

The basic function of the UJT is as follows:

1. If the triggering voltage is below the calculated value of V_P, the device remains "Off," with little conduction from base 1 to base 2.
2. When the voltage at the emitter reaches the proper value of voltage to trigger the device, the resistance between the two base leads drops very low and permits conduction through these leads. Also, conduction through the E-to-B_1 junction occurs.
3. When the voltage at the emitter falls below what is called the valley voltage (V_V), the UJT ceases to conduct (interrupts the B_1-to-B_2 current). Valley voltage should be below the value of peak voltage in practical applications.

Thus, much like the SCR and triac, the UJT produces an on/off control of current flow. Unlike these devices, there is control of the amount of triggering voltage needed to turn the device on and off. UJTs are often used to aid in triggering the gate leads of SCRs and triacs, as well as being used in nonsinusoidal oscillators and other special-purpose triggering circuits.

EXAMPLE 15.6

The circuit shown in Figure 15.25 uses a UJT to produce oscillations across R_2. Briefly describe how this occurs.

Solution:

See the waveforms in Figure 15.25.

FIGURE 15.25 a) UJT Oscillator

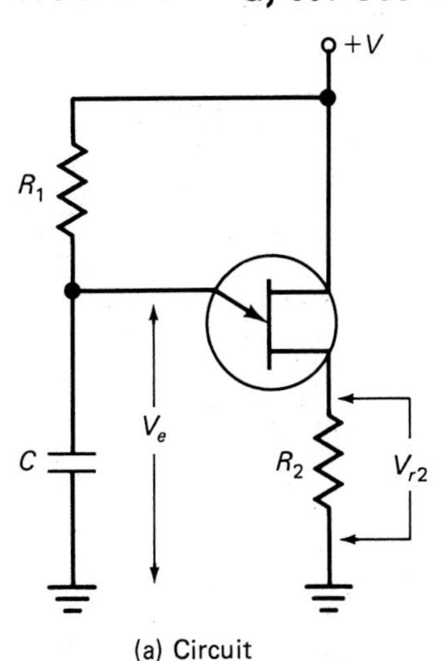

(a) Circuit

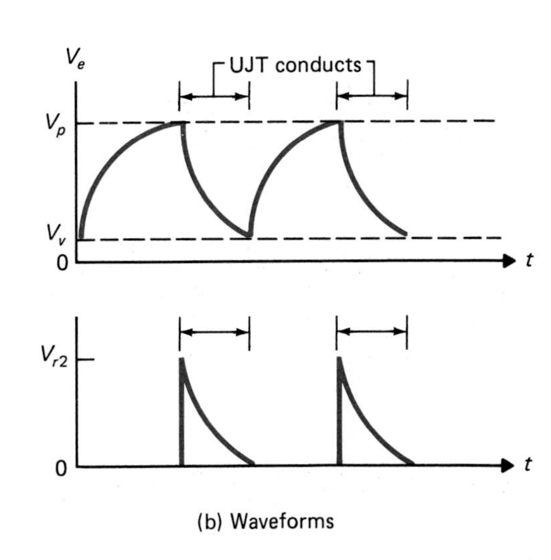

(b) Waveforms

1. The peak "triggering" voltage needed at the emitter lead of the UJT is determined by the supply voltage and the intrinsic standoff ratio for the UJT. The minimum holding voltage is also set by UJT specifications.
2. When the circuit first energizes, the UJT is nonconducting since emitter voltage is not at the proper voltage level. There is no voltage on R_2.
3. As the capacitor begins to charge ($R_1 \times C$ time constant circuit) the voltage on the emitter begins to rise to the value of V_P. Again the device is "off."
4. When the capacitor voltage reaches V_P the UJT begins to conduct. As it does, current flows (B_1 to B_2) and creates a drop on R_2. Also, emitter-to-B_1 current flows, discharging the capacitor. The voltage on R_2 follows the pattern of the capacitor's discharge voltage.
5. The capacitor continues to discharge, keeping the device "on" until the voltage on the capacitor drops below the minimum emitter voltage (V_V). At that time, the emitter current ceases to flow and the B_1-to-B_2 conduction ceases. Thus voltage to the resistor R_2 drops again to zero.
6. This process is repeated (charge/discharge of the capacitor), thus producing an oscillating (nonsinusoidal) waveform on the capacitor and a pulse-like waveform on the resistor R_2.

Programmable UJT

The programmable UJT (Figure 15.26) is physically similar to the UJT and operates somewhat like the UJT. However, the gate lead of the PUT is not used in the conventional triggering mode as seen in UJTs (or other triggered triacs, for that matter). The gate lead is used as a sensing point to determine the voltage level at which the PUT will begin conduction at its anode and cathode leads.

When the anode voltage is below the gate voltage (positive), the PUT is off and no conduction is seen. When the anode voltage rises to about 0.5 V above the level of voltage set at the gate lead, the PUT begins to conduct. Both anode and gate current flows at that time. (Gate voltage drops to about 0.5 V.) This conduction continues until the anode current falls below a minimum sustaining current (a device specification).

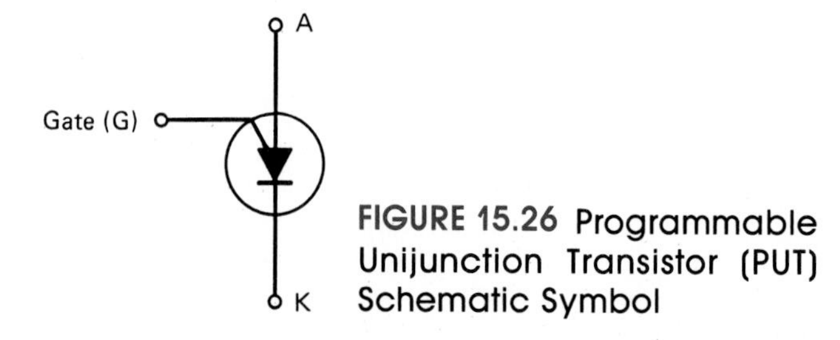

FIGURE 15.26 Programmable Unijunction Transistor (PUT) Schematic Symbol

The value of ''programmed'' voltage can be simply set by a voltage divider or by other conventional means. Again, the device will not conduct until that value of voltage is exceeded (by about 0.5 V) across the anode/cathode of the PUT.

[It is probably more logical to vision the PUT more as a programmable Shockley diode than as a UJT form of device. Its conduction from anode to cathode is more like that device. The voltage level set at the gate (''program voltage'') determines the threshold at which it will conduct. A variable ''overvoltage indicator'' circuit could be constructed by using the layout shown in Example 15.3. Variable control of the indicator lamp could be had by establishing a simple resistor/potentiometer (variable) voltage divider attached to a known voltage source (battery) and connecting this to the gate of the PUT.]

As can be seen, there are a wide variety of control devices available. Each has specific characteristics, making the group usable for a variety of control circuit applications.

REVIEW PROBLEMS

(1) Prepare a summary list of the names of the control devices listed at the beginning of this section, their schematic symbols, and a brief statement of their functions.

(2) Briefly contrast the function of the SCR to that of the triac. How are they similar? How do they differ?

SUMMARY

Both the circuitry and devices used for control functions were seen to differ dramatically from those used in amplification functions. Several new devices, each providing some new form of control function, were presented.

It was seen that the common inputs to control circuitry were either pulses or threshold voltages. These circuits were shown basically to provide an on/off function to control the current to a load device. Other control devices were seen to provide constant control functions and werre not of the triggered variety.

One of the most unique functions of most of the control devices was that they were triggered on by some minimum (or threshold) voltage level, but would switch off only when the controlled current dropped to a minimum level, and they thus could not be alternately triggered off.

KEY EQUATION

UJT voltage:

15.1 $V_P = \eta V_{BB} + V_D$

PROBLEMS

Section 15.1

15.1. Define the basic differences between control circuitry and amplification circuitry.

15.2. Illustrate a simple block diagram for both a closed-loop and an open-loop control system. Briefly contrast these two.

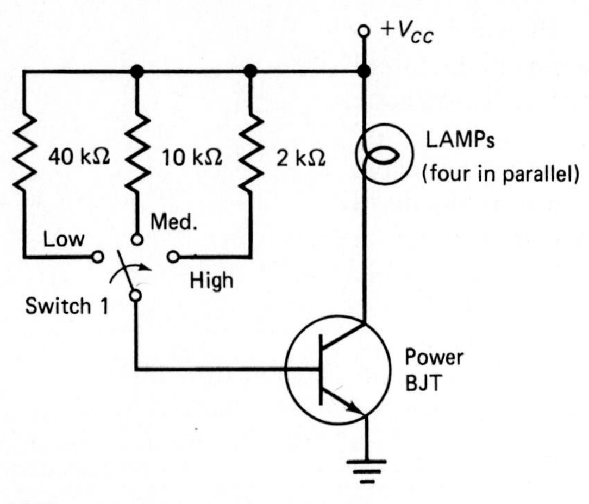

(a) Three-level brightness control for a clock radio dial light

(b) Photoresistive cell (see problem 15.4)

FIGURE 15.27 Control Circuit and Photoresistive Cell

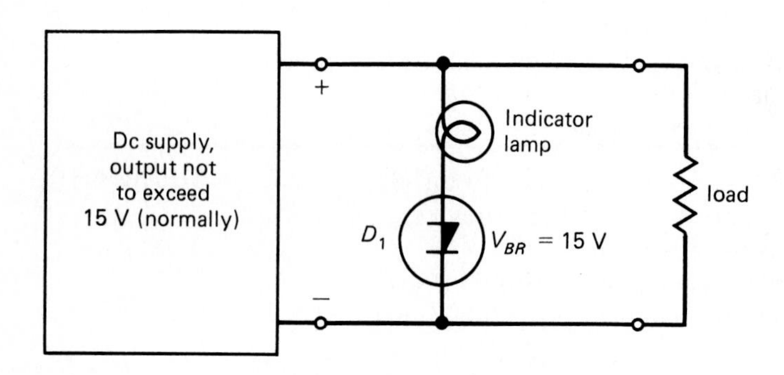

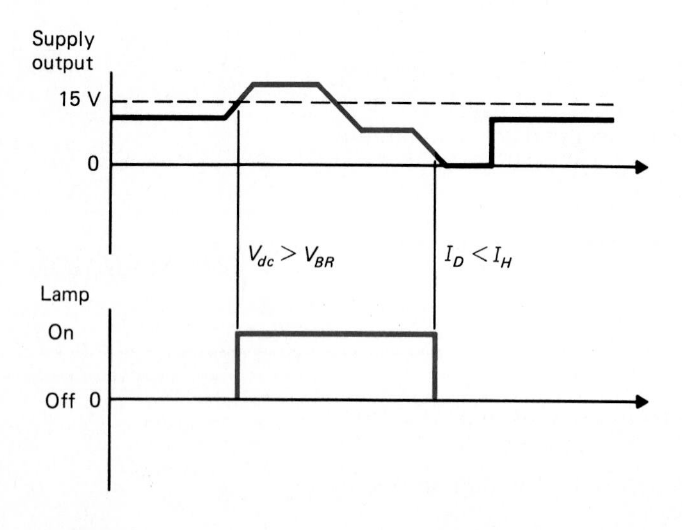

FIGURE 15.28

Section 15.2

15.3. Describe the basic operation of the control circuit shown in Figure 15.27. Why wouldn't this circuit be classified as an amplifier?

15.4. Convert the lamp control circuit shown in Figure 15.27(a) into an automatic lamp-level controller. The level of room lighting can be sensed by a photoresistive sensor [schematic symbol shown in Figure 15.27(b)]. The sensor has the following (inverse) resistance rating:

At full brightness: $R_{\text{sensor}} = 2\ \text{k}\Omega$
In darkness: $R_{\text{sensor}} = 45\ \text{k}\Omega$

15.5. Briefly describe thee difference between dc and pulsed control signals.

15.6. Briefly describe the difference between pulse rate and pulse-width control signals. (Use illustrations if they will aid in your description.)

15.7. Briefly describe the function of ac phase control. Be certain to include the concepts of phase shifting and threshold (pulse-like) triggering.

Section 15.3

15.8. Briefly compare the functions of the Shockley diode, the diac, and the silicon bilateral switch.

15.9. Briefly compare the functions of the silicon-controlled rectifier, the triac, and the silicon-controlled switch.

15.10. Briefly compare the UJT and PUT.

15.11. Which of the devices noted in problem 15.8 would be used to operate an ac overvoltage indicator as shown in Figure 15.28? Illustrate that device applied in a schematic such as that shown in Figure 15.28.

15.12. For the dc-operated SCR shown in Figure 15.29, what value must R_L be adjusted to for the SCR to turn off if it is once triggered by the pushbutton switch S_1?

15.13. If the load resistance in the circuit for Figure 15.29 were not adjustable, illustrate two circuit modifications that would allow the SCR to be reset once triggered.

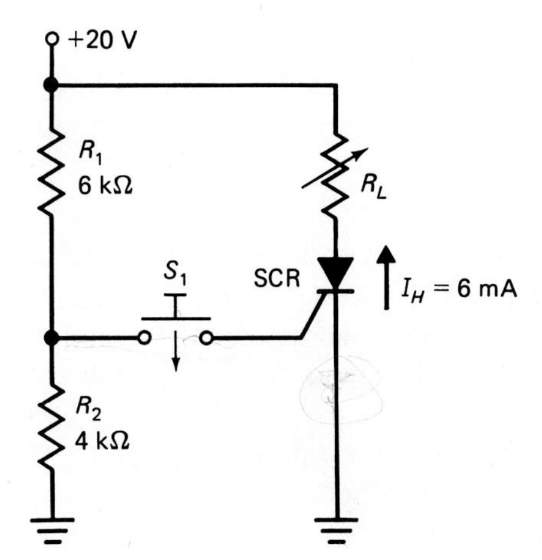

FIGURE 15.29 Pushbutton Control of SCR Latching Circuit

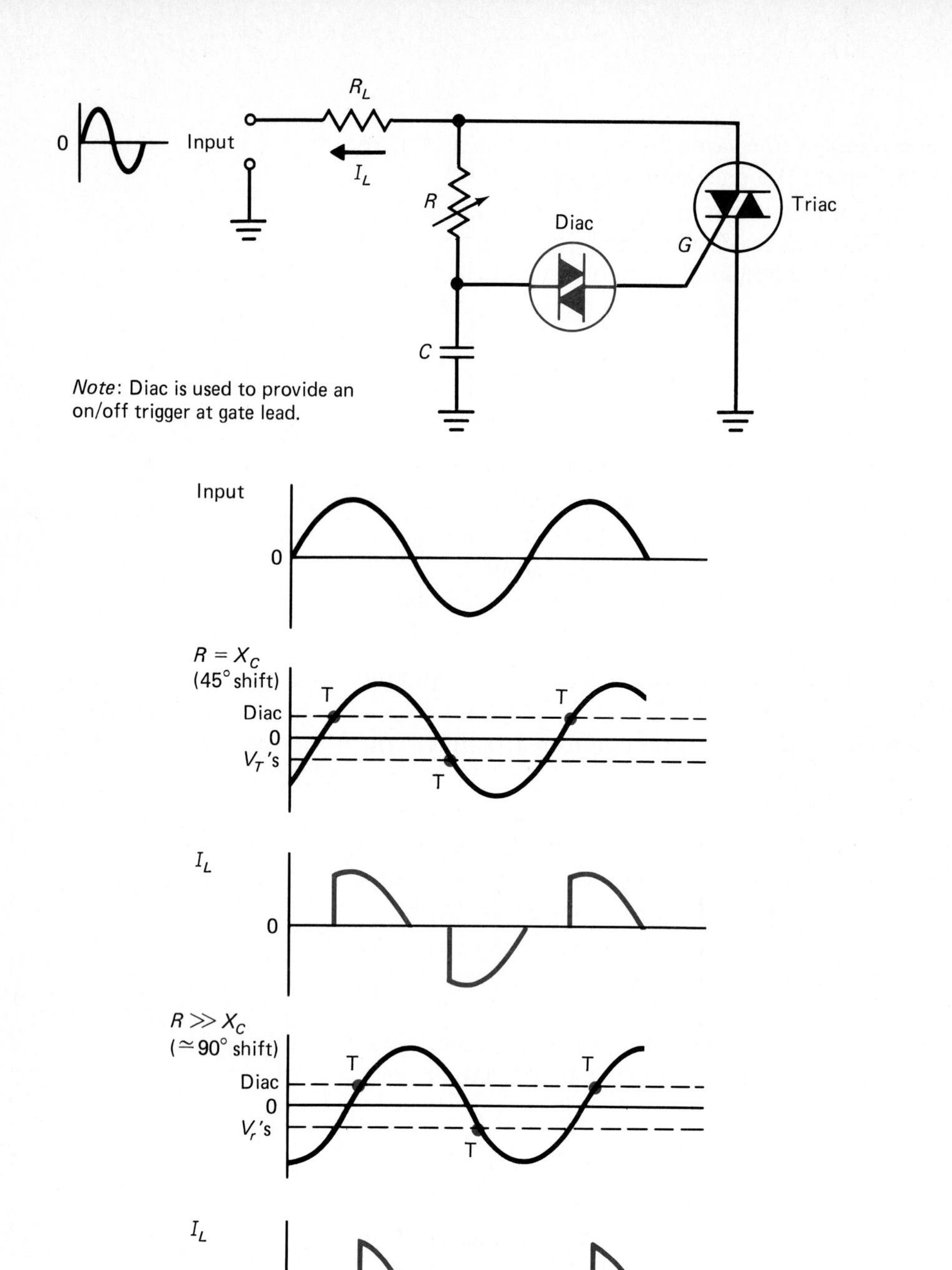

FIGURE 15.30

15.14. Briefly discuss the function of phase control as shown in Figure 15.30.

15.15. In Chapter 2 we discussed the development of a light dimmer circuit. Return to that chapter and describe, in your own words, the function of the circuit shown in Figure 2.29, relating it to the block diagram shown in Figure 2.28.

15.16. Discuss the function of the circuit (and devices) shown in Figure 15.31 (a control circuit used to operate a high current load).

15.17. Illustrate a circuit for a variable-frequency UJT oscillator.

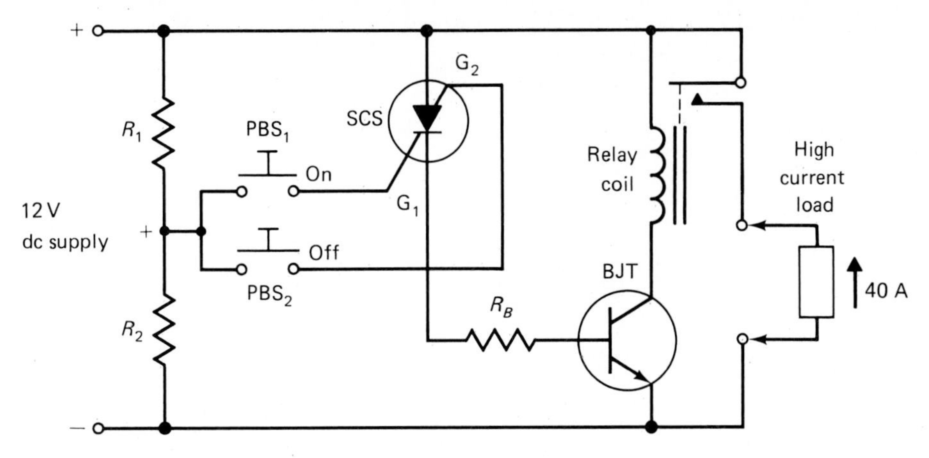

PBS1 and PBS2 are remotely located pushbutton switches.

FIGURE 15.31 Remote Pushbutton Power Supply "Connect/Disconnect" Circuit

ANSWERS TO REVIEW PROBLEMS

Section 15.1

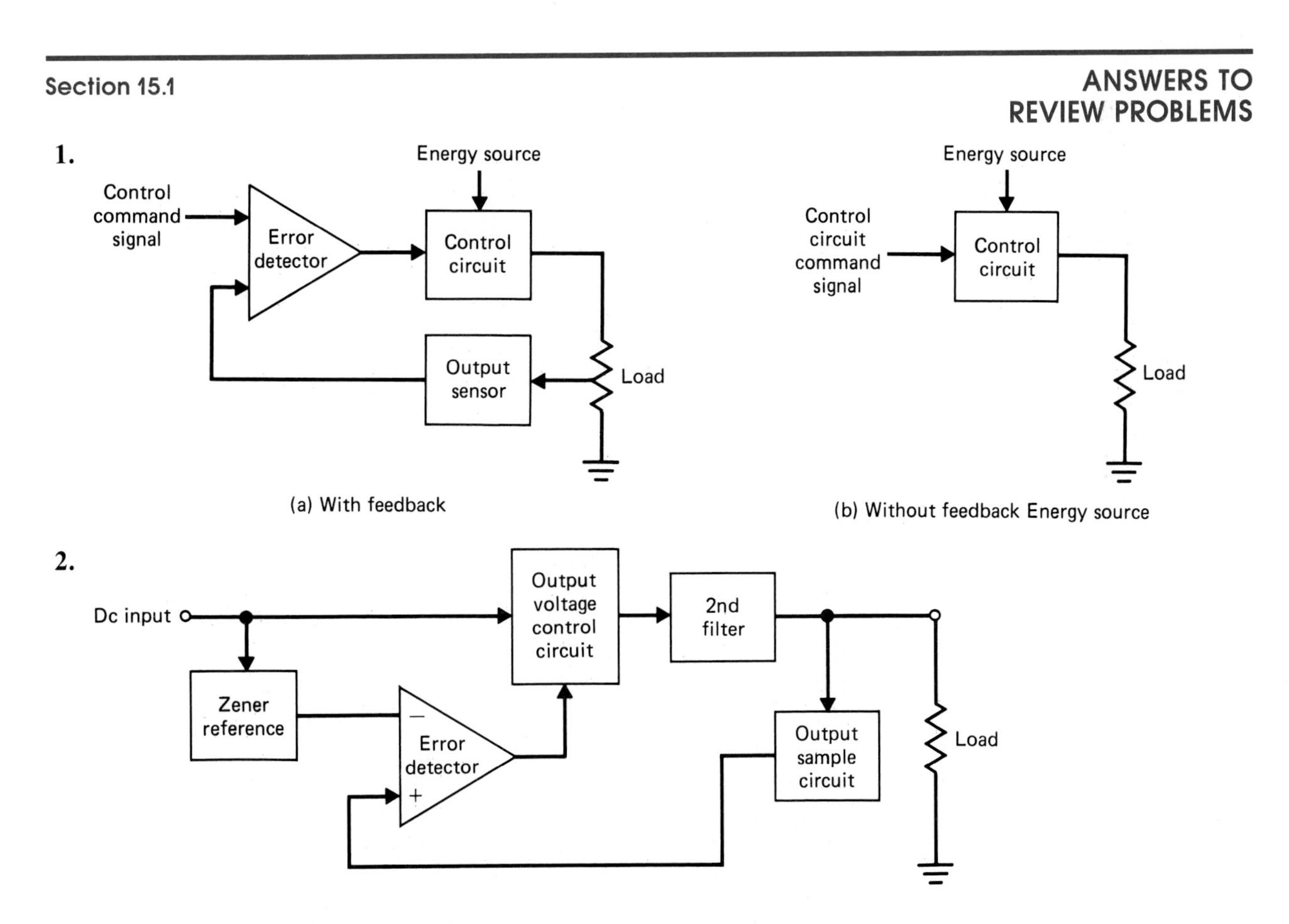

The switching regulator follows the basic function of a closed-loop control circuit. The output value is sensed and fed back to a comparator circuit. This value is compared to a reference voltage. Based on the error signal, the voltage to the second filter is controlled by the variable pulse-width oscillator and transistor circuit.

Section 15.2

1.

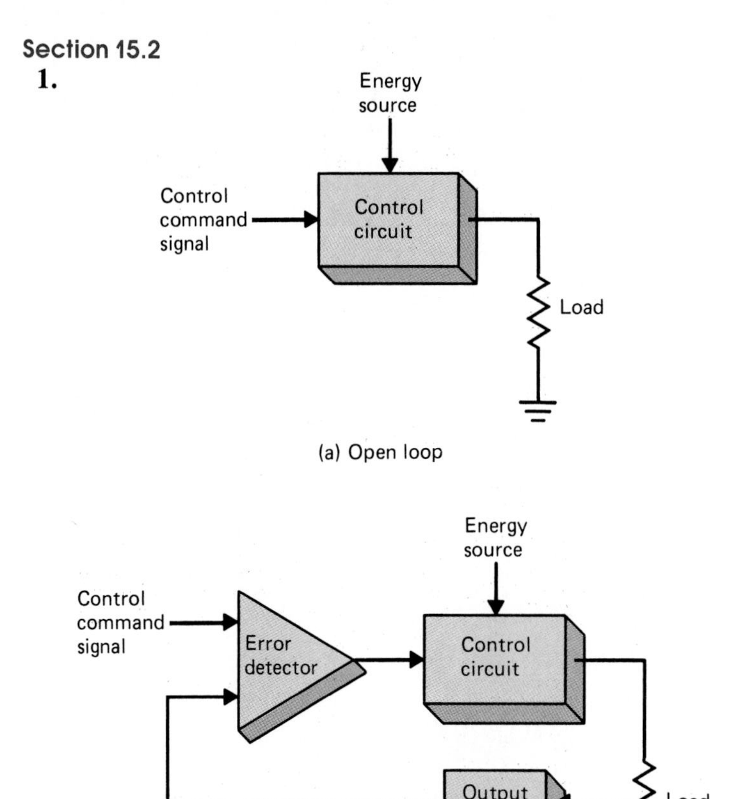

(a) Open loop

(b) Closed loop

2. Open-loop control produces control of the output quantity based on the value of a single control signal. Closed-loop control uses feedback from the controlled load which is compared to a preset control signal. The control device produces a response that is proportional to the amount of error (difference) between these two signals, thus attempting to maintain a match between the preset value and actual load function.
3. Dc amplitude, pulsed, and ac phase control signals. Dc amplitude control signals produce a continuous control output, the level of which determines the amount of energy delivered to the load. Pulsed control, unlike dc control, does not produce a continuous, constant level. Either a pulse rate or pulse width are adjusted to establish an average (over time) amount of energy delivered to the load. Ac phase control, although similar to pulse-width control, adjusts the duration of the ''on'' time of the control device based on a phase difference between two (same frequency) ac signals, thus adjusting the average amount of energy delivered to the load based on this phase difference.

Section 15.3

1. See the listing for each device in this section to compose a summary listing.
2. The SCR is similar to the triac in that the application of a triggering signal (of sufficient level) to its gate lead will cause conduction through the device. Also, the SCR will remain on, as does the triac, as long as load current (above a minimum) is drawn. The SCR differs from the triac in that it can conduct in only one direction (of current flow); a "reverse polarity" on the SCR, regardless of triggering, will not permit current flow. The triac can conduct current in either direction as long as proper triggering is present at the gate lead. Neither device can be "turned off" via a triggering signal at its gate lead.

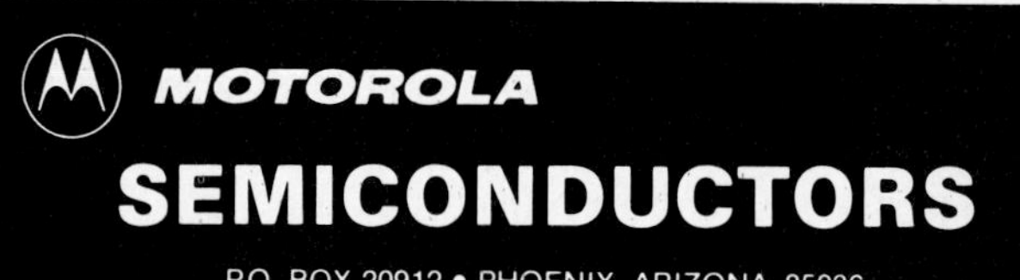

MKP9V120 MKP9V240
MKP9V130 MKP9V260
MKP9V270

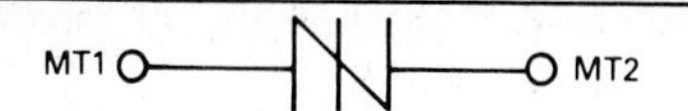

PLASTIC SIDAC HIGH VOLTAGE BILATERAL TRIGGER — HIGH VOLTAGE TRIGGERS

... designed for direct interface with the ac power line. Upon reaching the breakover voltage in each direction, the device switches from a blocking state to a low voltage on-state. Conduction will continue like an SCR until the main terminal current drops below the holding current. The plastic axial lead package provides high pulse current capability at low cost. Glass passivation insures reliable operation. Applications are:

- High Pressure Sodium Vapor Lighting
- Strobes and Flashers
- Ignitors
- High Voltage Regulators
- Pulse Generators

PLASTIC SIDAC HIGH VOLTAGE BILATERAL TRIGGER

0.9 AMPERE RMS
120 TO 130 VOLTS
240 TO 270 VOLTS

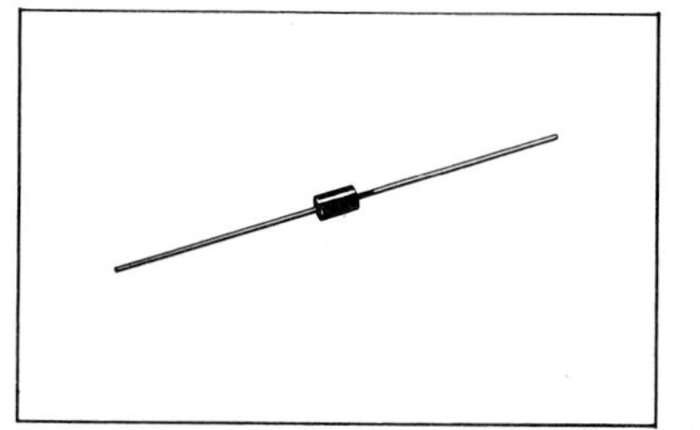

MAXIMUM RATINGS

Rating	Symbol	MKP9V120 MKP9V130	MKP9V240 MKP9V260 MKP9V270	Unit
Off-State Repetitive Voltage	V_{DRM}	±90	±180	Volts
On-State Current RMS (T_L = 80°C, LL = 3/8″, conduction angle = 180°, 60 Hz Sine Wave)	$I_{T(RMS)}$	0.9		Amps
On-State Surge Current (Nonrepetitive) (60 Hz One Cycle Sine Wave, Peak Value)	I_{TSM}	4.0		Amps
Maximum Rate of Change of On-State Current	di/dt	90		Amps/μs
Operating Junction Temperature Range	T_J	−40 to +125		°C
Storage Temperature Range	T_{stg}	−40 to +150		°C
Lead Solder Temperature (Lead Length ≥ 1/16″ from case, 10 seconds max)	—	230		°C

THERMAL CHARACTERISTICS

Characteristic	Symbol	Unit	
Thermal Resistance, Junction to Lead LL = 3/8″	$R_{\theta JL}$	40	°C/W

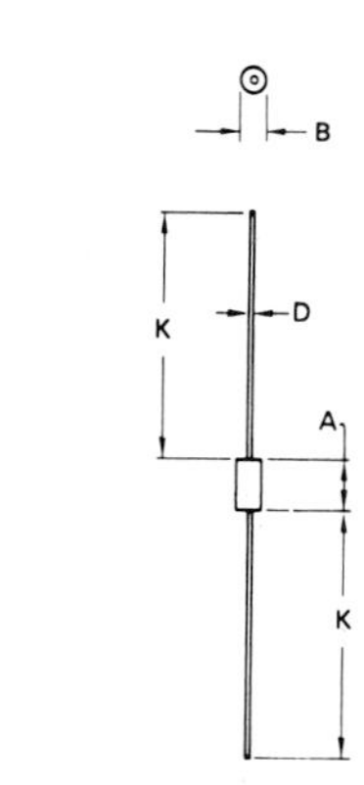

NOTES:
1. ALL RULES AND NOTES ASSOCIATED WITH JEDEC DO-41 OUTLINE SHALL APPLY.
2. POLARITY DENOTED BY CATHODE BAND.
3. LEAD DIAMETER NOT CONTROLLED WITHIN "F" DIMENSION.

DIM	MILLIMETERS MIN	MILLIMETERS MAX	INCHES MIN	INCHES MAX
A	5.97	6.60	0.235	0.260
B	2.79	3.05	0.110	0.120
D	0.76	0.86	0.030	0.034
K	27.94	–	1.100	–

CASE 59-04

DS3625

MKP9V120 • MKP9V130 • MKP9V240 • MKP9V260 • MKP9V270

ELECTRICAL CHARACTERISTICS (T_J = 25°C unless otherwise noted; both directions)

Characteristic	Symbol	Min	Typ	Max	Unit
Breakover Voltage	V_{BO}				
MKP9V120		110	—	125	
MKP9V130		120	—	135	
MKP9V240		220	—	250	Volts
MKP9V260		240	—	270	Volts
MKP9V270		250	—	280	Volts
Repetitive Peak Off-State Current (60 Hz Sine Wave, V = V_{DRM})	I_{DRM}	—	—	5.0	μA
T_J = 125°C		—	—	50	μA
Forward "On" Voltage (I_T = 1.0 A)	V_{TH}		1.3	1.5	Volts
Dynamic Holding Current (60 Hz Sine Wave)	I_H	—	—	100	mA
Switching Resistance	R_S	0.1	—	—	kΩ
Breakover Current (60 Hz Sine Wave)	I_{BO}			200	μA

FIGURE 1 — MAXIMUM LEAD TEMPERATURE

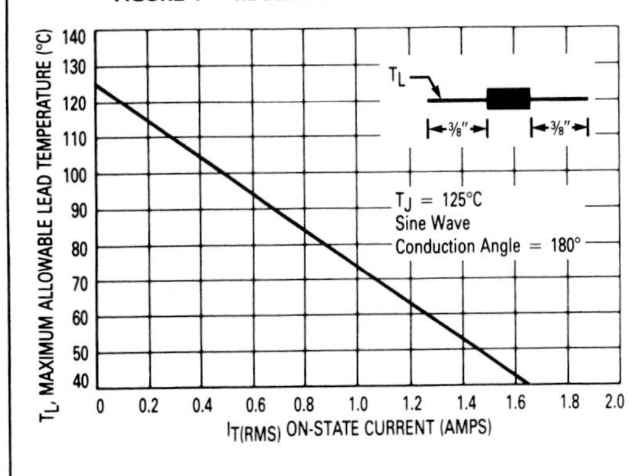

FIGURE 2 — MAXIMUM AMBIENT TEMPERATURE

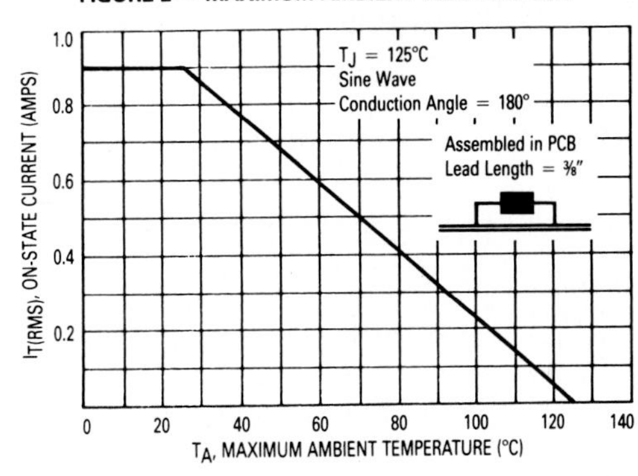

FIGURE 3 — TYPICAL ON-STATE VOLTAGE

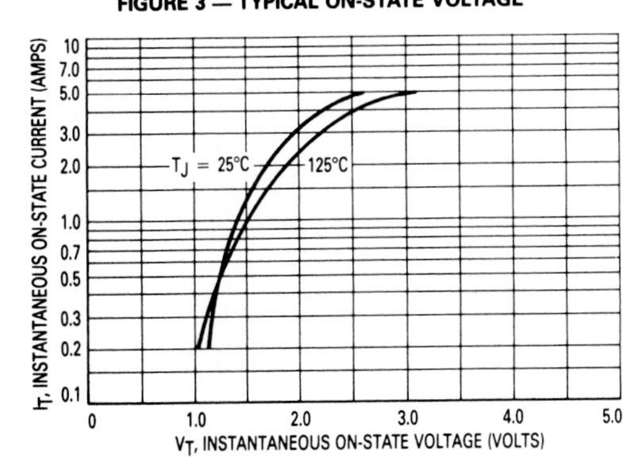

FIGURE 4 — POWER DISSIPATION

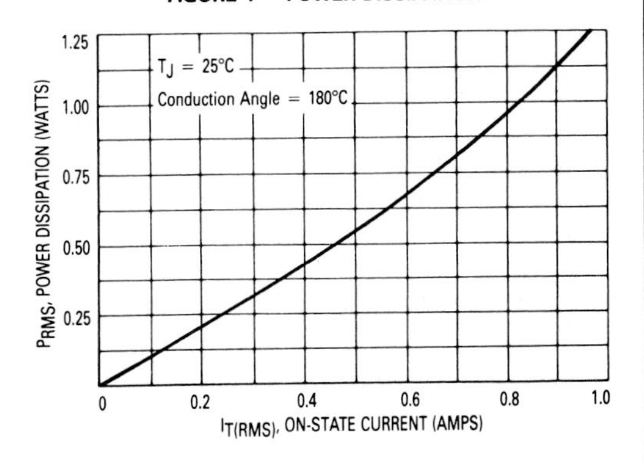

16 Special-Purpose Devices and Applications

Objectives

Upon completing this chapter, the reader should be able to:

- Identify, and describe applications for, various special semiconductor devices.
- Illustrate and describe various semiconductor circuits.
- Describe potential new applications/devices.

Introduction

In this chapter we present a collection of various semiconductor devices and applications and special circuit applications of more common semiconductor elements. A number of new devices and unique applications are shown. At the end of this chapter, guidelines are given as an aid to assisting in understanding the function of other new devices and circuits. ■

16.1 SPECIAL-PURPOSE SEMICONDUCTOR DEVICES

In this section several special-purpose semiconductor devices are investigated. The objective will be to investigate their characteristics and view typical applications. Since many of these devices are quite different from each other, they will be presented simply on a type-by-type basis.

Photodiodes

The photodiode (Figure 16.1) is a light-sensitive device. It is applied in a circuit in its reverse direction. It will be remembered that a standard diode exhibits a relatively constant value of low reverse leakage current. The reverse current of the photodiode is dependent on the amount of light (lumens) to which it is exposed. The photodiode has a transparent window covering the internal junction so that light may pass to the optically sensitive surface inside the device. As the diode is exposed to increasing intensities of light, greater reverse current will flow. The photodiode is similar in function to the photoresistive cell, except that it exhibits the single polarity sensitivity of regular diodes.

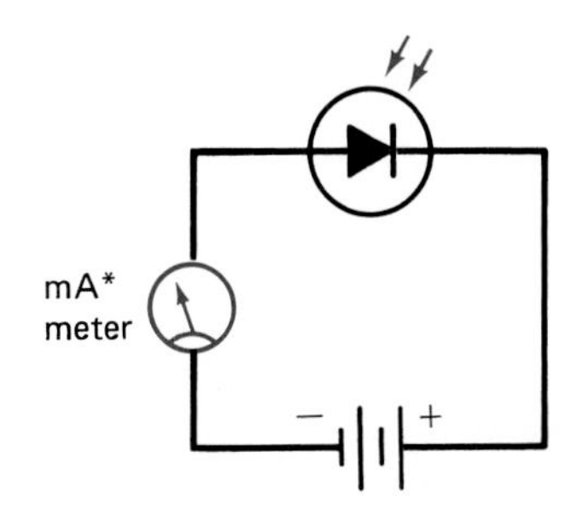

*May be recalibrated in Lumens

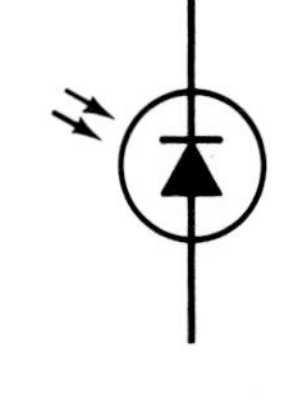

Photodiode

FIGURE 16.1 Photo Diode Circuit and Schematic Symbol

Phototransistors

The phototransistor exhibits characteristics similar to a standard BJT, but instead of depending on base current to control collector current, the phototransistor uses light as an input quantity. As light levels increase, the amount of collector current controlled is increased. Since the phototransistor produces "gain" (i.e., a greater energy output than is supplied at the input), it produces a more sensitive response to changes in light than would a passive device (photoresistive cell, photodiode).

In some cases, a phototransistor may be supplied with an electrical base lead. This lead is used (if necessary) to establish an average output current or to offset the collector flow. This electrical input can simply be considered in "series" with the optical input.

The simple light-activated relay circuit shown in Figure 16.2 operates from the output of the phototransistor. As that transistor conducts due to increased light levels, the output current is used to turn-on (saturate) transistor Q_2, thus activating the relay coil.

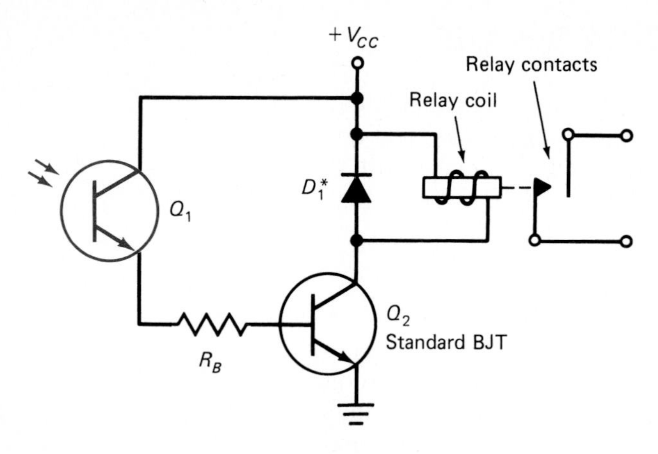

*D_1 is used to dissipate the inductive current produced when the relay is deenergized.

FIGURE 16.2 Optically Triggered Relay Circuit

Photo-Darlington Transistor

As will be remembered from the coverage of standard Darlington transistors, the Darlington is equivalent to two transistors connected in a series configuration (Figure 16.3). The gain of such a combination is typically in the thousands. As a photo-Darlington, this configuration produces much higher sensitivity to light levels than that of a standard phototransistor.

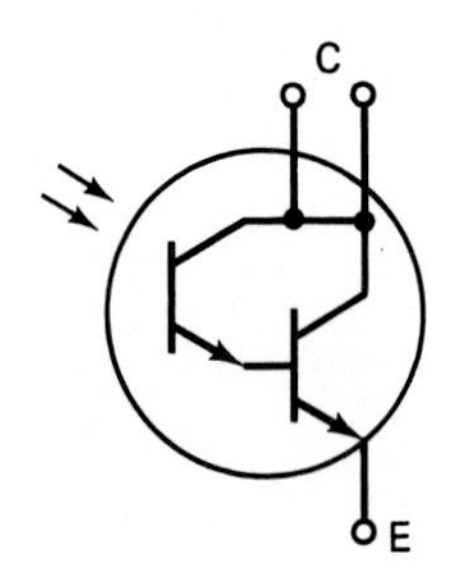

=

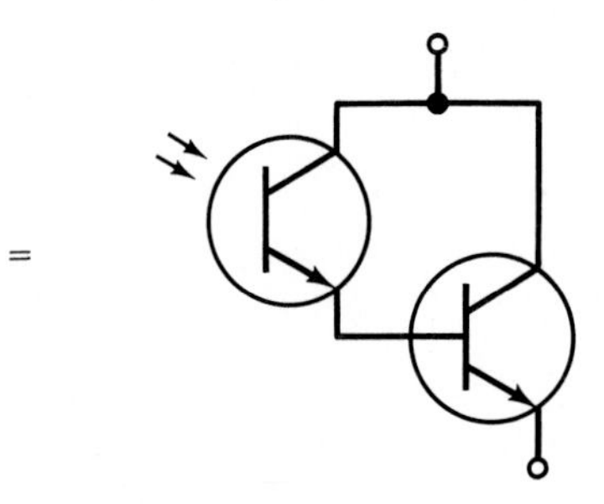

FIGURE 16.3 Photo-Darlington Circuit (IC and Discrete Configurations)

Light-Activated SCR

The light-activated SCR (LASCR) simply operates with an optical input to the gate lead, instead of an electrical input (Figure 16.4). Most LASCRs also have an electrical gate lead so that they can be alternately triggered with an electrical input if necessary. (There are also optical triacs, working in the same manner.) Minimum threshold light levels are needed to trigger the LASCR. (Turn-off is the same as for a standard SCR; anode current must fall to a minimum level.)

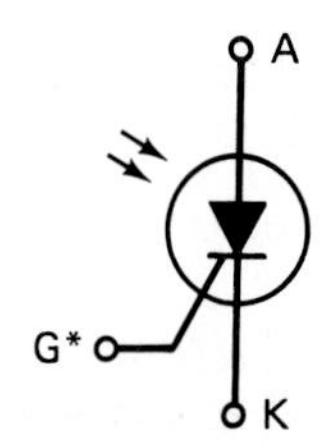

*Electrical gate may be omitted.

FIGURE 16.4 Light-Activated SCR (LASCR) Schematic Symbol

Special note about light-activated semiconductors: Most light-activated semiconductors have a limited spectral range of optical response. The spectral range coordinates with the wavelength of light applied to the device. (The wavelengths relate to the color of the light. For more details, consult a reference dealing with optics or physics.) The most common spectral range of response for semiconductor devices is the infrared range. This range is not visible to the human eye. The response of such devices to the light spectrum that is visible is often quite limited.

Light-Emitting Diodes

When forward current is passed through an LED (Figure 16.5), due to its special chemical composition inside, it emits light. This production of light is a direct effect of the current passing through the device. LEDs are not usable as standard diodes, and no effect is seen when they are reverse biased. It is necessary to limit current flow through the LED by the use of a series-limiting resistor. Specifications for LEDs include:

Maximum forward current
Forward voltage drop (different from standard diodes)
Emitted light color (typical colors: infrared, red, yellow, and green)

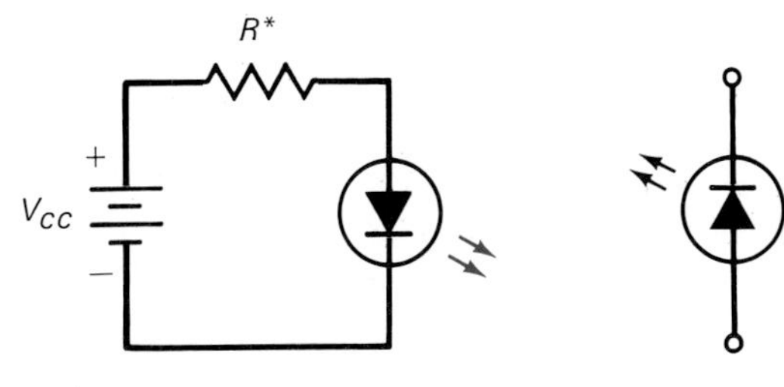

*Used to limit forward current.

FIGURE 16.5 Light-Emitting Diode (LED) Circuit and Schematic Symbol

The amount of light emitted from an LED is relatively small when compared to an incandescent light bulb of a similar physical size drawing the same current. A major advantage of the LED over a conventional lamp is that it is a solid-state device and therefore will not be subject to the mechanical wear that is common in a lamp. An additional advantage is that the spectrum of light emitted from an LED is usually limited. This spectrum can be matched to the spectrum that a photodiode or phototransistor will respond to. Therefore, an LED can be used to transmit information optically (with light) to a photodiode or phototransistor. The basic process of optical transmission of data is shown in Figure 16.6.

Optical Isolators (Opto-Isolators)

Optical isolators (sometimes called optical couplers; Figure 16.7) are used to provide total electrical isolation between one electronic circuit and another. Basically, the signal applied to the input of the coupler is converted to an optical (light) signal by a form of LED. This optical signal is passed to the

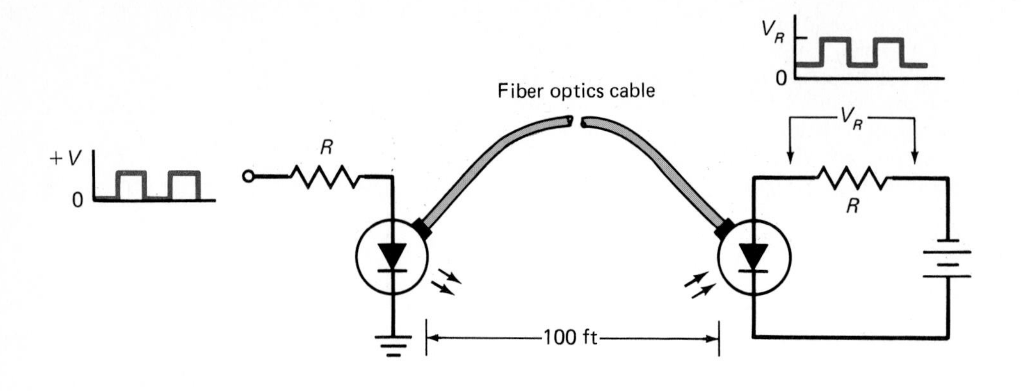

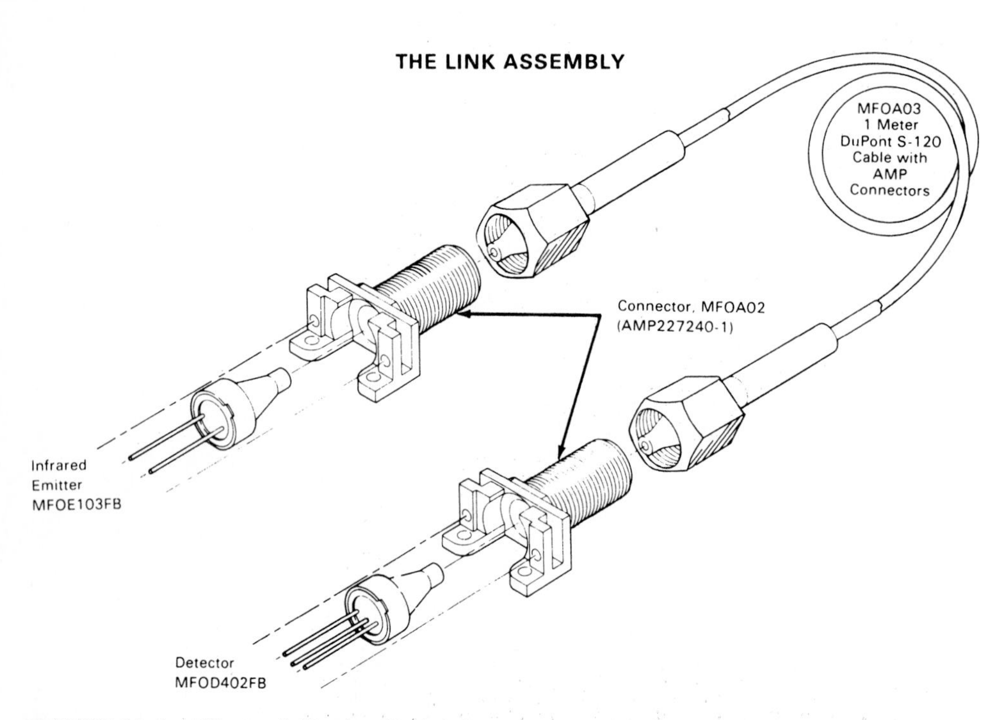

FIGURE 16.6 LED and Photodiode Coupled through Fiber Optics Cable (Reprinted by Permission of Motorola Semiconductor Products Inc.)

input of a phototransistor (or other optically sensitive device). The photodevice converts the light input into an electrical signal.

The reason that these devices are called isolators is that it is impossible for conditions occurring on the output device to affect the input. Therefore, there is no reflection of load resistance, load voltages, or any other quantity between the input and output of the isolator. Some isolator devices are designed to work as a pulse (or triggered) type of device, others can handle the passage of an analog signal from input to output. Optical isolators can be used with a variety of output devices, thus producing a variety of output functions. The optical isolator's circuit parts are built into one (usually DIP type) package, thus eliminating the chance for outside interference.

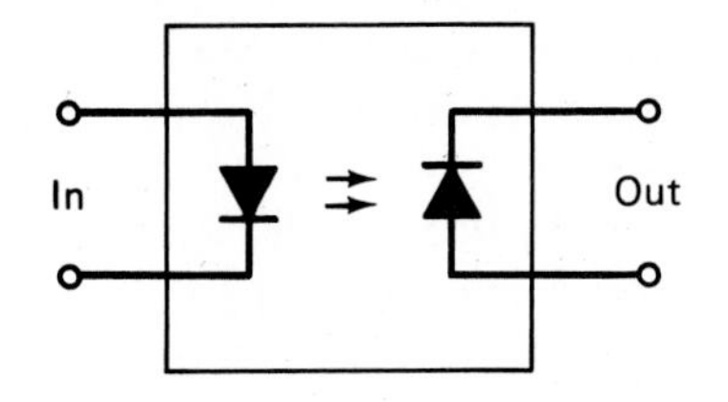

(a) Photodiode isolator (basic type)

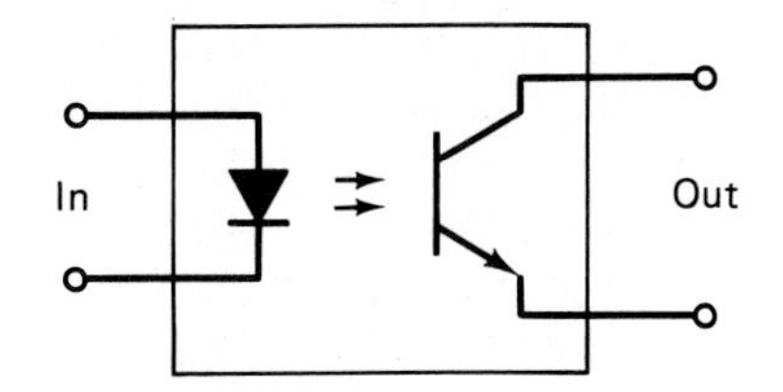

(b) Phototransistor output

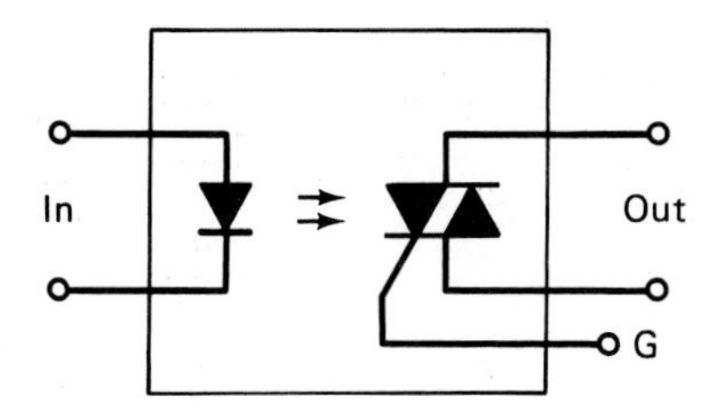

(c) Phototriac output (with electrical gate lead)

FIGURE 16.7 Three Forms of Opto-Isolators

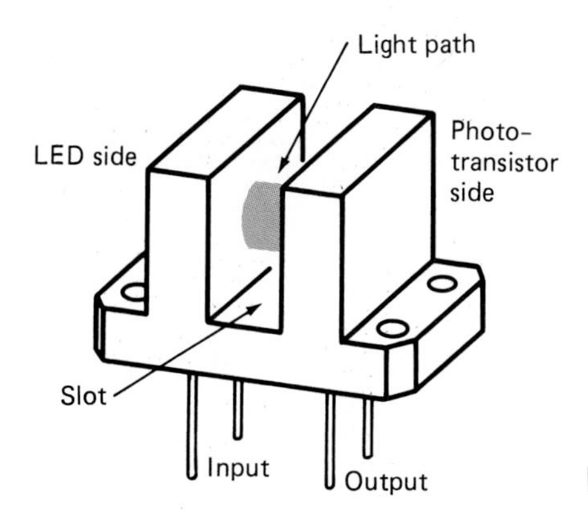

FIGURE 16.8 Optical Interruptor

Optical Interruptor (Special Form of the Isolator)

An optical interrupter is built in the same fashion as an optical isolator, except that the optical path between the LED and output device is mechanically exposed to the outside world. As such, the interrupter is used to sense the presence or absence of a mechanical object.

As shown in Figure 16.8, the opto-interrupter has a physical opening between the LED and opto-transistor output device. For typical operations, the LED is simply energized with a constant dc source. A light beam passes from the LED through the physical opening to the output device. If the beam is interrupted, the output reflects this interruption as a change in conduction for the output device. Opto-interrupters have a wide number of applications as input sensors for electronic circuits.

EXAMPLE 16.1

In Figure 16.9 an opto-interrupter is used to supply rotational information from the shaft of a motor. As the motor rotates it turns the disk which passes through the interrupter. As the regular surface of the disk passes through the interrupter it breaks the light beam from the LED to the phototransistor. The phototransistor is therefore off (cut-off) and no output current flows. When the slot in the disk passes through the interrupter, the beam of light passes on to the phototransistor, causing it to conduct. The output of the system is a series of pulses, one for each rotation of the disk (motor).

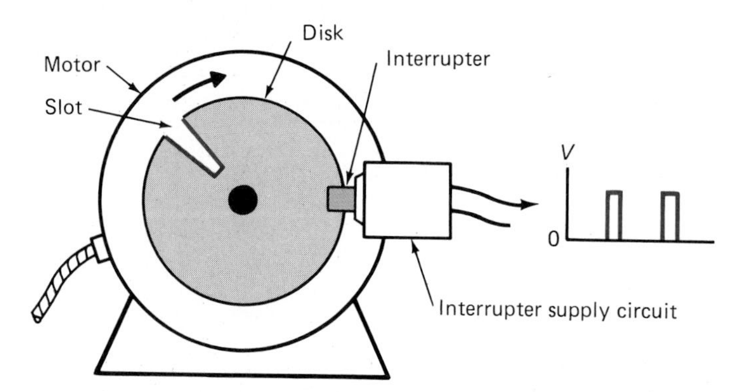

FIGURE 16.9 Interruptor Used to Monitor the Speed of a Motor

Varactor Diodes

A varactor diode (Figure 16.10) is a specially constructed diode used as a voltage-variable capacitor. This may seem unusual since the action of the diode was never described in terms of capacitance. It will be remembered from previous chapters that one concern about BJT and FET devices was their capacitance, a prime cause for limited high-frequency response. This capacitive effect is used as a by-product of the diode's function to produce the varactor diode effect.

The varactor diode is always used in its reverse-bias mode, where the capacitance figure is most predominant. As increasing amounts of dc reverse bias is applied to the device, the magnitude of capacitance changes. (Capacitance is inversely proportional to voltage.) Capacitances in the picofarad range are typical. Thus the capacitance of the varactor is inversely proportional to the reverse-bias voltage across the device.

Varactor diodes are used in many modern radio and television tuning (resonant) circuits. (Since they are low-capacitance devices, they are not usable in low-frequency circuits.) They have three distinct advantages over traditional variable capacitors. First, since they are voltage variable, using precision resistors or potentiometers to "tune" the resonant circuit via voltage dividers allows for easier "remote control" of tuning. Also, the varactor is

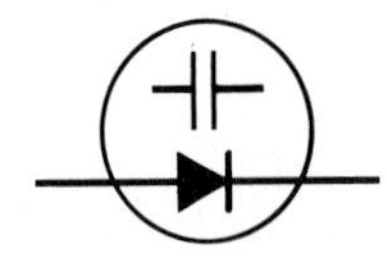

FIGURE 16.10 Varactor Diode Schematic Symbol

physically much smaller than a variable capacitor and thus can be used to produce smaller circuits (or be built into integrated circuits). Finally, since the varactor is a solid-state device, it is less sensitive to environmental conditions that could cause a variable capacitor to "drift" off its value.

Other Special-Purpose Diodes

There are a number of special-purpose diodes, designed to meet the needs of specialized applications. These diodes include the Schottky diode, tunnel diode, PIN diodes, IMPATT diodes, Gunn diodes, and various other forms of diodes and diode-like devices. The application of these devices can be found by referencing manufacturers' data sheets, application notes, and reference books.

Transient Suppressors

The transient suppressor is a special version of a zener diode (or dual zener diodes). The transient suppressor is a device designed to protect a system against the occasional voltage surges due to input or system faults and is usually used in power supply circuitry. The transient suppressor is used typically in parallel with the power supply lines. If the voltage in the circuit exceeds the rating of the suppressor, the suppressor dissipates the excess voltage internally (as heat) and maintains a constant voltage output (much like a zener diode).

Since the suppressor internally dissipates the overvoltage energy, it is not necessary to have a series resistance as is typical for the zener diode. An ac transient suppressor is like two zener diodes in series (see Figure 16.11), which will limit the positive or negative peak voltages applied across it. Transient suppressors are not used to hold a voltage at a constant level, such as a zener regulator does, but to suppress (consume) an occasional surge of voltage in a circuit. These surges may be the result of faults in an ac line input or could be caused in dc circuits by inductive transient voltages.

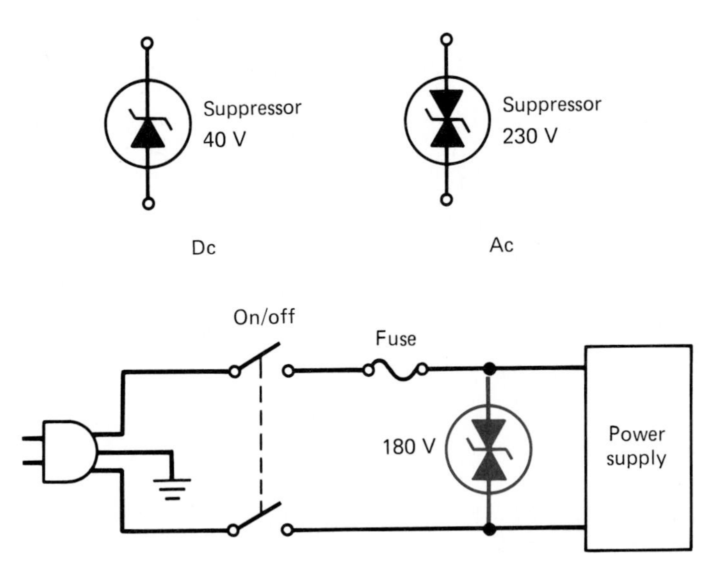

FIGURE 16.11 DC and AC Surge Suppressors and Application Circuit

Current Mode Op-Amps

The conventional op-amp operates as a voltage amplifier. Input voltages are sensed at the input of the device and are multiplied by the voltage gain of that device. Differential input voltages are handled as presented in past chapters. The current-mode op-amp (also called a current-mirror amplifier or Norton amplifier; Figure 16.12) deals with input currents in a special manner and responds to voltage inputs as well. The output of this form of op-amp is a voltage level that is proportional to the level of voltage at the input terminal. There are two major advantages of this form of amplifier over the traditional op-amp. First, current-mode op-amps operate on a single power supply and the output can have a dc offset much like that of a biased BJT amplifier.

This basic presentation may seem a bit unusual since it was stated that the amplifier is current operated at the input, but the gain is stated as a voltage-gain figure. Primarily, this amplifier is used in a differential mode (a nonground input at each terminal). The following is a step-by-step description of how this amplifier deals with input currents and then produces an output voltage when used as an inverting amplifier:

1. A constant-current value is input to the noninverting terminal of the device (a reference current).
2. The input current (signal) is applied to the inverting terminal of the device.
3. The device then mirrors (duplicates) the current applied to the noninverting terminal back to the inverting terminal. This process then sets the currents equal to each other. Note that the level of current at the inverting terminal is ''forced'' to be equal to the reference current at the noninverting terminal.
4. Voltage gain for the amplifier is established by multiplying the voltage at the inverting terminal by the A_v of the op-amp.

The actual application of this amplifier and how currents and voltages are related is still not clearly visible unless this device is placed in a typical application. In Figure 16.13 the current-mode op-amp is used as a typical inverting amplifier. The reference current is set by the value of R_3 and is flowing from the noninverting terminal. Due to current mirroring, the current flow out of the inverting terminal must be equal to this quantity. Since the capacitor (C_1) blocks dc current, the only path for this current must be through the output terminal of the op-amp, passing through R_2. Thus, without a signal input, the current through R_2 is equal to the current through R_3. The flow of current through R_2 causes a basic voltage drop on R_2, and this drop is reflected as a voltage potential on the output terminal. This potential remains constant regardless of the value of the signal (or of any gain values for the op-amp). When the signal is applied, it produces a voltage on the inverting terminal

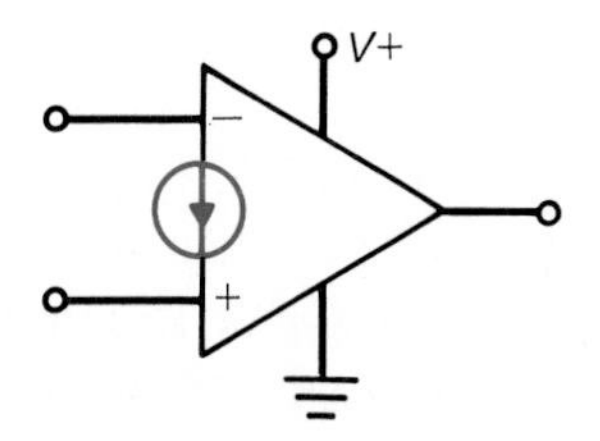

FIGURE 16.12 Norton (Current Mirror) Op-Amp

FIGURE 16.13 Application of Norton Amplifier

which is amplified by the voltage gain of the amplifier as an output. This output, then, is offset by the R_2 voltage drop. The voltage gain of this amplifier is simply set by the R_2-to-R_1 ratio ($A_v = -R_2/R_1$) as would be done for a standard op-amp.

If desirable, the current-mode amplifier may be used as a noninverting amplifier (but the circuitry is more complex). This amplifier can also be used where accurate sensing of current levels are needed. Some sensing or input devices are current sensitive, and thus this form of amplifier would be preferred over a standard op-amp in such an application.

> *Final note:* For the sake of clarity one functional feature of this current-mode op-amp was omitted from the presentation above. At each input terminal there is a standard "forward-bias" drop. This typically 0.7-V loss must be included in any calculations of bias potentials or current levels.

There are more forms of special-purpose semiconductor devices currently on the market. It can be assumed that there will be even more developed in the future. In this section we have summarized some of the more popular, currently used special-purpose semiconductors. Also, by reading descriptions of these devices, the reader should be able to form similar evaluations of devices made available in the future.

REVIEW PROBLEMS

(1) Prepare a list of the devices presented in this section which includes the device names, their schematic symbols, and a short note on the typical application of such a device.
(2) Compare the function of a LASCR to a standard SCR.
(3) Indicate how a phototransistor and LED could be used to make a smoke detector. (The buzzer in a smoke detector simply operates on a dc voltage.) What single device could be substituted for the phototransistor/LED combination?

16.2 SPECIAL APPLICATIONS OF STANDARD SEMICONDUCTOR DEVICES

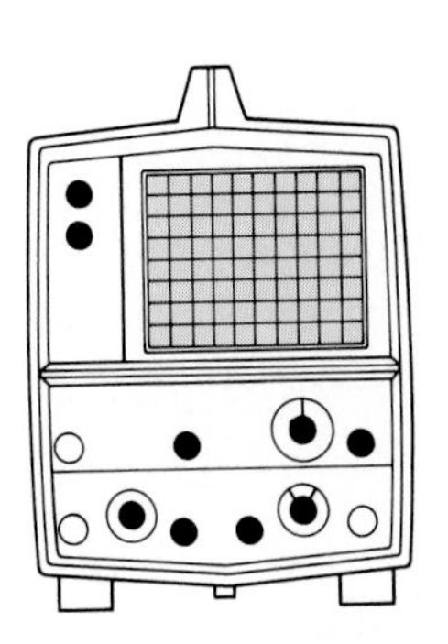

So far in this book relatively common applications of standard semiconductor devices have been presented. In this section several unique circuits using these devices are presented. Since these applications are distinct, a simple list of them will be presented.

Using an FET as a Voltage-Variable Resistor

Remember that an FET accepts a voltage on its gate lead and permits a current flow through the drain/source circuit in proportion to that voltage. As such, the FET can be applied as a voltage-variable resistor. See Figure 16.14, where an input voltage is applied to the gate lead of the FET. The conductivity through the drain/source leads is inversely proportional to that voltage. [This application is valid only for relatively low values of drain-to-source voltages, though (typically 1 V or below). At higher levels of drain-to-source voltage, the VVR function is invalid.]

The use of an FET as a VVR (voltage-variable resistor) is seen when attempting to convert voltage-producing sensors which have little current capacity to resistive like sensor which can handle some current levels. A second application is to use an FET as a VVR in a resonant circuit to adjust the bandwidth of the resonant circuit using a dc voltage level. (The effect of resistance in a resonant circuit as it relates to bandwidth via the Q of the circuit is explained in textbooks dealing with circuit analysis.) See also the discussion of AGC circuits in this section.

Clippers and Clampers

A clipper is a circuit that is used to limit the level of a voltage or signal above or below some predetermined value. A clamping circuit is used to offset an ac voltage with a dc level. A clamper is also often called a dc restorer.

The circuits and waveforms shown in Figure 16.15 illustrate clipper circuits and their relative outputs. Each of the circuits function by using the diode's forward/reverse conduction effect along with a "bias" battery. In cases where the diode is reverse biased by the battery (through the load resistor), the diode will conduct only when the input signal exceeds the bias potential (plus the normal diode drop). In cases where the diode is forward biased by the battery, a constant dc level is seen at the output unless the input signal exceeds the dc bias on the diode and thus reverse biases it. In either case, if the dc supply were made variable, the clipping effect would also be variable. Clipping is usable when it is necessary to place a limit on the level of an ac signal.

A clamping circuit, as shown in Figure 16.16, uses the rectifying action

FIGURE 16.14 Voltage-Variable Resistance (VVR) Curve for a JFET and Identification of Significant JFET Values

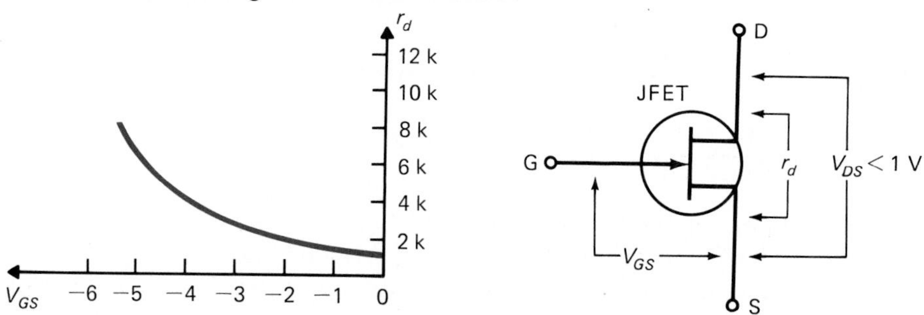

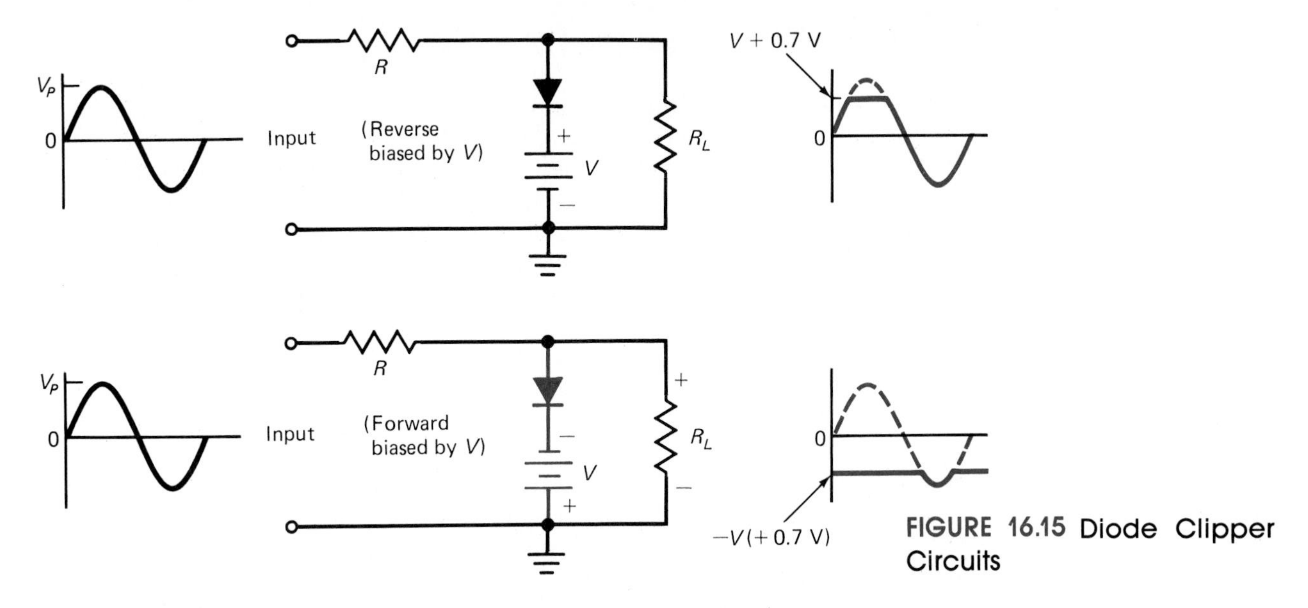

FIGURE 16.15 Diode Clipper Circuits

of the diode to charge a capacitor to the peak level of the ac input wave (minus the forward drop on the diode). Operating somewhat like a standard power supply rectifier/filter combination, the diode causes the capacitor to charge (approximately) to the peak voltage of the input signal. The capacitor holds this charge level (as a dc voltage) during the portion of the ac cycle that reverse biases the diode. This charge on the capacitor, since it is in series with the input voltage, is added to the input signal. Thus an output voltage which is an ac signal offset by (approximately) the peak level of that signal is produced at the load resistor. The diode does not affect the output voltage (rectify it) once the capacitor is charged to the peak level. The load device must be sufficiently high in resistance so as not to discharge the capacitor substantially (a common load would be an FET which has high input resistance).

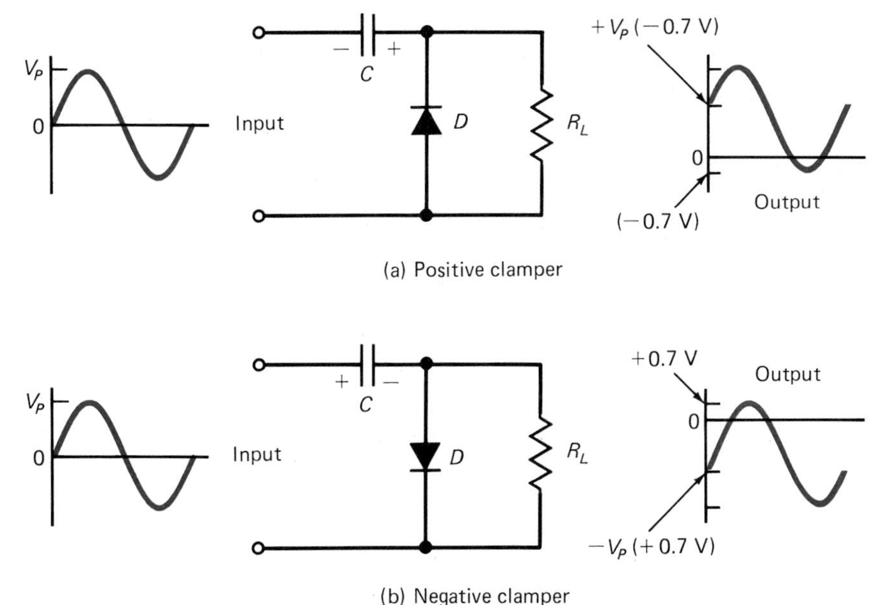

FIGURE 16.16 Clamper Circuits

Voltage Doublers/Triplers (Multipliers)

A voltage doubler or tripler (dc only; Figure 16.17) uses the basic clamping action seen in the presentation above to produce an output level which is twice or three times the peak level of an ac input. Each capacitor/diode combination functions as a peak voltage storage combination in these circuits. By tapping the output across the set of capacitors, the output voltage can be two or three times the value of the peak input voltage. Typically, the load current that can be supplied by these circuits (to avoid heavily discharging the capacitors) is relatively small.

Saturated Switches

In some cases it is necessary to use an input voltage to a transistor to produce a switched (on/off) output. Such applications are widely used in digital circuitry, but also have some application in control-type circuitry. Basically, the way to produce a saturated switch is either to use an input to the switching device (transistor or op-amp) that is sufficiently high or to have a gain for the device that is high enough that the output is driven to maximum current when the input begins to forward bias the device. Figure 16.18 illustrates an op-amp that is used as a saturated switch. The op-amp is used with an open-loop gain.

FIGURE 16.17 Voltage Doubler and Tripler Circuits

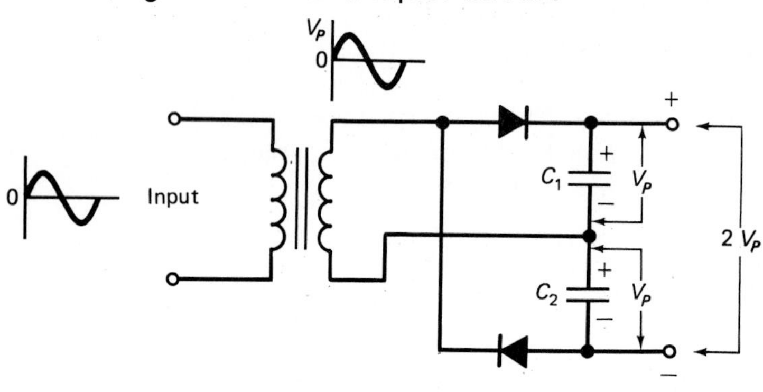

(a) Voltage doubler

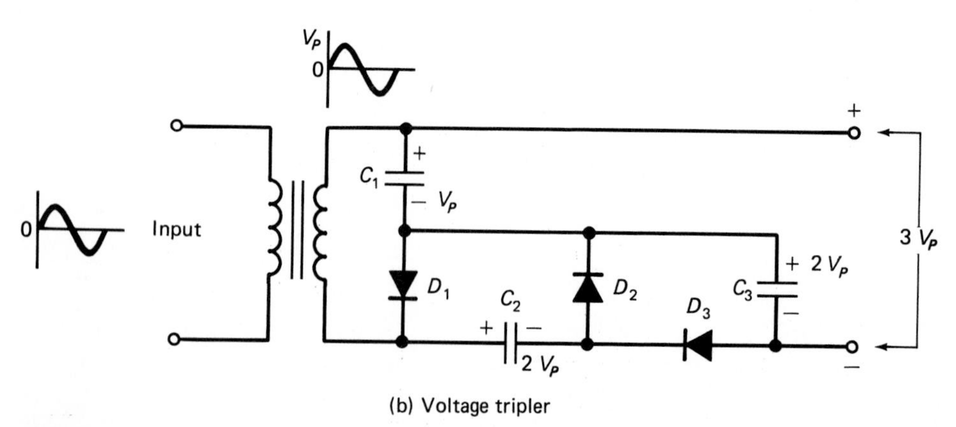

(b) Voltage tripler

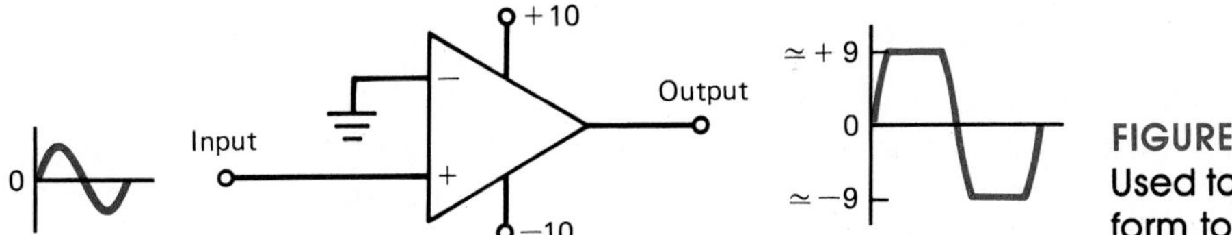

FIGURE 16.18 Saturated Op-Amp Used to Convert Sinusoidal Waveform to Square Wave

As a signal is applied to the noninverting terminal, the output will rise (positively or negatively in sync with the input) to the maximum output level of the amplifier. Such circuits can be used to convert sine (or other shaped waves) into square waves or as triggering devices that respond to small voltage or current levels at the input.

Dual-Polarity Power Supplies

For some circuits, especially op-amp circuits, it is necessary to produce power supplies that produce both a positive (+) and a negative (−) output relative to a common ground. Such a supply is illustrated in Figure 16.19. The basic function of the supply is as follows. (Note the "A, B, C, D" markings of the waveforms.)

1. Wave segment A is directed through diode D_1 to the positive side of bleeder resistor R_1. (Ground is negative with respect to A.)
2. At the same time, the negative wave segment C is conducted through diode D_4 to the negative side of bleeder resistor R_2. (Ground is positive with respect to C.)
3. When the top wave cycles to segment B, that negative voltage is conducted through diode D_3 to the bottom of resistor R_2. (Ground is now positive with respect to B.)
4. At the same time the bottom wave cycles positive (segment D). This voltage is directed to the top of resistor R_1 through diode D_2. (Ground is negative with respect to D.)

FIGURE 16.19 Dual-Voltage Power Supply Schematic

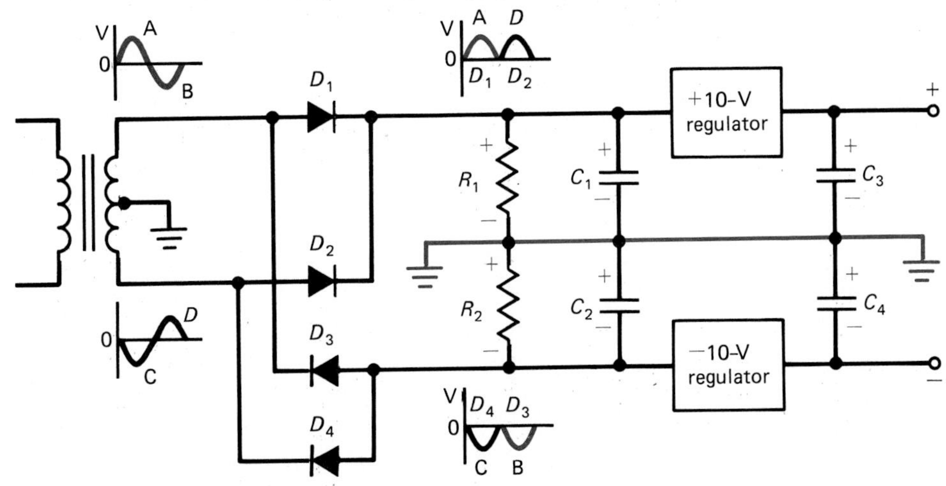

These voltages (+ on R_1, − on R_2 with respect to ground) are filtered by capacitors C_1 and C_2. The positive regulator stabilized that voltage to +10 V (with respect to ground) and the negative regulator stabilizes its voltage to −10 V. The output, then, is two voltages (+10 V and −10 V) with respect to ground. (Note that a measurement of the voltage from the $+V$ output to the $-V$ output would register 20 V.)

Instrumentation Amplifier

In many measurement situations, the sensors used to take measurements may be quite distant from the instrument (meter, scope, etc.) used to display the measured quantity. As such, the leads used to connect the sensor are easily subject to picking up stray electrical noise. Since the output of the sensor may be a relatively low electrical quantity in the first place, this introduced noise may be of a magnitude that interferes with the signal being measured (see Figure 16.20). If this signal, information plus noise, is applied to the input of an amplifier, both quantities would be amplified and a false reading would result. To help eliminate this problem, a special op-amp configuration has been developed. This special device is called an instrumentation op-amp. The symbol (and coordinating circuit) for an instrumentation amplifier is shown in Figure 16.21.

First, internally, the instrumentation op-amp blends three op-amps to achieve a highly controlled op-amp function. The CMRR (common-mode rejection ratio) and input impedance of such an amplifier is quite high. This amplifier functions basically in the following manner:

1. The measured signal (plus noise) is applied to the two input terminals. On this input is a high level of noise, common to both input lines and the initial signal (as a potential difference between each line).
2. The differential amplifier function of the inverting versus noninverting terminals, coupled with a high CMRR, causes a rejection of the common signal (noise) and an amplification of the differential signal (information from the sensor).

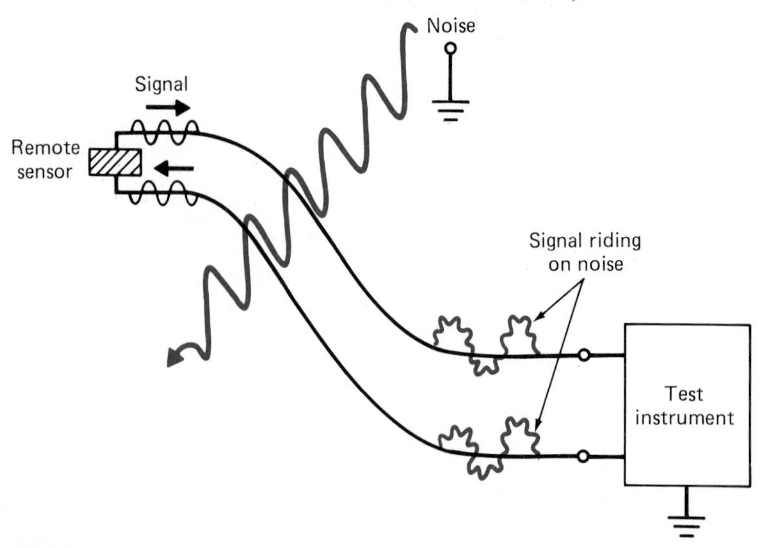

FIGURE 16.20 Noise Introduced in Connecting Cables

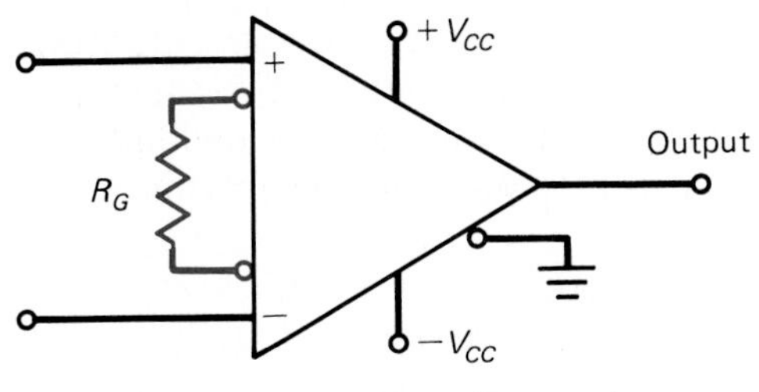

(a) Symbol

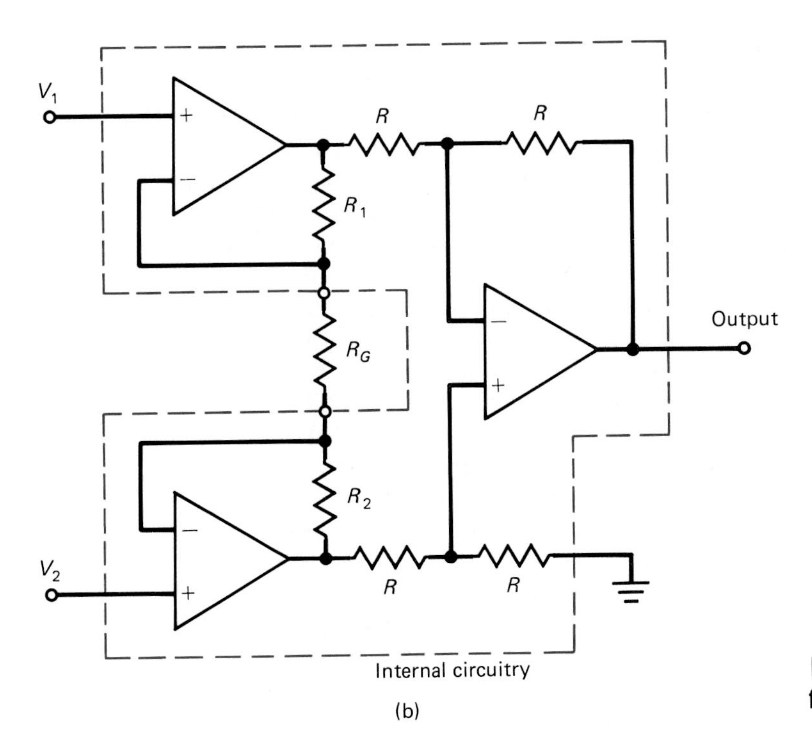

(b)

FIGURE 16.21 Instrumentation Amplifier (Op-Amp Type)

3. Unlike a common op-amp, gain is set by a single resistor (R_G). Although total open-loop gains are lower than most op-amps, there is sufficient gain to produce a highly usable, low-noise output.

The final gain of the instrumentation amplifier may either be based on the ratio of the gain-setting resistor to an internal feedback resistance, or the selection of this gain-setting resistor may be designed by gain information. In either case, manufacturers' application notes will describe the typical method of calculating these quantities. Naturally, it is assumed that this amplifier is also a very low (internal)-noise type and that terminals for offset adjustment and compensation are also provided.

Automatic Gain Control in Amplifiers

The process of producing an automatic gain control (AGC) in an amplifying circuit differs from the process of using feedback for gain stability. An example of automatic gain control can be seen in many modern portable cassette recorders. Many recorders have the ability to maintain a constant recording level (on the tape) given a variable level of input from the microphone. Therefore, if the person speaking into the microphone speaks loudly or softly, the average recording level is held nearly constant. (Sometimes this feature is called automatic volume control.)

The process of producing AGC differs, as noted above, from the process of using feedback gain control. In standard feedback gain control the gain of the system (or amplifier) is set to a fixed value; if the input is a high signal level or low signal level, it is reproduced with the same gain, although amplified it still has a high or a low level. The process of AGC actually adjusts the gain of the system (or amplifier) in order to maintain the same average output level for either high- or low-level inputs.

To achieve an AGC function, the average value of the output voltage is used to set the gain level of a previous stage. In the circuit shown in Figure 16.22, the output signal is first rectified and applied to a simple capacitor filter. This process converts the ac output to a dc level. The amount of dc stored on the capacitor, over a period of time, is proportional to the level of the ac output (a decrease in the ac causes a decrease in the dc level). This voltage is then applied to a voltage divider so that the proper level of dc is used in the feedback process.

In this circuit the special VVR (voltage-variable resistance) effect of the FET is used to control the gain of the first stage of the amplifier. It will be remembered that the gain of this circuit is given as $-R_L'/R_E$. In this case the ac emitter resistance is the value of the fixed R_E in parallel with the VVR. As the average level of the output increases, the value of the dc at the gate of the FET increases. As this voltage goes up, so does the resistance (drain to source). This then reduces the gain of this stage, which will tend to drive the output down in level. Thus a rather constant output is maintained.

It is not necessary to use the VVR to achieve the AGC process. Some

FIGURE 16.22 Use of Multistage Signal Feedback Which is Converted to a DC Level Which Is Used to Control an FET Applied as a VVR which in Turn Affects the Gain of the Input Amplifier

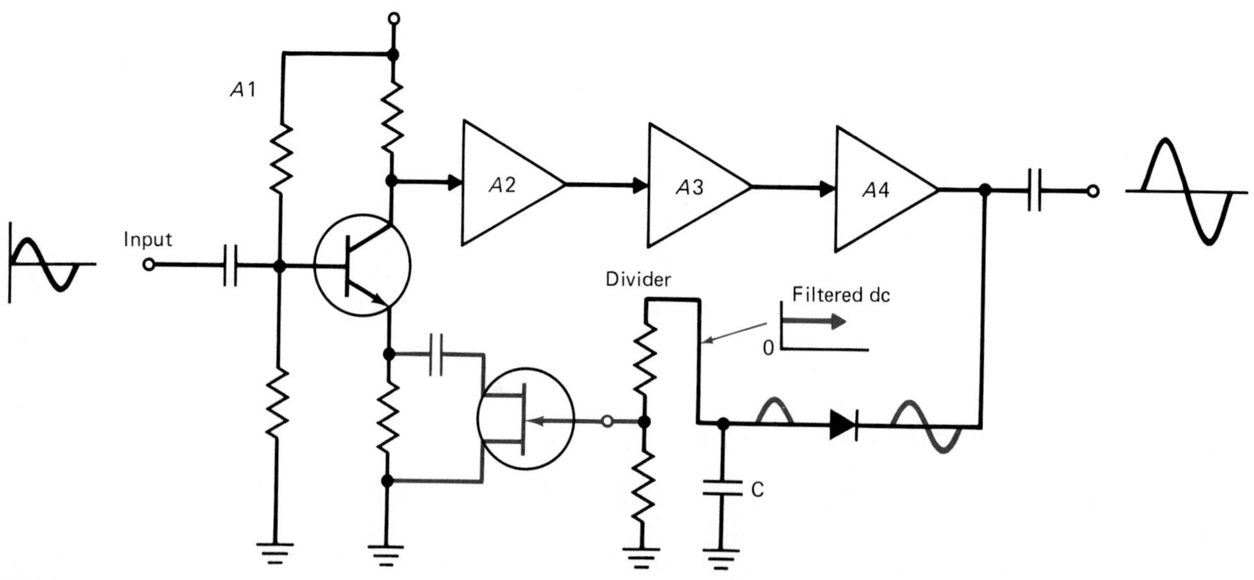

circuits use the AGC voltage directly to affect the bias on a previous stage, and thus affect the gain of that stage. The major key to the use of AGC is the production of the average dc voltage based on the ac output.

The following example is presented not so much because of its uniqueness, but because it demonstrates several principles in one composite system.

Complete Audio Power Amplifier Circuit

Figure 16.23 is a schematic for a complete power amplifier system. The amplifier consists of a preamplifier stage, a driver stage, and a complementary-symmetry power amplifier. The preamplifier (a small-signal amplifier) is capacitively coupled to a medium-power driver stage, which also contains a distortion volume control for the system. The following is an analysis of the function of each of the elements in the circuit, grouped by amplifier stage.

Preamplifier (Active Device Q_1)

1. Resistors R_1, R_2, R_3, and R_4 form a traditional beta-independent bias circuit around transistor Q_1.
2. Capacitor C_1 isolates this amplifier from the input signal source.
3. Capacitor C_2 feeds the output of this stage to the following driver amplifier.
4. The ac load for this amplifier is the parallel combination of R_3, R_5, and R_6 (the potentiometer) and the input resistance of transistor Q_2. Since this control affects this load resistance, it also affects the gain of this stage. A decrease in the R_6 resistance will decrease the gain of this amplifier.
5. Capacitor C_3 is a bypass capacitor for the emitter resistor R_4.

FIGURE 16.23 Multistage Power Amplifier

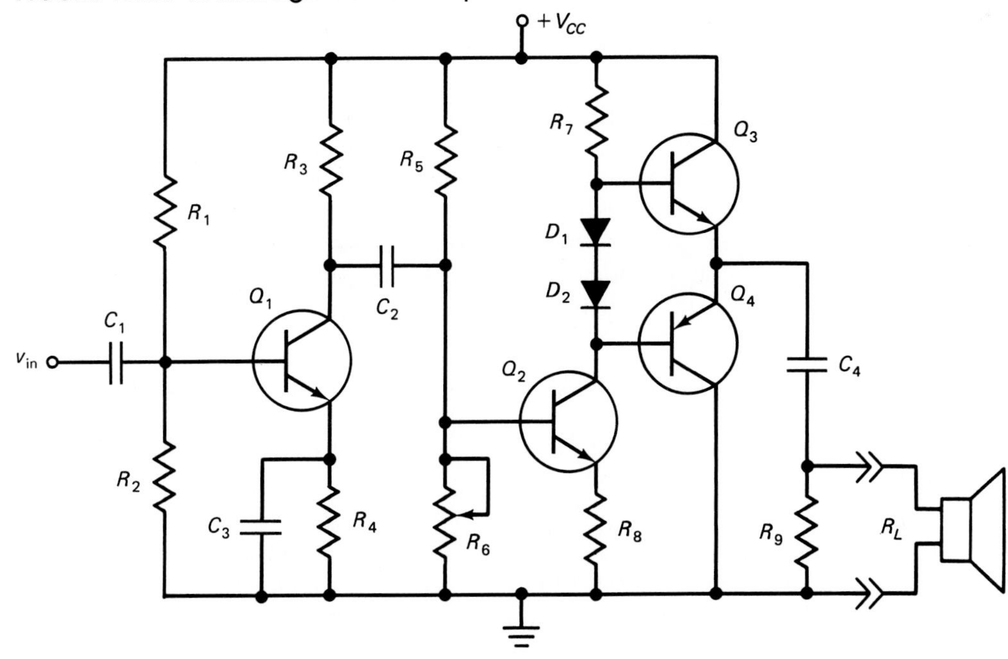

Driver (Medium-Power) Stage (Active Device Q_2)

1. Resistors R_5, R_6, R_7, and R_8 provide another beta-independent bias circuit for Q_2 (un-bypassed emitter resistor R_8). (Diodes D_1 and D_2 will be discussed later.)
2. The output of the preamplifier feeds the input of this stage via capacitor C_3.
3. The output of this amplifier is directly coupled to the following stage.
4. The ac load for this amplifier is R_7 in parallel with the input resistance of the next stage [which is basically the beta of either Q_3 or Q_4 multiplied by the load resistance to the amplifier (load in parallel with R_9)].
5. Potentiometer R_6 will adjust the quiescent (Q) point of the amplifier and will affect its current gain. Basically, this potentiometer is adjusted to produce the least distortion of the signal. In doing so it adjusts the Q point for this stage and, by its effect on the voltage on R_8, adjusts the operating point on Q_3 and Q_4. (This affects Q_3 and Q_4 since it and the dc drop on Q_2 provide the dc bias for the bases of these two amplifiers.)

Power Amplification Stage (Active Devices Q_3 and Q_4)

1. Transistors Q_3 and Q_4 form a complementary-symmetry class B power amplifier. Each transistor will amplify one-half of the ac input signal. Their outputs are directly coupled together.
2. The input signal is directly coupled from the output of the driver stage Q_2.
3. Diodes D_1 and D_2 are provided to slightly forward bias each device (Q_3, Q_4). The 1.4-V (average) drop across these two diodes is enough to provide the 0.7-V (average) forward potential for Q_3 (V_{BE}) and the 0.7-V potential (V_{BE}) for Q_4 as well. Without this small bias difference, some of the ac signal would be lost, as it would be used to forward bias each device. Thus the output would become distorted. Also, diodes are used so that their characteristics will match more closely the forward-bias characteristics of the transistors. Thus if the transistors would exhibit changes due to thermal (or other) conditions, the diodes would tend to duplicate these changes and rebias the device accordingly. If the transistors Q_3 and Q_4 were mounted on heat sinks, it would be common practice to mount diodes D_1 and D_2 to the same heat sinks; thus they would "track" the transistor changes more effectively.

Without going through all of the calculations here, this circuit could accept a 150-mV input (input power of approximately 22 μW) and produce an output of about 15 V at about 9.4 W of power (a power gain of about 4.27×10^5).

These past circuits have been presented to expose the reader to a wider variety of device applications and circuit configurations. It is suggested that these be reviewed in detail to realize possible applications fully and to comprehend the literally thousands of other potential applications.

REVIEW PROBLEMS

(1) Refer back to each of the schematics presented in this section and briefly describe the function of each circuit and what role is played by the solid-state device in that circuit.

(2) If the output of the power amplifier circuit (Figure 16.23) was a distorted ac wave, describe possible faults which may cause that problem. (Note that the output is present, but distorted.) There are at least four possible faults that could cause such distortion.

16.3 INVESTIGATING NEW DEVICES AND APPLICATIONS

In this section we offer some basic recommendations on how to evaluate and work with either new devices or new device applications. Throughout the past chapters, currently popular devices and circuit applications have been described and analyzed. Many of the same procedures that were used there can be applied to investigating new devices or circuits.

Information Resources

It is seldom possible to explore new devices or applications without using some form of written or specified information which describes these devices or systems. The following is a general list of the types of resources that are typically used to understand new devices or applications.

New Device Information Resources To obtain the necessary information about a new device, the following types of resources are commonly used. (Specific publication names are not given, as there usually are a variety of publications available in any one category.)

1. *Manufacturers' data sheets.* These publications list the electrical and functional specifications of a specific device. Often, recommended applications are given as examples of use.
2. *Manufacturers' application notes.* These publications are usually more informative than data sheets, as they include several suggested applications and often specific equations or circuits for use.
3. *Other manufacturers' publications.* Often, manufacturers publish special bulletins or even magazine-like periodicals that detail recent device or circuit developments. Most manufacturers publish books that deal with the applications of certain categories of devices (e.g., a publication detailing applications of all the BJT devices a company manufactures).
4. *Professional journals.* There are a large number of publications dedicated to the electronics industry. Most both describe new products and have feature articles that describe applications and new developments.
5. *Other text or reference books.* There are a large number of other text and reference books dealing with electronic devices. Often, companies have small in-house libraries which maintain such references.

6. *Seminars and trade shows.* Many companies and professional organizations provide training or informational seminars. These may deal with broad applications, new developments in a specific field, or specific new components (e.g., a seminar that highlights new devices for use with fiber optics.) There are also quite a few trade shows, which are gatherings where many manufacturers display their newest products and present application-oriented seminars.
7. *Computer access data banks.* More and more information is being made available via computer access data banks. A company subscribes to the service and when information is needed, accesses the system via a computer terminal. This process is becoming quite popular since it is easy to feed in a set of device specifications relative to an application and have the data bank search out a group of devices which can service that need.
8. *Electronic parts distributors.* In many cases, electronic parts are purchased through a distributor. These distributors publish catalogs (or use manufacturers' catalogs) which briefly describe specific devices. Often, the distributor can assist in describing a new device or help in locating a specific device to suit a need.

Since it is in the parts manufacturers' best interest to keep potential users informed, many of the services and information listed above may be available directly (free or at a limited cost) from the manufacturer.

New Circuitry or Systems Information Gathering information about new circuitry or systems depends on how the circuits are developed. If they are manufactured or developed by other companies, information is available from much the same resources as are listed above for device information. If new circuits or systems are developed by the company you work for, often, special in-house publications are made, training seminars are presented, or discussions are had with the design engineers who developed the circuits or systems. In any case, good documentation of circuits and products is essential.

Investigating New Devices

When working with a new form of device, whether discrete or integrated, certain facts must be known about that device. It is essential to investigate the basic application of the device. It is common to attempt to relate the device with known forms.

Application In which of the following categories does the device seem to be usable?

1. Linear amplification
   (a) Small signal or large signal
   (b) Low frequency or high frequency
   (c) Which amplification model does it fit (voltage, current, transconductance, transresistance) and what is its gain value, input impedance, and output impedance?

2. Control circuits
   (a) Low power or high power
   (b) Dc or ac (or both) applications
   (c) Triggered (as an SCR) or nontriggered (as a diac)
3. Power conversion (power supply or power supply control)
   (a) Ac to ac
   (b) Ac to pulsed dc
   (c) Regulation or other form of level control (fixed or variable)
   (d) Other forms of conversion
4. Other applications
   (a) Related to known applications (e.g., optical diode)
   (b) Totally new application

Device Correlation Which of the following categories does the device fit into (often stated in specifications)?

1. BJT types
   (a) NPN or PNP
   (b) Standard or Darlington
   (c) Special effects (e.g., optical, high current, etc.)
2. FET types
   (a) N-type or P-type
   (b) JFET, MOSFET (depletion or enhancement)
   (c) Other forms or special features
3. Op-amp types
   (a) Standard op-amp form
   (b) Norton (current) amplifier
   (c) Special-purpose or special characteristics (e.g., instrumentation amplifier, single power supply, high slew rate, etc.)
4. Special-purpose discrete type
   (a) Does it relate to other known applications (e.g., diode)?
   (b) It is designed for use in a unique application? (Must know the specifics of that application)
5. Special-purpose integrated-circuit types
   (a) Does it relate to known other applications (e.g., regulator)?
   (b) Is it designed for use in a unique application? (Must know the specifics of that application)

Once these fundamental questions have been asked, the following details should be investigated to obtain a clear understanding of the device and its applications:

Dc Operating Conditions

1. What are its bias requirements?
   (a) Is it polarity sensitive (e.g., diodes, BJTs, etc.)?
   (b) What are its bias limitations (maximum ratings)?
   (c) What are its normal bias conditions?
   (d) Must it be externally biased (e.g., BJT), or is it internally biased (e.g., op-amp), or (as a diode) is it customary not to provide external bias?

(e) Are there a variety of bias arrangements or one specific arrangement that must be used?

2. Does it normally function with dc inputs or signal inputs?
   (a) Does it function primarily with dc inputs (such as a regulator)?
   (b) Is it used mainly with ac inputs (e.g., BJT)?
   (c) Can it be typically used for either dc or ac inputs (e.g., op-amps)? (*note:* Since devices that are biased with dc can also usually accept dc input levels and produce an output in response, this category should be used for devices that are specially described as functioning with both dc and ac inputs.)

Ac Operating Conditions

1. Does the device function with ac inputs? (For example, a voltage regulator won't accept a sine-wave input.)
2. How does the device respond to an ac input?
   (a) Does it amplify?
   (b) Does it produce control functions?
   (c) Is it a triggering or fixed response device? (For example, a diode only rectifies.)
3. Are there special restrictions on the signal input?
   (a) Low or high level (small signal, large signal)
   (b) Frequency limitations
   (c) Is the input to be a triggering potential (e.g., SCR)?
   (d) Is the input to be in a special form (e.g., optical transistor)?

Special Integrated-Circuit Investigations

In many cases, linear integrated circuits will have one or a limited number of applications. In the case of such a device, it is necessary to investigate carefully all of the limits and applications of that device. If the device is for use as a single function, the manufacturer will often specify not only that application but also the appropriate circuitry needed to support that application. If the device can be used for multiple applications, the specifications will usually indicate typical uses along with enough details to customize the application.

EXAMPLE 16.2

Using the classification methods noted above, investigate the 2N6762 device listed in Appendix A.

Solution:

1. *Applications*. Control function, switching, high power, triggering function or switching regulator.
2. *Device type*. FET (discrete). MOSFET (''T''MOS may have to be checked with manufacturer for more information about what makes it special. Looking at the footnote at the bottom of the page it says that the ''TMOS'' name is a trademark of Motorola Inc. and thus

may be a specially defined form of MOSFET.) It is said to have a fast switching speed (related to high frequency). It includes an extra diode across the D/S leads. This is noted as being an advantage when using an inductive load.

3. *Dc conditions*. Its polarity sensitive (as an FET). Its maximum ratings indicate that it can handle high voltages and currents and up to 75 W of power. (See the data sheet for actual numbers.) In terms of normal bias, the second page of the sheets indicate the normal ON and OFF bias conditions and even indicate a typical bias condition. They also indicate that the typical input is a pulsed signal. Thus it seems to operate on a cross between constant dc and ac.
4. *Ac operation*. Since it is a pulsed device, it does not specify gain values (g_m) but does indicate switching conditions (ON/OFF specifications).

From this analysis it can be assumed that the device could be usable in a control function either as an electronic switch or like an SCR, but seems to be triggerable with a pulse-wave form. Since it is an FET it would have high gate resistance. From the specifications it is seen to have very little output leakage current in the off mode and rather low resistance in the on mode.

EXAMPLE 16.3

Using the same techniques as in Example 16.2, classify the MC13001XP listed in Appendix A.

Solution:

1. *Application*. Totally new application. Black-and-white TV. Subsystem.
2. *Device type*. Special-purpose IC.

The dc and ac classifications are skipped since it is a special-purpose device. A sample application circuit is listed with the specifications. Use of this circuitry or circuitry custom selected to support the rest of the TV circuitry will need to be done. It would be common to make direct contact with the manufacturer to discuss special applications or variations on the suggested circuit.

A Unique Situation

It is possible that in the course of working with circuitry one may encounter a device for which the specifications are not easily available. It may be from a manufacturer whose specification sheets are not readily available, or it may be a product that was discontinued and the specification sheet may not be available. There is one resource that can be used which will help to classify the device and roughly describe its electrical parameters. Manufacturers usually publish a cross-reference or substitution guide. These manuals list many devices made by many other manufacturers and then indicate a device made by that manufacturer that can be used as a substitution. By looking up the unknown device in this guide and then looking up the specifications of the substitute device, a general description can be had of the function of the unknown device.

The exact specifications for the substitute device should not be assumed for the unknown device, though! To be safe in recommending the substitution, the manufacturer selects a device that meets or exceeds the specifications of the coordinated devices from other manufacturers. For example, if a diode has a PIV rating of 75 V dc, the manufacturer would typically select a diode with a PIV of 80 V dc, just to be safe. But the general description (e.g., NPN BJT) can be used as well as an estimate of the unknown device's parameters can be had by using the data for the substitute device.

In this section some of the fundamental steps that are used to classify new devices and circuitry have been shown. Naturally, they must be applied with flexibility since a device may not be able to be strictly categorized. It is quite probable that a device can be described as fitting several categories—that just proves that it is usable in a wider variety of applications. In any case, when using a device, it should first be determined if that device will meet the needs of the intended application. Then more detailed specifications can be used to establish the necessary support circuitry to make that device totally functional.

REVIEW PROBLEMS

(1) Indicate at least four resources that may be used to determine the specifications of a device.

(2) Assume that you have encountered a circuit in a schematic and find that its function and operation are not clear. Indicate several steps that can be taken to obtain more details about that circuit.

(3) Select any device from Appendix A (except those used in Examples 16.2 and 16.3) and write a basic classification and description of that device using the guidelines presented in this section.

(4) (If available) Use a semiconductor substitution guide (or cross-reference) and find a substitute for the 2N2218,A listed in the Appendix A. Look up the specifications of that substitute and compare them to the specifications of the 2N2218.A.

SUMMARY

A number of new devices and circuits have been presented in this chapter. Many of these have quite limited and specialized applications. In addition, a basic procedure for more easily understanding new devices and applications has been indicated. It is expected that in the future, even as you read this book, new and unique devices and circuits are being designed and manufactured. Only by blending a knowledge of basic device function and the pros and cons of all applications with a logical investigation of these new developments can you expect to keep abreast of the rapid expansion of the electronics industry.

The future of electronics is certain to change and evolve new and more unique applications. It is the task of the electronics professional to be flexible enough to accept these changes and to be willing to keep informed of the most current developments. It is quite possible that the future of the discrete device (BJT, FET, and even op-amp) will be one of obsolescence. As the manufacturing process of integrated circuits becomes less costly and as more multifunction ICs are made available, the task of application will be more one of programming or selecting the desired function (block or blocks in a diagram) than of producing step-by-step functions from individual circuitry. Quite possibly the only application of discrete devices will be in very high power circuitry, where the problem of thermal power dissipation remains the only reason for having bulky, discrete devices. The reader is therefore encouraged to become skillful not only at understanding the function of currently produced devices and circuits, but to develop good analytical methods to use when the next waves of new devices or integrated processes come to popular use.

PROBLEMS

Section 16.1

16.1. Describe the basic function of the optically sensitive (photo) semiconductors noted in this chapter. Generally, how do they differ from other similar nonoptical devices?

16.2. Illustrate the schematic symbol for the following devices.
(a) Optical diode (b) LED (c) Phototransistor
(d) Photo-Darlington (e) LASCR (f) Optical coupler

16.3. Modify the illustration in Figure 16.24 to use a photo-Darlington device as a receiver.

16.4. Look up the specifications for the MFOE1201 device listed in Appendix A. From the information presented, write a brief response to the following questions.
(a) What form of device is it?
(b) What are some of the recommended uses?
(c) What does the (Figure 1 on that sheet) graph represent?

16.5. Repeat problem 16.4 for the MFOD1100 listed in Appendix A.

16.6. For the devices detailed in problems 16.4 and 16.5, illustrate a possible application using these together. Compare the information found in problems 16.4 and 16.5

FIGURE 16.24

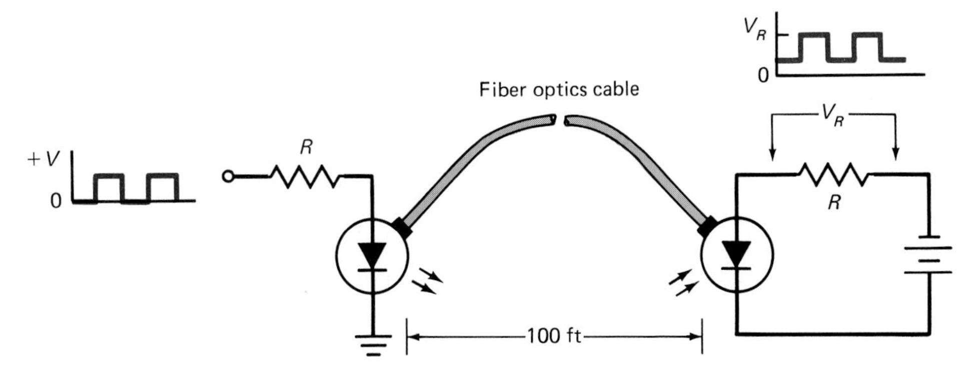

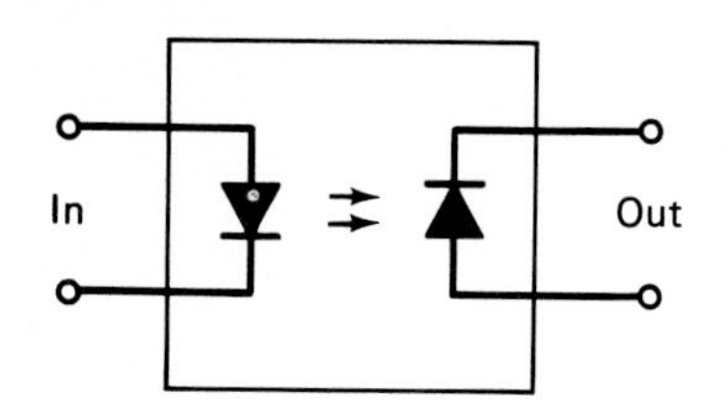

(a) Photodiode isolator (basic type)

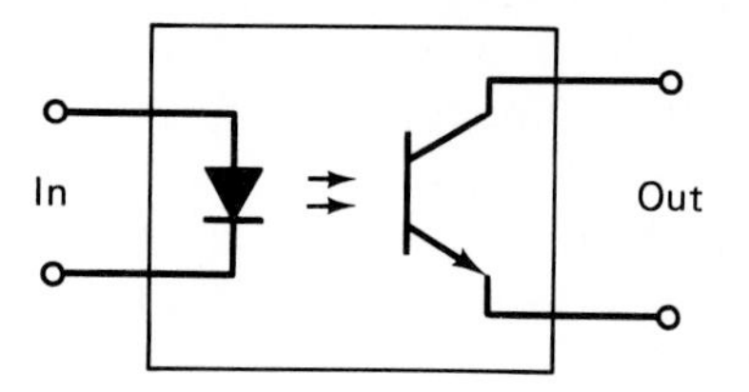

(b) Phototransistor output

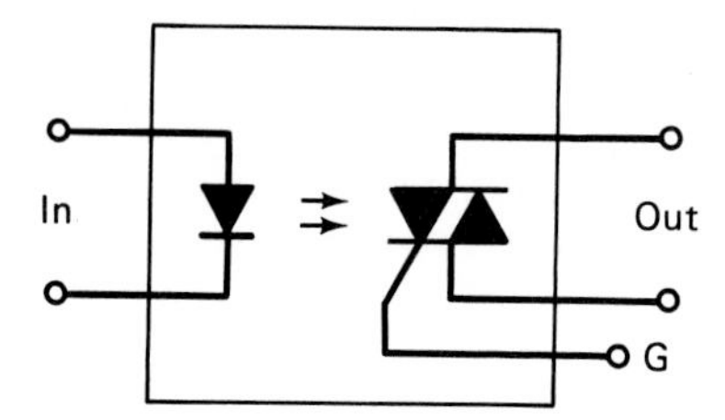

(c) Phototriac output (with electrical gate lead)

FIGURE 16.25

16.7. Briefly discuss the application of the MOC3009 device listed in Appendix A. Compare it to the devices listed in Figure 16.25.

16.8. Complete the circuit shown in Figure 16.26 using the MOC3009. Use a simple block symbol for the device, indicating pin numbers along the side (see the device spec sheet). Be sure to include a current-limiting resistor for the "emitter" side of the device. Calculate a value to limit the current to one-half of the maximum rating shown in the spec sheets.

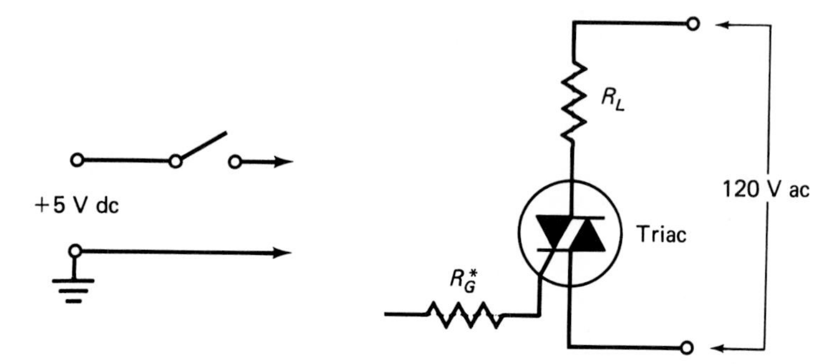

*Gate current limiter for the triac.

FIGURE 16.26 Triac Circuit

16.9. Illustrate an additional use (of your own invention) for an optical interruptor.

16.10. In the resonant circuit shown in Figure 16.27, substitute a varactor diode for the C_2 capacitor. Connect it to the variable dc supply shown in this illustration (be careful of bias polarity). (*Note:* Leave C_1 in place so that the dc supplied to the varactor will not be shorted by the inductor L.)

16.11. Look up the specifications for the MR2520L device in Appendix A. Is this for use in ac or dc circuitry? What are its voltage specifications? Illustrate a simple circuit that could use this device to protect a car radio from accidental surges in the supply lines (similar to Figure 15.11).

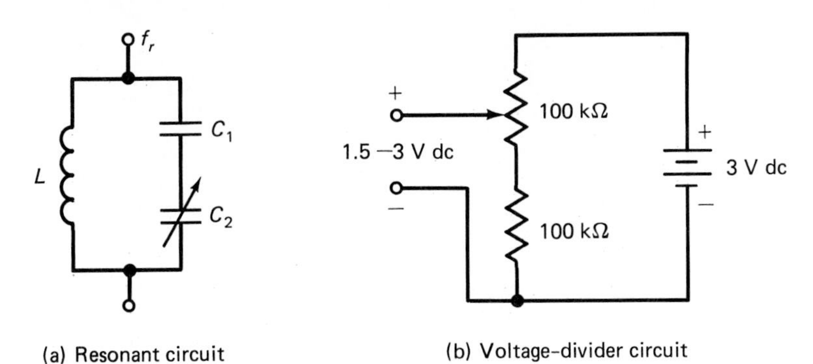

FIGURE 16.27 Resonant and Voltage-Divider Circuits

16.12. What makes the current-mode (or Norton) op-amp different from a standard op-amp?

Section 16.2

16.13. What does it mean to use a JFET as a VVR?

16.14. Illustrate a clipper circuit that is used to limit the amplitude of a sine wave to +6.7 V. Indicate the value of the voltage source used in this illustration.

16.15. Repeat problem 16.14 using a circuit to maintain a +6.7-V level and only pass on ac variations above the 6.7-V level. (Be cautious of voltage and diode polarities.)

16.16. Illustrate the schematic for a basic positive voltage clamper circuit.

16.17. Using information from Chapter 11 and the function of a saturated switch, illustrate a schematic using op-amps to achieve the tasks noted for the function generator shown in Figure 16.28. [Do not list component values, but identify each section's function. Also, do not include the effect of loading from one stage to the other (although in a real circuit it would require buffer amplifiers).]

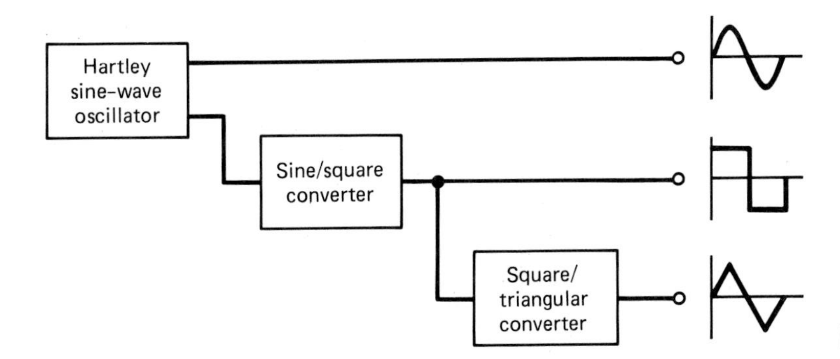

FIGURE 16.28 AC Function Generator

16.18. Illustrate the schematic for a dual-polarity power supply which produces + and − 5 V.

16.19. Briefly describe why instrumentation amplifiers are needed when the amplification of remote sensor values is needed. Then describe how the instrumentation amplifier differs from a standard op-amp.

16.20. How does automatic gain control differ from using signal feedback to control the gain of an amplifier (or amplifier system)? Explain both what it controlled and the form of the control signal (feedback).

16.21. The MC1306P device listed in Appendix A could be substituted for the function of the circuit listed in Figure 16.29. Briefly discuss the troubleshooting methods used if this device produces a distorted output in contrast to the circuit shown in Figure 16.29.

16.22. Briefly describe the function of each of the components shown in Figure 16.29.

16.23. If the following circuit defects were to occur in the circuit shown in Figure 16.29, first list what would happen to the output signal, then indicate one, two, or three dc voltage measurements that could be taken to find this defect. (Assume that reference voltages are supplied with the schematic.)

(a) Base–emitter of Q_3 opens.
(b) D_1 opens.

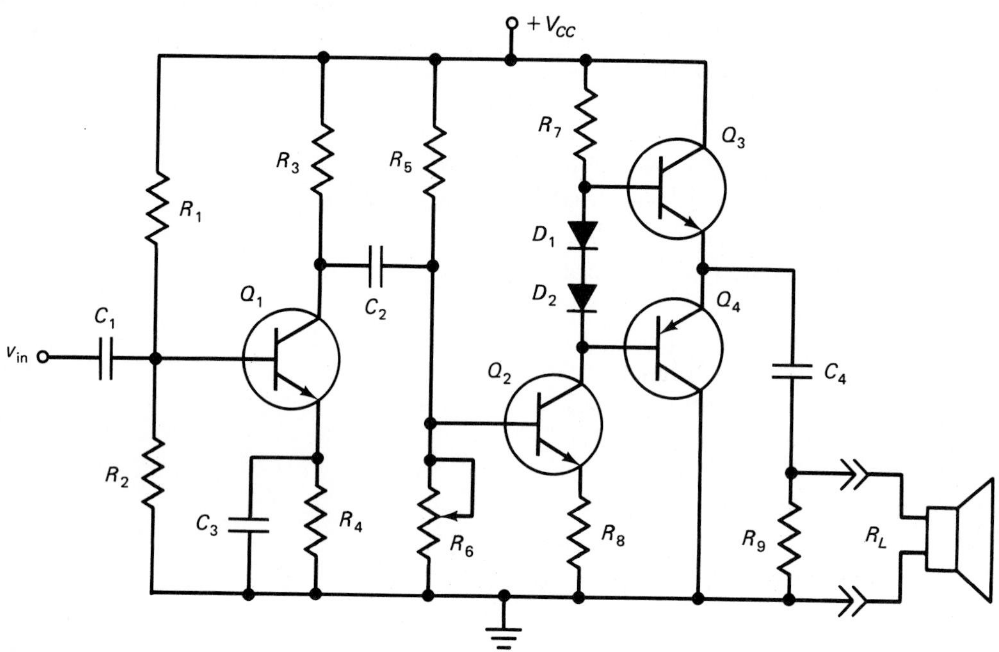

FIGURE 16.29

(c) R_4 raises to a high value (but not open).
(d) R_6 is improperly adjusted.
(e) C_3 develops leakage (begin passing dc current).
(f) The beta of Q_2 drops very low (or a low beta device is accidentally substituted).

ANSWERS TO REVIEW PROBLEMS

Section 16.1

1. The list should be a composite of information in Section 16.1.
2. The LASCR is an SCR where the gate lead is triggered by light levels, not electrical levels.
3. The LED could be made to shine on the phototransistor with a gap between them through which the smoke could pass. The phototransistor

would be connected to a "latching" circuit which operated the horn. If smoke passed between the LED and phototransistor, it would interrupt the light to the phototransistor and thus cause it to trigger the latching circuit. (A manual or time-delay reset could be attached to the latching circuit.)

Section 16.2

1. Figure 16.15 is of two biased clipper circuits. The role of the diode in each is to turn on (conduct) when a voltage above or below a specific level is seen at the input. Figure 16.16 shows two clamper circuits. The diode in each of these circuits is used to create a single-polarity dc charge on the capacitor. Figure 16.17 shows a voltage doubler and tripler. The diodes in each of these circuits are used to maintain a single-polarity charge on each of the associated capacitors in these circuits. Figure 16.18 is an illustration of a saturated op-amp circuit. The application of the op-amp is to produce a nearly square-wave output from a sine-wave input. Figure 16.19 is of a dual-polarity power supply. Diodes D_1/D_2 and D_3/D_4 form two full-wave rectifiers. The two regulators are used to stabilize the output to a constant level. Figure 16.22 is a multistage amplifier using automatic gain control. The diode is used as a simple half-wave rectifier; the FET is used as a voltage-variable resistor and the transistor is used as a basic common-emitter amplifier. Figure 16.23 is a power amplifier circuit. Transistor Q_1 is used as a simple common-emitter amplifier. Q_2 is used as a common-emitter amplifier. Transistors Q_4 and Q_5 form a complementary-symmetry amplifier. Diodes D_1 and D_2 provide forward-bias potentials for these transistors.
2. The input to the amplifier could be distorted; test it first; R_6 could be improperly adjusted or defective; Q_3 could be defective; Q_4 could be defective; Q_3 and Q_4 could be improperly "matched"; Q_3/Q_4 could be overheating; diode D_1 or D_2 could be shorted; the power supply could be introducing noise or may be defective, producing too little output voltage; capacitors C_1 to C_4 could be "leaky"; the load (R_L) could be too low, thus saturating Q_3/Q_4 (although unlikely, any of the fixed resistors could be defective).

Section 16.3

1. Manufacturers' data sheets, other manufacturers' publications, professional journals, text or reference books, seminars or trade shows, electronics parts distributors.
2. Check with associates who may have knowledge of the function; check resources listed in answer 1; look for in-house publications that explain the circuit; attend training seminars; discuss the circuit with the design engineers who worked with the circuit.
3. The answer is based on the device selected.
4. The answer is based on the substitute device found.

MC13001XP
MC13002XP

Advance Information

MONOMAX BLACK-AND-WHITE TV SUBSYSTEM

The MONOMAX is a single-chip IC that will perform the electronic functions of a monochrome TV receiver, with the exception of the tuner, sound channel, and power output stages. The MC13001XP and MC13002XP will function as drop-in replacements for MC13001P and MC13002P, but some external IF components can be removed for maximum benefit. IF AGC range has been increased, video output impedance lowered, and horizontal driver output current capability increased.

- Full Performance Monochrome Receiver with Noise and Video Processing — Black Level Clamp, DC Contrast, Beam Limiter
- Video IF Detection on Chip — No Coils, No Pins, except Inputs
- Noise Filtering on Chip — Minimum Pins and Externals
- Oscillator Components on Chip — No Precision Capacitors Required
- MC13001XP for 525 Line NTSC and MC13002XP for 625 Line CCIR
- Low Dissipation in All Circuit Sections
- High-Performance Vertical Countdown
- 2-Loop Horizontal System with Low Power Start-Up Mode
- Noise Protected Sync and Gated AGC System
- Designed to work with TDA1190P or TDA3190P Sound IF and Audio Output Devices
- Reverse RF AGC Types are Available: MC13008XP, MC13009XP

MONOMAX BLACK-AND-WHITE TV SUBSYSTEM

SILICON MONOLITHIC INTEGRATED CIRCUITS

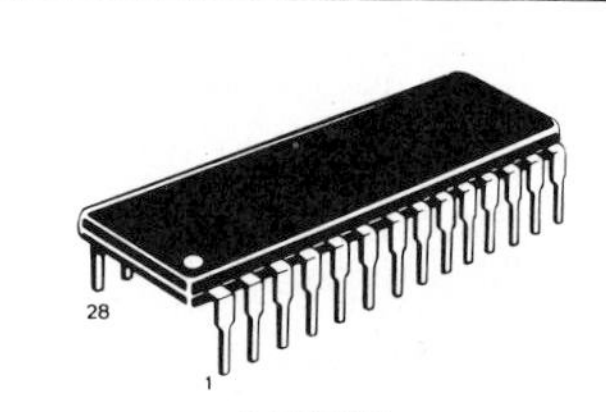

P SUFFIX
PLASTIC PACKAGE
CASE 710-02

FIGURE 1 — BASIC ELEMENTS OF THE SYSTEM

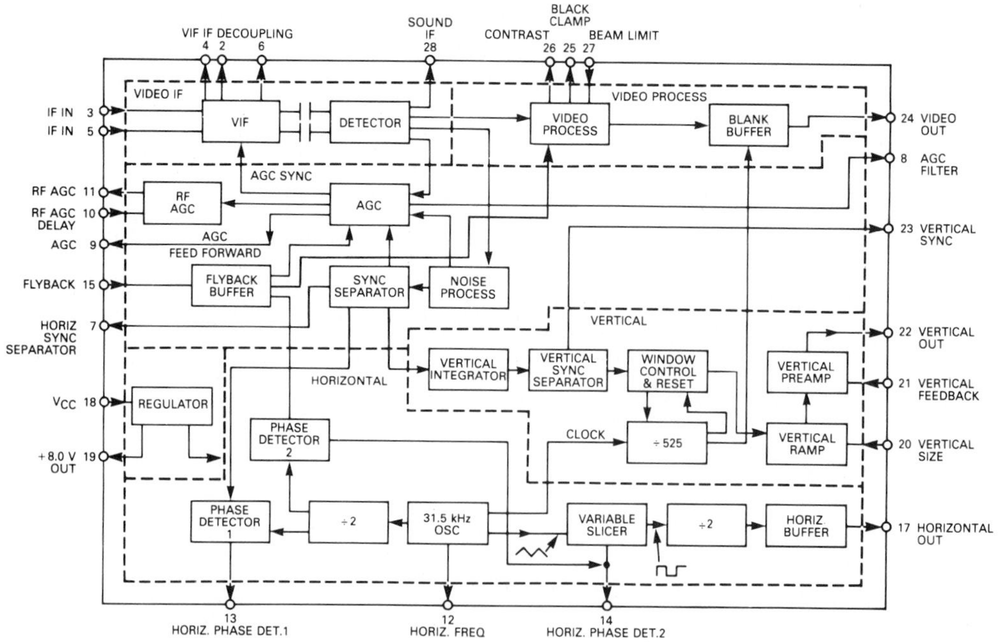

This document contains information on a new product. Specifications and information herein are subject to change without notice.

ADI-1081

MFOE1201
MFOE1202

Designer's Data Sheet

FIBER OPTICS
HIGH-POWER
AlGaAs LED

AlGaAs FIBER OPTIC EMITTER

. . . designed for fiber optic applications requiring high power and fast response time. It is spectrally matched to the first window minimum attenuation region of most fiber optic cables. Motorola's package fits directly into standard fiber optic connector systems. Applications include CATV, computer and graphics systems, industrial controls, military and others.

- Fast Response — Digital Data to 200 Mbaud (NRZ)
- Guaranteed 100 MHz Analog Bandwidth
- Hermetic Package
- Internal Lensing Enhances Coupling Efficiency
- Complements All Motorola Fiber Optic Detectors
- Compatible with AMP #228756-1, Amphenol #905-138-5001 and Radiall #FO86600380 Receptacles Using Motorola Alignment Bushing MFOA06 (Included)

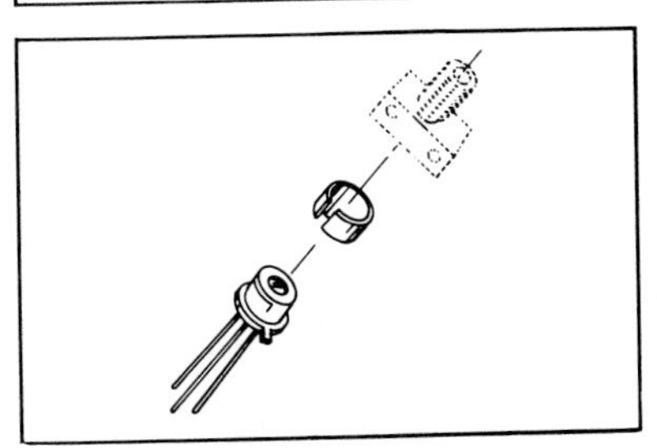

Designer's Data for "Worst Case" Conditions

The Designer's Data Sheet permits the design of most circuits entirely from the information presented. Limit data — representing device characteristics boundaries — are given to facilitate "worst case" design.

MAXIMUM RATINGS

Rating	Symbol	Value	Unit
Reverse Current	I_R	1.0	mA
Forward Current-Continuous	I_F	150	mA
Total Device Dissipation @ $T_A = 25°C$ Derate above 25°C	P_D	250 2.5	mW mW/°C
Operating Temperature Range	T_A	−65 to +125	°C
Storage Temperature Range	T_{stg}	−65 to +150	°C

THERMAL CHARACTERISTICS

Characteristics	Symbol	Max	Unit
Thermal Resistance, Junction to Ambient	θ_{JA}	400 225*	°C/W

*Installed in compatible metal connector housing with Motorola alignment bushing.

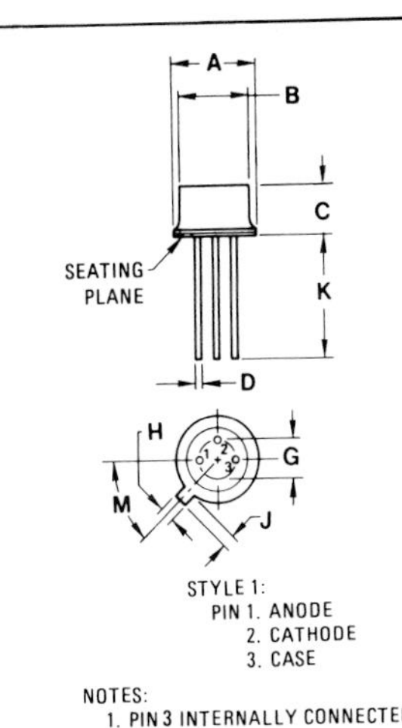

STYLE 1:
PIN 1. ANODE
2. CATHODE
3. CASE

NOTES:
1. PIN 3 INTERNALLY CONNECTED TO CASE.
2. LEAD POSITIONAL TOLERANCE AT SEATING PLANE: ⌖ ⌀ 0.36 (0.014) Ⓜ T A Ⓜ H Ⓜ
3. DIMENSIONS A AND H ARE DATUMS AND T IS A DATUM PLANE.

DIM	MILLIMETERS MIN	MILLIMETERS MAX	INCHES MIN	INCHES MAX
A	5.31	5.84	0.209	0.230
B	4.65	4.70	0.183	0.185
C	3.12	3.28	0.123	0.129
D	0.41	0.48	0.016	0.019
G	2.54 BSC		0.100 BSC	
H	0.99	1.17	0.039	0.046
J	0.84	1.22	0.033	0.048
K	12.70	–	0.500	–
M	45° BSC		45° BSC	

CASE 210A-01

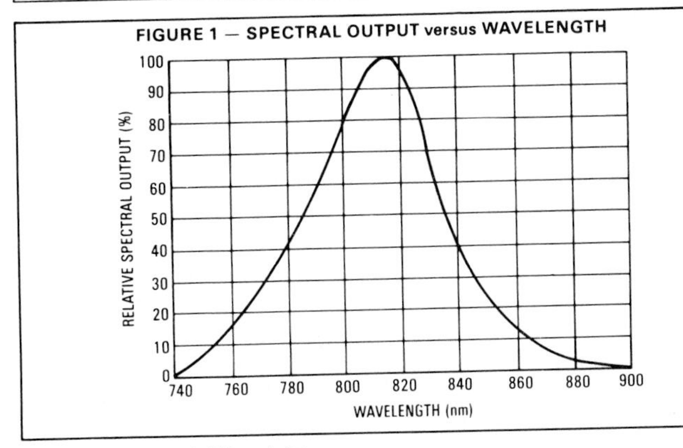

FIGURE 1 — SPECTRAL OUTPUT versus WAVELENGTH

Designer's is a trademark of Motorola Inc.

DS2689 R1

ELECTRICAL CHARACTERISTICS (T_A = 25°C)

Characteristic	Symbol	Min	Typ	Max	Unit
Reverse Breakdown Voltage (I_R = 100 μA)	$V_{(BR)R}$	2.0	4.0	—	Volts
Forward Voltage (I_F = 100 mA)	V_F	1.5	1.9	2.2	Volts
Total Capacitance (V_R = 0 V, f = 1.0 MHz)	C_T	—	70	—	pF
Crossmodulation Distortion, Figure 17 (80 mA bias, 50% depth of modulation)	XMOD	—	-50	—	dB
Intermodulation Distortion, Figure 17 (80 mA bias, 50% depth of modulation)	IMOD	—	-60	—	dB
Electrical Bandwidth, Figure 13 (I_F = 80 mAdc, measured 10 MHz to 110 MHz)	BWE	100	—	—	MHz

OPTICAL CHARACTERISTICS (T_A = 25°C)

Characteristic	Symbol	Min	Typ	Max	Unit
Total Power Output (I_F = 100 mA, λ ≈ 815 nm) MFOE1201	P_o	—	1500 (1.76)	—	μW(dBm)
MFOE1202		—	2400 (3.80)	—	
Power Launched, Figure 2 (I_F = 100 mA) MFOE1201	P_L	90 (-10.5)	—	180 (-7.4)	μW(dBm)
MFOE1202		150 (-8.2)	—	300 (-5.2)	
Numerical Aperture of Output Port (at -10 dB), Figure 7 (250 μm [10 mil] diameter spot)	NA	—	0.30	—	—
Wavelength of Peak Emission @ 100 mAdc	λ	—	815	—	nm
Spectral Line Half Width	—	—	50	—	nm
Optical Rise and Fall Times, Figure 12 (I_F = 100 mAdc)	t_r	—	2.8	4.0	ns
	t_f	—	3.5	6.0	

FIGURE 2 — LAUNCHED POWER (P_L) TEST SET

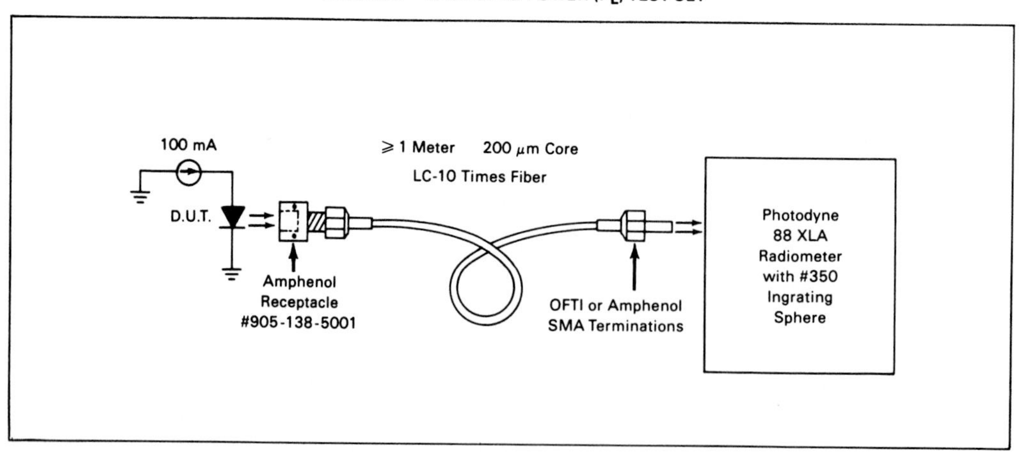

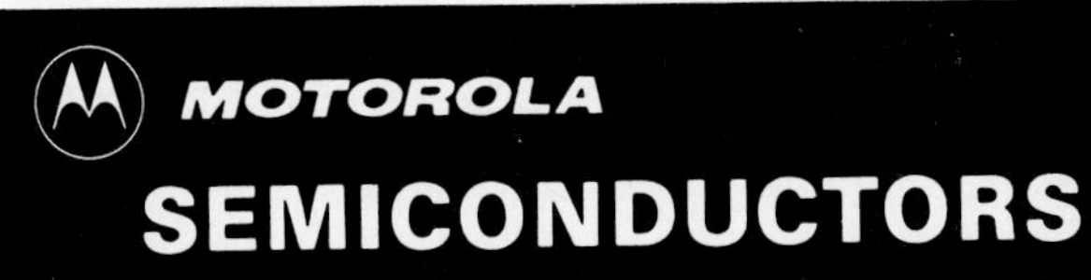

MFOD1100

FIBER OPTICS
PIN PHOTO DIODE

PIN PHOTO DIODE FOR FIBER OPTIC SYSTEMS

. . . designed for infrared radiation detection in high frequency Fiber Optic Systems. It is packaged in Motorola's hermetic TO-206AC (TO-52) case, and it fits directly into standard fiber optic connectors. The metal connectors provide excellent RFI immunity. Major applications are: CATV, video systems, M68000 microprocessor systems, industrial controls, computer and peripheral equipment, etc.

- Fast Response — 1.0 ns Max @ 5.0 Volts
- Analog Bandwidth (−3.0 dB) Greater Than 250 MHz
- Performance Matched to Motorola Fiber Optic Emitters
- TO-206AC (TO-52) Package — Small, Rugged, and Hermetic
- Compatible with AMP #228756-1, Amphenol #905-138-5001 and Radiall #F086600380 Receptacles Using Motorola Plastic Alignment Bushing MF0A06 (Included)

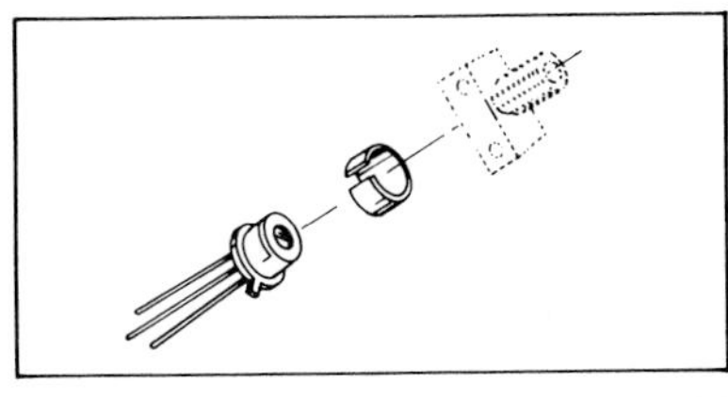

MAXIMUM RATINGS (T_A = 25°C Unless otherwise noted)

Rating	Symbol	Value	Unit
Reverse Voltage	V_R	50	Volts
Total Device Dissipation @ T_A = 25°C Derate above 25°C	P_D	50 0.50	mW mW/°C
Operating Temperature Range	T_A	−65 to +125	°C
Storage Temperature Range	T_{stg}	−65 to +150	°C

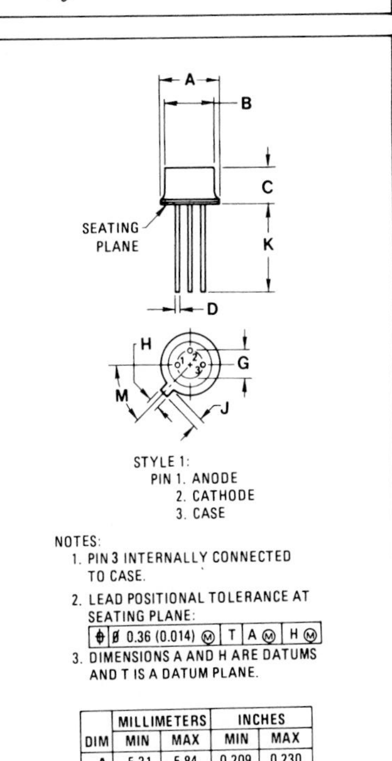

STYLE 1:
PIN 1. ANODE
2. CATHODE
3. CASE

NOTES:
1. PIN 3 INTERNALLY CONNECTED TO CASE.
2. LEAD POSITIONAL TOLERANCE AT SEATING PLANE: ⌖ Ø 0.36 (0.014) Ⓜ | T | A Ⓜ | H Ⓜ
3. DIMENSIONS A AND H ARE DATUMS AND T IS A DATUM PLANE.

DIM	MILLIMETERS MIN	MILLIMETERS MAX	INCHES MIN	INCHES MAX
A	5.31	5.84	0.209	0.230
B	4.65	4.70	0.183	0.185
C	3.12	3.28	0.123	0.129
D	0.41	0.48	0.016	0.019
G	2.54 BSC		0.100 BSC	
H	0.99	1.17	0.039	0.046
J	0.84	1.22	0.033	0.048
K	12.70	–	0.500	–
M	45° BSC		45° BSC	

CASE 210A-01

FIGURE 1 — RELATIVE SPECTRAL RESPONSE

RELATIVE RESPONSE (%): 0, 10, 20, 30, 40, 50, 60, 70, 80, 90, 100

λ, WAVELENGTH (μm): 0.2, 0.3, 0.4, 0.5, 0.6, 0.7, 0.8, 0.9, 1.0, 1.1, 1.2

DS2694R1

ELECTRICAL CHARACTERISTICS (T_A = 25°C)

Characteristic	Symbol	Min	Typ	Max	Unit
Dark Current (V_R = 5.0 V, R_L = 1.0 M, H ≈ 0, Figure 2)	I_D	—	—	1.0	nA
Reverse Breakdown Voltage (I_R = 10 μA)	$V_{(BR)R}$	50	—	—	Volts
Forward Voltage (I_F = 50 mA)	V_F	—	0.70	1.0	Volts
Total Capacitance (V_R = 5.0 V, f = 1.0 MHz)	C_T	—	—	2.5	pF
Noise Equivalent Power	NEP	—	50	—	fW/√Hz

OPTICAL CHARACTERISTICS (T_A = 25°C)

Characteristic	Symbol	Min	Typ	Max	Unit
Responsivity @ 815 nm (V_R = 5.0 V, P = 10 μW, Figure 3,5)	R	0.30	0.35	—	μA/μW
Response Time @ 815 nm (V_R = 5.0 V)	t_r, t_f	—	0.5	1.0	ns
Effective Input Port Diameter (Figure 4)	—	—	300 0.012	—	Microns Inches
10 dB (90%) Numerical Aperture of Input Port (Figure 4)	NA	—	0.40	—	—

TYPICAL CHARACTERISTICS

FIGURE 2 — DARK CURRENT versus TEMPERATURE

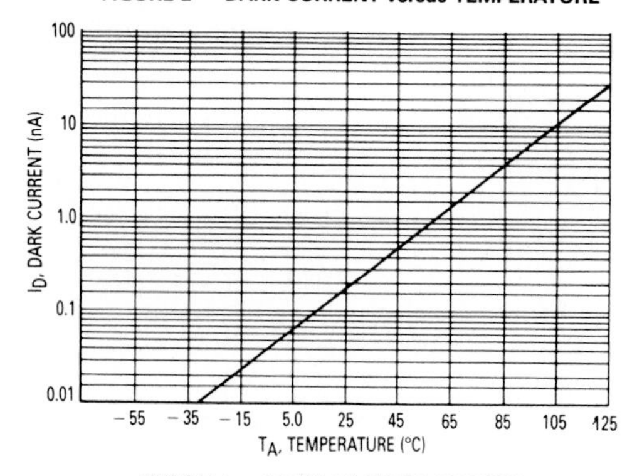

FIGURE 3 — RESPONSIVITY TEST CONFIGURATION

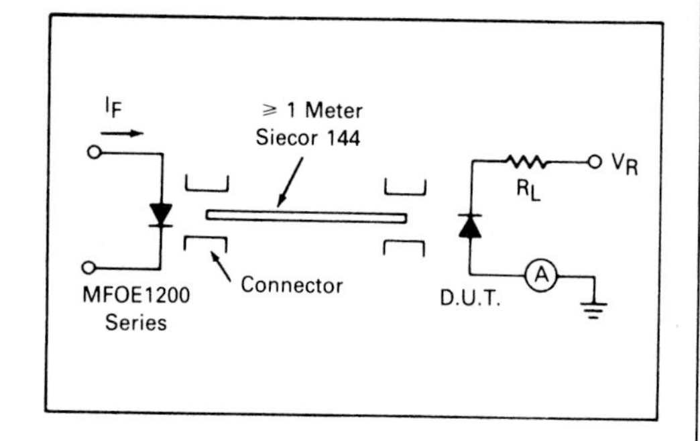

FIGURE 4 — PACKAGE CROSS SECTION

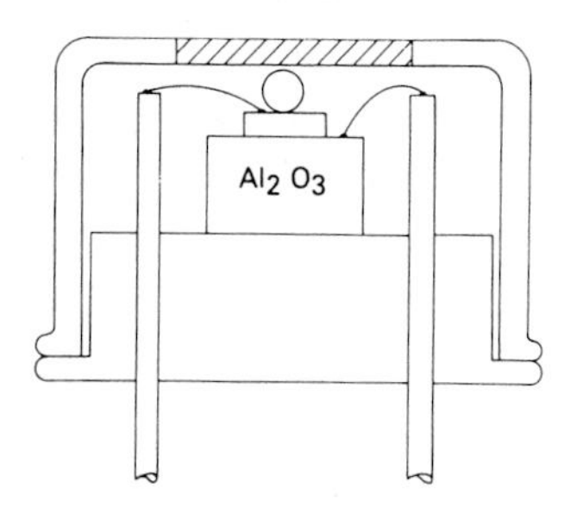

FIGURE 5 — DETECTOR CURRENT versus FIBER LENGTH

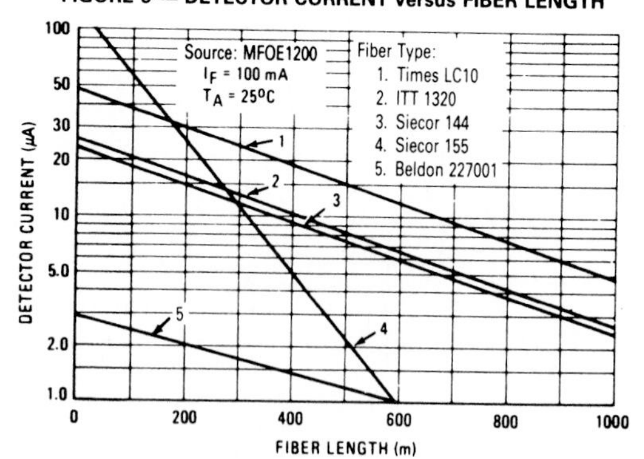

MOTOROLA *Semiconductor Products Inc.*

BOX 20912 • PHOENIX, ARIZONA 85036 • A SUBSIDIARY OF MOTOROLA INC.

16230 3 PRINTED IN USA (9/83) MPS 12M

MOC3009
MOC3010
MOC3011
MOC3012

OPTICALLY ISOLATED TRIAC DRIVERS

These devices consist of gallium-arsenide infrared emiting diodes, optically coupled to silicon bilateral switch and are designed for applications requiring isolated triac triggering, low-current isolated ac switching, high electrical isolation (to 7500 V peak), high detector standoff voltage, small size, and low cost.

- UL Recognized File Number 54915

OPTO COUPLER/ISOLATOR

PHOTO TRIAC DRIVER OUTPUT

250 VOLTS

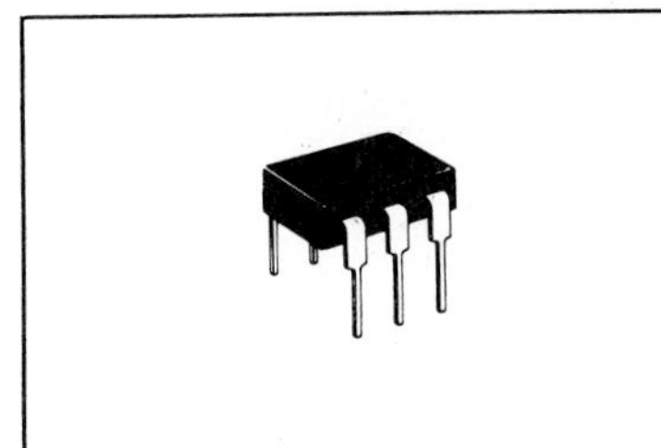

MAXIMUM RATINGS (T_A = 25°C unless otherwise noted)

Rating	Symbol	Value	Unit
INFRARED EMITTING DIODE MAXIMUM RATINGS			
Reverse Voltage	V_R	3.0	Volts
Forward Current — Continuous	I_F	50	mA
Total Power Dissipation @ T_A = 25°C Negligible Power in Transistor Derate above 25°C	P_D	100 1.33	mW mW/°C
OUTPUT DRIVER MAXIMUM RATINGS			
Off-State Output Terminal Voltage	V_{DRM}	250	Volts
On-State RMS Current T_A = 25°C (Full Cycle, 50 to 60 Hz) T_A = 70°C	$I_{T(RMS)}$	100 50	mA mA
Peak Nonrepetitive Surge Current (PW = 10 ms, DC = 10%)	I_{TSM}	1.2	A
Total Power Dissipation @ T_A = 25°C Derate above 25°C	P_D	300 4.0	mW mW/°C
TOTAL DEVICE MAXIMUM RATINGS			
Isolation Surge Voltage (1) (Peak ac Voltage, 60 Hz, 5 Second Duration)	V_{ISO}	7500	Vac
Total Power Dissipation @ T_A = 25°C Derate above 25°C	P_D	330 4.4	mW mW/°C
Junction Temperature Range	T_J	-40 to +100	°C
Ambient Operating Temperature Range	T_A	-40 to +70	°C
Storage Temperature Range	T_{stg}	-40 to +150	°C
Soldering Temperature (10 s)	—	260	°C

(1) Isolation Surge Voltage, V_{ISO}, is an internal device dielectric breakdown rating.

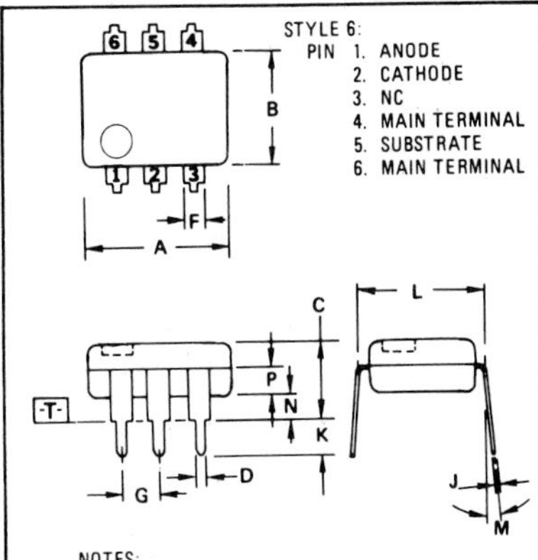

NOTES:
1. DIMENSIONS A AND B ARE DATUMS.
2. -T- IS SEATING PLANE.
3. POSITIONAL TOLERANCES FOR LEADS: ⌖ Ø 0.13 (0.005) Ⓜ T A Ⓜ B Ⓜ
4. DIMENSION L TO CENTER OF LEADS WHEN FORMED PARALLEL.
5. DIMENSIONING AND TOLERANCING PER ANSI Y14.5, 1973.

DIM	MILLIMETERS MIN	MILLIMETERS MAX	INCHES MIN	INCHES MAX
A	8.13	8.89	0.320	0.350
B	6.10	6.60	0.240	0.260
C	2.92	5.08	0.115	0.200
D	0.41	0.51	0.016	0.020
F	1.02	1.78	0.040	0.070
G	2.54 BSC		0.100 BSC	
J	0.20	0.30	0.008	0.012
K	2.54	3.81	0.100	0.150
L	7.62 BSC		0.300 BSC	
M	0°	15°	0°	15°
N	0.38	2.54	0.015	0.100
P	1.27	2.03	0.050	0.080

CASE 730A-01

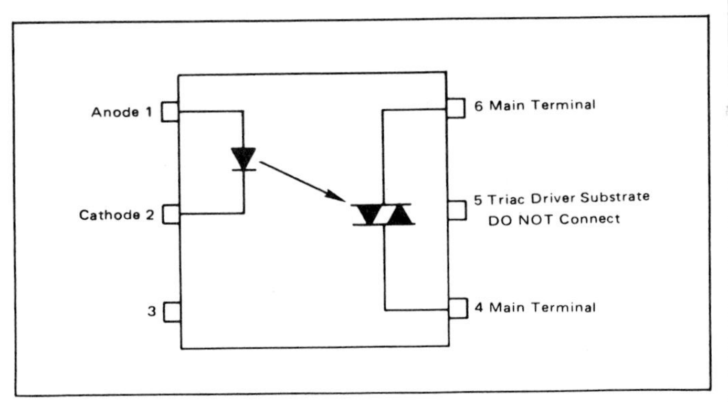

DS2535 R3

ELECTRICAL CHARACTERISTICS (T_A = 25°C unless otherwise noted)

Characteristic		Symbol	Min	Typ	Max	Unit
LED CHARACTERISTICS						
Reverse Leakage Current (V_R = 3.0 V)		I_R	–	0.05	100	μA
Forward Voltage (I_F = 10 mA)		V_F	–	1.2	1.5	Volts
DETECTOR CHARACTERISTICS (I_F = 0 unless otherwise noted)						
Peak Blocking Current, Either Direction (Rated V_{DRM}, Note 1)		I_{DRM}	–	10	100	nA
Peak On-State Voltage, Either Direction (I_{TM} = 100 mA Peak)		V_{TM}	–	2.5	3.0	Volts
Critical Rate of Rise of Off-State Voltage, Figure 3		dv/dt	–	2.0	–	V/μs
Critical Rate of Rise of Commutation Voltage, Figure 3 (I_{load} = 15 mA)		dv/dt	–	0.15	–	V/μs
COUPLED CHARACTERISTICS						
LED Trigger Current, Current Required to Latch Output (Main Terminal Voltage = 3.0 V)	MOC3009	I_{FT}		15	30	mA
	MOC3010		–	8.0	15	
	MOC3011		–	5.0	10	
	MOC3012		–	–	5.0	
Holding Current, Either Direction		I_H	–	100	–	μA

Note 1. Test voltage must be applied within dv/dt rating.

2. Additional information on the use of the MOC3009/3010/3011 is available in Application Note AN-780.

TYPICAL ELECTRICAL CHARACTERISTICS
T_A = 25°C

FIGURE 1 – ON-STATE CHARACTERISTICS

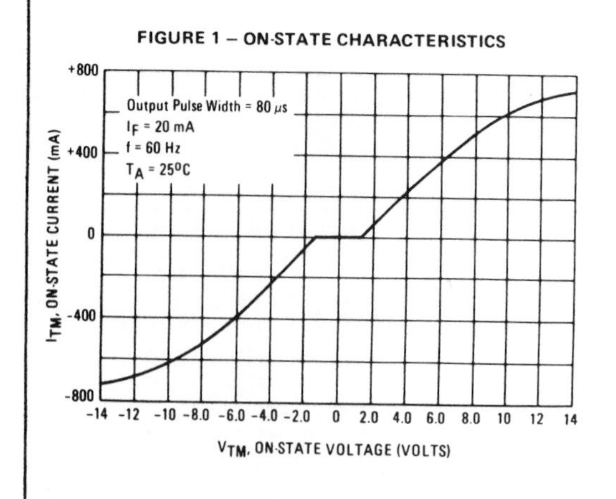

FIGURE 2 – TRIGGER CURRENT versus TEMPERATURE

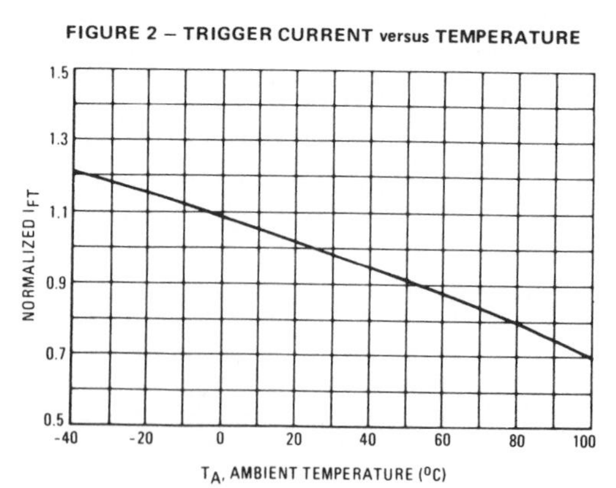

Order this data sheet by MR2520L/D

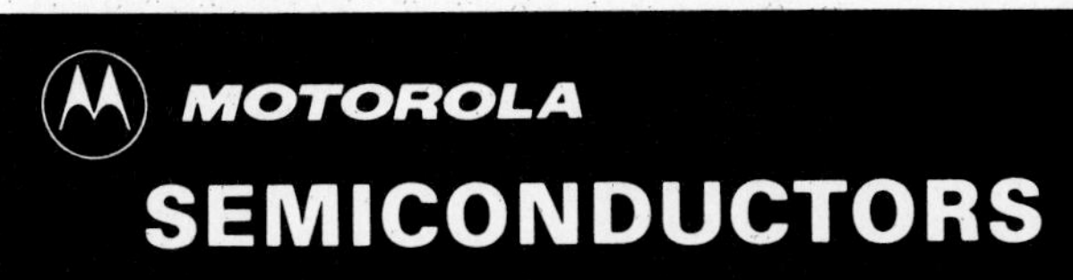

MR2520L
MR2525L

OVERVOLTAGE TRANSIENT SUPPRESSORS

. . . designed for applications requiring a diode with reverse avalanche characteristics for use as reverse power transient suppressors. Developed to suppress transients in the automotive system, these devices operate in the forward mode as standard rectifiers or reverse mode as power zener diodes and will protect expensive mobile transceivers, radios and tape decks from over-voltage conditions.

- High Power Capability
- Economical
- Increased Capacity by Parallel Operation

OVERVOLTAGE TRANSIENT SUPPRESSORS

MAXIMUM RATINGS

Rating	Symbol	Value	Unit
DC Peak Repetitive Reverse Voltage Working Peak Reverse Voltage DC Blocking Voltage	V_{RRM} V_{RWM} V_R	23	Volts
Repetitive Peak Reverse Surge Current MR2520L MR2525L (Time Constant = 10 ms, Duty Cycle ≤ 1.0%, T_C = 25°C)	I_{RSM}	 68 110	Amp
Operating and Storage Junction Temperature Range	T_J, T_{stg}	−65 to +175	°C

ELECTRICAL CHARACTERISTICS

Characteristic	Symbol	Min	Max	Unit
Reverse Current (V_R = 20 Vdc, T_C = 25°C) (V_R = 20 Vdc, T_C = 100°C)	I_R	 — —	 50 300	μAdc
Breakdown Voltage (1) (I_R = 100 mAdc, T_C = 25°C)	$V_{(BR)}$	24	32	Volts
Breakdown Voltage (1) MR2525L only (I_R = 40 Amp, T_C = 85°C)	$V_{(BR)}$	—	40	Volts

(1) Pulse Test: Pulse Width ≤ 300 μs, Duty Cycle ≤ 2.0%.

THERMAL CHARACTERISTICS

Characteristic	Lead Length	Symbol	Max	Unit
Thermal Resistance, Junction to Lead @ Both Leads to Heat Sink, Equal Length	1/4″ 3/8″ 1/2″	$R_{\theta JL}$	7.5 10 13	°C/W

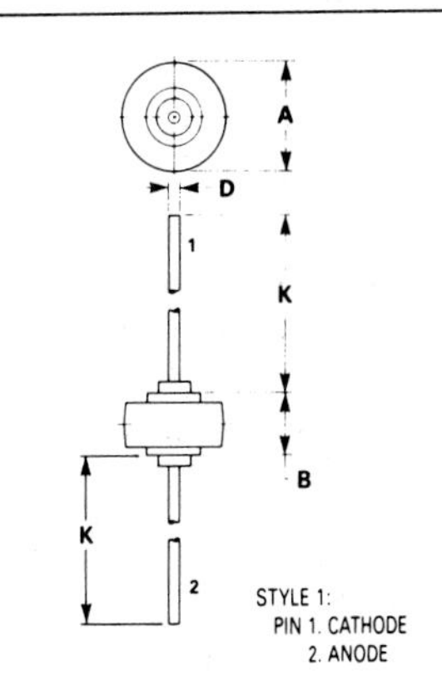

NOTE:
1. CATHODE SYMBOL ON PKG

DIM	MILLIMETERS MIN	MILLIMETERS MAX	INCHES MIN	INCHES MAX
A	10.03	10.29	0.395	0.405
B	5.94	6.25	0.234	0.246
D	1.27	1.35	0.050	0.053
K	25.15	25.65	0.990	1.010

CASE 194-01 (MR2525L)

DIM	MILLIMETERS MIN	MILLIMETERS MAX	INCHES MIN	INCHES MAX
A	8.43	8.69	0.332	0.342
B	5.94	6.25	0.234	0.246
D	1.27	1.35	0.050	0.053
K	25.15	25.65	0.990	1.010

CASE 194-04 (MR2520L)

MECHANICAL CHARACTERISTICS:

CASE: Transfer Molded Plastic

MAXIMUM LEAD TEMPERATURE FOR SOLDERING PURPOSES: 350°C 3/8″ from case for 10 seconds at 5 lbs. tension

FINISH: All external surfaces are corrosion-resistant, leads are readily solderable

POLARITY: Indicated by diode symbol or cathode band

WEIGHT: 2.5 Grams (approx.)

 DS6134R1

Glossary

Note: Numbers in parentheses refer to text sections where basic use of the defined term will be found. Turn to that section for further information.

Active device A component such as a transistor which is able to amplify voltage and/or current. (1.1)

Active region That part of the characteristic curves for an active device where typical operation is achieved. *See also* Saturation *and* Cutoff. (8.4)

AGC (automatic gain control) A process where the gain of a system is controlled automatically, usually via feedback. (16.2)

Alpha The gain figure for a common-base BJT amplifier; the ratio of collector-to-emitter currents in a BJT. (12.3)

Amp The abbreviation for amplifier (when not used to indicate amperes). (3.3)

Amplification The process of producing an output voltage or current which is higher in magnitude that the input. (1.5)

Amplifier An electronic circuit that is capable of producing amplification and/or input-to-output isolation. (1.6)

Anode The terminal of a semiconductor diode which must be biased positively to produce forward current flow. (5.2)

Attenuation The reduction of a voltage or current level.

Automatic gain control *See* AGC.

Barkenhausen criterion The necessary circuit functions which will maintain oscillation (unity system gain + positive feedback). (11.6)

Battery backup When a battery is used to provide power automatically should a power supply fail to operate. (13.5)

Beta The dc gain figure for a BJT transistor (when used in the common-emitter mode); the ratio of collector current to base current in a BJT amplifier circuit. (5.4)

Beta-independent biasing A bias arrangement for a BJT common-emitter amplifier where the gain of the circuit is not controlled by the gain of the device used. (6.5)

Bias The application of dc constant voltage or current levels to any active (or other) device to maintain a constant mode of operation (often prior to application of the device). (3.5)

Bipolar junction transistor *See* BJT.

BJT (bipolar junction transistor) One of several semiconductor active devices; a transistor that produces a current function (found in NPN and PNP forms). (3.3 and 5.4)

Bleeder resistor (power supply) A resistor used to deplete the charge on power supply capacitors when the power supply is turned off (for safety purposes). (14.4)

Block diagram A graphic sketch of the basic steps necessary to achieve a function, indicating the order of operation or flow of the related process. (1.2)

Bode plot A plot of the frequency response of an amplifier using a logarithmic axis for frequency and dB figures for gain. (10.3)

Break frequency The upper or lower frequency at which an amplifier's gain is −3 dB below the midband gain value. (10.4)

Breakover voltage Voltage necessary to cause conduction in one of several thyristor devices. (15.3)

Bridge rectifier A special four-diode arrangement that produces a full-wave rectified output using a simple transformer input. (14.3)

Buffering The process of matching the output of one system to the input of another system, often involving impedance matching or isolation relative to the first system. (11.6)

Butterworth filter A special arrangement of capacitive and resistive elements to provide low- or high-pass filtering, often used in an active filter. (11.5)

Bypass capacitor Any capacitor used to pass ac current instead of having it flow through a resistive (or other) device; a capacitor in parallel with another device. (6.7)

Capacitive coupling A process by which the flow of ac current from one amplifier is passed to another amplifier through the low impedance of a capacitor, but where dc bias current from one circuit is prevented from flowing into the other circuit by that same capacitor. (13.3)

Cathode The terminal of a semiconductor diode that must be biased negatively to produce forward current flow. (5.2)

Characteristic curve(s) A graph that illustrates the input-to-output relationships for an active device. (5.2)

Clamper A circuit that produces a dc level from an ac input and then adds that level to the ac [often producing a pulsed dc output (single polarity)]. (16.2)

Class A amplifier An amplifier that amplifies 100% of an input signal. (12.1)

Class B amplifier An amplifier that amplifies 50% of an input signal (half wave). (12.1)

Clipper A circuit that limits the amplitude or polarity of its output given a variable amplitude input. (16.2)

Closed-loop configuration (op-amp) An op-amp circuit that uses negative feedback to limit the gain of the circuit. (9.2)

Closed-loop control A control system that uses a sample of the output function to produce the control function. (15.1)

CMRR [common-mode rejection ratio (op-amp)] The rating of an op-amp's ability to eliminate common-mode signals. (Common-mode signals are those which are identical in frequency, phase, and shape.) (9.4)

Colpits oscillator An oscillator circuit using two capacitors and one inductor in a unique configuration. (11.6)

Common-base amplifier (BJT) A configuration of a BJT amplifier in which the base terminal of the BJT is common to both the input and output signals. (12.3)

Common-collector amplifier (BJT) A configuration of a BJT amplifier in which the collector terminal of the BJT is common to both the input and output signals. (12.3)

Common-emitter amplifier (BJT) A configuration of a BJT amplifier in which the emitter terminal of the BJT is common to both the input and output signals. (6.2)

Common-mode rejection ratio *See* CMRR.

Common-source amplifier (FET) A configuration of an FET amplifier in which the source terminal of the FET is common to both the input and output signals. (5.5)

Comparator Any of several circuits designed to produce an output based on two input signals. (1.7)

Coupling A category of means for feeding the output signal from one amplifier to the input terminal of a following amplifier or load. (13.3)

Current amplifier Amplifiers that accept an input current and produce a corresponding output current. (4.3)

Current mirror (or mode) op-amp A special form of op-amp which senses input current levels and produces an output proportional to these currents. (16.1)

Cutoff The point where an amplifier cannot produce an output change if the input increases any further. (6.5 and 8.4)

Darlington transistor A special form of semiconductor device blending the effects of two BJT devices, one driving the next, in one package. Often, the Darlington has quite high current-gain statistics. (8.5)

dB (Decibel) A special form of rating signal levels using a logarithmic scaling. (10.3)

Decimal gain Gain that is less than unity (1), a loss of signal level. (4.1)

Diac A semiconductor device that can conduct in either direction when a threshold voltage level is exceeded. (15.3)

Differential amplifier An amplifier that produces an output which is proportional to the difference between two input signals. (9.3)

Differentiator A circuit configuration that will produce the mathematical function of subtraction. (12.2)

Digital A category of electronic functions which are designed to manipulate finite mathematical and logic equations. (1.4)

Diode A semiconductor device that conducts current in only one direction. (5.2)

DIP (dual-in-line package) A form of packaging for integrated circuits and other electronic devices. (5.6)

Direct coupling The process of directly wiring the output of one amplifier or system to the input of another. Any dc or ac levels present in the first system will be passed on to the other system. (13.3)

D-MOSFET Depletion type of MOSFET. (6.5 and 8.5)

Dual-in-line package *See* DIP.

Emitter A terminal on a bipolar junction transistor. (5.4)

E-MOSFET Enhancement-type MOSFET. (5.5 and 8.5)

Family of curves *See* Characteristic curves.

Feedback Any process that extracts a sample of an output and uses that sample to control the function of the system. (2.4 and 4.7)

FET (field-effect transistor) A category of active semiconductor devices which produces a transconductance gain figure. (3.3 and 5.5)

Field-effect transistor *See* FET.

Filter Any frequency-selective electronic circuit. (2.3)

Filters (power supply) A circuit (often using a capacitor) used to convert the pulsed dc output of a rectifier into a constant dc output level. (14.3)

Forward bias The condition in a semiconductor device when conduction is had; any voltage level that specifically will produce such conduction. (5.2)

Frequency response A charting (or graph) of a system's ability to reproduce, at its output, any frequency. A notation of the limits to a system's ability to reproduce any frequency (e.g., a frequency response from 15 Hz to 32 kHz).

Full-wave rectifier An ac-to-pulsed dc conversion circuit that will produce an uninterrupted output with a variable (pulsed) level of a single polarity. (14.2)

Gain The ratio of output level to input level for an amplifier. (2.1)

Gain–bandwidth product A rating for (usually) op-amps which is a product of the maximum open-loop gain and (usually) maximum frequency that can be applied and still maintain a gain of unity (1) (rated in megahertz). (9.2)

GBW Gain–bandwidth product. (9.2)

Girator A circuit that simulates the voltage/current relationships seen in an inductor, but does not use an inductor. (11.5)

Half-power points (frequency response) The upper and lower frequencies at which the power output of an amplifier is reduced to one-half of its midband output level. This coordinates with voltage or current outputs that are 0.707 of the midband level as well. (10.3)

Half-wave rectifier An ac-to-pulsed dc conversion circuit which will produce an output only on either the positive or negative cycle of the ac input, thus producing an interrupted and variable (pulsed) level of a single polarity. (13.2)

Hartley oscillator An oscillator using two inductors and one capacitor in a unique configuration. (11.6)

Heat sink Any mechanical device used to draw heat away from the body of a semiconductor device. (12.1)

Holding current The minimum current necessary to maintain conduction in any of several thyristor devices. (15.3)

***h*-parameters** The hybrid parameters (characteristics) of a device which rate forward gain, reverse gain, input impedance, and output impedance. (7.3)

Hybrid parameters *See* *h*-parameters.

Ideal amplifier Any of four configurations of amplifiers which will produce ideal input-to-output relationships under any operating conditions. (2.4 and 4.1)

Impedance matching The process (or circuit) of matching the output impedance (resistance) of one amplifier to the input impedance (resistance) of another amplifier or a load device. (3.2)

Instrumentation amplifier (op-amp) A special form of op-amp circuit which displays a high value of CMRR and is especially useful when taking delicate measurements where common-mode noise could produce problems. (16.2)

Integrated circuit A collection of active and passive devices included in one common package, producing a circuit that can achieve at least one electronic function. (1.5)

Integrator A circuit configuration that will perform the mathematical function of integration. (12.3)

Inverting amplifier An amplifier that produces an output which is opposite in polarity from the input in the case of a dc input and is 180° out of phase with the input in the case of an ac input. [Inversion is indicated by the use of a negative (−) sign in front of the gain statistic (e.g., $A_v = -100$.)] (9.3)

JFET Junction field-effect transistor. *See* FET. (5.5)

Junction capacitance An undesirable capacitance effect occurring as a by-product of the internal connections within semiconductor (and other) electronic devices. (10.4)

Junction potential The minimum voltage necessary to cause forward conduction in a semiconductor device. The constant voltage drop seen across such a device when it is conducting. (5.2)

Large-scale integration *See* LSI.

LASCR (light-activated silicon-controlled rectifier) An SCR that uses an optically (light) triggered gate lead. *See also* SCR. (16.1)

Latch(ing) circuit An electronic circuit which operates in such a way that if an appropriate input signal (often pulsed) is applied, the output will assume a fully conductive or nonconductive mode and stay in that mode until "triggered" out of that mode with a special input or by other means. (2.5)

LED (light-emitting diode) A semiconductor device that produces light as a result of forward current flow. (16.1)

Light-activated SCR *See* LASCR.

Light-emitting diode *See* LED.

Linear system A system or amplifier which has an output that is directly proportional to an input quantity (e.g., a 5% change in the input will cause a 5% change in the output). (1.4)

Load Any element tied to the output of a system or amplifier which uses or senses either voltage, current, or both. (6.4)

Load line A line drawn on the characteristic curves of a device which represents the effect of a load (resistance) on that device; where the load line intersects with sections of the characteristic curves graphically illustrates the operational voltage and/or current values for that circuit. (8.4)

Load resistor *See* Load.

LSI (large-scale integration) Integrated circuitry which incorporates many elements (often over 100) and often completes several electronic functions in one single package. (3.3 and 5.7)

Metal-oxide semiconductor field-effect transistor *See* MOSFET.

Midband gain The gain of an amplifier or system at frequencies which are in the center of that amplifier's operational bandwidth. Sometimes called the amplifier's "normal" gain. (10.3)

Miller capacitance The effective input and/or output capacitance of an amplifier or system derived by using both true capacitance figures and capacitance effects seen through the function (usually voltage amplification) of the amplifier. (10.4)

Mini-DIP An eight-pin version of the popular integrated-circuit package called the dual-in-line package. (5.7)

Mixer An amplifier or system which accepts more than one input and produces an output that is a mathematical blend of the inputs. (1.7 and 12.2)

Mixing amplifier *See* Mixer.

Modular subsystems (electronic) Individual circuits or integrated circuits which perform a complete function as an integral part of a larger system. Often these subsystems are considered discrete blocks in the system and are not subject to internal repair (e.g., defective modules are discarded rather than repaired). (3.5)

MOSFET (metal-oxide semiconductor field-effect transistor) A special category of field-effect transistors. *See also* D-MOSFET and E-MOSFET. (5.5 and 8.5)

Multistage amplifier An amplifying system containing more than one amplifier section (or stage). (2.1 and Chapter 13)

Multistage feedback The application of feedback where the feedback potential (voltage or current) is returned to a stage other than the one at which the feedback sample is obtained. (4.7)

Negative feedback The use of a feedback quantity which is opposite in polarity from the input (dc) or is 180° out of phase with the input (ac). (*See also* Feedback.) (2.4 and 4.7)

Negative gain (statistic) When an amplifier or system produces both gain and a 180° phase inversion of the input signal, the gain statistic is listed with a negative sign (e.g., $A_v = -24$ indicates both a gain of 24 and inversion, as contrasted to an $A_v = +24$, which is the same gain but with no inversion). (4.1)

Noninverting amplifier (op-amp) An op-amp circuit which produces an output that is in phase with the input signal. (9.3)

Nonlinear distortion Distortion of the output waveform of an amplifier or system, when compared to the input, due to internal components which do not produce a linear (constant) gain statistic for all input conditions. (8.4)

Norton amplifier (op-amp) *See* Current mirror (or mode) op-amp.

NPN transistor One of two forms of bipolar junction transistors. (5.4)

N-type (JFET) (also called N-channel) One of two versions of the junction field-effect transistor. (5.5)

Null To set a quantity to zero. *See also* Nulling. (9.1)

Nulling (op-amp) To adjust the output of an op-amp to zero voltage when identical inputs are applied to both the inverting and noninverting inputs. (9.1)

Offset A (usually) dc quantity used to raise or lower the average voltage of (usually) an ac quantity. (6.2 and 9.1)

Op-amp (operational amplifier) A category of integrated circuits which provide linear amplification using a minimum of external components for support. (3.3 and 5.7)

Open-loop configuration (op-amp) An op-amp circuit that does not use feedback to control the gain of the circuit, therefore providing the maximum functional gain from the device. (9.2)

Open-loop control A control circuit that does not use a feedback process to sense the level of the output quantity and use that level to adjust the output. (15.1)

Operational amplifier *See* Op-amp.

Optical interruptor One of several optically coupled devices where the optical path can be interrupted (broken) by a mechanical device. (16.1)

Optical isolator Any of several devices where the electrical input is converted to a light beam which passes to a light-sensitive device that produces an electrical output. Isolation is achieved since electrical quantities seen at the output of the device cannot affect quantities produced at the input of the device via the optical path. (16.1)

Optical triac *See* LASCR.

Opto-isolator *See* Optical isolator.

Oscillator A circuit that generates an ac signal internally. Most oscillators are amplifiers that employ positive feedback and bandpass filters to produce a single-frequency output (no input signal is used). (11.6)

Output impedance (also called output resistance) The Thévenin or Norton equivalent resistance of an amplifier or system as viewed from the output terminals of that amplifier or system. [In this book output impedance usually refers to total nonreactive (capacitive or inductive) resistance.] (4.1)

Passive device Any electronic device that does not have the capacity to amplify or produce electronic control (usually resistors, capacitors, and inductors). (1.1 and 1.5)

Peak-point voltage (UJT) The triggering voltage for a UJT as applied between the emitter and base 1 lead. (15.3)

Phase control Using a phase-shifted ac waveform to produce a time delay in the turn-on function of an SCR or triac. (15.2)

Phase-shift oscillator An oscillator that uses a frequency selective phase-shifting network to produce positive feedback. (A 180° shift is usually achieved by using three filter sections which each provide 60° of phase shift at the selected frequency.) (11.6)

Photo *See* Optical.

Photo-darlington A Darlington transistor configuration which uses an optical (light) input to control output current. (16.1)

Photodiode A diode that uses an optical (light) input to provide various levels of conduction. (16.1)

Phototransistor A BJT transistor that uses an optical (light) input to control output current. (16.1)

PIV (peak inverse voltage) The maximum reverse-bias voltage that can be applied to a semiconductor device before it will break down and conduct in that direction. Usually used when rating diodes. (14.4)

Pinch-off For an FET, the point where increased gate voltage will not produce changes in drain current. (5.5)

PNP transistor One of two forms of bipolar junction transistors. (5.4)

Positive feedback The use of a feedback quantity which is identical in polarity to the input (dc) or is in phase with the input (ac). *See also* Feedback. (2.4 and 11.6)

Power amplifier Any of several amplifier configurations that is designed to produce power gain; usually accompanied by higher voltage and current levels necessary to drive an output load device (e.g., speaker). (1.7, 5.4, and 12.1)

Power conversion The conversion of one source of power to a different level or form (typically, an ac-to-dc power supply conversion). (1.5)

Power derating curve A curve that indicates a limit to power-handling capacity which is superimposed on the characteristic curves for a device. (12.1)

Power supply Any of several sources of (usually) dc or ac energy which is to be used by an amplifier or control device. *See also* Power conversion. (3.1)

Preamplifier Any small-signal amplifier used to increase the level of a signal prior to using a power amplifier. (1.7)

Programmable unijunction transistor *See* PUT.

P-type (JFET) (also called P-channel) One of two versions of the junction field-effect transistor. (5.5)

Pulsed dc Any waveform that has a variable amplitude but a constant polarity. (3.2)

Pulse rate control The control of an output quantity based on the rate (frequency) of occurrence of a sequence of pulses. (15.2)

Pulse-width control The control of an output quantity based on the width (duration) of the "on" portion of a pulsed control signal. (15.2)

PUT (programmable unijunction transistor) A unijunction transistor for which the trigger voltage can be set (programmed) by the use of an external voltage level. (15.3)

***Q*-point (quiescent point)** The voltage and current levels at which an amplifier (or other) system will function based on both the characteristics of the device and the bias conditions set on that device by external components. (8.4)

Radio-frequency choke *See* RFC.

Rectifier Either a device or circuit that can be used to convert ac waveforms to pulsed dc waveforms. (Diodes are often called rectifiers.) (14.2)

Reflected resistance (in this book) The Thévenin or Norton equivalent resistance seen at the input or output of an amplifier (or device) which is the result of both discrete resistances and the voltage or current amplification effect on other resistances. (13.4 and 13.5)

Regenerative feedback *See* Positive feedback.

Regulation The process of maintaining a constant output voltage and/or current under various operational and loading conditions (usually related to power supplies). (14.2)

Regulator A device or integrated circuit used to provide regulation. *See also* Zener diode. (14.2)

Remote regulators Regulators that are associated with specific circuit sections in a system rather than using one common regulator for all circuits within the system. (14.4)

Resonant circuit amplifier An amplifier using either bandpass or bandstop filtering at the input or output and thus providing amplification only at a select band of frequencies. (11.3 and 11.4)

Resonant circuit coupling Using a frequency-selective circuit to couple the output of one amplifier to the input of another (or to a load). (13.3 and 13.4)

Reverse bias The electrical polarity on a device which will not produce current flow. *See also* Forward bias. (5.2)

Reverse breakdown *See* PIV.

Reverse leakage current The small amount of current flow that occurs when a semiconductor device is reverse biased. (5.2)

RF Radio frequencies. (11.3)

RFC (radio-frequency choke) An inductive element used to block the passage of high (usually radio-frequency)-frequency signals. (11.3)

Ripple The amount of pulsed dc voltage still remaining on the output of a dc power supply after filtering has been applied (usually rated as a peak-to-peak level). (14.3)

Ripple factor The ratio of ripple voltage to the dc level outputted from a dc power supply (either as a decimal value or converted to a percentage). (14.3)

Ripple voltage *See* Ripple.

Roll-off The decline in output level of an amplifier (usually) at lower and upper frequency levels; the statistical rating of this decline (e.g., roll-off = −20 dB/decade). (10.3)

Saturation In an amplifier circuit, the condition where increased levels of input will not produce corresponding changes in the output level; a case where the maximum level of the output is produced. (8.4)

SBS (silicon bilateral switch) Functionally equivalent to the diac. *See also* Diac. (15.3)

Scaling amplifier A mixing amplifier where each input is not amplified using the same gain statistic. (12.2)

SCR (silicon-controlled rectifier) A control semiconductor device that uses a control input to trigger conduction of the output quantity. The SCR conducts only in one forward direction. (15.3)

Semiconductor A solid-state device, usually either silicon or germanium, whose conductivity is variable from that of an insulator to that of a low-resistance conductor. (5.1)

Series feedback A feedback process that introduces the feedback current in series with the input current. (4.7)

Series regulator A regulator circuit where the regulating device (often a transistor) is in series with the output load terminals. (14.4)

Shockley diode A semiconductor device that has a threshold (or triggering device) voltage level below which conduction is prevented. Above that potential conduction is maintained as long as adequate current is drawn through the device. This device conducts only in one forward direction. (15.3)

Shunt feedback A feedback process that introduces the feedback voltage in shunt (parallel) with the input voltage. (4.7)

Shunt regulator A regulator circuit where the regulating device is in parallel (shunt) with the output terminals. (14.4)

Silicon bilateral switch *See* SBS.

Silicon-controlled rectifier *See* SCR.

Silicon-controlled switch A control semiconductor that can be triggered into conduction using either positive or negative triggering potentials. (14.3)

Slew rate The maximum rate at which the output of an op-amp can respond to a square-wave or pulse input (units: volts per microsecond). (9.4)

Source follower amplifier (FET) An FET amplifier where the drain lead is common to both the input and output circuitry (also called a common-drain amplifier). (12.3)

Specification sheet *See* Spec sheet.

Spec sheet (specification sheet) A list of the electrical parameters of a device. (8.2)

Stability (in reference to amplifiers) The ability of an amplifier (or device) to maintain a constant-gain statistic under various electrical and environmental conditions. (3.2)

Stability biasing Providing a bias condition for an active device which assists in maintaining the stability of the circuit under conditions where the active device is not fully stable (usually by the use of negative feedback). (6.5)

Summing amplifier *See* Mixer.

Superposition (analysis) The process of analyzing a circuit with multiple sources by calculating the voltage and current values produced by each source independently and then preparing a composite algebraic sum of each of the solutions. (7.2)

Surge resistor (power supply) A resistor used to limit the current drawn by filtering elements from the rectifier in a power supply. (14.4)

Switching regulator A power supply regulator that uses a pulsed control signal to turn the flow of energy to a secondary filter circuit alternately on and off, based on the value of voltage maintained by that filter circuit. (14.2)

Thermal resistance (related to heat-sink function) The ability of either a power device or heat sink to transfer heat. Higher thermal resistance indicates a poorer ability to conduct heat; lower thermal resistance indicates a greater ability to conduct heat. (12.1)

Transconductance amplifier An amplifier that produces variations in output current based on variations in input voltage (gain statistic rated in amperes per volt or siemens). (4.5)

Transfer curve A graph that illustrates the ratio of input quantity to output quantity for an active device, commonly used for a JFET. (8.4).

Transformer coupling The process of coupling the output of one amplifier or system to the input of another using a transformer. (13.3)

Transient suppressors A circuit or device designed to prevent voltage or current from exceeding a predetermined level. (16.1)

Transistor Any of a group of solid-state semiconductor three-terminal (or more) devices designed to produce a control (gain) function. *See also* BJT and FET. (3.3)

Transresistance amplifier An amplifier that produces variations in output voltage based on variations in input current (gain statistic rated in volts per ampere or ohms). (4.5)

Triac A control semiconductor device that uses a control input to trigger conduction of the output quantity. The triac can conduct in either of two "forward" directions. (15.3)

Tuned circuit amplifier *See* Resonant circuit amplifier.

UJT (unijunction transistor) A control semiconductor device which is triggered to conduct based on the relationship of the triggering (control) voltage to the supply voltage. (15.3)

Unijunction transistor *See* UJT.

Unity gain An amplifier with a gain of exactly 1. (11.6).

Varactor diode A diode that exhibits predictable capacitance values in relationship to reverse-bias voltage. Used almost exclusively in resonant circuits. (16.1)

Virtual ground Any terminal or point in a circuit which has the exact electrical potential as ground but is not directly connected to ground by a conductor. (6.3)

Voltage amplifier Any amplifier that produces variations in output voltage in direct response to input voltage potentials (gain is represented as a V_{out}/V_{in} ratio and is unitless). (4.2)

Voltage doubler (tripler, etc.) A circuit that produces a voltage which is double (triple, etc.) the peak voltage inputted to the circuit. (16.2)

Voltage-variable resistor *See* VVR.

VVR (voltage-variable resistor) A special application of an FET where changes in input voltage produce a linearly responding change in output resistance. (16.1 and 16.2)

0-dB reference A process that rates other gain values in reference to a standard gain figure. This gain is considered to be 0 dB. Other gains are computed by the formula: 0-dB gain (in decibels) = 20 log (analog gain/reference gain); for power gains 10 is used instead of 20 in this formula. (10.3)

Zener diode A diode that can permit a moderate level of reverse current flow when the reverse breakdown voltage is reached or exceeded. Commonly used to provide a regulated voltage level. (14.4)

Zener regulator A regulator circuit using a zener diode to provide the regulation function. (14.4)

Zener voltage The reverse breakdown voltage (and thus regulation voltage) for a zener diode. Occasionally, the PIV for any diode is referred to as the zener voltage. (14.4)

A Data Sheets

The following Data Sheets
are Reprinted by Permission of
Motorola Semiconductor Products Inc.

MC34071, MC34072
MC35071, MC35072
MC33071, MC33072

Advance Information

HIGH PERFORMANCE SINGLE SUPPLY OPERATIONAL AMPLIFIERS

SILICON MONOLITHIC INTEGRATED CIRCUIT

HIGH SLEW RATE, WIDE BANDWIDTH, SINGLE SUPPLY OPERATIONAL AMPLIFIERS

A standard low-cost Bipolar technology with innovative design concepts are employed for the MC34071/MC34072 series of monolithic operational amplifiers. These devices offer 4.5 MHz of gain bandwidth product, 13 V/μs slew rate, and fast settling time without the use of JFET device technology. In addition, low input offset voltage can economically be achieved. Although these devices can be operated from split supplies, they are particularly suited for single supply operation, since the common mode input voltage range includes ground potential (V_{EE}). The all NPN output stage, characterized by no deadband crossover distortion and large output voltage swing, also provides high capacitive drive capability, excellent phase and gain margins, low open-loop high frequency output impedance and symmetrical source/sink ac frequency response.

The MC34071/MC34072 series of devices are available in standard or prime performance (A Suffix) grades and specified over commercial, industrial/vehicular or military temperature ranges.

- Wide Bandwidth: 4.5 MHz
- High Slew Rate: 13 V/μs
- Fast Settling Time: 1.1 μs to 0.10%
- Wide Single Supply Operating Range: 3.0 to 44 Volts
- Wide Input Common Mode Range Including Ground (V_{EE})
- Low Input Offset Voltage: 1.5 mV Maximum (A Suffix)
- Large Output Voltage Swing: −14.7 V to +14.0 V for $V_S = \pm 15$ V
- Large Capacitance Drive Capability: 0 to 10,000 pF
- Low T.H.D. Distortion: 0.02%
- Excellent Phase Margins: 60°
- Excellent Gain Margin: 12 dB

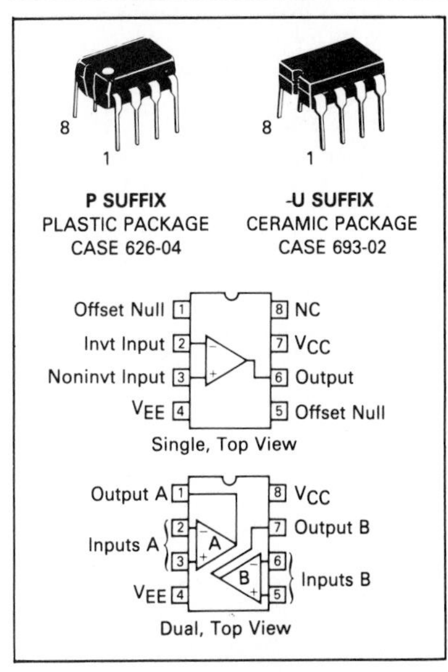

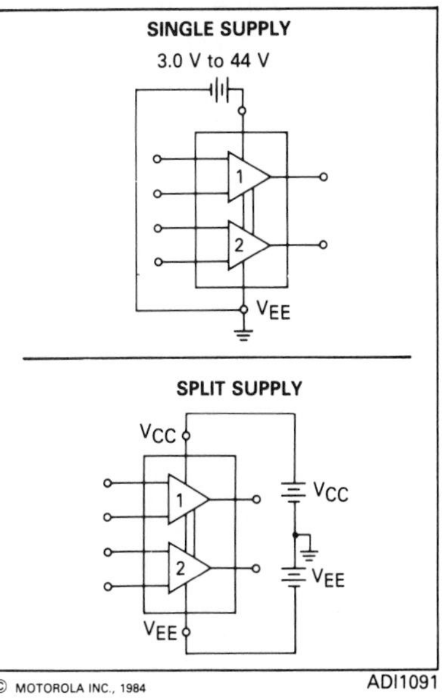

ORDERING INFORMATION			
Op Amp Function	**Device**	**Temperature Range**	**Package**
Single	MC35071U,AU	−55 to +125°C	Ceramic DIP
	MC33071U,AU	−40 to +85°C	Ceramic DIP
	MC33071P,AP	−40 to +85°C	Plastic DIP
	MC34071U,AU	0 to +70°C	Ceramic DIP
	MC34071P,AP	0 to +70°C	Plastic DIP
Dual	MC35072U,AU	−55 to +125°C	Ceramic DIP
	MC33072U,AU	−40 to +85°C	Ceramic DIP
	MC33072P,AP	−40 to +85°C	Plastic DIP
	MC34072U,AU	0 to +70°C	Ceramic DIP
	MC34072P,AP	0 to +70°C	Plastic DIP
Quad	MC34074 Series	Refer to MC34074 Data Sheet	

This document contains information on a new product. Specifications and information herein are subject to change without notice.

ADI1091

DC ELECTRICAL CHARACTERISTICS (V_{CC} = +15 V, V_{EE} = −15 V, R_L connected to ground, T_A = T_{low} to T_{high} [Note 3] unless otherwise noted)

Characteristic	Symbol	MC3507_A/MC3407_A/ MC3307_A			MC3507_/MC3407_/ MC3307_			Unit
		Min	Typ	Max	Min	Typ	Max	
Input Offset Voltage (V_{CM} = 0)	V_{IO}							mV
V_{CC} = +15 V, V_{EE} = −15 V, T_A = +25°C		—	0.5	1.5	—	1.0	3.5	
V_{CC} = +5.0 V, V_{EE} = 0 V, T_A = +25°C		—	0.5	2.0	—	1.5	4.0	
V_{CC} = +15 V, V_{EE} = −15 V, T_A = T_{low} to T_{high}		—	—	3.5	—	—	5.5	
Average Temperature Coefficient of Offset Voltage	$\Delta V_{IO}/\Delta T$	—	10	—	—	10	—	μV/°C
Input Bias Current (V_{CM} = 0)	I_{IB}							nA
T_A = +25°C		—	100	500	—	100	500	
T_A = T_{low} to T_{high}		—	—	700	—	—	700	
Input Offset Current (V_{CM} = 0)	I_{IO}							nA
T_A = +25°C		—	6.0	50	—	6.0	75	
T_A = T_{low} to T_{high}		—	—	300	—	—	300	
Large Signal Voltage Gain V_O = ±10 V, R_L = 2.0 k	A_{VOL}	50	100	—	25	100	—	V/mV
Output Voltage Swing	V_{OH}							V
V_{CC} = +5.0 V, V_{EE} = 0 V, R_L = 2.0 k, T_A = +25°C		3.7	4.0	—	3.7	4.0	—	
V_{CC} = +15 V, V_{EE} = −15 V, R_L = 10 k, T_A = +25°C		13.7	14	—	13.7	14	—	
V_{CC} = +15 V, V_{EE} −15 V, R_L = 2.0 k, T_A = T_{low} to T_{high}		13.5	—	—	13.5	—	—	
	V_{OL}							
V_{CC} = +5.0 V, V_{EE} = 0 V, R_L = 2.0 k, T_A = +25°C		—	0.1	0.2	—	0.1	0.2	
V_{CC} = +15 V, V_{EE} + −15 V, R_L = 10 k, T_A = +25°C		—	−14.7	−14.4	—	−14.7	−14.4	
V_{CC} = +15 V, V_{EE} = −15 V, R_L = 2.0 k, T_A = T_{low} to T_{high}		—	—	−13.8	—	—	−13.8	
Output Short-Circuit Current (T_A = +25°C) Input Overdrive = 1.0 V, Output to Ground	I_{SC}							mA
Source		10	30	—	10	30	—	
Sink		20	47	—	20	47	—	
Input Common Mode Voltage Range	V_{ICR}							V
T_A = +25°C		V_{EE} to (V_{CC} − 1.8)			V_{EE} to (V_{CC} − 1.8)			
T_A = T_{low} to T_{high}		V_{EE} to (V_{CC} − 2.2)			V_{EE} to (V_{CC} − 2.2)			
Common Mode Rejection Ratio (R_S ≤ 10 k)	CMRR	80	97	—	70	97	—	dB
Power Supply Rejection Ratio (R_S = 100 Ω)	PSRR	80	97	—	70	97	—	dB
Power Supply Current (Per Amplifier)	I_D							mA
V_{CC} = +5.0 V, V_{EE} = 0 V, T_A = +25°C		—	1.6	2.0	—	1.6	2.0	
V_{CC} = +15 V, V_{EE} = −15 V, T_A = +25°C		—	1.9	2.5	—	1.9	2.5	
V_{CC} = +15 V, V_{EE} = −15 V, T_A = T_{low} to T_{high}		—	—	2.8	—	—	2.8	

NOTES: (continued)

3. T_{low} = −55°C for MC35071,A/MC35072,A
 = −40°C for MC33071,A/MC33072,A
 = 0°C for MC34071,A/MC34072,A

T_{high} = +125°C for MC35071,A/35072,A
 = +85°C for MC33071,A/33072,A
 = +70°C for MC34071,A/34072,A

MAXIMUM RATINGS

Rating	Symbol	Value	Unit
Supply Voltage (from V_{CC} to V_{EE})	V_S	+44	Volts
Input Differential Voltage Range	V_{IDR}	Note 1	Volts
Input Voltage Range	V_{IR}	Note 1	Volts
Output Short-Circuit Duration (Note 2)	t_S	Indefinite	Seconds
Operating Ambient Temperature Range MC35071,A/MC35072,A MC33071,A/MC33072,A MC34071,A/MC34072,A	T_A	 −55 to +125 −40 to +85 0 to +70	°C
Operating Junction Temperature	T_J	+150	°C
Storage Temperature Range Ceramic Package Plastic Package	T_{stg}	 −65 to +150 −55 to +125	°C

NOTES:

1. Either or both input voltages must not exceed the magnitude of V_{CC} or V_{EE}.
2. Power dissipation must be considered to ensure maximum junction temperature (T_J) is not exceeded.

EQUIVALENT CIRCUIT SCHEMATIC (EACH AMPLIFIER)

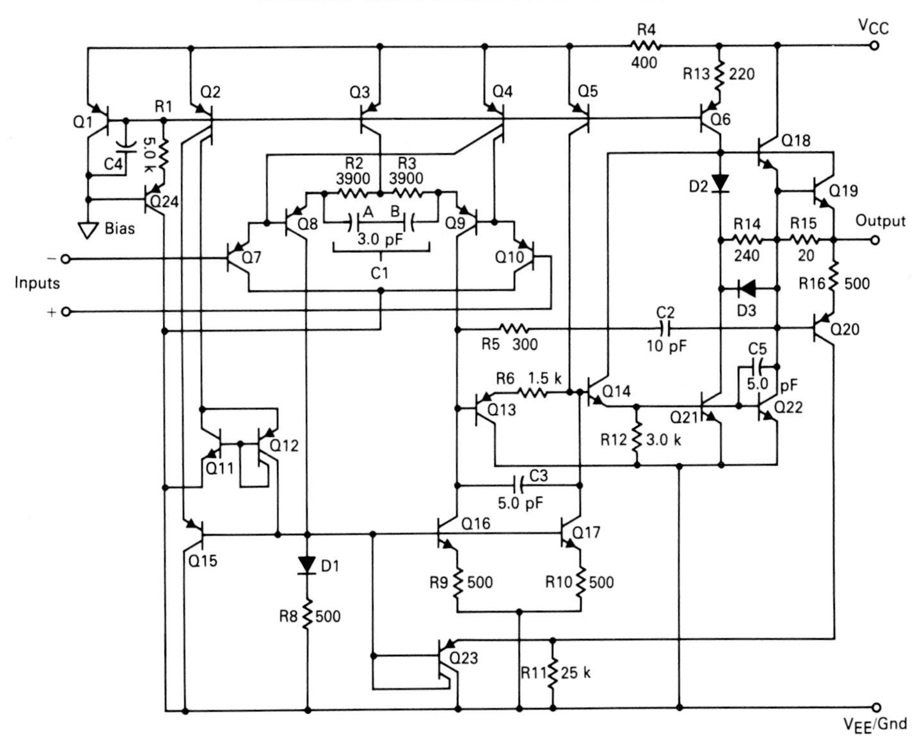

AC ELECTRICAL CHARACTERISTICS (V_{CC} = +15 V, V_{EE} = −15 V, R_L connected to ground, T_A = +25°C unless otherwise noted)

Characteristic	Symbol	MC3507_A/MC3407_A/ MC3307_A			MC3507_/MC3407_/ MC3307_			Unit
		Min	Typ	Max	Min	Typ	Max	
Slew Rate (V_{in} = −10 V to +10 V, R_L = 2.0 k, C_L = 500 pF)	SR							V/μs
A_V + 1.0		8.0	10	—	—	10	—	
A_V − 1.0		—	13	—	—	13	—	
Settling Time (10 V Step, A_V = −1.0)	t_s							μs
To 0.10% (±1/2 LSB of 9-Bits)		—	1.1	—	—	1.1	—	
To 0.01% (±1/2 LSB of 12-Bits)		—	2.2	—	—	2.2	—	
Gain Bandwidth Product (f = 100 kHz)	GBW	3.5	4.5	—		4.5	—	MHz
Power Bandwidth A_V = +1.0, R_L = 2.0 k, V_O = 20 V_{p-p}, THD = 5.0%	BWp	—	200	—	—	200	—	kHz
Phase Margin	ϕm							Degrees
R_L = 2.0 k		—	60	—	—	60	—	
R_L = 2.0 k, C_L = 300 pF		—	40	—	—	40	—	
Gain Margin	A_m							dB
R_L = 2.0 k		—	12	—	—	12	—	
R_L = 2.0 k, C_L = 300 pF		—	4.0	—	—	4.0	—	
Equivalent Input Noise Voltage R_S = 100 Ω, f = 1.0 kHz	e_n	—	32	—	—	32	—	nV/ $\sqrt{Hz}$
Equivalent Input Noise Current (f = 1.0 kHz)	I_n	—	0.22	—	—	0.22	—	pA/ $\sqrt{Hz}$
Input Capacitance	C_i	—	0.8	—	—	0.8	—	pF
Total Harmonic Distortion A_V = +10, R_L = 2.0 k, 2.0 ≤ V_O ≤ 20 V_{p-p}, f = 10 kHz	THD	—	0.02	—	—	0.02	—	%
Channel Separation (f = 10 kHz, MC34072,A Only)	—	—	120	—	—	120	—	dB
Open-Loop Output Impedance (f = 1.0 MHz)	z_o	—	30	—	—	30	—	Ω

For typical performance curves and applications information refer to MC34074 series data sheet.

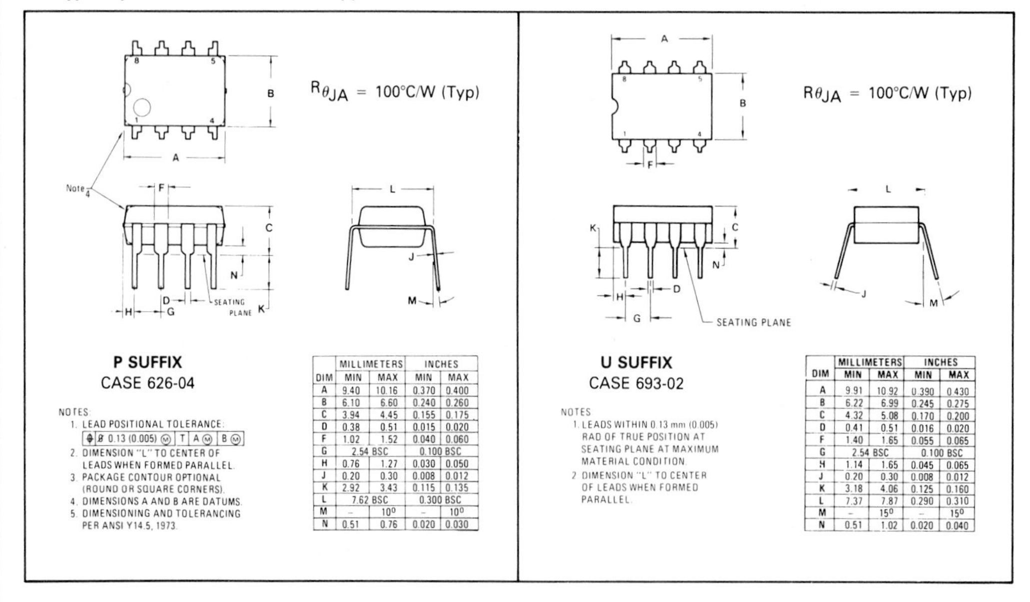

P SUFFIX
CASE 626-04

NOTES:
1. LEAD POSITIONAL TOLERANCE: ⌖ Ø 0.13 (0.005) Ⓜ | T | A Ⓜ | B Ⓜ
2. DIMENSION "L" TO CENTER OF LEADS WHEN FORMED PARALLEL.
3. PACKAGE CONTOUR OPTIONAL (ROUND OR SQUARE CORNERS).
4. DIMENSIONS A AND B ARE DATUMS.
5. DIMENSIONING AND TOLERANCING PER ANSI Y14.5, 1973.

DIM	MILLIMETERS MIN	MILLIMETERS MAX	INCHES MIN	INCHES MAX
A	9.40	10.16	0.370	0.400
B	6.10	6.60	0.240	0.260
C	3.94	4.45	0.155	0.175
D	0.38	0.51	0.015	0.020
F	1.02	1.52	0.040	0.060
G	2.54 BSC		0.100 BSC	
H	0.76	1.27	0.030	0.050
J	0.20	0.30	0.008	0.012
K	2.92	3.43	0.115	0.135
L	7.62 BSC		0.300 BSC	
M	–	10°	–	10°
N	0.51	0.76	0.020	0.030

U SUFFIX
CASE 693-02

NOTES:
1. LEADS WITHIN 0.13 mm (0.005) RAD OF TRUE POSITION AT SEATING PLANE AT MAXIMUM MATERIAL CONDITION.
2. DIMENSION "L" TO CENTER OF LEADS WHEN FORMED PARALLEL.

DIM	MILLIMETERS MIN	MILLIMETERS MAX	INCHES MIN	INCHES MAX
A	9.91	10.92	0.390	0.430
B	6.22	6.99	0.245	0.275
C	4.32	5.08	0.170	0.200
D	0.41	0.51	0.016	0.020
F	1.40	1.65	0.055	0.065
G	2.54 BSC		0.100 BSC	
H	1.14	1.65	0.045	0.065
J	0.20	0.30	0.008	0.012
K	3.18	4.06	0.125	0.160
L	7.37	7.87	0.290	0.310
M	–	15°	–	15°
N	0.51	1.02	0.020	0.040

MOTOROLA *Semiconductor Products Inc.*

BOX 20912 • PHOENIX, ARIZONA 85036 • A SUBSIDIARY OF MOTOROLA INC.

17857 PRINTED IN USA 5-84 IMPERIAL LITHO C22510 12,000

AD11091

2N2218,A/2N2219,A
2N2221,A/2N2222,A
2N5581/82

JAN, JTX, JTXV AVAILABLE

2N2218,A
2N2219,A
CASE 79-02
TO-39 (TO-205AD) STYLE 1

2N2221,A
2N2222,A
CASE 22-03
TO-18 (TO-206AA) STYLE 1

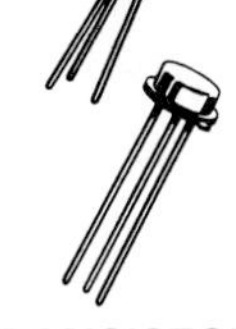

2N5581
2N5582
CASE 26-03
TO-46 (TO-206AB) STYLE 1

GENERAL PURPOSE TRANSISTOR

NPN SILICON

4

MAXIMUM RATINGS

Rating	Symbol	2N2218 2N2219 2N2221 2N2222	2N2218A 2N2219A 2N2221A 2N2222A	2N5581 2N5582	Unit
Collector-Emitter Voltage	V_{CEO}	30	40	40	Vdc
Collector-Base Voltage	V_{CBO}	60	75	75	Vdc
Emitter-Base Voltage	V_{EBO}	5.0	6.0	6.0	Vdc
Collector Current — Continuous	I_C	800	800	800	mAdc
		2N2218,A 2N2219,A	**2N2221,A 2N2222,A**	**2N5581 2N5582**	
Total Device Dissipation @ T_A = 25°C Derate above 25°C	P_D	0.8 4.57	0.4 2.28	0.6 3.33	Watt mW/°C
Total Device Dissipation @ T_C = 25°C Derate above 25°C	P_D	3.0 17.1	1.2 6.85	2.0 11.43	Watts mW/°C
Operating and Storage Junction Temperature Range	T_J, T_{stg}	−65 to +200			°C

ELECTRICAL CHARACTERISTICS (T_A = 25°C unless otherwise noted.)

Characteristic		Symbol	Min	Max	Unit
OFF CHARACTERISTICS					
Collector-Emitter Breakdown Voltage (I_C = 10 mAdc, I_B = 0)	 Non-A Suffix A-Suffix, 2N5581, 2N5582	$V_{(BR)CEO}$	 30 40	 — —	Vdc
Collector-Base Breakdown Voltage (I_C = 10 μAdc, I_E = 0)	 Non-A Suffix A-Suffix, 2N5581, 2N5582	$V_{(BR)CBO}$	 60 75	 — —	Vdc
Emitter-Base Breakdown Voltage (I_E = 10 μAdc, I_C = 0)	 Non-A Suffix A-Suffix, 2N5581, 2N5582	$V_{(BR)EBO}$	 5.0 6.0	 — —	Vdc
Collector Cutoff Current (V_{CE} = 60 Vdc, $V_{EB(off)}$ = 3.0 Vdc)	 A-Suffix, 2N5581, 2N5582	I_{CEX}	—	10	nAdc
Collector Cutoff Current (V_{CB} = 50 Vdc, I_E = 0) (V_{CB} = 60 Vdc, I_E = 0) (V_{CB} = 50 Vdc, I_E = 0, T_A = 150°C) (V_{CB} = 60 Vdc, I_E = 0, T_A = 150°C)	 Non-A Suffix A-Suffix, 2N5581, 2N5582 Non-A Suffix A-Suffix, 2N5581, 2N5582	I_{CBO}	 — — — —	 0.01 0.01 10 10	μAdc
Emitter Cutoff Current (V_{EB} = 3.0 Vdc, I_C = 0)	 A-Suffix, 2N5581, 2N5582	I_{EBO}	—	10	nAdc
Base Cutoff Current (V_{CE} = 60 Vdc, $V_{EB(off)}$ = 3.0 Vdc)	 A-Suffix	I_{BL}	—	20	nAdc
ON CHARACTERISTICS					
DC Current Gain (I_C = 0.1 mAdc, V_{CE} = 10 Vdc)	 2N2218,A, 2N2221,A, 2N5581(1) 2N2219,A, 2N2222,A, 2N5582(1)	h_{FE}	 20 35	 — —	—
(I_C = 1.0 mAdc, V_{CE} = 10 Vdc)	2N2218,A, 2N2221,A, 2N5581 2N2219,A, 2N2222,A, 2N5582		25 50	— —	
(I_C = 10 mAdc, V_{CE} = 10 Vdc)	2N2218,A, 2N2221,A, 2N5581(1) 2N2219,A, 2N2222,A, 2N5582(1)		35 75	— —	
(I_C = 10 mAdc, V_{CE} = 10 Vdc, T_A = −55°C)	2N2218A, 2N2221A, 2N5581 2N2219A, 2N2222A, 2N5582		15 35	— —	
(I_C = 150 mAdc, V_{CE} = 10 Vdc)(1)	2N2218,A, 2N2221,A, 2N5581 2N2219,A, 2N2222,A, 2N5582		40 100	120 300	

ELECTRICAL CHARACTERISTICS (continued) (T_A = 25°C unless otherwise noted.)

Characteristic		Symbol	Min	Max	Unit
(I_C = 150 mAdc, V_{CE} = 1.0 Vdc)(1)	2N2218,A, 2N2221,A, 2N5581		20	—	
	2N2219,A, 2N2222,A, 2N5582		50	—	
(I_C = 500 mAdc, V_{CE} = 10 Vdc)(1)	2N2218, 2N2221		20	—	
	2N2219, 2N2222		30	—	
	2N2218A, 2N2221A, 2N5581		25	—	
	2N2219A, 2N2222A, 2N5582		40	—	
Collector-Emitter Saturation Voltage(1)		$V_{CE(sat)}$			Vdc
(I_C = 150 mAdc, I_B = 15 mAdc)	Non-A Suffix		—	0.4	
	A-Suffix, 2N5581, 2N5582		—	0.3	
(I_C = 500 mAdc, I_B = 50 mAdc)	Non-A Suffix		—	1.6	
	A-Suffix, 2N5581, 2N5582		—	1.0	
Base-Emitter Saturation Voltage(1)		$V_{BE(sat)}$			Vdc
(I_C = 150 mAdc, I_B = 15 mAdc)	Non-A Suffix		0.6	1.3	
	A-Suffix, 2N5581, 2N5582		0.6	1.2	
(I_C = 500 mAdc, I_B = 50 mAdc)	Non-A Suffix		—	2.6	
	A-Suffix, 2N5581, 2N5582		—	2.0	

SMALL-SIGNAL CHARACTERISTICS

Characteristic		Symbol	Min	Max	Unit
Current-Gain — Bandwidth Product(2)		f_T			MHz
(I_C = 20 mAdc, V_{CE} = 20 Vdc, f = 100 MHz)	All Types, Except		250	—	
	2N2219A, 2N2222A, 2N5582		300	—	
Output Capacitance(3) (V_{CB} = 10 Vdc, I_E = 0, f = 100 kHz)		C_{obo}	—	8.0	pF
Input Capacitance(3)		C_{ibo}			pF
(V_{EB} = 0.5 Vdc, I_C = 0, f = 100 kHz)	Non-A Suffix		—	30	
	A-Suffix, 2N5581, 2N5582		—	25	
Input Impedance		h_{ie}			kohms
(I_C = 1.0 mAdc, V_{CE} = 10 Vdc, f = 1.0 kHz)	2N2218A, 2N2221A		1.0	3.5	
	2N2219A, 2N2222A		2.0	8.0	
(I_C = 10 mAdc, V_{CE} = 10 Vdc, f = 1.0 kHz)	2N2218A, 2N2221A		0.2	1.0	
	2N2219A, 2N2222A		0.25	1.25	
Voltage Feedback Ratio		h_{re}			$X\ 10^{-4}$
(I_C = 1.0 mAdc, V_{CE} = 10 Vdc, f = 1.0 kHz)	2N2218A, 2N2221A		—	5.0	
	2N2219A, 2N2222A		—	8.0	
(I_C = 10 mAdc, V_{CE} = 10 Vdc, f = 1.0 kHz)	2N2218A, 2N2221A		—	2.5	
	2N2219A, 2N2222A		—	4.0	
Small-Signal Current Gain		h_{fe}			—
(I_C = 1.0 mAdc, V_{CE} = 10 Vdc, f = 1.0 kHz)	2N2218A, 2N2221A		30	150	
	2N2219A, 2N2222A		50	300	
(I_C = 10 mAdc, V_{CE} = 10 Vdc, f = 1.0 kHz)	2N2218A, 2N2221A		50	300	
	2N2219A, 2N2222A		75	375	
Output Admittance		h_{oe}			μmhos
(I_C = 1.0 mAdc, V_{CE} = 10 Vdc, f = 1.0 kHz)	2N2218A, 2N2221A		3.0	15	
	2N2219A, 2N2222A		5.0	35	
(I_C = 10 mAdc, V_{CE} = 10 Vdc, f = 1.0 kHz)	2N2218A, 2N2221A		10	100	
	2N2219A, 2N2222A		25	200	
Collector Base Time Constant (I_E = 20 mAdc, V_{CB} = 20 Vdc, f = 31.8 MHz)	A-Suffix	$r_{b'}C_c$	—	150	ps
Noise Figure (I_C = 100 μAdc, V_{CE} = 10 Vdc, R_S = 1.0 kohm, f = 1.0 kHz)	2N2222A	NF	—	4.0	dB
Real Part of Common-Emitter High Frequency Input Impedance (I_C = 20 mAdc, V_{CE} = 20 Vdc, f = 300 MHz)	2N2218A, 2N2219A 2N2221A, 2N2222A	$Re(h_{ie})$	—	60	Ohms

(1) Pulse Test: Pulse Width ≤ 300 μs, Duty Cycle ≤ 2.0%.
(2) f_T is defined as the frequency at which $|h_{fe}|$ extrapolates to unity.
(3) 2N5581 and 2N5582 are Listed C_{cb} and C_{eb} for these conditions and values.

4

ELECTRICAL CHARACTERISTICS (continued) (T_A = 25°C unless otherwise noted.)

Characteristic		Symbol	Min	Max	Unit
SWITCHING CHARACTERISTICS					
Delay Time	(V_{CC} = 30 Vdc, $V_{BE(off)}$ = 0.5 Vdc, I_C = 150 mAdc, I_{B1} = 15 mAdc) (Figure 14)	t_d	—	10	ns
Rise Time		t_r	—	25	ns
Storage Time	(V_{CC} = 30 Vdc, I_C = 150 mAdc, I_{B1} = I_{B2} = 15 mAdc) (Figure15)	t_s	—	225	ns
Fall Time		t_f	—	60	ns
Active Region Time Constant (I_C = 150 mAdc, V_{CE} = 30 Vdc) (See Figure 12 for 2N2218A, 2N2219A, 2N2221A, 2N2222A)		T_A	—	2.5	ns

FIGURE 1 – NORMALIZED DC CURRENT GAIN

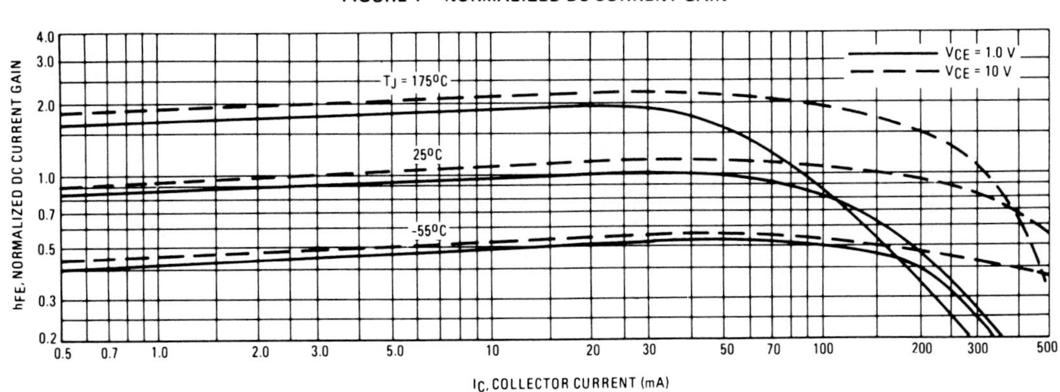

FIGURE 2 – COLLECTOR CHARACTERISTICS IN SATURATION REGION

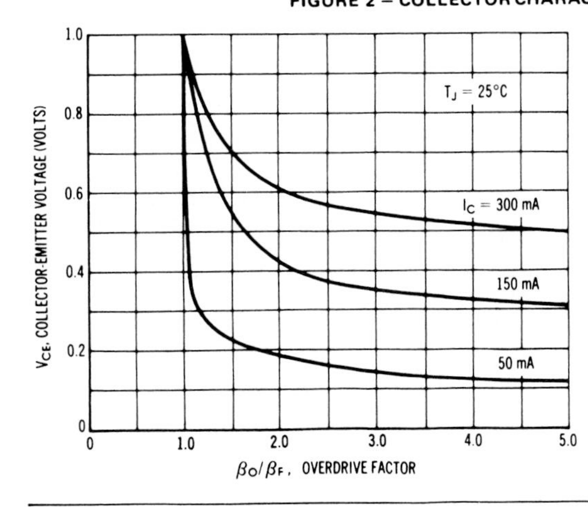

This graph shows the effect of base current on collector current. β_o (current gain at the edge of saturation) is the current gain of the transistor at 1 volt, and β_F (forced gain) is the ratio of I_C/I_{BF} in a circuit.

EXAMPLE: For type 2N2219, estimate a base current (I_{BF}) to insure saturation at a temperature of 25°C and a collector current of 150 mA.

Observe that at I_C = 150 mA an overdrive factor of at least 2.5 is required to drive the transistor well into the saturation region. From Figure 1, it is seen that h_{FE} @ 1 volt is approximately 0.62 of h_{FE} @ 10 volts. Using the guaranteed minimum gain of 100 @ 150 mA and 10 V, β_o = 62 and substituting values in the overdrive equation, we find:

$$\frac{\beta_o}{\beta_F} = \frac{h_{FE} @ 1.0\,V}{I_C/I_{BF}} \qquad 2.5 = \frac{62}{150/I_{BF}} \qquad I_{BF} \approx 6.0\ mA$$

MFQ5460P

P-CHANNEL

QUAD DUAL-IN-LINE
JUNCTION FIELD-EFFECT
TRANSISTORS

Type A

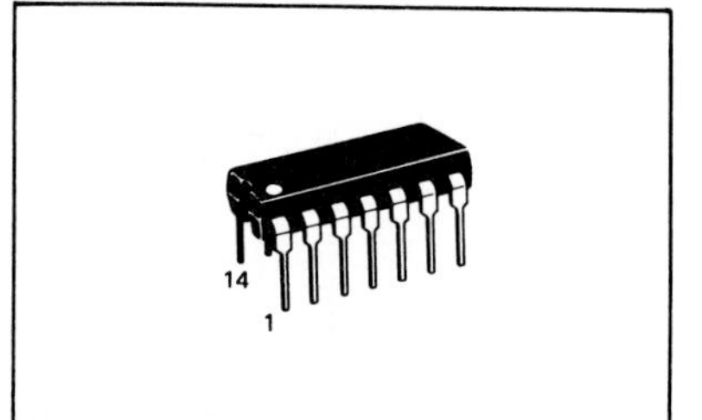

QUAD DUAL-IN-LINE
P-CHANNEL JUNCTION FIELD-EFFECT TRANSISTORS

. . . depletion mode (Type A) junction field-effect transistors designed for use in general-purpose amplifier applications.

- High Gate-Source Breakdown Voltage — $V_{(BR)GSS}$ = 40 Vdc (Min)
- Low Noise Figure — NF = 1.0 dB (Typ) @ f = 100 Hz
- Low Reverse Transfer Capacitance — C_{rss} = 2.0 pF (Max)
- Refer to 2N5460 Data Sheet for Performance Graphs

MAXIMUM RATINGS

Rating	Symbol	Value		Unit
Drain-Source Voltage	V_{DS}	40		Vdc
Drain-Gate Voltage	V_{DG}	40		Vdc
Reverse Gate-Source Voltage	V_{GSR}	40		Vdc
Drain Current	I_D	20		mAdc
Forward Gate Current	I_{GF}	10		mAdc
		Each Transistor	Total Device	
Total Device Dissipation @ T_A = 25°C Derate above 25°C	P_D	0.5 2.86	1.5 8.58	Watts mW/°C

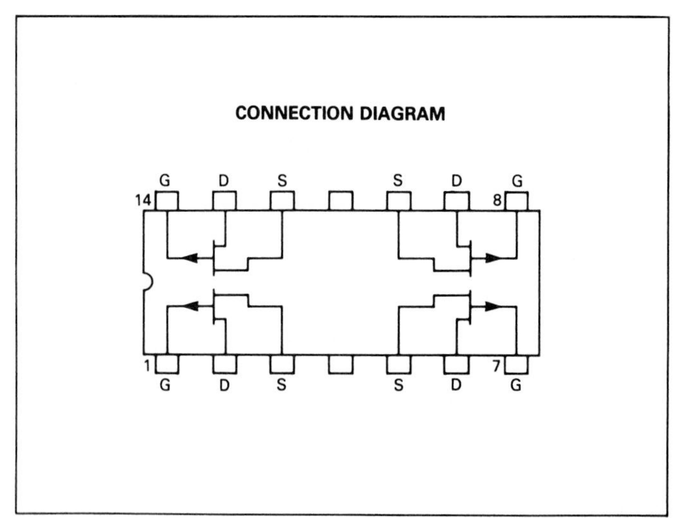

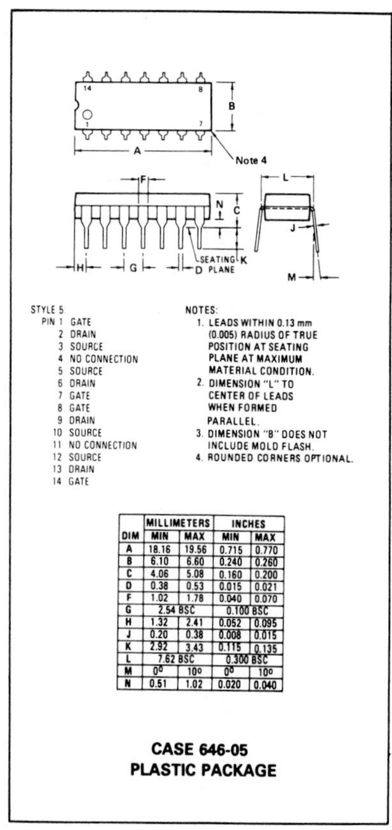

STYLE 5:
PIN 1 GATE
2 DRAIN
3 SOURCE
4 NO CONNECTION
5 SOURCE
6 DRAIN
7 GATE
8 GATE
9 DRAIN
10 SOURCE
11 NO CONNECTION
12 SOURCE
13 DRAIN
14 GATE

NOTES:
1. LEADS WITHIN 0.13 mm (0.005) RADIUS OF TRUE POSITION AT SEATING PLANE AT MAXIMUM MATERIAL CONDITION.
2. DIMENSION "L" TO CENTER OF LEADS WHEN FORMED PARALLEL.
3. DIMENSION "B" DOES NOT INCLUDE MOLD FLASH.
4. ROUNDED CORNERS OPTIONAL.

	MILLIMETERS		INCHES	
DIM	MIN	MAX	MIN	MAX
A	18.16	19.56	0.715	0.770
B	6.10	6.60	0.240	0.260
C	4.06	5.08	0.160	0.200
D	0.38	0.53	0.015	0.021
F	1.02	1.78	0.040	0.070
G	2.54 BSC		0.100 BSC	
H	1.32	2.41	0.052	0.095
J	0.20	0.38	0.008	0.015
K	2.92	3.43	0.115	0.135
L	7.62 BSC		0.300 BSC	
M	0°	10°	0°	10°
N	0.51	1.02	0.020	0.040

CASE 646-05
PLASTIC PACKAGE

 DS4635

ELECTRICAL CHARACTERISTICS (T_A = 25°C unless otherwise noted)

Characteristic	Symbol	Min	Typ	Max	Unit
OFF CHARACTERISTICS					
Gate-Source Breakdown Voltage (I_G = 10 μAdc, V_{DS} = 0)	$V_{(BR)GSS}$	40	—	—	Vdc
Gate-Source Cutoff Voltage (V_{DS} = 15 Vdc, I_D = 1.0 μAdc)	$V_{GS(off)}$	0.75	—	6.0	Vdc
Gate Reverse Current (V_{GS} = 20 Vdc, V_{DS} = 0)	I_{GSS}	—	—	5.0	nAdc
(V_{GS} = 20 Vdc, V_{DS} = 0, T_A = 100°C)		—	—	1.0	μAdc
ON CHARACTERISTICS					
Zero-Gate Voltage Drain Current (V_{DS} = 15 Vdc, V_{GS} = 0)	I_{DSS}	1.0	—	5.0	mAdc
Gate-Source Voltage (V_{DS} = 15 Vdc, I_D = 0.1 mAdc)	V_{GS}	0.5	—	4.0	Vdc
SMALL-SIGNAL CHARACTERISTICS					
Forward Transadmittance (V_{DS} = 15 Vdc, V_{GS} = 0, f = 1.0 kHz)	$\|y_{fs}\|$	1000	—	4000	μmhos
Output Admittance (V_{DS} = 15 Vdc, V_{GS} = 0, f = 1.0 kHz)	$\|y_{os}\|$	—	—	75	μmhos
Input Capacitance (V_{DS} = 15 Vdc, V_{GS} = 0, f = 1.0 MHz)	C_{iss}	—	5.0	7.0	pF
Reverse Transfer Capacitance (V_{DS} = 15 Vdc, V_{GS} = 0, f = 1.0 MHz)	C_{rss}	—	1.0	2.0	pF
Common-Source Noise Figure (V_{DS} = 15 Vdc, V_{GS} = 0, R_G = 1.0 MΩ, f = 100 Hz, BW = 1.0 Hz)	NF	—	1.0	—	dB
Equivalent Short-Circuit Input Noise Voltage (V_{DS} = 15 Vdc, V_{GS} = 0, f = 100 Hz, BW = 1.0 Hz)	e_n	—	60	—	nV/√Hz

MOTOROLA *Semiconductor Products Inc.*

BOX 20912 • PHOENIX, ARIZONA 85036 • A SUBSIDIARY OF MOTOROLA INC.

17443 PRINTED IN USA (3/84) MPS 12M

MOTOROLA
SEMICONDUCTORS
P.O. BOX 20912 • PHOENIX, ARIZONA 85036

MRF2628

The RF Line

15 W 136–220 MHz

RF POWER TRANSISTOR

NPN SILICON

NPN SILICON POWER TRANSISTOR

Designed for 12.5 volt VHF large-signal power amplifiers in commercial and industrial FM equipment.

- Compact .280 Stud Package
- Specified 12.5 V, 175 MHz Performance
  Output Power = 15 Watts
  Power Gain = 12 dB Min
  Efficiency = 60% Min
- Characterized to 220 MHz
- Load Mismatch Capability at High Line and Overdrive

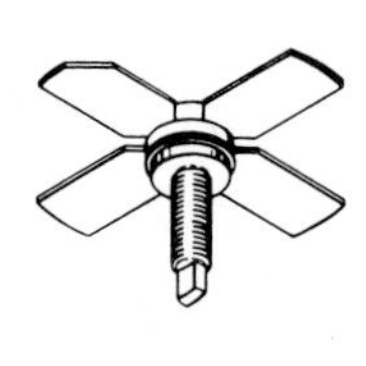

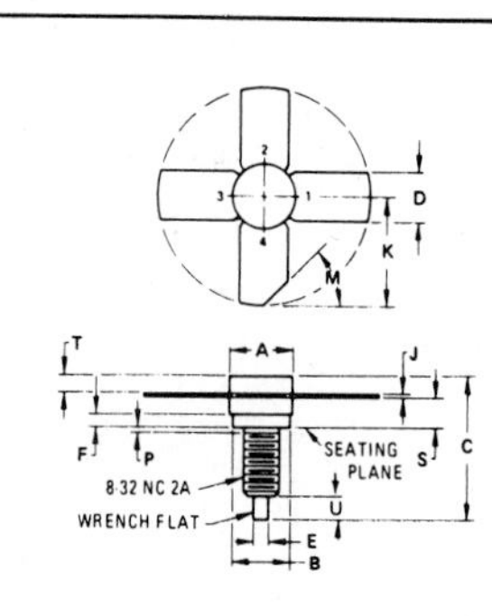

STYLE 1:
PIN 1. EMITTER
2. BASE
3. EMITTER
4. COLLECTOR

DIM	MILLIMETERS MIN	MILLIMETERS MAX	INCHES MIN	INCHES MAX
A	7.06	7.26	0.278	0.286
B	6.20	6.50	0.244	0.256
C	14.99	16.51	0.590	0.650
D	5.46	5.97	0.215	0.235
E	1.40	1.65	0.055	0.065
F	1.52	–	0.060	–
J	0.08	0.18	0.003	0.007
K	11.05	–	0.435	–
M	45° NOM		45° NOM	
P	–	1.27	–	0.050
S	3.00	3.25	0.118	0.128
T	1.40	1.78	0.055	0.070
U	2.92	3.68	0.115	0.145

CASE 244-04

MAXIMUM RATINGS

Rating	Symbol	Value	Unit
Collector-Emitter Voltage	V_{CEO}	18	Vdc
Collector-Base Voltage	V_{CBO}	36	Vdc
Emitter-Base Voltage	V_{EBO}	4.0	Vdc
Collector-Current — Continuous	I_C	2.5	Adc
Total Device Dissipation @ T_A = 25°C Derate above 25°C	P_D	40 0.23	Watts W/°C
Storage Temperature Range	T_{stg}	−65 to +150	°C
Junction Temperature	T_J	200	°C

THERMAL CHARACTERISTICS

Characteristic	Symbol	Max	Unit
Thermal Resistance, Junction to Case	$R_{\theta JC}$	4.0	°C/W

DS5881